Wireless Technologies

Circuits, Systems, and Devices

Wireless Technologies

Circuits, Systems, and Devices

Edited by
Krzysztof Iniewski

CRC Press
Taylor & Francis Group
Boca Raton   London   New York

CRC Press is an imprint of the
Taylor & Francis Group, an **informa** business

CRC Press
Taylor & Francis Group
6000 Broken Sound Parkway NW, Suite 300
Boca Raton, FL 33487-2742

© 2008 by Taylor & Francis Group, LLC
CRC Press is an imprint of Taylor & Francis Group, an Informa business

No claim to original U.S. Government works
Printed in the United States of America on acid-free paper
10 9 8 7 6 5 4 3 2 1

International Standard Book Number-13: 978-0-8493-7996-3 (Hardcover)

Library of Congress Cataloging-in-Publication Data

Iniewski, Krzysztof.
   Wireless Technologies : circuits, systems, and devices / Krzysztof Iniewski.
      p. cm.
   Includes bibliographical references and index.
   ISBN-13: 978-0-8493-7996-3 (alk. paper)
   ISBN-10: 0-8493-7996-2 (alk. paper)
   1. Wireless communication systems--Equipment and supplies--Design and construction. 2. Metal oxide semiconductors, Complementary--Design and construction. 3. Radio circuits--Design and construction. I. Title.

TK5103.2.I515 2007
621.384--dc22   2007011165

Visit the Taylor & Francis Web site at
http://www.taylorandfrancis.com

and the CRC Press Web site at
http://www.crcpress.com

Contents

Preface vii

Editor ix

Contributors xi

PART I Circuits for Emerging Wireless: A System Perspective

Chapter 1   RF Building Blocks for the Next-Gen Wireless Systems 3

Ali M. Niknejad

Chapter 2   Insights into CMOS Wireless Receivers toward a Universal Mobile Radio 53

Massimo Brandolini, Paolo Rossi, Danilo Manstretta, and Francesco Svelto

Chapter 3   Ultra Wide Band (UWB) Technology 81

Domine M. W. Leenaerts

Chapter 4   Design Considerations for Integrated MIMO Radio Transceivers 107

Yorgos Palaskas, Ashoke Ravi, Stefano Pellerano, and Sumeet Sandhu

Chapter 5   Cognitive Radio Spectrum-Sharing Technology 131

Danijela Cabric and Robert W. Brodersen

Chapter 6   Short-Distance Wireless and Its Opportunities 159

Jan M. Rabaey, Y. H. Chee, David Chen, Luca de Nardis, Simone Gambini, Davide Guermandi, Michael Mark, and Nathan Pletcher

Chapter 7   Ultra-Low-Power RF Transceivers 185

Emanuele Lopelli, Johan D. van der Tang, and Arthur H. M. van Roermund

Chapter 8   Human++: Emerging Technology for Body Area Networks 221

Bert Gyselinckx, Raffaella Borzi, and Philippe Mattelaer

Chapter 9   Progress toward a Single-Chip Radio 241

Ken K. O, Paul Gorday, Jau-Jr. Lin, Changhua Cao, Yu Su, Zhenbiao Li, Jesal Mechta, Joe E. Brewer, and Seon-Ho Hwan

PART II Chip Architectures and Circuit Implementations

Chapter 10 Digital RF Processor (DRP™) 265

Robert Bogdan Staszewski

Chapter 11   Low Noise Amplifiers 305

                Leonid Belostotski and James Haslett

Chapter 12   Design of Silicon Integrated Circuit W-Band Low-Noise Amplifiers 329

                *Sean T. Nicolson, Keith W. Tang, T. O. Dickson, P. Chevalier,
                B. Sautreuil, and Sorin P. Voinigescu*

Chapter 13   Power Amplifier Principles and Modern Design Techniques 349

                Vladimir Prodanov and Mihai Banu

Chapter 14   Phase-Locked Loop–Based Integer-N RF Synthesizer 383

                Vikas Choudhary and Krzysztof (Kris) Iniewski

Chapter 15   Frequency Synthesis for Multiband Wireless Networks 427

                John W. M. Rogers, Foster F. Dai, and Calvin Plett

Chapter 16   Design of a Delta-Sigma Synthesizer for a Bluetooth® Transmitter 455

                Jan-Wim Eikenbroek

Chapter 17   RFIC Parametric Converters: Device Modification, Circuit Design,
                Control Techniques 487

                *Sebastian Magierowski, Howard Chan, Krzysztof (Kris) Iniewski,
                and Takis Zourntos*

PART III   Device and Process Technology for Wireless Chips

Chapter 18   CMOS Technology for Wireless Applications 521

                John J. Pekarik

Chapter 19   Distributed Effects and Coupling in RF Integrated Circuits 543

                Calvin Plett

Chapter 20   Substrate Noise Coupling from Digital to Analog Circuits
                in Mixed-Signal Integrated Circuits 567

                *Piet Wambacq, Charlotte Soens, Geert Van der Plas,
                Mustafa Badaroglu, and Stéphane Donnay*

Chapter 21   Microelectromechanical Resonators for RF Applications 589

                Frederic Nabki, Tomas A. Dusatko, and Mourad N. El-Gamal

Chapter 22   Membrane-Supported Millimeter-Wave Circuits Based on Silicon
                and GaAs Micromachining 629

                Alexandru Müller, Dan Neculoiu, George Konstantinidis, and Robert Plana

Index 655

Preface

Advanced concepts for wireless communications present a vision of technology that is embedded in our surroundings and practically invisible, but present whenever required. From established radio techniques like the global system for mobile communications (GSM), 802.11, or Bluetooth to more emerging like ultra wide band (UWB) or smart dust motes, a common denominator for future progress is underlying integrated circuit technology. Although the use of deep-submicron CMOS processes allows for an unprecedented degree of scaling in digital circuitry, it complicates implementation and integration of traditional radio frequency (RF) circuits. The explosive growth of standard cellular radios and radically different new wireless applications make it imperative to find architectural and circuit solutions to these design problems.

Two key issues for future silicon-based systems are scale of integration and low power dissipation. The concept of combining digital, memory, mixed-signal, and RF circuitry on one chip in the form of system-on-chip (SoC) has been around for a while. However, the difficulty of integrating heterogeneous circuit design styles and processes onto one substrate still remains. Therefore, the system-in-package (SiP) concept seems to be gaining acceptance as well. While it is true that heterogeneous circuits and architectures originally developed for their native technologies cannot be effectively integrated "as is" into a deep-submicron CMOS process, one might ask the question whether those functions can be ported into more CMOS-friendly architectures to reap all the benefits of the digital design and flow. It is not predestined that RF wireless frequency synthesizers be always charge-pump-based phase-locked loops (PLLs) with voltage controlled oscillators (VCOs), RF transmit up-converters be I/Q modulators, and receivers use only Gilbert cell or passive continuous-time mixers. Performance of modern CMOS transistors is nowadays good enough for multi-GHz RF applications.

Low power has always been important for wireless communications. With new developments in wireless sensor networks and wireless systems for medical applications, power dissipation is becoming the number one issue. Wireless sensor network systems are being applied in critical applications in commerce, healthcare, and security. These systems have unique characteristics and face many implementation challenges. The requirement of long operating life for a wireless sensor node under limited energy supply imposes the most severe design constraints. This calls for innovative design methodologies at the circuit and system levels to address this rigorous requirement.

Wireless systems for medical applications hold a number of advantages over wired alternatives, including ease of use, reduced risk of infection, reduced risk of failure, reduced patient discomfort, enhanced mobility, and lower cost. Typically, applications demand expertise in multiple disciplines, varying from analog sensors to digital processing cores, suggesting opportunities for extensive hardware integration.

The book addresses state-of-the-art CMOS design in the context of wireless communication for emerging applications: 3G/4G cellular telephony, wireless sensor networks, and wireless medical applications. New exciting opportunities in body area networks, medical implants, satellite communications, automobile radar detection, and wearable electronics are discussed. The book is a must for anyone serious about future wireless technologies.

The book is written by top international experts on wireless circuit design representing both the integrated circuit (IC) industry and academia. The intended audience is practicing engineers in wireless communication field with some integrated circuit background. The book can also be used as a recommended reading and supplementary material in graduate course curriculum.

The book is divided into three different parts. Part I provides a wireless system perspective. In Chapter 1, Ali Niknejad from Berkeley provides a broad introduction to various aspects of wireless circuit design. In Chapter 2, researchers from the University of Pavia describe challenges in

CMOS design for multistandard radios. Emerging new UWB technology and its CMOS/BiCMOS implementations are discussed by Domine M. W. Leenaerts from NXP/Philips in Chapter 3. Yorgos Palaskas and Ashoke Ravi from Intel introduce novel circuit solutions for multiple-input multiple-output (MIMO) technology in Chapter 4. Danijela Cabric and Robert W. Brodersen from Berkeley reach even further out into the future in Chapter 5 by discussing possible architectures of cognitive radios.

Jan M. Rabaey and his group from Berkeley present an exciting perspective of short-reach wireless opportunities in Chapter 6. This is followed by a comprehensive review of wireless transceivers for short-reach applications given by a group from Eindhoven University of Technology in Chapter 7. Exciting applications of wireless technology to human body are discussed by a group from IMEC in Chapter 8. To conclude the first part of the book, a research group from the University of Florida presents efforts toward integrating on-chip antennae on CMOS substrate in Chapter 9.

The second part of the book deals with chip architectures and circuit implementation issues. Robert Bogdan Staszewski from Texas Instruments discusses revolutionary Digital Radio Processing™ chip architecture in Chapter 10. Low-noise amplifiers (LNAs) are introduced by Leonid Belostotski and James Haslett from the University of Calgary in Chapter 11. Design of LNAs for W-band applications is covered by a group from the University of Toronto in Chapter 12. In Chapter 13, Vladimir Prodanov and Mihai Banu from MHI Consulting provide an extensive treatment of power amplifiers.

Vikas Choudhary from PMC-Sierra and Krzysztof (Kris) Iniewski from CMOS cover integer PLLs in Chapter 14. Fractional PLLs for multiband synthesis are discussed by a group from Carleton University in Chapter 15. Jan-Wim Eikenbroek from Bruco describes interesting design considerations for delta-sigma PLL used in Bluetooth applications in Chapter 16. Finally, to close the second part of the book, Sebastian Magierowski and his collaborators describe RFIC parametric converters in Chapter 17.

The third part of the book deals with devices and technologies used to fabricate wireless integrated circuits (ICs). John J. Pekarik from IBM presents a broad overview of CMOS technology for wireless ICs in Chapter 18. Calvin Plett from Carleton University describes distributed effects in RF CMOS chips in Chapter 19. In Chapter 20, a group from IMEC provides deep insight into substrate coupling effects that frequently limit chip performance. Finally, the last two chapters are devoted to microelectromechanical systems (MEMS), an emerging technology that is bound to modify the way RF integrated circuits are built. Mourad N. El-Gamal and his collaborators from McGill University describe integration of CMOS and MEMS technologies in Chapter 21, while Alexandru Müller and his collaborators focus on membrane-based MEMS systems in Chapter 22.

I would like to thank all contributors for their hard work and carving out some precious time from their busy schedules to write their valuable chapters. I would also like to thank the reviewers, editorial staff at CRC Press, and my colleagues who have reviewed portions of the manuscript. Despite some challenges in integrating the material, there were over 70 contributors altogether, putting together this book was one of the most exciting projects in my life.

Krzysztof (Kris) Iniewski
Vancouver

Editor

Krzysztof (Kris) Iniewski is a founder and president of CMOS Emerging Technologies Inc., a high-tech consulting firm in Vancouver, Canada. His interests are in advanced CMOS devices, circuits, and systems for emerging applications in medical imaging, wireless communication, and optical networking. In addition to his consulting duties, Krzysztof (Kris) Iniewski serves as an adjunct professor at the Simon Fraser University, the University of Calgary, and the University of British Columbia (UBC). He has over 20 years of technical experience in semiconductor and communication IC industry.

From 2004 to 2006, Dr Iniewski was an associate professor at the University of Alberta in Edmonton performing research on ultra-low-power CMOS circuits. From 1995 to 2003, he was with PMC-Sierra and held various technical and management positions in research and development. During his tenure with PMC, he led numerous chip development from early architectural design stage to high volume production. Prior to joining PMC-Sierra, from 1990 to 1994 he was an assistant professor at the University of Toronto's Electrical Engineering and Computer Engineering Department.

Dr Iniewski has published over 100 research papers in international journals and conferences. For number of years he served on IEEE Custom Integrated Circuit Conference (CICC) technical committee and was invited to be a special 2006 CICC issue editor for *IEEE Journal of Solid-State Circuits*. He holds 18 international patents granted in the United States, Canada, France, Germany, and Japan. In 1988, he received his PhD in electronics (honors) from the Warsaw University of Technology (Warsaw, Poland). He can be reached at iniewski@ieee.org.

Contributors

Mustafa Badaroglu
AMI Semiconductor
Oudenaarde, Belgium

Mihai Banu
MHI Consulting LLC
Murray Hill, New Jersey

Leonid Belostotski
University of Calgary
Calgary, Alberta, Canada

Raffaella Borzi
IMEC Inc.
San Jose, California

Massimo Brandolini
Broadcom Corporation
Irvine, California

Joe E. Brewer
Department of Electrical
   and Computer Engineering
University of Florida
Gainesville, Florida

Robert W. Brodersen
Berkeley Wireless Research Center
University of California
Berkeley, California

Danijela Cabric
Berkeley Wireless Research Center
University of California
Berkeley, California

Changhua Cao
Department of Electrical
   and Computer Engineering
University of Florida
Gainesville, Florida

Howard Chan
University of Calgary
Calgary, Alberta, Canada

Y. H. Chee
University of California
Berkeley, California

David Chen
University of California
Berkeley, California

P. Chevalier
STMicroelectronics
Crolles, France

Vikas Choudhary
PMC-Sierra Inc.
Vancouver, BC, Canada

Foster F. Dai
Auburn University
Auburn, Alabama

Luca de Nardis
University of California
Berkeley, California

T. O. Dickson
University of Toronto
Toronto, Ontario, Canada

Stéphane Donnay
IMEC
Leuven, Belgium

Tomas A. Dusatko
McGill University
Montreal, Quebec, Canada

Jan-Wim Eikenbroek
Bruco B.V.
Borne, The Netherlands

Mourad N. El-Gamal
McGill University
Montreal, Quebec, Canada

Simone Gambini
University of California
Berkeley, California

Paul Gorday
Motorola Labs
Plantation, Florida

Davide Guermandi
University of California
Berkeley, California

Bert Gyselinckx
IMEC Netherlands/Holst Centre
Eindhoven,
The Netherlands

James Haslett
University of Calgary
Calgary, Alberta, Canada

Seon-Ho Hwan
Department of Electrical
    and Computer Engineering
University of Florida
Gainesville, Florida

Krzysztof (Kris) Iniewski
CMOS Emerging Technologies
Vancouver, BC, Canada

George Konstantinidis
FORTH-IESL-MRG
Crete, Greece

Domine M. W. Leenaerts
NXP Semiconductors, Research
Eindhoven, The Netherlands

Zhenbiao Li
Department of Electrical
    and Computer Engineering
University of Florida
Gainesville, Florida

Jau-Jr. Lin
Department of Electrical
    and Computer Engineering
University of Florida
Gainesville, Florida

Emanuele Lopelli
Eindhoven University of Technology
Eindhoven, The Netherlands

Sebastian Magierowski
University of Calgary
Calgary, Alberta, Canada

Danilo Manstretta
University of Pavia,
Pavia, Italy

Michael Mark
University of California
Berkeley, California

Philippe Mattelaer
IMEC Netherlands/Holst Centre
Eindhoven, The Netherlands

Jesal Mechta
Department of Electrical
    and Computer Engineering
University of Florida
Gainesville, Florida

Alexandru Müller
National R&D Institute
    for Microtechnologies
Bucarest, Romania

Frederic Nabki
McGill University
Montreal, Quebec, Canada

Dan Neculoiu
University of Bucharest
Bucarest, Romania

Sean T. Nicolson
University of Toronto
Toronto, Ontario, Canada

Ali M. Niknejad
University of California
Berkeley, California

Ken K. O
Department of Electrical
    and Computer Engineering
University of Florida
Gainesville, Florida

Yorgos Palaskas
Intel Corporation
Hillsboro, Oregon

John J. Pekarik
IBM Corporation
Essex Junction, Vermont

Stefano Pellerano
Intel Corporation
Hillsboro, Oregon

Robert Plana
LAAS CNRS
Toulouse, France

Nathan Pletcher
University of California
Berkeley, California

Calvin Plett
Carleton University
Ottawa, Ontario, Canada

Vladimir Prodanov
MHI Consulting LLC
Murray Hill, New Jersey

Jan M. Rabaey
University of California
Berkeley, California

Ashoke Ravi
Intel Corporation
Hillsboro, Oregon

John W. M. Rogers
Carleton University
Ottawa, Ontario, Canada

Paolo Rossi
Marvell Semiconductor,
Pavia, Italy

Sumeet Sandhu
Intel Corporation
Santa Clara, California

B. Sautreuil
STMicroelectronics
Crolles, France

Charlotte Soens
IMEC
Leuven, Belgium

Robert Bogdan Staszewski
Texas Instruments
Dallas, Texas

Yu Su
Department of Electrical
    and Computer Engineering
University of Florida
Gainesville, Florida

Francesco Svelto
University of Pavia
Pavia, Italy

Keith W. Tang
University of Toronto
Toronto, Ontario, Canada

Geert Van der Plas
IMEC
Leuven, Belgium

Johan D. van der Tang
Holst Center/Stichting IMEC-NL
Eindhoven, The Netherlands

Arthur H. M. van Roermund
Eindhoven University of Technology
Eindhoven, The Netherlands

Sorin P. Voinigescu
University of Toronto
Toronto, Ontario, Canada

Piet Wambacq
Vrije Universiteit Brussel
    and IMEC
Leuven, Belgium

Takis Zourntos
Texas A&M University
College Station, Texas

Part I

**Circuits for Emerging Wireless:
A System Perspective**

1 RF Building Blocks for the Next-Gen Wireless Systems

Ali M. Niknejad

CONTENTS

The Evolution of Wireless Transceivers 4
    Component Count Explosion in a Multimode Radio..5
    New Wireless Standards..5
    Universal Software-Defined Radio...6
    Dynamic Radio Concept 6
    A Cognitive Radio 8
CMOS Technology Scaling 9
    Passive Devices...9
    Active Devices 11
High Dynamic Range Front-End Receiver 14
    Receiver Specifications... 14
    Wideband LNA Design 15
    Shunt Feedback LNA.. 18
    Noise and Distortion Cancellation...20
    Broadband Low $1/f$ Noise Mixers 21
Power Amplifiers for the Future 24
    PA Target Specification 25
    Multimode Power Amplifiers 25
    PA Architecture 25
    Power-Combining Challenges 26
    Dynamic Biasing Power Amplifier...30
    Universal Frequency Synthesizer 30
    Wide Tuning Range VCO 31
    Analog and Digital Baseband 32
Microwave and mm-Wave CMOS 33
    Microwave CMOS 33
    mm-Wave CMOS..35
    mm-Wave Active Elements 35
    mm-Wave Passive Elements 37
    Key mm-Wave Building Blocks 41
    mm-Wave Radio Architecture 45
        Antenna Array 45
        Transceiver Architecture..48
Conclusion..49
Acknowledgments...50
References..50

Wireless devices have become ubiquitous, from cordless to cellular phones and wireless local area networks (WLANs). And yet, this is still the tip of the iceberg. New applications such as sensor networks, 3G and 4G multistandard radios, metropolitan area wireless data networks such as worldwide interoperability for microwave access (WiMAX), personal area networks (PANs) such as ultra wideband (UWB), multiantenna radios, 60 GHz, to name a few, promise even more choices and mobility.

Every wireless device needs a transceiver to drive and extract power from the antenna to enable two-way communication. Today, most wireless devices contain one, or at best a few, transceivers. Typically each radio is optimized for a particular application, trading off data rate versus range, or battery life versus performance. There is a real challenge in moving into the future of universal wireless devices, integrating many radios on the same device, capable of universal operation.

One of the biggest bottlenecks is the radio front-end. In particular, the low-noise amplifier (LNA), mixer, analog filters, and power amplifier (PA) are usually standard and frequency specific and do not scale with technology. This chapter will explore new radio building blocks that are inherently more broadband and linear and amenable to integration with scaled digital CMOS process.

THE EVOLUTION OF WIRELESS TRANSCEIVERS

The wireless revolution has created a great demand for low-cost, mass-producible, and reliable circuits for communications. Not surprisingly, the past decade of research and development in both industry and academia has been concentrated in this area, particularly work associated with employing integrated circuits to replace bulky and expensive discrete and multichip solutions. Major advancements have been made and these efforts have focused primarily on voice communications over cellular networks and short-range WLAN data communication. The Internet revolution has created an appetite for high-speed data communication, and users worldwide demand an always-on link to the Internet. Today a wireless device needs to communicate in a heterogeneous environment, including current and next-generation cellular metropolitan area networks (MAN), medium-range high-speed WLANs, and short-range PANs. The common element in all future wireless communication systems is the requirement for spectral efficiency, low power, and multistandard interoperability.

Cellular communication networks are now ubiquitous. Unfortunately, many disparate standards and frequencies are used and thus global connectivity is hard to achieve. The global system for mobile communication (GSM) standard is almost universal in Europe, and GPRS and enhanced data for GSM evolution (EDGE) have added medium data rate capability to the standard. Code division multiple access (CDMA) described by IS-95 and US-TDMA based on IS-136 are also common in the United States and abroad. These networks are the so-called 2G networks intended primarily for voice communication. They are also circuit-switched networks and therefore make very poor utilization of bandwidth for bursty data, typical of Internet browsing. These networks have been allocated spectrum in the range of 800 MHz to 2 GHz. Unfortunately, a given standard is often implemented in many different bands in different countries and regions, requiring multiband operation in the transceiver.

Medium-range WLAN networks based on 802.11a/b/g have grown exponentially in the past few years. It is now possible to build a high-speed WLAN network in an office or residential environment using off-the-shelf inexpensive parts. Data rates vary from 1 to 108 Mb/s. Fortunately, the frequency bands for WLANs are more standard and occur mostly in the 2.4 GHz industrial science medical (ISM) bands for 802.11b/g.

The Bluetooth standard is an example of a short-range PAN network. In theory, such a network can fulfill many tasks, such as providing connectivity to mice/keyboards, headphones, personal digital assistants (PDAs), and printers. One potential issue with Bluetooth is that it also operates in the 2.4 GHz band, with the possibility of interfering with the WLAN standards. With the adoption of the multiband orthogonal frequency division multiplexing (OFDM) UWB standard, Bluetooth can also provide high data rate short-range communication and utilize less crowded frequencies between

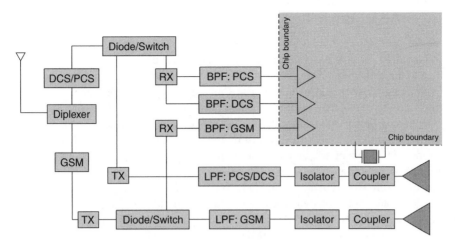

FIGURE 1.1 Typical present-day multiband transceiver off-chip components.

6 and 10 GHz. The Zigbee alliance has also created renewed interest in a reliable, cost-effective, low-power wireless network for monitoring and controlling applications based on an open standard.

COMPONENT COUNT EXPLOSION IN A MULTIMODE RADIO

Even inside a so-called "single-chip" cellular phone, there are hundreds of bulky components in addition to the core radio implemented in Si technology. Shown schematically in Figure 1.1, a simple triple-band cell phone requires band-specific filters to relax the requirements of the radio blocks on-chip. Switches are needed to change from band to band or from transmitter to receiver, and this introduces loss and complexity in the design. The PA on the transmitter drives the antenna through additional off-chip filtering. To implement power control and load isolation, directional couplers and circulators are often used. This radio is only a multiband radio operating with a single standard. Using conventional methods, it is very difficult and expensive to build a multistandard radio since all the various off-chip components must be duplicated for each new standard. Even today, there are several competing standards for cellular phones and short-range communication. To cover the important bands and standards available today globally would require a dozen or more parallel radio paths.

NEW WIRELESS STANDARDS

The explosion in the number of new wireless standards is a key motivation for the implementation of a multistandard radio. There are two major new wideband CDMA (W-CDMA) 3G cellular/data standards that require significant new radio design and architecture but in return offer high data rate communication. It is also clear that the 3G networks will coexist with the 2G and even 1G networks, therefore necessitating a more flexible radio architecture.

For higher data rates, there are proposals to use new frequency bands, such as WiMAX 802.16, and emerging standards at 24 GHz and 60 GHz. These higher frequencies require more advanced technology and new front-ends, but it is desirable again to share as much of the radio front-end with the existing 2.4/5 GHz radios. For instance, the voltage-controlled oscillator (VCO) and synthesizer can be shared [1] using frequency triplers and injection locking. Or alternatively, the core VCO can run at a much higher frequency and frequency dividers can select the appropriately divided frequency. Furthermore, if multistage down-conversion is employed, the radio can be directly integrated into an existing front-end, provided that enough bandwidth is available in the baseband. This requires a new flexible architecture for baseband filtering and the analog-to-digital converter (ADC).

UNIVERSAL SOFTWARE-DEFINED RADIO

Because of the explosion in standards and modulation schemes, many people dream of a software-defined radio (SDR), or an all digital and fully programmable radio. In such a radio, the entire spectrum up to the bandwidth of the antenna is digitized by an extremely high-resolution converter and all processing is done by a computer or a custom digital signal-processing unit.

Although this represents the ultimate in flexibility, unfortunately it cannot be implemented in today's technology for mobile applications. To receive signals up to 3 GHz, for instance, would require an ADC running at 6 GHz (the Nyquist rate) with more than 100 dB of dynamic range. This requires more than 16 bits of effective resolution in the ADC converter simply to meet the signal-to-noise ratio (SNR) requirements.

To relax the severe requirements on an SDR, some analog processing is needed on the incoming signal to limit its bandwidth and dynamic range. A universal radio with a wideband IF can perform this task. For instance, an IF of 100 MHz is a sufficiently large bandwidth to cover all major applications while remaining realizable. A 200 MHz ADC is thus required with a relaxed linearity of 10 bits. The VGA and filter are key components in the receiver as they compress the dynamic range of the ADC.

The universal radio has most of the advantages of the SDR, but it is realizable using today's technology. Unfortunately, it would consume an exorbitant amount of power and is thus not amenable to mobile battery-operated applications. For instance, if such a radio is implemented in 0.18 μm CMOS process, it is estimated to burn close to 1 W of power in the front-end continuously (40 mW/LNA, 80 mW/image-reject mixer and LO buffer, 400 mW/VGA and filter, 40 mW/VCO and synthesizer, 325 mW/ADC with 50 dB SNR).

DYNAMIC RADIO CONCEPT

Since the above specifications are only needed in a worst-case scenario, a dynamic radio can greatly reduce the power consumption requirements. A future wireless transceiver will be operating in multiple modes that require different transmit power levels and different receiver sensitivity levels. For instance, in many situations the receiver will be operating in a building or room with high-speed short-range wireless capability, such as Bluetooth or 802.11a/b/g WLAN. In such a situation, the transmit power level will average about 10 mW and the required receiver sensitivity might only be about −60 dBm (10 MHz bandwidth). On the contrary, in a cellular wireless network, due to the increased distance from the base station, the required power levels may increase to 100 mW and the required sensitivity to around −80 dBm. The same user might enter a vehicle and drive off to an isolated and distant location from the base station, thus increasing the transmit power levels to 1 W, and the required sensitivity may now be more than −100 dBm. Furthermore, the motion of the vehicle introduces new requirements due to time-varying multipath fading. Clearly, a cognitive dynamic radio should sense its environment and adapt its characteristics to save power for the mobile user.

Today a typical radio is optimized to operate with a fixed radio standard and the performance is optimized for the worst-case scenario, which embodies power levels of watts at the transmitter and maximum receiver sensitivity. This stems from the so-called "near-far" problem, shown in Figure 1.2. This scenario requires extremely good receiver sensitivity (and thus noise figure (NF)) since a distant transmission is being received. Very low phase noise and high linearity is required because a nearby interfering signal is jamming the band and thus causing phase noise from the VCO to reciprocally mix down into baseband. The interferer also compresses the gain of the LNA/mixer and thus compromises the overall sensitivity. Furthermore, since the blocker can travel through the receive chain with little attenuation, to prevent the ADC from clipping, the VGA would lower the gain and thus severely impact the overall sensitivity.

The dynamic radio architecture can result in significant power savings. For the receiver, the highest power consumption occurs in this worst-case scenario. This simultaneous requirement of

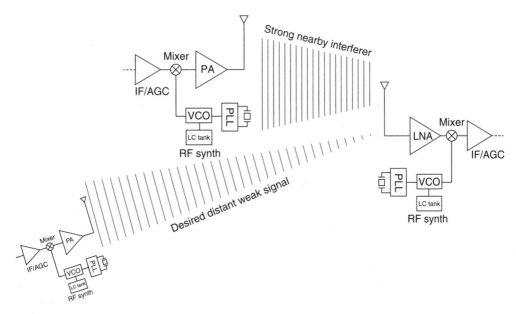

FIGURE 1.2 A strong interferer or "blocker" can jam a receiver. In the classic "near-far" scenario, a distant signal is jammed by a nearby interferer, setting the difficult requirement of high linearity and low noise.

low-noise operation and high linearity requires a large amount of supply power. However, it is important to note that this is a rare event and can be mitigated by dynamic receiver adjustments, resulting in great power savings.

Consider that a 3 dB increase in the required NF of an LNA can result in 10× power savings in the LNA. This is highly encouraging given that under normal operating conditions, the signal strength may be 20–30 dB above the required receiver reference sensitivity. Dynamic operation can be achieved by monitoring several signals in a transceiver, the bit-error-rate (BER) at the baseband, and the RF-received signal strength indicator (RSSI) with the ability to detect the presence of a jammer.* The VCO power can also be reduced if the requirements for low-phase-noise operation are relaxed. In a future wireless network with positioning capability at the base station (a 911 requirement in the United States), the base station can assist in the detection of nearby interferers and can guide the transceiver phase noise specification dynamically. Alternatively, the receiver can save power by moving to a new radio band to move away from the blocker. This is a major advantage of a dynamic and agile universal radio, since it minimizes the need for simultaneous high-linearity and low-noise operation.

Another potential power-saving measure is to replace the VGA at baseband with fixed gain and an attenuator. The attenuator is preferred over the VGA since the power consumption of an attenuator is much lower. Commercial VGAs with sufficient linearity and attenuation range consume 100 mW of power. An attenuator in CMOS as proposed by Dogan et al. [2] can consume as little as 2 mW of power while providing very high bandwidth and linearity. For instance, it is estimated that an attenuator with 40 dB peak attenuation, input third-order intercept point (IIP3) of +40 dBm, and 10 GHz bandwidth can be realized in an area of 4 mm^2 and consume only 2 mW. If such an attenuator is placed before the mixer, the linearity requirements completely fall onto the LNA.

---

*The RSSI is often a misleading indicator since a jammer can overpower a distant weak signal. When such a condition is detected, the linearity of a receiver should be maximized to process the incoming signal.

A COGNITIVE RADIO

A cognitive radio is aware of its environment and can dynamically adjust its operating frequency, bandwidth, sensitivity, linearity, and modulation scheme to best suit the environment [3]. One approach is to design a multistandard cognitive universal dynamic radio (COGUR) shown schematically in Figure 1.3. A few broadband receiver and transmitter blocks operate in parallel to cover the major frequency bands. For instance, most cellular phones occupy 1–2 GHz, and short-range data communication such as WLAN 802.11b/g and Bluetooth occur in the range of 2–3 GHz. A third broadband amplifier from 3 to 10 GHz can cover other WLAN bands such as 802.11a and new standards such as UWB. Each amplifier block is broadband and generic and can be shared. The signals are conditioned in the analog domain and mixed down to a lower intermediate frequency. To save power and improve the performance of each block, dynamic operation is employed. High power is only used in the rare circumstance when the radio is communicating with a distant base station in the presence of a strong interferer, requiring low-noise and high-linearity operation.

By eliminating much of the front-end off-chip filtering, the radios are now generic and no longer standard-specific. This simple front-end architecture should be compared with the complexity of a simple single-standard radio front-end shown in Figure 1.4. Even in a modern low-IF architecture, shown in Figure 1.5, a significant amount of band-specific signal processing occurs at RF and standard-specific analog processing occurs at baseband. In the COGUR architecture, channel selection will be done digitally in the baseband circuitry, while the analog front-end will perform image rejection and blocker attenuation through analog filtering. Without making such a receiver dynamic, the required linearity and noise performance would have to satisfy the worst-case requirement and consume an inordinate amount of power. But by dynamically adjusting the sensitivity, linearity, phase noise, filter bandwidth, and ADC resolution, significant power savings can be achieved.

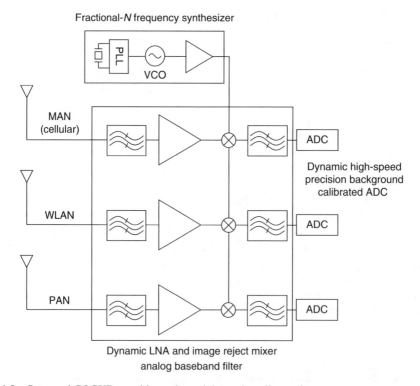

FIGURE 1.3 Proposed COGUR cognitive universal dynamic radio receiver.

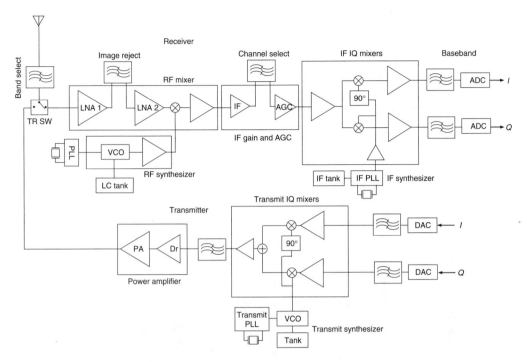

FIGURE 1.4 Classical super-heterodyne front-end architecture for a single-band transceiver.

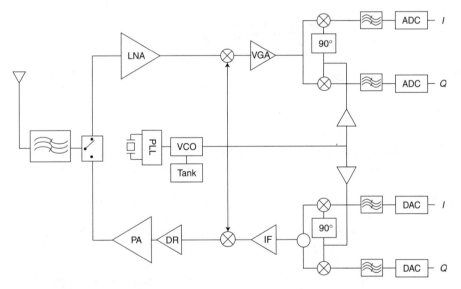

FIGURE 1.5 Modern low-IF front-end architecture.

CMOS TECHNOLOGY SCALING

PASSIVE DEVICES

The cross section of a typical modern CMOS process is shown in Figure 1.6. It is noteworthy that the back-end process (the metallization and dielectrics) has evolved substantially in the past decade, with increasing number of metal layers (5–8 is not uncommon), copper metal, and one or more

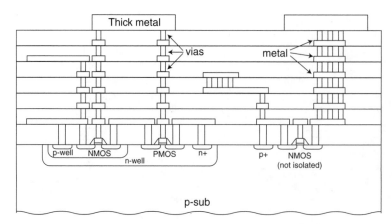

FIGURE 1.6   Cross-section of a typical modern CMOS process.

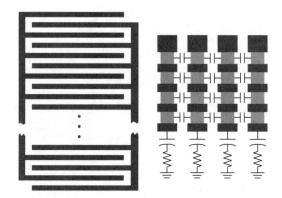

FIGURE 1.7   Top and side view of a high-density MIM capacitor constructed with native interconnect metal layers.

thick top metal layers. This has been driven by the increasingly important role of the interconnect delay in large digital circuits. This is of course a boon to analog and radio frequency (RF) circuitry since high-Q passive components are now within the reach of a digital CMOS process. It is now fairly typical to employ moderately conductive substrates with resistivity $\rho \sim 10\,\Omega\,\text{cm}$, sufficiently resistive so that magnetically induced eddy currents play an insignificant role in the quality factor of inductors and transformers. High-quality dense metal-insulator-metal (MIM) capacitors are an option in many analog flavors of CMOS technology, employing a high-K dielectric to boost the capacitance. But the availability of many dense metal layers allows the employment of finger capacitors, shown in Figure 1.7, with nearly the same density as MIM capacitors. These capacitors come without any extra processing options, but the lower density and resulting increase in chip area should be compared carefully when considering the cost reduction.

Depending on the process details, traditional spiral inductors, such as those shown in Figure 1.8a, can achieve relatively high-quality factors between 10 and 15 in the 1–5 GHz band. This is possible due to the availability of thick metals, or multiple layers strapped together, and the increased distance to the lossy substrate. Unfortunately, these inductors do not scale with process technology and occupy an increasing fraction of the die area of a modern RFIC. Given the increases in manufacturing costs in the 130 and 90 nm technology nodes, it is difficult to justify application of these inductors if other circuit techniques can be developed with competitive performance. This is in fact an important point when comparing broadband LNA topologies using inductors or distributed circuits

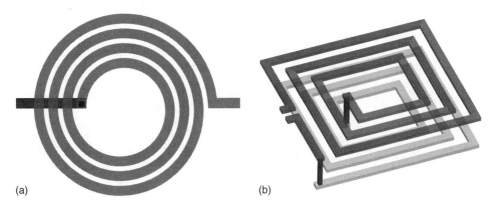

(a)   (b)

FIGURE 1.8 (a) A standard spiral inductor using thick top metal to achieve high-quality factor. (b) A multi-layer series-connected spiral inductor (the spacing between the layers has been exaggerated).

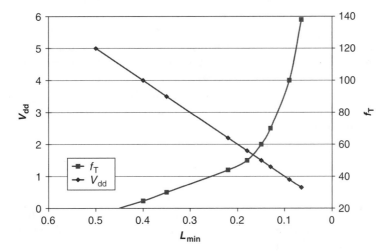

FIGURE 1.9 Unity gain frequency and maximum voltage of CMOS technology versus process node.

to simpler and compact feedback topologies. Inductor layouts that utilize the multiple metal layers to achieve higher density, as shown in Figure 1.8b, have an important advantage in this regard. In the case of perfect magnetic coupling between the spiral inductors on each metal layer, or $k \to 1$, the increase in inductance is proportional to n^2, where n is the number of spirals connected in series, resulting in substantial area reduction. Applications requiring only a modest Q inductor, such as a shunt-peaking amplifier, benefit tremendously.

ACTIVE DEVICES

With the continuous and virtually unabated improvement in CMOS technology, the RF microwave GHz regime came within the grasp of CMOS in the early 1990s [4,5] and now even the mm-wave band operation with CMOS is possible [6–8]. This trend is usually displayed by plotting the device-intrinsic gain f_T versus technology node (drawn minimum channel length L_{min}), as shown in Figure 1.9. Since analog circuits have a gain bandwidth product limited by f_T, this is a good way to judge the maximum frequency of operation in a given technology. But it is important to note that as f_T improves, the maximum supply voltage decreases due to oxide breakdown. In a short-channel transistor, drain control, due to finite junction depth, competes with gate control, reducing the performance of the device. To keep gate control in a short-channel transistor requires scaling T_{ox} to the extreme atomic limits, or the introduction of new high-K dielectrics.

This lower supply voltage has a dramatic impact on the power dissipation of analog circuits. Since dynamic range is proportional to supply voltage squared, for discrete-time analog circuits,

$$\text{DR} \propto \frac{V_{dd}^2}{kT/C} \tag{1.1}$$

To keep DR constant requires a quadratic increase in the sampling capacitor C, which results in a quadratic increase in current to operate at a given unity gain frequency

$$\omega_u \propto \frac{g_m}{C} \tag{1.2}$$

since $g_m \propto I/V_{dsat}$. This means that technology scaling is not necessarily good for analog circuits. For RF applications, the raw speed determined by f_T is an insufficient (but necessary) indicator. This is because the device-parasitic capacitances can be tuned out by inductors forming resonant circuits. Clearly, power gain cannot be obtained at an arbitrary frequency by simply choosing a sufficiently small inductance to tune out the transistor parasitics. This is because the device has additional losses not captured by f_T. These effects can be captured by power gain and the maximum frequency of activity, determined by f_{max}. In other words, beyond the frequency f_{max}, a transistor ceases to operate as an active device, or equivalently active power gain (greater than unity) cannot be obtained beyond this frequency. Since oscillators require power gain, this is also the maximum frequency of oscillation. For a bipolar device,

$$f_{max} \approx \sqrt{\frac{f_T}{8\pi r_b C_\mu}} \tag{1.3}$$

It is clear that though f_{max} improves with f_T, the improvement is more gradual and limited by device-parasitic base resistance r_b and feedback C_μ. In a metal oxide semiconductor (MOS) transistor, a better approximation to f_{max} is given by [6]

$$f_{max} \approx \frac{f_T}{2\sqrt{R_g(g_m C_{gd}/C_{gg}) + (R_g + r_{ch} + R_s)g_{ds}}} \tag{1.4}$$

This shows a proportional increase in f_{max} with improvement in f_T, which improves nearly linearly with decreasing channel length L in the velocity-saturated regime. Thus if the device parasitics can be kept under control, CMOS devices can operate at increasingly higher frequencies. Measurements on a 130 nm device show an f_{max} of about 135 GHz, when the device $f_T = 70$ GHz, dispelling the myth that f_{max} cannot exceed f_T.

Operation at 1–10 GHz in a modern CMOS technology is therefore not gain limited, allowing lower power operation. Other RF metrics include noise and linearity. For any two-port device, the spot noise frequency can be specified in terms of F_{min}, the minimum achievable NF, R_n, the noise sensitivity parameter, and Y_{opt}, the optimum source impedance

$$F = F_{min} + \frac{R_n}{G_s}|Y - Y_{opt}|^2 \tag{1.5}$$

The best NF is obtained by transforming the source impedance into Y_{opt}, whereas the optimal source impedance for gain is given by a power match Y_{in}^*. With lossless matching networks, this would result in an amplifier NF of F_{min}. By proper sizing and biasing of the transistor, the appropriate conditions can be engineered to achieve simultaneous noise and power match. The F_{min} of a CMOS transistor can be calculated using the Paspiesalski noise model [9]:

$$F_{min} = 1 + 2\left(\frac{f}{f_T}\right)\sqrt{g_m R_g \frac{\gamma}{\alpha}} \tag{1.6}$$

Here the gate-induced noise is captured through the NQS gate resistance $R_g = R_{poly} + R_{nqs}$. The device channel resistance is given by g_m/α and the excess drain noise is captured by the well-known term γ, which is 2/3 for long-channel devices and typically about 2 for short-channel devices (this is somewhat controversial [10,11]). For simplicity, assume that the device layout consists of multiple short fingers to reduce the physical external polysilicon gate resistance. Then the R_g term is dominated by the NQS resistance, $R_{nqs} = 1/5g_m$ [12], resulting in the following lower bound for F_{min}

$$F_{min} > 1 + 2\left(\frac{f}{f_T}\right)\sqrt{\frac{1}{5}\frac{\gamma}{\alpha}} \tag{1.7}$$

The term under the root is simply a technology constant, showing that the achievable noise is simply determined by the normalized frequency of operation. A plot of NF_{min} versus f_T at 2, 5, and 10 GHz is shown in Figure 1.10. Three current CMOS technology nodes are annotated, showing that sub-dB low-noise operation is possible from an intrinsic CMOS transistor in current 90 nm technology at 10 GHz. At this noise level, the dominant factors contributing to NF will be external parasitics, pads, electrostatic discharge (ESD), and matching network component Q, not the transistor. This also opens the door to circuit architectures that do not require explicit noise matching.

Finally, the linearity of CMOS devices is also an important metric for RF amplifiers. In RF amplifiers, the DR range is determined by the noise floor and the maximum signal amplitude. Typically, this amplitude is lower than V_{dd} because the distortion products must be sufficiently small by a

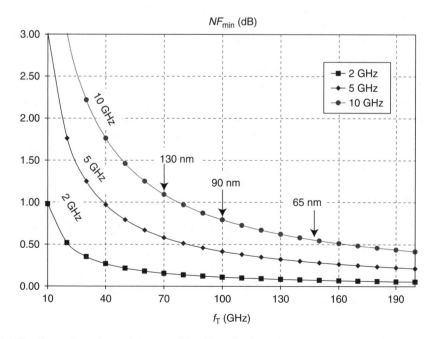

FIGURE 1.10   Lower bound on minimum achievable noise figure NF_{min} versus process f_T.

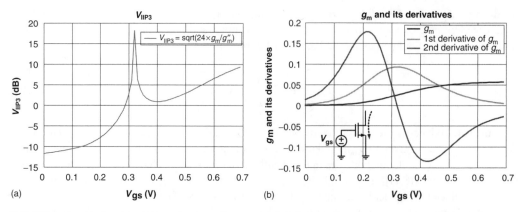

FIGURE 1.11 (a) Simulated V_{IIP3} of a short-channel CMOS device. Note the characteristic "sweet spot" V_{GS} bias point. (b) The simulated derivatives, g_{m}, g'_{m}, and g''_{m}, of the drain current versus V_{GS}.

factor of the required signal to noise and distortion ratio (SNDR). Since operation at an input power IIP3 theoretically implies equal signal and distortion powers,* amplifiers must back off from this power level (IM3 improves 2 dB for each 1 dB of back-off).

A plot of the low-frequency V_{IIP3} versus gate bias for a MOS device is shown in Figure 1.11. This curve is calculated from the derivatives of the *I–V* relation

$$V_{\text{IIP3}} = \sqrt{\frac{24 g_{\text{m}}}{g''_{\text{m}}}} \tag{1.8}$$

For small gate voltages, the device is in subthreshold and the IIP3 is determined by the exponential current–voltage relationship. In strong inversion, for a square law FET, $g''_{\text{m}} \approx 0$, which results in infinite IIP3. But due to velocity saturation and other nonideal effects, a real FET has finite IIP3 for large gate bias voltages. In the transition between weak and strong inversion, there is an interesting "sweet spot" where the device behaves very linearly. This stems from the opposite sign of g'_{m} in weak and strong inversion versus V_{g}, the g_{m} increasing initially with V_{g} but then decreasing. This sweet spot is also a function of technology scaling, and the ability to bias near the peak IIP3 performance is increasingly difficult owing to the narrowness in the curve and increasing process variation. Nevertheless, clever multitransistor circuits operating in different bias regions have been devised to exploit this feature [13]. By using two devices in parallel with different operating points, one in strong inversion and one in weak inversion, it is in theory possible to have a net G_{m} that has zero g''_{m} due to cancellation since the sign of g_{m} is different in the regions.

HIGH DYNAMIC RANGE FRONT-END RECEIVER

RECEIVER SPECIFICATIONS

The COGUR receiver front-end is shown in Figure 1.12. To meet the stringent specifications of a cellular system, the LNA will be designed with 15 dB of gain, 3 dB of NF, and an IIP3 of about 0 dBm. In practice the LNA is often preceded by surface acoustic wave (SAW) filters and switches, and thus the LNA NF should be about 2 dB to accommodate this extra loss at the input of the receiver. In the future, microelectromechanical system (MEMS) technology will replace many of the bulky external filters allowing more margin for the LNA NF.

---

*In practice, other distortion products begin to dominate at higher power levels, so the IIP3 is an extrapolated point.

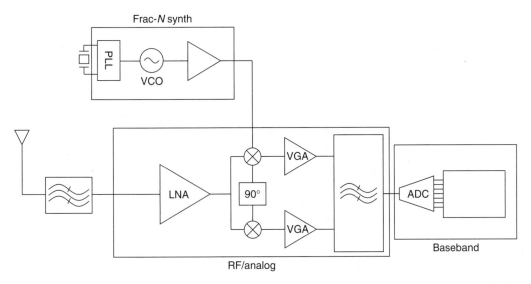

FIGURE 1.12   COGUR receiver architecture.

The image-reject mixer will have a gain of 10 dB, an NF of 12 dB, and an IIP3 of 10 dBm with 30 dB of image rejection. To enable direction conversion, the $1/f$ noise of the mixer needs to be very low, in the sub-50 kHz range, and IIP2 of 70 dBm. While achieving such a high IIP2 directly is difficult, a low-frequency calibration loop can greatly enhance the IIP2. The baseband analog filter needs to provide 40 dB of adjacent channel attenuation (fifth-order filter). The VGA should have a gain range of 10–70 dB, an NF of 5 dB, and an IIP3 of −10 dBm in high gain mode and +10 dBm in low-gain mode.

To accommodate multiple standards, the frequency synthesizer should have a channel spacing of 2.5 kHz, a phase noise better than −116 dBc/Hz at 600 kHz offset, and a settling time better than 150 μs. The phase noise specifications are set by blocker reciprocal mixing. The ADC needs about 10 bits of resolution, a 200-MHz sampling rate. Background digital calibration can save power and simplify the design. These specifications are derived by a combination of system-level analysis and simulation, with constraints based on power consumption and technology limitations based on simulations and previous research results.

WIDEBAND LNA DESIGN

The broadband CMOS LNA, shown in Figure 1.13, was proposed for UWB 3–10 GHz applications. The topology is very similar to a standard inductively degenerated LNA, utilizing L_s series feedback to obtain a real component in the input impedance of the amplifier. Neglecting C_{gd},

$$Z_{in} = j\omega L_g + j\frac{1}{\omega C_{gs}} + \omega_T L_s \tag{1.9}$$

In a narrowband tuned LNA, the reactive components of the input impedance are resonated out with an extra gate inductance L_g and AC coupling capacitor, to obtain a match at a given frequency. The voltage gain can be calculated by noting that the voltage v_{gs} is the voltage across a series resonant capacitor, which means that it is Q times as large as the voltage across the source resistor. For an input match, $v_{R_s} = \frac{1}{2}v_s$, so the voltage gain at resonance is given by

$$v_o = -g_m R_L v_{gs} = -g_m R_L Q \times \frac{v_s}{2} \tag{1.10}$$

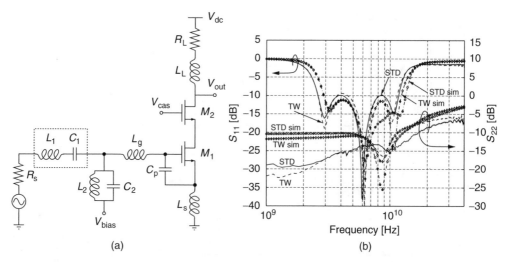

FIGURE 1.13   (a) Broadband LNA schematic. The input network can be simplified by eliminating the boxed series *LC* circuit at the cost of reduced bandwidth. (b) Measured and simulated input and output match of amplifier. (From Bevilacqua, A. and Niknejad, A. M., *IEEE International Solid-State Circuits Conference Digest of Technical Papers*, February 2004. © 2004 IEEE.)

or

$$A_v = -\frac{1}{2} g_m Z_L Q \tag{1.11}$$

The bandwidth of matching stage of the inductively degenerated amplifier is usually set by the Q factor of the input. Since the source impedance is fixed, there is little freedom in controlling the Q factor of the input stage. In most designs, the Q is fairly low and thus the input stage is relatively wideband. But many applications require larger bandwidth. A good example is a UWB amplifier that needs to cover the 3–10 GHz band.

The key observation in designing a wideband match is that the input impedance in fact appears as a series *RLC* circuit, and this resonator can be embedded as a filter section. In the particular case shown in Figure 1.13a, a third-order Chebyshev filter section is employed. The calculation of the amplifier gain is nearly identical to the simple inductively degenerated amplifier. Note that the amplifier input impedance is now given by

$$Z_{in} = \frac{R_S}{W(s)} \tag{1.12}$$

The function $W(s)$ is the filter transfer function. In-band, is $|W(s)| \approx 1$, whereas out of band $|W(s)| \approx 0$. The truth of these statements depends on the in-band ripple and the steepness of the filter skirt. Thus the input current generates an output voltage

$$\frac{v_o}{v_{in}} = \frac{v_o}{i_{in}} \times \frac{i_{in}}{v_{in}} = \frac{g_m}{sC_{gs}} Z_L(s) \times \frac{1}{R_S W(s)} \tag{1.13}$$

The shunt-peaking load compensates for the current gain roll-off. The above amplifier can be optimized to generate optimal power gain and noise performance by proper design. A prototype UWB LNA, fabricated in a 180 nm process, has been characterized [14]. The measured gain and NF are shown in Figure 1.14. The layout, shown in Figure 1.15, is dominated by the area of the spiral inductors.

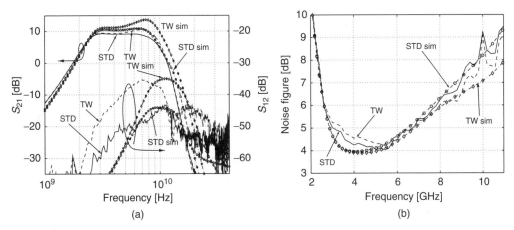

FIGURE 1.14   (a) Measured and simulated power gain of broadband UWB LNA. (b) Measured and simulated NF of LNA. (From Bevilacqua, A. and Niknejad, A. M., *IEEE International Solid-State Circuits Conference Digest of Technical Papers*, February 2004. © 2004 IEEE.)

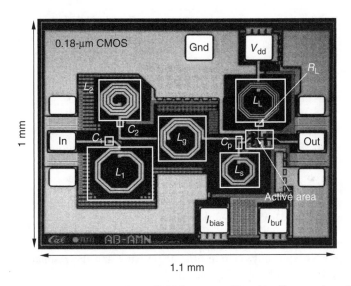

FIGURE 1.15   Layout of UWB LNA in 180 nm CMOS process. (From Bevilacqua, A. and Niknejad, A. M., *IEEE International Solid-State Circuits Conference Digest of Technical Papers*, February 2004. © 2004 IEEE.)

This architecture can also be used in a broadband multistandard LNA. By eliminating the series inductors on the input, the attenuation due to the filter is reduced and thus the NF improves. The simulated power gain and NF are shown in Figure 1.16a. Although the gain flatness is sacrificed, it is important to note that a multistandard LNA still operates with predominantly narrowband signals, and thus there is minimal distortion due to the gain variation. The dynamic operation of the LNA is shown in Figure 1.16b, as the input matching and NF are shown as a function of bias current. Wideband dynamic operation is achieved with a modest power consumption. It is important to note that the NF increases by only 3 dB when operating at a bias current 10× lower than the nominal value.

The techniques employed to generate a broadband input match can also be applied to a concurrent LNA design [15]. By designing a filter to provide the impedance match at two or more distinct frequencies, rather than in a broad range of frequencies, the amplifier can concurrently amplify multiple bands. Contrast a concurrent LNA with a tuned LNA with switchable resonant frequency.

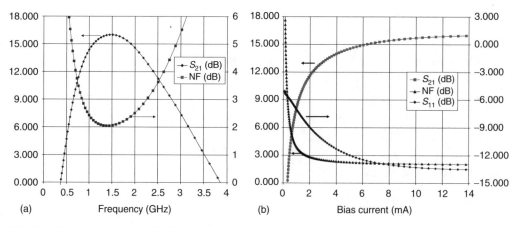

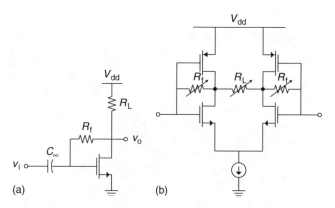

FIGURE 1.16 (a) Power gain of LNA (S_{21} dB), NF, and input match of LNA running at 15 mA demonstrate broadband performance. (b) Simulated power gain and NF of LNA as a function of dynamic bias.

FIGURE 1.17 (a) A classic shunt feedback CMOS amplifier. (b) A fully differential CMOS inverter shunt feedback amplifier with variable load and feedback resistance.

In such a case, the LNA can only operate in a single band and the resonant frequency must be switched by using varactors or by shorting inductor turns.

SHUNT FEEDBACK LNA

The shunt feedback amplifier topol ogy shown in Figure 1.17a is a popular approach for wideband impedance matching. It is not widely used in CMOS RF amplifiers due to its intrinsically higher NF (compared to an inductively degenerated amplifier) and the low gain due to feedback. But as transistors scale and become increasingly faster, it is now viable to obtain reasonable noise performance with very wideband capability. Working with the simplified Y matrix of the transistor,

$$Y_{\text{fet}} = \begin{bmatrix} j\omega C_{gs} & 0 \\ g_m & G_o + j\omega C_{ds} \end{bmatrix} \tag{1.14}$$

The feedback element has a Y matrix

$$Y_f = G_f \begin{bmatrix} +1 & -1 \\ -1 & +1 \end{bmatrix} \tag{1.15}$$

and so the overall amplifier Y matrix is given by

$$Y = \begin{bmatrix} G_f + j\omega C_{gs} & -G_f \\ g_m - G_f & G_f + G_o + j\omega C_{ds} \end{bmatrix} \tag{1.16}$$

The stability factor for the shunt feedback amplifier is given by

$$K = \frac{2G_f(G_o + G_f) - G_f(G_f - g_m)}{G_f |g_m - G_f|} = \frac{g_m + G_f}{g_m - G_f} = \frac{g_m R_f + 1}{g_m R_f - 1} > 1 \tag{1.17}$$

with the assumption that $g_m R_f > 1$. The choice of R_f and g_m is governed by the current consumption, power gain, and impedance matching. For a biconjugate match

$$G_{max} = \left| \frac{Y_{21}}{Y_{12}} \right| \left(K - \sqrt{K^2 - 1} \right) \tag{1.18}$$

$$= \frac{g_m - G_f}{G_f} \left(\left(\frac{g_m R_f + 1}{g_m R_f - 1} \right) - \sqrt{\left(\frac{g_m R_f + 1}{g_m R_f - 1} \right)^2 - 1} \right) = \left(1 - \sqrt{g_m R_F} \right)^2 \tag{1.19}$$

The input admittance is calculated as follows:

$$Y_{in} = Y_{11} - \frac{Y_{12} Y_{21}}{Y_{22} + Y_L} \tag{1.20}$$

$$= j\omega C_{gs} + G_f + \frac{G_f(g_m - G_f)(G_o + G_f + G_L - j\omega C_{ds})}{(G_o + G_f + G_L)^2 + \omega^2 C_{ds}^2} \tag{1.21}$$

At lower frequencies, $\omega < 1/(C_{ds} R_f \| R_L)$ and (neglecting G_o)

$$\Re(Y_{in}) = G_f + \frac{G_f(g_m - G_f)}{G_f + G_L} = \frac{1 + g_m R_L}{R_F + R_L} \tag{1.22}$$

$$\Im(Y_{in}) = \omega \left(C_{gs} - \frac{C_{ds}}{1 + (R_f/R_L)} \right) \tag{1.23}$$

Thus, a resistive source provides the optimum termination as long as $\omega C_{gs} \ll 1/R_S$. Relating this to the device ω_T,

$$\frac{\omega}{\omega_T} \ll \frac{1}{g_m R_S} \tag{1.24}$$

or within a multiplicative factor

$$\omega \ll A_v^{-1} \omega_T \tag{1.25}$$

For a conservative LNA voltage gain of 20 dB and a 100 GHz device f_T, this implies operating below 10 GHz to satisfy this condition, or precisely in the frequency range of most wireless transceivers. Thus the shunt feedback amplifier is a promising topology for scaled CMOS devices. Because of the absence of any inductors, the layout is compact and compatible with a process that offers low-parasitic resistors. In a pure digital process, the poly-gate material can be used to create this resistor. A lower bound for the NF of this topology is easily calculated

$$F > 1 + \frac{\gamma}{\alpha} \frac{1}{g_m R_S} \qquad (1.26)$$

For instance, if $g_m R_f \gg 1$, then to obtain an input match from Equation 1.22

$$G_S \approx \frac{g_m}{1 + (G_L/G_f)} \qquad (1.27)$$

Substituting this condition into the NF equation

$$F > 1 + \frac{\gamma}{\alpha} \frac{1}{1 + (G_L/G_f)} \qquad (1.28)$$

Assuming a short-channel value of $\gamma/\alpha \sim 2$, with $G_L = 2G_f$, the result is an NF of about 2.2 dB. An example design incorporating a shunt feedback topology is shown in Figure 1.17b. A differential NMOS and PMOS CMOS pair are used to realize the g_m to save power. The bias point is established with a current source, although the current source can be removed since the shunt feedback CMOS inverters will self-bias to the trip point of the amplifier ($V_{in} = V_{out}$). In this implementation, the feedback and load resistors are implemented by a switched resistor to vary the gain. Both resistors are varied together to maintain the input match. Using 3 bits for the feedback resistor and 2 bits for the load resistor, a gain variation of 13:1 is achieved while the input match is always better than 15 dB up to 6 GHz. The simulated prototype has a peak $S_{21} = 13$ dB, below 2 dB NF up to 6 GHz, and an $IIP_3 = +5$ dBm in low-gain mode.

Other promising feedback topologies include the voltage feedback amplifier discussed in Ref. 16. This topology has the nice property of automatically presenting the output-tuned load to the amplifier input, thus allowing one to tune the output load and still maintain an input match. This topology will be covered in a separate chapter as a candidate architecture for multistandard applications.

NOISE AND DISTORTION CANCELLATION

Even though it is possible to design relatively low-noise shunt feedback amplifiers, even better performance can be obtained by using the noise cancellation approach [17]. The Bruccoleri LNA, shown in Figure 1.18, is a two-parallel-path amplifier. The first path is a shunt feedback stage optimized to obtain an input match but at the expense of a poor NF. The second stage is designed to cancel the noise of the primary stage. To see that this is possible, simply observe that the dominant channel noise of a shunt feedback amplifier leaks to the input through R_f to produce an input noise in phase with the output noise. This is in contrast to the signal that experiences a phase inversion when passing from input to output. This is exploited by gaining this noise and subtracting it at the output-summing node of the amplifier.

The same idea can be utilized in a common-gate LNA. As shown in Figure 1.19, the noise cancellation is achieved in a similar vein as the shunt feedback amplifier. The common-gate amplifier signal experiences no inversion, whereas the noise at the input is antiphase with the output noise. A contour of constant NF as a function of the device biasing is shown in Figure 1.19b. It is interesting to note that the optimum point does not occur for perfect noise cancellation, but only partial noise cancellation.

The same observation holds for the distortion generated by the primary device. This distortion will leak back into the input with equal phase, and thus the auxiliary amplifier can also cancel this distortion. In practice it is difficult to achieve distortion cancellation because the nonlinearity of

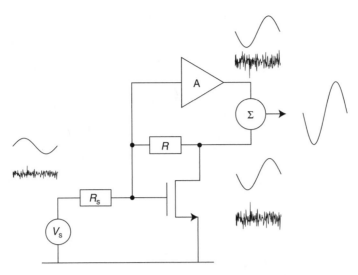

FIGURE 1.18 The Bruccoleri noise-canceling LNA.

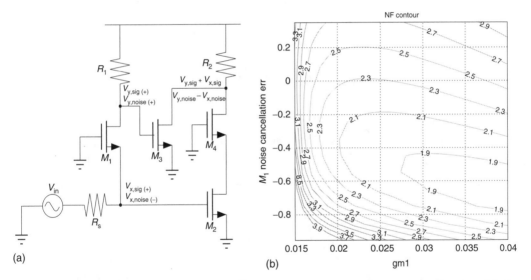

FIGURE 1.19 (a) A common-gate noise-canceling LNA. (b) Contours of constant NF as a function of device parameters.

the auxiliary amplifier can dominate the distortion. But by incorporating an ultra linear auxiliary amplifier, both noise and distortion can be canceled.

Broadband Low 1/f Noise Mixers

Passive ring mixers employing MOS switches, shown in Figure 1.20a, are a viable candidate for a multimode broadband front-end [18]. The advantage is the inherent high dynamic range and broadband operation. The disadvantage is the required LO drive signal and the conversion loss. A broadband transformer can be used to drive the mixers. The conversion loss can be overcome by preceding the mixer with sufficient gain to overcome the noise and by increasing the gain at IF or baseband. By codesigning the LNA and mixer, a careful choice can be made to allow optimal performance over a wideband. Another highly linear topology is shown in Figure 1.20b [19]. Here,

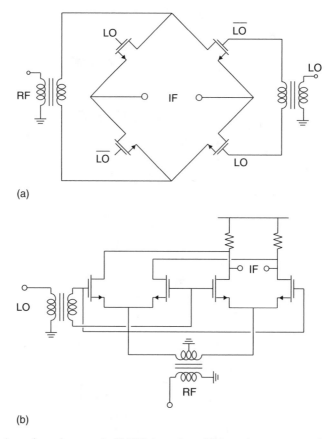

FIGURE 1.20   (a) A passive-voltage-mode CMOS ring mixer. (b) A passive-current-mode CMOS ring mixer.

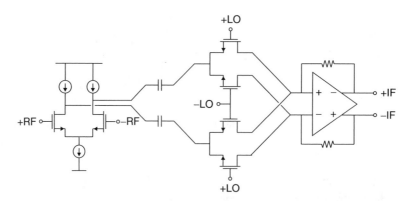

FIGURE 1.21   Broadband highly linear low/zero-IF CMOS mixer.

the input transconductor is replaced by a broadband transformer, thus improving the linearity of a common Gilbert cell mixer. In practice, the transconductance gain lost must be absorbed into the driver stage, and thus this architecture is favored when the LNA is single ended.

Current-mode passive mixers [20] are another promising alternative. A schematic of the mixer is shown in Figure 1.21. Here, a differential CMOS inverter style G_m stage generates the currents, which are switched by a passive Gilbert switching quad. By running the switching quad in zero DC current mode, the $1/f$ noise of the mixer is reduced dramatically, ideal for direct conversion or low-IF applications.

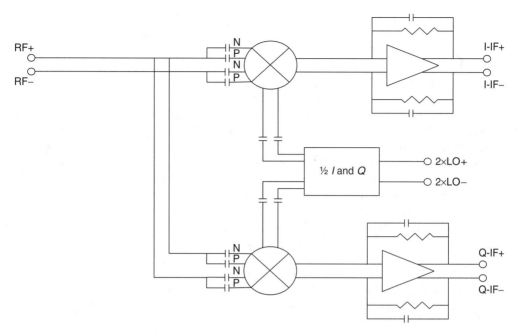

FIGURE 1.22 Block diagram of mixer chip fabricated in 130 nm CMOS.

Furthermore, by driving the output into a virtual ground, the voltage swing at the output of the mixer is minimized, thus allowing the linearity to be set by the G_m stage, rather than by the voltage headroom. This requires a low-noise, low-current, and highly linear op-amp, a challenging task on its own. But due to the low frequency of operation, traditional analog circuit design techniques can be applied to improve the op-amp. In particular, the slewing rate of the amplifier is important for achieving good performance. The LO path of such a mixer is likely to be the dominant source of power consumption.

A test chip based on the current commutating passive mixer including the LO path has been fabricated in a 130 nm process [21]. The block diagram of the chip is shown in Figure 1.22. Measurement results show that from 700 MHz to 2.5 GHz, the demodulator achieves 10 dB DSB NF with 9–33 kHz $1/f$ noise corner and 35 dB of voltage gain. The total chip draws 20–24 mA from a single 1.5 V supply.

A two-stage topology was chosen for the op-amp design to obtain both high-signal swing and low input-referred noise. The input-referred noise of the operational amplifier is given by [22]

$$\overline{v_{\text{in}}^2} = 8kT\gamma g_{\text{mn}}\left(1 + \frac{g_{\text{mn}}}{g_{\text{mp}}}\right) + \frac{2}{fC_{\text{ox}}}\left(\frac{K_{\text{N}}}{(WL)_{\text{N}}} + \frac{K_{\text{P}}}{(WL)_{\text{P}}}\frac{g_{\text{mn}}^2}{g_{\text{mp}}^2}\right)$$

(1.29)

To reduce op-amp noise contribution, the input NMOS transistors were sized to have a high W/L ratio with a long-channel length, whereas the PMOS had a low W/L ratio with a long-channel length. The output stage of the amplifier is simply a common-source stage and provides almost rail-trail output swing. The op-amp is designed to be able to handle the currents swing from the transconductance stage for the linear input range. For instance, the slew rate of the op-amp must be high enough to handle the current changes of

$$\frac{di_{\text{out}}}{dt} = \frac{2}{\pi}\omega_{\text{IF}}(g_{\text{mn}} + g_{\text{mp}})v_{\text{RF}}$$

(1.30)

The feedback resistors were chosen to be large to reduce the associated thermal current noise. The upper limit of the resistor value was set by linearity and voltage gain of the circuits. The feedback capacitors are large to attenuate the out-of-band blockers [23]. Although using big feedback capacitors creates a low-frequency gain roll-off at the IF output, this can be characterized and corrected in later stages as long as the NF is low and the gain is high enough for the IF frequency of interests. In practice, available chip area and gain of the circuit determine the upper limit of the capacitor value. The op-amps draw a total of 3.5 mA from the supply (both I and Q).

The LO generation path of the mixer uses two inverters in parallel that act as the input buffer to restore the high-frequency waveform distorted by the packaged pin, bond wire, and pad parasitics. A symmetric LO waveform is critical in ensuring the balanced switch operation so that the switching quad itself does not pose a substantial degradation to the mixer NF as well as the second-order intermodulation product [24,25]. A divide-by-two frequency scheme is employed to produce 50% LO duty cycle so as to minimize LO asymmetries. The LO frequency ranges from 0.7 to 2.5 GHz, while the divider operates at twice this frequency. This translates into higher power consumption and the need for a larger balun bandwidth.

The divide-by-two was implemented in current mode logic (CML) style. The core of the divider block consists of two CML flip-flops with the output cross-toggled back to its input. The CML divider draws constant current and has the advantage of generating less current spikes during its dynamic operation, which may propagate and appear as noise to other sensitive RF nodes. Because differential signaling is utilized in the CML divider, both I and Q LO outputs with good matching are available. Larger LO swing expedites the switch-quad transition and helps improve the mixer NF and second-order intermodulation product [24,25]. The mixer core design requires the LO differential swing of at least 1.5 Vpk-pk from a supply of 1.5 V. Two scaled inverters were cascaded in each path to provide sufficient drive capability. The driving inverters consume 4.1 mA and the CML circuits consume 1.43 mA (both I and Q).

POWER AMPLIFIERS FOR THE FUTURE

Although PA design today is dominated by the quest for high efficiency and higher power levels, the next generation of PAs deal with a multitude of issues. Foremost is the reduction in the distortion of the PA. Spectral regrowth leaks power into adjacent channels, necessitating wider interchannel spacing and thus reducing efficiency. Therefore, traditional high-efficiency class C or E PAs are no longer sufficient and linearization techniques, such as Cartesian feedback or polar modulation, must be employed.

Nonconstant envelope modulation schemes, such as quadrature amplitude modulation (QAM), achieve better spectral efficiency and hence are employed for high data rate wireless systems. Such systems require linear power amplification to preserve the integrity of transmitted data. The required linearity will be on the order of third-order intermodulation IM3 of −30 dBc to −50 dBc, depending on the standard (data rate and error vector magnitude [EVM] requirements). Such levels of linearity preclude class C and E architectures. Class A and B PAs, on the contrary, are capable of achieving such levels of linearity at the cost of greatly reduced efficiency levels. All wireless data systems employing OFDM (multitone) also require linear PAs due to the multicarrier envelope-varying signal. Digital and analog predistortion are strong candidates to enable operation at less power back-off while maintaining sufficient linearity.

The average power levels of future mobile standards are likely to drop significantly. Power levels will drop in crowded urban areas as the cell size decreases and also to reduce blocking and potential interference to nearby users. This will certainly relax the requirements on the PA design. But the requirement to regulate the output power level adds new complexities to the PA design. Closed feedback loops from the base station to the mobile unit regulate the power level of the transmitter to minimize interference and to maximize battery life. As previously mentioned, most PAs are designed to achieve peak efficiency only at peak power. Thus, if special measures are not taken in the design, the overall average efficiency of a PA will drop by an order of magnitude.

The ability to regulate the output power of the PA also greatly aids the typical near/far communication barrier shown in Figure 1.2. In many situations, the linearity and noise requirements of a front-end receiver are driven by the ability to reject a nearby jamming signal. In a cognitive radio, there exists the possibility of intercommunication between the jamming signal and the receiver to come to a compromise in the power levels transmitted. This of course requires that the transmitter can gracefully back off from its peak power.

Integration poses another challenge for the future PA. The PA will probably reside on the same die as several other key components of the wireless transceiver, such as a sensitive LNA, mixers, VCOs, and noisy digital blocks. Additionally, each component will consist of physically large inductors, capacitors, and transformers. These structures inject and receive signals into the oxide and substrate, transmitting signals from every block to every other block on the substrate. The signals propagate through the conductive substrates through electrical (conductive and displacement currents), magnetic (eddy currents), thermal, and electromagnetic mechanisms (radiation). Other sources of coupling are the large pads, ESD diodes, bond wires, and the package leads. In such an environment, every possible measure must be taken to minimize coupling, and differential operation is a natural choice. Reduced power levels of the PA are also critical in enabling such levels of integration.

PA Target Specification

The PA should ideally work from 600 MHz to 11 GHz to cover the many frequency bands of interest. In practice such a wideband PA would take the form of a distributed amplifier, but the overall efficiency would be too low for mobile applications. Instead, several parallel wideband PAs are needed. For 2G cellular bands, as much as 2 W of output power is needed to accommodate board losses, couplers, and other components between the PA and the antenna. Future wireless standards require lower power (below 1 W) but more linearity and power control.

Multimode Power Amplifiers

Currently, the wireless market is dominated by a few mobile standards: GSM and CDMA standards. GSM, which uses Gaussian minimum shift keying (GMSK), modulation, and CDMA, which uses variations of quadrature phase shift keying (QPSK), present virtually opposite requirements in PA specifications. While GMSK calls for high-efficiency, nonlinear PAs (such as class AB, E, or F), the linearity requirements of filtered QPSK demand linear PA operation (such as class A or B). With the convergence of wireless standards, the need for a multiple-mode PA, capable of switching between high-efficiency and high-linearity modes, becomes apparent.

Multimode PAs are capable of switching between high-efficiency class operation to allow backwards compatibility and linear class A/AB operation for 3G/WLAN wireless systems. In the simplest scheme, two PAs would be run in parallel and switch between the linear and the nonlinear PA depending on modulation scheme. At any given time, only one PA would be delivering power to the antenna. The difficulty in this implementation is that in "off" mode, a PA will contribute parasitics that must be accounted for in the PA that is in operation, unless bulky isolators or hybrids (or a quarter-wavelength line) are employed.

A dual-mode single amplifier solves this problem [26]. For instance, since class E/F and A/B amplifiers are similar in topology and differ primarily in the output resonator networks and bias, the core device can be shared between amplifiers. Techniques could be explored that would dynamically change the bias and output network to switch between a linear and a nonlinear PA.

PA Architecture

For CMOS PAs, a multistage differential building block serves as the core of the amplifier. Operating in class A naturally offers low distortion. Differential operation also minimizes the effects of package and substrate parasitics on the PA performance as well as elimination of even-order distortion. The interstage-matching network, for instance, can be made largely independent of bond wire and package inductance. Differential operation also offers enhanced gain, decreasing the number of stages and hence delay in the PA. Excessive delay can cause stability problems as output signals leak back into the input of the amplifier. The lower power level also means that the optimum load is closer to the 50 Ω load, which relaxes the requirements on the matching network, allowing a low-Q broadband match at lower insertion loss.

Differential operation requires a way to convert the differential output signal to a single-ended signal for compatibility with current transceivers (filters) and antennas. A discrete lumped LC–CL circuit could be employed to perform the conversion [27]. High-Q components are necessary to minimize the power loss. Other options are to design a board-level device such as a printed spiral transformer or coupler [28]. The main constraints preventing the integration of such a device on-chip are the losses of the balun and the need to supply high direct current. A possible solution is to realize the device within the package itself, using the leadframe to design a coupler. Such a device would need to withstand manufacturing tolerances. Another approach is to employ a transformer-based distributed power-combining scheme while simultaneously performing differential to single-ended conversion [29]. Transformers allow operation of the CMOS transistors at a low voltage on the primary while the secondary voltages are connected in series to drive the antenna. The efficiency of on-chip power combining is between 70 and 80% even when integrated into a CMOS process.

Another important issue is driving the output stage. Since the power gain at the last stage is low, it is important to drive this stage as efficiently as possible. The input impedance of the last stage requires careful modeling of MOS transistor parasitics such as gate resistance (distributed channel, poly, and interconnect).

POWER-COMBINING CHALLENGES

On-chip power combining with a transformer, shown schematically in Figure 1.23, is an effective solution for CMOS technology. Transformer power combining has been used successfully on-chip for interstage power and matching [27] and for output impedance matching and distributed active transformer (DAT) power combining [29]. If a simple architecture without the virtual grounds of the DAT is employed, shown in more detail in Figure 1.24, each stage can be driven independently and thus back off from peak power. Each stage, shown as a pseudodifferential CMOS stage, is in reality a differential cascode stage. A MIM cap at each driver tunes the input impedance to resonance for optimal power transfer. Each sub-PA drives the primary of a transformer with a low-supply voltage, circumventing problems with breakdown. The magnetic energy of several of such PAs is captured by a large secondary loop and delivered to the load. Note that the windings are in parallel at the primary, operating on a low-supply voltage. On the secondary, though, the windings are in series and add up to a large voltage to drive the off-chip antenna.

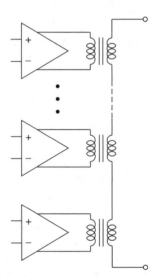

FIGURE 1.23 Transformer power combiner employing core 1:1 sections. For N stages, the impedance is transformed down by N for each driver.

To see that on-chip power combining is possible, and indeed efficient, consider the insertion loss of a simple 1:1 transformer. Since the insertion loss of a transformer highly depends on the source and load impedance, it is convenient to find the minimum insertion loss under ideal conditions, a biconjugate source and load match for optimal power transfer. Under such conditions, we know that for a passive device the gain of the transformer peaks at

$$G_{\max} = K_s - \sqrt{K_s^2 - 1}$$ (1.31)

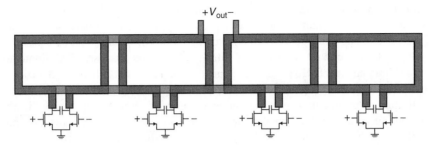

FIGURE 1.24 Simplified layout of transformer-power-combining circuit. A large secondary winding collects the power and delivers it to the antenna.

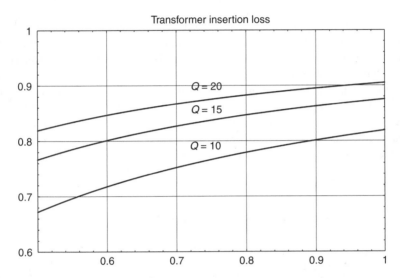

FIGURE 1.25 The maximum gain of a 1:1 transformer as a function of magnetic coupling factor K and winding quality factor Q.

The gain is thus related to the stability factor K_s. Ideally if $K_s = 1$, the insertion loss is 0 dB.*

One particularly simple case of interest is a 1:1 transformer where each winding has a resistance R_x. The maximum gain can be parameterized by the unitless $Q = \omega L / R_x$ factor and the transformer coupling factor $K = M/L_x$

$$G_{\max}(Q,K) = 1 + \frac{2}{Q^2 K^2} - 2\sqrt{\frac{1}{Q^4 K^4} + \frac{1}{Q^2 K^2}} \qquad (1.32)$$

A plot of $G_{\max}$ is shown in Figure 1.25, which shows that reasonably low insertion loss can be obtained with relatively modest Q factor and coupling factor K.[†] For instance, a $Q = 10$–15 is easily achieved in the 1–5 GHz frequency range. On-chip planar transformers have $K \approx 0.75$, which implies an insertion loss of about 0.8, or about 1 dB loss. Although the above result is derived for a simple 1:1 transformer, the result can be generalized if it is noticed simply that N windings driven in shunt on the primary and driven in series on the secondary will transform the impedance by a factor of N (not N^2

---

* Not to be confused with the transformer coupling factor K.

† The efficiency also depends on the magnetic coupling factor K since the amount of magnetic energy stored in the "leakage" inductance of a transformer depends on K. As $K \to 1$, very little magnetic energy is stored and consequently little is lost to the finite Q of the inductor.

since the currents in primary and secondary are equal), as shown in Figure 1.23. Furthermore, the insertion loss is independent of the number of sections. This is a powerful result because for LC matching networks, there is a strong dependence on the matching ratio. This is also true for transmission line quarter-wave matching sections. To test these simple calculations, extensive EM simulation using Agilent Momentum and Ansoft HFSS has been carried out. The results, though, confirm the simple analysis showing that an on-chip power combiner with an insertion loss of 1 dB is feasible.

The realization of an on-chip power combining and impedance-matching network is a powerful incentive to build such devices, but there is a further and equally important advantage in realizing the amplifier in this way. Since each driver sub-PA acts independently, these individual PAs can be turned off to lower the output power. This is a big advantage of this simple architecture, since the high efficiency of the overall PA becomes in essence independent of the output power level. In practice, each PA is turned off by forming a resonant circuit at the primary to minimize the loading of the PA. A detailed schematic of the driver stage is shown in Figure 1.26.

A 2.4 GHz PA test IC chip in 130 nm CMOS has been designed and fabricated [30]. The layout, shown in Figure 1.27, is highly scalable in power as more stages can be added to boost the output power. This test chip operates with a 1.2 V supply and utilizes fast thin-oxide device. The measured output power, plotted in Figure 1.28a, reaches 24 dBm with 25% efficiency at the 1 dB compression point. When driven into saturation, it transmits 27 dBm with 30% efficiency. When one stage of the PA is turned off, the output power drops by 2.5 dB, but the efficiency of 30% is maintained at back-off. This is in contrast to a conventional PA that has dramatically lower efficiency with back-off. A plot of the simulated efficiency over power back-off is shown in Figure 1.28b. A multistage transformer-coupled PA presents the optimum load impedance with back-off and therefore maintains nearly

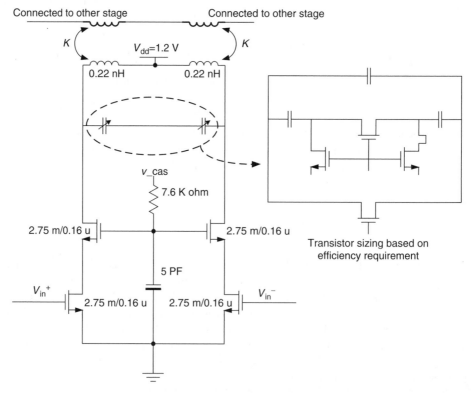

FIGURE 1.26 Schematic of the driver stage employing a cascode-tuned load amplifier with the capability to switch the resonance frequency for the "off" mode.

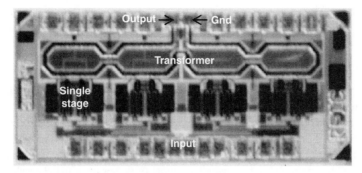

FIGURE 1.27 Layout of a prototype 130 nm CMOS PA test chip.

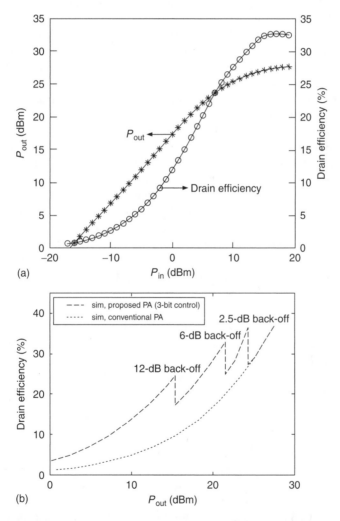

FIGURE 1.28 (a) Measured output power of prototype PA. (b) Simulated efficiency with power back-off. (From Liu, G., King, T.-J. and Niknejad, A. M., *Proceedings of the IEEE 2006 Custom Integrated Circuits Conference*, September 2006. © 2004 IEEE.)

optimum efficiency over a wide range [31]. The ability to transmit efficient power at back-off is also a characteristic of Doherty amplifiers, a modern CMOS implementation is found in Ref. 32.

DYNAMIC BIASING POWER AMPLIFIER

Power amplifiers universally suffer from reduced efficiency when operated below the maximum output power limited by the battery voltage, a condition known as power back-off. This problem is compounded by the use of high data rate standards as increasing degrees of amplitude modulation are being used to raise signal bandwidths. It is important to note the trends in peak–average power ratio (PAR) and the overall power control range. As higher order standards are implemented, transmitters are required to operate with increasing amounts of power back-off. Therefore, the trend for PAs is to operate with lower average efficiencies.

The idea of a dynamic class A PA has been proposed before [33–35]. Class A PAs are inherently the most linear PAs, and thus great candidates for linear modulation schemes. In fact, they are widely deployed in WLAN systems, especially OFDM 802.11.a/b/g. The efficiency of such PAs is at best 50%, and much lower in practice. To retain the linearity of the PA, they must be operated below saturation. The average efficiency of such PAs is very low in practice. If the modulation scheme has 10 dB of peak-to-average ratio, then in practice back off by at least 10 dB from the peak power point, resulting in dramatically lower efficiency.

One approach to maintain the efficiency at low power is to bias the supply voltage in a dynamic fashion. In other words, if track the envelope of the signal at the rate of the modulation, then reduce the supply just enough to prevent saturation. If the supply modulation is done with an efficient regulator, such as a DC–DC converter, then great efficiency enhancement is possible. It is therefore necessary to codesign the PA with the regulator. Many important challenges with such a scheme include the bandwidth of the regulator and the up-conversion of supply ripple into the RF spectrum.

If the bias current of the PA is also controlled dynamically, then nearly optimal conditions can be maintained over the entire output power level. To see this, consider that the load power is simply given by $P_L = (1/2)V_L \cdot I_L$. The DC power consumption is given by $P_{DC} = I_{av} \cdot V_{sup}$. Assuming large power gain, the efficiency approaches the drain efficiency

$$\eta = \frac{V_L I_L}{2 I_{av} V_{sup}} = \frac{1}{2} \frac{V_L}{V_{sup}} \frac{I_L}{I_{av}} = \frac{1}{2} \lambda_v \lambda_i \tag{1.33}$$

where λ_v and λ_i are the normalized voltage and current swings at the load. Since these terms approach unity for a class A PA at peak power, the efficiency peaks at 50%. But as the output power drops, the efficiency drops since λ_v and λ_i get smaller. If, on the contrary, I_{av} and V_{sup} are lowered, the average bias current and supply voltage, respectively, then near-unity λ_v and λ_i can be maintained and the efficiency at low power can be improved dramatically.

UNIVERSAL FREQUENCY SYNTHESIZER

The heart of every transceiver is the frequency synthesizer, a VCO phase-locked to a precision external reference signal. If multiple frequency band operation is needed, multiple VCOs must reside on-chip. Since the frequency of oscillation of each VCO is set by passive components, employing multiple resonators is bulky and expensive. This is true even if integrated on-chip spiral inductors and MIM capacitors are employed due to the large area consumed by these components. One solution is to employ a single wideband VCO inside a fractional-N frequency divider for flexible channel spacing and reduced VCO phase noise requirements.

In an integer-N frequency synthesizer, loop bandwidth and channel spacing are coupled, limiting the flexibility in the design of the synthesizer. A fractional-N architecture alleviates these contradictory requirements by introducing an extra degree of freedom. If the frequency division is done in

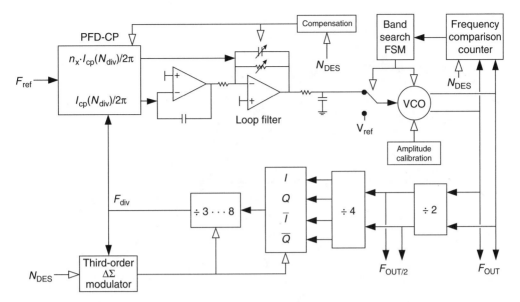

FIGURE 1.29 Block diagram of multiband frequency synthesizer.

fractional amounts, then the reference frequency can be increased relative to an integer-N architecture. This in turn allows wider phase-locked loops (PLL) bandwidths and faster settling and more relaxed requirements on the VCO phase noise.

The block diagram of a multimode frequency synthesizer is shown in Figure 1.29. A fractional-N architecture is chosen for the flexible channel spacing and relaxed VCO phase noise requirements. The drawback to fractional-N architectures is the spurious noise due to fractional division. Spurs occur since actual fractional division is achieved by periodically dividing by N and $N+1$ with the appropriate duty cycle to simulate dividing by $N+p$, with $p<1$. Dithering strategies reduce the fractional noise by randomizing the division. For example, application of a third-order $\Delta\Sigma$ modulator can produce noise shaping to reduce fractional spurs.

The design of the charge pump and loop filter is also challenging due to the large tuning range. To keep the phase noise specifications constant over the entire range, a different gain compensation is required for every frequency tuning word. The simplest solution of simply changing the charge-pump current is nonoptimal since the overall transfer function of the loop is strongly coupled to the charge-pump current. A better strategy is an active programmable op-amp RC loop filter that allows robust performance over frequency and process. System-level simulations show that good phase noise performance can be maintained across the entire tuning range (Figure 1.30). Full custom source-coupled logic (SCL) CMOS logic style is recommended in the programmable divider. Because of constant current and complementary switching action, SCL logic generates less substrate noise and should results in lower spurs in the synthesized spectrum.

WIDE TUNING RANGE VCO

Wide tuning range in the VCO can be obtained by employing a switched-capacitor topology. As shown in Figure 1.31a, the parallel combination of switched binary weighted capacitors and a MOS varactor can be used to tune over a wide frequency range. Since high-Q capacitors require small series resistance, aggressive scaling in CMOS technology is beneficial due to the increased on-conductance and lower parasitics. A recent demonstration of a CMOS VCO with over 1 GHz tuning range using this approach is found in Ref. 36. Because of the large tuning range, the VCO loop gain varies considerably from the low side to the high side of the tuning range. By employing digitally calibrated VCO current, shown in Figure 1.31b, the phase noise can be optimized over the entire

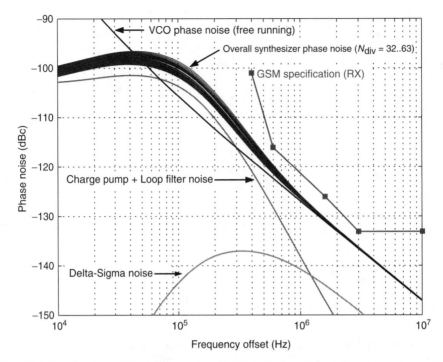

FIGURE 1.30 Simulated closed-loop phase noise of PLL including all known sources of noise. The contributions from various dominant sources are also shown for reference. The GSM RX mask falls above the simulated performance.

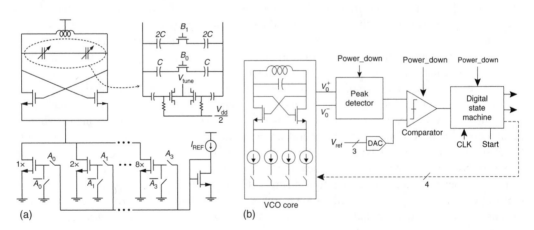

FIGURE 1.31 (a) Wideband tuning range VCO topology. (b) Digital amplitude calibration loop.

frequency range. Without this feedback, the power consumption is set by the lowest frequency of oscillation and the Q factor in this frequency range. Digital circuitry can also automatically select the proper VCO amplitude to meet the required phase noise specifications. This can result in power savings when operating with standards that do not have stringent phase noise requirements.

ANALOG AND DIGITAL BASEBAND

The simple RF front-end COGUR architecture needs a robust baseband to convert a large dynamic range analog signal into the digital domain. In fact, much of the programmability and flexibility of the radio comes about by implementing the baseband in software or programmable hardware. Because of the minimal amount of RF and IF filtering provided by the front-end, the ADC must handle a large dynamic range.

Recent research has resulted in several promising candidate architectures for the ADC [37,38]. These new architectures are compatible with standard CMOS technology, and unlike previous designs, they scale with technology by taking advantage of digital signal processing to overcome the analog impairments. It is projected that by consuming 500 mW of power, a sufficiently linear 12 bit ADC can be fabricated in 130 nm CMOS technology with over 500-Mb/s sampling rate and 100 dB of SFDR. This is sufficient performance to handle almost any wireless communication standard. This performance is achieved through background calibration. In fact, much power savings will be achieved if the sampling rate and DNL/INL are traded in a dynamic fashion. In most situation the full bandwidth and linearity of the ADC is not required.

MICROWAVE AND mm-WAVE CMOS

Despite the plethora of RF building blocks in the 1–5 GHz frequency range, there is a dearth of circuits in the 10–24 GHz range. Quite interestingly, there is perhaps more interest in the 60 and 77 GHz bands due to the availability of spectrum and many exciting applications. Moving CMOS circuits first into the 10–20 GHz microwave band and next into the mm-wave regime are explored in the new sections.

MICROWAVE CMOS

Many CMOS RF design techniques developed for 1–5 GHz map directly into the 10–20 GHz spectrum, albeit with some modifications in layout. This stems from the advancing speed of the technology and simple scaling of passive elements. As frequency is scaled, the desired inductance decreases, reaching hundreds of pH at 10 GHz. Since package and board parasitics are easily on the order of 1 nH, care must be exercised in designing the power/ground and bypass capacitors to provide a sufficiently low impedance on-chip. Unlike GaAs and other technologies with substrate through-vias, the lack of a ground plane in CMOS technology exacerbates the situation.

Consider for example the typical inductively degenerated LNA where a spiral inductor is used for both L_s to achieve an input match. The source inductance comprises the partial inductance of the spiral and the partial inductance of the leads, plus any partial mutual inductance.* Equivalently, the inductance is defined by the complete loop formed by the leads, the transistor, the spiral, and the ground plane. If there are multiple paths through the ground plane, it is somewhat unclear what path the current will take at any given frequency (without a full-wave simulation of the entire structure). This is compounded by the fact that mixed passive/active simulation is difficult, requiring patchwork, which could compromise the integrity of the simulation. Fortunately, at very high frequencies the path of least impedance is the dominant path of current flow, allowing simplifications to be made in the layout.

To see the importance of accounting for the entire current flow path, two LNAs were fabricated to operate 11 GHz, as shown in Figure 1.32. One LNA was designed with library inductors by simply ignoring the inductance of the parasitic path, whereas a second LNA was designed using ASITIC by assuming a dominant ground current flowing near the inductor. The results of the input match S_{11} are shown in Figure 1.33, clearly showing a mistuned input match for the first amplifier. One partial solution is to employ transmission lines instead of spiral inductors, since much smaller reactances can be synthesized by simply scaling the length of the transmission line. Furthermore, the ground return current of a transmission line flows along with the signal current, minimizing the impact of coupling to nearby structures and the substrate. At 11 GHz, though, transmission lines are somewhat too bulky to integrate on-chip, and spiral inductors are preferred to save area and cost.

---

* Technically, inductance is only defined for a closed loop. We can define the *partial* inductance of an open path [39].

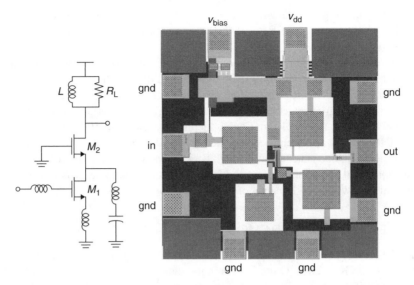

FIGURE 1.32 A classic inductively degenerated 11 GHz LNA employs spiral inductors at the gate and source of the input transistor.

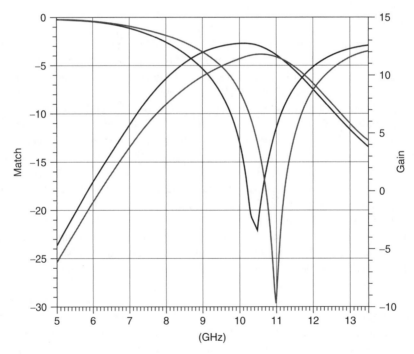

FIGURE 1.33 Input match of two versions of the LNA, designed with and without consideration of the return current through the ground plane.

Cascode transistor stages are the preferred building block for low-GHz amplifier. At lower frequencies, they exhibit higher gain, nearly unilateral performance, and nearly the same NF as a common-source amplifier. At high frequencies, though, the impedance at the cascode node drops and the noise of the cascode transistor begins to play a significant role. This is easily understood if we input refer the noise of the cascode device to the gate. At low frequencies, due to the large degeneration

$(r_o$ and $C_{db})$, this noise is rejected. A solution to this problem is to employ a resonant tank to increase the high-frequency impedance, as shown in Figure 1.32. Compared to a two-stage common-source topology with similar gain, this amplifier has comparable noise and better linearity.

mm-WAVE CMOS

The key motivation for moving to mm-wave is the availability of bandwidth that enables multi-Gb/s wireless communication. For instance, 7 GHz of unlicensed bandwidth is available in the 60 GHz spectrum. Although excessively high path loss at this frequency due to oxygen absorption precludes long-range communication, short-range local area networks actually benefit from the attenuation in providing extra spatial isolation and safety. Other applications for mm-wave CMOS include inexpensive automotive radar and imaging. Because of the small wavelength, antenna elements can be integrated on-chip or in the package, enabling a true integrated radio.

Employing inexpensive CMOS technology as opposed to SiGe, GaAs, or InP is a major departure from most mm-wave programs [40–42]. CMOS technology is constantly improving due to a fast-growing digital market. Today 130 nm bulk CMOS technology is capable of power gain in the 60 GHz band [6], whereas 90 nm technology offers higher gain and lower noise. Future bulk CMOS at the 65 nm and 45 nm nodes are expected to provide even more gain at less power.*

Naturally, the availability of inexpensive small footprint CMOS transceivers leads to the realization of dense antenna arrays, providing extra diversity, spatial power combining and power control, and electronic beam steering capability. A presumed disadvantage of the 60 GHz radio is the small antenna capture area compared to, say, 6 GHz WLAN systems. Since the received power P_r at a distance R is proportional to effective antenna area, and thus to wavelength, a factor of 20 dB loss is expected compared to a 6 GHz system. Assuming simple line-of-sight communication, the Friis propagation loss is given by

$$\frac{P_r}{P_t} = \frac{D_1 D_2 \lambda^2}{(4\pi R)^2} \propto \left(\frac{\lambda}{R}\right)^2 \tag{1.34}$$

where the received power P_r normalized to the transmitted power is seen to depend on the TX and RX antenna directivities D_1 and D_2, distance between RX and TX R, and wavelength λ. Although the directivity for a single antenna can be improved, it is more fruitful to increase the directivity of a device by employing an antenna array. The array provides extra diversity, automatic spatial power combining, and electronic beam steering capability. Therefore, it is in fact advantageous to move to a higher frequency, since the constant is total available area for a given application, and the antenna array gain actually improves with frequency.

mm-WAVE ACTIVE ELEMENTS

To ascertain the feasibility of a given technology for mm-wave operation, scattering parameters of active devices need to be measured and de-embedded. At mm-wave frequencies, the parasitics of the pad frame and leads can overwhelm the small transistor capacitance and inductance. Although any lossless reactance does not alter the maximum unilateral gain, the Si losses can dampen the gain significantly. The intrinsic MOSFET has several sources of loss that must be minimized. Transistors with many short fingers in parallel are employed to minimize the gate resistance of the transistor. Using short-finger transistors also allows substrate contacts to be placed close to the transistor to minimize the effects of substrate loss.

In a 130 nm process, measurements of single transistors and cascode device structures were carried out from DC to 65 GHz. Measurements of maximum stable gain (MSG) and unilateral gain are shown in Figure 1.34a. Standard open-short de-embedding is used and found to be adequate

---

* This is not a foregone conclusion as the performance of active devices at mm-wave frequencies is highly dependent on intrinsic and extrinsic device loss.

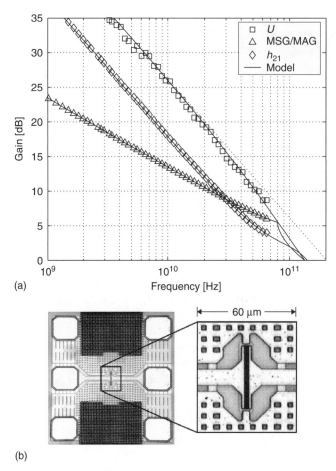

(a)

(b)

FIGURE 1.34 (a) Current gain and maximum power gain of a single transistor in the 130 nm technology node. Model extrapolated f_{max} is 135 GHz. (b) Optimized layout of NMOS transistor. (From Doan, C. H., Emami, S., Niknejad, A. M. and Brodersen, R. W., *IEEE J. Solid-State Circuits*, pp. 144–155, January 2005. © 2005 IEEE.)

up to 50 GHz. Beyond this frequency, an optimized model can predict performance. Custom layout transistors are fitted to models, and it is found that a modified lumped circuit sufficiently models small and large signal performance up to 65 GHz [6]. The model-extrapolated f_{max} curve shows a crossover at 135 GHz, and enough gain at 60 GHz to realize an amplifier.

It is noteworthy that the measured and modeled unilateral gain U shows close agreement, confirming the validity of the small-signal model. Mason's unilateral gain defined as

$$U = \frac{\left| y_{21} - y_{12} \right|^2}{4 \left(\Re(y_{11}) \Re(y_{22}) - \Re(y_{12}) \Re(y_{21}) \right)} \tag{1.35}$$

represents the maximum two-port gain when the device is unilaterized while embedded in a loss-less four-port. In practice, a device is unilaterized by applying series and shunt feedback around a two-port. In the special case that y_{12} is imaginary, then the device is unilaterized by employing neutralization, where a shunt feedback inductor resonates with the intrinsic C_{gd}. It can be shown that U has several important properties. First it is an invariant of the two-port in the sense that it is independent on the unilaterization circuitry. Also, if $U > 1$, the two-port is active, whereas if $U < 1$, the two-port is passive. U is also the maximum gain of a three-terminal device regardless of the

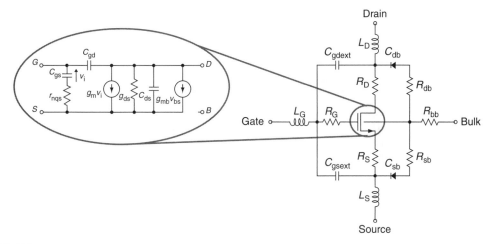

FIGURE 1.35 Modified BSIM3 model with enhanced parasitics is capable of predicting RF small signal and large signal performance up to 60 GHz.

common terminal. On the basis of these properties, note that U is very sensitive to device loss and a good way of testing the accuracy of both a model and measurement results. On the contrary, the MSG is relatively insensitive to device loss below the kink frequency. When the stability factor $K < 1$, the device is conditionally stable requiring termination loss to be added to stabilize the device. In this circumstance, the device parasitics can be buried in the large termination resistance.

In Figure 1.35, a modified BSIM3 model has been used to model an 80 finger device. S-parameters show a good match over the entire band of interest (Figure 1.36a). A plot of the transistor gain compression at 60 GHz is shown in Figure 1.36. Other linearity measures confirm the distortion behavior of the model [43]. It is important to note that the DC nonlinearity is used to predict high-frequency performance. This greatly simplifies extraction of the compact model parameters. This model is useful for the design of nonlinear and time-varying circuits, including mixers [44], dividers, multipliers, and VCOs.

The noise of the device is captured by employing Pospieszalski's noise model. The drain noise is captured as usual with $\gamma = 1.4$, while the gate noise is modeled through a noisy r_{nqs} resistance. Several devices were measured and modeled over bias to confirm the validity of the approach. A typical fit is shown in Figure 1.37. The results are encouraging and show that 130 nm CMOS has the potential to operate at reasonably low noise levels. The minimum achievable noise is found to be $F_{min} \sim 4$ dB. The optimum source impedance and noise sensitivity R_n are well captured by the model. The noise of a cascode transistor, modeled in the same way, also shows good agreement. The $F_{min} \sim 6.5$ dB indicates that the cascode device contributes significant noise at this frequency.

mm-Wave Passive Elements

The key passive building blocks for mm-wave circuit design include inductors, transformers, de-coupling or bypass capacitors, coupling DC-block capacitors, varactors, and transmission lines. Inductors are used extensively to tune out parasitics and take the form of spiral inductors or short sections of transmission lines. Capacitors can be process-specific MIM capacitors (using a high-K thin oxide) or simply realized using a finger structure incorporating multiple metal layers, as shown previously in Figure 1.7. Although lumped inductors are generally smaller than distributed circuit equivalents, the advantages of distributed circuit elements include better prediction, scalability, and improved isolation. The value of the inductance of a shorted transmission line depends only on the length, Z_0, and propagation constant $\gamma = \alpha + j\beta$. If a library of transmission lines is well characterized, then any arbitrary inductance can be synthesized by varying the length of the line.

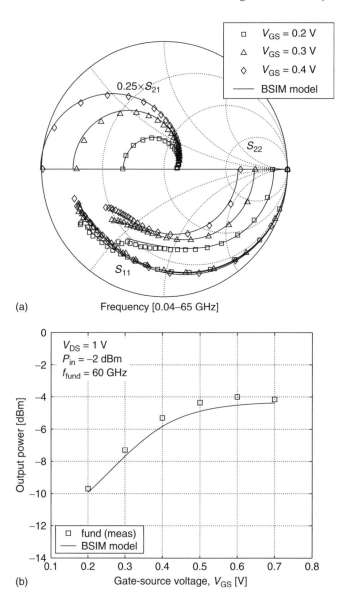

(a)   Frequency [0.04–65 GHz]

(b)   Gate-source voltage, V_{GS} [V]

FIGURE 1.36 (a) Measured (and scaled) S parameters of a typical NMOS transistor using the large signal model. (b) The output power versus gate bias point of a single transistor at 60 GHz. (From Doan, C. H., Emami, S., Niknejad, A. M., and Brodersen, R. W., *IEEE J. Solid-State Circuits*, pp. 144–155, January 2005. © 2005 IEEE.)

Furthermore, there are no leads that add parasitic inductance. Since ground return currents flow intrinsically in the transmission line, the lines radiate a dipole field pattern and display more isolation compared to ring and spiral inductors.

In many situations, though, a lumped inductor can be realized with higher quality factor. For typically small inductors with $L \sim 100\,\text{pH}$, simple ring inductors, as shown in Figure 1.38a, are sufficient. In fact, an analogy can be drawn between a ring inductor and a differential transmission line. To realize high Z_0 transmission line, increase the gap spacing to increase the inductance per unit length while reducing the odd-mode capacitance per unit length. But to short the line with low inductance, it is preferable to bend the end of the line and reduce the gap spacing. Furthermore, to present close leads to connect to a capacitor or transistor, it is preferable to also bend the front of the

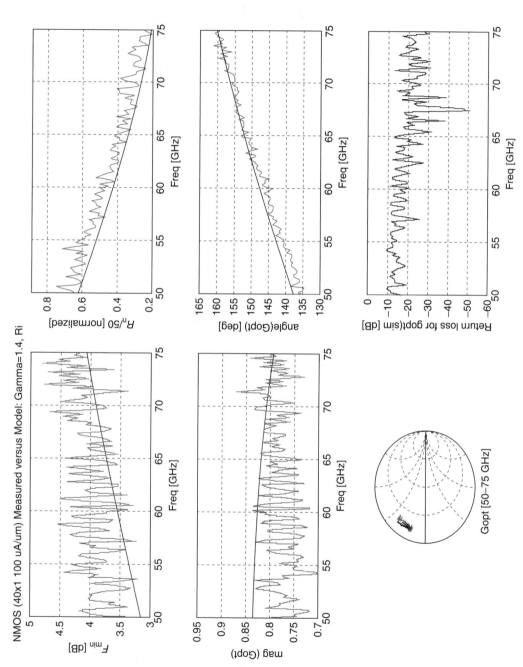

FIGURE 1.37 Measured and simulated minimum NF, noise sensitivity R_n, and optimum source impedance for a 130 nm NMOS device.

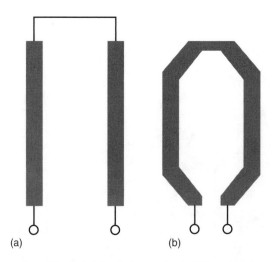

FIGURE 1.38   (a) Large spacing differential line inductor. (b) A differential transmission takes the form of a loop when practical leads are formed.

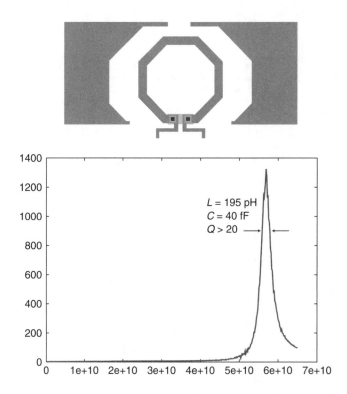

FIGURE 1.39   Measured resonance of an mm-wave tank formed by a ring inductor and MIM capacitors.

line, as shown in Figure 1.38b. Clearly, this structure very much resembles a ring inductor, showing a close connection between transmission lines and ring inductors.

A ring inductor has been designed and fabricated in an Si RF process. To accurately measure the Q factor, a parallel resonant circuit is formed with MIM capacitors. In this way, the Q factor is insensitive to small lead inductance and resistance, since at resonance, the circuit forms an open or large impedance circuit. The measured resonance curve is shown in Figure 1.39.

The resonator occupies an area of $150 \times 150\,\mu m^2$ and has a loaded measured quality factor over 20. The inductor has been simulated and measured separately and the Q factor is between 30 and 40, which shows that the MIM capacitors have comparable Q. An electric shield surrounds the structure to minimize the coupling to Si substrate. This open shield structure is used to minimize eddy current flow.

At mm-wave frequencies transmission lines are used extensively. Given the many metal layers offered by a modern CMOS process, there are many options for the realization of transmission lines. Microstrip lines are popular due to the self-shielding layout, preventing electric fields from penetrating the substrate. But owing to the fixed spacing between top metal and bottom metal, the Z_0 of the line can be controlled only by changing the signal conductor width W. To realize high Z_0 requires low W, and consequently high loss per unit length. For this reason, coplanar lines are preferred over microstrip lines due to higher obtainable inductive quality factors. This is important since transmission lines are used extensively to resonate MOS capacitors. Furthermore, since lateral dimensions are determined by lithography, the characteristic impedance of the coplanar line is more predictable and constant over process. Inductive quality factors as high as 25 have been measured at 50 GHz for coplanar lines. A measured and simulated plot of the inductive quality factor for a microstrip and coplanar line is shown in Figure 1.40, clearly demonstrating the benefit of the coplanar line.

Slow-wave differential lines and tapered lines also hold great promise for the realization of compact high-Q resonators [45,46]. By employing a dense array of filaments, as shown schematically in Figure 1.41a, the capacitance per unit length of a line is increased substantially while leaving the inductance per unit length unaltered. For a uniform transmission line, the product of LC is constant (by the speed of light), whereas for such nonuniform line, this no longer holds, and the waves traveling on these lines are slower, hence the name slow-wave transmission line. Nonuniform transmission lines can be optimized for standing waves. Since in a resonant quarter wave line, the current is ideally zero at the input of the line and maximum at the shorted end, the conductor width should be increased toward the end of the line to minimize the series resistive losses. Likewise, the voltage is maximum at the input and minimum at the end, and so the shunt-conductive losses dominate at the input and are negligible at the shorted end. These considerations would dictate a layout shown in Figure 1.41b. A more precise way to arrive at the optimal layout is discussed in Ref. 46.

Key mm-Wave Building Blocks

To realize gain close to the limits of process technology requires a careful trade-off between gain and noise for the LNA and gain and power efficiency for the PA. With process scaling, beginning with the 90 nm technology node, this trade-off is more manageable due to the inherently higher gain of the devices.

The schematic of a sample 60 GHz amplifier is shown in Figure 1.42. Simple gain stages consisting of NMOS common-source cascode amplifiers are used extensively for power gain. Inductive degeneration, common in low-GHz design, is avoided to maximize the gain. Transmission lines are used extensively to realize small inductive reactances in matching networks. The design of proper coupling and bypass capacitors are nontrivial at these frequencies due to self-resonance. If a bypass capacitor is oversized or designed incorrectly, it may self-resonate below 60 GHz and produce unwanted oscillations. Finger MIM capacitors are preferred to parallel-plate MIM capacitors for compatibility with a generic CMOS process. The measured gain is shown in Figure 1.43, showing an excellent match between measurement and simulation.

The Gilbert cell mixer is the workhorse of many low-GHz designs. Owing to the operation close to the limits of the process, simpler architectures may be preferred. Microwave diode ring mixers reign supreme in traditional microwave design. Low-loss diodes are not available in a traditional digital CMOS process, and so other alternatives such as dual-gate mixers or single-transistor mixers

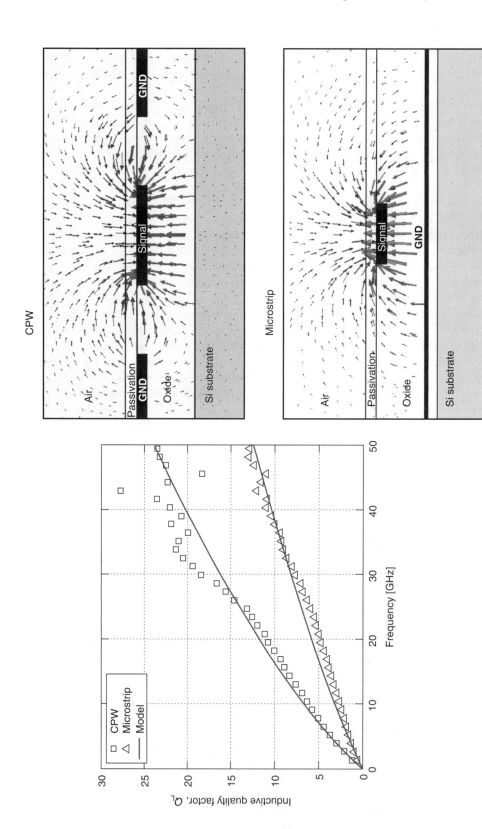

FIGURE 1.40   Measured and calculated inductive quality factor of a coplanar and microstrip transmission lines realized on the CMOS Si substrate. (From Doan, C. H., Emami, S., Niknejad, A. M., and Brodersen, R. W., *IEEE J. Solid-State Circuits*, pp. 144–155, January 2005. © 2005 IEEE.)

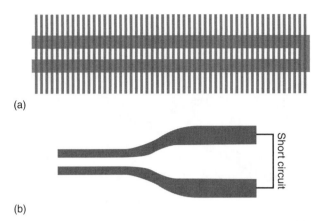

(a)

(b)

FIGURE 1.41 (a) A slow-wave transmission line employs orthogonal floating conductor strips that increase the capacitance per unit length without altering the inductance of the line. (b) A nonuniform standing-wave resonant transmission line optimized for low loss.

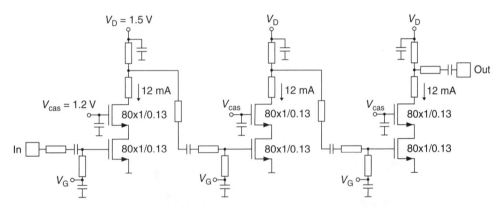

FIGURE 1.42 Schematic of an amplifier employing cascode gain stages. (From Doan, C. H., Emami, S., Niknejad, A. M., and Brodersen, R. W., *IEEE J. Solid-State Circuits*, pp. 144–155, January 2005. © 2005 IEEE.)

are explored. These architectures offer some conversion gain, which is desirable for rejecting the noise of the proceeding RF stages.

The pseudo-dual-gate mixer, shown in Figure 1.44, is a cascode device. The gate of transconductor is driven by an RF signal where its $g_m(t)$ is modulated by the LO signal applied to the gate of the cascode. This mixer is compact and provides a degree of isolation between RF and LO. The single-transistor mixer, shown in Figure 1.45, is even simpler but requires a hybrid coupler to combine an LO and RF signal. The transistor nonlinearity produces the desired mixing product, generated by the LO pumping of the transistor operating point. Because of the high frequency of operation, though, the hybrid can be easily and necessarily integrated on-chip. Dual-gate and quadrature-balanced mixers have been fabricated and measured in a 130 nm CMOS. The measured performance of these mixers shows low conversion loss with modest LO power below 0 dBm, a key requirement for a low-power integrated mixer.

CMOS VCOs have already been demonstrated in CMOS operating at 60 GHz [47,48]. Relatively high-Q (about 20) LC or transmission line resonators have been measured in these frequency bands. Demonstrated phase noise is around −85 dBc/Hz at a 100 kHz offset. Since the VCO is a shared block among the parallel transceiver sections, DC power consumption is not the primary

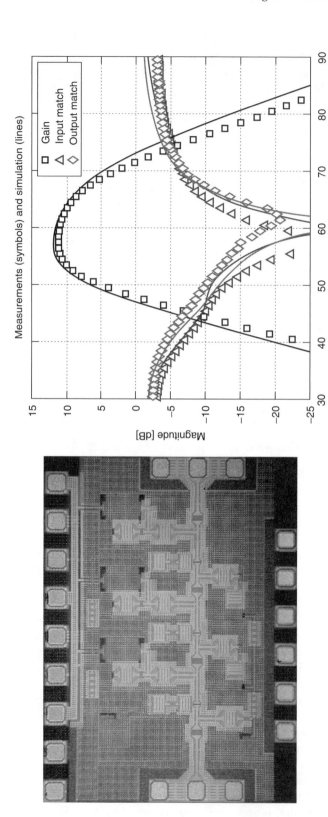

FIGURE 1.43 Die photo and measured performance of 60 GHz CMOS amplifier. (From Doan, C. H., Emami, S., Niknejad, A. M., and Brodersen, R. W., *IEEE J. Solid-State Circuits*, pp. 144–155, January 2005. © 2005 IEEE.)

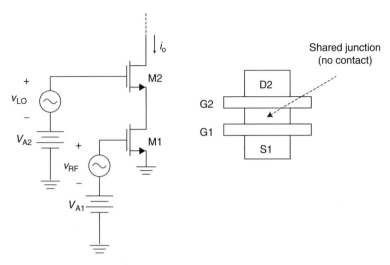

FIGURE 1.44 A dual-gate mixer employs a cascode device to modulate the g_m of the common-source input amplifier.

concern. As already discussed, generation of a stable frequency reference is a concern and may dictate a lower frequency LO.

The need for a PA is ameliorated by the spatial power combining shown in Figure 1.46. To transmit 100 mW, for instance, 10 amplifiers need only deliver 10 mW (or less due to antenna gain). The reduced power eases matching network design, a traditional source of loss in an integrated PA, and spatial power combining eliminates the need for power combining off-chip. Since power combiners are generally lossy and bulky, this greatly benefits the design and integration of the overall system. If needed, coarse power control is simply achieved by turning off parallel stages. But even generation of 10 mW of power in CMOS is nontrivial since optimal devices matched for gain do not have a low impedance. The performance of large-area devices suffers due to layout parasitics necessitating power combining or distributed architectures.

mm-WAVE RADIO ARCHITECTURE

Antenna Array

At mm-wave frequencies, antenna elements are small enough to be directly integrated into the package or on the Si substrate [49]. Beam forming improves the antenna gain while also providing spatial diversity and thus resilience to multipath fading. The main benefit of the multiantenna architecture used here is the increased gain that the directional antenna array pattern provides. This gain is needed to support data rates approaching 1 Gb/s at typical indoor distances. Additionally, the directive antenna pattern improves the channel multipath profile. By limiting the spatial extent of the antenna patterns to the dominant transmission path, the delay spread and Rician K-factor of an indoor wireless channel can be significantly improved.

Key elements of an antenna array are the RF phase shifters and voltage-controlled attenuators. An approach that eliminates the need for the RF phase shifters is to process each stage in parallel using independent RF/IF stages. Alternatively, the phase shifters and attenuators can be done directly at RF, and signals are combined at either RF or IF to reduce the required circuitry, power consumption, and linearity requirements. The accuracy of phase shifters and attenuators is greatly relaxed if one is only concerned with the location and magnitude of the peak in the antenna pattern, rather than the detailed response and the nulls. RF phase shifters based on an *I/Q* signal summing have been presented [50]. An alternative approach that eliminates the need for the RF phase shifters is to generate LO phases [51].

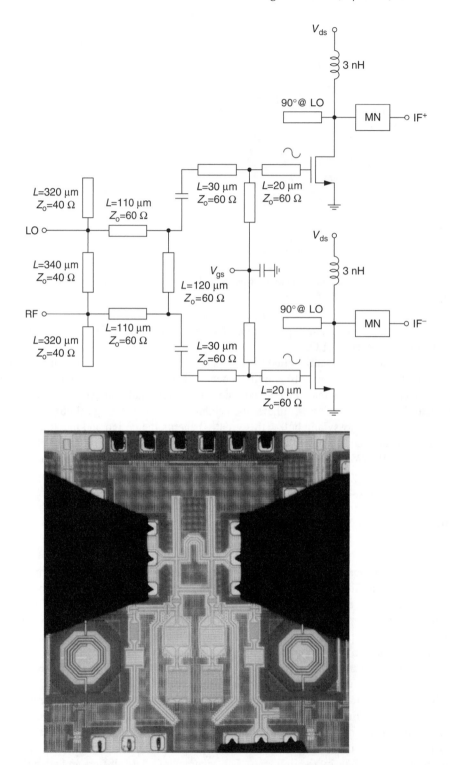

FIGURE 1.45 A quadrature-balanced mixer employing the nonlinearity of the transistor to generate the mixing products. A coupler is used to add the LO and RF signals. (From Doan, C. H., Emami, S., Niknejad, A. M., and Brodersen, R. W., *RFIC Digest of Papers*, pp. 163–166, June 2005. © 2004 IEEE.)

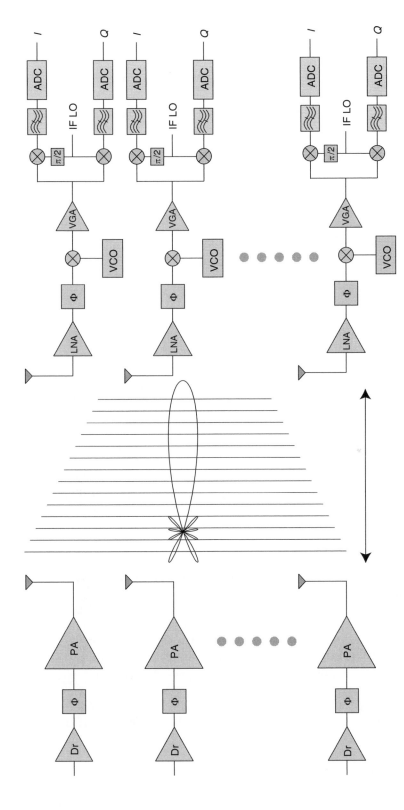

FIGURE 1.46 A multiple-antenna transceiver architecture employing antenna steering for maximum gain and power combining.

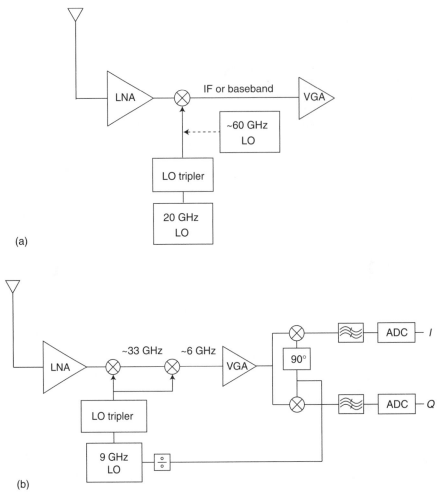

FIGURE 1.47   (a) A 60 GHz radio architecture based on a single conversion to an IF of 5 GHz or directly to baseband. (b) A multistage conversion architecture relaxes LO and frequency synthesizer design.

Transceiver Architecture

A simple receiver architecture for the 60 GHz radio front-end is shown in Figure 1.47. In a direct conversion architecture, the RF signal is directly down-converted to baseband in one step, but it requires a high-frequency LO. Although VCO design is not a major impediment at this frequency, a synthesizer would require frequency dividers to operate at this high frequency. Injection locking can be used to relax the power consumption of dividers. Alternatively, the 60 GHz LO signal can be obtained by a 20 GHz LO by employing a frequency tripler. Using a frequency multiplier has been the method of choice for most mm-wave systems [40]. The sliding IF architecture, on the contrary, uses double conversion. For instance, a 9 GHz LO is tripled and used to down-convert from 60 to 33 GHz and then reused at the fundamental mode to down-convert to IF. It is interesting to note that an image-reject mixer is not strictly needed for the 27 GHz LO since the image frequency is sufficiently distant. Since 60 GHz power gain comes at a DC power consumption premium, only enough gain is used to overcome the noise of subsequent stages.

A CMOS 60 GHz radio front-end has been fabricated in 130 nm technology. The die-photo, shown in Figure 1.48, implements the RF amplifier, mixer, and LO buffer in a single die. The measured RF conversion gain is 7 dB and the measured NF is about 11 dB in the 60 GHz band.

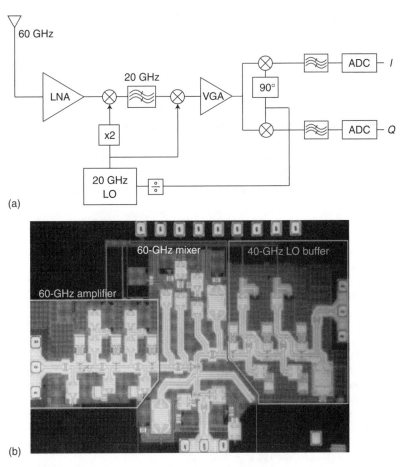

FIGURE 1.48   A 60 GHz front-end receiver in 130 nm CMOS technology.

The LO power is only −18 dBm due to the gain provided by the 40 GHz buffer. The NF of the receiver is dominated by the 60 GHz amplifier, which has been designed for power gain rather than low-noise performance. Measured $F_{min} = 3.5$ dB of a single transistor in this technology demonstrates that lower noise performance is possible by careful design of the LNA. The 60 GHz amplifier can double as a low-power transmitter, with a measured $P_{1dB} = 2$ dBm. The effective radiated power of an array can be orders of magnitude larger due to spatial power combining and antenna directivity. The total power consumption is 80 mW, determined primarily by the first stage amplifier and the LO buffer.

CONCLUSION

This chapter has explored building blocks for next generation of wireless transceivers. Broadband, linear, high-frequency, and low-noise RF building blocks are possible in nanoscale CMOS technology. A simplified front-end is proposed to realize a cognitive universal and dynamic front-end, backward and forward compatible with most wireless standards.

The feasibility of a CMOS wireless transceiver capable of 60 GHz operation has been demonstrated. By carefully developing passive and active device models from measured data, key RF building blocks such as amplifiers, mixers, and oscillators can be realized. By employing multiple antennas in a dynamic beam steering configuration, the extra path loss at 60 GHz can be tolerated.

ACKNOWLEDGMENTS

This work has been sponsored by our BWRC member companies and the DARPA TEAM project (DAAB07-02-1-L428). We thank ST Microelectronics and IBM for chip fabrication and support. Many companies have contributed through the UC Discovery and UC MICRO programs. We also thank Analog Devices, Broadcom, Conexant, Infineon, and Qualcomm for support.

Finally, most of the results presented in this chapter are a result of research carried out by students at UC Berkeley. In particular, I would like to acknowledge the work of Chinh Doan and Sohrab Emami on the mm-wave circuits, Gang Liu for the CMOS PA work, Axel Berny for the wideband VCO and frequency synthesizer work, Nuntachai Poobeupheun and Wei-Hung Chen for work on the broadband LNA and mixer, Andrea Bevilaque for the UWB LNA, and Ehsan Adabi, Bagher Afshar, Mounir Bohsali, and Babak Heydari for the 90 nm mm-wave effort at UC Berkeley.

REFERENCES

1. Ma, D. K., Long, J. R., and Hararne, D. L., "A subharmonically-injected quadrature LO generator for 17 GHz WLAN applications," *Proceedings of the IEEE 2003 Custom Integrated Circuits Conference*, pp. 567–570, September 2003.
2. Dogan, H., Meyer, R. G., and Niknejad, A. M., "A DC-2.5 GHz wide dynamic-range attenuator in 0.13 μm CMOS technology," in *Symposium VLSI Circuits Digest of Technical Papers*, pp. 90–93, June 2005.
3. "Facilitating opportunities for flexible, efficient, and reliable spectrum use employing cognitive radio technologies," FCC ET Docket No. 03-108, December 30, 2003.
4. Abidi, A. A., "RF CMOS comes of age," *IEEE J. Solid-State Circuits*, pp. 549–561, April 2004.
5. Rudell, J. C., Ou, J.-J., Cho, T. B., Chien, G., Brianti, F., Weldon, J. A., and Gray, P. R., "A 1.9 GHz wide-band if double conversion CMOS integrated receiver for cordless telephone applications," in *IEEE International Solid-State Circuits Conference (ISSCC) Digest of Technical Papers*, pp. 304–305, February 1997.
6. Doan, C. H., Emami, S., Niknejad, A. M., and Brodersen, R. W., "Millimeter-wave CMOS design," *IEEE J. Solid-State Circuits*, pp. 144–155, January 2005.
7. Razavi, B., "A 60-GHz CMOS receiver front-end," *IEEE J. Solid-State Circuits*, pp. 17–22, January 2006.
8. Chang, M.-C. F., Chien, C., Huang, D., Ku, T. W., Wang, N.-Y., Gu, Q., and Wong, R., "A 60 GHz CMOS differential receiver front-end using on-chip transformer for 1.2 volt operation with enhanced gain and linearity," in *Symposium VLSI Circuits Digest of Technical Papers*, pp. 144–145, June 2006.
9. Pospieszalski, M. W., "On the measurement of noise parameters of microwave two-ports," *IEEE MTT-S International Microwave Symposium Digest*, pp. 456–458, April 1986.
10. Abidi, A. A., "High-frequency noise measurements on FETs with small dimensions," *IEEE Trans. Electron. Devices*, pp. 1801–1805, November 1986.
11. Scholten, A. J., et al., "Compact modeling of drain and gate current noise for RF CMOS," *Proceedings of IEDM*, pp. 5.6.1–5.6.4, 2002.
12. Tsividis, Y., *Operation and Modeling of the MOS Transistor*, McGraw-Hill, New York, 1987.
13. Kim, B., Ko, J.-S., and Lee, K., "A new linearization technique for MOSFET RF amplifier using multiple gated transistors," *IEEE Microwave and Guided Wave Letters*, pp. 371–373, September 2000.
14. Bevilacqua, A. and Niknejad, A. M., "An ultra wideband CMOS low noise amplifier for 3.1–10.6 GHz wireless receivers," in *IEEE International Solid-State Circuits Conference (ISSCC) Digest of Technical Papers*, pp. 382–383, February 2004.
15. Hashemi, H. and Hajimiri, A., "Concurrent multiband low-noise amplifiers–theory, design, and applications," *IEEE Trans. Microwave Theory Tech.*, pp. 288–301, January 2002.
16. Rossi, P., Liscidini, A., Brandolini, M., and Svelto, F., "A variable gain RF front-end, based on a voltage–voltage feedback LNA, for multistandard applications," *IEEE J. Solid-State Circuits*, pp. 690–697, March 2005.
17. Bruccoleri, F., Klumperink, E. A. M., and Nauta, B., "Noise cancelling in wideband CMOS LNAs," in *IEEE International Solid-State Circuits Conference (ISSCC) Digest of Technical Papers*, pp. 406–407, February 2002.

18. Shahani, A. R., Shaeffer, D. K., and Lee, T. H., "A 12-mW wide dynamic range CMOS front-end for a portable GPS receiver," *IEEE J. Solid-State Circuits*, pp. 2061–2070, December 1997.

19. Long, J., et al., "Low voltage silicon bipolar RF front-end for PCN receiver applications," *IEEE International Solid-State Circuits Conference*, pp. 140–141, February 1995.

20. Crols, J. and Steyaert, M., "A 1.5 GHz highly linear CMOS downconversion mixer," *IEEE J. Solid-State Circuits*, pp. 736–742, July 1995.

21. Poobuapheun, N., Chen, W.-H., Boos, Z., and Niknejad, A. M., "A 1.5V 0.7–2.5 GHz CMOS quadrature demodulator for multi-band direct-conversion receivers," in *Proceedings of the IEEE Custom Integrated Circuits Conference*, pp. 797–800, September 2006.

22. Razavi, B., *Design of Analog CMOS Integrated Circuits*, McGraw-Hill, New York, 2001.

23. Aparin, V., Kim, N., Brown, G., Wu, Y., Cicalini, A., Kwok, S., and Persico, C., "A fully-integrated highly linear zero-IF CMOS cellular CDMA receiver," in *IEEE International Solid-State Circuits Conference (ISSCC) Digest of Technical Papers*, pp. 324–325, February 2005.

24. Darabi, H. and Abidi, A. A., "Noise RF-CMOS mixers: a simple physical model," *IEEE J. Solid-State Circuits*, pp. 15–25, January 2000.

25. Manstretta, D., Brandolini, M., and Svelto, F., "Second-order inter-modulation mechanisms in CMOS downconverters," *IEEE J. Solid-State Circuits*, pp. 394–406, March 2003.

26. Deltimple, N., Kerherve, E., Deval, Y., and Jarry, P., "A reconfigurable RF power amplifier biasing scheme," *The 2nd Annual IEEE Northeast Workshop on Circuits and Systems, 2004. NEWCAS 2004*, pp. 365–368, June 2004.

27. Simburger, W., Wohlmuth, H. D., Weger, P., and Heinz, A., "A monolithic transformer coupled 5-W silicon power amplifier with 59% PAE at 0.9 GHz," *IEEE J. Solid-State Circuits*, pp. 1881–1892, December 1999.

28. Tsai, K.-C. and Gray, P. R., "A 1.9 GHz, 1-W CMOS class-E power amplifier for wireless communications," *IEEE J. Solid-State Circuits*, pp. 962–970, July 1999.

29. Aoki, I., Kee, S. D., Rutledge, D. B., and Hajimiri A., "Distributed active transformer—A new power-combining and impedance-transformation technique," *IEEE Trans. Microwave Theory Tech.*, pp. 316–331, January 2002.

30. Liu, G., King, T.-J., and Niknejad, A. M., "A 1.2V, 2.4 GHz Fully Integrated Linear CMOS Power Amplifier with Efficiency Enhancement," in *Proceedings of the IEEE Custom Integrated Circuits Conference*, pp. 141–144, September 2006.

31. Liu, G., *Gang's Thesis Title*, PhD Dissertation, UC Berkeley.

32. Wongkomet, N., Tee, L., and Gray, P. R., "A 1.7 GHz 1.5W CMOS RF Doherty power amplifier for wireless communications," in *IEEE International Solid-State Circuits Conference (ISSCC) Digest of Technical Papers*, pp. 486–487, February 2006.

33. Saleh, A. A. and Cox, D. C., "Improving the power added efficiency of FET amplifiers operating with varying-envelope signals," *IEEE Trans. Microwave Theory Tech.*, pp. 51–56, January 1983.

34. Niknejad, A. M. and Meyer, R. G., "*The Design of an Efficient RF Power Amplifiers for CDMA Wireless Communication*," UC Berkeley ERL Research Summary, 1998.

35. Hannington, G., Chen, P.-F., Asbeck, P., and Larson, L., "High-efficiency power amplifier using dynamic power-supply voltage for CDMA applications," *IEEE Trans. Microwave Theory Tech.*, pp. 1471–1476, August 1999.

36. Berny, A. D., Niknejad, A. M., and Meyer, R. G., "A 1.8 GHz LC VCO with 1.3 GHz tuning range and mixed-signal amplitude calibration," in *Symposium VLSI Circuits Digest Technical Papers*, pp. 54–57, June 2004.

37. Chee, Y.-H., Rabaey, J., and Niknejad, A., "Ultra low power transmitters for wireless sensor networks," UC Berkeley Technical Report No. UCB/EECS-2006-57.

38. Murmann, B. and Boser, B. E., "A 12-bit 75-MS/s pipelined ADC using open-loop residue amplification," *IEEE J. Solid-State Circuits*, pp. 2040–2050, December 2003.

39. Ruehli, A. E., "Inductance calculations in a complex integrated circuit environment," *IBM J. Res. Develop.*, 16, pp. 470–481, September 1972.

40. Ohata, K., Maruhashi, K., Ito, M., Kishimoto, S., Ikuina, K., Hashiguchi, T., Takahashi, N., and Iwanaga, S., "Wireless 1.25 Gb/s transceiver module at 60 GHz-band," in *IEEE International Solid-State Circuits Conference (ISSCC) Digest of Technical Papers*, pp. 298–468, February 2002.

41. Floyd, B. A., Reynolds, S. K., Pfeiffer, U. R., Zwick, T., Beukema, T., and Gaucher, B., "SiGe bipolar transceiver circuits operating at 60 GHz," *IEEE J. Solid-State Circuits*, pp. 156–167, January 2005.

42. Hashemi, H., Xiang Guan, G., Komijani, A., and Hajimiri, A., "A 24-GHz SiGe phased-array receiver-LO phase-shifting approach," *IEEE Trans. Microwave Theory Tech.*, pp. 614–626, February 2005.

43. Emami, S., Doan, C. H., Niknejad, A. M., and Brodersen, R. W., "Large-signal millimeter-wave CMOS modeling with BSIM3," *RFIC Digest of Papers*, pp. 163–166, June 2004.

44. Doan, C. H., Emami, S., Niknejad, A. M., and Brodersen, R. W., "A 60-GHz down-converting CMOS single-gate mixer," *RFIC Digest of Papers*, pp. 163–166, June 2005.

45. Huang, D., Hant, W., Wang, N.-Y., Ku, T., Gu, Q., Wong, R., and Chang, M. F., "A 60 GHz CMOS VCO using on-chip resonator with embedded artificial dielectric for size, loss, and noise reduction," in *IEEE International Solid-State Circuits Conference (ISSCC) Digest of Technical Papers*, pp. 1218–1227, February 2006.

46. Andress, W. and Ham, D., "Standing wave oscillators utilizing wave-adaptive tapered transmission lines," *IEEE J. Solid-State Circuits*, pp. 638–651, March 2005.

47. Tiebout, M., Wohlmuth, H.-D., and Simburger, W., "A 1V 51 GHz fully-integrated VCO in 0.12 μm CMOS," in *IEEE International Solid-State Circuits Conference (ISSCC) Digest of Technical Papers*, pp. 238–239, February 2002.

48. Franca-Neto, L. M., Bishop, R. E., and Bloechel, B. A., "64 GHz and 100 GHz VCOs in 90 nm CMOS using optimum pumping method," in *IEEE International Solid-State Circuits Conference (ISSCC) Digest of Technical Papers*, pp. 444–538, February 2004.

49. Lin, J.-J., Li, G., Sugavanam, A., Xiaoling, G., Ran, L., Brewer, J. E., and O, K. K., "Integrated antennas on silicon substrates for communication over free space," *IEEE Electronic Device Letters*, pp. 196–198, April 2004.

50. Alalusi, S. H., *A 60 GHz Adaptive Antenna Array in CMOS*, PhD Dissertation, UC Berkeley, 2005.

51. Guan, X. and Hajimiri, A., "A 24-GHz CMOS front-end," *IEEE J. Solid-State Circuits*, pp. 368–373, February 2004.

2 Insights into CMOS Wireless Receivers toward a Universal Mobile Radio

Massimo Brandolini, Paolo Rossi, Danilo Manstretta, and Francesco Svelto

CONTENTS

Introduction..54
Global System for Mobile Communications...56
    GSM Receiver Requirements 56
    GSM Receivers: Evolution and State-of-the-Art 58
    GSM Direct Conversion Receiver: Building Blocks 59
    GSM: Conclusions 60
Universal Mobile Telecommunication System 60
    UMTS Receiver Specifications..61
    UMTS Receiver: State-of-the-Art 62
    UMTS Direct Conversion Receiver: Building Blocks................................63
    UMTS: Conclusions 64
Bluetooth..64
    Bluetooth Receiver Specifications and State-of-the-Art 64
    Bluetooth Receiver: The Proposed Solution...65
    Bluetooth: Conclusions 65
Wireless Local Area Networks 66
    IEEE 802.11/.11b..66
    IEEE 802.11a 68
    IEEE 802.11g 69
    Wireless LAN Receivers: State-of-the-Art..69
    Wireless LAN Direct Conversion Receiver: Building Blocks 70
    Wireless LANs: Conclusions 71
Toward a Multistandard Receiver 71
    Multimode and Multistandard Receivers: State-of-the-Art......................71
    A Proposed Architecture for Multistandard Applications 72
        Antenna and RF Filter 72
        LNA and Quadrature Mixer 73
        Baseband Filter and VGAs 74
        Analog-to-Digital Converter..74
        Frequency Synthesizer and VCO 74
Conclusions...74
References...75

INTRODUCTION

The widespread deployment of wireless communication infrastructures is turning the dream of being always connected to reality. The increasing demand for wireless services has been the driving force leading to the explosive growth of several wireless mass markets: cellular phones (global system for mobile communications [GSM], wideband code division multiple access [W-CDMA]), wireless local area networks [WLANs] (IEEE 802.11a/b/g), wireless personal area networks (Bluetooth), etc. A clear breakthrough in the wireless world, then dominated by broadcast services, came with the introduction of cellular networks. The simple concept of arranging an area in a cellular structure and reusing the same radio frequency (RF) carrier in a set of nonadjacent cells allowed serving a large number of users with a limited dedicated spectrum. Three generations of wireless phone networks have been deployed so far, starting with the first analog generation, through the second-generation digital systems, dominated by the worldwide success of GSM, to the third generation also known as W-CDMA within the universal mobile telecommunication system (UMTS) standard. A revolution of the same magnitude has affected the world of wired communications and personal computing with the introduction of the Internet. Bridging the gap between mobility and connectivity of personal computer networks was the goal of wireless-fidelity (Wi-Fi) technologies, including all IEEE 802.11 standards. Other technologies have also received worldwide acceptance. In the field of low power, low data rate wireless personal area networking (WPAN), Bluetooth is now the prevailing solution.

The remarkable growth of wireless communications owes much to the impressive reduction in mobile equipment costs brought about by the continuous advances in microelectronics. Progress in semiconductor process technology has led to the emergence of CMOS as RF-capable low-cost integrated circuit (IC) technology. The introduction of lightly doped CMOS substrates and thick top-metal layers significantly improved the quality of on-chip inductors and, coupled with the continuous scaling of transistor feature size, opened the way to high-performance RF circuits in CMOS [1]. Innovative simulation techniques and analysis methods improved the understanding of critical building blocks, such as mixers and oscillators [2–4]. Several techniques, some of which had been known for decades, were put to work. This completely changed the way radio terminals were built, increased the overall level of integration, and significantly reduced the bill-of-materials (BOM). Polyphase filters were rediscovered [5] and used at RF and intermediate frequency (IF) to implement on-chip image rejection [6]. This enabled the introduction of highly integrated low-IF receiver architectures, eliminating the need for expensive IF surface acoustic wave (SAW) filters, used in the classical super-heterodyne architecture, at the expense of a higher chip area. Further investigation of the mechanisms limiting critical performances of integrated CMOS receivers [7,8] ($1/f$ noise and second-order input intercept point [IIP2] of the down-conversion mixers) and the application of digital calibration techniques [9] have finally made possible to fulfill even the most stringent standard specification using a fully integrated direct conversion receiver, further reducing the costs [10,11]. The use of more digital-intensive approaches in the baseband part of the receiver has enabled more area-conscious designs and still promises to give significant innovations [12]. On the transmitter side, innovative solutions have appeared that are slowly replacing the once-dominant Cartesian modulator architecture. Fractional-N frequency synthesizers were used as direct modulation transmitters for constant envelope modulations, eliminating the need for post-power amplifier (PA) filtering [13]. An even more compact synthesizer solution later emerged, replacing the analog phase locked loops (PLL) with an all-digital PLL [14]. Techniques for efficient and linear amplification of variable envelope signals have been known for decades but have until recently remained only a theoretical essay due to various implementation obstacles. Thanks to the extensive use of digital calibrations, a polar modulator has become a real product for GSM-enhanced data for GSM evolution (EDGE) transmitters [15].

To fully exploit the wide variety of existing communication infrastructures, new mobile terminals are needed, capable to recognize the environment and adapt to various radio transmission and access management techniques, for voice and data communications, depending on the availability and convenience. To realize global roaming, for both voice and data applications, all of the already mentioned standards need to be included in a multiservice radio terminal. In fact, GSM and UMTS represent the present and the future of the vocal and mixed voice/data cellular service, respectively. 802.11a/b/g WLANs are the dominant standards for high data rate wireless Internet access, and Bluetooth enables the terminal to be air-connected with other peripherals to exchange data at low rates. This framework has also been depicted by the standardization bodies and a global scenario has been described, for example, by the 3GPP where a set of possible interworking solutions have been foreseen, defining for each some network requirements to support interoperability through seamless handover among heterogeneous systems [16]. In the longer term, next-generation wireless standards will incorporate an extended degree of flexibility to provide a variable capacity depending on the signal-to-interference-plus-noise ratio. For instance, next-generation cellular phones evolving from 3GPP-UMTS will have provisions for variable bit rate and channel bandwidth [17]. Furthermore, new services, such as location-aware services, based on a multiplicity of radio access technologies may become available [18]. This is motivating the trend toward the convergence of multiple communication devices (phone, video-game console, personal digital assistant [PDA] digital camera, web browser, e-mail, etc.) into a wireless device that can be used anywhere in the world. In this scenario multistandard radio terminals offer the possibility to integrate multiple devices into one, reducing form factor and cost and improving access to wireless services [19]. To achieve this goal, highly integrated systems must work across multiple standards and achieve maximum hardware reuse at minimum power consumption. The choice of receiver, transceiver, and building blocks architectures that are amenable to be efficiently reconfigured is therefore a crucial issue. To optimize power consumption in each operating mode, detailed knowledge of the system specifications and the underlying trade-offs is required. Moreover, multistandard terminals must follow procedures for switching between different standards that provide additional constraints on the reconfigurability of the underlying hardware in terms of time available to switch from one standard to another both in scan mode and in active mode.

The ultimate solution for the radio of such a multifunctional terminal would be a multistandard radio, built in a very low-cost CMOS technology, programmable to operate according to all major communications standards. A high degree of configurability can be achieved in different ways. One is the software-defined radio (SDR) approach [20]. The original idea shifted all signal processing to the digital domain. In addition to the advantage of being able to support different standards with a unified platform, SDR also allows to introduce new services and features with software upgrades. From a technological point of view, SDR is plagued by severe issues: first of all the unrealistic requests for the receiver RF front-end. However, the general concept of implementing as many signal processing tasks of a communication transceiver as possible by means of programmable devices remains a key implementation strategy for the efficient implementation of a multistandard radio.

For the design of a multistandard terminal, different aspects need to be considered, starting from the antenna down to the digital signal processing (DSP) and MAC interfaces. In this chapter, the possibility of integration of one of the most challenging parts of the receiver is evaluated: the RF front-end. This is typically a crucial part, requiring high dynamic range (low noise, large signal handling capability, and high linearity), especially in fully integrated reconfigurable solutions. To arrive at an efficient multistandard receiver definition each of the four modes of operations (GSM, UMTS, WLAN, and Bluetooth) must be first considered separately: GSM and UMTS, for cellular telephony, are treated in the sections "Global System for Mobile Communications" and "Universal Mobile Telecommunication System," respectively; Bluetooth, for short-range communications, in the section "Bluetooth"; and 802.11b-g, 802.11a, and HiperLAN2, for WLAN access, in the section "Wireless Local Area Networks." For each standard, the specifications of the receiver are derived and used to investigate the optimal architecture for a CMOS implementation. In each section, both commercial products and published research solutions are reviewed to better explain the different possible approaches and the various trade-offs involved. The section "Toward a Multistandard Receiver" formulates a proposal for a multistandard RF front-end. Bluetooth, operating simultaneously with other received signal, would have a dedicated processing path while all other standards would share the same path in a direct conversion solution. The reconfiguration of both RF and analog baseband blocks is the key point to keep consumption within acceptable levels. The final section draws the conclusions.

GLOBAL SYSTEM FOR MOBILE COMMUNICATIONS

Leading actor of the wireless telecommunications market for more than 10 years, GSM accounts today for 78% of the world's digital mobile market. Besides a financial turnover exceeding $500 billion in 2005, and over 1 billion users worldwide [21], GSM has been recently improved with the introduction of the air interface EDGE, providing multimedia services at the moderately high 384 kb/s data rate. Several frequency bands have been allocated to spread the GSM all over the world, enabling global roaming. The original version of the GSM standard, developed by the European Telecommunication Standards Institute (ETSI), was operative in the 900 MHz band and used throughout Europe. Today, as shown in Figure 2.1, the GSM standard is adopted and uses several different carriers. Enhanced GSM (E-GSM) and digital cellular communication system (DCS) are used in Europe, while personal communication services (PCS) has been deployed in the United States. Quad-band GSM cell phones operating in the most diffused bands are today widespread in the market. The most relevant characteristics of the GSM signals are summarized in Table 2.1.

GSM RECEIVER REQUIREMENTS

For the GSM standard [22], a bit error rate (BER) lower than 10^{-4} (or equivalently a 9 dB signal-to-noise ratio [SNR] at the demodulator input) is specified with a -102 dBm minimum received signal, in presence of additive white Gaussian noise. For the 200 kHz-wide GSM signal, a 9 dB maximum noise figure (NF) is allowed at the antenna connector. Consequently, the equivalent antenna-referred noise floor has to be kept below -111 dBm. NF requirement is extremely critical in direct conversion CMOS implementations due to $1/f$ noise in the narrow 100 kHz baseband channel.

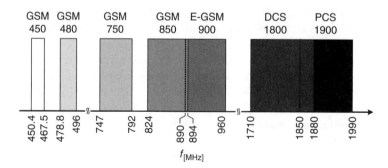

FIGURE 2.1   GSM frequency plan.

TABLE 2.1
GSM Signal Characteristics

Modulation	Gaussian-MSK
Channel bandwidth	200 kHz
Bit rate	270 kb/s
Spectral efficiency	1.3 b/s/Hz

The intermodulation test specifies that a GMSK signal 3 dB above sensitivity level has to be detectable in presence of a −49 dBm continuous wave and a −49 dBm GMSK-modulated interferer at 800 and 1600 kHz frequency offset from the desired signal, respectively. Therefore, the required third-order input intercept point (IIP3) is −18 dBm.

Because of the large allowed power of nearby users, one of the most stringent requirements set by the GSM standard is the local oscillator (LO) phase noise (PN). Figure 2.2 shows the in-band blocking profile. In this test case, −99 dBm GMSK desired signal is received together with a continuous wave blocker whose power is a function of the frequency offset from the desired signal. The reciprocal mixing has to be kept below the noise floor, or equivalently:

$$\mathrm{PN}_{\Delta f} + P_{\mathrm{CW}} + 10\log(200\,\mathrm{kHz}) < \mathrm{Noise\,floor} \qquad (2.1)$$

where $\mathrm{PN}_{\Delta f}$ is the phase noise at Δf frequency offset from oscillator carrier and P_{CW} is the interferer power. The worst condition is set by the 3 MHz blocker, giving a PN requirement of −141 dBc/Hz.

GSM sets a very stringent IIP2 specification. Although the constant-envelope GMSK modulation would not generate wideband second-order intermodulation (IM2) products by its own, the standard specifies an AM suppression test [22]. This test was introduced to avoid receiver desensitization in presence of a GMSK pulse jammer, as produced by the on/off switching signal on another carrier. Figure 2.3 highlights the most critical trade-off for the CMOS direct conversion receiver.

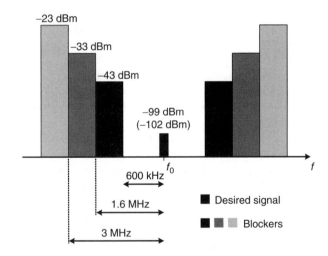

FIGURE 2.2   E-GSM blocking mask.

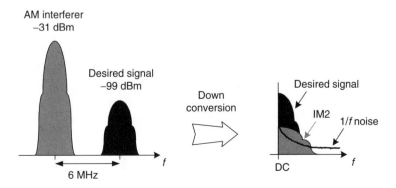

FIGURE 2.3   E-GSM "AM suppression test."

TABLE 2.2
E-GSM Receiver Specifications

Specification	Requirement
Noise figure	6.8 dB
IIP2	46 dBm
IIP3	−21 dBm
LO phase noise	−141 dBc/Hz @ 3 MHz

Both IM2 product generated in the AM suppression test and $1/f$ noise are superimposed onto the down-converted GSM signal, lying from DC to 100 kHz. According to the standard, a −99 dBm desired signal has to be correctly demodulated in presence of a −31 dBm AM interferer. The IIP2 requirement can be calculated as follows:

$$\text{IIP2} > 2P_{AM} - \text{Noise floor} \tag{2.2}$$

where P_{AM} is the AM interferer power. As a result, an IIP2 higher than +49 dBm has to be guaranteed. This is a very challenging specification because with a typical 16–18 dB low-noise amplifier (LNA) gain the IIP2 requirement for the down-converter exceeds +65 dBm. The fundamental limits of down-converters' second-order nonlinearity have been analyzed in details, for both CMOS [8] and BiCMOS [23] implementations. The most diffused technique employed to enhance the down-converter IIP2 is calibration [24–27]. The basic idea is to introduce a rather complicated digital circuitry setting a deterministic asymmetry in the down-converter (typically in the load resistors), to compensate the differential second-order distortion. On the contrary, the solution proposed in Ref. 10 directly addresses the intermodulation sources of the down-converter, leading to a measured IIP2 higher than +78 dBm without calibrations.

Table 2.2 summarizes the GSM receiver specifications, calculated at the antenna connector. BER of 10^{-4} shall be maintained with input power level up to −40 dBm, and can increase to 10^{-3} with input power up to −15 dBm. Notice that taking into account the attenuation profile of a commercial RF filter [28], the maximum allowed NF drops to 6 dB, when referred at the receiver input.

GSM Receivers: Evolution and State-of-the-Art

The very narrowband GSM signal (200 kHz) is extremely susceptible to $1/f$ noise and DC offsets. This is the reason why the earliest chipset solutions for GSM were based on super-heterodyne [29] or, more recently, on single [30] or double [31] conversion to a very low intermediate frequency (∼ 100 kHz). A pioneering investigation on the feasibility of a direct conversion solution was done in 1991 by Sevenhans et al. [32] in a 9 GHz silicon bipolar technology. The receive path in their study used an external LNA to reduce to 6 dB the NF of the IC, comprising quadrature mixers, LO generation circuitry, and postmixer amplifiers.

The extensive research activity and the technology advances have resulted, today, in several BiCMOS implementations of direct conversion receivers, covering multiple GSM bands, employing a number of techniques to overcome the inherent limitations of this architecture. Specifically, state-of-the-art SiGe and BiCMOS technologies lend themselves to excellent $1/f$ noise performances and optimum device matching. DC offset problem is solved by measuring and compensating offset in idle mode. Time-varying DC offsets are removed by dynamic calibration and DSP techniques [33]. IIP2 is usually improved by means of rather complex calibration techniques [27,34,35]. For example, in Kim et al.'s study [27] the second-order distortion (IM2) is reduced by about 15 dB by means of an 8-bit calibration block, composed of digitally controlled resistors.

Several companies, such as Texas Instruments [26], Samsung [27], Skyworks [35], Philips [36], and Analog Devices [37], are selling high-performance GSM direct conversion chipsets that adopt advanced SiGe or BiCMOS technologies. On the contrary, CMOS implementations are still extremely rare, mainly due to the worse performance in terms of $1/f$ noise and device matching. Most CMOS commercial products are still based on low-IF receiver topology [38], where both $1/f$ noise and DC offsets are minor concerns. At present, only GCT semiconductors [39] and Infineon [40] demonstrated direct conversion solutions. But in Song et al.'s study [39] the reported 37 dBm IIP2 is far from meeting the 49 dBm requirement. The solution offered by Bonnaud et al. [40], realized in 130 nm CMOS, shows very good NF and IIP2 performances. However, to reach the 50 dBm IIP2 spec, mixers are artfully driven by a large square wave, realized by an LO chain supplied at 2.1 V, while the core circuits are at 1.5 V.

State-of-the-art implementations demonstrate that the major bottleneck in direct conversion GSM CMOS receivers is within the down-converter design. In GSM not only IIP2 but also, and eventually even more, $1/f$ noise is a concern. This is the main reason why direct conversion GSM receivers are still shyly making inroads.

GSM Direct Conversion Receiver: Building Blocks

The proposed direct conversion receiver block diagram is shown in Figure 2.4. The overall seventh-order filtering reduces all the blockers to a level below the desired signal at the ADC input, that is, the ADC dynamic range is dominated by the useful signal. The adopted building blocks' 3 dB cut-off frequencies are reported in Table 2.3. Notice that the capacitors allowing this low-frequency filtering

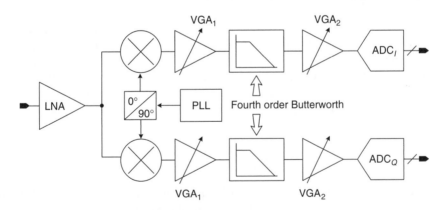

FIGURE 2.4   The proposed E-GSM CMOS direct conversion receiver.

TABLE 2.3
E-GSM Receiver Building Blocks Specifications

Block	Gain [dB]	Input Referred Noise	IIP2 [dBm]	IIP3 [dBm]	3 dB Cut-Off Frequency [kHz]
LNA	22	2 dB NF	–	−5	–
Mixer	10	6 nV/$\sqrt{\text{Hz}}$	70	5	300
VGA$_1$	20	10 nV/$\sqrt{\text{Hz}}$	45	6	250
Filter	20	25 nV/$\sqrt{\text{Hz}}$	25	6	120
VGA$_2$	30	30 nV/$\sqrt{\text{Hz}}$	−10	−5	200

can be in the order of few nanofarad, i.e., high-density structures should be employed in a fully integrated solution to restrain area occupation.

Table 2.3 shows the building block specs. The front-end design is extremely critical. The key idea is to achieve high gain in the LNA to ease the noise specification for the mixer and first VGA. Notice that even with a high 22 dB LNA gain, the mixer 6 nV/$\sqrt{\text{Hz}}$ requirement is still extremely critical in the 0–100 kHz band. However, a high LNA gain determines higher mixer linearity requirements. Also, VGA_1 deserves attention, because the noise and IIP2 requirements are still critical. After the VGA_1, the 6 MHz offset AM interferer is already sufficiently filtered and is no more a concern for the filter and VGA_2 design. The analog-to-digital converter (ADC) minimum required resolution is 6 bits. The receiver-cascaded gain is 102 dB.

GSM: CONCLUSIONS

The design of a CMOS direct conversion GSM receiver is very challenging, especially for the 1/f noise contribution in the narrowband 100 kHz-wide baseband channel and for the very high IIP2 requirement. However, a fully integrated direct conversion transceiver has been recently presented [40] in 130 nm CMOS. Furthermore, the 90 nm front-end supplied at a voltage as low as 750 mV as proposed in Ref. 11 showing 3.5 dB NF, 15 kHz 1/f noise corner, and 51 dBm minimum IIP2 opens the road for direct conversion GSM solutions also in future subscaled 65 and 45 nm nodes.

UNIVERSAL MOBILE TELECOMMUNICATION SYSTEM

The third generation (3G) of global wireless systems provides voice and information services with various data rates using W-CDMA. The European 3G version named UMTS–terrestrial radio access (UTRA) [41] specifies data rates up to 384 kb/s for outdoor applications and up to 2 Mb/s indoors. The information-bearing signal is modulated by a pseudorandom sequence (spreading code). Through this spreading process, the signal bandwidth (varying from 8 to 384 kHz) is broadened up to nearly the code bandwidth (fixed to 3.84 MHz). The resulting wideband signal modulates the carrier with quadrature phase shift keying (QPSK), in the downlink process. At the receiver, the de-spreading process uses the same code applied in the transmitter to recover the original spectrum of the data signal, whose power spectral density increases by an amount given by the spreading factor (SF):

$$SF = 10 \log \frac{T_{\text{bit}}}{T_{\text{chip}}} \tag{2.3}$$

where T_{bit} and T_{chip} are the bit and chip durations, respectively. The result of the spreading and de-spreading operations on the signal bandwidth is illustrated in Figure 2.5. The code generated in the receiver must be well synchronized to the desired signal because timing errors result in SNR loss.

UMTS is a continuously transmitting and receiving frequency division duplexing (FDD) system. At the user equipment side, the TX band is located between 1920 and 1980 MHz while the RX

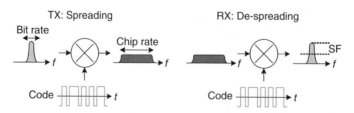

FIGURE 2.5   UMTS: Spreading and de-spreading processes. The proposed Bluetooth CMOS low-IF receiver. (From Brandolini, M., Rossi, P., Manstretta, D., and Svelto, F., *IEEE Trans. Microwave Theory Tech.*, 53, 1026–1038, 2005. With permission. © [2005] IEEE.)

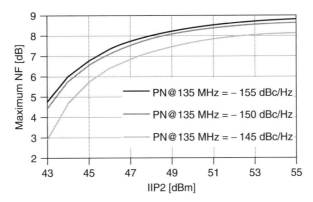

FIGURE 2.6 UMTS: Maximum allowed NF as a function of IIP2 with LO phase noise as a parameter. (From Brandolini, M., Rossi, P., Manstretta, D., and Svelto, F., *IEEE Trans. Microwave Theory Tech.*, 53, 1026–1038, 2005. With permission. © [2005] IEEE.)

band is between 2110 and 2170 MHz. The minimum frequency TX–RX spacing is thus 135 MHz. The channel spacing is 5 MHz. As shown in the next section, because of the duplexing technique adopted by UMTS, a highly linear receiver is required to reject the transmitter leakage into the RX section.

UMTS Receiver Specifications

The receiver specifications heavily depend on commercial duplexer performances. The following receiver requirements are obtained assuming typical commercially available duplexer with 1.8 dB insertion loss (IL) and 54 dB isolation between TX and RX sections.

In the sensitivity test, considering only antenna and receiver noise, 9 dB NF and −99 dBm antenna-referred maximum noise powers should not exceed 10^{-3} BER [42]. IM2 products, due to transmitter ($IM2_{TX}$), and reciprocal mixing of TX leakage, due to oscillator PN (RM_{TX}), also need to be considered in the sensitivity test and added to the receiver noise [43]. Consequently, the following relation holds:

$$-99\,dBm > N_R + IM2_{TX} + RM_{TX} \tag{2.4}$$

where N_R is the antenna-referred receiver thermal noise. Considering a class III UMTS transmitter [41], the TX leakage is −30 dBm. In Figure 2.6, the trade-off given by Equation 2.4 can be evaluated. The reciprocal mixing product is made negligible by setting the PN specification at −150 dBc/Hz at 135 MHz offset. To keep the NF specification within acceptable limit, an IIP2 requirement of at least +46 dBm has to be met. The resulting noise target is a 6 dB NF (referred at receiver input).

The transmitter leakage is also responsible for intermodulation products due to third-order nonlinearity, together with an out-of-band continuous wave interferer. In the worst case these two interfering signals are placed 135 and 67.5 MHz apart from the receive band, respectively. Because the desired signal power is 3 dB above the sensitivity level in presence of out-of-band interferers and the noise level is −99 dBm, the upper limit for the third-order intermodulation (IM3) product is set to −99 dBm as well. The required out-of-band IIP3 ($IIP3_{out-of-band}$), evaluated at the receiver input, is given by

$$IIP3_{out-of-band} = \frac{1}{2}T_{leak}[dBm] + P_{CW}[dBm]$$

$$-\frac{1}{2}(IM3[dBm] - IL[dB]) = -4.6\,dBm \tag{2.5}$$

where T_{leak} is the transmitter leakage power at the receiver, P_{CW} the continuous wave power (-40 dBm at the receiver input), and IM3 the power of the IM3 product.

The standard sets an in-band intermodulation test [41], where the following interferers are considered: a -46 dBm continuous wave and a -46 dBm W-CDMA–modulated interfering signal, placed 10 and 20 MHz apart from the desired signal carrier frequency, respectively. The resulting antenna referred in-band IIP3 requirement is -17 dBm [42]. Notice that, while the in-band requirement demands high linearity throughout the receiver (including the baseband circuits), the out-of-band specification mainly affects the RF front-end because the interferers can be strongly attenuated at the down-converter output. Finally the standard defines the maximum received signal power as -25 dBm. Table 2.4 summarizes the UMTS receiver specifications referred to receiver input.

UMTS Receiver: State-of-the-Art

The zero-IF architecture is the most promising solution for highly integrated UMTS receivers [42–45]. In fact, a low-IF approach would lead to a higher image rejection and ADC dynamic range requirements [42]. At the same time, the typical drawbacks of the zero-IF architectures, such as DC offset and, in CMOS technology, $1/f$ noise, are of minor concern due to the wideband nature of the signal. Moreover in the zero-IF architecture the image is the signal itself, and due to the low SNR required, 23 dB of I–Q accuracy is sufficient. Nevertheless, the TX leakage in the receiver path leads to extremely high linearity requirements hard to meet by the direct down-converter. State-of-the-art solutions employ a modified direct conversion architecture with an external filter [46–50]. As reported in Figure 2.7, an external interstage band-pass RF filter placed between two LNAs in the front-end section attenuates the TX leakage in the receive path, strongly relaxing IM2 requirements [44,45,51].

TABLE 2.4
UMTS Receiver Specifications

Specification	Requirement
Noise figure	6 dB
IIP2	46 dBm
IIP3$_{out-of-band}$	-4.6 dBm
IIP3$_{out-of-band}$	-18.8 dBm
LO phase noise	-150 dBc/Hz @ 135 MHz

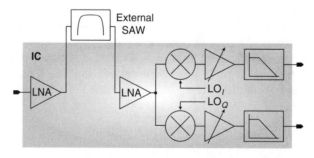

FIGURE 2.7 Typical commercial UMTS zero-IF receiver, employing an external SAW to suppress the transmitter signal leakage.

An important step toward fully integrated direct conversion solutions for UMTS is the high IIP2 mixer (up to 80 dBm) for UMTS [10] allowing the elimination of bulky and expensive interstage SAW filters.

UMTS DIRECT CONVERSION RECEIVER: BUILDING BLOCKS

The proposed zero-IF solution for a CMOS UMTS receiver is shown in Figure 2.8. The RF front-end consists of a low-noise amplifier and an active down-converter while the baseband is made of two VGAs, a fourth-order Butterworth filter and a 6-bit ADC. Table 2.5 summarizes the requirements of the receiver building blocks. The receiver has to be very linear in terms of IIP2 and IIP3. The mixer IIP2 requirement is very tough and specific design solutions to achieve high linearity are mandatory. The mixer first-order filtering is not sufficient to reduce the TX leakage; therefore, a quite stringent IIP2 requirement is also set for the first VGA.

Both front-end and baseband are required for high IIP3. The out-of-band intermodulation, resulting from TX leakage, together with the interferer 67.5 MHz apart from the receive band, determines the third-order linearity requirements of the RF front-end and of VGA_1. After the cascade of two first-order filtering, given by mixer and VGA_1, the out-of-band interferers are completely filtered out. On the contrary, the two interfering RF signals 10 and 20 MHz from received signal are slightly affected by the first stage filter; therefore, in-band intermodulation test mainly determines the IIP3 requirement for the fourth-order Butterworth filter.

The front-end noise performance is stringent. A 2 dB LNA NF is required to account for a typical 0.5–1 dB balun insertion loss; the estimated input-referred noise density for the mixer is 4.5 nV/√Hz. With about 33 dB of maximum front-end gain, the baseband noise contribution can be made negligible in the

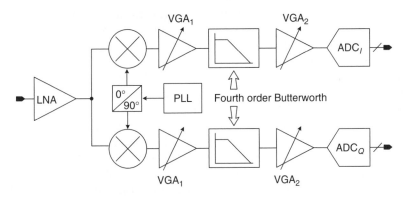

FIGURE 2.8   The proposed UMTS CMOS zero-IF receiver.

TABLE 2.5
UMTS Receiver Building Blocks Specifications

Block	Gain [dB]	Input Referred Noise	IIP2 [dBm]	IIP3 [dBm]	3 dB Cut-Off Frequency [MHz]
LNA	18	2 dB NF	–	0	–
Mixer	15	4.5 nV/√Hz	64	12	6
VGA_1	24	10 nV/√Hz	40	5	10
Filter	20	25 nV/√Hz	25	0	2
VGA_2	24	30 nV/√Hz	0	0	10

overall noise budget. The voltage-controlled oscillator (VCO) minimum tuning range is set to 60 MHz. A high-pass filter removes the DC offset. With the baseband signal going from DC to 1.92 MHz, a high-pass cut-off frequency of several kHz does not lead to a significant SNR degradation [42].

UMTS: Conclusions

UMTS sets stringent requirements in terms of noise and linearity for a zero-IF CMOS receiver. Careful design allows meeting linearity specification without the use of an additional external surface acoustic wave (SAW) filter between LNA and mixer. The resulting IIP2 and IIP3 are strongly dependent on the duplexer TX-to-RX isolation performance and on the class of transmitter used. Thanks to the large signal bandwidth, DC offset and $1/f$ noise are of minor concern. Therefore, zero-IF is the most suited architecture for a fully integrated solution.

BLUETOOTH

Bluetooth has definitely established itself as the short-range solution for WPAN. Mobile phones are dominating Bluetooth-enabled product shipments, and they are currently helping the penetration of this technology into other products, such as PDAs, notebook PCs, headsets, automobiles, and portable digital music players.

The Bluetooth standardization started at the beginning of 1998. A Special Interest Group (SIG) was formed by nine major telecom and PC companies (IBM, Intel, Nokia, Toshiba, Ericsson, Microsoft, Agere Systems, 3Com, and Motorola) to develop a standard for short-range radio connectivity between electronic devices [51]. The SIG at the end of 1999 released the first version of the Bluetooth protocol. The objective of these companies was to build a low-cost and low-power radio frequency technology for short-range communications. This enables to replace the special wires or cables used today to interconnect electronic devices together. Moreover, Bluetooth technology can be used to provide *ad hoc* networks or data/voice access points.

In this scenario, Bluetooth technology was standardized to allow a global operation range and a low-cost single-chip implementation [52,53]. The choice of the frequency allocation in the 2.4 GHz license-free industrial-scientific-medical (ISM) band enables a worldwide diffusion. Frequency-hopping spectrum spreading (FHSS) has been adopted to improve interference immunity from neighboring sources such as microwave ovens and cordless phones. Spectrum spreading is accomplished by pseudorandom hopping sequence through the RF channels, displaced by 1 MHz and centered at $2402 + k$ MHz with $k = 0–78$. The nominal hop rate is 1600 hops/s. Bluetooth channels are binary-FSK modulated, Gaussian shaped (BT = 0.5 and the modulation index between 0.28 and 0.35) with frequency deviations of ± 160 kHz around the carrier to code binary 1 and 0.

As Bluetooth chips are intended for portable, battery-driven equipments, such as cellular phones and personal digital assistants, provisions for saving power consumption are mandatory. Time division duplexing (TDD) eliminates the need for separate TX and RX oscillators and for expensive duplex filters. Moreover, at the transmitter side, the choice of GFSK constant-envelope signals allows the use of efficient transmitter architectures, on the basis of VCO or synthesizer direct modulation. The standard specifies three TX classes of 0, 4, and 20 dBm maximum powers, with an expected range of 10–50 m. At the receiver side, a high sensitivity level (−70 dBm) is specified [53] to allow the choice of inexpensive RF filters and CMOS single-chip implementations.

Bluetooth Receiver Specifications and State-of-the-Art

The −70 dBm required sensitivity level translates into 44 dB input-available SNR. The specified 10^{-3} maximum BER can be achieved with a 21 dB SNR. As a result, NF as high as 23 dB is allowed. In addition, the linearity requirement is relaxed. The intermodulation test specifies −39 dBm interferers with −64 dBm received signal [53], yielding an IIP3 of −15 dBm. The blockers mask sets a PN specification of −109 dBc/Hz at 1 MHz offset from the carrier, a much simpler requirement compared to cellular phone standards. The bandwidth that the VCO covers spans 2.4–2.48 GHz. The maximum mean power received at the antenna port is specified as −20 dBm.

The most attractive architecture for a Bluetooth receiver is low-IF. In this case, the required 29 dB image rejection [54] can be easily obtained without particular effort and power consumption. This compares to an image-rejection requirement of about 34 dB for direct conversion. Moreover, the Gaussian-shaped BFSK signal has most of its energy in about 200 kHz (3 dB bandwidth) so that a zero-IF CMOS solution would require some effort to address $1/f$ noise and DC offset removal.

Most of the Bluetooth commercial solutions employ inexpensive, fully integrated CMOS low-IF receivers [54–62] with an intermediate frequency of 1 or 2 MHz. The image suppression is achieved in the analog domain by polyphase or complex tunable filters. Low-IF receivers are strongly preferred over direct conversion [63] or super-heterodyne [64] versions because the former easily allow very good sensitivity performance with very low power consumption.

Moreover, to minimize the power consumption, analog demodulation is often preferred over power-hungry ADCs and digital demodulators [54,57,62,65–68].

BLUETOOTH RECEIVER: THE PROPOSED SOLUTION

The relaxed receiver specifications allow the designer to investigate low-power, small-area solutions without great concern on performance. Therefore, the proposed solution is a 2 MHz low-IF with analog demodulator, similar to that in Ref. 54, as shown in Figure 2.9.

In the RF front-end, a 5 dB NF LNA is assumed, with a gain of 18 dB and an IIP3 of −5 dBm. The mixer has to down-convert the received signal to a 2 MHz IF, where a fourth-order complex active filter performs the channel selection and image suppression. The baseband filter has 1 MHz bandwidth. A multistage limiter adjusts the desired signal to an adequate voltage level for the following processing. A differential analog demodulator recovers the signal binary data directly from the GFSK-modulated signal, centered at 2 MHz. The frequency demodulator is shown in Figure 2.10 [54]. It is composed of differentiators, multipliers, subtractor, and LPF. I and Q signals are differentiated, multiplied by Q and I signals, respectively, and then subtracted. The demodulated signal is then filtered, converted to digital bits by a slicer, and presented to the DSP.

BLUETOOTH: CONCLUSIONS

The Bluetooth technology asks for low-cost, low-power, moderate performance transceivers to allow a wide diffusion in portable electronic devices. Therefore, a fully integrated CMOS solution

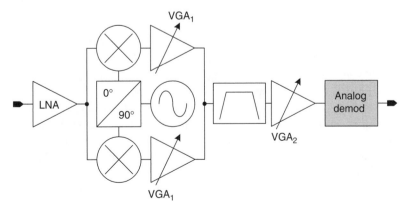

FIGURE 2.9   The proposed Bluetooth CMOS low-IF receiver. (From Darabi, H., et al., *IEEE J. Solid-State Circuits*, 36, 2016–2024, 2001. With permission. © [2006] IEEE.)

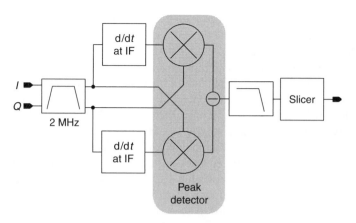

FIGURE 2.10   The Bluetooth analog demodulator.

is optimal. A CMOS receiver, using 2 MHz low-IF architecture with analog demodulator, seems to represent the best solution.

WIRELESS LOCAL AREA NETWORKS

The purpose of the IEEE 802.11 standard [69], formalized in 1999, was to allow high-rate wireless network connectivity between personal computers or workstations, avoiding the use of expensive and bulky wires. Today, WLANs are used worldwide in work environments, at home, and in "hot spots" at airports, hotels, and other public places for delivering high-speed Internet access. The original IEEE 802.11 standard provides a maximum data rate of 2 Mbps and allows radio implementations with frequency hopping (FH) or direct-sequence spread spectrum (DSSS), in the 2.4 GHz license-free ISM band. The need for more speed determined the definition of new standards, named IEEE 802.11a [70] (harmonized with ETSI HiperLAN2 [71]) and IEEE 802.11b [72]. The goal of the former was to standardize a high-rate (up to 54 Mbps) WLAN in the 5 GHz band, while the latter's objective was to extend the throughput of the original 802.11 standard to data rates higher than 2 Mbps (reaching 11 Mbps). Finally, the IEEE 802.11g committee has recently drafted a high-speed wireless LAN standard [73] in the 2.4 GHz band that is backward-compatible with 802.11b and supports data rates up to 54 Mbps.

IEEE 802.11/.11b

The IEEE 802.11 standard specifies two radio transmission schemes for wireless networking in the 2.4–2.4835 GHz ISM frequency range, with data rates up to 2 Mbps, on the basis FHSS and DSSS. In the latter case, the radio channel of about 14 MHz is modulated by differential binary PSK (DBPSK) and differential quadrature PSK (DQPSK) to provide 1 and 2 Mbps data rates, respectively. Then, the 802.11b committee extended the DSSS transmission scheme, and adopting complementary code keying (CCK) enables operation at 5.5 and 11 Mbps, employing the same signal bandwidth. The 11 Mbps operation mode, with a spectral efficiency of 0.78 b/s/Hz, is the most challenging for the receive section of the 802.11b mobile terminal. The adopted modulation scheme theoretically requires an SNR of about 9.5 dB at analog output to provide 8% maximum frame error rate (FER) allowed by the standard. However, taking into account the effects of the multipath channel, and the finite accuracy of the transmitted signal (specified as error vector magnitude [EVM]), as shown in Figure 2.11, the SNR requirement raises to about 11.5 dB.

Because the standard specifies −76 dBm sensitivity level, the maximum allowed NF can be easily calculated as follows:

$$NF = -76\,\text{dBm} - [-173.8\,\text{dBm} + 10\log(14\,\text{MHz})] - 11.5\,\text{dB} = 14.8\,\text{dB} \qquad (2.6)$$

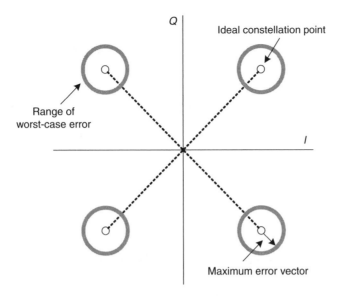

FIGURE 2.11 IEEE 802.11b: Maximum allowed deviation from the ideal constellation point. (From Brandolini, M., Rossi, P., Manstretta, D., and Svelto, F., *IEEE Trans. Microwave Theory Tech.*, 53, 1026–1038, 2005. With permission. © [2005] IEEE.)

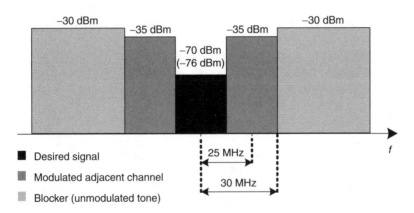

FIGURE 2.12 The IEEE 802.11b considered blocking mask. (From Brandolini, M., Rossi, P., Manstretta, D., and Svelto, F., *IEEE Trans. Microwave Theory Tech.*, 53, 1026–1038, 2005. With permission. © [2005] IEEE.)

Intermodulation test is not specified in IEEE 802.11 standard. Therefore, receiver third-order linearity requirement reflects 1 dB compression point requirement, determined by adjacent channel or blocking signals, as specified in Ref. 74. The possible cases are represented in the blocking mask depicted in Figure 2.12. The −70 dBm desired signal can be received together with −35 dBm modulated adjacent signal, or with a −30 dBm continuous wave interferer. The latter sets the receiver 1 dB compression point to about −26 dBm, taking into account 4 dB safety margin. The −30 dBm interferer also sets the LO PN requirement to −103 dBc/Hz at 1 MHz offset. As in the case of Bluetooth, the VCO tuning range is 80 MHz.

The IEEE 802.11b standard specifies a −10 dBm maximum received signal. In this case a more critical requirement is set in the 2 Mbps situation, where a −4 dBm maximum signal has to be received and correctly demodulated. The receiver low-gain-case 1 dB compression level is therefore set to about 0 dBm. The main IEEE 802.11/.11b receiver requirements are summarized in Table 2.6.

TABLE 2.6
Summary of Wireless LAN Receiver Specifications

Specification	802.11b	802.11a	802.11g
Noise figure	14.2 dB	7.5 dB	7.5 dB
1 dB C.P. (high gain)	−26 dBm	−26 dBm	−26 dBm
1 dB C.P. (low gain)	0 dBm	−20 dBm	−10 dBm
LO phase noise (1 MHz offset)	−101 dBc/Hz	−102 dBc/Hz	−102 dBc/Hz

IEEE 802.11a

In September 1999, the 802.11a task group approved the 5 GHz wireless LAN standard. The United States allocated the frequency spectrum universal networking information infrastructure (UNII) band and divided it into three portions: the UNII-1 and UNII-2 (from 5.15 to 5.35 GHz) are intended for indoor and outdoor use, while the UNII-3 (from 5.725 to 5.85 GHz) is for outdoor applications only. Each UNII band provides four nonoverlapping 16.6 MHz-wide OFDM-modulated radio channels. The allowed throughput varies from 6 to 54 Mbps.

The choice of this type of modulation is due to its high spectral efficiency and reduced multipath intersymbol interference (ISI). Each radio channel is composed of 48 low-rate modulated orthogonal subcarriers and 4 pilot tones, used for synchronization. Each subcarrier can be modulated by BPSK (with 6 or 9 Mbps throughput, depending on the coding rate), QPSK (12–18 Mbps), 16-QAM (24–36 Mbps), or 64-QAM (48–54 Mbps), and the actual bit rate is set by the channel interference conditions.

In Europe, the 5 GHz spectrum wireless LAN has been standardized by the ETSI task group broadband radio access networks (BRAN). This standard, named HiperLAN2, defines the wireless access in a very similar way as the IEEE 802.11a, adopting the same OFDM modulation with up to 54 Mbps throughput. The physical layer specifications are very similar, the major difference being the allocated spectrum, which ranges from 5.15 to 5.35 GHz for indoor use, and from 5.47 to 5.725 GHz for outdoor applications.

The most stringent receiver requirements are set in the 54 Mbps mode. In this case, to meet the specified packet error rate (<10%), a 20.5 dB SNR has to be guaranteed at the ADC output. For a 50 ns delay spread radio channel, the requirement grows to about 26.5 dB. The maximum transmitter EVM being −25 dB (not considering fading), the signal-to-receiver-noise ratio requirement becomes about 28 dB. Owing to the very high SNR requirement, even in the minimum signal condition, interferences play a significant role. Lumping various imperfections (quadrature accuracy, integral PN, finite ADC word-length, etc.) into a 1 dB implementation loss, and because the 54 Mbps sensitivity level is −65 dBm, the maximum allowed NF is set to

$$\text{NF} = -65\,\text{dBm} - [-173.8\,\text{dBm} + 10\log(16.6\,\text{MHz})] - 29\,\text{dB} = 7.5\,\text{dB} \tag{2.7}$$

To derive the linearity and LO PN requirements, it has to be noted that the IEEE 802.11a standard specifies only the adjacent channel (−63 dBm at ±20 MHz offset) and alternate channel (−47 dBm at ±40 MHz offset) power levels. Similar to the 802.11b case, interferers at more than 40 MHz frequency offset should be considered to ensure operation in a practical environment. In the 5 GHz band, the ETSI HiperLAN2 standard specifies the power level for these blockers, which is set to −30 dBm at frequency offset larger than 50 MHz [71]. If this scenario is also assumed for IEEE 802.11a, the continuous wave −30 dBm interferers set both the linearity and LO PN requirements. Therefore, the receiver 1 dB compression point is set to −26 dBm, while the oscillator spectral purity is set to −102 dBc/Hz (at 1 MHz offset). The maximum received signal is −30 dBm; therefore, the

1 dB compression point in the receiver low-gain mode is set to about $-20\,$dBm, taking into account 10 dB back-off. The IEEE 802.11a receiver requirements are summarized in Table 2.6.

IEEE 802.11g

To improve the data throughput in the 2.4 GHz ISM band, the IEEE 802.11g workgroup has drafted in January 2001 a high-speed wireless LAN [73] that is backward compatible with 802.11b (the same spectrum is used) and allows data rate up to 54 Mbps, so far reaching the 802.11a maximum speed. The idea is to use the same 802.11b modulation techniques to achieve 1, 2, 5.5, and 11 Mbps to allow interoperability with older cards, while also allowing the OFDM modulation to arrive at 802.11a data throughputs.

Therefore, 802.11g receivers have to meet the most critical requirements set by 802.11b and 802.11a. The main difference is the maximum received signal power level, set to $-20\,$dBm [73], which translates into a 1 dB compression point requirement of $-10\,$dBm in the receiver low-gain mode. The IEEE 802.11g receiver requirements are summarized in Table 2.6.

WIRELESS LAN RECEIVERS: STATE-OF-THE-ART

State-of-the-art fully integrated receivers for 802.11b standard mostly adopt the zero-IF architecture. Examples of transceivers exploiting this architecture by using SiGe and BiCMOS technology can be seen in Refs. 75, 76. The wideband channel and the spread nature of the data in the channel can be exploited to enable the use of CMOS technology. In fact, DC offsets can be easily removed by high-pass filtering during the long training sequence, while the $1/f$ noise impact has to be integrated in the whole channel bandwidth, resulting in a minor concern. On the contrary, prohibitively high image rejections are required for low-IF architectures. Up to date, 802.11b CMOS zero-IF transceivers have been presented [77,78].

For the 802.11a standard, the best choice is less straightforward. In fact, considering both zero-IF and low-IF as the most attractive alternatives, OFDM modulation sets very stringent requirements. Not only has the RF front-end to achieve low NF, more stringent than in all other standards considered here, and at more than twice the operating frequency, but due to the high SNR and large bandwidth required for 54 Mbps, the baseband/IF chain is also very challenging.

In the zero-IF case, the major concerns are DC offset, $1/f$ noise, and quadrature paths accuracy. The first two issues are related to the 20 ppm crystal oscillator accuracy allowed by the standard [70]. This means that a frequency offset as high as 240 kHz could exist between transmitter and receiver LOs. As a result, the DC offset and $1/f$ noise tail can fall very close to the first channel subcarrier. Therefore, a low $1/f$ noise corner and a low high-pass frequency are mandatory to avoid signal corruption. At the same time, the 802.11a training sequence is much shorter than that in 802.11 b, leaving only 8 μs for the DC offset cancellation settling. Moreover, quadrature accuracy as high as 38 dB has to be guaranteed because the required output SNR is very high [79].

In the low-IF case, the DC offset cancellation is not a major problem because the edges of the channels do not carry information. The image suppression required is lower than in the direct conversion case because the adjacent channel power level is 1 dB lower than that of the desired signal and it is specified with a signal 3 dB above sensitivity. The image rejection requirement is set to about 35 dB. The main disadvantage of this solution lies in the signal being centered on 10 MHz, leading to a higher maximum signal frequency and, therefore, higher power consumption in the IF chain.

Several fully integrated 802.11a zero-IF CMOS transceivers have been implemented [80–82]. The most interesting solution is presented in Ref. 81. To eliminate the frequency-offset problem, a fast digital estimation of the frequency error is performed. Then, the error is corrected by multiplying a digitally generated tone at the error frequency with the LO through a linear quadrature mixing stage.

The IEEE 802.11g is the most recent wireless LAN standard and up to date many semiconductor companies have presented solutions and some are even shipping products. The great advantage with respect to 802.11a is that the operating frequency is less than half. This leads to two important differences: first, for a given technology, achieving the same NF is easier, both for thermal and $1/f$ noise considerations. Second, the problems arising from the frequency offset are much less because now the frequency error can be at most 125 kHz. This, together with the fact that operation in the CCK mode (802.11b) needs to be guaranteed for backward compatibility, pushed virtually all solutions toward the zero-IF architecture.

The trend is to build dual-band/triple-mode radios, working in every wireless LAN environment. For example, in Ref. 83, a CMOS fully integrated zero-IF solution is presented that features dual-band/triple-mode capabilities.

Wireless LAN Direct Conversion Receiver: Building Blocks

The receiver proposed here is a CMOS fully integrated zero-IF dual-band triple-mode solution, shown in Figure 2.13. The primary goal is silicon area saving and baseband blocks reuse, tuning the filtering profile according to the standard. The most straightforward way is using the same architecture for every standard. As suggested by Behzad et al. [81], solutions to the problems arising from frequency offset can be found.

The main specifications of the receiver building blocks are summarized in Table 2.7. The multiband RF front-end is followed by a seventh-order filtering, fourth-order Butterworth filter, and two variable-gain amplifiers that together with mixers also perform a filtering function. The ADCs have 10-bit minimum resolution.

The gain of each block is defined considering the sensitivity level (−65 dBm) in maximum data rate condition (54 Mbps) for 802.11a and 802.11g. The noise performance is derived taking into account RF filters for the 2.4 GHz [84] and the 5 GHz [85] bands showing a 2 dB worst-case insertion loss

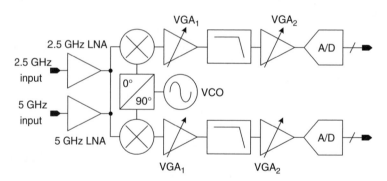

FIGURE 2.13 Wireless LAN proposed receiver solution. (From Brandolini, M., Rossi, P., Manstretta, D., and Svelto, F., *IEEE Trans. Microwave Theory Tech.*, 53, 1026–1038, 2005. With permission. © [2005] IEEE.)

TABLE 2.7
Wireless LAN Receiver Building Blocks Specifications

Block	Gain [dB]	Input Referred Noise	IIP2 [dBm]	1 dB C.P. [dBm]	3 dB Cut-Off Frequency [MHz]
LNA	18	2.5 dB NF	−	−15	−
Mixer	12	4 nV/$\sqrt{Hz}$	60	−5	15
VGA$_1$	12	10 nV/$\sqrt{Hz}$	65	0	20
Filter	0	20 nV/$\sqrt{Hz}$	65	5	11
VGA$_2$	11	30 nV/$\sqrt{Hz}$	65	0	20

and a typical 0.5–1 dB balun insertion loss. The resulting NF requirement for the LNA is quite critical considering carrier frequencies in the 5–6 GHz range.

WIRELESS LANs: CONCLUSIONS

A CMOS fully integrated receiver that enables operation in the 2.4 and 5 GHz bands can be realized in zero-IF architecture although the requirements are very challenging.

TOWARD A MULTISTANDARD RECEIVER

The need for global roaming and high-speed wireless data transfer is increasing the telecom companies' interest for a multistandard handset. Today, the concept of multimode terminal is already widespread. As an example, in Europe most GSM phones already support both the 900 and 1800 MHz versions of the standard. Many cellular phone manufacturers have developed tri-band cellular phones, which can be used both in the United States and in Europe. The next step for the wireless terminal is the introduction of UMTS, working together with GSM. In fact, because of the planetary success of second-generation cellular phones (GSM), the full transfer to the third generation will be carried out after a quite long period of coexistence, if ever.

On the contrary, the increasing demand for wireless services other than voice, such as high data rate Internet access and short-range radio connectivity, is raising the interest for a multistandard radio terminal, capable of satisfying both voice and data services.

The simplest multistandard terminal is realized by means of several transceivers, one for each standard, operating in parallel paths. However, this is not a cost-efficient solution, and with the increasing number of received standards it is, ultimately, unfeasible. To lower the terminal total cost, key aspects are sharing as much hardware as possible, increasing the level of integration and limiting the power consumption. These goals can be met only by careful system and design planning. To arrive at a multistandard terminal, selection of the receiver architecture is key to maximize hardware sharing, thus saving silicon area. Also, technology choice plays a fundamental role. Owing to the improvements in RF passives (high-Q inductors, linear capacitors), greater integration level and lower cost, ultra-scaled CMOS technology should be preferred over BiCMOS [42,86–88]. In the next sections a brief review of the available multistandard or multimode receivers is presented, and a system-level proposal for a future multistandard receiver is drawn.

MULTIMODE AND MULTISTANDARD RECEIVERS: STATE-OF-THE-ART

Most recently published receivers for multimode applications propose the implementation of multiband cellular phones for the GSM standard (900, 1800, and 1900 MHz). Examples of these applications are provided in Refs. 35, 36, 40, where the chosen architecture is zero-IF, for both BiCMOS [35,36] and CMOS [40] implementations. The signal coming from a multiband antenna [89–91] passes through one of the dedicated SAW filters and is then fed to a tuned LNA (one for each band). The mixers are the first blocks, in the receive path, to be shared. Actually, in Refs. 35, 40, there are two mixers, one for the 900 MHz band and the other for the 1800 and 1900 MHz bands. In Ref. 36, only one mixer is used for the three bands. Because the signal bandwidth is the same for each receive band and the RX specifications are very similar, the whole baseband is fully shared, thus saving silicon area.

In WLAN area, there are many examples of multistandard solutions at present. In Ref. 83, a dual-band (2.4 and 5 GHz) multistandard (IEEE 802.11a/b/g) CMOS chipset is presented. In this case the two signal paths, one for the 2.4 GHz band and the other for the 5 GHz band, share only the last baseband blocks of the zero-IF chain. Reference 92 is another example of all-CMOS dual-band radio chip for IEEE 802a/b/g applications, employing a dual-IF architecture. Darabi et al. [93] propose a multimode system that is capable of integrating 802.11b and Bluetooth. Both standards use the unlicensed 2.4 GHz ISM band, therefore, the RF path is shared, while two parallel baseband paths are used to perform channel selection.

UMTS and WLANs are appearing more and more as complementary for high data rate communications rather than as competitors. A receiver radio front-end for the two standards, showing a high level of hardware sharing, is reported in Ref. 94. In the chosen zero-IF architecture, the two dedicated 2.1 and 5.8 GHz LNAs are followed by a common broadband mixer.

A Proposed Architecture for Multistandard Applications

To realize global roaming, for both voice and data applications, GSM, W-CDMA, Bluetooth, and WLAN need to be included in a multiservice radio terminal. In fact, GSM and UMTS represent the present and the future of the vocal and mixed voice/data cellular service, respectively. 802.11a/b/g WLANs are the dominant standards for high data rate wireless Internet access, and Bluetooth enables the terminal to be air-connected with other peripherals, to exchange data at low rates. More precisely, the standards covered by the proposed architecture are two GSMs (1800 and 1900 MHz, for roaming in Europe and the United States), UMTS, and two WLANs (802.11g at 2.4 GHz and 802.11a at 5 GHz).

To arrive at the definition of the radio architecture the following need to be considered:

- The key point for a multistandard fully integrated receiver is minimum area consumption, meaning hardware share maximization.
- The receiver architecture has to enable building blocks reuse.
- All the considered standards except Bluetooth do not need to be covered at the same time, that is, when an application is active, the others can be switched off or kept in idle mode, to save power and reuse hardware resources.
- Bluetooth needs to operate concurrently to each of the other standards. In fact, the wireless link between the radio terminal and other peripherals cannot be halted when voice or data communications are active (for example, Bluetooth allows the use of headphones during a phone call).

The immediate consequence of the last need is that a stand-alone receive path has to be dedicated to Bluetooth, for which the optimum receiver architecture, exploiting analog demodulation, can be used. This section will analyze the possibility of hardware reuse in the other receive path, configurable to cover each of the four remaining standards.

From the analysis carried out in the previous sections, it is evident that the zero-IF is the most suitable architectures for a multistandard receiver in CMOS technology. In fact, it has been widely demonstrated that zero-IF is the most suitable solution for UMTS and WLANs. Furthermore, recent proposals [11,40] enable CMOS direct conversion for GSM applications as well. As a consequence, the block diagram for the multistandard receiver can be figured as the one shown in Figure 2.14.

The requirements of the considered communication standards differ over a very wide range in terms of central frequency, signal bandwidth, NF, and linearity. This will have its impact on the specifications of all building blocks of the radio. The most straightforward solution would be to satisfy the most critical requirements of each receive band. However, with this approach the cascaded performance would be too demanding leading to a nonefficient solution. All building blocks must be versatile and flexible, capable of reconfiguration to satisfy the specifications in different operative conditions. When the receiver is set to receive a given standard (by a digital control) in terms of frequency tuning and band selection, all building blocks must be reconfigured to meet the standard requirements. In other terms, the digital section, which plays a key role to control the receiver settings, can be adequately used to change the single-circuit performance, once a particular standard is selected.

In the following sections a review of the available reconfigurable receiver building blocks is presented.

Antenna and RF Filter

Major hurdles to take in the flexible radio design are associated with the antenna and the RF filter. Currently, antenna performances are generally optimized for narrowband operation and maximum

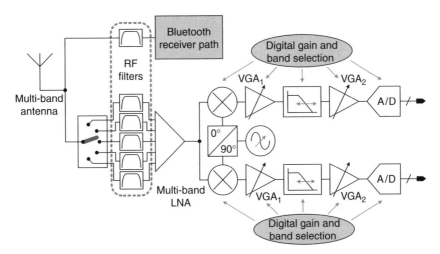

FIGURE 2.14 The proposed multistandard architecture. (From Brandolini, M., Rossi, P., Manstretta, D., and Svelto, F., *IEEE Trans. Microwave Theory Tech.*, 53, 1026–1038, 2005. With permission. © [2005] IEEE.)

blocking of out-of-band interferers. In a multistandard terminal, carrier frequencies vary from 800 MHz up to 6 GHz. The use of several antennas, each tuned to a specific standard, is straightforward, but it results in an unacceptably large form factor.

Recently, antennas for multiband cellular applications have been proposed. Examples of these solutions are provided in Refs. 95, 96, where dual- and tri-band antennas, developed for applications on vehicles, enable GSM and UMTS reception. In Ref. 97, the limitations in portability of the previous solutions are solved with a small volume design, optimized for handheld devices. Other interesting results can be achieved by means of tunable antennas, as reported in Ref. 98.

RF filters suffer from the same problems. It is important to notice that every standard needs a dedicated SAW to attenuate the strong interferers that surround the band of interest. By removing these filters, the linearity requirements of the receiver increases up to unacceptable levels. SAW filters are bulky and expensive, so several devices in parallel should be avoided. To reduce area and costs, Yoshimoto [99] proposes a multiband module encapsulated in a compact surface mount plastic package. Definitely more interesting should be the frequency-tunable RF filter. Examples are presented in Ref. 100, even if the operating frequency range is limited. A wider band is covered by Ref. 101, employing a varactor-tuned filter. Other promising solutions may be obtained by means of emerging technologies, such as radio frequency, microelectromechanical systems (RF MEMS) [102,103]. By combining mechanical and electrical properties, micromechanical filters on silicon substrate with high quality factor and wide tuning range can be obtained [104,105].

LNA and Quadrature Mixer

Present LNA implementations mostly employ an inductively degenerated common source input stage, to provide impedance matching, and a tuned LC output stage, to provide frequency-selective gain [106]. Some multiband extension of this topology has been proposed. For example, in Ref. 107, a four-band inductively degenerated LNA is proposed. The amplifier has four distinct input pins and large silicon/board area is occupied by inductor and external SMD. A different approach is given by LNA topologies able to continuously cover a broadband [108,109]. Signals belonging to different standards are processed by the same amplifier. The benefit of reduced occupied area is counteracted by loss of frequency selectivity that increases the RX sensitivity to out-of-band interferers. Narrowband tunable approaches are proposed by Rossi et al. [110] and Liscidini et al. [111]. Exploiting feedback action, these LNA topologies can be reconfigured to select different standards by switching one single resonant network.

In the proposed multistandard receiver topology, as shown in Figure 2.14, the multiband LNA is followed by a quadrature mixer. From the analysis carried out in the previous sections, mixer IIP2 requirement is really tough for UMTS and GSM applications. To avoid complex calibration algorithms or external filtering stages, the design as proposed by Brandolini et al. [10,11] seems mandatory.

Baseband Filter and VGAs

In the multistandard receive chain of Figure 2.14, all standards share the whole analog baseband circuitry, from the mixer to the analog-to-digital converter. To simplify the receiver reconfigurability, the same filtering strategy must be chosen for every considered standard. UMTS, GSM, and WLAN may share a common baseband chain made up of a fourth-order Butterworth filter, to perform the channel selection, and two VGAs, to provide the desired signal gain.

In multistandard systems, baseband signals vary widely in terms of signal level and bandwidth, so reconfigurable filter and amplifiers are needed. VGAs with high-gain variability have been already demonstrated [112]. Moreover, filtering stage with variable bandwidth can be realized by varying the value of capacitors, resistors, and transconductors, by means of a digital word. In literature many examples of programmable multimode baseband filters have been described. Alzaher et al. [113], Hollman [114], and Elwan et al. [115] propose channel select filters for dual-band applications, while Pavan et al. [116] and Chamla et al. [117] extend the filter programmability for multiple standard applications.

A further step toward digital programmability is taken in Ref. 118. After coarse analog filtering, following the down-conversion mixer, the signal is processed by a sampled data filter based on a cascade of three switched capacitors' stages with decreasing sampling frequency. This approach is characterized by wide programmability and has been shown to satisfy both GSM and 802.11g specifications.

Analog-to-Digital Converter

In the baseband section of the receiver the ADC must be able to quantize signals belonging to various standards, thus respecting different sample rate, dynamic range, and linearity requirements, depending on the application. In this field, Sumanen and Halonen [119] present dual-mode pipeline ADCs designed for a GSM–UMTS direct conversion receiver. Other interesting solutions are realized by means of oversampling structures, as described in Ref. 120.

However, to maximize the receiver efficiency and to minimize the power consumption, a higher reconfigurability is required to data converters. In Ref. 121, the ADC functionalities are extended with architectural reconfigurability. The converter is in delta-sigma mode for resolution higher than 12 bits, while for lower resolution it is in a pipeline mode. Moreover in both configurations, input bandwidth and resolution can be programmed. The ADC achieves a bandwidth range from 0 to 10 MHz over a resolution range from 6 to 16 bits.

Frequency Synthesizer and VCO

In a multiband receiver also the frequency synthesis is a critical function. For a flexible frequency generation, an integer-N synthesizer may not be able to satisfy the stringent multistandard specifications, while a fractional-N synthesizer can satisfy the fine frequency resolution and settling time requirements. Moreover, the voltage-controlled oscillator must be multiband and must satisfy all of the single-standard PN requirements. As examples, Refs. 122–124 present solutions for a multiband LC VCO. Other convenient solutions may arise from both ultra-wideband tuning range VCO [125] and the concept of frequency multiplication and harmonic mixing, as described in Refs. 126, 127.

CONCLUSIONS

In this chapter, a review of the most diffused standards for wireless applications was presented. The GSM, UMTS, Bluetooth, and WLAN standards were analyzed to derive the main specifications for the receive section of the mobile terminal. For each standard, the state-of-the-art of the realized integrated receivers was analyzed, to evaluate the optimal receiver architecture and to properly size the building blocks. Particular care was devoted to the RF front-end and highly integrated implementations, either in zero-IF or low-IF architectures, were focused. In the last section the trend toward a multistandard mobile terminal was identified, and a solution for a highly integrated multistandard receiver architecture was proposed. Bluetooth would be processed in parallel while all the other standards would share the same direct conversion path. As shown, the strategy for the reconfiguration of the analog baseband is the key point to keep silicon area and power consumption within acceptable levels.

REFERENCES

1. Abidi, A., "RF CMOS comes of Age," *IEEE J. Solid-State Circuits*, vol. 39, pp. 549–561, April 2004.
2. Kundert, K. S., "Introduction to RF simulation and its application," *IEEE J. Solid-State Circuits*, vol. 34, pp. 1198–1319, September 1999.
3. Hajimiri, A. and Lee, T. H., "A general theory of phase noise in electrical oscillators," *IEEE J. Solid-State Circuits*, vol. 33, pp. 179–194, September 1998.
4. Terrovitis, M. T. and Meyer, R. G., "Noise in Current-Commutating CMOS Mixers," *IEEE J. Solid-State Circuits*, vol. 34, pp. 772–783, June 1999.
5. Gingell, M. J., "Single sideband modulation using sequence asymmetric polyphase networks," *Elect. Commn.*, vol. 48, pp. 21–25, 1973.
6. Crols, J. and Steyaert, M., "An analog integrated polyphase filter for a high performance low-IF receiver," *Symposium on VLSI Circuits Digest of Technical Papers,* pp. 87–88, Kyoto, 8–10 June 1995.
7. Darabi, H. and Abidi, A., "Noise in RF-CMOS mixers: A simple physical model," *IEEE J. Solid-State Circuits*, vol. 35, pp. 15–25, January 2000.
8. Manstretta, D., Brandolini, M., and Svelto, F., "Second-order intermodulation mechanisms in CMOS downconverters," *IEEE J. Solid-State Circuits*, vol. 38, pp. 394–406, March 2003.
9. Darabi, H., Hea Joung Kim, Chiu, J., Ibrahim, B., Serrano, L., "An IP2 improvement technique for zero-IF down-converters," *IEEE ISSCC Digest of Technical Papers*, vol. 49, pp. 464–465, San Francisco, CA, February 2006.
10. Brandolini, M., Rossi, P., Sanzogni, D., Svelto, F., "A CMOS direct downconversion mixer with +78 dBm minimum IIP2 for 3G Cell-Phones," *IEEE International Solid-State Circuits Conference (ISSCC)*, vol. 48, pp. 320–321, San Francisco, February 2005.
11. Brandolini, M., Sosio, M., and Svelto, F., "A 750 mV, 15 kHz 1/f noise corner, 51 dBm IIP2, direct conversion front-end for GSM in 90 nm CMOS," *International Solid-State Circuit Conference (ISSCC) Digest of Technical Papers*, vol. 49, pp. 470–471, San Francisco, February 2006.
12. Jantzi, S., Martin, K., and Sedra, A., "A quadrature band-pass $\Sigma\Delta$ modulator for digital radio," *IEEE ISSCC Digest of Technical Papers*, vol. 46, pp. 216–217, San Francisco, February 1997.
13. Hegazi, E. and Abidi, A., "A 17-mW transmitter and frequency synthesizer for 900-MHz GSM fully integrated in 0.35 µm CMOS," *IEEE J. Solid-State Circuits*, vol. 38, pp. 782–792, May 2003.
14. Staszewski, R. B., Leipold, D., Muhammad, K., Balsara, P. T., "Digitally controlled oscillator (DCO)-based architecture for RF frequency synthesis in a deep-submicrometer CMOS Process," *IEEE Trans. Circuits Sys. II: Analog Digital Signal Process.*, vol. 50, no. 11, pp. 815–828, November 2003.
15. McCune, E. W., "Multi-mode and multi-band polar transmitter for GSM, NADC, and EDGE," *Wireless Commn. Networking, 2003. WCNC 2003. 2003 IEEE*, vol. 2, pp. 812–815, 16–20 March 2003.
16. Third Generation Partnership Project (3GPP), "Feasibility study on 3GPP system to WLAN interworking," 3GPP TR 22.934 v6.2.0, 2003.
17. Third Generation Partnership Project (3GPP), "Physical layer aspects for evolved UTRA," 3GPP TR 25.814 v1.2.2, 2006.
18. Mitola, J. and Maguire, G. O., "Cognitive radio: Making software radios more personal," *IEEE Personal Communications*, vol. 6, no. 4, pp. 13–18, August 1999.
19. Agnelli, F., Albasini, G., Bietti, I., Gnudi, A., Lacaita, A., Manstretta, D., Rovatti, R., Sacchi, E., Savazzi, P., Svelto, F., Temporiti, E., Vitali, S., Castello, R., "Wireless multi-standard terminals: system analysis and design of a reconfigurable RF front-end," *IEEE Circuits Syst. Mag.*, vol. 6, pp. 38–59, 2006.
20. Mitola, J., "Software radios-survey, critical evaluation and future directions," *National Telesystems Conference 1992*, pp. 13/15–13/23, Washington, DC, May 19–20, 1992.

21. On-line, www.gsmworld.com.
22. "Digital Cellular Telecommunications System (Phase 2), Radio Transmission and Reception (GSM 05 05)," Sophia Antipolis, Cedex, France.
23. Kang, S. and Kim, B., "Second order nonlinearity analysis of Gilbert mixer," *IEEE Radio Frequency Integrated Circuits Conference*, pp. 559–562, June 2003.
24. Kivekas, K., Parssinen, A., and Halonen, K., "Characterization of IIP2 and DC-offset in transconductance mixers," *IEEE Trans. Circuits Sys. II*, vol. 48, no. 11, pp. 1028–1038, November 2001.
25. Dow, S., Ballweber, B., Ling-Miao Chou, Eickbusch, D., Irwin, J., Kurtzman, G., Manapragada, P., Moeller, D., Paramesh, J., Black, G., Wolischeid, R., Johnson, K., "A dual-band, direct-conversion/VLIF transceiver for 50 GSM/GSM /DCS/PCS," *International Solid-State Circuit Conference*, vol. 1, pp. 230–462, San Francisco, CA, February 2002.
26. Duvivier, E., Puccio, G., Cipriani, S., Carpineto, L., Cusinato, P., Bisanti, B., Galant, F., Chalet, F., Coppola, F., Cercelaru, S., Vallespin, N., Jiguet, J. C., and Sirna, G., "A fully integrated zero-IF transceiver for GSM-GPRS quad-band application," *IEEE J. Solid-State Circuits*, vol. 38, pp. 2249–2257, December 2003.
27. Kim, Y. J., et al., "A GSM/EGSM/DCS/PCS direct conversion receiver with integrated synthesizer," *IEEE Trans. Microwave Theory Tech.*, vol. 53, no. 2, pp. 606–613, February 2005.
28. Sanyo TSM942AW9B (E-GSM SAW filter). On-line, http://www.sanyo.com.
29. Stetzler, T. D., Post, I. G., Havens, J. H., Koyama, M., "A 2.7–4.5 V single chip GSM transceiver RF integrated circuit," *IEEE J. Solid-State Circuits*, vol. 30, no. 12, pp. 1421–1429, December 1995.
30. Steyaert, M. S. J., Janssens, J., de Muer, B., Borremans, M., Itoh, N., "A 2-V CMOS cellular transceiver front-end," *IEEE J. Solid-State Circuits*, vol. 35, no. 12, pp. 1895–1907, December 2000.
31. Tadjpour, S., Cijvat, E., Hegazi, E., Abidi, A. A., "A 900-MHz dual-conversion low-IF GSM receiver in 0.35-μm CMOS," *IEEE J. Solid-State Circuits*, vol. 36, no. 12, pp. 1992–2002, December 2001.
32. Sevenhans, J., Vanwelsenaers, A., Wenin, J., Baro, J., "An integrated Si bipolar transceiver for a zero IF 900 MHz GSM digital mobile radio front-end of a hand portable phone," CICC, pp. 7.7.1–7.7.4, San Diego, CA, May 1991.
33. Loke, A. and Alì, F., "Direct conversion radio for digital mobile phones—Design issues, status, and trends," *IEEE Trans. Microwave Theory Tech.*, vol. 50, no. 11, pp. 2422–2435, November 2002.
34. Tanaka, S., Yamawaki, T., Takikawa, K., Hayashi, N., Ohno, I., Wakuta, T., Takahashi, S., Kasahara, M., Henshaw, B., "Circuit techniques for GSM/DCS 1800 direct conversion receiver," *IEEE European Solid-State Circuits Conference (ESSCIRC)*, pp. 494–497, Villach, Austria, September 2001.
35. Magoon, R., Molnar, A., Zachan, J., Hatcher, G., Rhee, W., "A single-chip quad-band (850/900/1800/1900 MHz) direct conversion GSM/GPRS RF transceiver with integrated VCOs and fractional-N synthesizer," *IEEE J. Solid-State Circuits*, vol. 37, no. 12, pp. 1710–1720, December 2002.
36. Le Guillou, Y., Gaborieau, O., Gamand, P., Isberg, M., Jakobsson, P., Jonsson, L., Le Deaut, D., Marie, H., Mattisson, S., Monge, L., Olsson, T., Prouet, S., Tired, T., "Highly integrated direct conversion receiver for GSM/GPRS/EDGE with on-chip 84-dB dynamic range continuous-time $\Sigma - \Delta$ ADC," *IEEE J. Solid-State Circuits*, vol. 40, no. 2, pp. 403–411, February 2005.
37. Analog Devices, Part Number AD6548, Othello-G Complete GSM/GPRS Transceiver, On-line, http://www.analog.com.
38. Silicon Laboratories, GSM/GPRS Transceiver, http://www.silabs.com/pdfs/FinalAeroPB.pdf.
39. Song, E., Koo, Y., Jung, Y.-J., Lee, D.-H., Chu, S., Chae, S.-I., "A 0.25 μm CMOS quad-band GSM RF transceiver using an efficient LO frequency plan," *IEEE J. Solid-State Circuits*, vol. 40, no. 5, pp. 1094–1106, May 2005.
40. Bonnaud, P., Hammes, M., Hanke, A., Kissing, J., Koch, R., Labarre, E., Schwoerer, C., "A fully integrated SoC for GSM/GPRS in 0.13 μm CMOS," *International Solid-State Circuits Conference (ISSCC)*, pp. 482–483, San Francisco, CA, February 2006.
41. "3rd Generation Partnership Project, Technical Specification Group Radio Access Networks, UE Radio Transmission and Reception (FDD)," Release 1999, V3.13.0 (2003-03).
42. Springer, A., Maurer, L., and Weigel, R., "RF system concepts for highly integrated RFICs for W-CDMA mobile radio terminals," *IEEE Trans. Microwave Theory Tech.*, vol. 50, no. 1, pp. 254–267, January 2002.
43. Gatta, F., Manstretta, D., Rossi, P., and Svelto, F., "A fully integrated 0.18 μm CMOS direct conversion receiver front-end with on chip LO for UMTS," *IEEE J. Solid-State Circuits*, vol. 39, pp. 15–23, January 2004.
44. Ryynanen, J., Kivekas, K., Jussila, J., Parssinen, A., Halonen, K. A. I., "A dual-band RF front-end for WCDMA and GSM applications," *IEEE J. Solid-State Circuits*, vol. 36, no. 8, pp. 1198–1204, August 2001.

45. Brunel, D., Caron, C., Cordier, C., Soudee, E., "A highly integrated 0.25 μm BiCMOS chipset for 3G UMTS/WCDMA handset RF sub-system," *IEEE Radio Frequency Integrated Circuits Symposium (RFIC)*, pp. 191–194, June 2002.

46. Rogin, J., Koucev, I., Brenna, G., Tschopp, D., Huang, Q., "A 1.5V 45 mW direct conversion WCDMA receiver IC in 0.13 μm CMOS," *IEEE J. Solid-State Circuits*, vol. 38, no. 12, pp. 2239–2248, December 2003.

47. Waite, H., Ta, P., Chen, J., Li, H., Gao, M., Chang, C. S., Chang, Y. S., Redman-White, W., Charlon, O., Fan, Y., Perkins, R., Brunel, D., Soudee, E., "A CDMA2000 zero IF receiver with low-leakage integrated front-end," in *Proceedings ESSCIRC 2003*, pp. 433–436, Estoril, Portugal, September 2003.

48. Maxim semiconductors, MAX2390-MAX2393/MAX2396/MAX2400, data sheet, www.maxim-ic.com.

49. Qualcomm, MSM6250 UMTS Chipset solution, www.qualcomm.com.

50. Texas Instruments, TRF6302 RF Transceiver, www.ti.com.

51. Haartsen, J. C. and Mattisson, S., "A new low-power radio interface providing short-range connectivity," *Proceedings of the IEEE*, vol. 88, no. 10, pp. 1651–1661, October 2000.

52. Wang, H., "Overview of Bluetooth technology," technical report.

53. Specification of the Bluetooth system, Version 1.1, February 2001.

54. Darabi, H., Khorram, S., Hung-Ming Chien, Meng-An Pan, Wu, S., Moloudi, S., Leete, J. C., Rael, J. J., Syed, M., Lee, R., Ibrahim, B., Rofougaran, M., Rofougaran, A., "A 2.4-GHz CMOS transceiver for Bluetooth," *IEEE J. Solid-State Circuits*, vol. 36, no. 12, pp. 2016–2024, December 2001.

55. Eynde, F. O., Schmit, J.-J., Charlier, V., Alexandre, R., Sturman, C., Coffin, K., Mollekens, B., Craninckx, J., Terrijn, S., Monterastelli, A., Beerens, S., Goetschalckx, P., Ingels, M., Joos, D., Guncer, S., Pontioglu, A. "A fully-integrated single-chip SOC for Bluetooth," *International Solid-State Circuit Conference (ISSCC) Digest of Technical Papers*, vol. 1, pp. 196–197, San Francisco, CA, February 2001.

56. Filiol, N., Birkett, N., Cherry, J., Balteanu, F., Gojocaru, C., Namdar, A., Pamir, T., Sheikh, K., Glandon, G., Payer, D., Swaminathan, A., Forbes, R., Riley, T., Alinoor, S.M., Macrobbie, E., Cloutier, M., Pipilos, S., Varelas, T., "A 22 mW Bluetooth RF transceiver with direct RF modulation and on-chip IF fi ltering," *International Solid-State Circuit Conference (ISSCC) Digest of Technical Papers*, vol. 1, pp. 202–203, San Francisco, CA, February 2001.

57. Van Zeijl, P., Eikenbroek, J.-W. T., Vervoort, P.-P., Setty, S., Tangenherg, J., Shipton, G., Kooistra, E., Keekstra, I. C., Belot, D., Visser, K., Bosma, E., Blaakmeer, S. C., "A Bluetooth radio in 0.18 um CMOS," *IEEE J. Solid-State Circuits*, vol. 37, no. 12, pp. 1679–1687, December 2002.

58. Cheah, J., Kwek, E.-H., Eng Chuan Low, Chee Kwang Quek, Yong, C., Enright, R., Hirbawi, J., Lee, A., Hongyu Xie, Longyin Wei, Le Luong, Jianping Pan, Shih-Tsung Yang, Lau, W. F. A., Wai-Lim Ngai, "Design of a low-cost integrated 0.25 μm CMOS Bluetooth SOC in 16.5 mm 2 silicon area," *International Solid-State Circuit Conference*, vol. 1, pp. 90–449, San Francisco, CA, February 2002.

59. Ishikuro, H., Hamada, M., Agawa, K.-I., Kousai, S., Kobayashi, H., Duc Minh Nguyen, Hatori, F., "A single-chip CMOS Bluetooth transceiver with 1.5MHz IF and direct modulation transmitter," *International Solid-State Circuit Conference*, vol. 1, pp. 94–480, San Francisco, CA, February 2003.

60. Ajjikuttira, A., Leung, C., Ee-Sze Khoo, Choke, M., Singh, R., Tian-Hwee Teo, Ban-Chuan Gheong, Jin-Hui See, Hwa-Seng Yap, Poh-Boon Leong, Choon-Tiong Law, Itoh, M., Yoshida, A., Yoshida, Y., Tamura, A., Nakamura, H., "A fully-integrated CMOS RFIC for Bluetooth applications," *International Solid-State Circuit Conference (ISSCC)*, vol. 1, pp. 198–199, San Francisco, CA, February 2001.

61. Komurasaki, H., Sano, T., Heima, T., Yamamoto, K., Wakada, H., Yasui, I., Ono, M., Miwa, T., Sato, H., Miki, T., Kato, N., "A 1.8V-operation RF CMOS transceiver for 2.4-GHz-band GFSK applications," *IEEE J. Solid-State Circuits*, vol. 38, no. 5, pp. 817–825, May 2003.

62. Philips, Part Nr. BGB100, available at www.semiconductors.philips.com

63. Lee, S. W., et. al., "A single-chip 2.4-GHz direct conversion CMOS transceiver with GFSK modem for Bluetooth application," *Symposium on VLSI Circuits Digest of Technical Papers*, pp. 245–246, June 2001.

64. Ugajin, M., Yamagishi, A., Kodate, J., Harada, M., Tsukahara, T., "A 1-V CMOS/SOI Bluetooth RF transceiver for compact mobile applications," *Symposium on VLSI Circuits Digest of Technical Papers*, pp. 123–126, Kyoto, Japan, June 2003.

65. Hyun, S.-B., Tak, G.-Y., Kim, S.-H., Kim, B.-J., Ko, J., Park, S.-S., "A dual-mode 2.4-GHz CMOS transceiver for high-rate Bluetooth systems," *ETRI J.*, vol. 26, no. 3, pp. 229–240, June 2004.

66. Pilsoon Choi, Hyung Chul Park, Sohyeong Kim, Sungchung Park, Ilku Nam, Tae Wook Kim, Seokjong Park, Sangho Shin, Myeung Su Kim, Kyucheol Kang, Yeonwoo Ku, Hyokjae Choi, Sook Min Park,

Kwyro Lee, "An experimental coin-sized radio for extremely low-power WPAN (IEEE 802.15.4) application at 2.4GHz," *IEEE J. Solid-State Circuits*, vol. 38, pp. 2258–2268, December 2003.

67. Sangjin Byun, Chan-Hong Park, Yongchul Song, Sungho Wang, Conroy, C. S. G., Beomsup Kim, "A low-power CMOS Bluetooth RF transceiver with a digital offset canceling DLL-based GFSK demodulator," *IEEE J. Solid-State Circuits*, vol. 38, no. 10, pp. 1606–1619, October 2003.

68. Wenjun Sheng, Bo Xia, Emira, A. E., Chunyu Xin, Valero-Lopez, A. Y., Sung Tae Moon, Sanchez-Sinencio, E., "A 3-V, 0.35 μm CMOS Bluetooth receiver IC," *IEEE J. Solid-State Circuits*, vol. 38, no. 1, pp. 30–42, January 2003.

69. Wireless LAN medium access control (MAC) and physical layer (PHY) specifications, ANSI/IEEE Std. 802.11, 1999 Edition.

70. Wireless LAN medium access control (MAC) and physical layer (PHY) specifications—High-speed physical layer in the 5 GHz band, ANSI/IEEE Std. 802.11a, 1999 Edition.

71. ETSI TS 101 475, "Broadband radio access network (BRAN), HiperLAN Type 2, physical (PHY) layer," ver.1.3.1, 2001.

72. Wireless LAN medium access control (MAC) and physical layer (PHY) specifications—Higher-speed physical layer extension in the 2.4 GHz band, ANSI/IEEE Std. 802.11b, 1999.

73. Wireless LAN medium access control (MAC) and physical layer (PHY) specifications—Further higher-speed physical layer extension in the 2.4 GHz band, IEEE Std. 802.11g/D1.1, January 2002.

74. ETSI EN 300 440-1 Ver. 1.3.1, "Electromagnetic compatibility and radio spectrum matters (ERM), short range devices, radio equipment to be used in the 1 GHz to 40 GHz frequency range, Part 1: Technical characteristics and test methods," 2001.

75. Stroet, P. M., Mohindra, R., Hahn, S., Schuur, A., Riou, E., "A zero-IF single-chip transceiver for up to 22Mb/s QPSK 802.11b wireless LAN," *IEEE International Solid-State Circuit Conference*, vol. 1, pp. 204–205, San Francisco, CA; February 2001.

76. Maxim "MAX2820—2.4 GHz 802.11b zero-IF transceiver," data sheet, available at http://www.maxim-ic.com, 2002.

77. Chien, G., Weishi Feng, Hsu, Y. A., Tse, L., "A 2.4 GHz CMOS transceiver and baseband processor chipset for 802.11b wireless LAN application," *IEEE International Solid-State Circuit Conference*, vol. 1, pp. 358–499, February 2003.

78. Kluge, W., Dathe, L., Jaehne, R., Ehrenreich, S., Eggert, D., "A 2.4 GHz CMOS transceiver for 802.11b wireless LANs," *International Solid-State Circuit Conference (ISSCC)*, vol. 1, pp. 360–361, February 2003.

79. Banu, M., Prodanov, V., Kiss, P., Manstretta, D., "The challenges of fully integrated OFDM transceivers for wireless-LAN systems," *IEEE International Solid-State Circuit Conference (ISSCC)—GIRAFE workshop*, San Francisco, CA, February 2003.

80. Pengfei Zhang, Thai Nguyen, Lam, C., Gambetta, D., Soorapanth, C., Baohong Cheng, Hart, S., Sever, I., Bourdi, T., Tham, A., Razavi, B., "A direct conversion CMOS transceiver for IEEE 802.11a wireless LANs," *International Solid-State Circuit Conference (ISSCC)*, vol. 1, pp. 354–498, San Francisco, CA, February 2003.

81. Behzad, A., Lin, L., Shi, Z. M., Anand, S., Carter, K., Kappes, M., Lin, E., Nguyen, T., Yuan, D., Wu, S., Wong, Y. C., Fong, V., Rofougaran, A., "Direct-conversion CMOS transceiver with automatic frequency control for 802.11a wireless LANs," *International Solid-State Circuit Conference*, San Francisco, CA, pp. 356–499, February 2003.

82. RF micro devices, "RF5405—true zero-IF 5 GHz CMOS transceiver," available at http://www.rfmd.com, 2003.

83. RF micro devices, "RF5421—Dual-band tri-mode 802.11a/b/g wireless LAN solution," available at http://www.rfmd.com, 2003.

84. Murata, Part Nr. DFCH22G45HDHAA, available at http://www.murata.com

85. Johanson, Part Nr. 5515BP15B725, available at http://www.johansontechnology.com

86. Lin, J. C. H., Yeh, T. H., Lee, C. Y., Chen, C. H., Tsay, J. L., Chen, S. H., Hsu, H. M., Chen, C. W., Huang, C. F., Chiang, J. M., Chang, A., Chang, R. Y., Chang, C. L., Wang, S. H., Wu, C. C., Lin, C. Y., Chu, Y. L., Chen, S. M., Hsu, C. K., Liou, R. S., Wong, S. C., Tang, D., Sun, J. Y. C., "State-of-the-art RF/analog foundry technology," *IEEE Bipolar/BiCMOS Circuits and Technology Meeting*, pp. 73–79, 2002.

87. Isaac, R. D., "The future of CMOS technology," *IBM J. Res. Develop.*, vol. 44, no. 3, pp. 369–378, May 2000.

88. Lee, T. H. and Wong, S. S., "CMOS RF integrated circuits at 5 GHz and beyond," *Proceedings of the IEEE*, vol. 88, no. 10, pp. 1560–1571, October 2000.

89. Economou, L. and Langley, R. J., "Multi-band mobile phone antennas," *International Conference on Antennas and Propagation*, vol. 2, pp. 754–757, April 2001.

90. Zhou, G. and Yildirim, B., "A multi-band fixed cellular phone antenna," *IEEE International Symposium on Antennas and Propagation*, vol. 1, pp. 112–115, July 1999.

91. Chen, T. L., "Multi-band printed sleeve dipole antenna," *Electron. Lett.*, vol. 39, no. 1, pp. 14–15, January 2003.

92. Atheros Communications, "AR5112—2.4/5 GHz dual band radio-on-a-chip," available at http://www.atheros.com, 2003.

93. Darabi, H., Chiu, J., Khorram, S., Hea Joung Kim, Zhimin Zhou, Hung-Ming, Chien, Ibrahim, B., Geronaga, E., Tran, L. H., Rofougaran, A., "A dual-mode 802.11b/bluetooth radio in 0.35 um CMOS," *IEEE J. Solid-State Circuits*, vol. 40, pp. 698–706, March 2005.

94. Hotti, M., Kaukovuori, J., Ryynanen, J., Kivekas, K., Jussila, J., Halonen, K., "A direct-conversion RF front-end for 2-GHz WCDMA and 5.8-GHz WLAN applications," *IEEE Radio Frequency Integrated Circuits Symposium*, pp. 45–48, Philadelphia, PA, June 2003.

95. Economou, L. and Langley, R. J., "Multi-band mobile telephone antennas," *11th International Conference on Antennas and Propagation*, pp. 754–757, April 2001.

96. Economou, L. and Langley, R. J., "Dual band hybrid vehicular telephone antenna," *IEEE Proceedings Microwaves, Antennas and Propagation*, vol. 149, no. 1, pp. 41–44, February 2002.

97. Ying, Z. and Andersson, J., "Multi band, multi antenna system for modern mobile terminal," *6th IEEE International Symposium on Antennas, Propagation and EM Theory*, pp. 287–290, November 2003.

98. Rostbakken, O., Hilton, G. S., and Railton C. J., "An adaptive microstrip patch antenna for use in portable transceivers," *IEEE 46th Vehicular Technology Conference*, pp. 339–343, 1996.

99. Yoshimoto, S., "Multi-band RF SAW filter for mobile phone using surface mount plastic package," *IEEE Ultrasonics Symposium*, pp. 113–118, October 2002.

100. Smole, P., Ruile, W., Korden, C., Ludwig, A., Quandt, E., Krassnitzer, S., Pongratz, P., "Magnetically tunable SAW-resonator," *IEEE International Frequency Control Symposium, 17th European Frequency and Time Forum*, pp. 903–906, May 2003.

101. Brown, A. R. and Rebeiz, G. M., "A varactor-tuned RF filter," *IEEE Trans. Microwave Theory Tech.*, vol. 48, pp. 1157–1160, July 2000.

102. Blondy, P., Champeaux, C., Tristant, P., Mercier, D., Cros, D., Catherinot, A., Guillon, P., "Applications of RF MEMS to tunable filters and matching networks," *IEEE International Semiconductor Conference*, pp. 111–116, October 2001.

103. Izadpanah, H., Warneke, B., Loo, R., Tangonan, G., "Reconfigurable low power, light weight wireless system based on the RF MEM switches," *IEEE MTT-S Symposium on Technologies for Wireless Applications*, pp. 175–180, February 1999.

104. Nguyen, C. T. C., "RF MEMS for wireless applications," *IEEE Device Research Conference (DRC)*, pp. 9–12, June 2002.

105. Nguyen, C. T. C., "Micromechanical circuits for communication transceivers," *IEEE Bipolar/BiCMOS Circuits and Technology Meeting (BCTM)*, pp. 142–149, 2000.

106. Shaeffer, D. K. and Lee, T. H., "A 1.5-V, 1.5-GHz CMOS low noise amplifier," *IEEE J. Solid-State Circuits*, vol. 32, no. 5, pp. 745–759, May 1997.

107. Ryynanen, J., Kivekas, K., Jussila, J., Sumanen, L., Parssinen, A., Halonen, K.A.I., "A single-chip multimode receiver for GSM900, DCS1800, PCS1900, and WCDMA," *IEEE J. Solid-State Circuits*, vol. 38, no. 4, pp. 594–602, April 2003.

108. Bruccoleri, F., Klumperink, E. A. M., and Nauta, B., "Wide-band CMOS low-noise amplifier exploiting thermal noise canceling," *IEEE J. Solid-State Circuits*, vol. 39, no. 2, pp. 275–282, February 2003.

109. Adiseno, A., Ismail, M., and Olsson, H., "A wide-band RF front-end for multiband multistandard high-linearity low-IF wireless receivers," *IEEE J. Solid-State Circuits*, vol. 37, pp. 1162–1168, September 2003.

110. Rossi, P., Liscidini, A., Brandolini, M., Svelto, F., "A variable gain RF front-end, based on a voltage-voltage feedback LNA, for multistandard applications," *IEEE J. Solid-State Circuits*, vol. 40, no. 3, pp. 690–697, March 2005.

111. Liscidini, A., Brandolini, M., Sanzogni, D., Castello, R., "A 0.13 µm CMOS front-end for DCS1800/UMTS/802.11b-g with multi-band positive feedback low noise amplifier," *IEEE Symposium on VLSI Circuits*, pp. 406–409, June 2005.

112. Chang, P. J., Rofougaran, A., and Abidi, A., "A CMOS channel-select filter for a direct-conversion wireless receiver," *Symposium on VLSI Circuits*, pp. 62–63, June 1996.

113. Alzaher, H. A., Elwan, H. O., and Ismail, M., "A CMOS highly linear channel-select filter for 3G multistandard integrated wireless receivers," *IEEE J. Solid-State Circuits,* vol. 37, no. 1, pp. 27–37, January 2002.

114. Hollman, T., "A 2.7-V CMOS dual-mode baseband filter for PDC and WCDMA," *IEEE J. Solid-State Circuits,* vol. 36, no. 7, pp. 1148–1153, July 2001.

115. Elwan, H., Ravindran, A., and Ismail, M., "CMOS low power baseband chain for a GSM/DECT multistandard receiver," *IEE Proc. Circuits, Devices Sys.*, vol. 149, no. 56, pp. 337–347, October–December 2002.

116. Pavan, S., Tsividis, Y. P., and Nagaraj, K., "Widely programmable high-frequency continuous-time filters in digital CMOS technology," *IEEE J. Solid-State Circuits*, vol. 35, no. 4, pp. 503–511, April 2000.

117. Chamla, D., Kaiser, A., Cathelin, A., Belot, D., "A Gm-C low-pass filter for zero-IF mobile applications with a very wide tuning range," *IEEE J. Solid-State Circuits*, vol. 40, no. 7, pp. 1443–1450, July 2005.

118. Bagheri, R., Mirzaei, A., Chehrazi, S., Heidari, M., Lee, M., Mikhemar, M., Tang, W., Abidi, A., "An 800 MHz to 5 GHz software-defined radio receiver in 90 nm CMOS," *International Solid-State Circuits Conference (ISSCC)*, pp. 480–481, San Francisco, CA, February 2006.

119. Sumanen, L. and Halonen, K., "Dual-mode pipeline A/D converter for direct conversion receivers," *IEEE Electronic Lett.*, vol. 38, no. 19, pp. 1101–1103, September 2002.

120. van Veldhoven, R. H. M., "A triple-mode continuous-time Σ-Δ modulator with switched-capacitor feedback DAC for a GSM-EDGE/CDMA2000/UMTS receiver," *IEEE J. Solid-State Circuits*, vol. 38, no. 12, pp. 2069–2076, December 2003.

121. Gulati, K. and Lee, H. S., "A low-power reconfigurable analog-to-digital converter," *IEEE J. Solid-State Circuits*, vol. 36, no. 12, pp. 1900–1911, December 2001.

122. Fong, N. H. W., Plouchart, J.-O., Zamdmer, N., Duixian Liu, Wagner, L. F., Plett, C., Tarr, N. G., "Design of wide-band CMOS VCO for multiband wireless LAN applications," *IEEE J. Solid-State Circuits*, vol. 38, no. 8, pp. 1333–1342, August 2003.

123. Kucera, J. J., "Wideband BiCMOS VCO for GSM/UMTS direct conversion receivers," *IEEE Solid-State Circuits Conference (ISSCC)*, vol. 1, pp. 374–375, February 2001.

124. Yim, S. M. and O, K.-K., "Demonstration of a switched resonator concept in a dual-band monolithic CMOS LC-tuned VCO," *IEEE Custom Integrated Circuits Conference (CICC)*, pp. 205–208, May 2001.

125. Berny, A. D., Niknejad, A. M., and Meyer, R. G., "A 1.8-GHz LC VCO with 1.3-GHz tuning range and digital amplitude calibration," *IEEE J. Solid-State Circuits*, vol. 40, no. 4, pp. 909–917, April 2005.

126. Shin, H., Xu, Z., and Chang, M. F., "A 1.8-V 6/9-GHz reconfigurable dual-band quadrature LC VCO in SiGe BiCMOS technology," *IEEE J. Solid-State Circuits*, vol. 38, pp. 1028–1032, June 2003.

127. Guermandi, D., Tortori, P., Franchi, E., Gnudi, A., "A 0.83-2.5-GHz continuously tunable quadrature VCO," *IEEE J. Solid-State Circuits*, vol. 40, pp. 2620–2627, December 2005.

128. Brandolini, M., Rossi, P., Manstretta, D., and Svelto, F., "Toward multistandard mobile terminals—Fully integrated receivers requirements and architectures," *IEEE Trans. Microwave Theory Tech.*, vol. 53, pp. 1026–1038, March 2005.

3 Ultra Wide Band (UWB) Technology

Domine M. W. Leenaerts

CONTENTS

Introduction 81
UWB Transceiver Specifications 83
    Receiver Sensitivity, Noise Figure, and Signal-to-Noise Ratio 83
    Interferer Scenario 84
    Receiver Linearity and Filter Requirements 85
    Transmitter Requirements 86
    Synthesizer Requirements 86
RF Receiver Building Blocks 87
    Low-Noise Amplifier 87
    Down-Converter Mixer 89
    IF/Baseband Filter 92
RF Transmitter Building Blocks 94
Fast-Hopping Synthesizer 95
    Integer-N PLL Approach 95
    Using Two PLLs and One Single-Side-Band Mixer 96
    Using a Single PLL and SSB Mixer(s) 98
    Other Options 100
RF Transceivers for MB-OFDM UWB 100
Conclusions 103
Acknowledgment 103
References 104

INTRODUCTION

Short-range communication systems (known as wireless personal area network [WPAN] systems) with ranges of up to 10 m are becoming popular for replacing cables and enabling new consumer applications. However, systems such as Bluetooth and Zigbee, which operate in the 2.4 GHz ISM band, have a limited data rate, typically about 1 Mbps, which is insufficient for many applications, such as fast transfer of large files (e.g., wireless USB) and high-quality video streaming. To increase the data rate to several hundreds of Mbps, a higher bandwidth is preferred over a larger signal-to-noise ratio (SNR). This became possible when the FCC released frequency spectrum for ultra wide band (UWB) in the United States spanning from 3.1 to 10.6 GHz with an average transmit power level of only −41.3 dBm/MHz [1]. Since then, several proposals have been presented to realize a short-range high data rate communication link. At present, both direct-sequence impulse communication and multiband orthogonal frequency division multiplexing (MB-OFDM) UWB systems are under

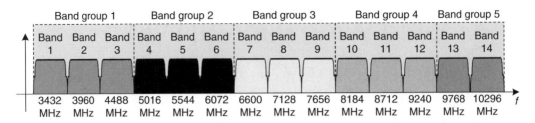

FIGURE 3.1   MBOA frequency bands and their partitioning.

consideration as a standard within the IEEE under IEEE802.15.3a. Industry has adopted MB-OFDM UWB for high data rates as the ECMA-368 standard [2].

In other proposals, an impulse radio UWB (IR-UWB) concept has been used [3]. The basic idea is to use a repetition of short pulses in time (and therefore wideband in frequency) to achieve moderate data rates, e.g., 10–100 Mbps. As other modulation schemes are deployed in such systems, power consumption can be low. These systems are, therefore, proposed as likely candidates for wireless sensor networks, for example. IR-UWB is expected to become an IEEE standard under IEEE802.15.4.

The ECMA-368 standard proposed by the MB-OFDM alliance (MBOA) is based on subdivision of the large available bandwidth into subbands of 528 MHz each (see Figure 3.1) [1,2]. Three subbands are grouped together in a band group, except for band group 5, which consists of two subbands only. Band group 1 is the mandatory mode of operation and spans from 3168 to 4752 MHz. The applied data mapping onto a complex constellation is QPSK for data rates up to 200 Mbps and dual carrier modulation (DCM) for data rates from 320 Mbps and higher. An orthogonal frequency division multiplexing (OFDM) technique using 128 subcarriers is applied to generate the discrete-time signal. Out of these 128 subcarriers, 100 carriers are used for data, 22 carriers are pilot tones, and the remaining carriers are null tones. A frequency-hopping (FH) scheme is implemented within each of the band groups, but not across band groups. For the first four band groups, seven time-frequency codes are defined, representing seven possible hopping scenarios. Band group 5 has only two possible hopping sequences. The permitted band-switching time is only 9.5 ns in all cases, whereas the symbol period is 312.5 ns.

Although five band groups have been defined, industry is currently focusing only on the mandatory mode (band group 1). Applications for this mode can be found in wireless computer peripheral communication, such as communication between one PC and a printer, scanner, or beamer. The advantage of only using the mandatory mode is that it is possible to use a selective band-pass antenna filter. This reduces not only interferers from the 2.4 GHz industrial science medical (ISM) band (e.g., IEEE802.11b/g, Bluetooth), but also interferers from the 5 GHz band (e.g., IEEE802.11a). The latter in particular are difficult to handle if band group 2 is also in use, as these interferers fall inside subband 4 and subband 5 (see Figure 3.1). However, even if a selective prefilter is used, interferers will pose a challenging linearity requirement on the receiver of the UWB radio.

The challenge of realizing a radio for UWB systems is already apparent from the name: ultra-wide bandwidth. Consider the mandatory mode. The bandwidth at radio frequency (RF) is over 1.5 GHz (three times 528 MHz), giving a bandwidth versus center frequency ratio of almost 38%. Compare this to the same ratio in the cellular standard GSM900 (50 MHz over 900 MHz), i.e., 6%, or to the ratio in the wireless local area network (WLAN) standard IEEE802.11a, yielding 3.8%. Such a ratio poses interesting design challenges for the RF circuitry with respect to linearity and noise. A similar property holds for the baseband frequencies. Even when a zero-intermediate frequency (IF) architecture is used in the receiver, the IF bandwidth is still 264 MHz, setting interesting design challenges for the IF filters and analog-to-digital (AD) converters. For comparison, the frequency bandwidth numbers are 100 kHz and 10 MHz for GSM900 and IEEE802.11a, respectively. One other interesting aspect of a UWB radio is the required FH. It has proven impossible to achieve a band-switching time of 9.5 ns with a traditional phase-locked loop (PLL) system. Other FH communication standards have a more relaxed switching time. For instance, the Bluetooth standard defines a FH of 200 μs.

This chapter aims to give an overview of the current status in the MB-OFDM UWB system. Various solutions proposed in literature for the design challenges mentioned are discussed. First of all, the most important system specifications are examined and considered within the context of UWB radio design. Transceiver specifications such as linearity, phase noise, noise figure (NF), and required filter transfer characteristic are discussed in the section "UWB Transceiver Specifications." The section "RF Receiver Building Blocks" highlights the progress made on receiver building blocks such as low-noise amplifiers (LNAs), down-converter mixers, and baseband amplifiers/filters. The section "RF Transmitter Building Blocks" focuses on the most important transmitter building block, i.e., the RF power amplifier (PA). Various design aspects related to the synthesizer are discussed in the section "Fast-Hopping Synthesizer." The first possible synthesizer architectures are discussed, followed by a detailed description of the most critical circuits. Several (fully) integrated transceivers are discussed in the section "RF Transceivers for MB-OFDM UWB," and, finally, some concluding remarks are given in the section "Conclusions."

UWB TRANSCEIVER SPECIFICATIONS

UWB receiver design is challenging because it requires simultaneously a low-noise density in a large bandwidth and a high linearity since large interferers can be present close to the frequency band used. An interferer scenario is required to determine the amount of filtering needed. On the transmit side, the challenge lies in achieving a tunable, flat gain response over a 1.584 GHz bandwidth. Probably the most challenging block is the synthesizer due to the fast-hopping requirement and the spectral purity required.

In this section, the most important system design requirements are derived from the UWB system specifications as defined in Ref. 2.

RECEIVER SENSITIVITY, NOISE FIGURE, AND SIGNAL-TO-NOISE RATIO

For a packet error rate (PER) of less than 8%, the minimum sensitivity in an arbitrary weighted Gaussian noise channel (AWGN) is -80.8 dBm for a data rate of 53.3 Mbps in band group 1. Here, a noise figure (NF) of 6.6 dB (referenced to the antenna), an implementation loss of 2.5 dB, and a margin of 3 dB have been assumed. This sensitivity number increases to -70.4 dBm when the data rate increases to 480 Mbps.

Any improvement in NF over the minimum requirement has a significant influence on the overall system performance. A lower NF gives a direct improvement in the ranging capabilities of the system, or, for a given range, allows for additional loss of the prefilter. A lower NF also allows for improved prefiltering, and thus a system with increased robustness to out-of-band interference. Hence, the target should be set for the lowest possible NF achievable, taking into account other requirements for the RF front-end. The maximum value allowed for the NF at the LNA input can be calculated from the following:

$$NF = NF_{system} - IL_{prefilter}(dB) = 6.6 - 2 = 4.6 \, dB \qquad (3.1)$$

where NF_{system} is the system NF requirement (assumed to be 6.6 dB) and $IL_{prefilter}$ the insertion loss of prefilter (2 dB seems a practical value). In reality, a lower NF will be required so that the NF requirements are met in all cases, e.g., over temperature and fabrication spread.

SNR is often discussed when considering noise requirements, particularly when the received signal is considered as a voltage, and voltage gain is defined rather than power gain. The SNR requirements vary according to the data rate of operation, with higher SNR levels needed for higher data rates. The NF can be taken to calculate the corresponding SNR at the radio output as follows:

$$SNR = reference \, sensitivity - No - NF - implementation \, loss + 10 \log\left(\frac{122}{100}\right) = -3.1 \, dB \qquad (3.2)$$

where the SNR (dB) is defined as the overall SNR and includes the signal power present in both the data and the pilot tones, and No is the thermal noise floor in the UWB signal bandwidth, assuming a 50 Ω impedance level (i.e., 174 dBm + 10 log(528 MHz) = −86.7 dBm). The conversion between SNR and energy per bit over noise (Eb/No) is as follows:

$$\frac{Eb}{No} = SNR - 10\log\left(\frac{data\ rate}{signal\ bandwidth}\right) + 10\log\left(\frac{100}{122}\right) = 6.8\,dB \qquad (3.3)$$

where Eb/No is defined as taking only the data energy per bit, and not taking account of the pilot tone energy.

INTERFERER SCENARIO

Clearly, there is a need for straightforward coexistence with other wireless technologies such as WLAN IEEE802.11 and WPAN IEEE802.15. Moreover, coexistence with present cellular communication devices operating under the global system for mobile communication (GSM), personal communication services (PCS), or code division multiple access (CDMA) mode is also almost mandatory. This requires an interferer-robust receiver.

Various types of interference need to be considered during analysis of a UWB system. When considering interferer signals where the wanted UWB signal is the victim, it is important to consider applicable application scenarios. To do this, determine where the interferer may be located with respect to the UWB receiver, for example, it may be located colocated in the same device, such as a laptop, or located in a different device but such that a UWB is operating in one device at the same time as another system is operating in a second device in the same vicinity. The interferer signals can be broken down further into in-band interferers, such as unwanted UWB interferers, and out-of-band interferers, such as WLAN 802.11 interferers and cellular interferers. These types of interferers need to be considered separately, since out-of-band interferers can be rejected more easily by filtering than in-band interferers. However, out-of-band interferers that lie close to the UWB band edge can also be problematic because of their proximity to the band, and thus the limitations on filtering. Another important consideration in the treatment of interference is the transmit levels of the different systems that generate interference, since systems that use a high transmit power level are more difficult to deal with.

The MB-OFDM alliance interference scenario recommendation is depicted in Figure 3.2. A wanted UWB transmitter is located at a distance of 10 m from the UWB receiver. The received

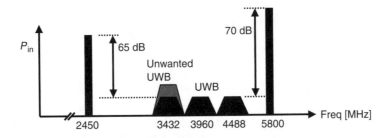

Interferer scenario: (MBOA recommendation)

• Distance wanted UWB:	10.0 m ➤	−73 dBm
• Distance WLAN interferer:	0.2 m ➤	−3 dBm
• Distance 2.4-GHz ISM interferer:	0.2 m ➤	−8 dBm
• Distance GSM1900 interferer:	1.0 m ➤	−8 dBm
• Distance unwanted UWB interferer:	2.0 m ➤	−60 dBm

FIGURE 3.2 Interferer scenario. The received interferer powers are indicated. (From Bergervoet, J., et al., *International Solid-State Circuits Conference (ISSCC)*, pp. 200–201, 2005.)

transmit power S at the antenna reference point (ARP) of the UWB receiver operating in subband 1 is -73 dBm, according to

$$S = \frac{1}{4\pi r^2} \cdot \frac{\lambda^2}{4\pi} \cdot P_{TX} \tag{3.4}$$

where r is the distance in meters, λ the wavelength in meters, and P_{TX} the transmit power, which is -10.3 dBm for an UWB signal. Here, a free-space loss is assumed. According to the interferer scenario, an unwanted UWB transmitter may be as close as 2 m, and if this transmitter operates in subband 2, the received transmit power at the ARP is -60 dBm, which is 13 dB stronger than the wanted UWB signal. In a similar way, calculate the received transmit powers of the out-of-band interferers, as they are mentioned in Figure 3.2. From the proposed interferer scenario and resulting power levels, observe that there can be a power difference of as much as 70 dB between the wanted signal and the interferer. Even with an antenna filter in front of the radio, this will set very stringent linearity requirements.

RECEIVER LINEARITY AND FILTER REQUIREMENTS

The linearity requirements for the receiver follow from the interferer scenario defined and the assumed attenuation of the antenna filter. The antenna filters currently available realize attenuation of about 20 dB in the 2.4 GHz and 5 GHz frequency bands.

The interference criteria define that the device is assumed to be operating 6 dB above sensitivity. The in-band noise floor No is equal to -86.7 dBm, leaving a total receiver noise of -80.1 dBm for the specified NF of 6.6 dB. If the signal is increased to 6 dB above sensitivity, then the sum of interferer and noise power can also increase by 6 dB to maintain the overall signal-to-noise and interferer ratio. Hence, the maximum allowable noise and interferer power is $(-80.1) + 6 = -74.1$ dBm. The allowable interferer level P_{IM} is then -75.4 dBm, to give a combined interferer and noise power of -74.1 dBm.

The system level iIP2 requirement for two interfering signals, referred to the ARP, is calculated as follows:

$$iIP2 = (Pin_1 + Pin_2) - P_{IM2} \tag{3.5}$$

where iIP2 (dBm) is the second-order intercept point (input referred to ARP), P_{IM2} (dBm) the power level of the second-order product (input referred), as defined above (-75.4 dBm), and Pin_1 (Pin_2) the first (second) interferer input power level at ARP. For a two-interferer scenario, where a second-order difference product falls in-band in the RF front-end, the maximum received interferer power level is -3 dBm for the IEEE802.11a UNII upper band for a first interferer signal, and -8 dBm for the PCS/GSM1900 band for a second interferer signal, according to Figure 3.2. This worst-case combination creates an in-band difference product in subband 2 in the RF front-end. If we assume 20 dB of prefiltering of the interferer signal level, then $Pin_1 = -23$ dBm and $Pin_2 = -28$ dBm, yielding an iIP2 of $(-23) + (-28) - (-75.4) = +24.4$ dBm.

In a similar way, the system level iIP3 requirement for two interfering signals, referred to the ARP, is calculated as

$$iIP3 = \left(Pin_1 + \frac{Pin_2}{2} \right) - \left(\frac{P_{IM3}}{2} \right) \tag{3.6}$$

where iIP3 (dBm) is the third-order intercept point (input referred to ARP), and P_{IM3} (dBm) the power level of the third-order product (input referred). For iIP3, two interferers from different systems/bands create the worst-case scenario. The maximum received interferer power level is -3 dBm for the

IEEE802.11a universal networking information infrastructure (UNII) upper band for a first interferer signal, and $-28\,$dBm for the IEEE802.11a UNII middle band for the second interferer signal. The calculated iIP3 is then $-9.3\,$dBm assuming 20 dB attenuation from the antenna filter.

Because of the strong interferers, there are stringent filter requirements at IF/baseband as well. Consider the case where the closest IEEE802.11a interferer is at a distance of 0.2 m while the wanted UWB signal is transmitted from a distance of 10 m. As the interferer is positioned at 5150 MHz while the edge of subband 3 is at 4752 MHz, the frequency spacing is only 398 MHz. In such a case, the filter has to provide more than 35 dB of attenuation relative to DC at 662 MHz offset (i.e., 264 MHz + 398 MHz, assuming a zero-IF or direct conversion receiver). In a similar way, for the upper band of IEEE802.11a, an attenuation of 46 dB is required at an offset of 1.3 GHz. Assume a roll-off of 6 dB/octave per filter order. Therefore, 35 dB attenuation is required in a frequency span of 1.4 octaves (frequency spacing between 662 and 264 MHz). This demands a filter order higher than 4.

TRANSMITTER REQUIREMENTS

A key requirement for a UWB transmitter is that the power spectral density P_{PSD} limit of $-41.3\,$dBm/MHz must be met. On the basis of this emission mask and the FH specification, the maximum transmit power can be calculated according to

$$P_{TX} = P_{PSD}(\text{dBm/MHz}) + 10\log_{10}(B/\text{MHz}) = -9.5\,\text{dBm} \qquad (3.7)$$

where P_{TX} is the transmit power and B represents the signal bandwidth determined by the 122 signal and pilot tones over three subbands. As the OFDM carrier spacing is 4.125 MHz, the bandwidth B is set to $3 \times 503.25\,\text{MHz} = 1509.75\,\text{MHz}$. Assuming a power loss of about 2.5 dB between the antenna and the power amplifier (PA), the power that needs to be generated is $-7.0\,$dBm. Study into the effect of nonlinearity on OFDM signals indicates that a back-off of 2–4 dB ensures acceptable degradation [3].

SYNTHESIZER REQUIREMENTS

As the radio must cover at least the lower three bands, as defined in the MB-OFDM UWB standard, and since a zero-IF architecture is most likely to be used, the synthesizer needs to provide quadrature carrier signals at the center frequencies of the mandatory subbands, i.e., at 3432, 3960, and 4488 MHz.

In the standard, FH between two subbands occurs once every symbol period of 312.5 ns. This period contains a 60.6 ns suffix, which is followed by a 9.5 ns guard interval in which the FH should be accomplished. Therefore, the synthesizer must be able to perform a frequency step and settle to the new frequency within these 9.5 ns, which is a challenging requirement.

The demands on the spectral purity of the generated carriers are stringent due to the presence of strong interferer signals. The spur requirement can be calculated from the interferer scenario according to

$$\text{Spur(dBc@offset freq.)} = P_{PSD\,wanted} - P_{PSD\,interferer} - \text{SNR} \qquad (3.8)$$

where P_{PSD} represents the power spectral density (in dBm/MHz). Following the interferer scenario and associated received power levels of the interferers, all spurious tones in the 5 GHz range must be below $-50\,$dBc. This spur-level requirement will avoid harmful down-conversion of strong out-of-band WLAN interferers into the wanted subbands. For the same reason, the spurious tones in the 2.4 GHz range should be below $-45\,$dBc to allow coexistence with the systems operating in the 2.4 GHz ISM band, such as IEEE802.11b/g and Bluetooth. The requirements for in-band spurious tones can be calculated in a similar way. The specified spurious tone level is $-35\,$dBc.

The system SNR can degrade due to intercarrier modulation. This mechanism sets an upper limit to the overall integrated phase noise of the local oscillator (LO) signal, used to drive the down-converter mixers. To ensure that the system SNR will not degrade by more than 0.1 dB, the overall integrated phase noise from DC to 50 MHz should not exceed 2° rms. This can be recalculated to a phase noise requirement of −100 dBc/Hz at 1 MHz offset from the carrier.

MB-OFDM UWB is based on FH, and therefore phase coherence is an issue. Phase coherence is not a problem for hopping from one frequency to another because the baseband will take care of it. However, when, after several frequency hops, the system returns to the old channel, the coherence of current phase and previous phase must be maintained, otherwise an unnoticed rotation in the constellation diagram can occur.

As the bit mapping used is quadrature phase shift keying (QPSK) (or DCM), the requirements on in-phase and quadrature phase (*I/Q*) mismatch are relaxed and must be lower than −30 dBc.

RF RECEIVER BUILDING BLOCKS

The signal bandwidth of a subband is 528 MHz and can be considered as really wideband. For this reason, most UWB radio architectures are based on a zero-IF concept. This means that the signal bandwidth at IF/baseband is reduced to 264 MHz, a bandwidth that can be handled by IF/baseband filters and AD converters. However, the LNA and input stage of the down-converter mixer must still be capable of handling a bandwidth of 1584 MHz.

This section examines important building blocks required to build a zero-IF MB-OFDM UWB receiver. An overview is given of possible circuit implementations for the LNA, down-converter mixer, and IF/baseband amplifiers.

LOW-NOISE AMPLIFIER

In addition to the receiver requirements mentioned with respect to noise and linearity, the LNA must provide broadband input matching and a broadband transfer. Several design options have been proposed in literature, such as applying band-pass input impedance matching, using distributed design techniques, or simply using feedback techniques. An overview of these techniques is given in this section.

Often an inductor is placed at the input of an LNA to cancel out the input capacitance of the LNA. The remaining capacitance is associated with the bond pad and electrostatic discharge (ESD) structures. This configuration has a low-pass filter behavior, which can easily be converted to a band-pass filter characteristic (see Figure 3.3). Here, Port 1 expresses the ARP and Port 2 reflects the input of the LNA. The conversion requires the use of an additional inductor and capacitor. However, this second inductor comes free because an inductively degenerated LNA is normally used. A conceptual

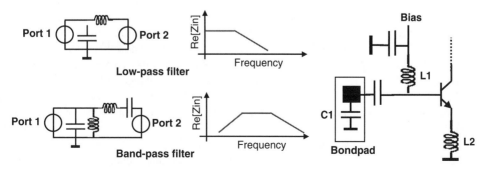

FIGURE 3.3 Low-pass-to-band-pass transformation (left) and a band-pass LNA implementation. (From Ismail, A. and Abidi, A., *International Solid-State Circuits Conference (ISSCC)*, pp. 384–385, 2004.)

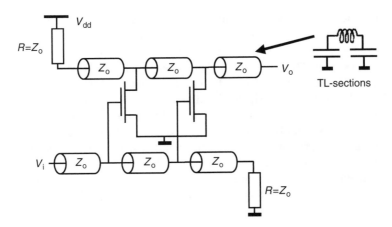

FIGURE 3.4 Distributed LNA.

example is shown in Figure 3.3, which is the basic idea behind the proposed LNAs in Refs. 5, 6. In Figure 3.3, inductor L1 together with capacitor C1 forms the shunt branch of the filter. The series branch of the filter is formed by (degeneration) inductor L2 together with the base-emitter capacitance of the transistor. The reactive part of the input impedance is now canceled over a wide frequency band. Implemented in a 0.18 μm silicon germanium (SiGe) bipolar and CMOS (BiCMOS) process, the LNA achieves an NF below 3 dB and an insertion gain above 20 dB over the mandatory band group 1.

Distributed amplifiers naturally achieve wideband behavior. Whereas in mm-wave design, coplanar wave guides or strip lines are used to implement the transmission lines, silicon implementations use integrated inductors and capacitors as the lumped element replacement circuits for the transmission line. An example is given in Figure 3.4, which shows a generic two-stage distributed amplifier. Although the resistive part of the inductors causes an increase in the NF, practical NF values around 3 dB are still achievable in 0.18 μm CMOS [7,8], similar to those achieved in SiGe BiCMOS technologies [9]. One drawback might be the large occupied die area caused by the silicon implementation of the inductors used.

An alternative CMOS LNA topology is presented in Ref. 10. Here, a common-gate (CG) input stage is loaded with three switched cascode devices with tanks resonating at the center frequency of each of the three subbands of band group 1. Note that the load switching must occur at the same speed as the hopping across the subbands, i.e., 9.5 ns. This means that the output node should settle within this time, which sets additional design constraints on the LNA. NFs between 5 and 7 dB and gains above 20 dB have been reported in literature.

Current feedback by means of a feedback resistor is also quite a commonly used method to broaden the bandwidth of the input match. As described in Ref. 11, a cascode topology including resistive feedback and a tuned load achieves an NF of 4 dB and a gain of 16 dB in a 0.18 μm CMOS process. Current feedback can be used in combination with voltage feedback. The output voltage can be fed back to the input using an integrated transformer, as is demonstrated in Ref. 12. Assume a degeneration inductor L_e placed at the emitter of the input transistor. The voltage across this inductor V_{ind} is equal to $I_e j\omega L_e$, where I_e is the emitter current. When the output voltage is fed back to this degeneration inductor by means of a transformer, the voltage V_{ind} changes into $I_e j\omega L_e(1 + k\sqrt{L_c/L_e})$, where L_c is the collector inductor and k the coupling factor. The increased voltage allows the degeneration inductor to be reduced. The extra term coming from the coupling factor means that the degeneration inductor can in fact be approximately three times smaller. The requirement in terms of series resistance (for low noise) remains the same, so the requirement on the Q-factor is three times less strict. Also take into account that the high level of degeneration reduces the current gain of the input transistor to hardly more than unity. This means that a resistor as collector load would contribute too much noise, so an inductive collector load would be necessary. In the transformer, this inductor is simply combined with the emitter degeneration inductor, thereby saving chip area again (in addition to the saving due to the smaller value of the emitter coil). A final advantage of the transformer feedback is that it flattens the frequency response. The LNA proposed in Ref. 12 is based on this principle and depicted in Figure 3.5. This LNA consists of a cascode input stage (Q1 and Q2), followed by a voltage buffer (Q3 and Q4) known as a White's emitter follower. There is voltage feedback by means of a transformer, formed by merging the collector coil and emitter degeneration coil of the input stage. In addition, there is current feedback formed by R1 and C1. This compound feedback mechanism gives high linearity, and also allows for matching of the input impedance to $50\,\Omega$ over the lower three bands, without the need for additional external matching components. The measured performance of the LNA is also provided in Figure 3.5, indicating an NF below 3 dB and wide band gain behavior. The input matching (i.e., S11) that takes into account the packaging and printed circuit board (PCB) is below $-15\,$dB for the mandatory band of interest.

Finally, to reduce area, the noise-canceling technique represents a promising wideband technique that, in principle, does not require any inductors. The CG stage is well known for its wideband input match and can realize wideband gain via a drain resistance (R_{CG}). However, transconductance g_m of the CG stage is fixed by the matching, whereas the next stage determines the capacitive load. As a result, only the resistance R_{CG} remains as a design variable, and a trade-off between gain (high R_{CG}) and bandwidth (low R_{CG}) exists. To remove this trade-off between gain and bandwidth—in other words, to increase the gain while maintaining the same bandwidth—an extra amplifier can be added. One disadvantage is that the additional amplifier will generate noise and consequently degrade the NF of the complete LNA circuit. However, an additional amplifier can be used to cancel the noise of the CG stage. By applying the LC technique, the gain-bandwidth of the LNA can be increased while maintaining a low overall NF. Figure 3.6 shows this concept conceptually. This idea is used by Chehrazi et al. [14] to realize an LNA using five inductors. A single-inductor implementation of the idea is demonstrated in Ref. 15. The LNA only uses one on-chip transformer, which takes about the same area as a single inductor. Moreover, this transformer is exploited to the maximum by being used simultaneously for biasing, source-impedance matching, and noiseless and powerless voltage amplification (see Figure 3.6). The LNA achieves a high voltage gain of 19 dB, a wideband input match, and an acceptable NF of 4–5.4 dB, while consuming 8 mW. This NF is acceptable because no balun with associated losses is needed as the circuit inherently converts the single-ended input into a differential output. The high voltage gain strongly reduces next-stage noise contributions. The LNA has been implemented in a 90 nm CMOS process.

DOWN-CONVERTER MIXER

In a zero-IF architecture, the LNA is in most cases followed directly by a (Gilbert) down-mixer. The mixer has a combined switching core for the in-phase and quadrature outputs. In fact, there are several ways to generate the two quadrature IF outputs:

- Two separate mixers for I and Q path
- Single understage with two merged switching cores and current-steering resistors for reducing current splitting noise
- Single understage with two merged switching cores and 25% LO duty cycle

The last option has been used by Bergervoet et al. [12]. In particular, the 25% duty-cycle drive gives conversion gain of about 3 dB more in the switching core, as can be easily seen from the Fourier expansion of the switching coefficient. This will not only lead to a better NF but also allow the use of an understage, which is required to provide 3 dB less current, thus directly reducing the current consumption.

In Figure 3.5, the subsequent mixer contains a combined common-emitter/common-base lower-stage, which is a well-known active balun structure [13]. The two bleeder resistors that go from

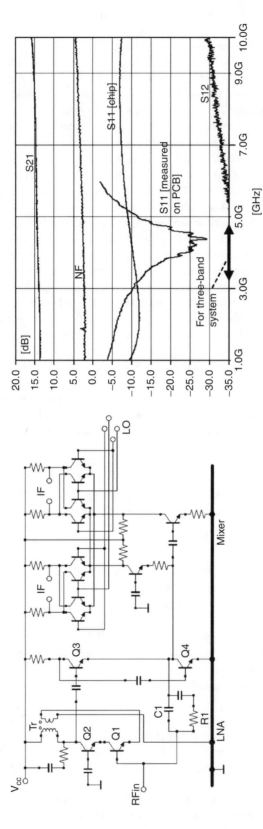

FIGURE 3.5 LNA and mixer design (left) and measured performance of LNA (right). (From Bergervoet, J., et al., *International Solid-State Circuits Conference (ISSCC)*, pp. 200–201, 2005.)

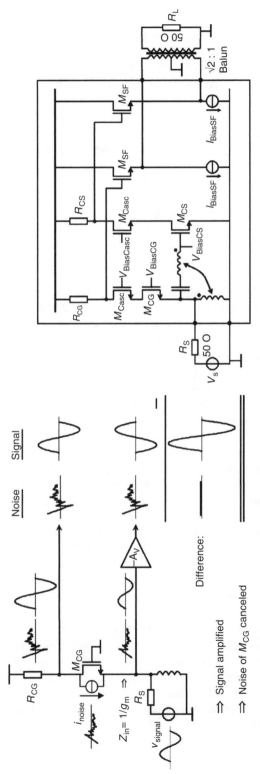

FIGURE 3.6 Noise-canceling technique applied to a CG-stage (left) and an implementation (right). (From Blaakmeer, S., et al., *Radio Frequency Integrated Circuits (RFIC)*, pp. 159–163, 2006.)

the switching core emitters directly to the supply voltage are there to reduce the DC current in the switching core. This is possible because, for linearity reasons, even a degenerated understage can only use about 50% of its available current swing. The advantage of removing DC current from the switching core is the noise reduction. A minor disadvantage is that the resistors will also remove a small fraction of the signal current. They will also act as an extra source of nonlinearity, since the signal current division between the linear bleeder resistors and the nonlinear base-emitter junction of the switching transistors is a nonlinear process. The understage is highly degenerated by emitter resistors to obtain the required linearity. A fully balanced eight-transistor switching core has been used, which creates both the in-phase and the quadrature phase baseband signals. Noise caused by cross-conduction is reduced to a minimum by appropriate shaping of the LO drive signals. These signals should ideally be sinusoidal signals, but as they are the output of frequency dividers, they also contain higher order harmonics. Therefore, (poly-phase) filtering is required especially to remove the third-order harmonic from the LO signal.

A mixer with a variable gain range is demonstrated in Refs. 3, 10. Here, the load resistor is decomposed into binary weighted segments so as to create dB-steps in the gain. Implemented in a 0.13 μm CMOS, a 30 dB gain is obtained over a large output bandwidth.

IF/BASEBAND FILTER

High-order filtering at IF/baseband is needed to achieve the sufficient attenuation of at least 35 dB at 662 MHz. This means that a roll-off of 35 dB is required within 662 MHz − 264 MHz = 398 MHz. Consequently, a fourth-order filter is the minimum requirement, but most UWB radios use a fifth-order filter to achieve sufficient attenuation. The large bandwidth in combination with high linearity involves a careful distribution of gain, filtering, and noise.

In Ref. 12, the baseband filter/variable gain amplifier (VGA) has been implemented as a fifth-order Chebyshev-like filter. A passive RC filter is used between the mixer outputs and the first active stage of the fifth-order filter, which is shown in Figure 3.7. This passive RC filter handles large out-of-band interferer signals linearly and filters them out considerably. The RC filter hardly influences the pass-band characteristic. The first active stage is a low-Q filtering stage, which makes a reasonable gain and has low noise. It is a modified Rauch filter and uses a simple two-stage operational amplifier (opamp) without compensation [16]. The Rauch filter realizes a complex conjugate pole pair and a real zero. The next stage of the filter is a passive attenuator, which is a network of resistors and switches, which direct the output current either to the input of the second active stage or to a differential ground node. The second active stage, which uses a structure similar to the first active stage, realizes the high-Q poles of the Chebyshev filter. The main challenge in realizing this stage has been to get the poles in the right location with a limited gain bandwidth of the opamp while ensuring the stability of the stage. With high-Q poles around the band edge, the gain of the stage will be high at these frequencies, resulting in a lower loop gain. However, with a low loop gain, the location of the poles is sensitive to spread in the opamp parameters. Therefore, the Q factor and the frequency of the poles have been made digitally tunable by means of independent tuning of the filter capacitances. The third active stage realizes a single real pole of the Chebyshev filter. This means that this stage has a high-frequency roll-off, leading to a lot of excess high-frequency loop gain, which ensures that it is highly linear and can drive the input of the analog-to-digital converter (ADC). For the same reason, the DC gain of this stage could be made relatively large without too much influence from the opamp parameters. The complete baseband filter provides a current gain ranging from 14 to 46 dB. This has been realized by the passive attenuator, as mentioned earlier, which can be controlled in steps of 6 dB. The bandwidth can be tuned in a range of 232–254 MHz. At 662 MHz offset, an attenuation of −57 dBr has been achieved (see Figure 3.7).

A fourth-order Sallen-and-Key filter has been used by Razavi et al. [10,17] and has been modified to a fifth-order channel selection filter by Aytur et al. [18]. Each filter section consists of a differential amplifier with a zeroing function and a current-mode digital-to-analog converter (DAC). In this amplifier configuration, offsets appear as a current applied to the resistive loads of the amplifier.

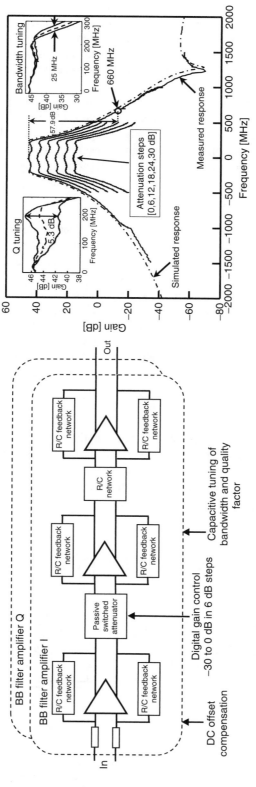

FIGURE 3.7 Block diagram of the filter implemented by Bergervoet et al. and its performance. (From Bergervoet, J., et al., *International Solid-State Circuits Conference (ISSCC)*, pp. 200–201, 2005.)

Offset cancellation is accomplished through corrective current provided by the DAC. Switched resistors across the output nodes allow for gain reduction.

A fifth-order elliptic channel filter has been used by Ismail and Abidi [19,20]. The on-chip filter is a passive LC filter, and it is therefore perfectly linear. Interestingly, the Gilbert mixer topology used has a low resistive load such that it can terminate directly the passive ladder filter, enabling the filter to be integrated in the mixer itself. Implemented as series-stacked spirals, the inductors are not too wasteful of area.

Finally, a third-order Gm-C followed by a third-order elliptic Gm-C filter has been implemented by Tanaka et al. [21]. The sixth-order filter has been realized in a 90 nm CMOS process with 1.1 V supply voltage. To enable operation at this low supply voltage, the use of stacked transistors should be minimized. Pseudodifferential amplifier techniques have been applied to ensure a robust gain stage. The filter provides over 50 dBr attenuation at 600 MHz offset.

RF TRANSMITTER BUILDING BLOCKS

A crucial aspect of a UWB transmitter is the need for power control to ensure that the transmitted level does not exceed the −41.3 dBm/MHz limit (−14 dBm across 528 MHz). Furthermore, as with WLAN systems, RF impairments (e.g., *I/Q* mismatch, phase noise, carrier feed-through) must be kept to a minimum.

The RF PA is in most cases based on an inductively loaded (cascode) transistor. An example is shown in Figure 3.8, where transistors M1 to M3 are used to implement a differential to single-ended structure [17]. Inductor L1 serves as a shunt peaking element for the cascode and as a series peaking element for the source follower. Transistor M4 delivers an output level of −10 dBm. The choice of the width of M4 matters as the output of the PA is directly coupled to the antenna and input of the LNA. The transistor therefore degrades the NF of the receiver. Careful design raises the NF of the receiver by 0.15 dB.

A straightforward approach has been used in Ref. 22. For the PA, a common-emitter amplifier has been used, which provides single-ended output. The signal is AC coupled, both at the input and at the output through a matching network. At the input, the PA is driven by one side of the differential output of the mixer through a matching circuit. The output-matching circuit consists of an on-chip pull-up inductor and a coupling capacitor. The output-matching circuit is also on-chip. For biasing the output transistor, a current-mirror configuration has been used as it provides a low-impedance source compared with the resistive-biasing network. The PA has a variable-bias current.

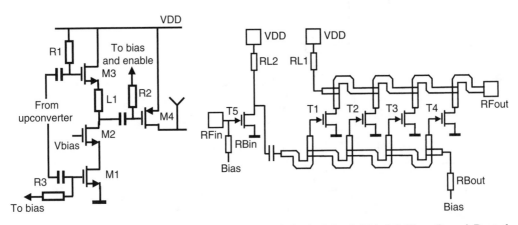

FIGURE 3.8 Two implementations of a CMOS RF PA: an inductively loaded PA (left [From Razavi, B., et al., *IEEE J. Solid-State Circuits*, vol. 40, pp. 2555–2562, 2005.]), and a distributed PA (right [From Grewing, C., et al., *Radio Frequency Integrated Circuits (RFIC)*, pp. MO4B-4, 2004.]).

By varying the bias, the gain of the PA can be varied. A 6 dB gain variation has been implemented in this stage. Another 6 dB of gain variation is obtained by changing the resistor values in a preceding *V*-to-*I* converter. The bias current and the resistors are programmed so as to provide gain steps of about 1.5 dB each in eight steps. The resulting output power is −7 dBm and the PA has a side-band suppression of 32 dB.

Again, the distributed amplifier has also been proposed. Reference 23 describes the implementation of a four-stage amplifier in a 0.13 μm CMOS process (see Figure 3.8), resulting in a compression point of +3.5 dBm. In this case, the transmission lines are implemented as microstrip lines.

At baseband, the required filters and gain stages are similar to those used in the receiver. The up-conversion mixers can be realized in several ways. Whereas Razavi et al. [10] have used an up-conversion circuit based on resistively degenerated passive mixers along with a current feedback amplifier, Aggarwal et al. [22] have used two single-side-band (SSB) Gilbert mixers. The voltage-to-current converter required as the understage for the Gilbert mixer core also implements a gain variation mechanism.

FAST-HOPPING SYNTHESIZER

A particularly challenging building block of the UWB receiver is the frequency synthesizer. The first challenge is the required fast frequency hop of 9.5 ns. The second challenge is the required spectral purity.

INTEGER-*N* PLL APPROACH

A classical integer-*N* PLL with programmable loop divider ratio seems a logical choice as only three LO signals need to be synthesized [24,25]. The greatest-common-divider frequency of the three mandatory LO signals is 264 MHz. This frequency can be the reference frequency in the integer-*N* PLL. Assuming that a realistic minimum of 20 reference cycles is needed before the PLL locks after a frequency hop, this is equivalent to 76 ns, which is much more than the required 9.5 ns. Straightforward implementation of an integer-*N* PLL is therefore not possible.

Because the symbol period is 312.5 ns, one could also argue that two integer-*N* PLLs are possible, each with a reference clock of 264 MHz. While one PLL is locked to the required LO signal, the other locks to the next required LO signal, as the band-switching sequence is known *a priori*. This idea has been exploited in Ref. 26 to cope with band groups 1 and 3. Here, a single 528 MHz reference clock has been used together with two quadrature LC oscillators generating frequencies in the range of 6–9 GHz (band group 3). Frequency division by 2 enables band group 1. A wide-loop bandwidth of 26 MHz results in a locking time of approximately 115 ns. Realized in a 0.18 μm CMOS technology, the complete synthesizer consumes 32 mA from a 1.8 V supply voltage. An issue with the large-loop bandwidth might be the total integrated phase noise, which is required to be below 2° rms.

Another option might be a fractional-*N* PLL synthesizer, allowing de-coupling of the loop bandwidth and reference frequency. In the integer-*N* PLL, they are linked to each other by the stability conditions of the feedback loop [25]. A locking time of 9.5 ns within 20 reference cycles requires a reference clock above 2 GHz. As the PLL stays locked for 312.5 ns, the output of the fractional-*N* PLL will settle to many different frequencies several times, generating an average output frequency equal to the wanted LO signal. This will most probably give rise to many spurious tones, and it is difficult to reconcile these with the stringent spurious-tone requirements. In open literature, no synthesizers for MB-OFDM UWB based on fractional-*N* PLL techniques are reported.

If only the mandatory mode is considered, a straightforward frequency synthesizer architecture would use three separate PLLs. Each PLL generates one of the three required carrier frequencies. In combination with a high-speed three-to-one multiplexer, the LO signal is selected. This option is only practical in those cases where RC ring oscillators can fulfill the requirements. Three LC-oscillator-based PLLs will raise issues with respect to frequency pulling and occupation of die area.

The option of using ring oscillators has been considered by Zheng et al. [3] and Razavi et al. [10] for a three-band UWB system in a 0.13 μm CMOS process, where each PLL consumes 15 mW from a 1.5 V supply voltage. The use of a 66 MHz reference frequency allows a loop bandwidth of about 5 MHz, thus suppressing the close-in phase noise of the oscillators considerably. Although injection pulling caused by magnetic coupling is of no concern when using RC oscillators, pulling via supply lines is still possible. However, as the three PLL frequencies are far from each other and not related by integer multiples, injection pulling is negligible.

USING TWO PLLS AND ONE SINGLE-SIDE-BAND MIXER

Most other proposed synthesizer concepts are based on frequency translation, where two frequencies can be added or subtracted by means of a SSB mixer (Figure 3.9).

Synthesizers using this method are also known as multitone generators. One PLL generates a fixed frequency of 3960 MHz, which is the carrier frequency of the second UWB subband in band group 1. When the carrier frequency of the first subband needs to be generated, this 3960 MHz is shifted down by 528 MHz to 3432 MHz; when the carrier frequency of the third subband is required, the 3960 MHz signal is shifted up in frequency by the same amount to give 4488 MHz. The shift frequency of 528 MHz is obtained from a second nonprogrammable PLL. Because both PLLs deliver a fixed frequency, their lock time (and, with that, bandwidth and reference frequency) is completely de-coupled from the 9.5 ns guard time.

The frequency translation has been realized by means of a full complex SSB mixer. This mixer generates the shifted output frequency while at the same time suppressing the image frequency. The direction (up or down) of the shift of the 3960 MHz signal can be reversed by changing the sign of the 528 MHz signal. It can be easily shown that this requires either the I and Q signals of the complex 528 MHz signal to be swapped or one of these signals to be inverted (shifted by 180°). Because in many designs, differential signaling is used, the second option is easily implemented by swapping the $+Q$ and $-Q$ signals. The inversion of the Q signal is represented in Figure 3.9 by the -1 gain block.

Rather than using a multiplexer at the SSB mixer's output to select either the fixed 3960 MHz PLL output or the shifted frequency generated by the SSB mixer, a multiplexer at the SSB mixer's input has been used to select from -528 MHz, $+528$ MHz, or DC. Selection of DC passes the 3960 MHz signal directly to the SSB mixer output, thus eliminating the need for a high-speed multiplexer.

The problem of SSB mixing lies in the inherently generated spurious tones, for example, due to nonlinear behavior of the mixer. In the scheme discussed, the third harmonic of the 528 MHz signal (at 1584 MHz) is particularly troublesome because after mixing with 3960 MHz, this harmonic will cause a spurious tone at either 3960 MHz + 1584 MHz = 5544 MHz or 3960 MHz − 1584 MHz = 2376 MHz. Both spurious tones are close to possible strong interferer signals (5 GHz and 2.4 GHz ISM bands, respectively), and this may result in UWB signal corruption.

One of the very first approaches using the concept outlined above was presented in Ref. 27. The complete block schematic of the frequency synthesizer is also shown in Figure 3.9. The PLL labeled "PLL8G" generates the quadrature 3960 MHz signals. The internal LC oscillator of this PLL generates a 7920 MHz signal that is fed through a static divide-by-2 frequency divider, which yields the desired quadrature signals and is also part of the PLL main divider. The PLL marked "PLL2G" generates the quadrature 528 MHz signals, using an internal oscillator of 2112 MHz and static frequency dividers; the voltage controlled oscillator (VCO) runs at four times rather than two times the required frequency to facilitate the use of an integrated LC oscillator. Both PLLs are fully integrated and derive their output from a common 44 MHz reference signal that is supplied externally and buffered internally, as shown in the block diagram of Figure 3.9 (see also Ref. [36]). Because the 528 MHz signal is the output of a static divide-by-2 circuit, its harmonic content will inevitably be strong. Because of the use of quadrature signals, the third harmonic of +528 MHz is located at −1584 MHz. An integrated two-stage polyphase filter at the divide-by-2 output (see Figure 3.9) was used to place a notch at this frequency. The zeroes are chosen on both sides of the third-harmonic

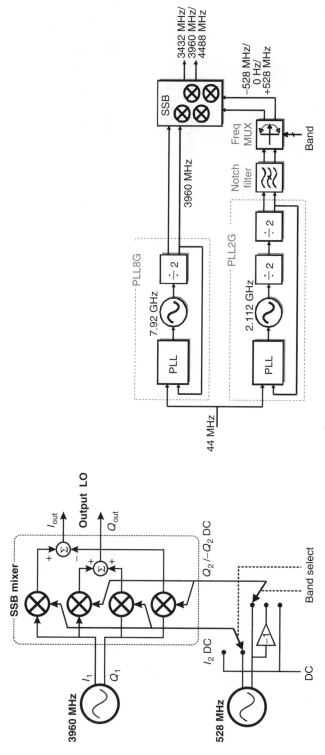

FIGURE 3.9 LO scheme based on SSB mixing (left) and a possible implementation (right). (From Leenaerts, D., et al., *International Solid-State Circuits Conference (ISSCC)*, pp. 202–203, 2005.)

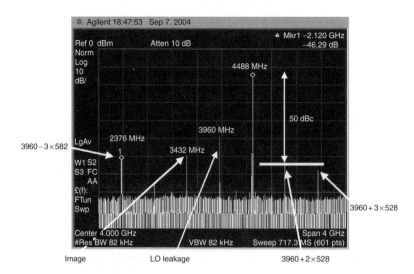

FIGURE 3.10 Measured spectral output of the synthesizer when generating the LO for band 3. (From Leenaerts, D., et al., *International Solid-State Circuits Conference (ISSCC)*, pp. 202–203, 2005.)

frequency rather than both exactly at this frequency to cope with process spread. The placement of two somewhat different zeroes widens the notch in the transfer function. In this way, all spurious tones in the 5 GHz band are below -50 dBc, as can be seen in Figure 3.10. The fully integrated 1 mm^2 synthesizer consumes 73 mW from a 2.7 V supply and achieves FH within 1 ns.

The same strategy has been applied in Refs. 28, 29. The 528 MHz signal has been generated from an RC ring oscillator rather than an LC VCO to save die area because a 0.13 µm CMOS technology has been used. A 2-bit current DAC controls the tail current of the RC oscillator, and therefore the frequency. The double quadrature SSB mixer, however, uses integrated inductors as load to boost the performance. Consequently, the overall die area is still considerable: 0.91 mm^2. In-band side-band suppression is better than the requirement; unfortunately, nothing is mentioned with respect to the out-of-band suppression. The synthesizer consumes 62 mW from a 1.5 V supply.

To cover seven bands, a different clock distribution is needed. One possible solution is provided in Refs. 30, 31. The center PLL generates the center frequencies 6864 MHz (subband 5) and 3432 MHz (subband 1), whereas the incremental PLL produces twice the incremental frequencies, 2112 MHz and 1056 MHz, for frequency addition and subtraction. A tri-mode divider then provides DC or quadrature output signals with different phase sequences of 528 and 1056 MHz. As a result, band groups 1 and 3 are covered. Side-band spurious tones are suppressed by inductively band-pass loading the SSB mixer, resulting in a -48 dBc out-of-band attenuation. The synthesizer dissipates 48 mW and has been realized in a 0.18 µm CMOS technology.

Using a Single PLL and SSB Mixer(s)

To eliminate the need for two PLLs, the 3960 MHz signal needs to be divided by 7.5 to derive a 528 MHz signal. The challenge lies in the design of this divider, especially because of the need for quadrature signals with a 50% duty cycle. In Ref. 32 this is accomplished by two modified versions of the Miller divider, one realizing divide-by-3 and the other divide-by-2.5. The regenerative loop naturally leads to quadrature outputs and 50% duty cycle. Realized in 0.18 µm CMOS, the image suppression of the divider is -20 dBc and it consumes 18 mW from a 1.8 V supply.

One other possibility is demonstrated in Ref. 33. The architecture of the synthesizer is shown in Figure 3.11. The VCO runs at a frequency of 7920 MHz, from which quadrature 3960 MHz signals are derived using a static divide-by-2 frequency divider. To obtain a frequency of 528 MHz, one could

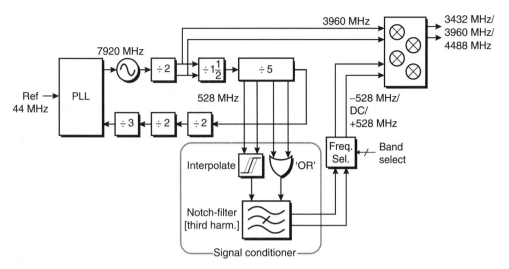

FIGURE 3.11 Single PLL, single SSB mixer synthesizer implementation. (From van de Beek, R., et al., *European Solid-State Circuit Conference (ESSCIRC)*, pp. 173–176, 2005.)

divide the 7920 MHz tone by 15. This, however, would lead to extra VCO loading and, with that, a power consumption penalty. Instead, the 3960 MHz signal is divided by 7.5, and this operation is split into a divide-by-1.5 and a divide-by-5 operation. An added benefit of this approach is the 50% duty-cycle signal delivered by the divide-by-1.5 circuit, which helps in generating the quadrature 528 MHz signals. The 528 MHz signal is further divided down to a 44 MHz signal that is locked to an external reference clock. To generate quadrature 528 MHz signals, a time resolution of one-quarter of the input signal is needed. This is not directly available from the outputs of the divider. An "OR" function is needed to generate the in-phase signal and an interpolation functionality is required to generate the quadrature signal. Again, filtering is necessary to attenuate the third-order harmonics of 528 MHz. The single PLL, single SSB-mixer concept consumes 52 mW from a 2.7 V supply and is implemented in a 0.25 μm SiGe BiCMOS process. Again due to additional filtering, out-of-band spurious tones are below −50 dBc. The integrated phase noise is below 2° rms and the measured hopping speed is well below the required 9.5 ns. The occupied die area is 0.44 mm², saving 60% compared with the two-PLL approach in Ref. 27.

Seven subbands can also be covered using a single PLL, two SSB mixers, and only divide-by-2 blocks [20]. The synthesizer is built up around a 16.896 GHz LC-VCO, which after frequency division provides 8.448, 4.224, 2.112, and 1.056 GHz signals. One SSB mixer is used to multiply 8.448 GHz with 4.224 GHz to realize 12.672 GHz, which is then divided in steps toward 6.336 and 3.168 GHz. The other SSB mixer has as inputs the 16.896 GHz signal and one of the three intermediate signals: 2.112, 1.056, and 3.168 GHz. The output of this SSB mixer is again divided into two steps. Consequently, the LO signal for subband 2 is generated from 16.896 GHz, multiplied by −1.056 GHz and then divided by 4.

The LC oscillator is built around two coupled oscillator cores with fully integrated LC resonators. The quadrature VCO draws 20 mA from a 2.7 V supply, and is realized in a 0.18 μm SiGe BiCMOS process. The remaining LO generator blocks consume 46 mA.

A single PLL approach to cover the first four band groups has been proposed in Ref. 34. A quadrature VCO operating at 8.448 GHz together with five divide-by-2 blocks and two SSB mixers synthesize all possible frequencies. A harmonics-suppressing filter is introduced at one of the SSB mixer outputs. The cut-off frequency of the filter is programmed to switch in accordance with the changing (sub) bands. The configuration dissipates 47 mW in a 90 nm CMOS process. Spurious tones are at −25 dBc level.

OTHER OPTIONS

A synthesizer for the mandatory band using a direct digital synthesizer (DDS) is demonstrated in Ref. 35. The principal idea is to add or subtract a low frequency, ± 264 or ± 792 MHz from a fixed frequency of 4224 MHz. This fixed frequency is locked by means of a PLL to an external reference frequency. The DDS generates the low frequencies. Inherently, the harmonics of these signals are much lower than if they had been generated by divider chains, SSB mixers, or both. The 4224 MHz signal is used to clock the read only memory (ROM) lookup tables and the 4-bit current-steering DACs. The sinusoidal I/Q waveforms are stored in the ROM tables. An SSB mixer is needed to mix the 4224 MHz signal with the output signal from the DDS. The measured phase error is less than $2°$ rms. FH occurs within 2 ns. The LO chain consumes 124 mA from a 1.5 V supply and has been implemented in a 0.13 μm CMOS technology.

RF TRANSCEIVERS FOR MB-OFDM UWB

As already mentioned, due to the wide channel bandwidth, the receiver and transmitter signal paths of UWB systems naturally employ direct conversion, i.e., zero-IF. The advantage of zero-IF conversion lies in the absence of the image signal. However, drawbacks include flicker noise, DC offset, and I/Q imbalance.

A direct-conversion three-band MB-OFDM UWB transceiver has been demonstrated in Refs. 10, 17. The complete system on chip, i.e., radio plus baseband, has been presented in Ref. 18. In receive mode, a 5-bit ADC digitizes the in-phase and quadrature phase signals at 528 MS/s. Signal de-mapping is performed after transformation by a 128 point fast Fourier transform (FFT). For transmission, the process is reversed, except that the signals are up-sampled to 1056 MS/s. To enable agile FH, the design utilizes three separate RF blocks for each operating frequency in band group 1. Each block consists of a tuned LNA, I/Q mixers, and integer-N PLL using an RC-oscillator. Voltage regulators have been employed to reduce LO leakage and other unwanted coupling. The transmitter section uses many of the same concepts as the receiver for implementing the FH. The modulator circuit employs switched inductive loading. The use of switched inductors allows each inductor to be optimized separately, allowing a wider operating range. The chip was implemented in a 0.13 μm CMOS technology and packaged in a 64-pin quad flat package no leads (QFN) package. The RF transceiver draws 100 mA in receive mode and 70 mA in transmit mode from a 1.5 V supply. The measured receiver sensitivity is -82 dBm at 53.3 Mbps and -71.3 dBm at 480 Mbps data rate at 8% PER using 200-byte packets.

A fully integrated transceiver front-end has been realized in a 0.25 μm SiGe BiCMOS technology with an NPN-f_T of 70 GHz [36,37]. The transceiver architecture is shown in Figure 3.12. The antenna signal is filtered by an external passive prefilter to reduce the level of out-of-band interferers. The direct conversion receiver consists of a wideband LNA and a quadrature mixer. The baseband filter provides both filtering and variable gain. The filtered baseband signal is digitized by an ADC, which is followed by the digital baseband processor. The transmitter is a direct up-conversion architecture, delivering -41.3 dBm/MHz output power to the antenna. The transmit chain features wideband elliptic baseband filters, a VGA with dynamic range of 12 dB, an up-conversion mixer, and an RF output stage with a power of -7 dBm. The PA is a modified (inductor less) version of the circuit presented in Ref. 22. The totem-pole-based PA is shown in Figure 3.13 and is designed such that the NF of the LNA is degraded by only 0.1 dB. Part of the transceiver's front-end is shown in Figure 3.13 too, indicating that the output of the PA shares the same pin as the input of the LNA. The LNA has been already depicted in Figure 3.5. Simulations indicate a transmitter output IP3 of +17 dBm. The synthesizer provides the FH quadrature LO signals. Dedicated LO buffers ensure high isolation between receive and transmit mixers while also providing the possibility to allow simultaneous transmission and reception, enabling a loop-back test mode. The chip with a total area of 4 mm^2 has been packaged in an heatsink very-thin quad flat-pack no leads (HVQFN) package and mounted on an flame resistant 4 (FR4) board. Digital control blocks for tuning the VCOs and

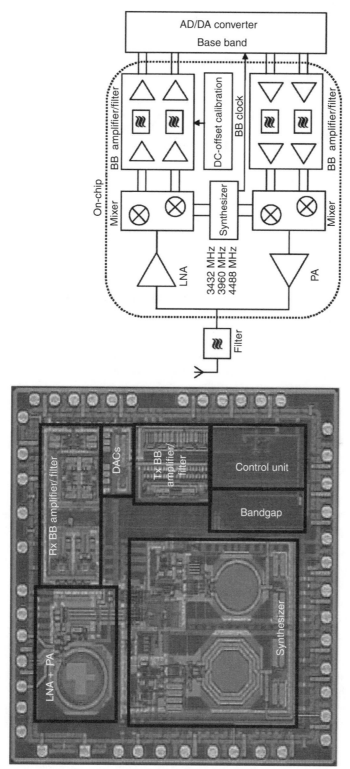

FIGURE 3.12 Chip photograph (left) and block diagram (right) of a fully integrated UWB transceiver. (From Bergervoet, J., et al., *Radio Frequency Integrated Circuits (RFIC)*, pp. 301–304, 2006.)

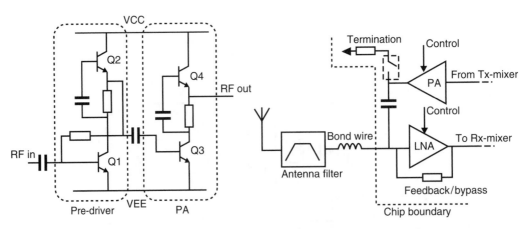

FIGURE 3.13   Totem-pole RF PA (biasing not shown) (left) and front-end design (right). (From Bergervoet, J., et al., *Radio Frequency Integrated Circuits (RFIC)*, pp. 301–304, 2006.)

TABLE 3.1
Measured Data for the Transceiver (Assuming 20 dB Attenuation by Prefilter)

Parameter	Performance	Measured	Info
Current consumption		47 mA @ 2.7 V	Receiver
		43 mA @ 2.7 V	Transmitter
		27 mA @ 2.5 V	Synthesizer
Noise figure	<6.6 dB[a]	4.5 dB	On PCB, center of IF band, LO is 3960 MHz
Input IP2	>+20 dBm	+25 dBm	f_{in1} : 5 GHz ISM, f_{in2} : GSM1900
Input IP3	>−9 dBm	−6 dBm	f_{in1} : 5 GHz ISM, f_{in2} : 5 GHz ISM
Maximum gain		59 dB	Power gain from RF input to baseband output
VCO phase noise	<−100 dBc/Hz	−104 dBc/Hz	At 1 MHz offset
Integrated phase noise	<3.5° rms	1° rms	Integrated from 0–50 MHz
In-band spurs	<−30 dBc	<−30 dBc	
Out-of band spurs	<−50 dBc	<−50 dBc	For 5 GHz ISM
	<−45 dBc	<−45 dBc	For 2.4 GHz ISM
Hopping speed	<−9.5 ns	<1 ns	For all permitted hopping sequences
EVM	<10%	8%	Loop-back mode test
Output power	−9.5 dBm	−6 dBm	
OIP3	10 dBm	12 dBm	

[a] Requirement is <4.6 dB, assuming a prefilter insertion loss of 2 dB.

Source: Bergervoet, J., et al., *International Solid-State Circuits Conference (ISSCC)*, pp. 200–201, 2005.

the IF filter as well as a band-gap unit have also been implemented. The error vector magnitude (EVM) is measured while the transceiver is set in a loop-back test mode. The resulting eye diagram and constellation diagram reveal an EVM below 8%. The measured performance is provided in Table 3.1, indicating that it is possible to achieve low NFs together with high linearity for complete UWB transceivers.

A four-band group MB-OFDM UWB transceiver has been demonstrated in Ref. 34. The zero-IF architecture has been implemented in 90 nm CMOS using thin oxide-only devices. The overall transmit output IP3 is 7.2 dBm and is flat over the complete spectrum up to 9.2 GHz, whereas the

TABLE 3.2
Comparison of Several Transceiver Performances

	[18]	[37]	[34]	[35]	[38]
Technology	0.13 μm CMOS	0.25 μm SiGe BiCMOS	90 nm CMOS	0.13 μm CMOS	0.18 μm CMOS
NF (dB)	6–7	4.5	6.9	4.1	4.7
iIP3 (dBm)	−15	−6	−16	−22	−0.8
iIP2 (dBm)		+25			+22
EVM (dB)	−19.5	−24	−28	−27	−28.6
P_{out} (dBm)		−6	−3.8	+5	−12.6
TX OIP3		12	8.6		11.8
Pdiss RX (mW)[a]	100 mA @ 3.3/1.5 V	199	224	237	412
Pdiss TX (mW)[b]	70 mA @ 3.3/1.5 V	190	131	284	397
Chip area (mm²)	2	4	4.5	6.6	16

[a] RX stands for receive mode.
[b] TX stands for transmit mode.

receiver linearity is −17 dBm. Overall NF is 6.3–7.8 dB. Operating from a 1.1 V supply, the transceiver dissipates 131 mW in the transmit mode and 224 mW in the receive mode.

A fully integrated 0.13 μm CMOS transceiver for the mandatory mode is demonstrated in Ref. 35. The measured EVM is −28 dBm up to −4 dBm output power. The NF is close to 4 dB in subband 3. An external band-pass filter is required to suppress all transmit output spurious tones below the −30 dBc level.

Finally, a dual-antenna phased array UWB transceiver for the mandatory mode is described in Ref. 38. To achieve extended range and improved attenuation of interferers, antenna diversity is implemented in a 0.18 μm CMOS process. The transceiver employs two transmit and two receive antennas. Selection diversity has been implemented on both transmit and receive sides by choosing the branch with the higher SNR, as determined by the accompanying baseband processor. The two received signals are combined with the optimal LO phase shift at the mixer output. Interferers can be suppressed by phase combining.

An overview of published transceiver performances for MB-OFDM UWB is provided in Table 3.2. As can be seen, a radio implemented in a dedicated RF process has improved linearity and noise behavior at lower dissipation levels compared with the CMOS counterparts. However, it is expected that very soon, UWB radios in standard deep submicron CMOS processes will deliver similar performance. These CMOS solutions will then be preferred on the basis of integration and cost aspects.

CONCLUSIONS

Several circuit design techniques for multiband UWB have been discussed. Challenging design aspects in UWB include the combination of wideband behavior at radio frequencies and baseband in combination with low NFs and high linearity, as well as the required fast LO hopping.

Currently, most UWB transceivers are realized in a BiCMOS technology. However, recently presented circuit techniques and achievements in CMOS indicate that CMOS transceivers will start competing with their BiCMOS counterparts.

ACKNOWLEDGMENT

The author would like to acknowledge the much-appreciated inputs from the NXP Semiconductors UWB-team, both in Eindhoven and in San Jose, CA.

REFERENCES

1. IEEE P802.15 Working Group for Wireless Personal Area Networks, March 2004, http://www.ieee802. org/15/pub/TG3a.html.
2. ECMA-368, High rate ultra wideband PHY and MAC standard, http://www.ecma-international.org/ publications/standards/Ecma-368.htm.
3. Zheng, Y., Tong, Y., Wei Ang, C., Xu, Y-P., Wooi, G., Lin, F. and Singh, R., A CMOS carrier-less UWB transceiver for WPAN applications, *International Solid-State Circuits Conference (ISSCC), IEEE*, vol. 49, pp. 116–117, 2006.
4. Razavi, B., Multiband UWB transceivers, *IEEE Custom Integrated Circuits Conference (CICC)*, pp. 141–148, 2005.
5. Ismail, A. and Abidi, A., A 3 to 10 GHz LNA using a wideband LC-ladder matching network, *International Solid-State Circuits Conference (ISSCC), IEEE*, vol. 47, pp. 384–385, 2004.
6. Bevilacqua, A. and Niknejad, A., An ultra-wideband CMOS LNA for 3.1 to 10.6 GHz wireless receivers, *International Solid-State Circuits Conference (ISSCC), IEEE*, vol. 47, pp. 382–383, 2004.
7. Heydari, P. Lin, D., Shameli, A. and Yazdi, A., Design of CMOS distributed circuits for multiband UWB wireless receivers, *Radio Frequency Integrated Circuits (RFIC), IEEE*, pp. 695–698, 2005.
8. Heydari, P. and Lin, D., A Performance Optimized CMOS Distributed LNA for UWB Receivers, *IEEE Custom Integrated Circuits Conference (CICC)*, pp. 337–340, 2005.
9. Tsai, M-D., Lin, K-Y. and Wang, H., A 5.4 mW LNA using a 0.35 μm SiGe BiCMOS technology for 3.1-10.6 GHz UWB wireless receivers, *IEEE Custom Integrated Circuits Conference (CICC), IEEE*, pp. 335–338, 2005.
10. Razavi, B., Aytur, T., Yang, F., Yan, R., Kang, H. Hsu, C. and Lee, C., A 0.13 μm CMOS UWB transceiver, *International Solid-State Circuits Conference (ISSCC), IEEE*, vol. 48, pp. 216–217, 2005.
11. Iida, S., A 3.1 to 5 GHz CMOS DSSS UWB for WPANS, *International Solid-State Circuits Conference (ISSCC), IEEE*, vol. 48, pp. 214–215, 2005.
12. Bergervoet, J., Harish, K., van der Weide, G., Leenaerts, D., van de Beek, R., Waite, H., Zhang, Y., Aggarwal, S., Razzell, C. and Roovers, R., An interference robust receive chain for UWB radio in SiGe BiCMOS, *International Solid-State Circuits Conference (ISSCC), IEEE*, vol. 48, pp. 200–201, 2005.
13. Chevallier, G. and Stikvoort, E. F., Transformer circuit, double-balanced mixer, U.S. patent 5825231.
14. Chehrazi, S., Mirzaei, A., Bagheri, R. and Abidi, A., A 6.5 GHz wideband CMOS low noise amplifier for multi-band use, *IEEE Custom Integrated Circuits Conference (CICC)*, pp. 801–804, 2005.
15. Blaakmeer, S., Klumperink, E., Leenaerts, D. and Nauta, B., A wideband noise-canceling CMOS LNA exploiting a transformer to realize low-noise, low-power, voltage gain, *Radio Frequency Integrated Circuits (RFIC)*, pp. 137–140, 2006.
16. Heinlein, W. E. and Holmes, W. H., *Active Filters for Integrated Circuits*, Prentice-Hall International Inc., London, 1974.
17. Razavi, B., Aytur, T., Yang, F., Yan, R., Kang, H., Hsu, C. and Lee, C., A UWB CMOS transceiver, *IEEE Journal of Solid-State Circuits*, vol. 40, pp. 2555–2562, 2005.
18. Aytur, T., Kang, H-C., Mahadevappa, R., Altintas, M., ten Brink, S., Dieo, T., Hsu, C., Shi, F., Yang, F-R., Lee, C., Yan, R. and Razzavi, B., A fully integrated UWB PHY in 0.13 μm CMOS, *International Solid-State Circuits Conference (ISSCC), IEEE*, vol. 49, pp. 124–125, 2006.
19. Ismail, A. and Abidi, A., A 3.1 to 8.2 GHz direct conversion receiver for MB-OFDM UWB communication, *International Solid-State Circuits Conference (ISSCC), IEEE*, vol. 48, pp. 208–209, 2005.
20. Ismail, A. and Abidi, A., A 3.1 to 8.2 GHz zero-IF receiver and direct frequency synthesizer in 0.18 μm SiGe BiCMOS for mode-2 MB-OFDM UWB communication, *IEEE Journal of Solid-State Circuits*, vol. 40, pp. 2573–2582, 2005.
21. Tanaka, A., Okada, H., Kodama, H. and Ishikawa, H., A 1.1 V 3.1 to 9.5 GHz MB-OFDM UWB transceiver in 90 nm CMOS, *International Solid-State Circuits Conference (ISSCC), IEEE*, vol. 49, pp. 120–121, 2006.
22. Aggarwal, S., Leenaerts, D., van de Beek, R., van der Weide, G., Harish, K., Bergervoet, J., Landesman, A., Zhang, Y., Razzell, C., Waite, H. and Roovers, R., A low power implementation for the transmit path of a UWB transceiver, *IEEE Custom Integrated Circuits Conference (CICC)*, pp. 149–152, 2005.
23. Grewing, C., Winterberg, K., van Waasen, S., Friedrich, M., Puma, M., Wiesbauer, G. and Sandner, C., Fully integrated distributed power amplifier in CMOS technology, optimized for UWB transmitters, *Radio Frequency Integrated Circuits (RFIC)*, pp. 87–90, 2004.

24. Leenaerts, D., van der Tang, J. and Vaucher, C., *Circuit Design for RF Transceivers*, Kluwer Academic Publishers, Dordrecht, 2001.

25. Vaucher, C., *Architectures for RF frequency synthesizers*, Kluwer Academic Publishers, Dordrecht, 2002.

26. Tak, G-Y., Hyun, S-B., Kang, T., Choi, B. and Park, S., A 6.3-9 GHz fast settling PLL for MB-OFDM UWB applications, *IEEE Journal of Solid-State Circuits*, vol. 40, pp. 1671–1679, 2005.

27. Leenaerts, D., van de Beek, R., van der Weide, G., Harish, K., Wiate, H., Zhang, Y., Razzell, C. and Roovers, R., A SiGe BiCMOS 1ns fast hopping frequency synthesizer for UWB radio, *International Solid-State Circuits Conference (ISSCC), IEEE*, vol. 48, pp. 202–203, 2005.

28. Chae, H-S., Park, E-C. and Cha, C-Y., A fast hopping frequency synthesizer for UWB systems in a CMOS technology, *Wireless Communication Systems 2005*, pp. 370–374, 2005.

29. Lee, J-E., Park, E., Cha, C., Chae, H-S., Suh, C-D., Koh, J., Lee, H. and Kim, H-T., A frequency synthesizer for UWB transceiver in 0.13 μm CMOS technology, *Silicon Monolithic Integrated Circuits in RF systems*, pp. 294–297, 2006.

30. Lee, J. and Chiu, D., A 7-band 3 to 8 GHz frequency synthesizer with 1ns band-switching time in 0.18 μm CMOS technology, *International Solid-State Circuits Conference (ISSCC), IEEE*, vol. 48, pp. 204–205, 2005.

31. Lee, J., A 3-to-8-GHz fast-hopping frequency synthesizer in 0.18 μm CMOS technology, *IEEE Journal of Solid-State Circuits*, vol. 41, pp. 566–573, 2006.

32. Lin, C. and Wang, C., A semi-dynamic regenerative frequency divider for mode-1 MB-OFDM UWB hopping carrier generation, *International Solid-State Circuits Conference (ISSCC), IEEE*, vol. 48, pp. 206–207, 2005.

33. van de Beek, R., Leenaerts, D. and van der Weide, G., A fast-hopping single-PLL 3-band UWB synthesizer in 0.25 μm SiGe BiCMOS, *European Solid-State Circuit Conference (ESSCIRC)*, pp. 173–176, 2005.

34. Tanaka, A., Okada, H., Kodama, H. and Ishikawa, H., A 1.1 V 3.1-to-9.5 GHz MB-OFDM UWB transceiver in 90 nm CMOS, *International Solid-State Circuits Conference (ISSCC), IEEE*, vol. 49, pp. 120–121, 2006.

35. Sandner, C., Derksen, S., Draxelmayr, D., Ek, S., Filimon, V., Leach, G., Marsli, S., Matveev, D., Mertens, K., Micli, F., Paule, H., Punzenberger, M., Reindl, C., Salerno, R., Tiebout, M., Wiesbauer, A., Winter, I. and Zhang, Z., A WiMedia/MBOA-compliant CMOS RF transceiver for UWB, *International Solid-State Circuits Conference (ISSCC), IEEE*, vol. 49, pp. 122–123, 2006.

36. Roovers, R., Leenaerts, D., Bergerviet, J., Harish, K., van de Beek, R., van der Weide, G., Waite, H., Zhang, Y., Aggarwal, S. and Razzell, C., An interference-robust receiver for ultra wide band radio in SiGe BiCMOS technology, *IEEE Journal of Solid-State Circuits, IEEE*, vol. 40, pp. 2563–2572, 2005.

37. Bergervoet, J., Kundur, H., Leenaerts, D., van de Beek, R., Roovers, R., van der Weide, G., Waite, H., and Aggarwal, S., A fully integrated 3-band OFDM UWB transceiver in 0.25 μm SiGe BiCMOS, *Radio Frequency Integrated Circuits (RFIC)*, pp. 301–304, 2006.

38. Lo, S., Sever, I., Ma, S., Jang, P., Zou, A., Arnott, C., Ghatak, K., Schwarz, A., Huynh, L. and Nguyen, T., A dual-antenna phased array UWB transceiver in 018 μm CMOS, *International Solid-State Circuits Conference (ISSCC), IEEE*, vol. 49, pp. 118–119, 2006.

4 Design Considerations for Integrated MIMO Radio Transceivers*

Yorgos Palaskas, Ashoke Ravi, Stefano Pellerano, and Sumeet Sandhu

CONTENTS

Introduction... 107
MIMO Theory 109
    Fading Wireless Channels 110
    Spatial Diversity 111
    Spatial Multiplexing 113
    Channel Knowledge at the Transmitter... 115
MIMO Implications on Radio IC Design 116
    MIMO Crosstalk.. 116
        Linear MIMO Crosstalk 116
        Nonlinear MIMO Crosstalk 118
    Shared LO Generation... 119
    VGA Gain Control... 119
    Transmitter Error Vector Magnitude (EVM) 120
Chip Architecture and Design 121
Testing Procedures and Measured Results 122
Future Directions 127
Acknowledgments... 128
References.. 129

INTRODUCTION

Recent years have witnessed an explosive growth in the demand for wireless connectivity. This is especially true for wireless local area networks (WLANs) [1,2], where complete systems-on-chip incorporating the radio frequency integrated circuits (RFICs) and the digital baseband have become available in just a few years [3,4]. Upcoming wireless applications, such as video, music, voice, and data, require higher data rates and improved network coverage and capacity to keep up with

---

* Portions reprinted, with permission, from Palaskas, Y., Ravi, A., Pellerano, S., Carlton, B. R., Elmala, M. A., Bishop, R., Banerjee, G., Nicholls, R. B., Ling, S., Dinur, N., Taylor, S. S., and Soumyanath, K., "A 5-GHz 108-Mb/s 2 × 2 MIMO RFIC transceiver with fully integrated 20.5-dBm P_{1dB} power amplifiers in 90-nm CMOS," *IEEE J. Solid-State Circuits*, December 2006 (INVITED) © 2006 IEEE.

corresponding wireline data rates. Wireless rates are governed by fundamental theoretical limits captured by Shannon's theorem as follows: $C = B \cdot \log(1 + \text{SNR})$, where C is the maximum possible rate, B the radio frequency (RF) bandwidth, and the signal-to-noise ratio (SNR) at the receiver. One way to increase the rate is to improve SNR at the receiver so that higher-order modulations such as 1024-QAM may be transmitted. This requires power-hungry power amplifier (PAs), sensitive low-noise amplifiers (LNAs), wider dynamic range analog-to-digital converter (ADCs) and digital-to-analog converter (DACs), and generally more expensive hardware at the transmitter and the receiver, which may not be cost effective for commercial products.

A more cost-effective method for increasing data rate is to use wider RF bandwidths. For example, ultra wide band (UWB) technology [5] uses wide channel bandwidths to achieve high data rates and transmits at low power levels to coexist with existing networks in the same band. Because of the low transmitted power, UWB achieves a limited range of coverage and it is inherently susceptible to interference from other wireless services coexisting in the same band. In general, wider bandwidths can only be allocated per user at the expense of other users in a network with fixed spectrum. Frequency spectrum is not cheap for service providers, especially in licensed bands used for WiMAX and other cellular services. Recently, there has been increased interest in exploiting underutilized spectrum at mm-wave frequencies (e.g., 24 or 60 GHz) to achieve high data rates [6]. The high operating frequency in this case creates challenges for the radio integrated circuit (IC) design and may result in a costly and power-hungry solution. Additionally, the high propagation loss might necessitate line-of-sight (LOS) communication and directional antennas, complicating the deployment of this technology.

The availability of multiple antennas at the receiver, the transmitter, or both, offers new possibilities for increased data rates and range in wireless systems [7,8]. Wireless rates for multiple input, multiple output (MIMO) systems with multiple antennas at both transmitter and receiver are governed by the following fundamental theoretical limit: $C \approx B \cdot M \cdot \log(1 + \text{SNR})$, where M is the number of antennas at the transmitter or receiver, whichever is lower. This linear scaling of rate holds in environments that are rich in multipath, e.g., indoor office environments. In a multipath environment, the transmitted signal is reflected off random scatterers giving rise to multiple delayed replicas of the signal. At each receive antenna these replicas can combine constructively or destructively resulting in significant variations of the received signal power. MIMO technology takes advantage of multipath fading, which has traditionally been a scourge for wireless communications.

MIMO can operate in two general modes: (1) spatial diversity (e.g., beam steering, MRC, Alamouti), where the same information is encoded in multiple antennas, thus introducing redundancy and improving the reliability of the link, and (2) spatial multiplexing, where the different antennas process different data streams resulting in improved spectral utilization. Details on MIMO algorithms and channel models will be provided in the next section.

MIMO processing has significant implications on the radio IC design. These implications are thoroughly analyzed for the case of MIMO spatial multiplexing and solutions are proposed. For example, it is shown that crosstalk between the multiple transceivers residing on the same die can degrade MIMO performance and has to be carefully minimized, especially when power amplifiers are integrated on the RFIC die. A shared LO can be used to reduce power dissipation and cost and also to improve phase noise immunity in a MIMO system. Techniques are proposed to prevent crosstalk between the multiple radio chains through this shared LO. A fully integrated 5 GHz 2×2 MIMO WLAN transceiver RFIC has been implemented in 90 nm CMOS to demonstrate these ideas. The fabricated MIMO receiver achieves a sensitivity of -63 dBm while receiving 108 Mb/s. This data rate is double of what is achievable using conventional WLAN technology and illustrates the increased spectral utilization associated with MIMO spatial multiplexing. The chip includes linearized 3.3 V, 5 GHz power amplifiers with $P_{1dB} = 20.5$ dBm that deliver an average power of $+13/+16$ dBm each, in MIMO/SISO modes, respectively. The measured radio performance demonstrates the effectiveness of the MIMO techniques used, e.g., with regard to crosstalk.

This chapter is organized as follows: The section "MIMO Theory" describes the basic principles of MIMO and following that is a section that discusses the implications of MIMO on the radio IC design. The section "Chip Architecture and Design" describes a 2×2 MIMO CMOS transceiver RFIC for 5 GHz WLAN. The section "Testing Procedures and Measured Results" presents measurement results from the fabricated prototype. The fabricated chip can support both MIMO modes, but the emphasis is on data rate increase through spatial multiplexing. The fabricated prototype uses two receive and two transmit antennas to achieve a data rate of 108 Mb/s in a 16 MHz RF channel. The resulting spectral efficiency is double that of a legacy 1×1 system based on 802.11a/g technology. The ideas presented in this chapter can be extended to MIMO systems with a larger number of antennas and correspondingly higher data rate and/or range of coverage. For example, the ideas presented would be applicable to a 3×3 MIMO system using a 40 MHz RF channel that can support >300 Mb/s (note that the final data rate also depends on other orthogonal frequency division multiplexing [OFDM] parameters such as code rate, guard interval duration, and number of subcarriers used). The same principles can be used to optimize a 2×3 system (two transmitters, three receivers) that can achieve the same data rate as a 2×2 system but with improved sensitivity and range owing to extra diversity associated with the third receiver. Within years from its inception, MIMO technology is already becoming a reality in consumer applications. The chapter concludes with a discussion of upcoming MIMO-inspired technologies that might shape the wireless systems of tomorrow.

MIMO THEORY

In general, MIMO technology uses multiple antennas and RF chains at both transmitter and receiver. Simpler versions of MIMO include single-input, multiple-output (SIMO), where only the receiver has multiple antennas, and multiple-input, single-output (MISO), where only the transmitter has multiple antennas. When both receiver and transmitter have single antennas, MIMO defaults to SISO (single-input, single-output). In a SISO system, a single spatial stream of baseband information symbols is mapped to the transmit antenna, upconverted to RF, and transmitted. In contrast, a MIMO system allows for transmission of multiple spatial streams [9–12]. For a MIMO system with M_t transmit antennas and M_r receive antennas, at most $M = \min(M_t, M_r)$ spatial streams are transmitted [13]. This is shown in Figure 4.1 for a 2×2 MIMO system ($M_t = M_r = 2$). Note that information symbols on M spatial streams must be carefully mapped to M_t transmit antennas. This mapping can be performed in many ways, and the next two sections describe the two popular mapping schemes previously mentioned in the introduction: spatial diversity and spatial multiplexing. To motivate these schemes, a brief background on fading channels is provided.

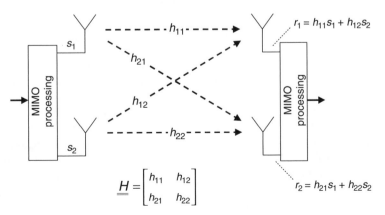

FIGURE 4.1   A 2×2 MIMO system. $\underline{H}$ represents the channel matrix.

FADING WIRELESS CHANNELS

Consider the 2×2 system from Figure 4.1. The baseband input–output equation on a single frequency tone in an OFDM system can be written as

$$\begin{bmatrix} r_1 \\ r_2 \end{bmatrix} = \begin{bmatrix} h_{11} & h_{12} \\ h_{21} & h_{22} \end{bmatrix} \cdot \begin{bmatrix} s_1 \\ s_2 \end{bmatrix} + \begin{bmatrix} n_1 \\ n_2 \end{bmatrix} \Leftrightarrow \underline{r} = \underline{H} \cdot \underline{s} + \underline{n} \qquad (4.1)$$

where $\underline{s}$ is the transmitted signal vector with total power E_s (for two transmit antennas each branch has a transmit power of $E_s/2$), $\underline{r}$ the received vector, $\underline{n}$ the receiver noise with power N_o per element, and $\underline{H}$ a 2×2 channel matrix describing the equivalent baseband channel between each pair of transmit and receive antennas. Similar equations can be written for all tones, for example, there are 48 data tones in a 64-point fast Fourier transform (FFT) in the IEEE 802.11a standard.

The transmitted signal undergoes multiple reflections as it travels through a richly scattering environment containing various objects in the path between the transmitter and the receiver. The superposition of reflections at the receiver can be modeled as a Gaussian random variable for a large enough number of reflections (called multipath), especially for a narrowband signal where the receiver cannot separate different reflections arriving at typical time of arrival delays. The amplitude of a complex Gaussian channel (e.g., h_{ij} in Equation 4.1) follows the Rayleigh distribution, hence the term *Rayleigh fading*. Wireless networks such as Wi-Fi and WiMAX are designed with narrowband signals that exhibit Rayleigh fading. Wi-Fi and WiMAX are also affected by shadowing due to large obstructions in the environment, and path-loss, which is the power fall-off with distance. Shadowing and path-loss are the dominant channel impairments in short-range systems, e.g., UWB and mm-wave. An example of detailed wireless channel models can be found in Ref. 14. Only fading is discussed next because shadowing and path-loss can be absorbed into the average SNR at the receiver.

Performance of the MIMO system in Equation 4.1 is maximized when the channel matrix $\underline{H}$ is full rank, that is, no row or column can be written as a linear combination of other rows or columns. This is explained intuitively for the case of MIMO spatial multiplexing using the concept of "noise amplification" in page 113. For a random matrix $\underline{H}$, maintaining full rank is likely to happen when elements of $\underline{H}$ are identical and independently distributed random variables. When antennas are spaced more than half a wavelength apart (see the section "Future Directions" for more details on antenna spacing), received signals tend to exhibit independent statistics and therefore optimal MIMO performance. In some cases, however, signals on multiple transmit or receive antennas are highly correlated, which leads to performance degradation. This is likely to happen when the transmitter and the receiver are within LOS of each other and the scattered reflections are insignificant as compared to strong direct signals, which may happen to be at unfavorable phases with respect to each other in the channel matrix leading to reduced channel rank. In summary, performance of MIMO systems is strongly dependent on properties of the fading channel matrix.

Finally, performance of MIMO systems depends on the amount of channel knowledge at the transmitter. The more the transmitter knows about the channel, the better it can steer its signal in the correct direction toward the receiver (see "Spatial Diversity"). The channel matrix is estimated at the receiver using preamble training sequences or data pilot tones in the packet. It can be fed back to the transmitter explicitly via feedback frames or learned implicitly by assuming channel reciprocity in time division duplex systems. (Channel reciprocity and transmit/receive chain calibration are described in the section "Channel Knowledge at the Transmitter.") If the channel information is fed back explicitly, it can only be used for a short duration until the channel changes due to mobility of the transmitter, receiver, or objects in the environment. Therefore, a trade-off exists between the overhead of channel knowledge feedback and performance gains, especially in highly mobile environments. In conclusion, MIMO algorithms must be designed for both cases: when the channel is known at the transmitter (closed-loop MIMO) and when the channel is unknown at the transmitter (open-loop MIMO).

In the next two sections, two well-understood schemes for mapping input spatial streams to multiple transmit antennas are discussed. Spatial diversity is a technique that spreads a small amount of information over a large number of space–time ("space" = antennas) resources to provide additional reliability. Spatial multiplexing is a technique that multiplexes a large amount of information in available space–time resources to provide additional data throughput.

SPATIAL DIVERSITY

Spatial diversity can be classified into two classes: receive diversity, e.g., SIMO systems, and transmit diversity, e.g., MISO systems. An example of receive diversity with one transmit and two receive antennas is:

$$\begin{bmatrix} r_1 \\ r_2 \end{bmatrix} = \begin{bmatrix} h_1 \\ h_2 \end{bmatrix} \cdot s + \begin{bmatrix} n_1 \\ n_2 \end{bmatrix} \Leftrightarrow \underline{r} = \underline{H} \cdot s + \underline{n} \qquad \text{Input – output equation}$$

$$\Leftrightarrow \hat{s} = \underline{H}^* \cdot \underline{H} \cdot s + \underline{H}^* \cdot \underline{n} = \begin{bmatrix} h_1^* & h_2^* \end{bmatrix} \begin{bmatrix} h_1 \\ h_2 \end{bmatrix} \cdot s + \tilde{n} = \left(|h_1|^2 + |h_2|^2 \right) s + \tilde{n} \qquad \text{Receiver processing}$$

$$\Leftrightarrow \text{SNR} = \left(|h_1|^2 + |h_2|^2 \right) \frac{E_s}{N_o} = \|H\|^2 \frac{E_s}{N_o}$$

where $\hat{s}$ is the estimate of the transmitted signal calculated at the receiver.

The optimal receiver performs maximal ratio combining (MRC) of two antennas [15,16], which maximizes the SNR. Intuitively, maximal ratio combining optimally aligns the channel phases on both antennas and weights their amplitudes in proportion to the channel amplitudes. The resulting SNR is proportional to the channel power, that is, the sum of the squares of channel amplitudes on all antennas.

Two examples of transmit diversity, both closed-loop and open-loop transmit diversity are provided next. In closed-loop transmit diversity, where the transmitter knows exact channel elements, beamforming is used at the transmitter to prealign the phases and amplitudes such that they add up constructively at the receiver. An example of 2×1 transmit diversity with signaling over a single symbol time is:

$$r = \begin{bmatrix} h_1 & h_2 \end{bmatrix} \cdot \begin{bmatrix} s_1 \\ s_2 \end{bmatrix} + n \Leftrightarrow r = \underline{H} \cdot \underline{S} + n \qquad \text{Input – output equation}$$

$$\Leftrightarrow r = \begin{bmatrix} h_1 & h_2 \end{bmatrix} \cdot \left(\begin{bmatrix} h_1^* \\ h_2^* \end{bmatrix} \cdot s / \|H\| \right) + n = \|H\| s + n \qquad \text{Transmitter processing}$$

$$\Leftrightarrow \text{SNR} = \|H\|^2 \frac{E_s}{N_o}$$

Note that performance of ideal closed-loop diversity with perfect channel knowledge at the transmitter is identical to receive diversity. This is in contrast to open-loop transmit diversity.

In open-loop transmit diversity, where the transmitter does not know the channel elements, specially designed mappings called space–time codes are used to spread input spatial streams onto multiple antennas and multiple symbol times. The most popular spatial diversity scheme for two transmit antennas is the Alamouti space–time code [17], shown in the next equation (Figure 4.2)

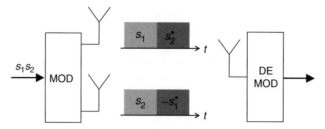

FIGURE 4.2 Spatial diversity scheme for the Alamouti space–time code.

over two symbol times (the equation represents a single OFDM tone over two symbol times t_1 and t_2 corresponding to successive OFDM frames):

$$\begin{bmatrix} r_{t1} & r_{t2} \end{bmatrix} = \begin{bmatrix} h_1 & h_2 \end{bmatrix} \cdot \begin{bmatrix} s_1 & s_2^* \\ s_2 & -s_1^* \end{bmatrix} + \begin{bmatrix} n_{t1} & n_{t2} \end{bmatrix} \qquad \text{Input – output equation}$$

$$\Leftrightarrow \hat{s}_i = \|H\|^2 s_i + \tilde{n}_i \quad i = 1, 2 \qquad \text{After receiver processing}$$

$$\Leftrightarrow \text{SNR} = \|H\|^2 \frac{E_s}{2N_o}$$

Note that postprocessing SNR of open-loop diversity is half that of closed-loop SNR. This loss is expected because open-loop systems have less channel knowledge at the transmitter compared to the closed-loop system. Furthermore, the Alamouti code is known to be optimal for a single receive antenna but loses capacity for two or more receive antennas [18,19]. Recently, perfect space–time codes based on algebraic number theory have been proposed that provide full diversity and full multiplexing gains [20].

Another simple form of transmit diversity is called cyclic delay diversity (CDD). This code transmits cyclically shifted versions of the time-domain transmit sequence on different antennas [21]. It provides less performance gain than the Alamouti code but is backward compatible in the sense that it can be used on two or more transmit antennas without requiring changes at the receiver. An example is provided for a frequency-flat fading channel where S_1 and S_2 are the time-domain inverse fast fourier transforms (IFFTs) of the frequency-domain information stream s_1 and s_2, i.e., $[S_1\ S_2] = \text{IFFT}$ $([s_1\ s_2])$ where FFT size $N = 2$.

$$\begin{bmatrix} R_1 & R_2 \end{bmatrix} = \begin{bmatrix} h_1 & h_2 \end{bmatrix} \cdot \begin{bmatrix} S_1 & S_2 \\ S_2 & S_1 \end{bmatrix} + \begin{bmatrix} N_1 & N_2 \end{bmatrix} \qquad \text{Time domain}$$

$$\Leftrightarrow \hat{\underline{s}} = \left(h_1 + e^{j2\pi/N} h_2 \right) \underline{s} + \tilde{\underline{n}} \qquad \text{Frequency domain } (N = 2 = \text{FFT size})$$

$$\Leftrightarrow \text{SNR} = \left(|h_1|^2 + |h_2|^2 + 2\,\text{Re}\left\{ h_1^* h_2 e^{j2\pi/N} \right\} \right) \frac{E_s}{2N_o}$$

The diversity schemes discussed in this section can be ordered in terms of SNR performance as follows:

$$\text{SNR}_{\text{receive\_MRC}} \geq \text{SNR}_{\text{transmit\_beamform}} \geq \text{SNR}_{\text{transmit\_Alamouti}} \geq \text{SNR}_{\text{transmit\_CDD}}$$

The examples considered so far examined SIMO and MISO systems with a single spatial stream (measured by the number of symbols transmitted per time). In general, for MIMO systems, spatial diversity transmits a small number of spatial streams, say m where $m < M = \min(M_t, M_r)$, using space–time codes to extract diversity gain from independent Rayleigh fading on different antennas. It can be combined with receive diversity when multiple receive antennas are present at the receiver. Both open-loop and closed-loop transmit diversity methods can be combined with receive diversity. For example, Ref. 22 presents an example of a single spatial stream (up to) 4×4 system employing beamforming at the transmitter and MRC at the receiver. Multiple spatial streams are discussed in detail in the next section.

SPATIAL MULTIPLEXING

In contrast to spatial diversity, spatial multiplexing transmits the maximum possible number of spatial streams, $m = M = \min(M_t, M_r)$ to maximize throughput rather than reliability. At high SNRs, spatial multiplexing performs better than spatial diversity [23]. At low SNRs, spatial diversity performs better than spatial multiplexing at the same rate.

Consider the spatial multiplexing input–output Equation 4.1. Simple linear operations for separating two spatial streams is described. The receiver applies the inverse of the channel matrix to the received data (channel equalization):

$$\hat{\underline{s}} = \underline{H}^{-1} \cdot \underline{r} = \underline{s} + \underline{H}^{-1} \cdot \underline{n} \tag{4.2}$$

The decomposition of Equation 4.2 allows transmission of different symbol streams within the same frequency channel. This is similar to establishing multiple parallel data pipes between the transmitter and the receiver, resulting in increased spectral efficiency [11]. For example, if each of the data streams has a data rate of 54 Mb/s (64-QAM OFDM modulation over a 16 MHz channel), the 2×2 MIMO system can achieve two times this rate or 108 Mb/s (Figure 4.3). The expression of Equation 4.2 is referred to as zero-forcing equalization.

The inversion of the channel matrix $\underline{H}$ required in Equation 4.2 is possible if there is significant multipath fading in the environment between transmitter and receiver antennas. As a result of multipath, the transmitted signal of TX_1 arrives at the two receiver antennas with random phases. This can be perceived as the signal of TX_1 arriving from a specific direction. Similarly, the signal of TX_2 can be perceived as originating from a direction that is generally different due to the random phases involved. MIMO processing can then distinguish the two transmitted signals according to their angles of arrival. Spacing the antennas far apart (see the section "Future Directions" for more details) ensures de-correlated phases that are required to generate the perception of different angles of arrival. In a LOS environment, both transmitted signals arrive from the same direction and the MIMO receiver might not be able to de-convolve them. Recent studies have demonstrated that the use of cross-polarized antennas can help in reducing the correlation even in LOS environments [24].

In certain degenerate cases the channel matrix $\underline{H}$ can be noninvertible (singular) [11]. This causes MIMO spatial multiplexing to fail and prevents the system from supporting multiple independent

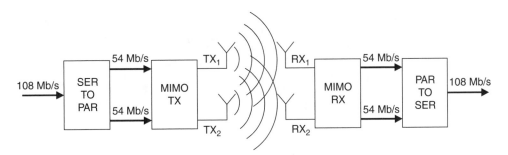

FIGURE 4.3 Increasing the data rate through spatial multiplexing in a 2×2 MIMO system.

bit streams and high data rates. If $\underline{H}$ is close to singular, the determinant $|\underline{H}|$ is close to zero, and the noise term $\underline{H}^{-1}\underline{n}$ in Equation 4.2 becomes significant, resulting in low SNR and degraded performance. This phenomenon is called noise amplification.

Noise amplification can be mitigated by employing more sophisticated equalization algorithms rather than the simple expression of Equation 4.2, for example, the minimum mean square error (MMSE) algorithm [11]:

$$\hat{\underline{s}} = \left(\underline{H}^*\underline{H} + \frac{1}{\text{SNR}}\underline{I}_N \right)^{-1} \underline{H}^* \cdot \underline{r} \tag{4.3}$$

where $\underline{H}^*$ denotes the complex conjugate transpose of the channel matrix, $\underline{I}_N$ a unit matrix, and the received SNR. The MMSE equalizer of Equation 4.3 tries to balance the errors due to imperfect channel equalization and noise amplification so as to achieve optimal overall performance. For high SNR values, the MMSE algorithm simplifies to the zero-forcing expression of Equation 4.2. The simulations and measurements reported in this work were performed using MMSE equalization according to Equation 4.3. To keep the following discussions simple, MIMO spatial multiplexing is described in terms of the zero-forcing equalizer of Equation 4.2. It is expected that the actual MMSE equalizer used follows similar trends as far as noise amplification is concerned, since MMSE equalization reduces noise amplification but cannot eliminate it.

In fact, the MMSE receiver is known to be optimal only for an orthogonal channel, that is, when the matrix H has full rank and all its eigenvalues are equal. In general, the optimal MIMO receiver requires computationally intensive nonlinear operations such as maximum likelihood (ML) decoding (e.g., exhaustive search over all possible transmitted symbol vectors). There are simpler forms of ML decoding such as lattice decoding; however, they are still rather complex for implementation [25]. If the channel is known at the transmitter (closed loop), performance of the MMSE receiver can be improved by diagonalizing the channel with precoding at the transmitter. In particular, the singular value decomposition (SVD) of the channel (Figure 4.4) can be written as $H = U\Sigma V$, where U and V are unitary matrices and Σ a diagonal matrix of positive eigenvalues. If the transmitter knows V, it can precode the transmitted vector as $x = V^*s$, where s is the information to be transmitted, as shown below. The receiver applies U^*, resulting in a one-to-one mapping between transmit symbols and received symbols with no cross terms:

$$\underline{r} = \underline{H} \cdot \underline{x} + \underline{n} \Leftrightarrow \underline{r} = \underline{U} \cdot \underline{\Sigma} \cdot \underline{V} \cdot \underline{x} + \underline{n}$$

$$\Leftrightarrow \underline{r} = \underline{U} \cdot \underline{\Sigma} \cdot \underline{V} \cdot \underline{V}^* \underline{s} + \underline{n} = \underline{U} \cdot \underline{\Sigma} \cdot \underline{s} + \underline{n} \quad \text{Transmitter processing}$$

$$\Leftrightarrow \hat{\underline{s}} = \underline{\Sigma} \cdot \underline{s} + \tilde{\underline{n}} \quad \text{After receiver processing}$$

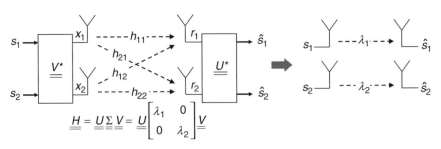

FIGURE 4.4 Implementation of spatial multiplexing using singular value decomposition (SVD) when the channel is known at the transmitter.

Furthermore, performance of closed-loop MIMO with SVD can be improved by using the water-pouring principle to optimize channel capacity and maximize SNR at the receiver. Details of the water-pouring algorithm are beyond the scope of this chapter and can be found in Refs. 10,12.

CHANNEL KNOWLEDGE AT THE TRANSMITTER

Multiplexing according to Figure 4.4 is very sensitive to precise channel knowledge at the transmitter. As was discussed earlier, this can be achieved by explicitly feeding back the channel information from the receiver to the transmitter. The channel matrix is estimated at the receiver using preamble training sequences or data pilot tones in the packet. The receiver can then perform SVD of the channel matrix and feed the right singular matrix V back to the transmitter. This feedback link represents an overhead that might reduce spectrum efficiency if appropriate compression techniques are not used.

The channel information at the transmitter can only be used for a short time duration until the channel changes due to mobility of the transmitter, receiver, or objects in the environment. Therefore, a trade-off exists between the overhead of channel knowledge feedback and performance gains, especially in highly mobile environments.

A second technique for achieving channel knowledge at the transmitter is by taking advantage of the channel reciprocity in a time division duplex system. Channel reciprocity is a basic principle in antenna theory and states that the channel coefficient from antenna A_1 to antenna B_1 is the same as the channel coefficient from antenna B_1 to antenna A_1 (see Figure 4.5). Because of channel reciprocity, station A can use frames transmitted from B to estimate the $A \rightarrow B$ channel and use this to beamform back to station B. For the technique to work, the channel should not vary too much between the two transmissions ($B \rightarrow A$ and $A \rightarrow B$). This technique is usually referred to as implicit feedback as opposed to the explicit feedback discussed in the previous paragraphs where the channel information is explicitly fed back to the transmitter. Implicit feedback can take advantage of existing medium access (MAC) packets sent from station B to station A and can, therefore, result in smaller overhead relative to the explicit feedback scheme. For the same reason, implicit feedback might be advantageous in highly mobile environments.

Even though the actual radio propagation channel is reciprocal, the overall MIMO system is usually not [26]. This is because, in addition to the radio channel, the overall MIMO channel also consists of RF, intermediate frequency (IF), and baseband circuits that are typically not reciprocal as explained in Figure 4.5. It can be seen that the effective channel coefficient from A_1 to B_1 is $\tilde{h}_{A_1 \rightarrow B_1} = \alpha \cdot h_{A_1 \rightarrow B_1} \cdot \delta$

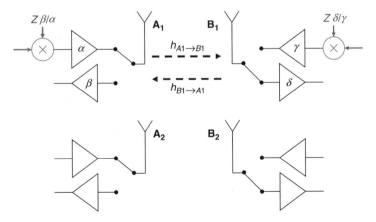

FIGURE 4.5 Implicit closed-loop channel feedback is based on the channel reciprocity assumption. Digital signal processing (gray part) can be used to correct for gain and phase mismatches to restore the reciprocity of the system.

while the corresponding coefficient for the reverse channel is $\tilde{h}_{B_1 \to A_1} = \gamma \cdot h_{B_1 \to A_1} \cdot \beta$ (α and β are the complex gains of the transmitter and receiver paths of station A_1, respectively, and similarly γ and δ for B_1). Even though the radio channel is reciprocal ($h_{A_1 \to B_1} = h_{B_1 \to A_1}$), the overall MIMO channel is not, unless $\alpha \cdot \delta = \beta \cdot \gamma$. This requirement is difficult to meet in practice, as it involves transmitter and receiver paths in two different stations.

The reciprocity of the overall MIMO channel can be restored by using the digital signal processing (DSP) techniques in the baseband [26]. A simple example is shown by the gray part of Figure 4.5. Here, the data of A_1 is premultiplied by $Z\beta/\alpha$ before transmission and similarly the data of B_1 is premultiplied by $Z\delta/\gamma$ (Z is a constant). It is easy to see that this DSP operation restores the reciprocity of the overall MIMO system.

The gain coefficients α, β, γ, and δ can be estimated during a calibration phase that might involve loop-back from the transmitter to the corresponding receiver [26] or on-the-air communication between different stations [27]. In the latter case, the overhead in spectral efficiency is usually small since the calibration tries to correct relatively static gain and phase errors and therefore does not have to be performed very frequently. This is an advantage over the explicit feedback case discussed earlier in this section where it is attempted to continuously feed back the varying channel information to the transmitter. By the same argument, implicit feedback might be advantageous for mobile applications. It should be noted, however, that a number of practical issues exist with the calibration scheme described. For example, temperature variations, accuracy of automatic gain control (AGC) or transmitter power control steps, or slow voltage controlled oscillator (VCO) phase/frequency drifts [22] might degrade the calibration accuracy and the performance of the closed-loop MIMO system.

MIMO IMPLICATIONS ON RADIO IC DESIGN

The previous section introduced the basic concepts of MIMO spatial multiplexing. This section discusses the implications of MIMO on the design of the radio IC. Many of these implications are a direct or indirect result of the unavoidable crosstalk between multiple RF transceivers implemented on the same silicon die.

MIMO CROSSTALK

It was mentioned in the previous section that when the channel matrix is close to singular ($|\underline{H}| \approx 0$) the performance of the spatial multiplexing MIMO system degrades due to noise amplification. This occurs when there is significant crosstalk between the ideally independent data pipes established between TX and RX. As an example, contrast the case of zero crosstalk ($\underline{H} = [1\ 0;\ 0\ 1]$) to the case of complete crosstalk ($\underline{H} = [1\ 1;\ 1\ 1]$) for which the channel matrix is not invertible (the notation $[h_{11}\ h_{12};\ h_{21}\ h_{22}]$ is used to describe a 2×2 matrix $\underline{H}$). This undesired MIMO crosstalk can occur in the radio channel, between the antennas, and also on the radio IC. Crosstalk in the channel can be quite small assuming a rich multipath environment (see the section "MIMO Theory"). Crosstalk between the antennas is minimized by spacing the antennas far from each other, i.e., at least half a wavelength distance. This section focuses on crosstalk that occurs at the radio IC level. This type of crosstalk can be linear or nonlinear as explained next. First linear crosstalk is discussed.

Linear MIMO Crosstalk

Linear crosstalk between the transmitters in a MIMO system can be described by the matrix $\underline{A}$, as shown in Figure 4.6. Similarly, linear crosstalk between the receivers can be described by a crosstalk matrix $\underline{B}$. For the MIMO system of Figure 4.6, the received signal vector can be expressed as

$$\underline{r} = (\underline{B} \cdot \underline{H} \cdot \underline{A}) \cdot \underline{s} + \underline{B} \cdot \underline{n} \tag{4.4}$$

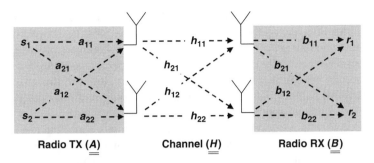

Radio TX ($\underline{\underline{A}}$) Channel ($\underline{\underline{H}}$) Radio RX ($\underline{\underline{B}}$)

FIGURE 4.6 Model of a MIMO system including the effect of the channel (matrix $\underline{\underline{H}}$), linear crosstalk between the transmitters (matrix $\underline{\underline{A}}$), and linear crosstalk between the receivers (matrix $\underline{\underline{B}}$). (From Palaskas, Y., Ravi, A., Pellerano, S., Carlton, B. R., Elmala, M. A., Bishop, R., Banerjee, G., Nicholls, R. B., Ling, S., Dinur, N., Taylor, S. S., and Soumyanath, K., *IEEE J. Solid-State Circuits*, 41, 2746–2756, 2006. With permission. © [2006] IEEE.)

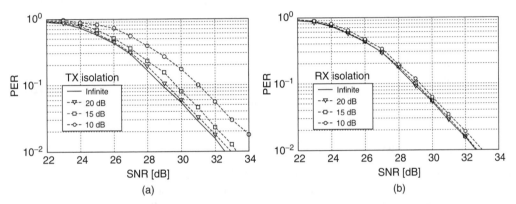

FIGURE 4.7 The effect of linear crosstalk on MIMO spatial multiplexing performance. (a) Transmitter crosstalk and (b) receiver crosstalk. It can be seen that MIMO is more tolerant to receiver crosstalk. The simulations were performed for a 75 ns Rayleigh fading channel. (From Palaskas, Y., Ravi, A., Pellerano, S., Carlton, B. R., Elmala, M. A., Bishop, R., Banerjee, G., Nicholls, R. B., Ling, S., Dinur, N., Taylor, S. S., and Soumyanath, K., *IEEE J. Solid-State Circuits*, 41, 2746–2756, 2006. With permission. © [2006] IEEE.)

where it has been assumed that most of the receiver crosstalk occurs after the channel noise and the receiver noise have been added (this point is further explained in item (c) that follows). Similarly, as in Equation 4.2, the transmitted signals can be estimated as

$$\hat{\underline{s}} = (\underline{\underline{B}} \cdot \underline{\underline{H}} \cdot \underline{\underline{A}})^{-1} \cdot \underline{r} = \underline{r} + \underline{\underline{A}}^{-1} \cdot \underline{\underline{H}}^{-1} \cdot \underline{n} \tag{4.5}$$

From Equation 4.5 the following observations are made:

a. Matrices $\underline{\underline{B}}$, $\underline{\underline{H}}$, and $\underline{\underline{A}}$ should *all* be invertible. Otherwise, the overall system matrix $(\underline{\underline{B}} \cdot \underline{\underline{H}} \cdot \underline{\underline{A}})$ would become singular and MIMO spatial multiplexing would fail.

b. Transmitter crosstalk degrades MIMO performance due to noise amplification (last term in Equation 4.5). This can be seen in Figure 4.7a that depicts simulated waterfall curves (packet error rate [PER] versus SNR) for a 2×2 MIMO system in the presence of transmitter crosstalk. The TX isolation requirements are seen to be rather benign, since a moderate isolation of 20 dB results in negligible performance degradation with respect to the ideal case. This low-isolation requirement is a somewhat surprising result, since 20 dB of isolation between the two uncorrelated data streams would result in about 20 dB of SNR for each one of them, which is worse than the ~30 dB SNR required for good MIMO performance (e.g., for PER = 10%, see Figure 4.7a). The reason for this apparent inconsistency is that MIMO equalization according to Equations 4.2 or 4.5 inherently tries to eliminate crosstalk between the ideally independent bit streams; this is true regardless of where the crosstalk occurs, be it in the channel or in the radio IC. However, this crosstalk cancellation inherent to MIMO equalization comes at the expense of noise amplification that eventually limits the performance.

c. The MIMO system is not very sensitive to receiver crosstalk, since the receiver crosstalk matrix $\underline{B}$ cancels in Equation 4.5 and does not introduce noise amplification ($\underline{B}$ still has to be invertible according to observation (a) as just described). This is verified by the simulated waterfall curves of Figure 4.7b where RX crosstalk is seen to have a minimal effect on MIMO performance. It should be noted that this result is based on the assumption that the crosstalk occurs after the noise has been added to the system. This assumption might not be accurate for certain noise sources and crosstalk mechanisms. For example, the noise of the down-conversion mixers occurs after the coupling between the LNA input traces. In such cases, the performance degradation will be comparable to the TX case since the noise is introduced after the crosstalk. For some of the receiver noise sources, the sensitivity to RX crosstalk is smaller than in the transmitter, whereas for others it is at most comparable to the transmitter sensitivity. Therefore, the transmitter crosstalk requirement of 20 dB is adequate for the receiver too.

In summary, to minimize performance degradation due to linear crosstalk, the MIMO radio has to be designed to maintain at least 20 dB linear isolation between the transmitters and 20 dB between the receivers. If priorities have to be made (e.g., during IC floorplanning), transmitter isolation should be favored.

Nonlinear MIMO Crosstalk

Nonlinear crosstalk can occur, for example, through the common local oscillator (LO) signal used to drive the multiple transceivers in a MIMO system (the reasons for sharing the LO are explained in the section "Shared LO Generation"). This is shown in Figure 4.8, where the output of TX_1 is seen to couple to the common LO signal, which is no longer a pure sinusoid. When this corrupted LO signal is used to upconvert the baseband signal BB_2, an undesired nonlinear crosstalk component will appear at the output of TX_2. This component corresponds to the convolution between the upconverted BB_1 signal, coupled from TX_1 output to the LO, and the BB_2 signal. A similar

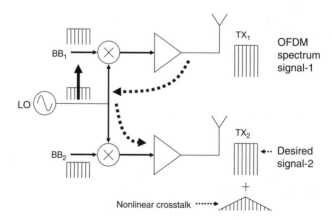

FIGURE 4.8 Nonlinear MIMO crosstalk can occur if the PA output disturbs the LO signal used in the upconversion mixers. (From Palaskas, Y., Ravi, A., Pellerano, S., Carlton, B. R., Elmala, M. A., Bishop, R., Banerjee, G., Nicholls, R. B., Ling, S., Dinur, N., Taylor, S. S., and Soumyanath, K., *IEEE J. Solid-State Circuits*, 41, 2746–2756, 2006. With permission. © [2006] IEEE.)

effect occurs even in a non-MIMO system (only one TX) because of coupling from the TX output to the LO line. In this case, the nonlinear component corresponds to the convolution between the upconverted baseband signal coupled from the TX output to the LO line and the baseband signal itself. This self-coupling effect is shown more clearly in the measurements of Figure 4.16a and discussed in the section "Testing Procedures and Measured Results." Nonlinear coupling effects become particularly problematic when the PAs are integrated on the same die with the rest of the radio transceiver as is the case of this system. The appendix in Ref. 32 presents a detailed mathematical analysis of nonlinear crosstalk through the LO. Nonlinear crosstalk can occur through other mechanisms too, i.e., shared supplies or the substrate.

MIMO spatial multiplexing processing is a linear equalizer and cannot mitigate nonlinear crosstalk. To ensure minimal performance degradation, the system has to be designed such that nonlinear crosstalk is well below the error vector magnitude (EVM) floor so that it is innocuous to begin with. In the fabricated system, the nonlinear crosstalk was specified to be smaller than $-35\,dB$ with respect to the desired signal, so that it does not raise the $-27\,dB$ EVM floor appreciably. Nonlinear crosstalk is not expected to occur between the receivers; therefore, transmitter isolation has to be prioritized over receiver isolation similar to the linear crosstalk case discussed earlier.

SHARED LO GENERATION

It was mentioned earlier that a common LO can be used to drive all transmitters and all receivers in a MIMO system. This minimizes the VCO/synthesizer power dissipation and chip area and prevents pulling between multiple VCOs on the same die. Sharing the LO can also improve the accuracy and stability of the calibration used in closed-loop MIMO systems as explained in Ref. 22. Additionally, a common LO relaxes the phase noise requirements of the system by improving the pilot-assisted phase noise tracking at the receiver as explained with the aid of Figure 4.9. The figure shows the phasor representation of the pilot tones at the two receivers, where $\Delta\varphi_1$ and $\Delta\varphi_2$ are due to phase noise. If the LO is shared between the two transmitters, and the same holds for the receivers, then $\Delta\varphi_1 = \Delta\varphi_2$. An additional uncorrelated noise component (e.g., from thermal noise) corrupts the phase estimate. Averaging out the pilot estimates reduces the effect of these uncorrelated components, thus improving the accuracy of the phase noise tracking. The same advantage can be obtained to some extent by sharing a reference signal between multiple synthesizers.

VGA GAIN CONTROL

Baseband variable gain amplifiers (VGAs) are typically used in wireless receivers to amplify the received signal to the full range of the ADC. This reduces the ADC resolution requirements resulting in reduced power dissipation and chip area for the overall system. In a MIMO system, the power

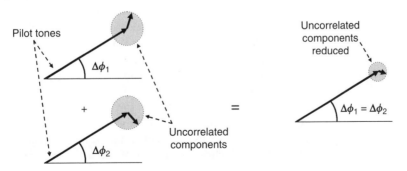

FIGURE 4.9 Sharing the LO signal between the transmitters and between the receivers in a MIMO OFDM system improves the pilot-assisted phase tracking, resulting in increased immunity to VCO phase noise. (From Palaskas, Y., Ravi, A., Pellerano, S., Carlton, B. R., Elmala, M. A., Bishop, R., Banerjee, G., Nicholls, R. B., Ling, S., Dinur, N., Taylor, S. S., and Soumyanath, K., *IEEE J. Solid-State Circuits*, 41, 2746–2756, 2006. With permission. © [2006] IEEE.)

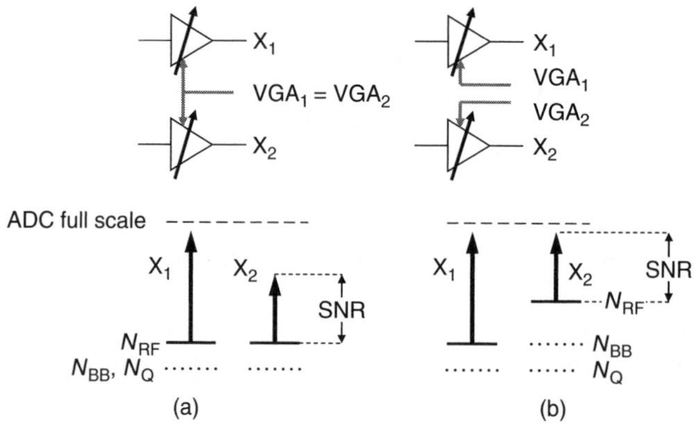

FIGURE 4.10 The strength of the received signals in a MIMO receiver can vary significantly depending on channel conditions. (a) The case where the two VGAs are set to the same gain which simplifies the digital MIMO equalizer. (b) The situation for independent VGA controls. Using the same VGA gain as shown in (a) should not change the SNR appreciably assuming the latter is dominated by RF thermal noise (N_{RF}). (From Palaskas, Y., Ravi, A., Pellerano, S., Carlton, B. R., Elmala, M. A., Bishop, R., Banerjee, G., Nicholls, R. B., Ling, S., Dinur, N., Taylor, S. S., and Soumyanath, K., *IEEE J. Solid-State Circuits*, 41, 2746–2756, 2006. With permission. © [2006] IEEE.)

of the two (or more) received signals might be significantly different from each other depending on channel conditions. Following the SISO common practice described previously the VGA gains could be set independently to drive both ADCs to their full range. This is the approach taken in the MRC receiver of the digital baseband processor presented in Ref. 15. Independent setting of the VGA gains, however, might result in additional computations in the digital MIMO equalizer and therefore higher power dissipation. The next few paragraphs show that in many practical MIMO systems it might be acceptable to use the same gain setting in both VGAs allowing for a simpler, lower power MIMO equalizer.

This is explained in Figure 4.10 where X_1 and X_2 are the two received signals, N_{RF} is the thermal noise of the RF front-end, N_{BB} the thermal noise of the baseband circuits and the ADC, and N_Q the quantization noise of the ADC. All signal and noise levels are referenced to the input of the ADC. Normally a wireless receiver is designed so that the performance is limited by RF thermal noise rather than baseband thermal noise or quantization noise ($N_{RF} \gg N_{BB}, N_Q$), since this usually results in the most economical implementation in terms of power dissipation and chip area.

In Figure 4.10a both VGAs are set to the same gain and this gain is chosen so that the stronger of the two received signals X_1 spans the full range of the ADC. This results in suboptimal utilization of the ADC dynamic range by the weakest signal X_2. In Figure 4.10b, the VGA gains are set independently so as to drive both signals to the full range of the corresponding ADCs. VGA_2 amplifies the weakest signal X_2 but at the same time amplifies the RF thermal noise resulting in approximately the same SNR as in Figure 4.10a; X_2 is much stronger than the corresponding quantization and baseband noise, but this is immaterial since the performance was limited by RF thermal noise. In such a case, it might be acceptable to use a common VGA gain setting as in Figure 4.10a, thereby simplifying the design of the digital MIMO equalizer.

It should be noted that during testing of the fabricated prototype the proposed arrangement of Figure 4.10a was not used. The independent setting of the VGA gains as in Figure 4.10b was used. This was done because the analog baseband circuits contributed significant noise (i.e., N_{BB} in Figure 4.10 was not negligible compared to N_{RF}), which necessitated the highest possible gain for any given signal power to minimize the system noise figure [28].

TRANSMITTER ERROR VECTOR MAGNITUDE (EVM)

The transmitter EVM was specified to be better than -27 dB. Simulations of the complete MIMO system showed that this level of EVM results in 1 dB loss in the link budget, that is, the receiver sensitivity in the presence of TX imperfections is 1 dB worse than when using an ideal MIMO transmitter. These simulations were performed using a fairly idealized receiver with lossless channel estimation and synchronization. If a more realistic digital baseband receiver is used, simulations show that the EVM requirement becomes about −29 dB. Transmitter measurements reported in the section "Testing Procedures and Measured Results" were performed at an EVM of −27 dB.

CHIP ARCHITECTURE AND DESIGN

Figure 4.11 shows the architecture of the direct-conversion 2 × 2 MIMO transceiver. The design is based on the WLAN transceiver presented in Ref. 29. The system consists of two receivers, two transmitters, and a shared LO generation circuit. Each of the two receivers consists of an inductively degenerated differential LNA, a pMOS current-commutating down-conversion mixer, and a sixth-order Gm-C elliptic filter interleaved with a five-stage VGA. The cutoff frequency of the filter can be programmed using switched capacitor banks to accommodate different channel bandwidths [28]. The VGA has 54 dB of

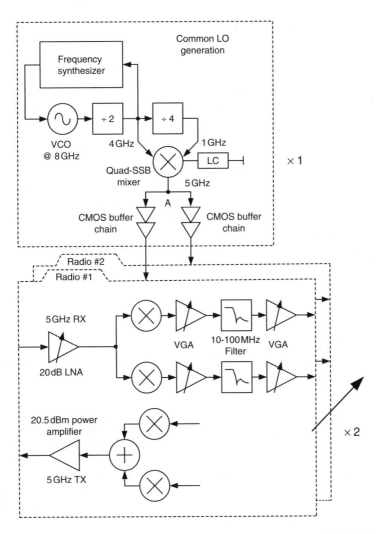

FIGURE 4.11 The architecture of the fabricated 2 × 2 MIMO transceiver. The chip includes two receive chains, two transmit chains with integrated power amplifiers with $P_{1dB} = 20.5$ dBm, and a shared LO generation and distribution scheme specifically optimized for MIMO. (From Palaskas, Y., Ravi, A., Pellerano, S., Carlton, B. R., Elmala, M. A., Bishop, R., Banerjee, G., Nicholls, R. B., Ling, S., Dinur, N., Taylor, S. S., and Soumyanath, K., *IEEE J. Solid-State Circuits*, 41, 2746–2756, 2006. With permission. © [2006] IEEE.)

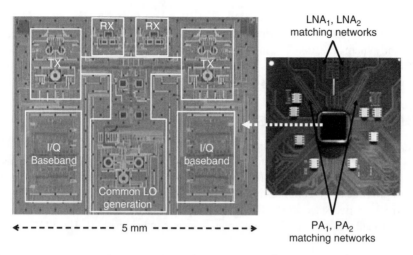

FIGURE 4.12 Die shot (left) and package photo (right). (From Palaskas, Y., Ravi, A., Pellerano, S., Carlton, B. R., Elmala, M. A., Bishop, R., Banerjee, G., Nicholls, R. B., Ling, S., Dinur, N., Taylor, S. S., and Soumyanath, K., *IEEE J. Solid-State Circuits*, 41, 2746–2756, 2006. With permission. © [2006] IEEE.)

gain, variable in 2 dB steps. Each of the transmitters consists of a Gilbert upconversion mixer and an integrated, class AB power amplifier with $P_{1\,dB} = 20.5$ dBm. A varactor in parallel with the inductive load of the upconversion mixer is used to introduce a signal-dependent phase shift to counteract the AM–PM distortion of the PA [30]. The varactor is controlled by the amplitude of the baseband IQ data through an appropriate lookup table. The matching networks of the PAs and the LNAs are realized as microstrips on the top layer of a multilayer flip-chip package to get a high Q and minimize external component count.

To minimize detrimental MIMO coupling (see the section "MIMO Implications on Radio IC Design"), the two transmitters and the associated matching networks are placed at the maximum possible distance on the die and the package, respectively (see Figure 4.12). The distance between the two receivers is somewhat smaller, which is acceptable on the basis of the reduced sensitivity of the receiver to MIMO crosstalk (see the section "MIMO Implications on Radio IC Design"). Several other isolation techniques are used to minimize MIMO coupling: the supplies of the power amplifiers are separated from the rest of the transceiver; extensive de-coupling is used on supplies and bias lines; lightly doped, high-resistivity regions are used around the different blocks to minimize coupling through the substrate; large numbers of substrate contacts close to the devices are used for the same reason; ground shields are used to protect sensitive lines; and differential signaling is employed throughout the chip.

The 5 GHz LO is generated from an 8 GHz synthesizer by division and single side–band mixing [31], as shown in Figure 4.11. Operating the VCO at a frequency nonharmonically related to the RF frequency minimizes VCO pulling by the integrated PAs, thus reducing nonlinear MIMO crosstalk and self-coupling. CMOS scaling is exploited and rail-to-rail CMOS buffer chains used to drive the mixers. CMOS inverter buffers have larger swings and present a lower output impedance than low-swing LC-loaded buffers, thereby minimizing nonlinear coupling through the LO. Two separate chains of inverter buffers are used to drive the two transceivers, as shown in Figure 4.11. The two buffer chains exhibit significant reverse isolation minimizing undesired crosstalk from TX_1 and TX_2 to the common LO node A. If a common LO distribution buffer chain had instead been used, there would have been no isolation between the LO ports of the two upconversion mixers. Extensive de-coupling capacitors and shielding techniques are used to protect the sensitive RX circuits from the switching noise of the inverter buffers.

TESTING PROCEDURES AND MEASURED RESULTS

The 2 × 2 MIMO transceiver RFIC was fabricated in a 90 nm CMOS process and occupies a total die area of 18 mm² (Figure 4.12). The transceiver was tested using a custom MIMO modem implemented

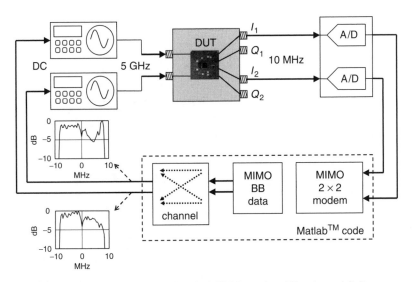

FIGURE 4.13 The test setup used to characterize the MIMO receiver. The channel fading was emulated in a software. Multiple different channels were used for statistical analysis. (From Palaskas, Y., Ravi, A., Pellerano, S., Carlton, B. R., Elmala, M. A., Bishop, R., Banerjee, G., Nicholls, R. B., Ling, S., Dinur, N., Taylor, S. S., and Soumyanath, K., *IEEE J. Solid-State Circuits*, 41, 2746–2756, 2006. With permission. © [2006] IEEE.)

in a software, as shown in Figure 4.13. The software generates MIMO OFDM signals and passes them through an appropriate model of a 2 × 2 radio channel. Random Rayleigh channels with 25 ns rms delay spread, typical of office environments, were used during the testing. The resulting RF signals upconverted by two vector signal generators (VSGs) are then applied to the two receivers and the outputs are digitized and passed through timing and frequency synchronization routines, MMSE channel equalization, and Viterbi decoding. The MIMO demodulation was performed at low-IF (10 MHz) due to the limited number of matched inputs available at the digitizer (two instead of four). Zero-IF demodulation for the legacy 1 × 1 mode degrades the sensitivity by 1.5 dB due to flicker noise.

For this test setup, particular care has been taken to synchronize the two VSGs responsible for upconverting the baseband signals. In principle, this would not be necessary, since the MIMO algorithm should be able to equalize any delay mismatch between the channels. However, there is a limit on the maximum delay mismatch that can be tolerated and this limit depends on the duration of the cyclic prefix in the packet [12]. Therefore, a time skew between the two VSGs comparable to the cyclic prefix duration can reduce the immunity of the receiver to the delay mismatch introduced by the channel, corrupting the accuracy of the measurement. In the test setup of Figure 4.13, the VSGs were synchronized within a few nanoseconds.

Figure 4.14 shows measured waterfall curves obtained from the fabricated prototype. All measurements were performed at low-IF to allow for direct comparison. Two hundred random channels per data point were used to obtain statistically significant results. In legacy 54 Mb/s mode, each receiver achieves a sensitivity of −76 dBm (PER = 10%) in the presence of additive white Gaussian noise (AWGN). Because of multipath fading, this sensitivity degrades to −68.5 dBm for a 25 ns Rayleigh channel. In the 108 Mb/s 2 × 2 MIMO spatial multiplexing mode, the corresponding sensitivity is −63 dBm. This matches closely the MIMO sensitivity as estimated from theory and simulations:

$$kT + 10\log(BW) + SNR + Implem\_Loss + NF$$
$$\approx -174 + 72 + 29 + 3 + 7.5 = -62.5 \, dBm$$

where

BW (= 16 MHz) is the channel bandwidth

SNR is the average SNR required for PER = 10% in a 25 ns Rayleigh channel (this is estimated using MATLAB simulations of the complete MIMO system)

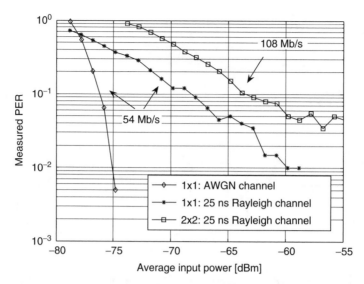

FIGURE 4.14 Measured packet error rate (PER) as a function of average received power (per channel). The 2×2 MIMO receiver achieves a sensitivity of −63 dBm while receiving 108 Mb/s in the presence of multipath fading (25 ns Rayleigh channel). The 1×1 AWGN results are reported for easy comparison with the literature. Receiver measurements were performed at low-IF (10 MHz) instead of zero-IF due to limitations of the test setup. (From Palaskas, Y., Ravi, A., Pellerano, S., Carlton, B. R., Elmala, M. A., Bishop, R., Banerjee, G., Nicholls, R. B., Ling, S., Dinur, N., Taylor, S. S., and Soumyanath, K., *IEEE J. Solid-State Circuits*, 41, 2746–2756, 2006. With permission. © [2006] IEEE.)

Implem_Loss is the loss due to channel estimation, pilot tracking, and other synchronization routines

NF is the effective noise figure of the receiver

Note that the noise figure (NF) for this calculation is higher than the 6 dB noise figure reported in Table 4.1. This is because at the MIMO sensitivity level of −63 dBm the noise figure of the baseband chain is somewhat worse than at higher gains. The MIMO sensitivity would be significantly better had a third receive chain been used (2×3 MIMO) due to extra diversity gain associated with the third receiver.

MIMO to SISO performance is compared next. To make a meaningful comparison an estimation is made of what the SISO sensitivity would have been at 108 Mb/s. To double the rate from 54 to 108 Mb/s in the 1×1 mode while using the same 16 MHz bandwidth, Shannon's theorem dictates that the SNR at the receiver must be doubled in decibel. The SNR required for 54 Mb/s in 1×1 mode is about 18 dB. Therefore, the SISO sensitivity for 108 Mb/s is estimated to be −68.5 + 18 = −50.5 dBm. From these calculations, the conclusion is that the 108 Mb/s 2×2 MIMO transceiver has a 12.5 dB SNR advantage over a conventional 1×1 transceiver. This would also translate to 12.5 dB improvement in range assuming that the total transmitted power is the same in both cases.

Figure 4.15 shows the 5 GHz transmitter constellation and spectral mask in SISO 1×1 and MIMO 2×2 modes. In legacy 1×1 mode, the TX delivers an average power of +16 dBm (EVM = −25 dB) with an efficiency of 7%. The AM–PM linearization allows the PA to operate at a small backoff from the $P_{1\,dB}$ of 20.5 dBm [30]. In the 2×2 mode, each PA delivers an average power of +13 dBm while meeting the more stringent EVM of −27 dB required for MIMO.

Turning off one of the transmitters in MIMO mode improves the EVM of the other transmitter by 1 dB (from −27 to −28 dB), which implies that there is some residual crosstalk between the two. Figure 4.16 presents several experiments that can be used to quantify MIMO crosstalk. Figure 4.16a shows the output spectrum of transmitter TX_2 when TX_1 is off. The input of TX_2 is a complex sinusoid at frequency $f_2 = 7$ MHz. The output contains the desired tone at $f_{LO} + f_2$ and a nonlinear self-coupling

TABLE 4.1
Summary of Transceiver Performance

5 GHz RX Noise Figure (IF = 10 MHz/DC)	6 dB/7.5 dB
SISO 5 GHz RX sensitivity[a] (54 Mb/s, IF = 10 MHz)	−76 dBm (AWGN channel)
	−68.5 dBm (25 ns Rayleigh)
MIMO 5 GHz RX sensitivity[a] (108 Mb/s, IF = 10 MHz)	−63 dBm (25 ns Rayleigh)
RX IIP3	−12 dBm
RX power dissipation	1 × 1 Mode: 170 mW (1.4 V)
	2 × 2 Mode: 280 mW (1.4 V)
5 GHz TX average power (64-QAM)	1 × 1 Mode: +16 dBm (EVM: −25 dB)
	2 × 2 Mode: +13 dBm (EVM: −27 dB)
5 GHz TX power dissipation	1 × 1 Mode: 840 mW (1.4/3.3 V)
	2 × 2 Mode: 1400 mW (1.4/3.3 V)
MIMO RX-RX isolation	30 dB
MIMO TX-TX isolation (linear)	37 dB
MIMO TX-TX isolation (nonlinear)	> 39 dB
Process/area	90 nm CMOS/18 mm^2

[a] Zero-IF would degrade receiver performance by 1.5 dB.

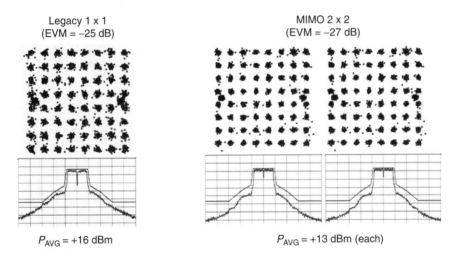

Legacy 1 x 1
(EVM = −25 dB)

MIMO 2 x 2
(EVM = −27 dB)

P_{AVG} = +16 dBm

P_{AVG} = +13 dBm (each)

FIGURE 4.15 The transmitter constellation and spectral mask in legacy 1 × 1 mode and MIMO 2 × 2 mode. In MIMO mode the average linear power of each PA is 3 dB smaller than in the SISO case (13 dBm versus 16 dBm) because of the tighter MIMO EVM requirement and some residual crosstalk between the transmitters. (From Palaskas, Y., Ravi, A., Pellerano, S., Carlton, B. R., Elmala, M. A., Bishop, R., Banerjee, G., Nicholls, R. B., Ling, S., Dinur, N., Taylor, S. S., and Soumyanath, K., *IEEE J. Solid-State Circuits*, 41, 2746–2756, 2006. With permission. © [2006] IEEE.)

component at $f_{LO} + 2f_2$. This component is generated when the desired output at $f_{LO} + f_2$ couples to the LO line which is then used to upconvert the baseband signal with frequency f_2. Figure 4.16b shows the output spectrum of TX_2 when the corresponding baseband signal is turned off and a complex sinusoid with frequency $f_1 = 9$ MHz is applied to the transmitter TX_1. The tone at $f_{LO} + f_1$ at the output of TX_2 is due to linear coupling from TX_1. Even though TX_1 couples a disturbance to

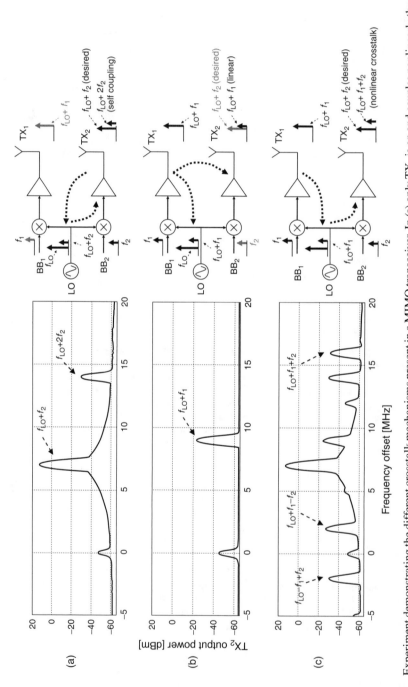

FIGURE 4.16 Experiment demonstrating the different crosstalk mechanisms present in a MIMO transceiver. In (a) only TX_2 is on and couples nonlinearly through the LO to itself. Case (b) demonstrates linear crosstalk from TX_1, which is on, to TX_2, which is off. In (c) the output of TX_1 couples nonlinearly to TX_2 through the LO. The spectrum in (c) also contains the crosstalk terms of (a) and (b) and some additional nonlinear terms. (From Palaskas, Y., Ravi, A., Pellerano, S., Carlton, B. R., Elmala, M. A., Bishop, R., Banerjee, G., Nicholls, R. B., Ling, S., Dinur, N., Taylor, S. S., and Soumyanath, K., *IEEE J. Solid-State Circuits*, 41, 2746–2756, 2006. With permission. © [2006] IEEE.)

the LO line, this does not create a nonlinear crosstalk component at the output of TX_2 because the corresponding baseband signal is zero. In Figure 4.16c, both transmitters are on. The output spectrum of TX_2 contains the crosstalk components of Figures 4.16a and 4.16b and additional nonlinear crosstalk terms. For example, the tone at $f_{LO} + f_1 + f_2$ is generated when the output of TX_1 couples a $f_{LO} + f_1$ disturbance on the LO line, which is then used to upconvert the f_2 baseband signal of TX_2. A detailed derivation of all the coupling components shown in Figure 4.16c is presented in Ref. 32.

The crosstalk products of Figure 4.16c add up to a total power of about -33 dBc, which explains the 1 dB degradation in EVM from -28 to -27 dB when both transmitters are on. This calculation assumes that the nonlinear coupling is the same for the crosstalk experiment of Figure 4.16 and the EVM measurement of Figure 4.15. To ensure this, the average power of the PAs was set to the same value of 13 dBm in both cases. The MIMO EVM measurements of Figure 4.15 were performed using a commercial SISO demodulator. Alternatively, a MIMO demodulator could be used to estimate the EVM of the two transmitters in MIMO mode. A MIMO demodulator would eliminate the linear crosstalk between the two transmitters as explained in conjunction with Figures 4.6 and 4.7. From the measurements of Figure 4.16c it is estimated that this would improve the EVM by about 0.5 dB.

On-the-air tests were performed using the fabricated chip at the receiver and the transmitter side of an actual wireless MIMO link. This experiment demonstrated the functionality of the complete MIMO spatial multiplexing system in a real-life environment. The transceiver RFIC performance is summarized in Table 4.1.

FUTURE DIRECTIONS

The demand for higher data rates in the future will lead to higher order MIMO systems, as MIMO capacity can theoretically be increased by increasing the number of transmit and receive antennas. However, several limitations exist in practice [33]. First of all, cost (including die area) and complexity of the system will likely increase linearly with the number of TX/RX antenna pairs. Power consumption will also increase but less than linearly if the total transmitted power is kept constant (due to government regulations). Moreover, some blocks, such as the LO synthesizer, can be shared, thus their power consumption gets averaged over the multiple radios. However, there is a more fundamental limit to the maximum achievable capacity of a MIMO system, which is the physical volume available to deploy antennas. This becomes a critical issue especially in portable devices such as handhelds where the form factor is constrained. When antennas are placed too close together, they start to interact and the channel capacity can get significantly smaller, deviating from theoretical results. As stated in Ref. 34, the fundamental question to be answered is how small can the antenna array be made without losing a significant amount of capacity. Several studies based on ray tracing simulations and indoor measurements [34–37] have demonstrated that increasing the antenna spacing beyond $\lambda/2$ does not lead to significant increase in the channel capacity in environments with uniformly distributed scatterers. In Ref. 38, the same conclusions are drawn by using a completely different approach, i.e., the laws of electromagnetism and the Nyquist sampling theorem in the spatial domain. There are two main factors that affect channel capacity when antennas are placed too close and experience mutual coupling: correlation between the fades and radiation efficiency. In Ref. 39, it is shown that the reduction in radiation efficiency caused by mutual coupling is largely responsible for channel capacity degradation. The correlation between received signals at antenna ports has a less significant effect. Different matching networks [40,41] and antenna configurations [42] that maximize the channel capacity for a given array size have also been investigated.

A new technology being developed to overcome space limitations on handheld wireless devices is distributed or virtual MIMO [43]. In this scenario, different autonomous wireless devices can transmit or receive on the same band. Collaboration comes from the fact that idle devices can choose to help other devices in transmission and reception. Diversity gain is achieved through collaborative communication between clusters of closely located nodes. However, for collaboration to happen, some information has to be shared between the nodes. In Ref. 43, it is shown that it is possible to

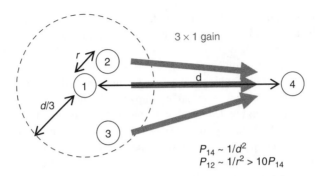

FIGURE 4.17 Cluster size for an achievable 3×1 space–time gain with three transmit nodes and one receiving node.

reduce the overhead due to the intracluster negotiation so that almost full diversity gain of traditional space–time coding can be achieved. Analysis demonstrates that if the channel path-loss between the collaborators is 10 dB better than that of the targeted receiver, almost full diversity gain can be achieved. From a geometric point of view, this translates into a transmit cluster whose radius can be as large as one-third of the distance between the source and the destination nodes over a free-space channel with path-loss exponent of 2 (Figure 4.17) [44]. Similar collaboration schemes can be used between multiple receivers. These collaboration ideas might be particularly important to heavily dispersed, ubiquitous wireless sensor networks. Practical challenges for distributed MIMO include synchronization of independent devices in time and frequency, network layer algorithms for selection of optimal cooperators, and novel billing methodologies to provide incentive for cooperation.

Another interesting research direction is the use of multiple antennas to mitigate cochannel interference [45,46]. This allows for increased frequency reuse and improved performance for the overall network. The receiver in this case employs multiple antennas that receive the superposition of a desired signal plus a number of interfering signals. The desired and interfering signals generally appear to arrive at the receiver from different directions due to the different physical locations of the corresponding transmitters and also due to multipath fading (see the section "Spatial Multiplexing"). The receiver can then optimally combine the signals of the individual antennas in such a way as to enhance the desired signal while minimizing the interfering signals. Intuitively, this is similar to "steering" the receiver antenna array toward the perceived angle of arrival of the desired signal and away from the interference. Directionality can also be used to minimize out-of-channel interference [16], which might relax the requirements of the radio, e.g., the required number of bits in the ADC. The combination of MIMO ideas with actual or perceived directionality is an area of active research today.

MIMO technology has already begun to transform wireless networks. Ongoing research efforts, building on the original MIMO concepts, are opening up new possibilities for wireless connectivity. Examples include distributed communications, interference cancellation, and wireless sensor networks. The portable device of tomorrow will be based on the convergence of computing, data storage, and communications in a compact form factor. MIMO and its derivatives will continue to play a key role in enabling always-on connectivity at reasonable cost. Continuous innovation at the radio IC level will be required to realize the full potential of these upcoming wireless technologies.

ACKNOWLEDGMENTS

The authors would like to thank B. R. Carlton, M. A. Elmala, R. Bishop, G. Banerjee, R. B. Nicholls, S. Ling, N. Dinur, S. S. Taylor, K. Soumyanath, D. Steele, Minnie Ho, K. Holt, P. Seddighrad, A. Biran, S. Somayazulu, H. Alavi, Qiang Li, D. Martin, R. Wang, N. Yaghini, Jian Gong, C. Zhan, John Vu, D. Trammo, A. Crouch, and K. Kahn for their contributions.

REFERENCES

1. IEEE Std 802.11a-1999, High-speed physical layer in the 5 GHz band.
2. Zargari, M., Su, D. K., Yue, C. P., Rabii, S., Weber, D., Kaczynski, B. J., Mehta, S. S., Singh, K., Mendis, S., and Wooley, B. A., "A 5-GHz CMOS transceiver for IEEE 802.11a wireless LAN systems," *IEEE J. Solid-State Circuits*, vol. 37, no. 12, pp. 1688–1694, December 2002.
3. Khorram, S., "A fully integrated SOC for 802.11b in 0.18 µm CMOS," *IEEE J. Solid-State Circuits*, vol. 40, no. 12, pp. 2492–2501, December 2005.
4. Mehta, S. S., Weber, D., Terrovitis, M., Onodera, K., Mack, M. P., Kaczynski, B. J., Samavati, H., Jen, S. H.-M., Si, W. W., Lee, M-L., Singh, K., Mendis, S., Husted, P. J., Ning Z., McFarland, B., Su, D. K., Meng, T. H., and Wooley, B. A., "An 802.11g WLAN SoC," *IEEE J. Solid-State Circuits*, vol. 40, no. 12, pp. 2483–2491, December 2005.
5. Razavi, B., Aytur, T., Lam, C., Yang, F-R., Li, K-Y., Yan, R-H., Kang, H-C., Hsu, C.-C., and Lee, C-C., "A UWB CMOS transceiver," *IEEE J. Solid State Circuits*, vol. 40, no. 12, pp. 2555–2562, December 2005.
6. Natarajan, A., Komijani, A., and Hajimiri, A., "A fully integrated 24-GHz phased-array transmitter in CMOS," *IEEE J. Solid-State Circuits*, vol. 40, no. 12, pp. 2502–2514, December 2005.
7. Foschini, G. J., "Layered space–time architecture for wireless communication in a fading environment when using multiple antennas," *Bell Labs Tech. J.*, vol. 1, pp. 41–59, 1996.
8. Telatar, E., "Capacity of multi-antenna Gaussian channels," *Bell Labs Technical Memorandum*, pp. 1–28, October 1995. Also published in *Eur. Trans. Telecommun.*, vol. 10, no. 6, pp. 585–595, November/December 1999.
9. Paulraj, A. J., Gore, D. A., Nabar, R. U., and Bolcskei, H., "An overview of MIMO communications—A key to gigabit wireless," *Proc. IEEE,* vol. 92, no. 2, pp. 198–218, February 2004.
10. Gesbert, D., Shafi, M., Shiu, D., Smith, P. J., and Naguib, A., "From theory to practice: An overview of MIMO space-time coded wireless systems," *IEEE J. Sel. Areas Commun.*, vol. 21, no. 3, pp. 281–302, April 2003.
11. Paulraj, A., Nabar, R., and Gore, D., "Introduction to space-time wireless communications," Cambridge University Press, Cambridge, 2003.
12. Jankiraman, M., "Space-time codes and MIMO systems," Artech House, Norwood, MA, 2004.
13. IEEE P802.11n™/D1.02, Part 11: Wireless LAN medium access control (MAC) and physical layer (PHY) specifications, July 2006.
14. IEEE 802.11-03/940r4, IEEE p802.11 Wireless LANs TGn channel models, May 2004.
15. McFarland, W., Choi, W-J., Tehrani, A., Gilbert, J., Kuskin, J., Cho, J., Smith, J., Dua, P., Breslin, D., Ng, S., Zhang, X., Wang, Y-H., Thomson, J., Unnikrishnan, M., Mack, M., Mendis, S., Subramanian, R., Husted, P., Hanley, P., and Zhang, N., "A WLAN SoC for video applications including beamforming and maximum ratio combining," *IEEE International Solid-State Circuits Conference*, pp. 452–609, February 2005.
16. Paramesh, J., Bishop, R., Soumyanath, K., and Allstot, D. J., "A four-antenna receiver in 90-nm CMOS for beamforming and spatial diversity," *IEEE J. Solid-State Circuits*, vol. 40, no. 12, pp. 2515–2524, December 2005.
17. Alamouti, S., "A simple transmit diversity technique for wireless communications," *IEEE J. Selected Areas Commn.*, vol. 16, pp. 1451–1458, October 1998.
18. Sandhu, S. and Paulraj, A., "Space-time block codes: a capacity perspective," *IEEE Commn. Lett.,* vol. 4, pp. 384–386, December 2000.
19. Hassibi, B. and Hochwald, B., "High-rate codes that are linear in space and time," *IEEE Trans. Inf. Theory*, vol. 48, pp. 1804–1824, July 2002.
20. Oggier, F. E., Rekaya, G., Belfiore, J.-C., and Viterbo, E., "Perfect space time block codes," *IEEE Trans. Inf. Theory*, vol. 52, pp. 3885–3902, September 2006.
21. Gore, D., Sandhu, S., and Paulraj, A., "Delay diversity codes for frequency selective channels," *IEEE International Conference on Communications*, vol. 3, pp. 1949–1953, 2002.
22. Rahn, D. G., Cavin, M. S., Dai, F. F., Fong, N. H. W., Griffith, R., Macedo, J., Moore, A. D., Rogers, J. W. M., and Toner, M., "A fully integrated multiband MIMO WLAN transceiver RFIC," *IEEE J. Solid-State Circuits*, vol. 40, no. 8, pp. 1629–1641, August 2005.
23. Zheng, L. and Tse, D., "Diversity and multiplexing: A fundamental tradeoff in multiple-antenna channels," *IEEE Trans. Inf. Theory*, vol. 49, pp. 1073–1096, May 2003.

24. Nabar, R. U., Bolcskei, H., Erceg, V., Gesbert, D., and Paulraj, A. J., "Performance of multiantenna signaling techniques in the presence of polarization diversity," *IEEE Trans. Signal Process*, vol. 50, pp. 2553–2562, October 2002.

25. Gamal, H. E., Caire, G., and Damen, M., "Lattice coding and decoding achieve the optimal diversity-multiplexing tradeoff of MIMO channels," *IEEE Trans. Inf. Theory*, vol. 50, pp. 968–985, June 2004.

26. Bourdoux, A., Come, B., and Khaled, N., "Non-reciprocal transceivers in OFDM/SDMA systems: Impact and mitigation," *IEEE Radio and Wireless Conference*, Boston, MA, USA, pp. 183–186, August 2003.

27. Nanda, S., Walton, R., Ketchum, J., Wallace, M., and Howard, S., "A high-performance MIMO OFDM wireless LAN," *IEEE Commn. Mag.*, vol. 43, no. 2, pp. 101–109, February 2005.

28. Elmala, M., Bishop, R., and Soumyanath, K., "A highly linear filter and VGA chain with novel DC-offset correction in 90 nm digital CMOS process," *VLSI Circuits Symposium*, pp. 302–303, June 2005.

29. Ravi, A., Carlton, B. R., Palaskas, Y., Banerjee, G., Bishop, R. E., Elmala, M. A., Nicholls, R. B., Rippke, I. A., Lakdawala, H., Franca-Neto, L. M., Taylor, S. S., and Soumyanath, K., "A 1.4V, 2.4/5 GHz, 90 nm CMOS system in a package transceiver for next generation WLAN," *VLSI Circuits Symposium*, pp. 294–297, June 2005.

30. Palaskas, Y., Taylor, S. S., Pellerano, S., Rippke, I., Bishop, R., Ravi, A., Lakdawala, H., and Soumyanath, K., "A 5-GHz, 20-dBm power amplifier with digitally-assisted AM-PM correction in a 90 nm CMOS process," *IEEE J. Solid-State Circuits*, vol. 41, pp. 1757–1763, August 2006.

31. Zhang, P., Nguyen, T., Lam, C., Gambetta, D., Soorapanth, T., Cheng, B., Hart, S., Sever, I., Bourdi, T., Tham, A., and Razavi, B., "A 5-GHz direct-conversion CMOS transceiver," *IEEE J. Solid-State Circuits*, vol. 38, no. 12, pp. 2232–2238, December 2003.

32. Palaskas, Y., Ravi, A., Pellerano, S., Carlton, B. R., Elmala, M. A., Bishop, R., Banerjee, G., Nicholls, R. B., Ling, S., Dinur, N., Taylor, S. S., and Soumyanath, K., "A 5GHz 108Mb/s 2 × 2 MIMO transceiver with fully integrated 20.5 dBm P1dB power amplifiers in 90 nm CMOS," *IEEE J. Solid-State Circuits*, vol. 41, no. 12, pp. 2746–2756, December 2006.

33. Foschini, G. and Gans, M. J., "On limits of wireless communications in a fading environment when using multiple antennas," *Wireless Personal Commn.*, vol. 6, pp. 311–335, 1998.

34. Burr, A. G., "Channel capacity evaluation of multi-element antenna systems using a spatial channel model," *Millennium Conference on Antenna and Propagation*, Davos, Switzerland, April 2000.

35. Steinbauer, M., Molisch, A., Burr, A., and Thomä, R., "MIMO channel capacity based on measurement results," *European Conference on Wireless Technology*, pp. 52–55, October 5–6, 2000.

36. Pohl, V., Jungnickel, V., Haustein, T., and von Helmolt, C., "Antenna spacing in MIMO indoor channels," *IEEE 55th Vehicular Technology Conference*, pp. 749–753, May 2002.

37. Jungnickel, V., Pohl, V., and von Helmolt, C., "Capacity of MIMO systems with closely spaced antennas," *IEEE Commn. Lett.*, vol. 7, pp. 361–363, August 2003.

38. Loyka, S., "On the relationship of information theory and electromagnetism," *IEEE 6th International Symposium on Electromagnetic Compatibility and Electromagnetic Ecology*, pp. 100–104, 2005.

39. Kildal, P. S. and Rosengren, K., "Correlation and capacity of MIMO systems and mutual coupling, radiation efficiency, and diversity gain of their antennas: Simulations and measurements in a reverberation chamber," *IEEE Commn. Mag.*, vol. 42, pp. 104–112, December 2004.

40. Wallace, J. W. and Jensen, M. A., "Mutual coupling in MIMO wireless systems: A rigorous network theory analysis," *IEEE Trans. Wireless Commn.*, vol. 3, pp. 1317–1325, July 2004.

41. Lau, B. K., Ow, S. M. S., Kristensson, G., and Molisch, A. F., "Capacity analysis for compact MIMO systems," *IEEE 61st Vehicular Technology Conference*, vol. 1, pp. 165–170, 30 May–1 June 2005.

42. Waldschmidt, C., Schulteis, S., and Wiesbeck, W., "Complete RF system model for analysis of compact MIMO arrays," *IEEE Trans. Vehicular Technol.*, vol. 53, pp. 579–586, 2004.

43. Mitran, P., Ochiai, H., and Tarokh, V., "Space-time diversity enhancements using collaborative communications," *IEEE Trans. Inf. Theory*, vol. 51, pp. 2041–2057, June 2005.

44. Tarokh, V., "Cognition, collaboration and competition in space and time: From theory to implementation," *MSRI Workshop: Mathematics of Relaying and Cooperation in Communication Networks*, April 10–12, 2006.

45. Winters, J. H., "Optimum combining in digital mobile radio with cochannel interference," *IEEE Trans. Vehicular Technol.*, vol. 33, no. 3, pp. 144–155, August 1984.

46. Dai, H., Molisch, A. F., and Poor, H. V., "Downlink capacity of interference-limited MIMO systems with joint detection," *IEEE Trans. Wireless*, vol. 3, pp. 442–452, June 2003.

5 Cognitive Radio Spectrum-Sharing Technology

Danijela Cabric and Robert W. Brodersen

CONTENTS

Introduction... 131
Spectrum Sharing 132
Underlay Approach: UWB... 133
    UWB System Architecture... 134
    UWB Impulse Radio Architecture.. 135
    UWB Impulse Detection 136
Overlay Approach: Cognitive Radios 137
    Cognitive Radio Systems Architecture 139
    Wideband Spectrum-Sensing Radio Architecture 141
        Frequency Agility Regimes 141
        Dynamic Range Reduction Techniques.. 144
    Signal Processing for Spectrum Sensing.. 146
        Optimal Coherent Detection: Matched Filter 146
        Suboptimal Noncoherent Detection: Radiometer.................................. 148
        Feature Detection... 149
    Frequency-Agile Wideband Transmission... 153
Conclusions 156
References.. 156

INTRODUCTION

A major shift in radio design is now just beginning, which attempts to share spectrum in a fundamentally new way. This situation arises from the fact that spectrum is actually poorly utilized in many bands, in spite of the increasing demand for wireless connectivity. In addition, current crowding of unlicensed spectra has pushed the regulatory agencies to be more aggressive in providing new ways to use spectra. To obtain spectra for unlicensed operation, new sharing concepts have been introduced to allow use by secondary users under the requirement that they limit their interference to preexisting primary users.

Two basic approaches to spectrum sharing have been identified. One is an underlay approach with severe restrictions on transmitted power levels with a requirement to operate over ultra wide band (UWB) widths and the other is an overlay approach based on avoidance of higher priority users through the use of spectrum sensing and adaptive allocation cognitive radios (CRs). Both of these techniques are a major shift from the long-standing approach that once a frequency band is assigned, no interference is allowed to the primary user, in spite of the fact that it is clearly not possible to ensure absolutely no interference.

This spectrum access paradigm shift also has major implications on radio architectures, since traditional narrowband radio design techniques are not applicable. Spectrum sharing required in UWB and CR over wide bands implies frequency agility and significant dynamic range improvements to accommodate the in-band primary users. In addition, new radio functions are required, which involve high-sensitivity sensing, signal processing, and protocols that can exploit this sensing information to minimize interference. Even more interesting is that entirely new signaling strategies may be employed, since the combination of wide bandwidth availability and severe power and dynamic range restrictions were not present before. Finally, new application areas may be facilitated.

Major opportunities and challenges of this new era in CMOS radio design are presented in this chapter by focusing on some of the issues raised by these new sharing strategies. Radio architectures that address the unique new requirements of these radios will be discussed including the analog and digital circuit partitioning, and the issues involved in signal processing and protocols.

SPECTRUM SHARING

In the past, the approach for spectrum allocation was based on specific band assignments designated for a particular service, as illustrated by the Federal Communications Commission's (FCC) frequency allocation chart [1]. This spectrum chart contains overlapping allocations in most frequency bands and seems to indicate a high degree of spectrum scarcity. To overcome the bandwidth limitations, several great advances in communications system design, such as sophisticated modulations, coding, and multiple antennas, have greatly improved the spectrum efficiency of some commercial radio systems (e.g., cellular and WiFi bands). However, these systems are now faced with increased self-interference that limits network capacity and scalability as inevitable consequence of spectrum crowding. On the contrary, some bands are poorly utilized. Measurements taken in downtown Berkeley (Figure 5.1) reveal a typical utilization of roughly 30% below 3 GHz, and 0.5% in the 3–6 GHz frequency band. This view is supported by recent studies of the FCC's Spectrum Policy Task Force that reported [2] vast temporal and geographic variations in the use of allocated spectrum with utilization ranging from 15 to 85%.

To promote more flexibility in spectrum sharing, the FCC has provided new opportunities for unlicensed spectrum use with fewer restrictions on radio parameters. Three new opportunities in spectrum access have thus been introduced: (1) an underlay approach with severe restrictions on transmitted power levels with a requirement to operate over (UWBs); (2) an opening of

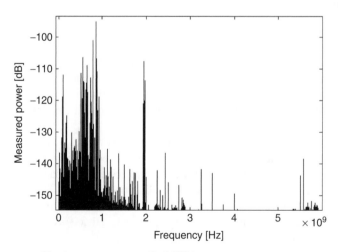

FIGURE 5.1 Spectrum utilization measurement (0–6 GHz).

TABLE 5.1
Potential System-Level Specifications Consistent with FCC Regulations and IEEE Standards

Radio Type	UWB radio	60 GHz radio	Cognitive radio
Spectrum Access	Underlay	Unlicensed	Overlay
Carrier	0–1, 3–10 GHz	57–64 GHz	0–1, 3–10 GHz
Bandwidth	>500 MHz	>1 GHz	>500 MHz
Data Rates	100–500 Mb/s	>1 Gb/s	~10–1000 Mb/s
Spectrum Efficiency	~0.1–1 b/s/Hz	~1 b/s/Hz	~0.1–10 b/s/Hz
Range	1–10 m	~10 m	1 m to 30 km

7 GHz of unlicensed spectrum at millimeter-wave frequencies (around 60 GHz) where oxygen absorption limits long-distance interference; and (3) an overlay approach based on avoidance of higher-priority users through the use of spectrum sensing CRs.

The potential opening of these new spectra introduces new opportunities for vastly more wireless connectivity. As indicated in Table 5.1, these three radio systems are (or should be) allowed to operate in 500 MHz or wider spectrum. Therefore, the design of high-throughput radios with data rates from 100 Mb/s to even 1 Gb/s is achievable at moderate-to-low spectrum efficiencies. The power limitations and wireless channel propagation characteristics for these bands dictate the range capability that extends from 1 m to 30 km, so that a wide variety of communication modes can be supported with these three new wireless radio technologies.

Although opening of the 60 GHz band is another spectrum allocation for unlicensed use, both UWB and CR are introduced as sharing concepts with interference being the main regulation concern. Furthermore, they are two fundamentally opposite ways of managing the inter-system and intra-system interference. The underlay UWB approach allows transmission in previously used bands using power levels that are sufficiently low so that the interference is felt to be not harmful. The overlay strategy through CRs has more dynamic spectrum access where radios must vacate the spectrum sufficiently fast to only yield no interference.

UNDERLAY APPROACH: UWB

In spectrum sharing, choosing the radio transmit power that causes minimal interference to primary users presents the crucial challenge. In principle, the transmit power constraint cannot be globally set so that it meets interference requirements at any location and time for arbitrary primary users. The extreme and quite conservative strategy was used to regulate the UWB transmission. It relies on the fact that if the bandwidth is increased, then reliable data transmission can occur even at power levels so low that primary radios in the same spectral bands are not affected. Figure 5.2 illustrates the transmitter power spectrum density profiles in UWB underlay spectrum-sharing approach. To overcome the transmit power limitations, systems are allowed to use a very large bandwidth so that they can trade off the data rate for robustness. This strategy is suitable for noise-like channels where signal-to-noise ratio (SNR) can be improved by spreading or coding. However, spreading transmission power equally across a wide bandwidth could be largely suboptimal in case of strong in-band interference.

As a conservative protection margin, the maximum transmission level was set to be equivalent to the previously allowed unintentional transmission from electronic devices such as a personal computer (PC) (i.e., <40 μW/500 MHz). Since the only difference from the PC noise emission was to allow data to be modulated onto the emission, it seemed reasonable to allow this new underlay use. Of course, even this conservative approach caused much concern, as any change in spectrum policy

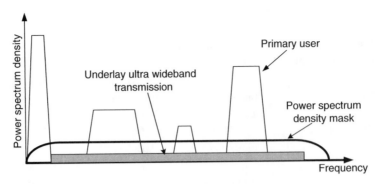

FIGURE 5.2 Underlay approach for sharing spectrum with primary users.

always does, and after inclusion of protection for the particularly sensitive global positioning system (GPS) receivers, this new approach opened bands 0–960 MHz, 23.6–24 GHz, and 3.1–10.6 GHz to a variety of new uses in 2002. UWB transmission is presently legal only in the United States, and regulations are underway in Europe, Japan, and China.

To encourage new technical approaches, the FCC has restricted the minimum occupied bandwidth to a 500 MHz minimum bandwidth [3] and identified applications in three main areas as appropriate for the three bands that were (1) imaging systems in 0–960 MHz, and 1.99–10.6 GHz, (2) vehicular radar (23.6–24 GHz), and (3) communication and measurement (3.1–10.6 GHz). These applications take advantage of the large bandwidth to allow transmission at very high data rates or reduced implementation cost by facilitating integration and low power consumption.

As the FCC has only specified the spectral mask for UWB, leaving the implementation details open, there is an enormous, unexplored space for UWB radio architectures, standards, and applications. Two exciting directions of UWB research are (1) ultra low power consumption and (2) high-rate, low-transmit-power communication.

UWB System Architecture

Although the large bandwidth required for UWB operation suggests high throughput, the low transmit power drastically limits transmit distances for high-rate communication, making this approach suitable primarily for short-range applications such as wireless personal area networks (WPAN). Present commercial attention is focused on high-data-rate consumer applications, as expressed in the IEEE 802.15.3a standard; targeting 110, 200, and 480 Mbps communication links over 1–10 m [4]. Since the FCC did not specify a modulation scheme and access mechanism, the development of this standard has been quite contentious. Main reason for the argument is that transmission over UWB channel can be addressed by time or frequency domain signaling. As a consequence, two competing approaches have been introduced: one utilizing frequency-hopping (FH) orthogonal frequency division multiplexing (OFDM) and the other employing the impulse radio technique with direct-sequence coding.

The approach based on OFDM operates over the minimum 500 MHz bandwidth with the addition of FH to provide a way to avoid large interfering signals and allow higher transmit power levels. The advantage of this approach is that OFDM modulation scheme is well understood, since it is already widely used in IEEE 802.11a and 802.11g standards. In addition, the frequency diversity provided by spreading the data over many subchannels gives additional robustness against strong in-band primary users and multipath, together with potential power allocation for optimizing channel capacity. However, a major challenge of this approach is that the overall complexity is on the order of present 802.11 systems, which means that opportunities for dramatic cost and power reductions are unlikely.

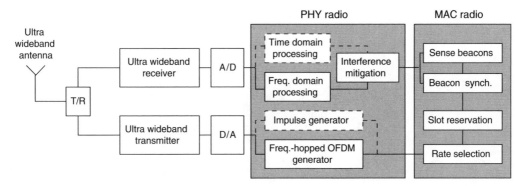

FIGURE 5.3 UWB radio transceiver block diagram.

The second approach deploys a novel type of signaling that makes use of impulses to carry information. Time domain processing of impulses modulated with direct sequence pseudonoise (PN) codes allows optimal combining for multipath processing and coding gains via spreading and repetition. Furthermore, data throughput is simply controlled by the PN sequence length. However, impulses are not as robust to narrowband in-band interference as hopping signals. The main advantage of impulse radio approach is that simpler architecture can be implemented with power consumption and cost expected to be lower owing to the duty-cycled nature of pulse communication and high levels of integration possible.

In addition to these short-range, high-throughput applications, UWB is also being considered for lower rate, medium-range systems such as sensor networks, which require ultra low power communication with scalable, but low data rates. Another IEEE 802.15.4a standard targets a physical layer radio suitable for ranges up to 30 m maximum with data rates from a few kbps to 1–10.

Mbps rate desires the ability for ranging with accuracy from 1 m to around tens of centimeters [5]. The processing of short impulses allows measurement of distances and locations so that a high-accuracy ranging may be implemented simply by using time-of-flight with fine time resolution of impulse-UWB. Therefore, an impulse-based UWB design is specified as an alternative physical layer for this standard.

Owing to short range and small geographic distribution, UWB systems tend to have an *ad hoc* configuration. As a result, UWB network deploys a simple medium access (MAC) mechanism using beacons to access the channel [6]. Since the high throughput needs to be maintained, a combination of carrier sense and time-division access manages the time slot reservation. At the startup, a device scans for beacons, and if no beacons are found, it sends its first beacon. If other beacons are received, then the device finds an available time slot. The main purpose of beacons is device discovery, time synchronization, exchanging reservations, and interference mitigation. It is envisioned that with this scheme, up to 127 devices can be supported by a single host.

Regardless of the application, it is possible to derive common building blocks and functionalities of UWB radio in the form of a block diagram. Figure 5.3 presents the architecture of a generic UWB radio including radio front-end, physical layer, and MAC layer functionalities. In the following sections, the approach to realize the architecture of UWB radio utilizing time domain impulse-based signaling is presented.

UWB IMPULSE RADIO ARCHITECTURE

By using pulse-based UWB signaling, it is possible to approach a fully digital and hence fully integrated radio; taking advantage of the technology scaling of CMOS to reduce both the transmit power and receiver's analog complexity beyond what may be achieved through simply scaling a narrowband transceiver. By moving analog-to-digital (A/D) conversion as close to the antenna as is

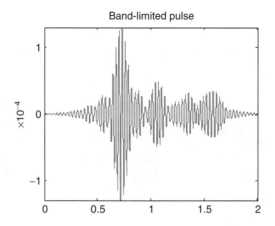

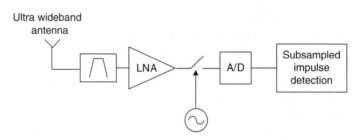

FIGURE 5.4   Low-complexity impulse UWB radio front-end.

feasible (post matching and gain), the signal is directly sampled and processed digitally, as shown in Figure 5.4. This architecture eliminates the need for frequency translation and synthesis and removes external filtering and components as would be needed for a narrowband radio. Also, due to the pulse-based nature of signaling, further power may be saved by duty-cycling the analog circuitry: activating only when a pulse is received. The most popular modulation schemes for impulse communication are antipodal signaling and pulse position modulation, which can dramatically reduce linearity requirements at the expense of increased timing sensitivity.

When operating in 3–10 GHz band, instead of the conventional heterodyne topology utilizing one or two mixing stages to down-convert the passband signal, the proposed receiver directly subsamples the incoming signal after amplification. This is accomplished by sampling at a rate below the Nyquist rate of the radio frequency (RF) signal, but at or above the Nyquist rate of the data itself. An interesting consequence of impulse signaling is that the A/D conversion precision may be kept low, whereas the conversion rate is high, on the order of a GigaSample/s. Because transmit power is limited, and relatively large interferers are likely to be present, the resulting channel is interference dominated, as opposed to thermal-noise dominated. In such environments, there is no benefit in sampling with high resolution, since it is not necessary to cancel interference. The analysis in Ref. 7 indicates that adequate throughput is still possible even with a 1-bit A/D converter. The A/D simplification saves substantial power, without severely degrading system performance. The proposed system avoids wideband analog processing with increased digital processing [8], which results in a more efficient solution than conventional UWB implementations and results in significant increases in power dissipation.

UWB IMPULSE DETECTION

The main challenge in UWB impulse detection is to recover the short impulse in a very low SNR regime. The optimal receiver is matched filter in the form of RAKE combiner, which requires

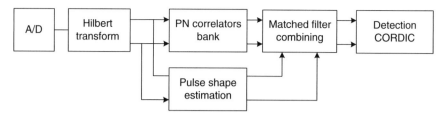

FIGURE 5.5 Digital backend processing for UWB impulse detection.

perfect time and phase synchronizations of impulse multipath replicas. Most conventional narrow-band receivers perform synchronization comprising timing-recovery and carrier-frequency compensation using synchronization loops. Since there is no sinusoidal carrier in an impulse radio, the only issue becomes the timing recovery. A timing recovery is typically done via over-sampling or interpolation. However, the cost of these approaches in UWB radio is very high due to the large bandwidths and high sampling frequency requirements.

The suboptimal approach to impulse detection is to collect the energy over the duration of the impulse using square-law device and integrator. However, the energy detection suffers the 3 dB SNR loss with respect to matched filter due to noncoherent processing. The inferiority of energy detector becomes even more significant in the presence of narrowband and multiuser interference, which are dominant issues in UWB channels. To gain partial coherence, it is possible to perform the correlation of the received impulse with the estimated pulse shape, i.e., autocorrelation (Figure 5.5). This approach requires a training sequence to learn the received pulse shape.

If pulses are processed as real signals, the perfect estimation of an impulse phase is not possible. Commonly, the phase is extracted from the analytic signal composed of the in-phase and quadrature components. In narrowband receiver, the analytic signal is obtained via mixing with sine and cosine at the intermediate frequency (IF) stage of the receiver. In case of wideband, it can be obtained by performing a Hilbert transform on the received real signal after A/D conversion. Note that this approach effectively does not require I and Q mixers; thus, only one A/D converter is sufficient. The Hilbert transformers can be implemented in digital domain as an finite impulse response (FIR) or fast Fourier transform (FFT). The real and imaginary parts of the analytic signal are orthogonal, and the phase information can be studied on the Euler plane. A coordinate rotation digital computer (CORDIC) block can be used to calculate the phase and magnitude of the complex signal, which are used for impulse detection. For antipodal signaling the constellation plot allows threshold-based detection, similar to binary phase shift keying (BPSK) detection [8].

In severe interference channels, autocorrelation receiver cannot maintain the throughput without bit error rate (BER) degradation. However, the throughput can be traded for robustness by applying the PN spreading codes across impulses. For 2^N spreading sequence length, the throughput is scaled down by 2^N, but SNR is improved by $3N$ dB. The increase in SNR achieved by spreading gain can be used for range extension or BER improvement. To support spreading, the digital backend needs to include a bank of template filters whose outputs are correlated against a programmable spreading sequence. A key advantage of CMOS integration is that digital signal processing can be used to assist the analog circuits. As a result, the low-cost and low-power UWB impulse front-end is feasible due to robust impulse detection via sophisticated signal processing.

OVERLAY APPROACH: COGNITIVE RADIOS

After enabling UWB transmission and understanding its limitations, regulatory domains also realized the need for new technologies to use other available spectral resources more efficiently. Recent studies by the FCC Spectrum Policy Task Force have reported vast temporal and geographic variations in the use of allocated spectrum [2]. To utilize these white spaces, the FCC has issued a Notice

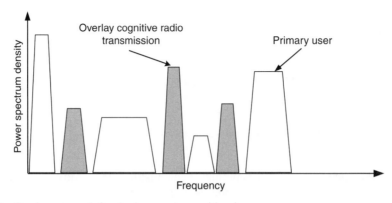

FIGURE 5.6 Overlay approach for sharing spectrum with primary users.

of Proposed Rule Making [9] advancing CR technology as a candidate to implement negotiated or opportunistic spectrum sharing. Opposite to UWB, cognitive approach does not necessarily limit the transmission power, but rather attempts to share the spectra through a dynamic avoidance strategy and would use higher transmit power in white spaces to maximize capacity. Figure 5.6 illustrates the transmitter power spectrum density profiles in CR overlay spectrum-sharing approach. Note that CRs are allowed to transmit increased transmit power levels but must not cause cochannel and adjacent-channel interference to primary users in the vicinity.

Unfortunately, there has not been a clear definition of what actually constitutes a CR. A relatively conservative definition would be that a CR is a network of radios that coexists with higher priority primary users, by sensing their presence and modifying its own transmission characteristics in such a way that they do not yield any harmful interference. It is this sensing function and ability to rapidly modify their transmitted waveform that is the unique characteristic and challenge of CR implementation. More generically speaking, a CR is an environment-aware radio, that is, it has awareness of the RF channel, location, user profiles, etc. to which it is capable of adapting. However, we shall constrain our attention to the spectrum awareness and adaptation aspects (spectrum agility) that impact the physical layer of a CR.

Although at present, there are no CR networks in commercial deployment, the basic ideas have already been demonstrated. In the middle of the 5 GHz band of 802.11a, there are two approximately 200 MHz-wide bands (5150–5350 MHz, 5470–5725 MHz, or both) that are shared with sensitive aeronautical navigation radars. Although these radars have limited geographic distribution, it was deemed necessary to protect them from 802.11a transmissions. A technique called dynamic frequency selection (DFS) is employed, which involves sensing if a radar signal is present and then avoiding those frequency channels to prevent interference. Although the relatively simplistic strategy of abandoning the impacted 802.11a frequency channels is far from optimal in terms of exploitation of the spectrum, it does indicate that the regulators are willing to accept that spectrum can be shared through sensing and avoidance.

There is also evidence that regulators are willing to go even further and actually experiment with CR operation by allowing shared use in the digital television (DTV) bands from 400 to 800 MHz [10]. This is hotly contested now by the incumbent users (TV broadcasters), but it does seem clear that some limited form of CR operation may be allowed. The primary application being promoted is to use the long-range capability of the TV carrier frequencies to provide Internet access in rural areas. Furthermore, given the static TV-channel allocations, the timing requirements for spectrum sensing are very relaxed. Another natural band to allow CR operation is the same 3–10 GHz band that is already in use by UWB radios. This band is sufficiently wide that the true advantages of CR operation could be demonstrated and the simultaneous availability of two approaches to sharing could actually be mutually supportive. It is important to realize that CR approach to spectrum sharing is a fundamentally new paradigm and, if regulations allow it, could be virtually applied to a large number of currently deployed and future wireless systems.

COGNITIVE RADIO SYSTEMS ARCHITECTURE

CRs are considered lower priority or secondary users of spectrum allocated to a primary user, but they have the ability to greatly exploit unused spectrum: through sensing and adaptation. Their fundamental requirement is to avoid interference to potential primary users in their vicinity. Spectrum sensing has been identified as a key enabling functionality to ensure that CRs would not interfere with primary users, by reliably detecting primary user signals. In addition, reliable sensing creates spectrum opportunities for capacity increase of cognitive networks. The first application of spectrum sensing is studied under IEEE 802.22 standard group [11] to enable secondary use of ultra high frequency (UHF) (400–800 MHz) spectrum for a fixed wireless access. In addition, there are a number of indoor applications in which spectrum sensing would increase spectrum efficiency and utilization. For example, it could improve coexistence of wireless local area networks (WLANs) (802.11) or create new spectrum opportunities for sensor and *ad hoc* networks. Regardless of application, sensing requirements are based on primary user modulation type, power, frequency, and temporal parameters. For example, the actual primary signal used for sensing could be the regular data transmission signal. Alternatively, it could be a special permission or denial signal to use the spectrum, in the form of a pilot or a beacon.

Spectrum sensing is often considered as a detection problem, which has been extensively researched since early days of radar. Since primary user signal decays with distance, spectrum sensing for faraway radios becomes increasingly difficult. Therefore, the key challenge of spectrum sensing is the detection of weak signals in noise with a very small probability of missing detection, which requires better understanding of very low SNR regimes [12]. This requirement must be applied over all available degrees of freedom (time, frequency, and space) to identify frequency bands currently available for transmission. In addition, spectrum sensing is a cross-layer design problem in the context of communication networks. If benefits of higher power transmission are to be exploited together with minimum interference, then reliable sensing becomes the most critical functionality of CRs.

CR spectrum-sensing requirements are set by the primary user systems and are dependent on the quality of primary receivers and their type of service. For example, current requirements set by TV broadcasters allow maximum of 2 s of interference time. Minimum signal level of −116 dBm with reliability of 90% detection and 10% false alarm must be met, which translates into 32 dB better sensitivity than DTV receivers. Effective time spent for sensing should be minimized while maintaining reliability, which translates into a dynamic and periodic operation. Spectrum sensing is best addressed as a cross-layer design problem. CR sensitivity can be improved by enhancing radio RF front-end sensitivity, exploiting digital signal processing (DSP) gain for specific primary user signal, and ensuring network cooperation where users share their spectrum-sensing measurements. On a radio design level, the implementation of the spectrum-sensing function also requires a high degree of flexibility, since the radio environment is highly variable because of different types of primary user systems, propagation losses, and interference.

Because of its vital functionality for cognitive operation, sensing receiver is, in essence, a separate receiver within a CR, as depicted in Figure 5.7. Therefore, it requires an independent antenna, and constantly active radio front-end, sampling the band of interest and processing the signals to find white spaces. Sensing receivers should have enhanced sensitivity to detect weak signal but at the same time must filter out the strong blocking signals that can reduce the sensitivity.

After reliable detection, a CR should use transmission schemes that provide the best spectrum utilization and capacity. There are several unique requirements that a modulation scheme should provide. First, spectrum bands available for transmission could be spread over a wide frequency range, with variable bandwidths and band separations. The unoccupied spectrum distribution is a function of geographic location and time of use, and it is updated after every spectrum-sensing period. Second, for optimal spectrum and power efficiency, every CR estimates the quality of unoccupied frequencies to provide higher layers with signal-to-noise measurements to be used for power and bit allocation. Last, different applications might require selection of frequency bands based on propagation characteristics or interference measurements. These requirements translate into dynamic frequency allocation and frequency-agile modulation schemes.

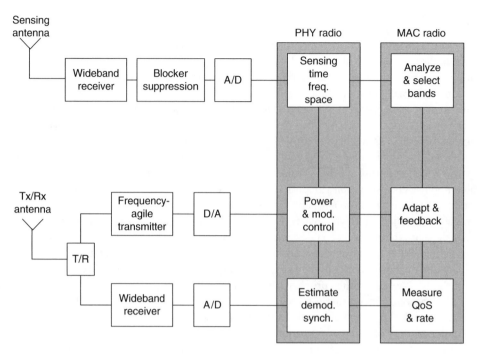

FIGURE 5.7 Cognitive radio transceiver block diagram.

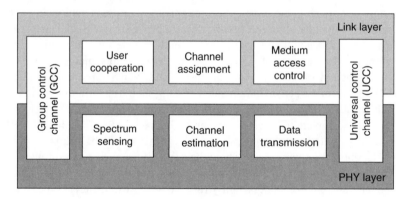

FIGURE 5.8 Cognitive radio physical and link layer functionalities.

CR networks could be either centrally organized (base station/access point mode) or distributed (*ad hoc* mode), but in any case radio cooperation within the network is required to jointly perform reliable sensing and coordinate communication. Besides frequency-agile opportunistic channel use for data transmission, cognitive networks need dedicated channels for the exchange of control and sensing information to accommodate dynamic frequency band operation. Two different kinds of control channels [13], a universal control channel (UCC) and group control channels (GCCs), need to be maintained (Figure 5.8). The UCC is globally unique, known to every CR *a priori*, whose main purpose is to allow coexistence of multiple cognitive networks. The need for the UCC arises from the requirement that while sensing the primary user band, no additional noise or interference should be added by nearby cognitive users to achieve good sensitivity measurement. In addition, mutual interference and bandwidth allocation should be coordinated through agreements on sharing strategies (e.g., time-division multiple access, or carrier sense mechanisms) among different groups. As a result, the UCC channel must cover larger distances, but its throughput requirement is fairly low. On the contrary, the

GCC is set up to exchange sensing information, perform channel allocation, and link maintenance. It is local within one group and thus has a shorter range but higher throughput.

Note that the UCC and the GCCs are logical concepts, which might even be mapped to a single physical channel. An implementation using UWB transmission is especially attractive if we are considering use of the 3–10 GHz band. UWB control channels would be unlicensed with low impact on other types of communication and with the possibility to operate independently using different spreading codes. There are severe limitations on the power of UWB emissions limiting its range, but the control channel requires very low data rates; so spreading gain will increase the range to be adequate for most applications (more than 10,000 times lower data rate than the commercial UWB systems being envisaged in this band). This example shows that UWB and CR can be complementary technologies.

WIDEBAND SPECTRUM-SENSING RADIO ARCHITECTURE

Frequency Agility Regimes

In conventional radios, receiver sensitivity is a quantitative measure that specifies the minimum signal level that can be detected for targeted modulation scheme and BER. However, the CR front-end is used to sense wide range of signals and modulations; thus such sensitivity metric is impossible to quantify. Spectrum sensing requires detection of weak signals whose power levels are often so low that they fall below the receiver noise floor. Therefore, it is a paramount requirement to condition a received signal from the CR antenna to detector circuits with minimal signal-to-noise degradation and signal distortion. The main design challenge is to define RF and analog architecture with right trade-offs between linearity, sampling rate, accuracy, and power so that DSP techniques can be utilized for spectrum sensing.

Different spectrum utilization regimes require different radio architecture designs. In essence, the amount of spectrum opportunities for secondary spectrum use dictates how much frequency agility the cognitive wideband radio should have. What is important to realize is that cognitive systems need to be designed with forward compatibility; thus the future of CR technology development must be looked at further. Three different regimes of spectrum scarcity, are distinguished here and also illustrated in Figure 5.9:

- *No-spectrum-scarcity regime*: In case bandwidth utilization by primary systems is low and there are limited temporal and spatial variations in spectrum use by the primary system, there is no spectrum scarcity for the CR network.
- *Medium-spectrum-scarcity regime*: Over a period of time, the progression of CR technology will increase the spectrum utilization. The temporal and spatial variations will also increase and the number of unused bands will become comparable to the number of cognitive networks sharing the same spectrum pool.
- *Significant-spectrum-scarcity regime*: Eventually, there will be multiple cognitive networks competing for the same spectrum so that spectrum will become truly scarce resource. In addition, there could be significant temporal and spatial variations due to dynamic allocation.

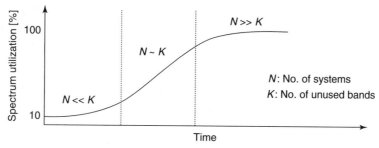

FIGURE 5.9   Three regimes of spectrum utilization enabled by cognitive radios.

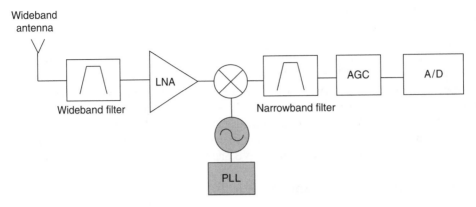

FIGURE 5.10   Spectrum-sensing front-end for sequential band processing.

No-Spectrum-Scarcity Regime

This regime corresponds to an early stage of a CR network deployment, which is a situation today with the abundance of spectrum available in some frequency bands. An example of such regime would be the VHF-UHF bands, for which current IEEE 802.22 standard is proposed. It has been shown that there are approximately 5–20 (6 MHz-wide) unoccupied TV channels available for cognitive operation in this band, especially in the rural areas. Because of continuous transmission and static geographic TV transmitter allocations, these bands show no temporal variation of primary user signals. Under these conditions, spectrum sensing can be done sequentially, on a moderate time scale (from 100 ms to 2 s), by searching one frequency band at the time through frequency sweeping using a tunable local oscillator (LO), as illustrated in Figure 5.10. The main challenge for this architecture is to design a wideband voltage-controlled oscillator (VCO) with the tuning range over a band of interest (in the order of 1 GHz). A phase-locked loop (PLL) controlling the VCO should guarantee short settling time and small phase noise. The biggest advantage of this architecture is that the baseband portion of the radio is still narrowband. As a result, a fixed narrowband–baseband filter for channel selection and low-resolution, low-speed A/D converter can be optimized for reduced cost and power.

CR antennas should provide wideband reception with omni-directional pattern and minimal signal loss. Conveniently, great progress has been made in the design of UWB antennas in 0–1 and 3–10 GHz bands and can be used for CRs as well. In contrast with low noise amplifiers (LNA) for UWB systems, where noise figure (NF) was not an issue and was traded for power dissipation, CRs require small NF in the order of 2–3 dB together with low power (~10 mW). The critical design specification for mixers and amplifiers is to maintain the linearity across large dynamic range and frequency range. This is also conflicting requirement with respect to power consumption. Potentially, the linearity requirements of the mixer could be relaxed by employing programmable gain in the LNA.

Medium-Spectrum-Scarcity Regime

Once the number of unoccupied bands and that of systems competing for them become comparable, CRs cannot afford sequential sensing because the probability of sensing the occupied band is increased. In addition, some bands (for example, public safety or radar bands) are very infrequently used, but still CRs must detect primary user reappearance and free the band. Therefore, the front-end circuitry should support simultaneous parallel sensing over several frequency bands, as illustrated in Figure 5.11. Inevitably, the number of components increases. For most practical solutions, the number of parallel paths is bounded to four or five. Even though there are multiple LOs, they are operating at fixed frequencies and therefore have relaxed requirements. To justify parallelization, the frequency planning should be sufficiently wide. As a result, the baseband portion of parallel radio channels has increased bandwidth and would require A/D converters with higher speed (100–300 MHz) and resolution (5–8 bits) than in the case of large spectrum opportunities.

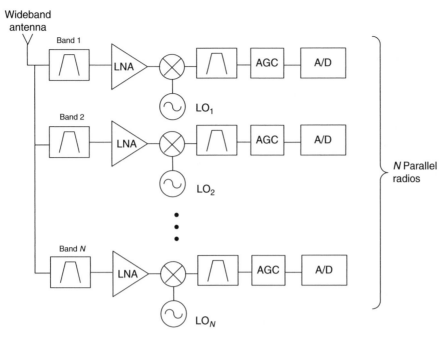

FIGURE 5.11 Spectrum-sensing front-end for parallel band processing.

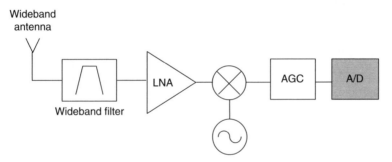

FIGURE 5.12 Spectrum-sensing front-end for wideband processing.

Significant-Spectrum-Scarcity Regime

If the radio spectrum environment becomes highly variable, because of different types of primary user systems, propagation losses, and competing cognitive networks, then the implementation of the spectrum-sensing function requires the highest degree of flexibility. Figure 5.12 depicts an architecture of a wideband RF front-end capable of simultaneous sensing of several-GHz-wide spectrum. This architecture is commonly proposed for software-defined radios except for significantly narrower bandwidths. For CRs, all front-end circuitry should be wideband, which further imposes high-speed sampling requirement for A/D converter. In addition, the RF signal presented at the antenna of such a front-end includes signals from close and widely separated transmitters and from transmitters operating at widely different power levels and channel bandwidths. As a result, the large dynamic range becomes the main challenge, since it requires high-linearity circuits and high-resolution (12 bits or more) A/D converters. The scaling of CMOS technology improves the transistor operating frequency, which could be used for increased sampling rate, but adversely affects the design of precision analog circuitry. With the reduction in supply voltage, which scales with the transistor features, the inherent transistor gain decreases, the voltage headroom decreases, and device noise and leakage increase, making it difficult to design high-resolution A/D converters [14]. Therefore, reducing the strong in-band primary user signals is necessary to receive and process weak signals.

Dynamic Range Reduction Techniques

A key advantage of CMOS integration is that DSP can be used to assist the analog circuits. This motivates research of mixed signal techniques that can relax challenging requirements for analog, specifically wide-band amplification and mixing and A/D conversion of over a GHz or more of bandwidth and enhance overall radio sensitivity. The main challenge in the wideband-sensing receiver design is to achieve good sensitivity in the presence of strong blocking signals requiring large dynamic range. However, measurements [15] show that there are typically a few strong in-band narrowband primary signals. Furthermore, these strong signals are of no interest to detect, since the spectrum sensing focuses on processing of weak signals. We explore three different approaches for reducing the dynamic range before sampling circuits by recognizing that strong signals can be resolved and removed in frequency, time, or space.

Frequency Domain Filtering

In conventional radios, fixed filters provide frequency discrimination, precondition input signal, and relax the requirements on the radio circuit components. However, CRs encounter scenarios where, at every location and time, different strong primaries fall in-band. Therefore, CRs require bandpass and bandstop filters with challenging specifications: high center frequency, narrowband with large out-of-band rejection, and tuning ability. Tunable bandpass filters are used for channel selection and reduction of out-of-band interference, whereas tunable bandstop filters are needed for isolation of transmit and receive paths and in-band blocker rejection.

Filters are external components and not favorable for integrated single-chip radios. In general, CMOS implementation and integration lead to low cost and power, so the number of external filters should be minimized. Furthermore, sharp roll-off RF filters need high Q, which leads to high power consumption and large circuit area to accommodate passive elements, inductors and capacitors. Nonideal filters cause leakage of the strong signal across bands and can degrade weak-signal-sensing performance.

Tunable RF filters require exploration of nonconventional filter architectures. Novel techniques for filtering like radio frequency-microelectro-mechanical systems (RF-MEMS) show some promise. However, MEMS filters suffer from insertion loss and are also hard to design for high frequencies. In addition, they require a long time to tune to the desired band. The majority of lumped-element MEMS-tuned filter designs uses fixed inductors and achieves tuning via adjustable MEMS capacitors, so their frequency is in general limited to 3 GHz [16]. Tuning bandwidth and center frequency is possible via switching distributed resonators, but again resolution is limited by the placement density and electrical sizes of resonators, since the design is discrete in nature. To improve resolution, the number of resonator elements must increase. It is required that on-going research efforts in RF-MEMS filters address these requirements and provide more filtering flexibility.

Time Domain Cancellation

A cancellation in time domain relies on the great result from a multiuser-detection theory [17] applicable to an interference channel. This result establishes that it is possible to decode a strong interfering signal first and subtract it so that weak signals can be decoded too. This result can be applied to a wideband channel that has a few strong primary narrowband signals. It is important to note that it is sufficient to attenuate signal, not perfectly cancel; thus estimation error can be tolerated. Figure 5.13 illustrates an RF front-end architecture with digitally assisted active cancellation. Active cancellation is achieved through the use of an adaptive linear prediction filter and reconstruction digital-to-analog (D/A) converter in a feedback loop [18].

There are several challenges in this feedback approach. First, the adaptive filter used to regenerate the interference has a time-varying input signal, and its estimation error due to noise, quantization, and prediction errors limits the performance of the interference cancellation. Second, since this is a closed-loop structure, it is difficult to guarantee its stability. Last, the key challenge in this approach is to perform analog subtraction in a closed loop with stringent timing constraints.

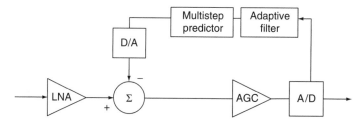

FIGURE 5.13   Feedback architecture for time domain digitally assisted analog interference cancellation.

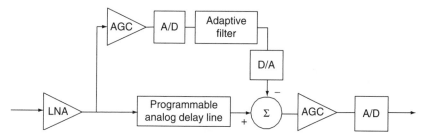

FIGURE 5.14   Feed-forward architecture for time domain digitally assisted analog interference cancellation system.

To overcome the limitations of the feedback architecture, alternative feed-forward architecture is proposed (Figure 5.14). The main idea for this approach is to use two low-resolution A/D converters with N and M bits to achieve a high-resolution A/D converter of $N + M$ bits. For example, under strong interference conditions with up to 50 dB dynamic range increase with respect to desired weak signals, it is possible to realize an equivalent 12-bit A/D converter using two 6-bit converters. This architecture effectively behaves as a two-stage pipelined A/D converter, where the first stage deploys sophisticated signal processing to provide reduced dynamic range to the second stage. The incoming signal goes through two paths; one branch deals with interference estimation and reconstruction, whereas the other is a programmable analog delay line used to align the signals in two branches so that proper cancellation timing is achieved. The first-stage A/D converter is the cancellation A/D and the A/D after subtraction is the residue A/D converter. In general, the trade-off between the number of bits in the cancellation A/D converter and residue A/D converter depends on the interference strength, that is, strong interference situations require more bits in cancellation A/D converter. Although the timing constraints of the feedback loop are avoided by this architecture, it still requires matching of the latency through the two paths using an analog delay line. Despite the fact that the active cancellation approach will consume significantly more hardware, it has the important advantage of ultimately being more flexible.

Spatial Filtering
An alternative approach for dynamic range reduction would be to filter the received signals in the spatial domain by using multiple antennas. This idea is inspired by recent theoretical work on multiple-antenna channels identifying that spatially received signals occupy a limited number of directions or spatial clusters [19]. Through beamforming techniques, signals can be selectively received or suppressed using antenna arrays. However, multiple-antenna processing must be done in the analog domain before the automatic gain control circuits used to properly amplify reduced dynamic range signal for the best utilization of the number of bits in the A/D converter. The architecture of a wideband RF front-end that is enhanced with an antenna array for spatial filtering is shown in Figure 5.15. This architecture could be implemented as a phased antenna array in which the antenna-array coefficients are computed in the digital backend and fed back to analog phase shifters, which then adjust the gains and

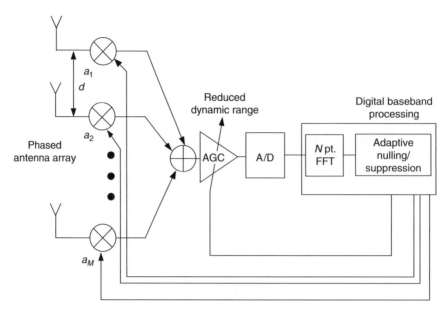

FIGURE 5.15   Wideband RF front-end with antenna array for spatial filtering.

phases of the antenna elements. The use of simple phase shifters is particularly attractive due to their very low latency needed for fast convergence of the desired array response.

A simple algorithm for the computation of optimal array coefficients could be derived by noticing that strong primary users occupy distinct frequency bands and have different directions of arrival [20]. By applying an FFT on a wideband signal at the output of the A/D, a power profile in frequency domain is measured. To obtain the estimate of angle of arrivals, the antenna-array coefficients must sweep through sufficient number of directions. Given M antenna elements, any set of $K > M$ independent array coefficients is sufficient to obtain the estimation of spatial distribution. After identifying the frequency-spatial location of M strongest primary users through least square estimation on these K measurements, array coefficients are set to attenuate their directions of arrivals. Effectively, optimal coefficients spatially equalize the received energy from all angles and frequencies. Figure 5.16 shows the outlined algorithm performance for the case of two strong primary users whose power is 30–40 dB larger than average power in other frequency bands. After the optimal coefficients are applied, dynamic range reduces by approximately 22 dB (saving 3–4 bits in A/D converter) using a four-element antenna array. This approach is a novel application of multiple antennas at the receiver front-end that uses spatial filtering techniques to relax requirements for the implementation of wideband sensing. In addition, this architecture could improve the sensing of weak signals through the received beamforming gain.

Signal Processing for Spectrum Sensing

In case of spectrum sensing, the need for signal processing is twofold: improvement of radio front-end sensitivity by processing gain and primary user identification based on knowledge of the signal characteristics. In this section, advantages and disadvantages of three candidate approaches for signal detection are discussed: matched filter, energy detector, and cyclostationary feature detector.

Optimal Coherent Detection: Matched Filter

Spectrum sensing addresses SNR regimes that are often much below the ones encountered in UWB channels. Similar to impulse detection, the optimal way for any signal detection is a matched filter (Figure 5.17), since it maximizes received SNR. However, a matched filter effectively requires demodulation of a primary user signal. This means that the CR has *a priori* knowledge of primary user signal at

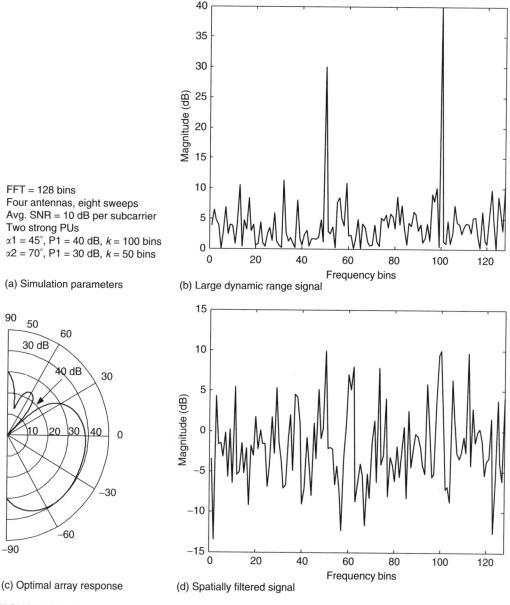

FFT = 128 bins
Four antennas, eight sweeps
Avg. SNR = 10 dB per subcarrier
Two strong PUs
$\alpha1 = 45°$, P1 = 40 dB, k = 100 bins
$\alpha2 = 70°$, P1 = 30 dB, k = 50 bins

(a) Simulation parameters

(b) Large dynamic range signal

(c) Optimal array response

(d) Spatially filtered signal

FIGURE 5.16 An example of dynamic range reduction using antenna arrays.

both physical layer (PHY) and MAC layers, e.g., modulation type and order, pulse shaping, packet format, etc. Such information might be prestored in a memory, but the cumbersome part is that for demodulation the CR has to achieve coherency with primary user signal by performing timing and carrier synchronization, even channel equalization. This is still possible, since most primary users have pilots, preambles, synchronization words, or spreading codes that can be used for coherent detection. For example, a TV signal has narrowband pilot for audio and video carriers; code division multiple access (CDMA) systems have dedicated spreading codes for pilot and synchronization channels; OFDM packets have preambles for packet acquisition. The main advantage of matched filter is that due to coherency, it requires minimal time to achieve high processing gain, since only $T \sim O(1/SNR)$ samples are needed to meet a given probability of detection constraint. However, a significant drawback of a matched filter is that a CR would need a dedicated receiver for every primary user class.

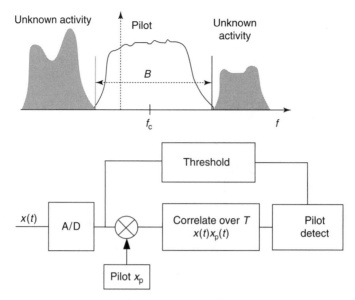

FIGURE 5.17   Spectrum sensing via matched filter (correlator).

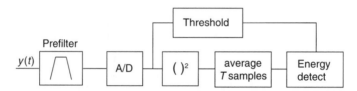

FIGURE 5.18   Energy detector implementation using analog prefilter and square-law device.

Suboptimal Noncoherent Detection: Radiometer

Again, one approach to simplify matched filtering approach is to perform noncoherent processing through energy detection. This suboptimal technique has been extensively used in radiometry and in principle can be applied to an arbitrary signal. In the context of spectrum sensing, it can serve as a practical solution in the situation when there are no known primary user signals available, or if the CR cannot establish synchronization to primary user signal due to very low SNR. A conventional implementation, presented in Figure 5.18, involves analog filtering, A/D converter, squaring circuit, and an integrator. Sensing performance is controlled by setting the averaging time T. One big constraint in this design is the fixed bandwidth of the analog filter, which if not narrow enough could allow more noise into the system than necessary. Note that over-sampling would correlate noise samples; thus, detection could be always reduced to Nyquist sampling.

The analog filtering implementation is quite inflexible, particularly in the case of narrowband signals and sine waves. An alternative approach could be devised by using a periodogram to estimate the spectrum via squared magnitude of the FFT, as depicted in Figure 5.19. This architecture also provides the flexibility to process wider bandwidths and sense multiple signals simultaneously. As a consequence, an arbitrary bandwidth of the modulated signal could be processed by selecting corresponding frequency bins in the periodogram. Furthermore, digital implementation provides more flexibility and additional postprocessing like windowing and filtering. Processing gain is proportional to FFT size N and observation/averaging time T. Increasing N improves frequency resolution, which helps narrowband signal detection. Also, longer averaging time reduces the noise power and thus

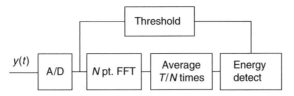

FIGURE 5.19 Energy detector implementation using periodogram FFT and averaging.

improves SNR. However, due to noncoherent processing, $T \sim O(1/\text{SNR}^2)$ samples are required to meet a given probability of detection constraint.

Unfortunately, an increased sensing time is not the only disadvantage of the energy detector. More important, there is a minimum input signal level below which detection is not possible. The main reason is that noise is an aggregation of various sources including not only thermal noise at the receiver and underlined circuits but also interference due to nearby unintended emissions, weak signals from transmitters far away, etc. A noise variance is not precisely known to the receiver; therefore, the threshold cannot be set optimally. How does the noise uncertainty affect detection of signals in low SNR? Essentially, setting the threshold too high on the basis of wrong noise variance would never allow the signal to be detected. If there is an x-dB noise uncertainty, then the detection is impossible below $\text{SNR}_{\text{wall}} = 10 \log_{10}[10^{(x/10)} - 1]\,\text{dB}$ [21]. For example, if there is a 0.5 dB uncertainty in the noise variance, then signal in -21 dB SNR cannot be detected using energy detector.

Detection of sine wave and 4-MHz-wide QPSK-modulated signals in noise were experimentally tested for signal levels down to -128 and -110 dBm, respectively. In the case of sine wave sensing, the scaling law $T \sim 1/\text{SNR}^{1.5}$ due to partial coherent processing gain of the FFT implementation (Figure 5.20) was observed. However, beyond -124 dBm level, the slope becomes increasingly steep and signals below -128 dBm cannot be detected. In the case of quadrature phase shift keying (QPSK) detection, we observe a noncoherent detection scaling law of $T \sim 1/\text{SNR}^2$, consistent with the theoretical prediction. The limit in QPSK detection happens at -110 dBm, which is significantly inferior to pilot detection.

In frequency-selective fading channels, sine wave detection becomes inferior due to multipath notches. Modulated signals with wider bands exploit frequency diversity and improve the detection. Therefore, robust solutions to sensing should employ both pilot and signal energy detectors. However, in some applications, energy detector would not perform well enough. Besides the minimum detectable levels set by uncertainty, energy detector does not differentiate between modulated signals, noise, and interference. Since it cannot recognize the interference, it cannot benefit from adaptive signal processing for canceling the interferer. Furthermore, spectrum policy for using the band is constrained only to primary users; so a cognitive user should treat noise and other secondary users differently. Therefore, more sophisticated signal-processing algorithms are needed for robust spectrum sensing.

Feature Detection

A natural question arises: Are there any detectors that perform somewhere between the optimal matched filter and the suboptimal energy detector? Some intuition can be gained by understanding what information is thrown away by an energy detector. In essence, the energy detector threats noise and signal in the same way as wide-sense stationary white processes. However, modulated signals are in general coupled with sine wave carriers, pulse trains, repeating spreading, hopping sequences, or cyclic prefixes, which result in built-in periodicity. Even though the data is a wide-sense stationary random process, these modulated signals are characterized as cyclostationary, since their statistics, mean, and autocorrelation exhibit periodicity. This periodicity is typically introduced intentionally in the signal format so that a receiver can exploit it for parameter estimation such as carrier phase, pulse timing, or direction of arrival. This information can then be used for detection of a random signal with a particular modulation type in a background of noise and other modulated signals [22].

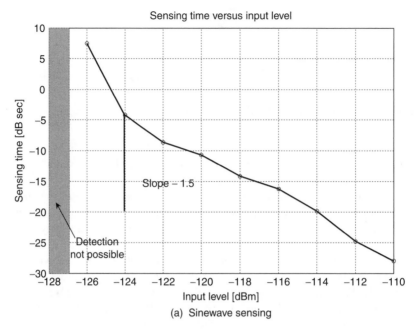

(a) Sinewave sensing

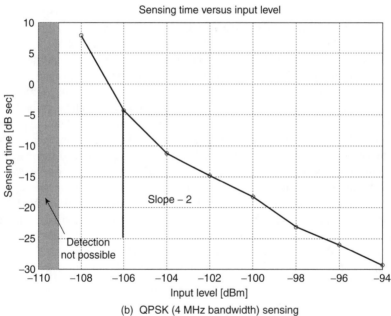

(b) QPSK (4 MHz bandwidth) sensing

FIGURE 5.20   Required sensing time versus signal input level for fixed P_d and P_{fa}.

Common analysis of wide-sense stationary random signals is based on autocorrelation function and power spectral density. On the contrary, cyclostationary signals exhibit correlation between widely separated spectral components due to spectral redundancy caused by periodicity. By analogy with the definition of conventional autocorrelation, one can define spectral correlation function (SCF):

$$S_x^\alpha(f) = \lim_{T \to \infty} \lim_{\Delta t \to \infty} \frac{1}{\Delta t} \int_{-\Delta t/2}^{-\Delta t/2} \frac{1}{T} X_T\left(t, f + \frac{\alpha}{2}\right) X_T^*\left(t, f - \frac{\alpha}{2}\right) dt$$

where finite-time Fourier transform is given by

$$X_T(t,v) = \int_{t-T/2}^{t+T/2} x(u)e^{-j2\pi vu}\,du$$

SCF is also termed as *cyclic spectrum*. Unlike power spectrum density, which is real-valued one-dimensional transform, the SCF is two-dimensional transform, in general complex valued, and the parameter α is called cycle frequency. Power spectral density is a special case of an SCF for $\alpha = 0$.

The distinctive character of spectral redundancy makes signal selectivity possible. Signal analysis in cyclic spectrum domain preserves phase and frequency information related to timing parameters in modulated signals. As a result, overlapping features in the power spectrum density are the nonoverlapping feature in the cyclic spectrum. Different types of modulated signals (such as BPSK, QPSK, and staggered-quadrature phase-shift keying (SQPSK)) that have identical power spectral density functions can have highly distinct SCFs. Furthermore, stationary noise and interference exhibit no spectral correlation.

The sufficient statistics used for the detection are obtained through nonlinear squaring operation; therefore, cyclostationary detectors also fall in the category of noncoherent detectors. Given N samples divided in blocks of T samples, SCF is estimated as

$$\tilde{S}_x^\alpha(f) = \frac{1}{N}\frac{1}{T}\sum_{n=0}^{N} X_T\left(n, f + \frac{\alpha}{2}\right) X_T^*\left(n, f - \frac{\alpha}{2}\right)$$

$X_T(n,f)$ is the T point FFT around sample n. To detect feature at cycle frequency α, the optimal processing is the projection of the estimated SCF onto the ideal SCF of a known primary user signal.

$$z(N) = \int_{-f_s/2}^{f_s/2} S_x^\alpha(f)^* \tilde{S}_x^\alpha(f)\,df$$

Because of this coherent projection in a domain of SCF, cyclostationary detectors have a performance gain over completely noncoherent energy detectors. The gain is proportional to the amount of spectral redundancy of underlying modulation scheme. For example, slow decay of pulse-shaping filters or double-sided modulations have larger amount of spectral redundancy.

The main advantage of feature detectors is that detection is not susceptible to variation in the noise power and accuracy of the noise estimate. Figure 5.21 illustrates the advantages of cyclostationary detection versus energy detection for continuous-phase 4-frequency shift keying (FSK)-modulated signals. In case of high SNR, both power spectral density and SCF allow reliable detection. In addition, SCF allows modulation recognition based on the pattern of cycle frequencies. For low SNR $= -20$ dB, distinct pattern of 4-FSK modulation in an SCF is preserved. Since cyclostationary feature detector operates in the region outside of $\alpha = 0$ where energy detector is limited by the large noise, its detection performance is more robust.

Cyclostationary detectors involve computation of a two-dimensional SCF on a bi-frequency plane. A detection of the specific feature requires sufficient over-sampling and minimum resolution both in frequency and in cycle domain. These requirements are more demanding for cycle domain because the signal has to be over-sampled such that Nyquist rate for maximum resolvable cycle frequencies is satisfied. For example, if the desired feature is a sine wave carrier of the modulated signal, then the sampling frequency needs to be at least twice the carrier. This requirement is met if the CR performs wideband sensing using high-speed A/D converters. The resolution in frequency and cycle domain is improved by observing the signal over a long period of time. Since the cycle resolution depends on signal periodicities, in case of long symbol times, total observation increases overall sensing time.

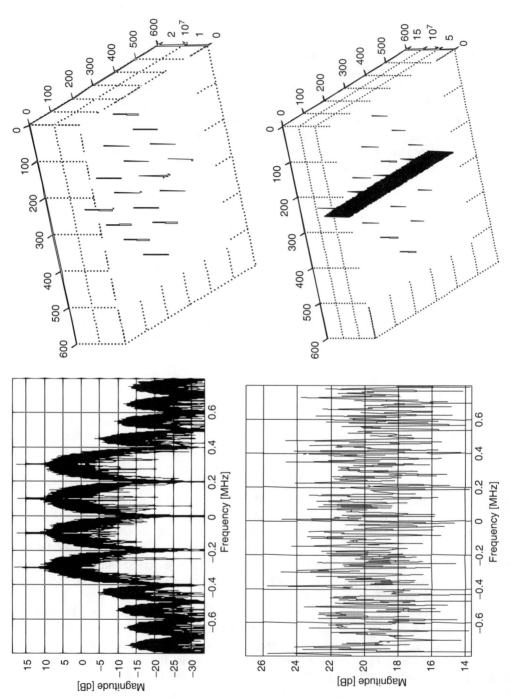

FIGURE 5.21   Detection of a continuous-phase 4-FSK using energy detection and feature detection.

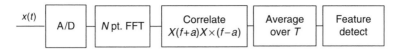

FIGURE 5.22 Implementation of a cyclostationary feature detector.

Digital frequency smoothing algorithms are the most attractive implementation solution. Direct algorithms first compute the spectral components of the data through FFT and then perform the spectral correlation directly on the spectral components (Figure 5.22). Estimates can be improved by applying windows for smoothing at the cost of additional processing. The computational complexity of an SFC estimator is easily estimated. For a stream of N samples, it requires a computation of N point FFT, which requires $N \log N$ multiplications, and N^2 multiplication for cross-multiplications. Note that this algorithm is extremely parallel so that the computation of the SCF can be organized across frequency or across cycle plane independently. Also, this method could be employed to compute SCF for smaller parts of bi-frequency plane and save the area and power.

However, some RF and analog circuit implementations might introduce impairments that could potentially hurt cyclostationary detectors. Because of the periodically time-varying nature of some circuit blocks, such as mixers, the noise that is generated and processed by the system has periodically time-varying statistics. There are two reasons why the mixer output noise has periodically time-varying statistics. First, the operating point of the device may vary with time, and second, the transfer function of the noise signal from the point at which it is generated to the output can have time-varying characteristics. Cycle frequencies will appear whenever there is a correlation between two different frequency components. There are two cases where the spectral correlation could be significant. If there is a strong interfering signal or a blocker at the input of a receiver, it can change the operating point of the devices and affect circuit noise performance. The noise generated by the circuit will acquire the cyclostationary characteristics with cycle equal to the blocker period. In case the blocker is not filtered or modulated to a different frequency, it acts as a common LO for successive blocks. The LO is periodically time–varying, which can introduce cyclostationary components into output noise of the mixer [15].

FREQUENCY-AGILE WIDEBAND TRANSMISSION

In opportunistic spectrum access, bands available for transmission could be spread over a wide frequency range with variable bandwidths and band separations. Besides the frequency agility, there is an implicit power constraint that requires minimum interference to primary users operating in the same frequency band and in the adjacent bands. In contrast with standard MAC mechanisms that are directed from the higher layers, here there is a necessary closed-loop interaction between PHY and MAC layers, as outlined in Figure 5.7.

To provide protection to primary user in adjacent bands, it is necessary to understand the performance of primary system receivers. Because of the inevitable need for frequency reuse, every primary receiver is designed to tolerate a limited amount of cochannel as well as adjacent-channel interference. There are so-called desired-to-undesired (D/U) ratios or reference interference ratios for cochannel and adjacent-channel interferers, usually specified for the reference sensitivity levels. For example, to reuse a vacant TV channel, a CR must satisfy transmission mask depicted in Figure 5.23. To control the interference, the relative power measurements are obtained from the spectrum sensing and used to create transmit power mask for the CR transmitter including additional margin.

The modulation scheme based on OFDM is a natural approach to frequency agility. OFDM has become the modulation of choice in many broadband systems due to its inherent multiple-access mechanism and simplicity in channel equalization, plus benefits of frequency diversity and coding. The transmitted OFDM waveform is generated by applying an inverse FFT (IFFT) on a vector of data, where the number of points N determines the number of subcarriers for independent channel use, and minimum resolution channel bandwidth is determined by W/N, where W is the entire frequency band

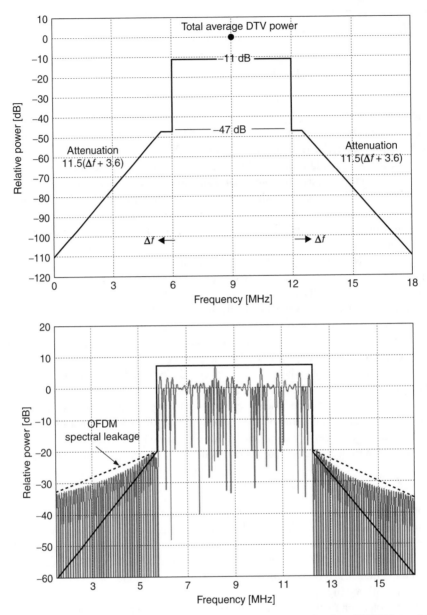

FIGURE 5.23 DTV transmitter spectrum mask (top) and power spectrum density of OFDM signal (bottom).

accessible by any cognitive user. The frequency domain characteristics of the transmitted signal are determined by the assignment of nonzero data to IFFT inputs corresponding to subcarriers to be used by a particular cognitive user. Similarly, the assignment of zeros corresponds to channels not permitted to use due to primary user presence or channels used by other cognitive users. The output of the IFFT processor contains N samples that are passed through a D/A converter producing the wideband waveform of bandwidth W. A great advantage of this approach is that the entire wideband signal generation is performed in the digital domain, instead of multiple filters and synthesizers required for the signal processing in analog domain.

From the cognitive network perspective, OFDM spectrum access is scalable while keeping users orthogonal and noninterfering provided the synchronized channel access. However, this conventional

OFDM scheme does not provide truly band-limited signals due to spectral leakage caused by sinc-pulse-shaped transmission that results from the IFFT operation. The slow decay of the sinc-pulse waveform, with first side lobe attenuated by only 13.6 dB, produces interference to the adjacent-band primary users, which is proportional to the power allocated to the cognitive user on the corresponding adjacent subcarrier (Figure 5.23). Therefore, a conventional OFDM access scheme is not an acceptable candidate for wideband CR transmission.

CRs operate under the assumption that primary users define their protection requirements and allowable interference margin. The protection is guaranteed with a certain probability of harmful interference. It is then the responsibility of a CR to dynamically manage the interference through spectrum sensing. There is an explicit trade-off between the radio sensitivity and allowable transmission power [23]. If a radio can meet specified probability of detection for very low sensitivity levels, then in principle, it should be allowed to transmit higher power levels because it is located far away from protected radius. Therefore, reliable sensing of very weak signals provides mechanism to adapt transmission power.

Even though a conventional OFDM access scheme is not an acceptable candidate for wideband CR transmission due to significant spectrum leakage to adjacent band, there are a number of spectrum-shaping techniques that can be used to enhance the OFDM transmitter and meet the out-of-band emission constraints. They can be divided into two categories dependent on the processing domain. One is the frequency domain processing that happens prior to subcarrier assignment, and the other is time domain processing realized after the IFFT operation.

The simplest frequency domain processing is to introduce the guard bands by assigning the outer subcarriers to zero. However, guard banding results in significant power loss and inefficient spectrum use, but it preserves user/channel orthogonality and conventional OFDM demodulation. The second option is to assign independent power constraints for each subcarrier and optimally fit the spectrum mask. In contrast to guard banding, this approach would preserve all benefits of OFDM transmission without sacrificing spectrum utilization. Another interesting approach was proposed in Ref. 24, which introduced the concept of cancellation carriers, similar to peak-to-average power reduction carriers in conventional OFDM transmitters. Out of N subcarriers, their subset located at the edges of the transmitted spectrum can be devoted to optimization of side-lobe power and reductions up to 30 dB can be achieved. However, these subcarriers no longer bear information data and like guard bands introduce loss in effective throughput. Additional power has to be allocated to these cancellation carriers thus further reducing power efficiency.

Time domain techniques include windowing and filtering. Windowing prefilters each subcarrier to reduce the side lobes but introduces power loss on all subcarriers. The main disadvantage of the windowing approach is the increased complexity needed on the receiver side due to loss of subcarrier orthogonality. On the contrary, filtering reduces side lobes independently but adds computational complexity and increases delay so that cyclic prefix must be extended. Figure 5.24 presents the

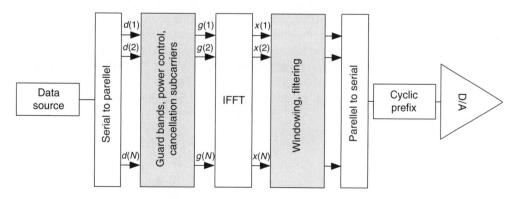

FIGURE 5.24 OFDM transmitter enhanced for out-of-band interference management.

envisioned architecture of the OFDM transmitter that combines different out-of-band interference optimization techniques based on processing before and after the IFFT block. It is expected that the optimal approach would be a hybrid of time and frequency domain techniques. Combining windowing and cancellation carriers proposed in [24] could achieve side-lobe rejection of up to 50 dB at the expense of 25% of total transmit power. Further research is needed to understand trade-offs and implications on the physical layer implementation, as well as higher layers.

CONCLUSIONS

Spectrum sharing is becoming a new paradigm in wireless communications. This transition from restricted spectrum access has major implications on radio architectures, since traditional narrowband radio design techniques are not applicable. Spectrum-sharing radios operate in much wider frequency bands and require frequency agility and significant dynamic range improvements to accommodate the in-band primary users. In addition, new radio functions are required, which involve high-sensitivity sensing and algorithms that can exploit this sensing information to minimize interference. Even more interesting is that entirely new signaling strategies may be employed, since the combination of wide bandwidth availability and interference power restrictions were not present before.

The underlay and overlay sharing strategies to increasing spectrum utilization, which have been discussed, represent only the first steps toward implementing such systems. In particular, interference to the secondary users, which must tolerate in-band, allocated users, is a challenging design task. Ways of coping with an "interference channel" as opposed to a Gaussian noise channel have little theoretical support, other than the information theory result that a large interfering signal does not significantly reduce capacity, since it can be detected and subtracted out. How such subtraction is actually implemented is of course not indicated by the information theoretic results, but it is comforting to the implementer that we are not working against fundamental limits.

In addition to mitigating interference, the new sharing radios will need to have awareness of the channel in not only frequency but time and space as well. This implies more complicated transmit signal generation and a more complicated receiver to process the incoming energy. The design trade-offs for channel measurements have not been fully explored yet regarding the amount of time and computation needed. Another problem that may arise is that of rapid acquisition. Quick identification and synchronization to a UWB signal in a low-SNR channel with large interferers may require an excessively large amount of computation if attempted in parallel, and a highly flexible CR needs to have a way to keep all users in the network synchronized while undergoing rapid changes in the signaling waveform.

As the characteristics of the physical layer of these sharing radios become better understood, issues at the higher levels of the protocol stack will also appear. For example, the trade-offs in multiple access are not understood, and the issues of setup and maintenance in very low signal-to-interference environments are expected to be particularly challenging. Probably of main interest in these recent developments to increase spectrum utilization is that the challenge of developing radio systems that coexist must only meet the requirement of not creating harmful interference as opposed to the need to produce no interference. This provides a great opportunity to find new ways such that wireless systems can coexist beyond simple frequency allocation, which has been essentially the only technique used since interference first became an issue. This fundamental change will no doubt be far-reaching and will ultimately make the spectrum allocation chart appear embarrassingly naïve to the future radio system designer.

REFERENCES

1. NTIA, "U.S. frequency allocations," available at http://www.ntia.doc.gov/osmhome/allochrt.pdf.
2. FCC, "*Spectrum Policy Task Force Report*," ET Docket No. 02-155, November 2002.

3. *First Report and Order*, Federal Communications Commission Std. FCC 02-48, February 2002.
4. http://www.ieee802.org/15/pub/TG3a.html.
5. http://www.ieee802.org/15/pub/TG4a.html.
6. Pavon, J. P., Sai Shankar, N., Gaddam, V., Challapali, K., and Chun-Ting Chou., "The MBOA-WiMedia specifications for ultra wideband distributed networks," *IEEE Commn. Mag.*, vol. 44, pp. 128–133, June 2006.
7. O'Donnell, I. D. and Brodersen, R. W., "A 2.3 mW baseband impulse-UWB transceiver front-end in CMOS," *Proceedings of the IEEE 2006 Symposium on VLSI Circuits*, vol. 1, pp. 200–201, Hawaii, June 2006.
8. Chen, M. S. W. and Brodersen, R., "A subsampling UWB radio architecture by analytic signaling," *Proceedings of the IEEE ICASSP'04*, vol. 4, pp. 533–536, Montreal, Canada, May 2004.
9. FCC, "*Notice of Proposed Rule Making and Order*," ET Docket No. 03-322, December 2003.
10. FCC, "*Notice of Proposed Rule Making (NPRM)*," Docket No. 04-113, released 25 May 2004.
11. Cordeiro, C., Challapali, K., Birru, D., and Sai Shankar, N., "IEEE 802.22: The first worldwide wireless standard based on cognitive radios," *Proceedings of the IEEE DySPAN'05*, vol. 1 pp. 328–337, Baltimore, MD, November 2005.
12. Sahai, A., Hoven, N., and Tandra, R., "Some fundamental limits on cognitive radio," *Allerton Conference on Communications, Control and Computing*, October 2004.
13. Cabric, D., Mishra, S. M., Willkomm, D., Brodersen, R. W., and Wolisz, A., "A cognitive radio approach for usage of virtual unlicensed spectrum," *14th IST Mobile and Wireless Communications Summit*, Dresden, Germany, June 2005.
14. Walden, R. H., "Analog-to-digital converters survey and analysis," *IEEE J. Selected Areas Commn.*, vol. 17, no. 4, 539–550, April 1999.
15. Terrovitis, M. T., Kundert, K. S., and Meyer, R. G., "Cyclostationary noise in radio-frequency communication systems," *IEEE Transactions on Circuits and Systems I: Fundamental Theory and Applications*, vol. 49, no. 11, pp. 1666–1671, November 2002.
16. Carey-Smith, B. E., Warr, P. A., Rogers, P. R., Beach, M. A., and Hilton, S. G., "Flexible frequency discrimination subsystems for reconfigurable radio front ends," *EURASIP J. Wireless Commn. Networking, Special Issue on Reconfigurable Radio for Future Generation Wireless Systems*, vol. 2005, no. 3, pp. 364–381, 2005.
17. Verdu, S., *Multiuser Detection*, Cambridge University Press, Cambridge, UK, 2003.
18. Cabric, D., Chen, M. S. W., Sobel, D. A., Yang, J., and Brodersen, R. W., "Future wireless systems: UWB, 60 GHz, and cognitive radios," *Proceedings of the 2005 Custom Integrated Circuits Conference*, vol. 1, pp. 788–791, San Jose, September 2005.
19. Poon, A., Brodersen, R. W., and Tse, D., "Degrees of freedom in spatial channels," *IEEE Transactions on Information Theory*, vol. 51, pp. 523–536, February 2005.
20. Cabric, D. and Brodersen, R. W., "Physical layer design issues unique to cognitive radio systems," *Proceedings of the 16th IEEE International Symposium on PIMRC*, vol. 2, pp. 759–763, Berlin, Germany, September 2005.
21. Tandra, R. and Sahai, A., "Fundamental limits on detection in low SNR," *Proceeding of 2005 International Conference on Wireless Networks, Communications and Mobile Computing*, vol. 1, pp. 464–469, Maui, June 2005.
22. Gardner, W. A., "Signal interception: performance advantages of cyclic-feature detectors," *IEEE Transactions on Communications*, vol. 40, no. 1, pp. 149–159, January 1992.
23. Hoven, N. and Sahai, A., "Power scaling for cognitive radio," *Proceeding of 2005 International Conference on Wireless Networks, Communications and Mobile Computing*, vol. 1, pp. 250–255, Maui, June 2005.
24. Brandes, S., Cosovic, I., and Schnell, M., "Reduction of out-of-band radiation in OFDM based overlay systems," *Proceedings of the 2005 IEEE DySPAN*, vol. 1, pp. 662–665, Baltimore, November 2005.

6 Short-Distance Wireless and Its Opportunities

Jan M. Rabaey, Y. H. Chee, David Chen, Luca de Nardis, Simone Gambini, Davide Guermandi, Michael Mark, and Nathan Pletcher

CONTENTS

Introduction...159
The Channel...160
    Electrically Small Antennas...160
    Reactive versus Radiative 161
    Capacitive versus Inductive 163
Classification of Short-Distance Wireless Transceivers 165
    Receiver 166
    Transmitter..166
Narrowband Sinusoidal Transmission 167
    Background..167
    State-of-the-Art...169
Wideband Capacitive Transmission...171
Wideband Inductive Transmission...172
Dense Networks..172
Passive Receivers 178
Implementation Comparisons...180
Conclusions and Opportunities..181
References...181

INTRODUCTION

The vast majority of the wireless transceivers designed over the past couple of decades have focused on medium (10–100 m) and long distance (100 m and more) transmission ranges. Only recently with the emergence of wireless sensor networks and radio-frequency identifications (RFIDs) has substantial attention been paid to the opportunities offered by creating wireless connectivity over very short distances (well below 1 m). Unfortunately, short distance often corresponds to ultra-small size, power, or cost, making the realization of such transceivers a major challenge. Continued technology scaling, combined with the emergence of novel microelectromechanical system (MEMS) components, advanced packaging techniques, and novel energy sources, may finally make the realization of such dust nodes possible. Before discussing the needs and requirements imposed on these nodes, it is worth exploring what applications may be enabled by their availability.

- Intelligent objects: Adding a set of connected sensors to common objects may dramatically extend their usefulness, functionality, or reliability. Imagine, for instance, tires that can measure their shape while driving through a set of sensors embedded in the surface. This could lead to a substantial enhancement in driving performance, road sensitivity, and safety. Embedding distributed sensors in user interfaces for gaming could lead to a far more engaging interaction.

- Health care (body area networks): It is generally acknowledged that health care costs could be dramatically reduced if affordable around-the-clock and unobtrusive monitoring were available. Distributed sensors also can improve the performance of prosthetics such as hearing aids or artificial limbs. Further down the road are systems that not only observe but also adjust (for instance, by injecting medication). In the long term, Arthur C. Clarke-like ideas of artificial antibodies roaming within the human body may even become reality [1,2].
- Seamless system assembly: Stacking silicon die or assembling small components together into a tiny package often requires expensive and unreliable wiring (through bumps, vias, or wire bonds). By realizing these connections in a wireless fashion, assembly cost, modularity, and even reliability can be improved. A number of researchers have started to explore the opportunities in this area, mostly to enable high data rate/high bandwidth connections between processors (in supercomputer-like applications) or between processors and memories [3–6].
- Smart surfaces and paintable electronics: In a sense, these are similar to the previous category; only the scale is larger. Larger numbers of small nodes are deposited on a surface using a pseudorandom assembly process (such as paint). The ensemble of nodes auto-configures into a distributed computational network that may implement a range of innovative functions, such as wall-sized displays, artificial skin, intelligent materials that respond to strain or stress, coatings with long-term memory, etc. [6,7].

In short, the abundant availability of ultra-low-power nodes that communicate over short distances can lead to a veritable plethora of exciting opportunities. The obvious question is how realistic it is to expect this to happen soon, or if at all. The power and size constraints imposed by some of these applications may be hard to meet. Even worse, pure physics bounds may eliminate the possibility of some wireless scenarios altogether. In this chapter, the general constraints and conditions of short-distance wireless are discussed first. This is followed by a classification of the different options open to the designer, each of which is illustrated with some examples of actual designs. The chapter is concluded with a summary and perspectives on what the future may hold.

THE CHANNEL

To appreciate the choices available to the designer of a short-distance link, it is quite essential that the behavior of the wireless channel under these conditions is understood. A thorough understanding of the channel will enable conscientious decisions about frequencies, impedances, and antenna sizes.

ELECTRICALLY SMALL ANTENNAS

Together with the energy source, the antenna is the primary component that defines the size of the ultra-small nodes required by most of the applications listed. Hence, keeping the antenna as small as possible is one of the primary design goals (unless very high carrier frequencies are used, which does not fit well with ultra-low power). Consider, for instance, the loop antenna shown in Figure 6.1a. This antenna is called electrically small if there is no significant phase shift over the electrical path length [8]. This is the case when the electrical path length is smaller than 1/10 of the wavelength λ. This leads to the following condition (with a being the radius of the loop and N the number of turns):

$$2\pi aN < 0.1\lambda \tag{6.1}$$

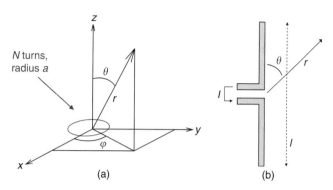

FIGURE 6.1 Electrically small antennas: (a) loop; (b) dipole.

For instance, at a carrier frequency of 800 MHz, an antenna is considered electrically small if its radius is 6 mm or less (for a single winding). Similarly, a dipole is considered electrically small when $l < 0.1\lambda$, where l is the length of the dipole (for 800 MHz, this translates to a length of 3.75 cm).

REACTIVE VERSUS RADIATIVE

The transmission behavior of an electrically small antenna can be described by a relatively simple set of equations. For instance, the electric field generated by the elementary electric dipole in Figure 6.1b of length l driven by a current I at its input port is given in Equation 6.2, with r the distance from the antenna and k the wave number ($2\pi/\lambda$) [8]:

$$E_r = \frac{U}{2\pi\varepsilon}\left(\frac{1}{r^3} + \frac{jk}{r^2}\right)\cos\theta$$

$$E_\theta = \frac{U}{4\pi\varepsilon}\left(\frac{1}{r^3} + \frac{jk}{r^2} - \frac{k^2}{r}\right)\sin\theta$$

$$\hspace{6cm}(6.2)$$

$$H_\phi = \frac{U}{4\pi\varepsilon\zeta}\left(+\frac{jk}{r^2} - \frac{k^2}{r}\right)\sin\theta$$

$$U = \frac{I \cdot l}{j\omega}$$

Observe that a dipole antenna is used as the reference for a capacitive link throughout this chapter. In reality, a patch antenna would be a better match. However, the equations for a patch are harder to formulate, and yield the same dependencies, hence the choice.

Because of the duality between electric and magnetic fields, the same equations hold for the field generated by a magnetic dipole if the electric dipole momentum U is replaced by the magnetic dipole momentum U_m, ε by μ, and E by H. (An elementary loop of radius a generates a field equivalent to that of a magnetic dipole of momentum $U_m = \mu\pi a^2 I$.) Use the previous equation to calculate the power flux on a sphere of radius r centered around the antenna by integrating the radial component of the Poynting vector $S = E \times H$. This yields

$$P_r = \frac{\pi}{3}\zeta \cdot I^2 \left(\frac{l}{\lambda}\right)^2 \left[1 - j\left(\frac{\lambda}{2\pi r}\right)^3\right]$$

$$\hspace{6cm}(6.3)$$

with $\zeta = \sqrt{\mu/\varepsilon}$, the characteristic (wave) impedance.

For distances $r < \lambda/2\pi = 1/k$, the imaginary part of P_r is larger than the real part, so that the reactive component of the energy generated by the antenna is dominant with respect to the radiative component. This condition defines a region in the range versus operating frequency (r, f) plane that will be called the near-field region. Figure 6.2 delineates the near-field and radiative regions in the range versus frequency plane. For example, at 800 MHz communication is predominantly reactive for distances below 6 cm.

From Equation 6.3, observe that the power flux of the reactive antenna (integrated over a sphere) decays with r^3, while remaining constant for radiative antennas, which is equivalent to the well-known result that the radiative power received over a fixed area decreases with r^2 in free space. While radiative (electromagnetic) communication obeys a law of power conservation, reactive communication (electrostatic or magnetic) is strongly dependent upon proximity and coupling between sending and receiving antennas. Information is transferred by modulating the standing wave of reactive power that encloses the proximity of the antenna, which is significantly larger than the power of the radiated wave. This has some fundamental impact on how transceivers are designed:

- In the radiative regime, optimizing the power transfer is the main design goal. Antennas are hence designed to match the characteristic impedance of medium.
- Maximizing the electrical (E) or magnetic (H) coupling is the primary design goal in the near-field regime. Antenna impedances are very different from what is typically observed in wireless systems.

The partitioning of the communication paradigms into different operational regions is demonstrated in Table 6.1 (with r, l, and λ as essential design parameters). Most wireless radio frequency (RF)

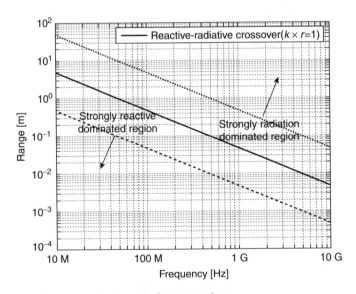

FIGURE 6.2 Regions of operation in the range-frequency plane.

TABLE 6.1
Operation Regions of RF Communication Systems

	Reactive ($r < \lambda/2\pi$)	Radiative ($r < \lambda/2\pi$)
Electrically small ($l < \lambda/10$)	$10l < \lambda; 2\pi r < \lambda$	$r/l > 1.6; 2\pi r > \lambda$
Electrically large ($l > \lambda/10$)	$r/l < 1.6; 2\pi r < \lambda$	$10l > \lambda; 2\pi r > \lambda$

communications deployed today are operating in the electrically large, radiative regime (lower right corner), with both the distance and the antenna size relatively large with respect to the wavelength. Electrically small antennas are very poor radiators, which makes operating them over long distances not attractive (upper right). Electrically large, reactive operation only rarely makes sense, as the antennas are approximately of the same size or larger than the communication distance. Most short-distance communication systems hence operate in the electrically small, reactive regime.

CAPACITIVE VERSUS INDUCTIVE

When near-field conditions hold and $Im(P_r) > 0$ at the distance of interest, the field is predominantly magnetic; otherwise, it is predominantly electric. Antennas for which $Im(P_r) > 0$ within the near-field region are therefore called magnetic antennas; otherwise, they are called electric antennas. The immediate result of this fact is that antennas appear to the circuit as almost ideal reactive elements; loop antennas behave inductively, while dipole antennas behave capacitively. For example, a loop antenna with a radius of 0.5 cm has a radiation resistance R_r of only 0.5 Ω at 1 GHz. Impedance matching to the antenna is not necessary at the transmitter or receiver side of the link, and is in fact detrimental to the link performance. In fact, a near-field radio link is better thought of as a transformer coupling, where no power is dissipated in the primary winding, unless the secondary winding is loaded. This creates the opportunity for extremely power-efficient transmitter design.

Consider, for instance, the case of links composed of loop antennas, for which the transformer analogy can be exploited in a literal sense to obtain the equivalent of Figure 6.3a. A link that exploits dipole antennas or any other electrical antenna can be represented by the π network of Figure 6.3b. Transmission channels, such as the ones displayed in Figure 6.3, are clearly energy conserving, since no dissipative elements (such as resistors) are present.

The values of L, M, C, and C_c can be calculated analytically only in simple cases. For any other case, electromagnetic simulation is necessary. For example, for two coplanar elementary loops find that $L \approx \mu a N^2 O$ and $M \approx (1/2)(\mu \pi a^4 N^2)/r^3$, where N is the number of turns of radius a in the loop. Similarly, when short dipoles with a common axis are used for transmitting and receiving antennas, find $C \approx 2\pi \varepsilon l/O$ and $C_c \approx 2\pi \varepsilon l^4/r^3 O^2$, where O is the thinness of the dipole (defined as log(2l/t), with l and t being the antenna length and diameter, respectively). Observe that both M and C_c decrease with distance cubed.

It is also possible to describe the link in terms of coupling factors instead of impedances. In fact, both representations can be derived from S-parameter measurements, which is most often done in practice. The derivation in terms of coupling factors can be done by applying Thevenin's theorem at the (receiver and transmitter) antenna terminals, obtaining $k = M/L \approx \pi a^3/2r^3 O$ for the loop and $k = C_c/(C_c + 2C) \approx l^3/2r^3 O$ for the dipole. To preserve the symmetry of the networks of Figure 6.3, a pair of voltage-controlled voltage sources should be used (one representing the impact of antenna A on antenna B, and vice versa). However, for small values of k, the return path can be ignored, leading to the equivalent circuits of Figure 6.4, which give a more intuitive understanding of the attenuation properties of the channel.

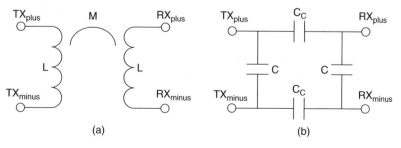

(a)　　　　　　　　　　　　　(b)

FIGURE 6.3　Equivalent circuits of near-field links composed of loop antennas (a) and electrical dipoles (b).

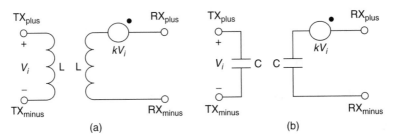

FIGURE 6.4 Equivalent circuits of near-field links composed of small loop antennas (a) and small electrical dipoles (b) after Thevenin transformation.

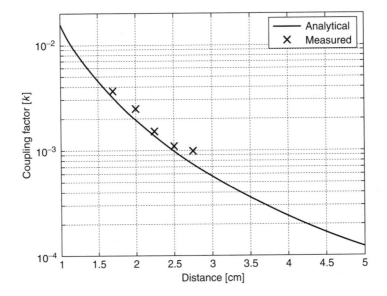

FIGURE 6.5 Coupling factor versus distance for a loop antenna of $r = 0.4\,\text{cm}$ and $N = 1$.

Using these equivalent models, it is possible to compare the effectiveness of dipoles and loop antennas, and to determine which choice is the best for a given size constraint and communication range. It is interesting to note that, if elementary dipoles are considered and $2a = l$, the following relation holds:

$$\frac{k_\text{m}}{k_\text{e}} = \frac{\pi}{4} \approx 0.8 \tag{6.4}$$

Therefore, at a fundamental level, the same levels of attenuation are achievable using either of the two topologies. The choice may therefore be deferred to more realistic antenna models, and is also largely influenced by circuit design considerations, which are the topic of a later section. To give the reader some idea of the actual value of k, Figure 6.5 plots both the analytical and measured channel attenuation of an inductive link with $a = 0.4\,\text{cm}$ and $N = 3$. Clearly, the fast roll-off of the near-field components results in a high attenuation even at short distances. For $r = 5\,\text{cm}$, an attenuation of approximately 80 dB is observed. The small antenna hence makes it necessary to adopt a low-noise, high-gain receiver to recover the signal, even for these very small ranges.

One positive point is that the effective area of the antenna from a radiative perspective is extremely small for a wide range of frequencies. This provides the system with an inherent robustness to the traditional RF interferences, which allows for the use of simple architectures in the transceivers.

CLASSIFICATION OF SHORT-DISTANCE WIRELESS TRANSCEIVERS

From the previous discussion, some of the essential choices available to the designer are already apparent. Given a communication distance r, the operational regime of the transceiver is determined by the choice of the center frequency and the size of the antenna. In addition, a number of important decisions remain open:

- Capacitive versus inductive antennas (in case of near field). As stated earlier, this decision is mostly guided by circuit and environmental conditions. Placement and orientation of the antennas have an important impact on the coupling factor.
- Narrowband versus ultra-wideband communications. Virtually all traditional RF wireless communication links are narrowband and based on modulation of sinusoids. With the advent of ultra wideband (UWB), interest in pulse-based communications has been renewed. This is especially the case in short-distance wireless links, where a number of recent implementations have opted for a pulse-based strategy. One of the major advantages of pulse-based transmission is that the transmit power can be extremely low since the amount of energy required to transmit a pulse is extremely small. On the negative side, the need for precise timing acquisition complicates the receiver design and impacts its power.
- Passive versus active. The extreme power limits often imposed on short-distance wireless links lead to the interesting trade-off of whether to spend the power at the transmitter or at the receiver. In the latter case, the transmitter modulates an incoming waveform (transmitted from the intended receiver) with its data using backscattering. Extending the concept further, the transmitter can be powered by energy extracted from the same waveform. This concept is used extensively in RFID tags but can be exploited in a far wider range of short-distance links.

Before going into a discussion of some detailed designs in different operational modes, some general considerations for circuit design in the near-field regime should be addressed. First, antenna impedance becomes an essential parameter in the near field. Figure 6.6 plots the absolute value of antenna impedance versus frequency for different antenna sizes. For a given size d, the input impedances of dipole and loop antennas become equal at a frequency f_0, such that $c/f_0 \approx 4d$. Since the small antenna regime is characterized by $f \ll f_0$, the input impedances of dipole and loop are significantly different in this region. Intuitively, the low impedance characteristic of inductive antennas makes them close

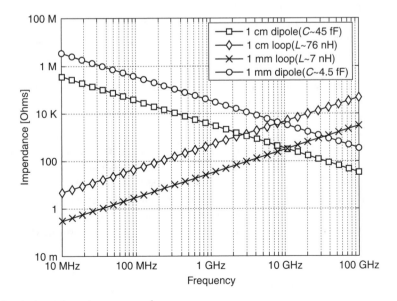

FIGURE 6.6 Antenna impedance versus frequency.

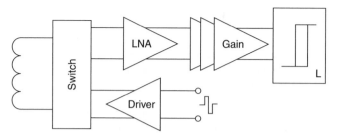

FIGURE 6.7   Architecture of a fully asynchronous inductive pulse-based transceiver.

to ideal receivers, but the low impedance become problematic when these antennas are driven by a transmitter. In a dual sense, electrical antennas have much higher impedance, which makes them very easy to drive, but leads to complications in the receiver design.

RECEIVER

A typical architecture of a multistage transceiver operating in the near field is shown in Figure 6.7. A low-noise amplifier picks up the signal at the antenna and feeds a cascade of several gain stages, which restore the signal amplitude to a level such that the latch can operate reliably. As already discussed, the design of the low-noise amplifier and the antenna driver is impacted by the impedance level of the chosen antenna, which is a strong function of the type of coupling used (magnetic or electric) and the operating frequency.

Inductively coupled systems do not suffer much from the parasitic capacitance at the receiver input, since such capacitance simply results in a resonance in the input circuit. In a broadband system, this resonance can either be designed to be well above the upper limit of system bandwidth, or can be placed in band after an appropriate reduction of its quality factor. Given the relatively low impedance of the inductors, both these choices can be pursued up to frequencies in the low-GHz range. For example, a three-turn inductor with a radius of 0.4 cm realized on a printed circuit board (PCB) substrate has a typical inductance of 76 nH. To avoid resonance in the 0–1 GHz band, the low-noise amplifier (LNA) input capacitance should be less than 330 fF. This value is realizable in modern integrated technologies. Furthermore, in narrowband systems, this capacitance can be used to offset the path loss by sizing it to resonate with the antenna inductance, obtaining a passive gain equal to the (loaded) quality factor of the antenna itself. This approach has the advantage of providing filtering at the receiver input. If the receiver is to be broadband, the quality factor of this resonance must be reduced by explicit resistive termination.

On the contrary, a dipole antenna of comparable size would have an input capacitance C around 45 fF. If a voltage mode amplifier is to be used as the sensing circuit, its input capacitance should be much lower than C to avoid increasing the effective path loss. This is a challenge when the range exceeds a few microns, and the receiver needs a low input–referred noise. Hence, the parasitic capacitance introduced by the LNA input devices, interconnect and pads, easily limits the performance. In addition, the low capacitance value virtually eliminates the option of building a resonant input circuit, as the value of the inductance required to tune out such a small capacitor is prohibitively large for frequencies up to several GHz. For instance, the inductance needed to tune out 45 fF at 1 GHz is 560 nH, which is not available in integrated form. These considerations make capacitive links not very practical for systems with ranges more than a few centimeters. One option is to use a current-input LNA to take advantage of the high-source impedance and to reduce the sensitivity to parasitics.

TRANSMITTER

When considering the transmitter design, inductive implementations present more difficulties than capacitive. Because of the high input impedance, a dipole (capacitor) is easily driven. Conversely,

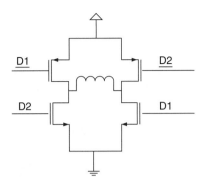

FIGURE 6.8   Bridge-based inductive driver.

the low input impedance of a loop antenna is problematic; the current flowing through the transmitter when on becomes prohibitively large. A wideband pulse-based scheme is often preferred in this setting, with the driver typically being either a bridge (Figure 6.8) [9] or a simple inverter [5]. Under these conditions, peak current becomes the dominant issue. If a voltage step of amplitude V_{DD} and duration T is applied to an inductor, a current $I_{DD} = (V_{DD}/R)(1 - \exp(-tR/L))$ is drawn from the supply, where R is the total resistance in the driving circuit (the on-resistance of the driver and the series resistance of the inductor). When $T \ll L/R$, the current increases linearly with time, to a peak of $I_{max} = (V_{dd}/L)T$. Assuming the 0.4 cm radius inductor described earlier (with $L = 76$ nH), a pulse duration T of 1.5 ns, and a voltage of 1 V, the peak current I_{max} is 20 mA, which is still acceptable due to the small duration. If L is reduced however, T must be reduced accordingly to keep the drawn current constant. This results in increased receiver bandwidth, and consequently increased power dissipation. For comparison, if a narrowband system with $f_c = 1/T$ is considered, the driver needs only to source and sink a peak current of $I_{max} = V_{dd}T/2QL$, which is roughly $2Q$ times lower (where Q is the passive quality factor).

A different option to limit current in the pulse-based transmitter is to operate the driver in the resistance-limited regime ($T \approx L/R$). In this case, significant droop is observed on the received waveform over the bit interval, again resulting in increased receiver bandwidth and dynamic range requirements, and hence an increase in power dissipation. Because of these disadvantages, the resistance-limited regime has been traditionally avoided in implementations.

In the following sections, the principles discussed are applied in some practical design examples and some useful guidelines for the design of short-range wireless links are derived.

NARROWBAND SINUSOIDAL TRANSMISSION

BACKGROUND

Most traditional wireless communication systems utilize narrowband architectures. In a narrowband communication system, the baseband information signal is modulated onto a higher frequency-carrier waveform so that the spectrum is restricted to a relatively small frequency range. In contrast to pulse-based communication, channelization in narrowband systems is accomplished in the frequency domain by defining frequency slots for each channel and assigning a specific channel to each communicating link. In this way, multiple systems can communicate simultaneously without interference.

At the block level, a typical narrowband communication system consists of three main types of circuit components: amplifier blocks, filters, and frequency conversion modules. A generalized direct-conversion transceiver block diagram is shown in Figure 6.9.

The transmitter takes the digital baseband data and modulates it onto the desired carrier frequency, while the power amplifier (PA) drives the antenna. The receiver amplifies the low-level input signal at the antenna and down-converts it to baseband to recover the original digital bits. In a narrowband system, the amplifiers only need gain over a relatively small frequency range. Therefore,

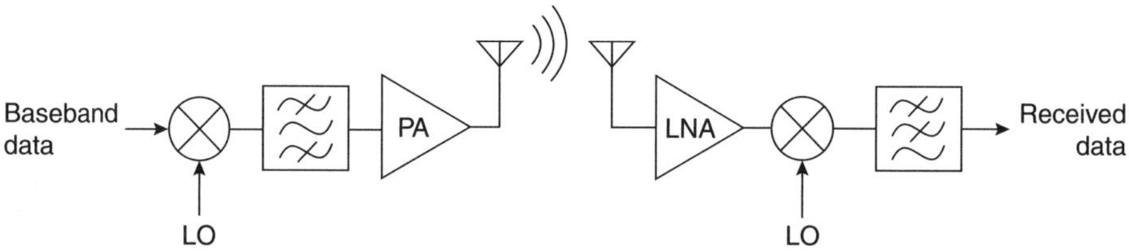

FIGURE 6.9   Block diagram of a generalized narrowband transceiver architecture.

resonant circuit elements, typically inductors and capacitors, are used to tune the amplifiers to a specific center frequency. To a first-order approximation, the gain-bandwidth product of an amplifier is constant and independent of center frequency. Accordingly, amplifiers using resonant circuits can achieve large gain, even at the high frequencies used in narrowband systems, by restricting the effective bandwidth. For example, consider a typical LNA design for a global system for mobile communication (GSM) receiver. The required gain is typically about 15 dB in the GSM receive band (935–960 MHz), leading to a gain-bandwidth product of about 30 GHz for a broadband amplifier design. However, gain is only required over a bandwidth of about 30 MHz, reducing the gain-bandwidth product for a tuned amplifier to 950 MHz, which is much more manageable.

Narrowband architectures are quite popular for all types of wireless communication, including short-distance, low-power systems. There are several reasons for the popularity of narrowband architectures. First, frequency spectrum is easily regulated and channels can be allocated to licensed users only, simplifying interference avoidance. In addition, as described above, gain at a tuned frequency is readily obtained using resonant circuits.

On the contrary, narrowband architectures also present some difficulties when designing highly integrated, low-power, short-distance links. First, they require a highly accurate frequency reference to ensure that transmitters and receivers are operating at the same frequency. This reference has traditionally been provided by off-chip quartz crystals, which are prohibitively large and expensive. Bulky external passive filtering components are sometimes also necessary to remove signals in adjacent frequency bands and ensure immunity to interference.

An additional challenge for narrowband systems used in power-constrained applications is how to provide sufficient RF gain with very little current. The gain of the generalized amplifier of Figure 6.10 is given by

$$A_V = \frac{V_{\text{out}}}{V_{\text{in}}} = g_\text{m}Z(s)\Big|_{s = j\omega_0} \tag{6.5}$$

where g_m is the device transconductance and $Z(j\omega_0)$ is the impedance of the load network. In power-constrained designs, the device transconductance is limited to fairly low values, meaning that a high load impedance is necessary to achieve gain. Unfortunately, it can be difficult to realize this large impedance at high frequencies.

Consider a fully integrated load network $Z(s)$ consisting of an on-chip inductance and capacitance, as is desirable for low-cost solutions. Unfortunately, the quality factor of integrated inductors is typically much lower than that of on-chip capacitors, especially in digital CMOS processes. Therefore, the load impedance is limited by the integrated inductor, whose equivalent resistance at resonance is given by Equation 6.6

$$R_{\text{parallel}} = QL\omega \tag{6.6}$$

Typical values of parallel resistance are plotted in Figure 6.10 for different values of inductance and on-chip Q factors of 5 and 10. The calculation assumes that the quality factor is independent of inductance, which is approximately true over the range of inductance plotted. While bondwire

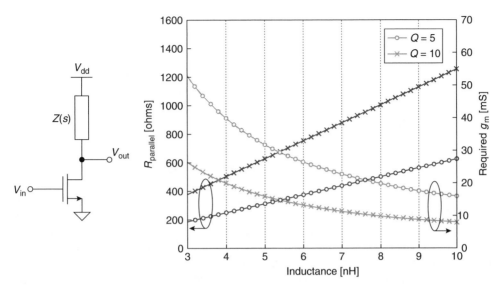

FIGURE 6.10 Generalized gain stage with resonant load impedance: equivalent resistance at resonance and required transconductance versus inductance.

inductors can be used to achieve higher quality factors, this solution is not compatible with highly integrated implementations. From Figure 6.10, it is evident that load impedances greater than about 1400 Ω are difficult to achieve with a fully integrated design. The plot also shows the required transconductance to achieve a gain of 10 using load impedance, as given by Equation 6.5. An amplifier using a 5 nH inductor with $Q = 10$ requires about 15 mS of transconductance to achieve a gain of just 10 at 2 GHz. With a device transconductance efficiency (g_m/I_d) of 15, 1 mA is needed to provide the required transconductance, which is prohibitively high for ultra-low-power systems. From this example, it is clear that the limited load impedance of integrated passives presents a significant impediment to the design of narrowband gain stages for short-range wireless systems.

STATE-OF-THE-ART

One solution to overcome the challenges described above is the use of emerging radio frequency microelectromechanical systems (RF-MEMS) technologies. RF-MEMS resonant elements can provide high quality factors (often greater than 1000) and have the potential for cheap integration with active circuits, breaking the trade-off between quality factor and size/integration. More importantly, the tighter manufacturing tolerances of MEMS compared to on-chip passives allow RF-MEMS components to serve as accurate frequency references. Although currently available RF-MEMS devices are not as accurate as traditional quartz crystal references, active RF-MEMS research continues to improve the devices. A recent transceiver implementation designed for 10–20 m links [10] illustrates the power reduction and high level of integration enabled by RF-MEMS. Both receiver and transmitter utilize micromachined bulk acoustic wave (BAW) resonators as frequency references [13]. These resonators contain a thin membrane of aluminum nitride piezoelectric material and are fabricated on silicon substrates in a low-temperature process, which has been shown to be compatible with CMOS [14]. The narrowband super-regenerative receiver achieves 10 kbps data rate using on/off keying (OOK) modulation at a carrier frequency of 1.9 GHz. The receiver sensitivity is −100 dBm for a power consumption of only 400 μW.

The high sensitivity of the receiver in Ref. 10 is not needed for short-range (a few meters or less) radiative communication systems where the path loss is less severe. For example, a narrowband communication system with a carrier frequency of 2 GHz using radiative propagation will incur only about 40 dB path loss for a 1 m link distance. For shorter distances, the path loss will be even less; within 10 cm, the propagation can no longer be assumed to be purely radiative, as described in the section "Introduction." The range of the wireless link determines the path loss, which together with the transmit power sets the required sensitivity of the receiver according to Equation 6.7:

$$\text{Sens}_{\text{Rx}} = P_{\text{Tx,out}} - 20 \log_{10}\left(\frac{4\pi d}{\lambda}\right) \tag{6.7}$$

Because of the short communication range, no multipath interference is assumed. For a carrier frequency of 2 GHz, Figure 6.11 plots the required receiver sensitivity for various transmit power levels over the communication range of interest (10 cm to 1 m).

Figure 6.11 shows that, even for very low transmit power levels, the receiver is not required to have high sensitivity. In this case, its architecture can be simplified substantially to reduce the power consumption. A block diagram of a possible architecture is shown in Figure 6.12, where the RF-MEMS component is used as a filtering element rather than as a frequency reference. The receiver is based on the tuned RF (TRF) architecture, which includes only RF amplification, filtering, and energy detection. The input signal is filtered at the input using the narrowband response of the BAW resonator. The OOK signal is then amplified before driving the envelope detector, which down-converts the modulated data to baseband. The local oscillator and mixer of a typical heterodyne or direct-conversion receiver are eliminated, allowing significant power savings.

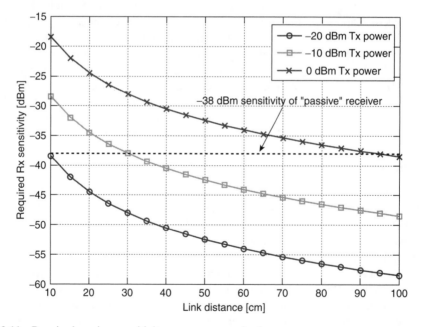

FIGURE 6.11   Required receiver sensitivity versus communication range.

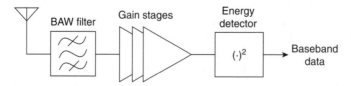

FIGURE 6.12   Tuned RF (TRF) receiver architecture.

The sensitivity of the TRF receiver is determined by the amount of RF gain in the receive chain before the energy detector. The energy detector can be implemented as a simple diode detector, or a subthreshold-biased field effect transistor (FET) in a CMOS process, with negligible power consumption. For the ultimate low-power receiver, the gain stages can be omitted completely, leaving just a bandpass filter and energy detector, reminiscent of the first amplitude modulation (AM) radio "passive" receivers. A passive receiver comprising a BAW filter input matching network and subthreshold MOSFET detector achieved a measured sensitivity of -38 dBm while consuming less than 1 µW of power. Again referring to Figure 6.11, such a passive receiver implementation may be feasible if the communication distance is sufficiently short.

From the above example, it is apparent that the overall energy consumption of this short-range system will be dominated by the transmitter. Several published transmitters with output power between -5 and 0 dBm [11,12,15] consume $1-3$ mW to transmit 300 kbps. Transmitters designed for even lower output power will ultimately be limited by the power necessary to generate the carrier frequency for a narrowband system (more than 100 µW). Thus, a 2 GHz narrowband TRF transceiver designed for communication in the range of 25 cm can be expected to achieve data transmission efficiency of approximately 1 nJ/bit.

Although narrowband architectures are the dominant choice for traditional wireless communication, it is clear that short-range systems require unique considerations. In the transmitter, very low output power is required (< 0 dBm), making power-efficient design difficult. In the receiver, a simple architecture may be used; if RF gain is required, however, the power consumption of the receiver will be increased dramatically. In the following sections, several alternatives to traditional narrowband architectures are presented, which are particularly attractive for short-range wireless systems.

WIDEBAND CAPACITIVE TRANSMISSION

One option to circumvent the constraints of narrowband transceivers is to use pulses instead of modulated carriers. These systems are wideband in nature, and concentrate the energy in narrow bursts. A number of examples of short-distance pulse-based wireless links can be found in the literature. Most of them are exploring the use of wireless to facilitate 2.5D integration of integrated circuits. Examples in literature exploit both capacitive [16,17] and inductive [4,5,18] coupling to transmit data and clock between two stacked dice. For this application, the most important metrics to be considered are the area, power efficiency, and communication bandwidth required by one communication channel. In this section, the potential and challenges of a capacitive link are analyzed first.

In Ref. 17, the transmitter and receiver electrical antennas are square pads built on the top metal layer of the process. The dice are stacked face to face, and pads couple capacitively through the passivation layer. Given that the communication links are never in contact with the environment, electrostatic discharge (ESD) structures can be omitted, saving area and parasitic capacitance. For this structure, derive that $C_c = 2\varepsilon_{Si}A/d$ and $C = \varepsilon_{Si}A/t$ (see Figure 6.13) with A, d, and t being the pad area, communication distance, and metal stack height, respectively. The coupling factor k when the output is not terminated is proportional to t/d and as such is only a weak function of process technology. Reduction in signal-to-noise ratio (SNR) due to misalignment of the pads is recognized in Ref. 16 as the most important limitation in link design, and counteracted by a calibration system that exploits two Verniers (X and Y axes) to measure misalignment, and a redundant transmitter array to compensate for it. The techniques enabled a net (including calibration overhead) pad pitch of 50 µm, using coupling structures of 35 µm on a side. While pad pitch is chosen to ensure robustness against channel-to-channel crosstalk, the size of the coupling structures themselves must be large enough that the parasitic loading due to the receiver is negligible.

A test chip demonstrates 16 channels communicating with a bit error rate (BER) of 10^{-12} at a data rate of 1.35 Gb/s/channel. In a 0.35 µm CMOS process, each channel consumes 3.6 mW static power, plus an additional 3.9 pJ/bit due to switching, which translates to 3.9 mW when clocked at 1 Gb/s. In a more advanced technology, the parasitic loading on the pads due to the receiver will be

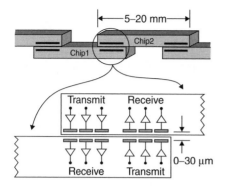

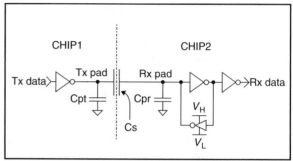

FIGURE 6.13   Capacitive inter-chip communication: topology and circuit architecture. (From Drost, R., et al., *IEEE J. Solid-State Circuits*, 39, 1529–1535, 2004. With permission. © [2004] IEEE.)

reduced, allowing smaller structures and smaller pitch to be used. Simultaneously reduced switched capacitance in the logic and voltage swings will therefore substantially decrease energy per bit, while increasing density and maximum bandwidth per channel. Because of the reduced parasitics, communication range will also increase, albeit more slowly.

WIDEBAND INDUCTIVE TRANSMISSION

A different approach is taken in Refs. 4, 5, 18. In Refs. 4, 5, on-chip inductors are used to obtain 1.25 Gb/s/channel at a distance of 240 μm. This technique has the advantage of not requiring the communicating chips to be mounted face to face; however, the demonstrated pad pitch increases to 300 μm because of the relatively large area of the inductors, and the transmit power amounts to 43 mW/channel. A solution that improves over the previous ones in terms of area and power efficiency is reported in Ref. 18. Magnetic coupling is again used, but now the clock is transmitted along with data from transmitter to receiver, by using a dedicated, higher SNR channel. This technique, combined with the use of antipodal modulation, allows the pad pitch to be reduced to 30 μm. However, wafer thinning to 10 μm is required, which adds cost and might also impair the mechanical robustness of the die. Power dissipation is also lowered to 6 and 2 mW/channel for the clock and data transmitters, respectively, part of which is due to migration to 0.18 μm technology (Figure 6.14).

One conclusion that emerges from this discussion is that pulse-based wideband links can be very energy efficient and may be superior from that perspective to narrowband links. This is obtained through a higher peak power consumption and accurate synchronization of transmitter and receiver system clocks to enable efficient duty cycling. The latter is the most challenging part in the design of pulse-based links. In the chip-to-chip communication system, described above, a special channel with a higher SNR is used to transmit the clock separately. Such a channel may not always be available. Hence, either accurate but expensive timing references such as crystals must be used, or the clock must be extracted from the incoming data stream. Such a system is described in the next section.

DENSE NETWORKS

With the development of new technologies such as body networks and wearable/paintable computer, a new model of wireless sensor network must be considered. Instead of a few nodes in a wide space, these new applications require hundreds of nodes close to each other, with negligible cost, area, and power consumption. The small distances (<5 cm) between nodes and the small size of the nodes (<1 cm) lead to a near-field condition with electrically small antennas.

In this particular case, similar to the one presented in Ref. 9 for chip-to-chip communications, coupled inductors will be considered. However, in contrast to the study by Mizoguchi et al. [9], the inductive loops are placed in the same plane and not stacked on top of each other (Figure 6.15).

Total bandwidth	1Tb/s
Number of data transceivers	1024
Channel pitch	30μm
Clock	1GHz, Inductive coupling
Power dissipation	3W@1.2V (Tx:2.2W, Rx:0.6W, Clock:0.2W)
Total area	Data:1mm² (Clock:1mm²)
Communication distance	15μm (Chip thickness:10μm, Adhesive:5μm)
Process	0.18μm CMOS
Signaling	Bi-phase modulation + 4-phase TDMA
Power/Bandwidth	3mW/Gb/s (50% of [5])
Area/Bandwidth	1mm²/Tb/s (40% of [5])

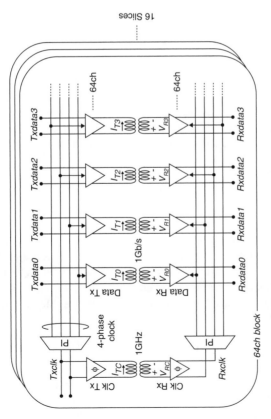

FIGURE 6.14 Inductive inter-chip communication: topology and link performance. (From Miura, N., et al., *International Solid-State Circuits Conference*, 2006. With permission. © [2006] IEEE.)

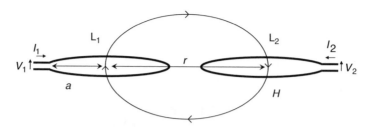

FIGURE 6.15   Inductive in-plane antennas.

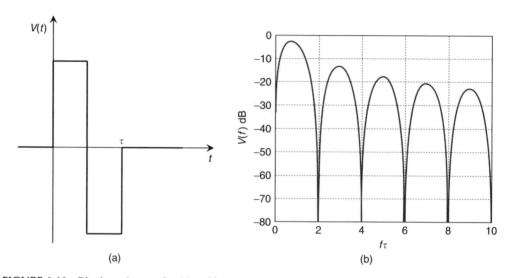

FIGURE 6.16   Bipolar voltage pulse (a) and its energy spectrum (b).

When an alternating current (AC) voltage V_1 is forced across the inductor L_1, a voltage $V_2 = k\, V_1$ is induced across the second inductor L_2, where k is the coupling factor. For inductors laid out in the same plane, it can be shown that the coupling factor scales inversely with the third order of the distance. In fact, the magnetic field at a distance r (smaller than the wavelength) on the same plane as the inductor can be written as

$$H \approx \frac{U_\mathrm{m}}{4\pi}\frac{1}{r^3} \tag{6.8}$$

where U_m is the magnetic momentum (which for a single loop coil equals $\pi a^2 I$).

The same relationship holds for the fluxes and hence the induced voltages. For example, the attenuation between two 3-turn coils with a 4 mm radius ($L = 76\,\mathrm{nH}$, self-resonant frequency $> 800\,\mathrm{MHz}$) equals 80 dB at a 5 cm separation. These results suggest that communication can only be established between inductors that are closely spaced, while distant inductors are automatically isolated because of the high attenuation.

In a low power and low data rate scenario, the efficiency of the system is related to the energy that is needed to transmit a single bit from the source to the destination. From the transmitter perspective, a high efficiency can be achieved by transmitting the bits over short voltage pulses. The power spectrum of a bipolar pulse of duration τ (Figure 6.16a) is shown in Figure 6.16b. Of the total energy of the pulse, 85% is allocated between the zero and $2/\tau$ frequencies, with the maximum at about $1/\tau$. Yet the overall energy of the pulse spreads over a wide bandwidth.

The duration of the pulse must be chosen such that $2/\tau$ is lower than the self-resonant frequency of the inductor, while at the same it must be long enough to limit the maximum current I_{max} in the inductor L, which is proportional to τ as demonstrated in Equation 6.9.

$$I_{max} = \frac{\tau}{2} \frac{V_O}{L}$$

(6.9)

On the other side of the channel, the receiver should amplify and detect the small and short pulse over the noise. Because of the wideband nature of the signal, noise limits the maximum communication range. Whereas with narrowband communication the SNR can be increased by means of high Q filter, this method is not applicable here because both signal and noise are broadband.

While the amplitude of the received signal only depends on the transmitter characteristic and the coupling factor k (which is a function of the distance and the antenna sizes), the noise is given by KT/C (where C is the input capacitance of the receiver and K is the Boltzmann factor). Limiting the bandwidth of the signal at $2/\tau$ fixes the minimum noise level compliant with the signal integrity (e.g., 10 dB at the receiver output for 10e−3 BER [19]), and hence the minimum value of k allowed for a given SNR. Simulations show that when using a resonant RLC filter with $Q = 0.8$ in the receiver, the optimal resonance frequency to maximize the ratio of peak signal to effective noise equals $0.61/\tau$. A Q of 0.8 is needed to dampen the oscillation induced in the tank by the pulse.

The only way to increase the SNR at a fixed k is to reduce the noise bandwidth, which means increasing C (and lowering L). This translates into increased power at the transmitter side because of the longer pulse and the higher peak current, according to Equation 6.8. Figure 6.17 shows the pulse duration τ and the peak current I_{max}, required for a SNR of 13 dB at the receiver side as a function of the coupling factor k, assuming a 1 V pulse at the transmitter. For example, for $k = 1e−4$, a 3 ns pulse and a 20 mA peak current are needed when using the previously discussed 76 nH inductors as antennas.

The SNR limitation can be relaxed by means of a coherent receiver [20], where the received signal is sampled and correlated with a local reconstruction of the pulse. However, this requires an accurate time base at a frequency higher than the pulse spectrum at the receiver side. With a pulse duration τ of 3 ns, and assuming five samples per pulse, a 1.66 GHz sampling clock is needed, which is not compatible with the stringent low-power constraints.

It is worth observing that a reactive communication link based on inductors is relatively insensitive to narrowband electromagnetic interference due to its poor antenna gain. Consider, for instance,

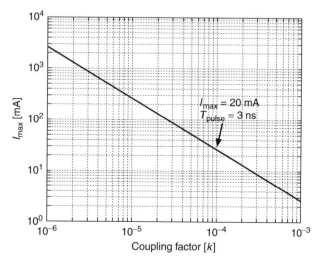

FIGURE 6.17 Peak current I_{max} and pulse duration τ as a function of the coupling factor k for an SNR of 13 dB and antenna inductance $L = 76$ nH.

an interfering electromagnetic (EM) wave at 300 MHz with 10 mV/m of electric field. This translates into a magnetic field B of $E/c = 33$ pT, which induces a voltage into a loop antenna (of equivalent area $A = 1$ cm^2) equal to $2\pi fAB = 12.5$ μV. By comparison, the voltage induced by the transmitter would be $k \times 1$ V $= 100$ μV. Magnetic coupling of electronic signals generated by the receiver itself (self-noise) may be a greater source of concern.

The properties of the inductive link discussed earlier in this section suggest that the channel resembles a serial line rather than a standard RF channel, as the main limitation stems from noise and not from interference. An energy detection scheme may hence be more appropriate for the receiver [19]. In a pure energy-detection scheme (incoherent or asynchronous receiver), the received signal is amplified and simply compared to a (set of) threshold(s) without any sampling. In principle, this kind of receiver is power hungry, as it must fully recover the channel attenuation over the full bandwidth without taking advantage of the positive feedback that would be introduced by sampling. For a low data rate link, however, the duration of the pulses is negligible with respect to the bit period. The power efficiency of the receiver can be dramatically improved by turning it off for most of the time and switching it on only when a pulse is expected. In contrast to the sampling-based system, where the clock accuracy must be of the order of a fraction of the pulse duration, accuracy must be just enough to ensure that the clock fits in the on window. The larger the windows, the lower the accuracy required, but the higher the average power consumption. Of course this windowing operation itself costs power and is only interesting as long as the power consumption of the local time base is smaller than the power saved by the duty cycling of the receiver.

The circuit diagram of a potential inductive transceiver for a dense network of inductively coupled nodes, designed on the basis of trade-offs discussed at the beginning of this section, is shown in Figure 6.18. For distances below 5 cm, energy levels as low as 30 pJ/bit for transmit and 500 pJ/bit for receive can be obtained using this scheme. A large fraction of the energy goes into the timing subsystem, which is a typical property of wideband transmission schemes.

One option to obtain the local time base is to equip every node with a precision crystal oscillator. Unfortunately, this choice increases both the size and the cost of the nodes in an undesirable manner. An alternative approach is to extract the time base from the received signal and to synchronize a local oscillator to it by means of a phase locked loop (PLL). That oscillator can then be used to drive the transmitter of the receiving node, and further propagate timing information (together with data) to its neighbors. In this kind of network, all the nodes are synchronized by simply tuning the frequencies of their own local oscillators to the frequency of the received signals. This simple principle results in a self-synchronized network; each node acts as a reference for its neighbor, while getting the reference from the neighbors themselves. The network can hence maintain global synchronization

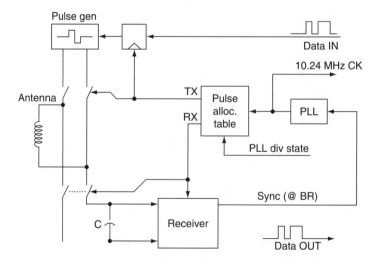

FIGURE 6.18   Schematic diagram of inductive transceiver for dense networks.

using only local asynchronous communications. Since each PLL adds some jitter to the clock, there is a maximum number of nodes through which the clock can be propagated depending on the PLL bandwidth and noise. This puts a limit on the size of the network that can be self-synchronized. The problem can be solved by putting in the network a certain number of synchronization nodes equipped with a stable crystal oscillator with low phase noise (or some other kind of global synchronization). In this way the clock signal is periodically (in space) realigned to remove the jitter introduced by the PLLs.

To analyze the feasibility of such an approach, a distributed synchronization algorithm where each node eventually locks to the closest reference clock was designed, so that each node experiences the minimum possible level of jitter. The performance of the algorithm was analyzed by means of computer simulations. A scenario was considered that composed of 500 nodes randomly distributed over an area of $0.5\,m^2$, and the jitter suffered by each node as a function of the number of nodes equipped with a crystal oscillator acting as reference nodes was determined. The simulator included a detailed model of the PLL circuit adopted in the transceiver presented in Figure 6.15. Results indicate that under realistic hypotheses on the stability of the PLL circuit (as determined by the maximum capacitance compatible with subcentimeter-size chips), a low number of reference nodes is sufficient to reliably synchronize the whole network. An example of the final result of the synchronization phase is presented in Figure 6.19, showing the case where three reference clocks are present in the network.

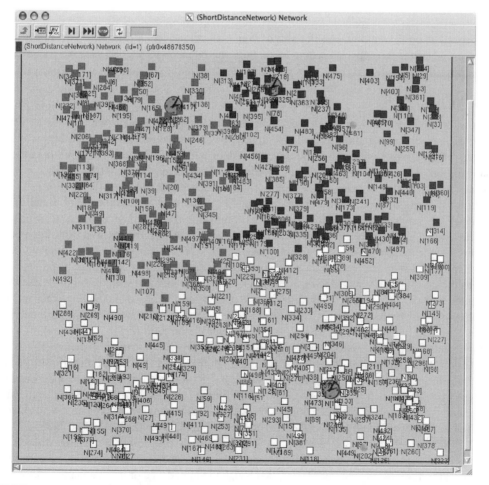

FIGURE 6.19 An example of dense network scenario considered in the analysis of the distributed synchronization algorithm. Large clocks represent active reference nodes, while squares represent active nodes locked to a clock, with different colors (white, black, and gray) indicating lock to different clocks.

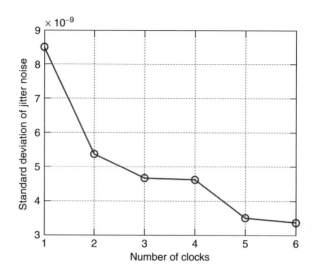

FIGURE 6.20 Standard deviation of jitter experienced by nodes in the simulated scenario as a function of the number of reference nodes in the network.

Depending on the requirements in terms of synchronization accuracy, additional reference nodes can be introduced in the network leading to a significant reduction of jitter, as shown in Figure 6.20.

PASSIVE RECEIVERS

For ultra-low-power communication over short distances, communication via backscattering is a potential alternative to traditional radiative approaches. Communication utilizing backscattering is often referred to as passive RF communication, since it allows a device to transmit data without generating any active radio signals itself. Currently, it is most often used in RFID systems, in which a powered reader extracts a simple ID number from a passive card or tag.

In most cases, the communication is initiated by a reader by emitting a signal at the desired frequency. This signal is detected by a passive data tag within the communication range, which then uses the energy of the received signal to wake up and power its electronics, and send the required data back to the reader. Passive tags exploit either near-field communication, which employs inductive coupling between the tag and the magnetic field of the reader, or far-field propagation, which utilizes the radiated electric field to receive energy from the reader and send the required data back. Depending on the spatial separation between the reader and the tag, one coupling mechanism or the other may apply. As previously discussed, devices that are less than $\lambda/2\pi$ apart are operating in the near field, while devices further apart operate in the far-field regime. In general, near-field communication is used by systems operating in low frequency (LF) and high frequency (HF) bands, whereas far-field communication is utilized by ultra-high frequency (UHF) and microwave systems.

The principle of transmitting data from the tag back to the reader, however, is independent of the type of field in which the devices operate. This principle of passive RF communication can be most clearly illustrated for near-field coupling by considering the analogy between near-field inductive coupling and a transformer introduced earlier in the chapter. The reader and the tag form a loosely coupled transformer with the tag antenna being the secondary coil. Just like any other transformer, changes in the load of the secondary coil are seen at the primary coil, which is in this case the reader antenna. Therefore, the tag chip only needs to change the impedance of its antenna to transmit data back to the reader, which is done by an internal circuit that translates the data stream into varying antenna impedance. The same principle is used in the far field, but now actual power is reflected back via backscattering, modulated by the changing antenna impedance. The only basic difference between near-field and far-field communication in a passive RFID system is the way the antennas are designed.

Since the entire electronic system of the passive tag runs from the power received from the radio signal of the reader device, two important aspects need to be considered when designing passive transponder circuits. First, the power consumption of the entire circuit needs to be kept as low as possible, often leading to simple circuits with very limited features. The second important consideration is the design of the rectifier, which is responsible for converting the received radio signal to a reliable power supply for the tag [21–24], and which has a significant impact on the energy efficiency of the receiver.

Another important design variable is the carrier frequency. Depending on the application, systems working with frequencies ranging from 135 kHz to 5.8 GHz have been proven to be feasible [25,27,28]. Higher frequencies usually lead to higher data rates and lower module sizes; however, lower frequencies usually provide better penetration in free space and the propagation characteristics degrade faster at higher frequencies, especially in the presence of water and metal. Another factor that influences the choice of frequency is the required data rate. A higher data rate usually requires a larger bandwidth, which might not be available in all frequency bands due to regulation restrictions. In general, unlicensed bands at higher frequencies provide more bandwidth than the bands at lower frequencies. While communication through backscattering is quite simple, it is also not very efficient; modulation schemes other than OOK or binary shift keying (BSK) are difficult to realize. This limits the efficiency of the link and typically translates to higher bandwidth requirements. Regulation also sets the amount of power that can be transmitted. In most bands, these regulations are quite severe, hence restricting the range of the link as well as the complexity of the receiver.

While most implementations of passive radios combine electromagnetic power scavenging with backscattering-based communication, this is not truly essential. For instance, it is possible to use a frequency signal in one band to provide enough power so that an active radio of the type discussed in the previous sections can be powered with it. This allows for individual optimization of the energy delivery and the communication channel, which is beneficial especially when higher data rates are needed.

Besides classical RFID applications, where only a fixed tag-ID is read out, passive RF communication is increasingly used for wireless sensor network–like applications. Several different sensor node architectures utilizing passive RF communication have been developed over the last several years. A pure passive sensor node uses the energy obtained from the radio link to power the sensor and to sense, convert, and transmit the data. Examples for such passive sensor nodes are presented in Refs. 26, 29. A different approach uses surface acoustic wave (SAW)-based transponders. In such systems SAW devices are used to reflect the incident radio signal. By using SAW devices having reflection characteristics that are sensitive to certain physical or chemical parameters, very simple purely passive sensor nodes can be built [32].

One drawback to the use of purely passive transmitters in sensing applications is that data can only be sensed while a radio signal is applied to the sensor node by the reader. In some applications, continuous collection of the data is required—while readouts occur in bursts. In general, the power obtained from the reader signal is not sufficient to power the sensor node for a long enough period of time. Tags with batteries for the sensor part of the device can be used under these conditions [25]. The read-out is done passively to remove the power needed for the radio transmission from the battery power budget. In principle, the batteries for the sensor part might be recharged wirelessly using the incident RF power, extending the lifetime of the sensor node and making it more independent from the restricted charge of the battery. Another option is to scavenge other sources of energy to power the electronics (such as motion, pressure, or acceleration) [31].

A typical example of the architecture of a passive transceiver for short distances is shown in Figure 6.21. This design by Hitachi [30] takes only 0.16 mm^2 in a 0.18 μm CMOS process. The maximum reading distance equals 40 cm for a 12.5 kbps data rate using amplitude-shift keying (ASK) modulation [30].

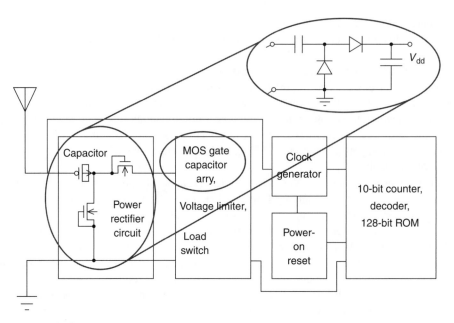

FIGURE 6.21 Architecture of passive short-distance transceiver link. (From Usami, M., et al., *Solid-State Circuits Conference*, 2003. With permission. © [2003] IEEE.)

TABLE 6.2
Energy Efficiency of Reported Transceivers

System	Range (m)	Peak Operating Power	Data Rate (Pulse Rate)	Energy/Bit (Energy/Pulse)	Notes
[33]	1	N.R.	20 MP/s	1.44 nJ/P	Carrier-based UWB
[34]	1	5 mW	1 MP/s	1 nJ/P	Baseband UWB
[10]	10	400 µW	20 Kb/s	20 nJ/b	Narrowband active
[15]	10	400 µW	100 Kb/s	3 nJ/b	Narrowband active
[35]	9.25	16.7 µW	250 Kb/s	60 pJ/b	Backscattering
[21]	12	2.7 µW	1 Mb/s	2.7 pJ/b	Backscattering

Note: For RFID systems, power consumed by the reader is not included, as it is usually not constrained. Including this power would increase the energy/bit to values in the order of 1 few µJ/bit.

IMPLEMENTATION COMPARISONS

In this chapter, the many trade-offs have been discussed that must be considered by the designer of a short-distance wireless link. Obviously, what strategy to use is a strong function of the application settings and the resulting constraints in terms of power availability, size, and cost. Yet, some universal metrics to help compare the various approaches are necessary. Table 6.2 compares a number of published transceivers in terms of range, peak operating power, data rate, and energy per bit. Obviously, other metrics (such as energy per useful bit) should be considered as well.

There are a few conclusions to be drawn from Table 6.2.

- The energy efficiency of nodes exploiting backscattering is outstanding due to the extremely low power consumption. Therefore, whenever an interrogator or a base station is available that does not have major power constraints, this should be the preferred choice. In many applications, however, the network topology imposed by this choice is not tolerable.

- The energy efficiency of UWB systems seems to be superior to that of narrowband systems. However, it comes at a higher peak power consumption, and further requires accurate synchronization of transmitter and receiver system clocks to enable efficient duty cycling. Also, Table 6.2 reports the energy/pulse, which does not take into account the need for most UWB systems to introduce spreading gain to cope with interference. In reactive links, as described earlier, interference is often only a minor concern due to the poor antenna efficiency, and spreading may not be necessary.

- Narrowband systems conversely achieve lower peak power consumption but, because of their high per-bit activity factor, lower energy efficiency. The rationale is that the narrowband impedance transformation through LC networks can be used to reduce power in such systems. Because impedances can be increased by a factor of Q, power savings of the order of $1/Q$ can be achieved using these techniques. However, the minimum achievable power dissipation is determined by the maximum value of parallel resistance obtainable in the process, as described earlier. A dominant source of power dissipation is the need for an (accurate) frequency reference.

CONCLUSIONS AND OPPORTUNITIES

Short-distance wireless links have the potential to revolutionize a wide range of applications, and may lead to entirely new concepts. However, major reductions in size, cost, and power are essential for many of these applications to become realistic. In this chapter, a range of potential solutions and options were evaluated. Yet, while the goals seem to be within reach for some applications, some major breakthroughs are still needed for the dream of others to come true. Novel technologies such as nanoscale RF-MEMS and novel computational paradigms such as self-organization and synchronization are needed to get there. The authors are convinced that the coming decade will see a wide range of new ideas emerging in each of these areas.

REFERENCES

1. Gyselinckx, B., Van Hoof, C., Ryckaert, J., Yazicioglu, R. F., Fiorini, P., and Leonov, V., "Human++: autonomous wireless sensors for body area networks," *Custom Integrated Circuits Conference*, pp. 13–19, September 2005.
2. Carmena, J. M., Lebedev, M. A., Crist, R. E., O'Doherty, J. E., Santucci, D. M., Dimitrov, D. F., Patil, P. G., Henriquez, C. S., and Nicolelis, M. A. L., "Learning to control a brain-machine interface for reaching and grasping by primates," *PLOS Biology*, vol. 1, no. 2, November 2003.
3. Drost, R., Forrest, C., Guenin, B., Ho, R., Krishnamoorthy, A. V., Cohen, D., Cunningham, J. E., Tourancheau, B., Zingher, A., Chow, A., Lauterbach, G., and Sutherland, I., "Challenges in building a flat-bandwidth memory hierarchy for a large-scale computer with proximity communication," *13th Symposium on High-Performance Interconnects*, pp. 13–22, 2005.
4. Miura, N., Mizoguchi, D., Inoue, M., Tsuji, H., Sakurai, T., and Kuroda, T., "A 195 Gb/s 1.2 W 3D-stacked inductive inter-chip wireless superconnect with transmit power control scheme," *2005 International Solid State Circuits Conference. Digest of Technical Papers*, pp. 142–143, February 2005.
5. Miura, N., Mizoguchi, D., Sakurai, T., and Kuroda, T., "Analysis and design of inductive coupling and transceiver circuit for inductive inter-chip wireless superconnect," *IEEE J. Solid-State Circuits*, vol. 40, pp. 829–837, April 2005.
6. Zimmerman, T., "Personal area networks: Near-field intra body communication," *M.I.T. Media Lab, Internal Report*, 1995.
7. Butera, W., "Programming a paintable computer," *PhD Thesis, M.I.T. Media Lab*, 2002.
8. Balanis, C., "*Antenna Theory, Analysis and Design*," 2nd Ed., John Wiley and Sons, New York, 1997.

9. Mizoguchi, D., Yusof, Y. B., Miura, N., Sakurai, T., and Kuroda, T., "A 1.2 Gb/s/pin wireless supercon-nect based on inductive inter-chip signaling (IIS)," *International Solid State Circuits Conference 2004. Digest of Technical Papers*, pp. 142–517, February 2004.

10. Otis, B., Chee, Y. H., and Rabaey, J. M., "A 400 μW Rx, 1.6 mW Tx super-regenerative transceiver for wireless sensor networks," *International Solid State Circuits Conference 2005. Digest of Technical Papers*, February 2005.

11. Chee, Y. H., Niknejad, A., and Rabaey, J. M., "A 46% efficient 0.8 dBm transmitter for wireless sensor networks," *2006 Symposium on VLSI Circuits Digest of Technical Papers*, June 2006.

12. Chee, Y. H., Niknejad, A. M., and Rabaey, J., "An Ultra-low power injection locked transmitter for wireless sensor networks," *J. Solid State Circuits*, vol. 41, no. 8, pp. 1740–1748, August 2006.

13. Ruby, R., Bradley, P., Larson III, J., Oshmyansky, Y., and Figueredo, D., "Ultra-miniature high-Q filters and duplexers using FBAR technology," *International Solid State Circuits Conference 2001. Digest of Technical Papers*, pp. 120–121, 438, February 2001.

14. Dubois, M. A., Billard, C., Muller, C., Parat, G., and Vincent, P., "Integration of high-Q BAW resonators and filters above IC," *International Solid State Circuits Conference 2005. Digest of Technical Papers*, pp. 392–606, February 2005.

15. Cook, B. W., Berny, A. D., Molnar, A., Lanzisera, S., and Pister, K. S. J., "An ultra-low power 2.4 GHz RF transceiver for wireless sensor networks in 0.13 μm CMOS with 400 mV supply and an integrated passive RX front-end," *International Solid State Circuits Conference 2006. Digest of Technical Papers*, pp. 370–371, 659, February 2006.

16. Drost, R., Hopkins, R. D., Ho, R., and Sutherland, I., "Proximity communication," *IEEE J. Solid-State Circuits*, vol. 39, pp. 1529–1535, September 2004.

17. Drost, R., Ho, R., Hopkins, D., and Sutherland, I., "Electronic alignment for proximity communication," *International Solid State Circuits Conference 2004. Digest of Technical Papers*, pp. 144–518, February 2004.

18. Miura, N., Mizoguchi, D., Inoue, M., Niitsu, K., Nakagawa, T., Tago, M., Fukaishi, M., Sakurai, M., and Kuroda, T., "1Tb/s 3 W inductive-coupling transceiver for inter-chip clock and data link," *International Solid State Circuits Conference 2006. Digest of Technical Papers*, pp. 424–425, 663, February 2006.

19. Tamtrakarn, A., Ishikuro, H., Ishida, K., Takamiya, M., and Sakurai, T., "A 1-V 299 μW flashing UWB transceiver based on double thresholding scheme," *2006 Symposium on VLSI Circuits. Digest of Technical Papers*, pp. 202–203, June 2006.

20. O'Donnell, I. D., Chen, M. S. W., Wang, S. B. T., and Brodersen, R. W., "An integrated, low-power, ultra-wideband transceiver architecture for low-rate, indoor wireless systems," *IEEE CAS Workshop on Wireless Communications and Networking*, September 2002.

21. Curty, J. P., Joehl, N., Dehollain, C., and Declercq, M. J., "Remotely powered addressable UHF RFID integrated system," *IEEE J. of Solid-State Circuits*, vol. 11, pp. 2193–2202, November 2005.

22. Curty, J. P., Joehl, N., Krummenacher, F., Dehollain, C., and Declercq, M. J., "A model for u-powered rectifier analysis and design," *IEEE Transactions on Circuits and Systems I*, vol. 52, pp. 2771–2779, December 2005.

23. Umeda, T., Yoshida, H., Sekine, S., Fujita, Y., Suzuki, T., and Otaka, S., "A 950-MHz rectifier circuit for sensor network tags with 10-m distance," *IEEE J. Solid-State Circuits*, vol. 41, pp. 35–41, January 2006.

24. Kocer, F., and Flynn, M. P., "A new transponder architecture with on-chip ADC for long-range telemetry applications," *IEEE J. Solid-State Circuits*, vol. 41, pp. 1142–1148, May 2006.

25. Claes, W., De Cooman, M., Sansen, W., and Puers, R., "A 136-μW/channel autonomous strain-gauge datalogger," *IEEE J. of Solid-State Circuits*, vol. 38, pp. 2280–2287, December 2003.

26. Aytur, T. S., Ishikawa, T., and Boser, B. E., "A 2.2-mm² CMOS bioassay chip and wireless interface," *2004 Symposium on VLSI Circuits. Digest of Technical Papers*, pp. 314–317, June 2004.

27. Usami, M., Sato, A., Sameshima, K., Watanabe, K., Yoshigi, H., and Imura, R., "Powder LSI: An ultra small RF identification chip for individual recognition applications," *International Solid State Circuits Conference 2003. Digest of Technical Papers*, pp. 398–501, February 2003.

28. Strassner, B., and Chang, K., "Passive 5.8-GHz radio-frequency identification tag for monitoring oil drill pipe," *IEEE Transactions on Microwave Theory and Techniques*, vol. 51, pp. 356–363, February 2003.

29. Stangel, K., Kolnsberg, S., Hammerschmidt, D., Hosticka, B. J., Trieu, H. K., and Mokwa, W., "A pro-grammable intraocular CMOS pressure sensor system implant," *IEEE J. Solid-State Circuits*, vol. 36, pp. 1094–1100, July 2001.

30. Hitachi Europe Limited, "Mu Chip Datasheet," http://www.hitachi-eu.com/mu/products/mu_chip_data_sheet.pdf

31. Roundy, S., Wright, P. K., and Rabaey, J. M., "Energy scavenging for wireless sensor networks," Kluwer Academic Publishers, Dordrecht, 2004.

32. Reindl, L., "Wireless passive SAW identification marks and sensors," *2nd International Symposium on Acoustic Wave Devices for Future Mobile Communication Systems*, Chiba University, March 2004.

33. Ryckaert, J., Badaroglu, M., De Heyn, V., Van Der Plas, G., Nuzzo, P., Baschirotto, A., D'Amico, S., Desset, C., Suys, H., Libois, M., Van Poucke, B., Wambacq, P., and Gyselinckxi, B., "A 16 mA UWB 3-to-5 GHz 20 Mpulses/s quadrature analog correlation receiver in 0.18 μm CMOS," *International Solid State Circuits Conference 2006. Digest of Technical Papers*, pp. 114–642, February 2006.

34. Terada, T., Yoshizumi, S., Muqsith, M., Sanada, Y., and Kuroda, T., "A CMOS ultra-wideband impulse radio transceiver for 1-Mb/s data communications and ±2.5-cm range finding," *IEEE J. Solid-State Circuits*, vol. 41, pp. 891–898, April 2006.

35. Karthaus, U., and Fischer, M., "Fully integrated passive UHF RFID transponder IC with 16.7-μW minimum RF input power," *IEEE J. Solid-State Circuits*, vol. 38, pp. 1602–1608, October 2003.

7 Ultra-Low-Power RF Transceivers

*Emanuele Lopelli, Johan D. van der Tang,
and Arthur H. M. van Roermund*

CONTENTS

Introduction 185
Energy-Scavenging Techniques 186
Low-Power Wireless Systems Trends 186
Recent Advances in Ultra-Low-Power Transceivers 187
    Research in Industries 187
    Research in Universities 187
System-Level Aspects for Wireless Microsensor Nodes 188
    System Architecture Trends 189
        Ultra Wideband Transceivers 190
        RFID, Subsampling, and Super-Regenerative Architectures 190
    Spread-Spectrum Systems 191
        DSSS versus FHSS 192
Frequency-Hopping Spread-Spectrum Synthesizers 194
    Predistortion-Based Hopping Frequency Synthesizers 195
        Deterministic Errors 196
        Stochastic Errors 197
        Data Recovery 200
FHSS Predistortion-Based Transmitter Design 201
    Experimental Results 204
Receiver Planning 205
    Receiver Architecture 207
    A 2.4 GHz Receiver Link Budget Analysis 209
        Propagation-Link Budget Analysis 209
        Link Budget Analysis of Discrete Parts 210
        Link Budget Analysis for Integrated Parts 210
    Simulation Results 215
    Intermodulation Distortion Characterization 216
Conclusions 218
References 219

INTRODUCTION

In recent years, a trend toward a world in which people will be surrounded by networked devices that are sensitive and adaptive to their needs has been witnessed. This trend has been expressed in a vision called ambient intelligence (AmI). It is possible to partition this world into three different classes of devices called Watt nodes, milli-Watt nodes, and micro-Watt nodes [1].

The Watt nodes and the milli-Watt nodes demand a further improvement in technology scaling to meet the low-power target. In contrast, the design of a micro-Watt node requires meeting the limit of miniaturization, cost reduction, and power consumption. Therefore, the complexity of this task does not lie in the number of transistors but in the ability to optimally combine technologies, circuit, and protocol innovation to obtain the utmost simplicity of the wireless node.

At the system level, there are several challenges engineers have to face to achieve such a reduction in terms of power consumption. Like in a paging channel, a low-power wake-up circuitry will be required together with received signal strength indication (RSSI) measurement circuits, so that energy expenditure can be optimally adjusted. Finally, advances in energy sources and scavenging techniques are mandatory to avoid battery replacement.

ENERGY-SCAVENGING TECHNIQUES

Several scavenging techniques have been studied in the recent years. However, it is unlikely that a single solution will satisfy the total application space. For example, a solar cell requires minimum lighting conditions, a piezoelectric generator requires sufficient vibration, and a Carnot-based generator requires a sufficient temperature gradient.

One of the most common scavenging techniques is to harvest energy from a radio frequency (RF) signal. An electric field of 1 V/m yields only 0.26 μW/cm^2, but such field strengths are quite rare [2]. This technique is generally used for radio frequency identification (RFID) tags, which have power consumption between 1 and 100 μW. Energy can be harvested by using solar cells. Although 1 cm^2 of standard solar cell produces around 100 mW under bright sun, it only generates no more than 100 μW in a typically illuminated office [2]. Also, thermoelectric conversion can be used as an energy-scavenging technique. Unfortunately, the Carnot cycle limits the use of this technique for small temperature gradients by squeezing the efficiency below 5% for about 15 degrees temperature difference* [2]. Another possible solution to the scavenging problem can be found using vibrational energy. If 1 cm^3 volume is considered, then up to 4 μW power can be generated from a typical human motion, whereas 800 μW can be harvested from machine-induced stimuli [2].

Considering the aforementioned state of the art in energy-scavenging techniques, researchers have plenty of room for improvement both in increasing energy harvesting efficiency and in reducing the overall power consumption of wireless nodes.

LOW-POWER WIRELESS SYSTEMS TRENDS

Increasing demands in wireless systems for sensing and monitoring applications have culminated in a high demand for low-power autonomous devices. Although AmI and the autonomous node concepts still are far from an everyday reality, much effort has been put in the elaboration of novel standards, which can better match the low-power trend. Following this novel trend, standards such as Bluetooth and Zigbee have been proposed in the recent years.

Whereas these standards have reduced power consumption at the expense of a lower data rate and quality of services (QoS), their protocol and their physical layer are still too complex to be implemented in an autonomous wireless device. Indeed, even though recently a simplified Bluetooth low-end extension (LEE) [3] has been released, the overall transceiver power consumption remains too high for an autonomous node. Although the data rate has been scaled down to less than 0.5 Mbps and other constraints have been simplified at both the physical and the media access control (MAC) layers, a state-of-the-art current consumption of 9 mA (at −2 dBm transmitted power) in transmit (TX) mode and 8 mA in received (RX) mode has been reported for a Bluetooth transceiver [3].

---

* The Carnot efficiency is $(T_H - T_L)/T_H$ where T_L and T_H are the low and the high temperatures in Kelvin. Going from the body temperature (36°C) to a cool room (20°C), the efficiency is only 5%.

In an effort to reduce the transceiver power consumption, the Zigbee standard has been recently released (preliminary standard by Philips in 2001, now updated to 802.15.4-2006). Whereas the data rate is reduced with respect to the Bluetooth standard, the overall transceiver power consumption remains high due to complex MAC and physical layers. A recent work [4] showed a Zigbee-compatible radio transceiver, with 21 mW power consumption in RX mode and 30 mW in TX mode (at 0 dBm output power). It follows that these standards cannot be used in the design of a micro-Watt node where an average power consumption of less than 100 µW, and therefore a peak power consumption lower than 10 mW, is required for duty cycles lower than 1%.

RECENT ADVANCES IN ULTRA-LOW-POWER TRANSCEIVERS

Starting with university research, the interest in ultra-low-power wireless devices has increasingly spread among companies as well. In the wide scenario of ultra-low-power devices, various pioneering investigations have been conducted to prove the feasibility of a micro-Watt node in terms of power consumption and robustness of the communication link.

RESEARCH IN INDUSTRIES

Several wireless products, which claim to be ultra-low power, are present on the market. Rarely these products can be used as core block for a micro-Watt node. Pioneering research toward the development of this kind of wireless nodes can be found in Refs. 5 and 6.

The Eco node [5] has been designed to monitor the spontaneous motion of preterm infants using the 2.4 GHz industrial scientific and medical (ISM) band at 1 Mbps data rate. While showing a good form factor ($648 \, \text{mm}^3 \times 1.6 \, \text{g}$), its power consumption is still far away from the minimum target required by an autonomous node. Indeed, even at 10 kbps and −5 dBm output power, it consumes 20.4 mW in TX mode and 57 mW in RX mode (at 1 Mbps) considering only the radio device. Whereas these nodes are designed for duty-cycled operation, as stated in the section "Low-Power Wireless Systems Trends," peak power consumption should not exceed 10 mW. Robustness of the link by frequency diversity is achieved by using a frequency-hopping spread spectrum (FHSS) technique.

The Telos node [6] complies with the ZigBee standard. As a result, while having a reduced data rate (250 kbps), it has an overall power consumption of around 73 mW from 1.8 V power supply at 0 dBm transmitted power.

RESEARCH IN UNIVERSITIES

Different universities are involved in pioneering research on ultra-low-power devices and networks. At Berkeley University, an ultra-low-power microelectromechanical system (MEMS)-based transceiver has been developed [7]. Despite using a 1.9 GHz carrier frequency and only two channels, the receiver power consumption is 3 mA from a 1.2 V power supply. The data rate is 40 kbps at 1.6 dBm output power. The low receiver power consumption is mainly obtained by using a high quality-factor (Q) MEMS resonator implemented as a thin-film bulk acoustic resonator (FBAR). If more channels are needed, such as in the case of an FHSS transceiver, the hardware requirement increases linearly with the number of channels, making this choice impractical from a low-power point of view. The transmitter part adopts direct modulation of the oscillator and MEMS technology, eliminating power-hungry blocks such as phase-locked loops (PLLs) and mixers, therefore reducing the overall power consumption. Two major drawbacks can be foreseen in the proposed architecture. While reducing the circuit and technological gap toward a micro-Watt node, it relies on nonstandard components (MEMS), which increase the cost and require higher driving voltage. Furthermore, it lacks in robustness due to the use of only two channels, while requiring a linear increase of the power consumption with the channel number if a more robust frequency diversity scheme has to be implemented.

At the CSEM Institute, the WiseNet [8] project aims to optimize both the MAC and the physical layers to obtain a robust, low-power solution for sensor networks. By using the 2.4 GHz ISM band available worldwide but the lower 433 MHz ISM band, it achieves a power consumption of only 1.8 mW from a minimum supply voltage of 0.9 V in RX mode. This result was achieved by a combination of circuit and system innovative techniques and the use of the low-frequency 433 MHz band, which reduces the power consumption of the most power-hungry blocks such as the frequency synthesizer. In TX mode, a high power consumption of 31.5 mW was reported mainly due to the choice of a high output power of 10 dBm. Data rate is 25 kbps with frequency-shift keying (FSK) modulation.

The proposed solution, while relying partially on the frequency band choice to reduce the power consumption, still requires external components such as high-Q inductors for the LC-tank circuit and external RF filters, which deteriorate form factor and power consumption at higher frequencies.

SYSTEM-LEVEL ASPECTS FOR WIRELESS MICROSENSOR NODES

Receiver sensitivity depends on the noise figure (NF), noise bandwidth, and the required signal-to-noise ratio (SNR) of the demodulator. The NF depends on the receiver architecture and technology used and in an asymmetric scenario can be considered to be smaller than 10 dB. The required SNR depends on the modulation scheme and so no improvement can be expected, given a certain bit error rate (BER) and a modulation scheme. For example, for a 10^{-3} BER and a 2-FSK noncoherent modulation scheme, the required SNR is about 12.5 dB. The only parameter left is the noise bandwidth, which ultimately affects the data rate. To meet the required ultra-low-power target, the transmitter node has to be duty cycled; it wakes up, transmits the required data, and falls back in an ultra-low-power mode called idle mode. Assume P_{idle} the power consumption in the idle mode, P_{TX} the power radiated from the antenna including the power amplifier (PA) efficiency, and P_{diss} the power used by the whole circuitry before the PA. Then, the average power consumption of the transmitter node can be approximated by the following equation:

$$P_d = P_{\text{TX}} \frac{T_{\text{TX}}}{T} + P_{\text{diss}} \frac{(T_{\text{TX}} + T_{\text{wu}})}{T} + P_{\text{idle}} \frac{(T - T_{\text{TX}} - T_{\text{wu}})}{T} \tag{7.1}$$

where T_{TX} is the time required for each transmission, T_{wu} the wake-up time of the transmitter (i.e., the time required to start up the circuitry and to acquire the synchronization), and T the time interval between two consecutive transmissions. The duty cycle of the system, d, can be defined as the ratio between the time required to transmit the data and the time between two consecutive transmissions. This time is the sum of the wake-up time, the transmission time, and the time for which the transmitter is in the idle mode. The transmission time depends on the packet length L_{pack} and the data rate D:

$$T_{\text{TX}} = \frac{L_{\text{pack}}}{D} \tag{7.2}$$

From these considerations, Equation 7.1 can be rewritten in the following way:

$$P_d = (P_{\text{TX}} + P_{\text{diss}}) \times d + P_{\text{idle}}(1 - d) + (P_{\text{diss}} - P_{\text{idle}}) \times d \frac{T_{\text{wu}}}{T_{\text{TX}}} \tag{7.3}$$

The required transmitted power is given by

$$P_{\text{TX}} = N_0 \times \frac{B_{\text{noise}}}{B_{\text{data}}} \times \frac{E_{\text{b}}}{N_0} \times \text{NF} \cdot D \cdot L_{\text{path}} \tag{7.4}$$

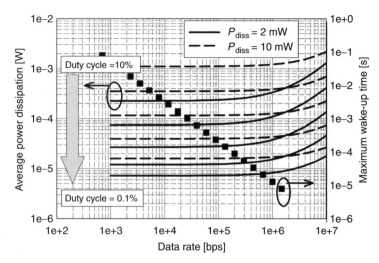

FIGURE 7.1 Average transmitter power consumption as a function of the data rate (packet length of 1000 bits).

where E_b is the energy per bit of information, N_0 the additive white Gaussian noise (AWGN) spectral density, B_{noise}* the noise bandwidth, B_{data}† the data bandwidth, and L_{path}‡ the path losses due to propagation. From Equation 7.3, it can be seen that the power consumption can be reduced by reducing the duty cycle and the data rate and by making the wake-up time small compared to the transmission time. This last requirement becomes difficult to achieve in an FHSS system at high data rate due to pseudorandom noise (PN) code synchronization.

Therefore, reducing the data rate will help relaxing the wake-up time for a given T_{wu}/T_{TX}. The transmitter average power consumption as a function of the data rate for different duty cycles is plotted in Figure 7.1. At high data rate, the average power consumption is dominated by the transmitted power. At data rates below a threshold value, the average power consumption is dominated by the pre-PA power. This threshold value depends on the pre-PA power dissipation and it is lower for lower values of the pre-PA power dissipation.

As shown in Figure 7.1, when the pre-PA power dissipation is 2 mW, this threshold value is around 100 kbps, whereas at pre-PA power of 10 mW, it is located around 1 Mbps. Furthermore, as expected at lower duty cycle, the average power consumption decreases. At higher data rate, the wake-up time has to decrease considerably to keep the contribution to the average power consumption negligible. Synchronization times below 1 ms are not easily achievable. Therefore, from the previous analysis, it is possible to conclude that a good strategy toward the reduction in the average transmitter power consumption consists in reducing the data rate and the synchronization time for a given node duty-cycle.

SYSTEM ARCHITECTURE TRENDS

Different radio architectures have been recently studied to reduce the power consumption. Some of these architectures comprise ultra-wideband (UWB) transceivers, superregenerative transceivers, subsampling transceivers as well as spread-spectrum-based transceivers (both frequency hopping [FH] and direct sequence [DS]).

---

* The 2-FSK noise bandwidth can be approximated by Carson's rule as $B_{noise} = 2(\Delta f + f_m)$, where $f_m = 2/T_s$, with T_s the symbol period and Δf the frequency deviation.

† For a 2-FSK modulated signal it equals four times the data rate.

‡ To calculate the path loss, the following expression has been used: $L_{path} = 27.56 - 20\log_{10}(f_c) - 20\log_{10}(d_0) - 10 \cdot n \cdot \log_{10}(d/d_0)$, where f_c is the carrier frequency expressed in megahertz, d the communication distance in meters, d_0 the reference distance for free-space propagation (unobstructed transmission distance, which is less than 2 or 3 m in an indoor environment) in meters, and n is the path-loss exponent. The carrier frequency is 915 MHz.

Ultra Wideband Transceivers

Among different architectures suitable for an ultra-low-power implementation, UWB-based systems are gaining more and more attention. FCC rules specify UWB technology as any wireless transmission scheme that occupies more than 500 MHz of absolute bandwidth.

The most important characteristic of UWB systems is their ability to operate in the power-limited regime. In this regime, the channel capacity increases almost linearly with power, whereas at high SNR it increases only as the logarithm of the signal power as shown by Shannon's theorem

$$C = BW \times \log_2\left(1 + \frac{P_S}{P_N}\right) \tag{7.5}$$

where P_S is the average signal power at the receiver, P_N the average noise power at the receiver, and BW the channel bandwidth. For low data rate applications (small C), it can be implied from Equation 7.5 that the required SNR can be very small, given an available bandwidth in excess of several hundred megahertz. A small SNR translates into a small transmitted power and, as a result, in a reduction of the overall transmitter power consumption.

Although UWB transceivers can have reduced hardware complexity, they pose several challenges in terms of power consumption. In Figure 7.2, a schematic block diagram of an UWB transceiver is shown. The biggest challenge in terms of power consumption is the analog-to-digital converter (ADC). If all the available bandwidth is used, the sampling rate has to be in the order of several gigasamples per second. Furthermore, the ADC should have a very wide dynamic range to resolve the wanted signal from the strong interferers. This implies the use of low-resolution full-flash converters. It can be proven [9] that a 4 bit, 15 GHz flash ADC can easily consume hundreds of milliwatts of power. Even if a 1 bit ADC at 2 gigasample per second is used, the predicted power consumption of the ADC remains around 5 mW [10]. Furthermore, the requirement on the clock generation circuitry can be very demanding in terms of jitter.

Besides these drawbacks, wideband low-noise amplifier (LNA) and antenna design are challenging when the used bandwidth is in excess of some gigahertz. The antenna gain, for example, should be proportional to the frequency [11], but most conventional antennas do not satisfy this requirement. LNA design appears quite challenging when looking at power consumption of state-of-the-art wideband LNAs [12]. A wideband LNA consumes between 9 and 30 mW making it very difficult to fulfill a constraint of maximum 10 mW peak power consumption for the overall transceiver. Although several successful designs are recently published showing the potential of UWB systems, their power consumption remains too high to be implemented in a micro-Watt node. In the design described in Ref. 12, the total power consumption is around 136 mW at 100% duty cycle, and in the one described in Ref. 13, a power consumption of 2 mW has been reported for the pulse generator only.

RFID, Subsampling, and Super-Regenerative Architectures

In the wide arena of low-power architectures, RFIDs represent a good solution when the applications scenario requires an asymmetric network. In this case, the micro-Watt node needs to transmit data

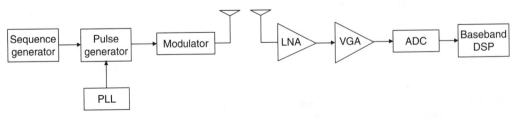

FIGURE 7.2 Building blocks of an impulse-based UWB transceiver. (From van Roermund, A. H. M., Casier, H., and Steyaert, M., *Analog Circuit Design* [AACD 2006], Springer, p. 385. November 2006.)

and to receive only a wake-up signal. The required energy is harvested from the RF signal coming from the interrogator. In the design described in Ref. 14, the interrogator operates at the maximum output power of 4 W, while generating 2.7 μW by inductive coupling. This power allows a backscattering-based transponder to send on/off keying (OOK)-modulated data back to the interrogator in a 12 m range using the 2.4 GHz ISM band. Unfortunately, the limited amount of intelligence at the transmitter side makes this architecture inflexible and only suitable for a highly asymmetric wireless scenario.

The Nyquist theorem has been explored in subsampling-based receiver to reduce the overall power consumption. The power consumption of analog blocks mainly depends on the operating frequency. Applying the theory of band-pass sampling [15], it can be proven that the analog front-end can be considerably simplified reducing the operating frequency. This has the potential to lead to a very-low-power-receiver implementation. Unfortunately, due to the noise aliasing, it can be proven that the noise degradation in decibels is

$$Dg = 10 \log_{10}\left(1 + \frac{2MN_P}{N_0}\right) \tag{7.6}$$

where M is the ratio between the carrier frequency and the sampling frequency, and N_p the filtered version of N_0. In this sense, the choice of the band-pass filter (BPF) as well as the choice of the sampling frequency becomes quite critical. Besides this, the phase noise specification of the sampling oscillator becomes quite demanding. Indeed, the phase noise is amplified by M^2 requiring a careful design of the voltage controlled oscillator (VCO). Consequently, when interferers are present, a poor phase noise characteristic can degrade the BER through reciprocal mixing considerably. Hence, till now, this architecture has been used mainly in interferer-free scenarios (space applications) [16].

Super-regenerative architectures date back to Armstrong, who invented the principle. Despite many years of development, they still suffer from poor selectivity and lack of stability, while having the potential to be low power. Furthermore, it is restricted to OOK modulation techniques only. In Ref. 17, bulk acoustic wave (BAW) resonators are used to reduce the power consumption and to provide selectivity. In spite of achieving an overall power consumption of 400 μW, it relies on nonstandard technologies (BAW resonators), which increase cost and form factor of the micro-Watt node.

In Ref. 18, a 1.2 mW receiver has been designed and fabricated in 0.35 μm CMOS technology. Even though the power consumption is very close to the requirements of a micro-Watt node, selectivity is quite poor. Indeed, to demodulate the wanted signal in the presence of a jamming tone placed 4 MHz away from the wanted channel with a BER of 0.1%, the jamming tone has to be no more than 12 dB higher than the desired signal. Generally, to achieve a reliable communication, the receiver should be able to handle interferers that have a power level 40 dB higher than the wanted signal with a BER smaller than 0.1%. This specification is very demanding for a superregenerative architecture, and it requires the use of nonstandard components such as BAW resonator to achieve a better selectivity.

SPREAD-SPECTRUM SYSTEMS

Among the various competing spread-spectrum (SS) techniques, the frequency-hopping (FH) system, in which the transmitting carrier frequency is changing according to a prescribed PN sequence, offers an attractive solution in terms of hardware complexity, system performance, and power consumption.

FH also enables the simplicity of using FSK modulation, which allows the possibility of employing a direct-conversion receiver architecture. In this way, a greater integration level and lower power consumption can be achieved. Furthermore, FSK modulation can be easily superimposed on the hopping carrier by simple digital techniques. Finally, the modulated output has a constant envelope and is amenable to nonlinear power amplification. A schematic block diagram of an FH transmitter is depicted in Figure 7.3a, whereas the distribution of the signal in the time–frequency plane is depicted in Figure 7.3b.

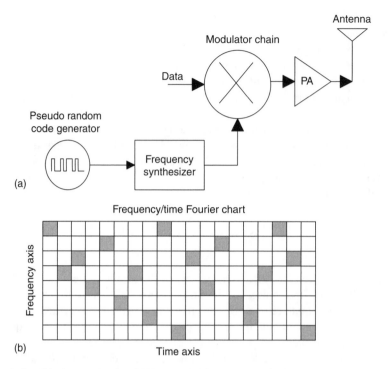

FIGURE 7.3 A simplified example of an FHSS transmitter: (a) block diagram of a frequency-hopping transmitter, (b) frequency-hopping signal. (From van Roermund, A. H. M., Casier, H., and Steyaert, M., *Analog Circuit Design* [AACD 2006], Springer, p. 388, November 2006.)

DSSS versus FHSS

Differently from a direct sequence spread-spectrum (DSSS) system, in an FHSS system, the spreading code is applied on the carrier frequency rather than on the modulated data. This choice has some advantages as well as some drawbacks. In terms of power consumption, the wake-up time required by the system is considerably lower. The wake-up time also takes into account the synchronization time. It can be proven that the average acquisition time for an SS system is

$$\overline{T_{sync}} = (N_{cell} - 1)T_{da}\left(\frac{2 - P_d}{2P_d}\right) + \frac{T_i}{P_d} \tag{7.7}$$

where T_i is the integration time for the evaluation of each cell in the time–frequency plane, P_d the probability of detection when the correct cell is being evaluated, T_{da} the average dwell time at an incorrect phase cell, and N_{cell} the total number of cells. Now, assuming that no frequency uncertainty is present, there is a time misalignment between the two PN sequences at the transmitter and receiver side equal to ΔT_i. Therefore, though for a DSSS, the system has to be synchronized within $T_C/2$ where T_C is the chip duration, an FHSS system needs to be synchronized within $T_s/2$. Because of the fact that in a DSSS system the processing gain is related to the ratio between the chip rate and the symbol rate, the chip period is at least an order of magnitude smaller than the symbol period. As a result, the number of cells that have to be evaluated in a DSSS system is considerably more than that in an FHSS system. As implied from Equation 7.7, the mean DSSS synchronization time is larger than that in the case of an FHSS system. This will increase the wake-up time and therefore the overall system power consumption.

   SS systems are generally affected by the so-called near-far problem. The near-far problem is the major limitation in DSSS systems and increases the complexity of the transceiver due to the

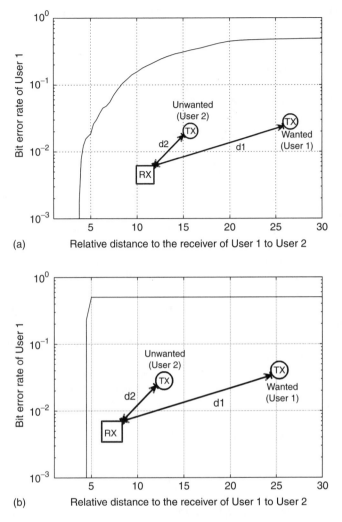

(a)   Relative distance to the receiver of User 1 to User 2

(b)   Relative distance to the receiver of User 1 to User 2

FIGURE 7.4 Near-far sensitivity comparison between FHSS and DSSS: (a) FHSS, (b) DSSS. (From van Roermund, A. H. M., Casier, H., and Steyaert, M., *Analog Circuit Design* [AACD 2006], Springer, p. 389, November 2006.)

need for power control circuitry. Without any power control, the performances of a DSSS system in an environment in which other DSSS systems are present can be heavily spoiled. As shown in Figure 7.4, in an FHSS system (Figure 7.4a), when User 1 is five times further away from the receiver compared to user 2, the BER is still below 5%. At the same relative distance, the DSSS system (Figure 7.4b) has already a BER of about 50%. The results shown in these plots have been obtained using Simulink® models, in which the users were interfering continuously with each other and only an AWGN channel was considered. In reality, the probability that two or more users will communicate simultaneously is in the order of a few percent. Consequently, the average BER should be scaled down accordingly. In conclusion, there is a high probability that in a DSSS system, an automatic gain control (AGC) should be used, whereas in an FHSS system, it can be easily avoided, reducing the complexity and the power consumption of the system.

Unfortunately, the requirements on the hopping synthesizer in terms of settling time and accuracy in the frequency synthesis are quite stringent. Therefore, the implementation of FHSS functionalities in a micro-Watt node requires a novel simplified architecture with power consumption an order of magnitude smaller than the state-of-the-art design.

FREQUENCY-HOPPING SPREAD-SPECTRUM SYNTHESIZERS

The main challenges in the implementation of an FHSS system result from the requirements for agile and accurate FH at very low power levels. Several ways to generate the hopping bins have been proposed in the recent years. Although widely used also for high data rate links, fractional-N synthesizers [19] and direct-digital frequency synthesizers (DDFS) [20] are relatively power hungry. As reported in Ref. 19, the synthesizer power consumption is 55 mW from a 2.5 V power supply, whereas according to Ref. 20, the synthesizer dissipates 40 mW from a 3 V power supply.

Fractional-N synthesizers have to deal with increased phase noise coming from the $\Sigma\Delta$ modulator. Indeed, the quantization noise is filtered by the PLL phase transfer function and converted into phase noise [21]. To conclude, a trade-off between phase noise and settling time exists. Phase noise can increase the system BER by reciprocal mixing. Therefore, it should be kept low by increasing the reference frequency or the order of the loop filter [22]. This increases the overall power consumption in both cases.

Recently, a new system concept based on analog-double quadrature sampling (A-DQS) has been proposed to relax the specifications on the synthesizer [23]. The channel selection originally performed by the synthesizer can be partitioned to the A-DQS. In this way, the step size of the PLL synthesizer can be doubled and the locking position of the local oscillator (LO) in the entire frequency band is halved. Nevertheless, the power consumption remains too high, mainly due to the ADC requirements [24].

In a DDFS, a sine wave is synthesized in the digital domain through the use of a simple accumulator, which produces, as an output, a ramp proportional to the desired frequency and a phase-to-sine amplitude converter. In the simplest case, this converter is a read-only memory (ROM). A digital-to-analog converter (DAC) and a low-pass filter (LPF) are used to convert the sinusoid samples into an analog waveform.

The main operation in a phase accumulator of N bit length is the N bit addition. As discussed in Ref. 25, the energy required for an addition by an arithmetic and logical unit (ALU) can be considered in the range of 300 pJ per addition (16 bit addition). If a bandwidth of 9.6 MHz (64 channels spaced by 150 kHz in the 915 MHz ISM band) should be synthesized using the architecture proposed in Ref. 20, then a 25 MHz reference clock is needed and the power consumption of the phase accumulator can be predicted by the following equation:

$$P_{\text{phase-acc}} = f_{\text{ref-clk}} \times E_{\text{add}} \tag{7.8}$$

where $f_{\text{ref-clk}}$ is the reference clock frequency and E_{add} the energy per N bit addition (300 pJ). From Equation 7.8, the predicted power consumption for the phase accumulator is evaluated as 7.5 mW. The second block in a DDFS, which to a great extent has significant power consumption, is the ROM.

Generally, even if a resolution of few hertz is needed to keep the spurious level low, practical words longer than 14 bits will lead to a very large ROM even if compression techniques are employed. Considering a truncated 14 bit phase word and a 12 bit word length for the amplitude mapping, the size of the ROM will be approximately 192 kbit. Splitting, for convenience, the ROM in three banks of approximately 64 kbit, it can be implemented by three $2^8 \times 2^8$ matrices.

Defining the storage array as a $2^n = 2^{n-k} \times 2^k$ matrix with 2^n memory cells, 2^{n-k} rows, and 2^k columns, then as described in Ref. 26, most of the power consumption in a ROM comes from the precharge or the evaluation of the 2^k memory cells. Therefore, approximate the total power consumption per operation by the following equation:

$$P_{\text{mem-cell}} = 3\frac{2^k}{2}\left(c_{\text{int}}l_{\text{column}} + 2^{n-k}C_{\text{tr}}\right)V_{\text{dd}}V_{\text{swing}} \tag{7.9}$$

where $P_{\text{mem-cell}}$ is the approximated power consumption of the ROM, the factor 3 takes into account that the total memory required has been split into three ROMs of smaller size, c_{int} is the capacitance of a unit wire length with minimum width, C_{tr} is the minimum size gate capacitance, V_{dd} is the power supply voltage, and V_{swing} is the voltage swing of each memory cell.

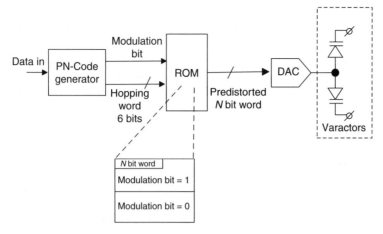

FIGURE 7.5   Proposed hopping frequency synthesizer architecture. (From Lopelli, E., van der Tang, J., and van Roermund, A. H. M., *IEEE ISCAS 2006*, p. 2310.)

Defining the memory cell as $d_m \times d_m$ square, the column interconnection length of the memory matrix is $l_{column} = 2^{n-k}d_m$.

Now, considering as an example the 0.18 μm CMOS technology, a 5.4 mW predicted power consumption for the ROM is obtained. As a result, the peak power consumption of the DDFS excluding DACs is higher than the 10 mW limit of a micro-Watt node.

PREDISTORTION-BASED HOPPING-FREQUENCY SYNTHESIZERS

A conceptual block diagram of the proposed architecture is depicted in Figure 7.5. The incoming data, together with the wanted hopping code, addresses a particular word cell in the ROM. The ROM has been split into two blocks depending on the modulation bit. In each memory cell, the predistorted word that will drive the DAC with a defined data bit is stored. The DAC then drives directly the varactor array changing the capacitance and therefore the VCO oscillation frequency according to the desired frequency bin and data.

The frequency of an LC-type oscillator and the tank capacitance are related by the following well-known relation:

$$f_{osc} = \frac{1}{2\pi\sqrt{LC}} \tag{7.10}$$

where L and C are the total capacitance and the inductance of the tank, respectively and f_{osc} the oscillation frequency. This, in practice, is realized by using a varactor diode, which has a capacitance that varies nonlinearly with its reverse voltage. Therefore, by applying the correct voltages to the varactor diodes, it is possible to synthesize all the required frequency bins with minimum hardware complexity (virtually only a voltage-controlled oscillator [VCO]).

In an FHSS system, the various frequency bins are addressed in a pseudorandom fashion. Pseudorandom codes are generated in the digital domain, whereas the varactor diodes require an analog control voltage. Consequently, a DAC is required as an interface between the digital world and the analog world. While passing from the digital world to the analog world, several nonidealities can affect the precision with which the frequency bins are generated. The main sources of error in this conversion process are the following:

- DAC quantization error (deterministic)
- Varactor nonlinearity (deterministic and stochastic)

- DAC integral non-linearity (INL) (stochastic)
- Square-root relation between frequency and tank capacitance (deterministic)

All these nonidealities in the transmitter chain can be divided into two groups. One group has a deterministic behavior and it does not vary due to process spread. The square-root nonlinear relation and the DAC quantization error fall in this category. The other group has a stochastic behavior. This means that it will depend on the process spread, and therefore a different behavior can be expected for different integrated circuits (ICs).

Deterministic Errors

The nonlinear relation between frequency and tank capacitance can be easily corrected by mapping the required frequencies in required capacitance values. From these values, knowing the varactor characteristic, a set of required voltages can be mapped and knowing the DAC specifications a set of digital words, which can be stored in a ROM, can be derived.

Therefore, neglecting for the moment the stochastic nature of the DAC linearity and of the C–V characteristic of the varactor, the problem simplifies in correctly choosing the DAC resolution to reduce the residual frequency error below a certain threshold.

The quantization error produces a nonlinear frequency error passing through the nonlinear C–V characteristic of the varactor and through the square-root relation between frequency and tank capacitance.

Looking at the square-root relation, when the overall capacitance is the smallest (at higher frequencies), an error on the capacitance due to quantization produces the largest error in the synthesized frequency. In this situation, the reverse voltage applied to the varactor is at its maximum value (for example, $-1.6\,\text{V}$). In this region, the varactor exhibits a highly linear behavior, contributing less to the overall frequency error. In these conditions, the frequency error is mainly caused by the quantization error passing through the square-root relation.

Close to the minimum reverse voltage (for example, $-0.2\,\text{V}$), the varactor characteristic is highly nonlinear. In this case, the frequency error is mainly caused by the quantization error passing through the varactor nonlinearity, whereas the square-root relation contributes minimally to it.

In both cases, a maximum capacitance error can be defined above which a frequency error larger than the required threshold is present at least in one of the synthesized frequency bins. In these two cases, the maximum errors ($\Delta C_{\sqrt{LC}}$ and ΔC_{var}) are defined as follows:

$$\Delta C_{\sqrt{LC}} = \frac{1}{[2\pi(f_{\max} + \Delta f)]^2 L} - C_{\min} \qquad (7.11)$$

$$\Delta C_{\text{var}} = \frac{1}{[2\pi(f_{\min} - \Delta f)]^2 L} - C_{\max} \qquad (7.12)$$

where $f_{\max}$ is the highest channel center frequency, $f_{\min}$ the lowest channel center frequency, Δf the maximum allowed frequency offset, $C_{\min}$ the capacitance at the highest channel frequency, $C_{\max}$ the capacitance at the lowest channel frequency, and L the LC-tank inductance value.

The maximum acceptable frequency error after predistortion can be derived by the following considerations. The orthogonality of PN codes has to be preserved. This means that the relative position of the channels along the frequency grid has to remain unchanged with respect to the ideal case. From this consideration, a maximum frequency error equal to the inter-channel spacing minus the channel bandwidth is allowed (for example, 100 kHz). If a 100 kHz maximum channel frequency shift is considered, then there would be the possibility that two channels get adjacent to each other (the interchannel spacing becomes zero). In this situation, if the specification on the oscillator phase noise remains unchanged, the amount of noise leaking in the adjacent channels increases, degrading the SNR in those channels.

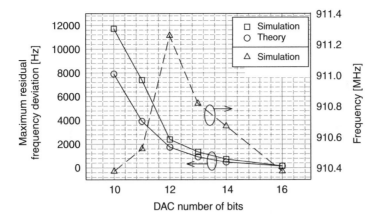

FIGURE 7.6 Maximum uncorrected frequency error and its position in the frequency band versus DAC number of bits.

To avoid degrading the BER in the adjacent channels considerably, a 0.5 dB maximum degradation on the phase noise has been considered.* Under this condition, a maximum uncorrected frequency error of 25 kHz can be tolerated [27].

Considering a 1.4 V swing on the varactor control voltage, an inductance value of 4.1 nH, and 64 channels placed around 915 MHz (ISM band), it can be found from Equations 7.11 and 7.12 that the largest error comes from the quantization error passing through the varactor nonlinear C–V characteristic. This is shown in Figure 7.6. The two solid-line curves represent the calculated (via Equations 7.11 and 7.12) and simulated maximum residual frequency error after predistortion is applied versus the DAC resolution. In this example, the DAC is considered linear (INL = 0). The dotted-line curve represents the position in the frequency range at which the aforementioned frequency error occurs. As can be seen, the largest frequency error due to the quantization error occurs at the lower portion of the frequency range.[†]

From the previous discussion, it has become evident that the quantization error translates into a nonlinear frequency error by passing through the varactor C–V curve and through the square-root relation. It has been shown that in the lower portion of the frequency band, the frequency error mainly depends on the effect of the varactor nonlinearity on the quantization error, whereas in the upper portion of the frequency band, the square-root function has the dominant effect. Looking now at Figure 7.6, in which both the nonlinearities are considered, it can be concluded that the frequency error coming from the DAC quantization error is mainly caused by the nonlinear mapping of this error in the frequency domain via the C–V characteristic of the varactor.

Given that the maximum residual frequency error has to be lower than 25 kHz, from Figure 7.6 it can be concluded that a 10-bit DAC is sufficient to achieve the required specification.

Stochastic Errors

In the previous analysis, the DAC has been considered linear and the spread on the varactor capacitance has been neglected. In the real case, they will affect the overall residual frequency error.

Given a certain DAC, its nonlinear behavior can be taken into account by applying a dedicated predistortion table. Unfortunately, the INL of each DAC will be different due to its statistical behavior. Therefore, this would require a different programmable look-up table per chip, which would be too

---

* The error probability of a noncoherent binary frequency-shift keying (BFSK)-modulated signal is given by $1/2e^{-E_b/2N_0}$.
   Considering a 0.1% initial BER, a 0.5 dB degradation in the phase noise translates in a 0.5 dB degradation in the SNR at the demodulator input and therefore in a BER close to 0.2%.
† Given 64 channels and a 150 kHz separation between adjacent channels, the minimum and maximum channel frequencies are 910.35 and 919.65 MHz, respectively.

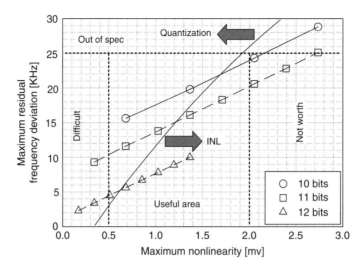

FIGURE 7.7 Effect of DAC INL on the residual frequency error.

costly for a system that aims to be very cheap. As a result, it is necessary to fulfill the required specifications on the residual frequency error even when the DAC statistical properties do change.

A DAC model has been built in Simulink, which also includes its nonlinear behavior [27]. The results are shown in Figure 7.7. Here both the quantization and the INL of the DAC are considered. Roughly four regions can be recognized. The upper region (*Out of spec*) does not fulfill the maximum residual frequency offset requirement. Then there are two regions, namely, *Difficult* and *Not worth*. The first region presents a difficult task for the designer due to harsh requirements in terms of maximum INL. The second one, while relaxing the INL requirements, necessitates the design of a higher resolution DAC than in the *Useful area* to fulfill the maximum residual frequency error specification. Designing one more bit of resolution generally requires more area and more power. So the design of the DAC is near optimum inside the *Useful area* shown in Figure 7.7.

Indeed, in the *Useful area*, the INL requirements are not too harsh and the residual frequency error due to the combined effect of both quantization error and DAC INL is smaller than the 25 kHz specification. This region is divided two parts by the bold line. This line represents the points at which the contributions of the quantization error and of the INL to the residual maximum frequency error are equal. Therefore, the right part of this area is dominated by the INL error, whereas the left part by the quantization error. Reducing the number of bits will reduce the chip area and in the end, the costs and the power consumption of the DAC. As a result, given the low-frequency operation of the DAC, a lower resolution DAC, which does not require an extremely small INL, can be chosen.

Among different DAC specifications that fulfill the maximum residual frequency error requirement, given the previous considerations and looking at Figure 7.7, a 10 bit DAC with an INL between 1 and 1.5 mV can be chosen as a near-optimum solution.

The last source of error is the variation in the *C–V* characteristic of the varactor due to the process spread. As for the DAC INL, this problem can be corrected by a dedicated predistortion look-up table. Although this is an effective solution, it can be costly.

Figure 7.8 shows the measured fine-tuning range of 20 IC samples, each calibrated to a common center frequency (915 MHz) via coarse tuning (namely, f_c in Figures 7.12a and 7.12b). The channel bin tuning voltages are generated by the DAC, driven by the baseband microprocessor. The look-up table in the ROM has not been changed and also the DAC is the same; therefore, the final effect is only due to the process spread on the varactor. Although the maximum frequency deviation of 20 ICs within the same batch is found to be no more than 220 kHz ($\sigma \approx 32$ kHz), interbatch spreads are larger.

In Figure 7.9a, the two extreme cases shown in Figure 7.8 are plotted. The effect of the varactor spread (in the two aforementioned extreme cases) on the position in the band of the frequency bins

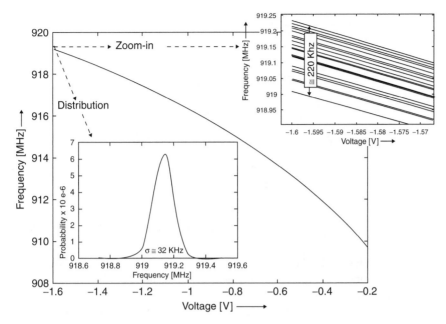

FIGURE 7.8 Measured fine-tuning range of 20 IC samples (same DAC).

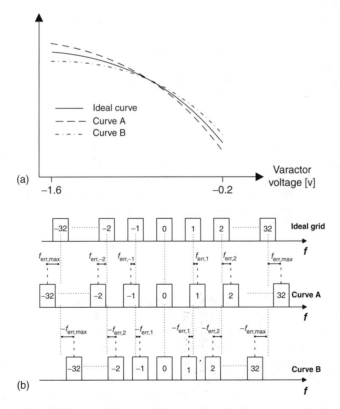

FIGURE 7.9 Effect of the C–V varactor characteristic spread on the frequency synthesis: (a) the two extreme cases shown in Figure 7.8, (b) effect of the varactor spread on the channel positions in the frequency band.

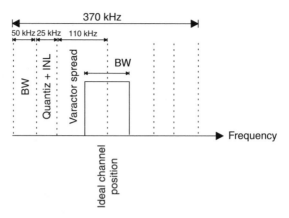

FIGURE 7.10   Minimum interchannel spacing when varactor spread and DAC INL are considered.

for a given predistortion table is illustrated in Figure 7.9b. It can be seen that due to the difference in the varactor C–V characteristic, there is a frequency offset accumulating while moving from channel 1 to channel 32 or from channel −1 to channel −32. The maximum error is present at the channels ±32 as can be seen also from the measured curves in Figure 7.8. Given the monotonicity of the C–V varactor characteristic, the amount of frequency error due to the different C–V characteristic between two adjacent channels is negligible compared with the frequency error due to the quantization error. Unfortunately if the predistortion table is kept the same for all the ICs in a batch, these phenomena will pose problem at the receiver side. Due to its statistical dependence from the process, the absolute position of the last channel with respect to the ideal position is known with a precision of ±110 kHz (see Figure 7.8). All the other channel positions are known with a better precision.

Therefore, at the receiver side, the channel bandwidth has to be as large as 370 kHz, whereas the interchannel spacing larger than 185 kHz. This is clearly shown in Figure 7.10. Indeed, the absolute position of the channel can spread 110 kHz (at 4σ) in both directions. Furthermore, the quantization error plus the INL of the DAC adds in the worst case 25 kHz more uncertainty in the channel position. Finally, the two channels should not overlap in the two worst cases, and therefore the center frequency of the adjacent channel has to be at least the bandwidth apart (in this case 50 kHz). Therefore, the baseband filter has to span two times 185 kHz, which makes 370 kHz.

If all the 64 channels are utilized, then the occupied bandwidth is around 12 MHz. Given that no crystal has to be used, then this poses some more strict requirements on the reference frequency accuracy. Indeed, the accuracy has to become better than 0.75% in this case, whereas in the implemented case, it can be relaxed to 1%. On the contrary, FCC rules demand for only 25 hopping channels for power levels below 0.25 W (which is generally the case for ultra-low-power wireless nodes). Therefore, another possibility is to still keep the 1% accuracy for the reference frequency but reduce the number of hopping channels.

Data Recovery

FSK modulation is very sensitive to frequency offsets. For example, in a correlator-type demodulator, the frequency offset has to be much smaller than the data rate for a BER lower than 1%. If a 1 kbps data rate is chosen, then the residual offset has to be no more than few hundred hertz. From Figures 7.6 and 7.7, it can be seen that a DAC with more than 14 bits is needed for a reasonable INL requirement.

Furthermore, a technological solution or a calibration circuitry would be needed at the transmitter side to correct for the statistical variation of the C–V varactor curve. Therefore, the choice of the demodulator at the receiver side is crucial to reduce the complexity at the transmitter side.

In Ref. 28, several demodulator topologies have been studied, with respect to their performances, in the presence of static frequency errors (these errors are caused by the statistical properties of the DAC and of the varactor and therefore are time invariant in first approximation). The short-time

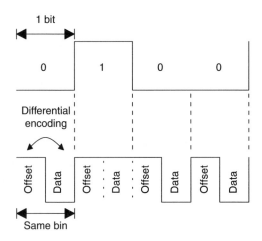

FIGURE 7.11   Residual offset cancellation technique.

discrete Fourier transform (DFT) (ST-DFT) algorithm, through differential encoding, shows a remarkable immunity against static frequency offsets. Furthermore, it can be applied in the digital domain, which has the potential to be low power. Indeed, if a zero-IF architecture is employed, due to the small signal bandwidth, the operating frequency of the ADC will be around 200k sample/s at the Nyquist rate (signal bandwidth equal to 50kHz).

As shown in Ref. 29, the capability of the ST-DFT algorithm to reject the frequency offset depends upon the condition that the offset is slowly varying. In other words, the frequency offset should vary at a rate smaller than the data rate. Therefore, the offset between two consecutive bits can be considered the same and it will be canceled out when differential encoding is applied. In this way also the frequency error due to temperature and power supply variations can be tracked and actively canceled out without requiring any additional circuitry.

Although differential encoding can cancel out, when applied to two consecutive bits, frequency errors induced by temperature, or power supply variations, it cannot cancel the frequency error induced by the statistical properties of the DAC and of the C–V varactor characteristic. Indeed, in the particular case of the proposed FHSS system, each bit is sent on a different channel, which is affected by a different offset due to the INL distribution of the DAC as well as the varactor C–V characteristic spread.

This means that two different bits will have two different offsets and, therefore, the simple differential encoding cannot cancel it out. Therefore, a straightforward way to cope with such a problem is to use again a dedicated look-up table for each IC. Although this approach guarantees an easy solution to the previous problem, it increases the cost of the final wireless node due to an increase in testing and calibration costs.

Another solution is to send at the beginning of each hopping information about the offset present at that particular hopping frequency. The principle is depicted in Figure 7.11. As shown in the figure, the offset is sent as a high-logic level, but it can be chosen to be a low-logic level as well. Because of the fact that the offset and the data are sent now on the same frequency bin, they are both affected by the same frequency error (due to the statistical properties of the DAC INL and the C–V characteristic of the varactor). The differential encoding, therefore, can provide the final cancellation in this case. The drawback of such a solution is an increase in the data rate. Given the large modulation index employed ($m > 5$) and the low data rate, there will be no severe drawback at the receiver side. Therefore, this technique has been chosen for the proposed implementation.

FHSS PREDISTORTION-BASED TRANSMITTER DESIGN

Two different transmitter RF front-ends have been realized in bipolar silicon-on-anything (SOA) technology. The front-end has to be as simple as possible while pushing the complexity in the digital domain. This is achieved by using the predistortion technique previously described. All the digital words stored in the ROM (see Figure 7.5) are calculated so that they take into account the DAC quantization error and its effect via a nonlinear transfer function (square root and varactor) on the frequency error. The INL of the DAC has to be compensated separately for each IC or can be left uncompensated if it is smaller than 1.5 mV for a 10 bit DAC (see Figure 7.7). The two architectures are depicted in Figure 7.12.

In Figure 7.12a, a common oscillator-divider front-end is shown. The system front-end consists of a resonator-based LC-VCO, a divider, and an output stage able to deliver −25 dBm power on a 50 Ω load. To minimize oscillator pulling, the VCO operates at 1.8 GHz and is divided down to the TX frequency. V_{f_c} is the coarse control voltage and is used to calibrate the hopping channels inside the ISM band. The second divider, connected to a second buffer, is used inside a frequency-locked loop (FLL) for the initial center frequency calibration. Once this calibration is performed, the FLL and related dividers are powered down, thus not contributing to the total power dissipation. V_{f_f} is the fine control voltage and is used to synthesize the 64 channel frequencies. The FHSS control, which is realized in

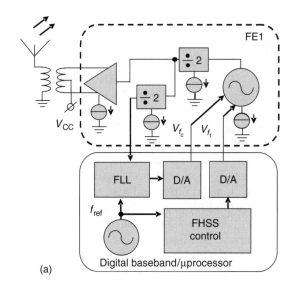

(a)

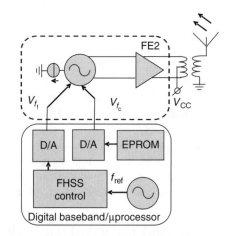

(b)

FIGURE 7.12 Transmitter RF front-end: (a) oscillator-divider-based front-end (From Lopelli, E., van der Tang, J., and van Roermund, A. H. M., *IEEE RFIC 2006*, p. 497.), (b) single building block front-end. (From van Roermund, A. H. M., Casier, H., and Steyaert, M., *Analog Circuit Design* [AACD 2006], Springer, p. 401, November 2006.)

baseband, generates the PN-hopping code and outputs the N bits predistorted digital words. These words are applied to the DAC to synthesize the required control voltage for the fine varactor bank.

Figure 7.12b shows a single building block RF front-end. It is a combination of a VCO and a PA, of which the coarse frequency calibration code can be factory set and stored in an erasable programmable read only memory (EPROM). This code will be converted into an analog voltage by a DAC. The output of the DAC will directly drive the coarse varactor on the power-VCO. The synthesis of the channel frequencies is obtained in the same way as for the architecture shown in Figure 7.12a. In this case, the oscillator center frequency is at 915 MHz instead of 1.83 GHz, which can reduce the system power consumption. Furthermore, no PA is required, but the VCO is directly coupled to the antenna through a balun.

The baseband part has been implemented using discrete components for both architectures. The microprocessor is a PIC18F627A, which consumes 12 µA at 32 kHz, whereas the DAC is an AD7392 [30], which draws 100 µA. The transistor-level schematics of the two front-ends are depicted in Figure 7.13.

The divider shown in Figure 7.13a is a traveling wave divider (TWD). In this design, the often-present external base resistors of the upper stage are not used, and the design is optimized by proper transistor dimensioning to have maximum divider sensitivity at 1.8 GHz and low power dissipation (200 µA). The output buffer (not shown) is a differential pair (2 mA). The oscillator core consumes 550 µA while achieving a −109 dBc/Hz phase noise at 450 kHz from the carrier.

The second architecture (Figure 7.13b) consists only of a directly modulated RF cascoded Colpitts power VCO. Cross-coupled oscillators (Figure 7.13a) have been preferred over other topologies for monolithic integrated circuit implementation because they are easily realized using CMOS technology and differential circuitry. Because of the use of the tail current source, cross-coupled oscillator

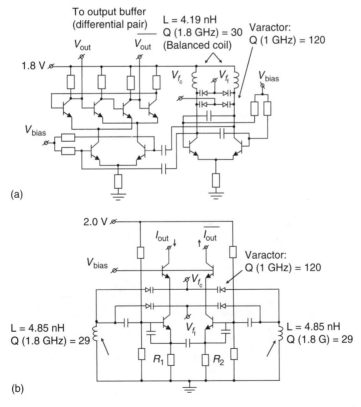

(a)

(b)

FIGURE 7.13 Transistor-level schematic of the realized front-ends: (a) FE$_1$ transistor-level schematic (From Lopelli, E., van der Tang, J., and van Roermund, A. H. M., *IEEE ISCAS 2006*, p. 2312.), (b) FE$_2$ transistor-level schematic. (From van Roermund, A. H. M., Casier, H., and Steyaert, M., *Analog Circuit Design* [AACD 2006], Springer, p. 403, November 2006.)

topologies present phase noise performances worse than classical types of oscillator with one of the active device ports grounded. In addition, classical type of oscillator topologies provide larger oscillation amplitudes for a given bias current because there is no voltage drop of the DC current across the current source element enabling, in this way, optimization of the power consumption.

One of these configurations is based on the Colpitts topology. Isolation problems can degrade VCO performances in terms of phase noise and frequency stability through phenomena like VCO pulling. This problem arises from changes in the load conditions and therefore requires a buffer stage between the VCO and the output stage, which increases the overall power consumption. To minimize the power consumption, the VCO and its buffer are connected in series to reuse the bias current between the two stages. For these reasons, a common collector Colpitts oscillator has been chosen together with a common-base buffer stage, arranged in a cascode configuration.

In this way, the current consumption is minimized, pulling is reduced due to better isolation between the tank and the load, and no PA is required; instead, the cascode stage can directly drive the antenna through a balun. As a result, a single-block RF front-end is obtained. A differential configuration has been chosen to reduce the second-order harmonic distortion term, which can degrade performances at the receiver side in a zero-IF architecture.

The bias current level largely depends on the required output power levels. It is desirable that phase noise requirements are met over a wide output power range. The output power can then be controlled by changing the bias current while meeting the BER specifications over the entire control range.

Experimental Results

The two RF frond-end ICs are shown in Figures 7.14a and 7.14b. The FE1 front-end occupies 2.8 mm², whereas the FE2 front-end 3.6 mm². Figure 7.15 shows the required bias current for the power VCO versus the output power and phase noise performance at 450 kHz away from the carrier for the front-end shown in Figure 7.14b. The required bias current ranges approximately between 1 and 2 mA to obtain an output power between −18 and −5 dBm. Simulated and measured results show a good agreement allowing minimization of the current consumption for a given set of output power and phase noise specifications. Indeed, in Figure 7.15, the phase noise varies between −102 and −115 dBc/Hz when the bias current changes between 1 and 2 mA, which is better than the required specification of −100 dBc/Hz at 450 kHz from the carrier.

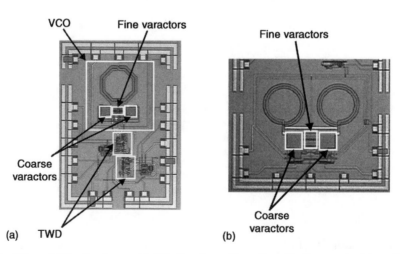

FIGURE 7.14 FE₁ and FE₂ die photos: (a) FE₁ die photo (From Lopelli, E., van der Tang, J., and van Roermund, A. H. M., *IEEE ISCAS 2006*, p. 2312.), (b) FE₂ die photo. (From van Roermund, A. H. M., Casier, H., and Steyaert, M., *Analog Circuit Design* [AACD 2006], Springer, p. 404, November 2006.)

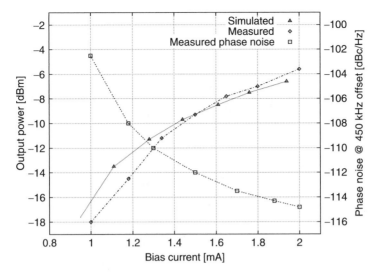

FIGURE 7.15 Output power and phase noise at 450 kHz offset versus bias current at 450 kHz offset. (From van Roermund, A. H. M., Casier, H., and Steyaert, M., *Analog Circuit Design* [AACD 2006], Springer, p. 405, November 2006.)

In Figure 7.16, the whole chain, from the look-up table in the microprocessor to the evenly spaced FH spectrum, is shown. In the measurement results shown in Figure 7.16, the 64 channels are addressed sequentially rather than in a pseudorandom fashion. As can be seen, the output of the DAC has a non-linear shape in time due to the predistortion algorithm. Moreover, the channels are equally spaced with a maximum frequency error between adjacent channels smaller than 5 kHz. This frequency error is smaller than theoretically predicted, due to the higher DAC resolution. Indeed, a 12 bit DAC has been used even if a 10 bit DAC was sufficient, as explained in the section "Predistortion-Based Hopping Frequency Synthesizers."

To demonstrate a reliable wireless link at the specified output power levels and phase noise specifications and to prove the predistortion concept, an FH transmitter has been realized. Two isotropic antennas placed 8 m apart were used in a non-line-of-sight (NLOS) configuration in a normal office environment.

The transmitted power has been set to −25 dBm, whereas the receiver employs a ST-DFT demodulation algorithm and a superheterodyne architecture. The measured received power is on an average −75 dBm. In this condition, the measured raw BER is lower than 1.1% at 1 kbps data rate and 1 khop/s hopping rate. The overall measured transmitter power consumption is 2.4 mW (for the front-end shown in Figure 7.14b at 100% duty cycle) from a 2 V power supply and 5.4 mW (including the buffer) for the front-end shown in Figure 7.14a from 1.8 V power supply. The baseband and mixed signal circuitry including the digital signal processor (DSP) dissipate 0.4 mW mostly consumed in the DACs.*

In Figure 7.17, the received spectrum before demodulation (Figure 7.17a) is shown together with a summary of the performances of the realized front-ends (Figure 7.17b).

RECEIVER PLANNING

A receiver front-end needs to achieve different objectives: amplification, mixing, filtering, and demodulation. For micro-Watt nodes, given the low mobility of the wireless nodes and the small channel bandwidth, a slow-fading condition is foreseen. Thus, diversity and equalization are not needed, simplifying, in this way, the demodulator.

---

* The baseband power consumption can be reduced to 0.2 mW if the coarse calibration DAC is switched off after coarse calibration has been achieved.

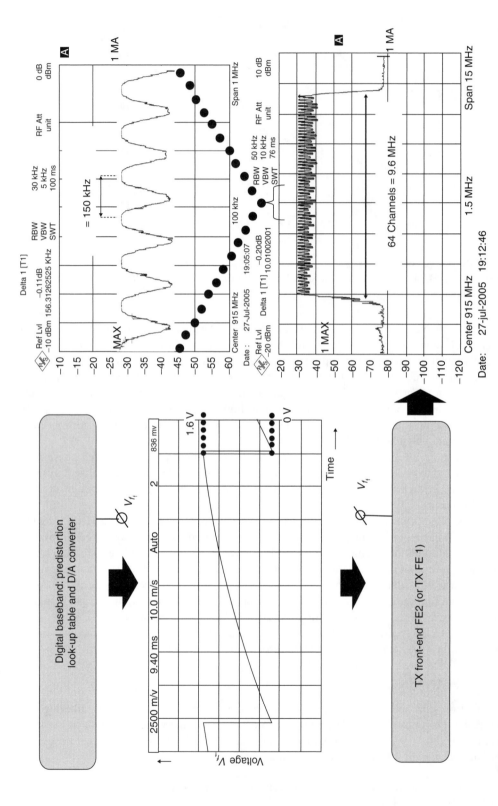

FIGURE 7.16 Predistortion chain with measured DAC predistorted output voltage and output spectrum of TX FE_2 (FE_1 has an identical spectrum). (From Lopelli, E., van der Tang, J., and van Roermund, A. H. M., *IEEE ISCAS 2006*, pp. 2311–2312.)

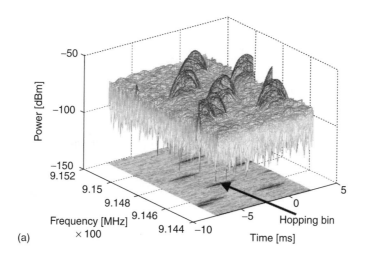

(a)

Parameter	FE1	FE 2 (1mA)	FE 2 (2 mA)	Unit
Technology	SOA	SOA		
Active chip area	2.8	3.6		mm^2
V_{cc}	1.8	2		V
Max. front-end current (20 samples)	2.8	1	2	mA
Max. front-end dissipation (20 samples)	5	2	4	mW
Baseband dissipation		0.4		mW
Total active FHSS TX dissipation	5.4	2.4	4.4	mW
Min. coarse tuning range (20 samples)	94.4	50		MHz
Min. fine tuning range (20 samples)	8.9	10		MHz
Worst case phase noise (20 samples)	−109	−102	−115	dBc/Hz @ 450 kHz
Output power	−25	−18	−5	dBm
Raw BER (8 meter distance, P_{out}=−25 dBm)		<1.1E−2		

(b)

FIGURE 7.17 Received spectrum and RF front-ends performance summary: (a) Received power spectrum (8 m distance, NLOS condition and −25 dBm transmitted power), (b) RF front-ends performance summary. (From van Roermund, A. H. M., Casier, H., and Steyaert, M., *Analog Circuit Design* [AACD 2006], Springer, p. 407, November 2006.)

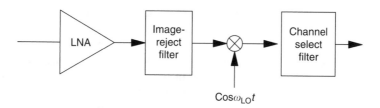

FIGURE 7.18 Simplified heterodyne receiver.

RECEIVER ARCHITECTURE

The transceiver architecture determines complexity, cost, power dissipation, and the number of external components. Two very common architectures are generally used in integrated transceivers: heterodyne and homodyne. In heterodyne architecture, the signal is down-converted to a lower IF frequency to relax the required Q needed at high frequencies to filter a narrowband signal from large interferers. Figure 7.18 shows a simplified block diagram of a heterodyne receiver.

The principal consideration in heterodyne architecture is the image frequency. This issue can be easily understood noting that all the signals at a distance equal to ω_{IF} from the LO frequency will be

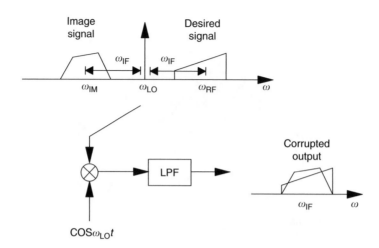

FIGURE 7.19 Image problem.

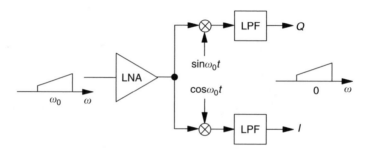

FIGURE 7.20 Proposed receiver architecture.

down-converted to the same IF frequency. This is illustrated in Figure 7.19. To suppress the image frequency, an image-reject mixer is often used. The choice of the IF is not trivial, and it entails a trade-off between sensitivity and selectivity. Generally, the image-reject mixer is realized using external components. This translates into a worst transceiver form factor and requires the LNA to drive a $50\,\Omega$ impedance. This finally translates into higher power consumption due to a larger required current.

The simplicity of the homodyne architecture makes this topology very attractive for ultra-low-power transceiver. The RF signal is directly down-converted to baseband. Figure 7.20 shows a simple homodyne receiver. In the general case of phase- or frequency-modulated signals, the information at positive and negative frequencies of the spectrum is different. Therefore, this topology requires quadrature down-conversion. Nevertheless, the homodyne receiver offers two big advantages over the heterodyne architecture. First, no image rejection filter is required, and second, the often-external IF filter is replaced by two simple LPFs, which can be easily integrated.

Unfortunately, homodyne receivers suffer from I/Q mismatch, even-order distortion, DC offset, and flicker noise. Flicker noise and DC offset can be overcome for low data rate application by employing wideband FSK modulation followed by a high-pass filter which can be sometimes a simple AC-coupling capacitance. This technique also solves the even-distortion problem partially in case the received signal contains some amplitude modulation. Indeed, if the information is carried by the frequency, the received signal can be hard-limited canceling any unwanted amplitude modulation. I/Q mismatch can be reduced by proper layout and matching techniques.

Given the aforementioned consideration, the proposed architecture is a zero-IF topology and is depicted in Figure 7.21.

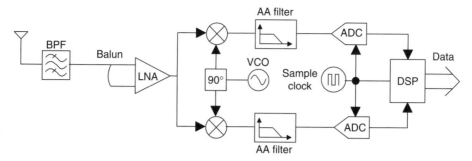

FIGURE 7.21   Homodyne receiver.

A 2.4 GHz RECEIVER LINK BUDGET ANALYSIS

Any channel and demodulator impose some boundary conditions on the RF front-end. These conditions can be specified in terms of SNR and can be translated into circuit concepts such as NF, gain and distortion. Next it is necessary to translate these boundary conditions into boundary conditions of the front-end subblocks.

Propagation-Link Budget Analysis

The first step consists in calculating the minimum received signal known as receiver sensitivity. The received signal P_{RX} can be expressed in terms of transmitted power P_{TX} and channel losses L_{path} by the following equation:

$$P_{RX} = P_{TX} - L_{path} \tag{7.13}$$

The losses in the channel can be expressed by the following equations [31].

$$L_{path,LOS} = -27.56\,dB + 20\log_{10} f_c + 20\log_{10} d_0 \tag{7.14}$$

$$L_{path} = L_{path,LOS} + 10 \cdot n \cdot \log_{10} \frac{d}{d_0} \tag{7.15}$$

where f_c is the carrier frequency expressed in megahertz, n the path loss exponent, which indicates how fast the path loss increases with distance, d_0 the reference distance in meters for free-space (unobstructed) propagation, $L_{path,LOS}$ the corresponding propagation loss of the line-of-sight (LOS) path, and d the distance in meters between transmitter and receiver.* Considering a 3 m reference distance, a $-6\,dBm$ transmitted power, and a maximum communication distance of 30 m with $n = 3$, the receiver sensitivity is about $-85.6\,dBm$ in the 2400 MHz ISM band.

The required SNR at the demodulator input largely depends on fading conditions. When a BFSK modulation technique is used, it can be proven that the required SNR is [31]

$$\Gamma = \frac{E_b}{N_0}\overline{\alpha^2} \tag{7.16}$$

---

* Isotropic antennas have been assumed at both the receiver and the transmitter sides. Indeed, in a sensor network, there is not, in general, a LOS condition and therefore isotropic antennas must be used.

where α is the gain of the channel with Rayleigh distribution. The term Γ represents, indeed, the average value of the normalized SNR. The error probability, for noncoherent BFSK modulation, is then given by [31]

$$P_{e,\text{NCFSK}} = \frac{1}{2+\Gamma}$$

(7.17)

For a 1% BER, the required SNR is 20 dB. Furthermore, this already large SNR has to be met also when interferences are present. Because these interferences are random in nature, the demodulator cannot differentiate them from the channel noise and will process them during the demodulation operation. In such situation, the SNR is reduced even more.

Link Budget Analysis of Discrete Parts

The first step is to filter out the noise and the interferences that are out of the band of interest. This is accomplished by the band-pass filter (BPF), as shown in Figure 7.21. The quality factor of this filter is generally quite high. For the band of interest, the required Q has to be $2400/83.5 \approx 29$. It is generally realized as an external LC network, but this approach increases the form factor and the cost of the wireless node. Therefore the integration of the BPF is highly desirable.

Several ways can be used in the design of integrated BPF. One possibility consists in using integrated passive components to realize the common LC ladder of a BPF. Unfortunately, on-chip inductors have very poor Q, limiting *de facto* the Q of the entire filter. The Q is related to the losses in the inductive element of the filters and, therefore, can be made very large if the losses are reduced. One way to accomplish this result is to use on-chip transformers [32]. The unloaded filter Q is given by

$$Q_{\text{unloaded}} \approx Q_{\text{inductor}} \frac{1+k}{1-\beta k}$$

(7.18)

where k is the coupling factor and β takes into account the losses in the transformer. Equation 7.18 shows that a small increase of k leads to a significant increase of the filter's unloaded quality factor. Starting from a $Q_{\text{inductor}} \approx 2$, a filter Q of about 10 has been achieved. To increase the unloaded Q of the filter, an active topology is needed [33]. Unfortunately, though a $Q \approx 3000$ is achieved, the required active components (CMOS transistors) consume around 10 mW while achieving 3.1 dB gain. Consequently, this topology cannot be used for ultra-low-power applications.

The last possibility lies on using FBAR. Their size is between five and eight times smaller than surface acoustic wave (SAW)-based filters. A 1.9 GHz FBAR BPF has been realized in Ref. 34. The total required volume is 1.0 mm × 1.0 mm × 0.7 mm. An insertion loss of 3.3 dB has been measured with 35 dB rejection in the upper stopband and 25 dB in the lower stopband.

To conclude, the two most promising ways to achieve integration of the BPF are the one based on on-chip transformers and the one based on FBAR resonators. Nevertheless, the Q of the first topology is still below the receiver specifications, whereas the second one cannot be still considered fully integrated. Therefore, there is a large margin of improvement in this field for researchers.

In this work, passive structures will be used for the BPF, the balun, and the TX–RX switch. The only required specification is the attenuation, which is then taken from available components off the shelf.

Link Budget Analysis for Integrated Parts

The receiver chain has to assure a certain SNR at the demodulator input. This SNR has been previously evaluated around 20 dB when worst-case fading is considered for a 1% BER. For noise calculation, the noise bandwidth needs to be calculated. The noise bandwidth is the smallest bandwidth

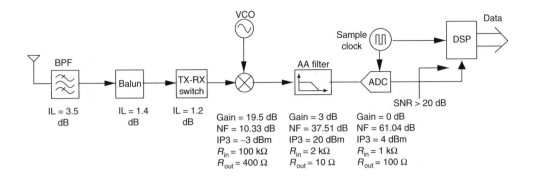

FIGURE 7.22   Receiver front-end blocks' specifications.

in the receiver chain, and it strongly depends on the channel bandwidth. Considering a 2 kbps data rate and a modulation index equal to 5, from Carson's rule, a 24 kHz signal bandwidth is required. Increasing the data rate increases the bandwidth and, therefore, reduces the available receiver NF. If the bandwidth has to remain the same to relax the receiver requirements, then the modulation index has to be reduced. This will make the effect of flicker noise and DC offset more severe. Therefore, the data rate can be increased by a combination of a smaller modulation index and larger signal bandwidth.

If a 24 kHz bandwidth is considered, then the specification on the antialiasing (AA) filter becomes prohibitive. This means that generally a certain oversampling ratio is needed at the ADC side to relax the AA filter specifications. In this case, an oversampling ratio of 4 has been considered. The choice of an oversampling ratio, though it appears spoiling the SNR by enlarging the noise bandwidth, does not have, in reality, any counterproductive effect. Indeed, at the DSP side, the SNR improves, through decimation, by a factor roughly equal to the oversampling ratio. For the moment, a Nyquist bandwidth of 96 kHz will be considered for noise calculations.

In Figure 7.22, the specifications for each block have been derived. Although it appears that no LNA is present, in reality it is merged with the mixer. To meet the low-power target, the simple cascaded LNA-mixer topology is not efficient. It requires two current sources while achieving a voltage gain and an NF, which are too close to the required specifications. As described in Ref. 35, the LNA-mixer front-end consumes 3.6 mA while achieving a 21.4 dB voltage gain and 13.9 dB of NF. For low-power applications, LNA and mixer should be merged in a single block, reusing the current. Following this approach in Ref. 36, a current-reused LNA-mixer front-end with 31.5 dB voltage gain and 8.5 dB NF at 500 μA current consumption (1.0 V power supply) in 0.18 μm technology has been recently proven. Comparing these figures with the specifications shown in Figure 7.22, it clearly appears that there is still room for a further reduction in the current consumption. The signal and the noise levels along the receiver chain are depicted in Figure 7.23a.

Although in principle the ADC can be replaced by a limiter (1 bit ADC), in reality this choice poses some drawbacks. Any radio system has to work in an environment full of other radios using the same propagation medium (the air) potentially interfering each other. Therefore, every radio must have a certain dynamic selectivity to cope with this scenario. Considering a simple second-order Butterworth filter as an AA filter, the signal and blockers levels throughout the chain are depicted in Figure 7.24. As can be seen, only around 10 dB carrier-to-interferer (C/I) ratio is achieved in the adjacent channel, whereas for the alternate and third and beyond channels, the situation is even more dramatic because the interferer has a higher power compared to the wanted signal. Therefore, either the selectivity of the analog filter has to improve by increasing the order of the AA filter or the ADC has to have enough dynamic range and linearity to convert in the digital domain the interferer as well. In this way, a digital postfiltering will achieve the required selectivity. Because of the fact that in the newer technologies the cost for such kind of digital techniques in terms of power is

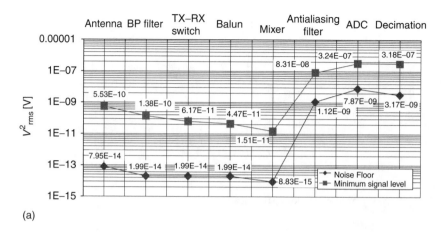

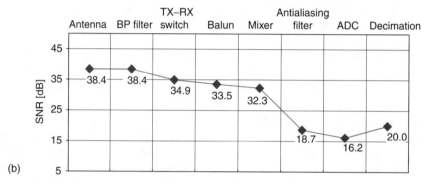

FIGURE 7.23   Receiver noise analysis: (a) signal and noise levels along the receiver chain at the input of each block, (b) SNR along the receiver chain at the input of each block.

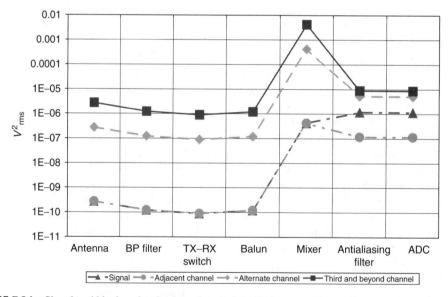

FIGURE 7.24   Signal and blockers levels along the receiver chain.

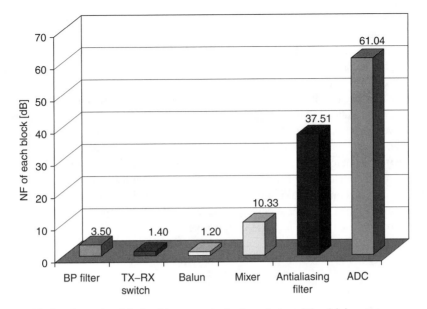

FIGURE 7.25   NF of each receiver block with respect to the impedance of the driving stage.

foreseen to decrease with time, it looks more promising from a power point of view a weak analog filtering followed by a strong digital postfiltering compared to a pure analog approach.

The values of signal, noise, and SNR given in Figure 7.23 have to be intended after decimation is performed. In Figure 7.23b, after decimation an SNR equal to about 20 dB is achieved.

Considering the noise factors, they have been calculated with respect to the source impedance driving the stage. The NF values for each block are depicted in Figure 7.25. The ADC NF is close to 61 dB with respect to the output impedance of the AA filter ($10\,\Omega$). It is important to notice that, due to the decimation, the new noise bandwidth at the ADC output is one-fourth of the Nyquist bandwidth. Therefore, the ADC noise power expressed in V_{rms}^2 is given by the following equation:

$$V_{rms,ADC}^2 = 10^{NF/10} \times kTR_s \frac{B_{Nyq}}{4} \tag{7.19}$$

where NF is the ADC noise factor expressed in decibels, k the Boltzmann constant, T the temperature expressed in Kelvin, R_s the source resistance of the driving stage, and B_{Nyq} is the Nyquist bandwidth (96 kHz in this case). The noise contribution of the ADC is then about 35.5 µV. Now supposing that the ADC full-scale range (FSR) is about 0.4 V, the required ADC SNR in decibels can be derived from the following equation:

$$SNR_{ADC} = 20 \log_{10} \left(\frac{\Delta V_{IN}}{2\sqrt{2V_{rms,ADC}^2}} \right) \tag{7.20}$$

where ΔV_{IN} is the ADC FSR and $2V_{rms,ADC}^2$ the ADC noise contribution in rms. From Equation 7.20 the required ADC SNR is evaluated as 72 dB. This corresponds to a required effective number of bits (ENOB) that can be calculated from the following equation:

$$ENOB = SNR_{ADC} - 1.76 - 10 \log_{10} \left(\frac{f_{sample}}{2B_{noise}} \right) \tag{7.21}$$

where f_{sample} is the ADC sampling frequency. For the system under analysis, this corresponds to 10.67 effective number of bits.

From Equations 7.20 and 7.21, it can be seen that the SNR due to quantization can be improved by oversampling the incoming signal. Taking oversampling to the extreme, it is possible to achieve any desired SNR with 1 bit sampling at a sufficient high rate. Unfortunately, oversampling places a bigger burden on the DSP because it increases the bit rate. It can be easily proven that if the SNR is kept constant, and bits of resolution are traded off for a higher sampling rate, the overall bit rate increases. This can pose a severe overhead in the DSP stage especially if the overall channel-filtering requirements are mainly shifted to the digital domain while leaving a simple low-power AA filter in the analog domain.

It is also important to notice that there is a trade-off between the requirements on the ADC SNR and the maximum gain achievable in the receiver chain without using an AGC. Indeed, the receiver gain strongly depends on the maximum signal the system needs to handle. When the transmitter is at its minimum distance from the receiver, the signal at the ADC input does not have to exceed the FSR. The maximum achievable receiver gain is given by

$$G_{\mathrm{rec}}^2 = \frac{\mathrm{FSR}^2}{2V_{\mathrm{mix\text{-}in,rms}}^2} \tag{7.22}$$

where the factor 2 converts the FSR from peak value to rms value, and $V_{\mathrm{mix\text{-}in,rms}}$ is the rms voltage at the mixer input which is equal to

$$V_{\mathrm{mix\text{-}in,rms}} = V_{\mathrm{in,max}} - L_{\mathrm{BPF}} - L_{\mathrm{TX\text{-}RX}} - L_{\mathrm{Bal}} \tag{7.23}$$

where $V_{\mathrm{in,max}}$ is the maximum signal at the receiver antenna and L_{BPF}, $L_{\mathrm{TX\text{-}RX}}$, and L_{Bal} are the respective attenuations due to the BPF, the TX–RX switch, and the balun. Given the values in Figure 7.22, the maximum voltage gain is limited to 34 dB. The gain along the receiver chain is shown in Figure 7.26.

The gain achieved till the ADC input is very close to the maximum value of 34 dB. A greater gain requires either a larger ADC FSR (with an increase in the required ENOB) or the use of an AGC system, which increases the gain when the signal is weak and reduces it when the signal is strong at the cost of decreased sensitivity in the latter.

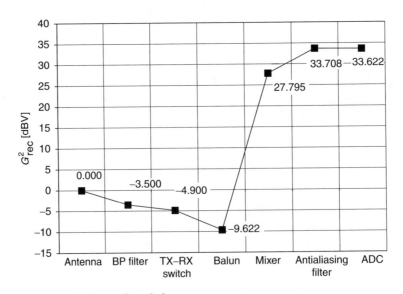

FIGURE 7.26   Gain through the receiver chain.

A good figure of merit (FOM) to characterize the ADC performances is given by the following expression:

$$\text{FOM} = \frac{P}{2^{\text{ENOB}} \times f_{\text{sample}}} \qquad (7.24)$$

where P is the power consumption and f_{sample} the sampling frequency. Supposing a target power consumption of 250 μW, 10.67 effective bits, and 192 k sample/s, we obtain an FOM equal to about 0.8 pJ/conversion, which is still feasible. It can be noticed that, given the previous FOM, and Equation 7.21, it is wise to keep the sampling rate low while increasing the ENOB to save power also at the ADC level. Besides this, a saving in the DSP power consumption is achieved given the lower generated bit rate.

Given the previous specifications for the ADC and supposing that no AGC system is used, a minimum distance of 40 cm between TX and RX nodes can be achieved without saturating the ADC.

SIMULATION RESULTS

The previous theoretical results have been verified by using ADS®. A linear model has been built on the basis of a two-port network. The comparison between simulated results and predicted results is shown in Figures 7.27 and 7.28. Figure 7.28 indicates a difference between the predicted SNR and the simulated one. Actually this is due to the fact that in ADS, no decimation has been performed and the SNR goes down due to the ADC noise. This is predicted theoretically also, and indeed, the gain due to the decimation process is less than the theoretical 6 dB and roughly equal to 4 dB. If no decimation is used, the predicted SNR becomes 15.5 dB, which correspond to the simulated value.

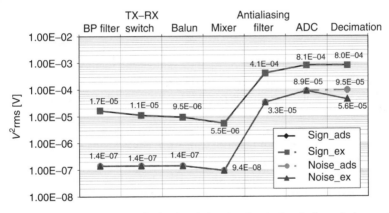

FIGURE 7.27 Theoretical (excel) and ADS-simulated signal and noise levels through the receiver chain.

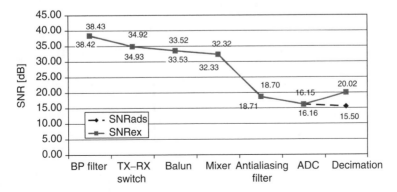

FIGURE 7.28 Theoretical (excel) and ADS-simulated SNR values through the receiver chain.

Intermodulation Distortion Characterization

In a real environment, there is a finite probability that in-band interferers can produce, due to the receiver nonlinearities, intermodulation products, which can fall in the band of interest. If the interferers are very strong, then it is possible that the resulting intermodulation product overwhelms the wanted signal.

Interferers can come from other nodes communicating on different channels or other wireless sources using the same bandwidth. The 2.4 GHz band is used also from standards like Bluetooth, Zigbee, and Wi-Fi, which can act as sources of interferers for the network. In the following analysis, only interferers coming from nodes of the same network will be considered. Indeed, if coexistence between different standards has to be taken into account, then the selectivity of the ultra-low-power node will increase considerably. This increment in the receiver selectivity is not consistent with the ultra-low-power target above a certain limit, and therefore an ultra-low-power network will be supposed to work properly mainly in a scenario in which other standards or wireless networks do not constitute the main sources of interference.

For the intermodulation distortion, the first passive blocks in the chain will not be considered. Therefore, the three blocks to be considered are the LNA-mixer block, the AA filter block, and the ADC. For the third-order intermodulation distortion calculation, the two interferers to be considered are placed in the adjacent and in the alternate channels. When they mix due to the nonlinearities in the receiver chain, they generate a third-order intermodulation product (IM_3), which falls in the wanted channel causing degradation in the SNR. The power levels of these two interferers along the receiver chain are depicted in Figure 7.29. Each of the aforementioned blocks can be described by its input third-order input intercept point (IIP_3). Then given the IIP_3, the third-order intermodulation product at the output of each block can be derived using the following equation:

$$IIP_3 = \frac{P_{out} - P_{IM_3}}{2} + P_{IN} \tag{7.25}$$

where P_{out} is the alternate channel interferer output power, P_{IM_3} the power of the third-order intermodulation product, and P_{IN} the adjacent channel interferer input power.

From Table 7.1 it is possible to highlight possible field of research. Concerning the mixer, given the required gain, NF, and IIP_3, the use of a merged mixer LNA-mixer topologies requires improvements either in the linearity or in reducing the power consumption. The work in [36] fulfills the gain and NF requirements but it has very poor linearity with a 1 dB compression point of only −31 dBm. This figure translates into an IIP_3 of about −21.3 dBm, which is very far from the required specification.

The work described in Ref. 37 meets NF and IIP_3, but it has a current consumption of 8 mA and a conversion gain of 23 dB. Given a 3.4 dB NF, there is a good margin to relax the NF and to

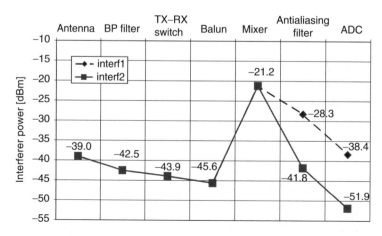

FIGURE 7.29 Adjacent and alternate channel interferer levels through the receiver chain.

TABLE 7.1
Theoretical and Simulated IM$_3$ Products

	IM$_3$ (dBm) (Theoretical)	IM$_3$ (dBm) (Simulated)	IM$_3$ (μV^2 rms)	IIP$_3$ (dBm) (Required)
Mixer	-100.41	-100.4	182	-3
AA filter	-88.67	-88.5	1360	20
ADC	-109.81	-109.8	91	4

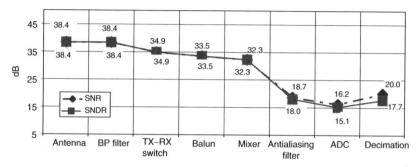

FIGURE 7.30   SNDR and SNR along the receiver chain with the input signal at the sensitivity level.

increase the gain while reducing the power consumption. The AA filter, as expected, has the largest IIP$_3$ requirement. Several configurations are feasible for baseband filters:

- g_m-C
- Switched capacitor
- Active RC

Active filter based on operational amplifier (OPAMP) are more linear than the one using open-loop active transconductors. The linearity is constrained by the linearity of the passive components in the OPAMP external feedback loop. A switched capacitor topology is also not suitable. Indeed, the filter must be able to handle signal at frequencies well above the channel frequency. Therefore, either the sampling rate becomes unacceptable or the order of the AA filter has to increase. Both choices lead to an increase in the filter power consumption.

To conclude, the best solution in terms of linearity and power consumption is the active-RC filter in which the OPAMP can be designed with transistors biased in weak inversion to reduce the power consumption. In a zero-IF architecture, special care has to be taken to avoid the degradation in filter performances due to flicker noise. Finally, a calibration is needed to correct for the spread in the position of poles and zeros of the filter due to process spread in resistances and capacitances. Nevertheless, this architecture assures very good linearity at reasonable cost in terms of power consumption.

Although the IIP$_3$ looks quite high, it can be achieved with less than 375 μA per pole of the filter transfer function [38]. Indeed, in this work, specifications are tighter than required. IIP$_3$ is higher than required, gain is 18 dB while only 6 dB are required in this work, and worst-case input-referred noise is 142 nV$_{rms}$/$\sqrt{Hz}$ for a 100 kHz bandwidth. The noise value is inline with the required specification of 106 nV$_{rms}$/$\sqrt{Hz}$. To avoid severe constraints on the linearity of the receiver, the SNR degradation due to the third-order intermodulation product has been fixed to less than 3 dB. This means that to keep the same SNR (20 dB), the signal cannot be at its minimum but 3 dB above. Generally, in low-power standards like Bluetooth, the signal is allowed to be 6 dB above the sensitivity level when intermodulation is considered. This gives a way to relax further the linearity requirement of the receiver chain. The SNR and the signal-to-noise-and-distortion (SNDR) are depicted in Figure 7.30. As can be seen from this figure, the presence of the third-order intermodulation product decreases

TABLE 7.2
FHSS Products Comparison

Company/Type	Operating Voltage (V)	Frequency Band	Current Consumption
Chipcon CC1050	2.1–3.6	300–1000 MHz	10 mA @ 868 MHz, −5 dBm
Analog Devices ADF7010	2.3–3.6	902–928 MHz	20 mA @ 0 dBm
Chipcon CC2550	1.8–3.6	2.4 GHz	12.8 mA @ −12 dBm
TI Dolphin	2.2–3.6	902–928 MHz	35 mA @ +7 dBm
Nordic nRF24E2	1.9–3.6	2.4 GHz	10.5 mA @ −5 dBm
Nordic nRF2402	1.9–3.6	2.4 GHz	10 mA @ −5 dBm
Nordic nRF24L01	1.9–3.6	2.4 GHz	7 mA @ −18 dBm
This work[a]	*1.8–2.0 V*	*902–928 MHz*	*1.1 mA @ −18 dBm*
			2.1 mA @ −5 dBm

[a] Coarse calibration DAC switched off.

the SNR after decimation by 2.5 dB. If the signal is allowed to be 6 dB above the sensitivity level, as in the Bluetooth standard, then the IIP_3 requirements for the LNA-mixer block and the AA filter can be relaxed to −7 and 17 dBm, respectively.

CONCLUSIONS

The implementation of a micro-Watt network, envisioned in the AmI concept, has a unique set of design constraints that focuses attention on small required bandwidth, low duty-cycle operation, short communication range, and the most challenging requirements in average and peak power demand. Nevertheless, a robust wireless link is mandatory, requiring spread-spectrum techniques implemented in a low-cost technology.

Reducing the transmission rate reduces the noise floor, giving the possibility to increase the noise added by the receiver. This translates into a smaller current consumption for the receiver part. Furthermore, decreasing the data rate for a given duty cycle allows for a longer synchronization time, which is an important parameter in SS radios.

Between the two most common SS techniques (DSSS and FHSS), an FHSS system has been found as the most robust and simplest architecture to be implemented in a micro-Watt node. The possibility to use the FSK modulation and a switching type PA, together with smaller wake-up time and no need for an AGC system, offers a unique possibility to lower the overall power consumption by an order of magnitude.

Problems arise due to the complexity of the hopping synthesizer in terms of accuracy and settling time in the frequency synthesis. A new architecture, based on the direct synthesis of frequency bins, is proposed in this chapter. The proposed architecture reduces the complexity of the hopping synthesizer, reducing by a factor 5 the power consumption with respect to state-of-the-art hopping synthesizers. The required accuracy in the synthesis of the frequency bins is obtained by a combination of digital predistortion and offset robust demodulation techniques. In this way, an absolute-accurate synthesizer is obtained and no feedback loop is used [39].

To demonstrate the feasibility of the digital predistortion concept, an FHSS transmitter has been realized in SOA technology. Two front-ends have been realized, one based on the common combination of VCO and divider [40] and a second one that reduces the RF front-end to a single block by combining the VCO and the PA.

Measurement results showed the possibility to achieve a BER smaller than 1.1% at −25 dBm transmitted power, non line-of-sight (NLOS) condition in a common office environment with a distance between TX and RX antennas of around 8 m. An overall power consumption for the complete

FHSS transmitter as low as 2.4 mW at −18 dBm transmitted power has been reported. This is five times less than the state-of-the-art FHSS transmitters (see Table 7.2*).

Ability to detect small signals in the presence of other interfering signals in a power-constrained situation is the main target in the design of an ultra-low-power receiver.

A zero-IF architecture is the most suitable architecture for full integration and avoids the IF section, which requires additional power. Due to the small signal bandwidth, the receiver NF is relaxed allowing reduction in the power consumption both in the LNA and in the mixer.

Considering interferences coming from other nodes of the same network, the IIP_3 requirements are also relaxed. Indeed, the zero-IF architecture concentrates the higher linearity requirements on the AA filter. In this block, noise is of no concern due to the high gain of the LNA-mixer block and therefore linearity is the only target. In the LNA-mixer, instead, the main target is the noise, though an 11 dB NF allows using a small current even at high voltage gain.

To conclude, if a low data rate is employed, NF can be extremely relaxed while linearity will not have severe degradation effect due to the low probability of collision. These architectural choices translate into an overall power optimization without spoiling the transceiver performances.

REFERENCES

1. H. D. Man, "Ambient intelligence: Gigascale dreams and nanoscale realities," in *International Solid-State Circuit Conference*, pp. 29–35, February 2005.
2. J. A. Paradiso and T. Starner, "Energy scavenging for mobile and wireless electronics," *IEEE Pervasive Computing*, vol. 34, January 2005.
3. M. Honkanen, A. Lappetelinen, and K. Kiveks, "Low end extension for Bluetooth," in *IEEE Radio and Wireless Conference*, pp. 199–202, September 2004.
4. P. Choi, H. C. Park, S. Kim, S. Park, I. Nam, T. W. Kim, S. Park, S. Shin, M. S. Kim, K. Kang, Y. Ku, H. Choi, S. M. Park, and K. Lee. "An experimental coin-sized radio for extremely low-power WPAN (IEEE 802.15.4) application at 2.4 GHz," *IEEE J. Solid-State Circuits*, vol. 38, pp. 2258–2268, December 2003.
5. C. Park, J. Liu, and P. H. Chou, "Eco: An ultra-compact low-power wireless sensor node for real-time motion monitoring," in *4th International Symposium on Information Processing in Sensor Networks*, pp. 398–403, April 2005.
6. J. Polastre, R. Szewczyk, and D. Culler, "Telos: Enabling ultra-low power wireless research," in *4th International Symposium on Information Processing in Sensor Networks*, pp. 364–369, April 2005.
7. B. P. Otis, Y. H. Chee, R. Lu, N. M. Pletcher, and J. M. Rabaey, "An ultra-low power MEMS-based two-channel transceiver for wireless sensor networks," in *Symposium on VLSI circuits*, pp. 20–23, June 2004.
8. C. Enz, A. El-Hoiydi, J.-D. Decotignie, and V. Peiris, "WiseNet: An ultralow-power wireless sensor network solution," *Computer*, vol. 37, August 2004.
9. P. Heydari, "A study of low-power ultra wideband radio transceiver architectures," in *Wireless Communications and Networking Conference*, pp. 758–763, March 2005.
10. I. D. O'Donnell and R. W. Brodersen, "An ultra-wideband transceiver architecture for low power, low rate, wireless systems," *IEEE Trans. Veh. Technol.*, vol. 54, pp. 1623–1631, September 2005.
11. R. H. T. Bates and G. A. Burrell, "Towards faithful radio transmission of very wideband signals," *IEEE Trans. Antennas Propagat.*, vol. AP-20, pp. 684–690, November 1972.
12. J. Ryckaert, C. Desset, A. Fort, M. Badaroglu, V. De Heyn, P. Wambacq, G. van der Plas, S. Donnay, B. van Poucke, and B. Gyselinckx. "Ultra-wide-band transmitter for low-power wireless body area networks: Design and evaluation," *IEEE Trans. Circuits Syst. I*, vol. 52, pp. 2515–2525, December 2005.
13. L. Stoica, A. Rabbachin, H. O. Repo, T. S. Tiuraniemi, and I. Oppermann. "An ultrawideband system architecture for tag based wireless sensor networks," *IEEE Trans. Veh. Technol.*, vol. 54, pp. 1632–1645, September 2005.
14. J.-P. Curty, N. Joehl, C. Dehollain, and M. J. Declercq, "Remotely powered addressable UHF RFID integrated system," *IEEE J. Solid-State Circuits*, vol. 40, pp. 2193–2202, November 2005.

---

* In the table for the transceiver products, only the power dissipation in TX mode has been considered for comparison.

15. R. G. Vaughan, N. L. Scott, and D. R. White, "The theory of bandpass sampling," *IEEE Trans. Signal Process.*, vol. 39, pp. 1973–1984, September 1991.

16. M. R. Yuce and W. Liu, "A low-power multirate differential PSK receiver for space applications," *IEEE Trans. Veh. Technol.*, vol. 54, pp. 2074–2084, November 2005.

17. B. Otis, Y. H. Chee, and J. Rabaey, "A 400 µW-RX, 1.6 mW-TX super-regenerative transceiver for wireless sensor networks," in *International Solid-State Circuit Conference*, pp. 396–397, February 2005.

18. A. Vouilloz, M. Declercq, and C. Dehollain, "A low-power CMOS super-regenerative receiver at 1 GHz," *IEEE J. Solid-State Circuits*, vol. 36, pp. 440–451, March 2001.

19. D. Theill, C. Durdodt, A. Hanke, S. Heinen, S. van Waasen, D. Seippel, D. Pham-Stabner, K. Schumacherl, "A fully integrated CMOS frequency synthesizer for Bluetooth," in *Radio Frequency Integrated Circuits Symposium*, pp. 103–106, May 2001.

20. G. Chang, A. Rofougaran, M.-K. Kuand, A. A. Abidi, and H. Samueli, "A low-power CMOS digitally synthesized 0-13 MHz agile sinewave generator," in *International Solid-State Circuit Conference*, pp. 32–33, Feb. 1994.

21. T. A. D. Riley, M. A. Copeland, and T. A. Kwasniewski, "Delta-sigma modulation in fractional-N frequency synthesis," *IEEE J. Solid-State Circuits*, vol. 28, pp. 553–559, May 1993.

22. N. Christoffers, R. Kokozinski, S. Kolnsberg, B. J. Hosticka, "High loop-filter-order $\Sigma\Delta$—fractional-N frequency synthesizers for use in frequency-hopping-spread-spectrum communication-systems," in *International Symposium on Circuits and Systems*, vol. 2, pp. 216–219, May 2003.

23. P.-I. Mak, C.-S. Sou, S.-P. U, and R. P. Martins, "Frequency-downconversion and IF channel selection A-DQS sample-and-hold pair for two-step-channel-select low-IF receiver," in *10th IEEE International conference on Electronics, Circuits and Systems*, vol. 2, pp. 479–482, December 2003.

24. P.-I. Mak, C.-S. Sou, S.-P. U, and R. P. Martins, "Frequency-downconversion and IF channel selection A-DQS sample-and-hold pair for two-step-channel-select low-IF receiver," in *International Symposium on Circuits and Systems*, vol. 1, pp. 1068–1071, May 2004.

25. V. Kalyanaraman, M. Mueller, S. Simon, M. Steinert, and H. Gryska, "A power dissipation comparison of ALU-architectures for ASIPs," in *ECCTD 2005*, vol. 2, pp. 217–220, September 2005.

26. D. Liu and C. Svensson, "Power consumption estimation in CMOS VLSI chips," *IEEE J. Solid-State Circuits*, vol. 29, pp. 663–670, June 1994.

27. E. Lopelli, J. van der Tang, and A. van Roermund, "Ultra-low power frequency-hopping spread-spectrum transmitters and receivers, in *Analog Circuit Design*, Springer, November 2006.

28. E. Lopelli, J. van der Tang, and A. van Roermund, "A FSK demodulator comparison for ultra-low power, low data-rate wireless links in ISM bands," in *ECCTD 2005*, vol. 2, pp. 259–262, September 2005.

29. S. Hara, A. Wannasarnmaytha, Y. Tsuchida, and N. Morinaga, "A novel FSK demodulation method using short-time DFT analysis for LEO satellite communication systems," *IEEE Trans. Veh. Technol.*, vol. 46, pp. 625–633, August 1997.

30. *Analog Devices*, "AD7392 data sheet" [Online]. Available: www.analog.com.

31. B. Leung, *VLSI for Wireless Communication*. Prentice Hall: Upple Saddle River, NJ,2002.

32. A. H. Aly, D. W. Beishline, and B. El-Sharawy, "Filter integration using on-chip transformers," in *Microwave Symposium Digest IEEE MTT-S Digest*, vol. 3, pp. 1975–1978, June 2004.

33. Y.-C. Wu and M. F. Chang, "On-chip spiral inductors and bandpass filters using active magnetic energy recovery," in *Custom Integrated Circuit Conference*, pp. 275–278, May 2002.

34. D. Shim, Y. Park, K. Nam, S. Yun, D. Kim, B. Ha, and I. Song, "Ultra-miniature monolithic FBAR filters for wireless application," in *Microwave Symposium Digest IEEE MTT-S Digest*, pp. 213–216, June 2005.

35. F. Beffa, R. Vogt, W. Bachtold, E. Zellweger, and U. Lott, "A 6.5-mW receiver front-end for Bluetooth in 0.18-µm CMOS," in *Microwave Symposium Digest IEEE MTT-S Digest*, vol. 1, pp. 501–504, June 2002.

36. T. Song, H.-S. Oh, S. Hong, and E. Yoon, "A 2.4-GHz sub-mW CMOS receiver front-end for wireless sensors network," *IEEE Microwave Wireless Compon. Lett.*, vol. 16, pp. 206–208, April 2006.

37. H. Sjöland, A. Karimi-Sanjaani, and A. Abidi, "A merged CMOS LNA and mixer for a WCDMA receiver," *IEEE J. Solid-State Circuits*, vol. 38, pp. 1045–1050, June 2003.

38. H. A. Alzaher, H. O. Elwan, and M. Ismail, "A CMOS highly linear channel-select filter for 3G multi-standard integrated wireless receivers," *IEEE J. Solid-State Circuits*, vol. 37, pp. 27–37, January 2002.

39. E. Lopelli, J. van der Tang, and A. van Roermund, "An ultra-low power predistortion-based FHSS transmitter," in *IEEE International Symposium on Circuits and Systems*, pp. 2309–2312, May 2006.

40. E. Lopelli, J. van der Tang, and A. van Roermund, "A sub-mA FH frequency synthesizer technique," in *IEEE Radio Frequency Integrated Circuits Symposium*, pp. 495–498, June 2006.

8 Human++: Emerging Technology for Body Area Networks

Bert Gyselinckx, Raffaella Borzi, and Philippe Mattelaer

CONTENTS

The Growing Economic Burden of Health Care Systems 221
    Demographics: Toward an Aging Society and Active Aging 222
    Epidemiological Transition: Chronic Diseases 222
    Rising Patient Expectations: Toward a Patient-Centric Approach 223
e-Health: Toward a Proactive and Connected Health 223
Body Area Networks: An Enabling e-Health Technology 224
Ambulatory Multiparameter Monitoring as Test Case 224
Wireless Communication 229
    UWB Pulse Generator 229
    UWB Analog Receiver 230
Micropower Generation and Storage 232
Sensor and Actuators 236
Integration Technology 237
Conclusions 240
References 240

THE GROWING ECONOMIC BURDEN OF HEALTH CARE SYSTEMS

Many national health services struggle in the face of financial resource constraints and shortages of skilled labor. The cost of health care delivery is steadily on an upward trend. A recent survey shows that by 2020 health care spending is projected to triple in dollars, consuming 21% of gross domestic products (GDP) in the United States and 16% of GDP in other organisation for economic co-operation and development (OECD) countries. As a result, the pressure on health systems to step up efforts in cost containment and efficiency improvement keeps growing. Consensus about the main determinants of expenditure is not complete but revolves generally around cost drivers such as rising income and patient expectations; demographic change, in particular the aging of population; and new technologies.

In wealthier nations, consumer demand increases, leading to a higher spend on health care. Statistics show that the cost-lowering effect of technology and automation is more than offset by the impact of an aging society, consumerism, biotechnology, and medical breakthroughs. This results in an overall increase in cost between 2 and 3% per year. As a result, alternative ways of increasing efficiency, productivity, and usability while controlling cost are being sought. One strategy that is gaining major attention consists of offloading health care institutions by shifting the health management outside the expensive formal medical institutions. Other strategies seek to improve the appropriateness of treatment or emphasize preventive care rather than treatment. For example, the field of chronic diseases is a vast domain in which the provision of real-time data from and to the patient anywhere and at any point of time may hold significant potential for cost reduction.

The supporting role of an adequate technology platform is critical here. e-Health technology, enabling wireless and mobile-based health care services, is increasingly coined as the revolutionizing enabler for the next decades to come. However, at this early stage today it is only through the many pilot projects on e-health ongoing in different countries around the world that evidence will be gathered to determine the economic viability.

There are three strong drivers for an e-health technology platform: demographics, the epidemiological transition, and the rising patient expectations.

DEMOGRAPHICS: TOWARD AN AGING SOCIETY AND ACTIVE AGING

Demographic development in the United States, Europe, and Japan is leading to a situation where a decreasing number of younger people will have to support an increasing number of older retired people. Close to 40% of the population will be older than 65 by 2050. This major development, better known as "population aging," is due to a fall in birthrates, following the baby boomer wave, as well as a progressive lengthening of life expectancy. Consequently, as people live longer and healthier lives than ever before, their habits and lifestyles will change accordingly. A vision, often captured by the notion of "active aging," emerges. It calls for a society where later life is as active and fulfilling as the earlier years, with older people participating in their families and communities.

Both the age wave and the steady movement toward active aging call for radical changes in how care will be provided for the elderly and how technology may assist. In addition, a society increasingly based on looser social structures, as is the case in many western countries, suggests further a need for solutions supporting an independent lifestyle.

The advent of self-care and self-management support programs, the boom of developing mobile solutions for personal health monitoring, personal health coaches obtaining immediate feedback, and health information from patients at home are strong indicators today for a silent revolution of an aging society reshaping the health care and well-being landscape.

EPIDEMIOLOGICAL TRANSITION: CHRONIC DISEASES

Chronic diseases are now catalogued as the major cause of death and disability worldwide. Health care systems were historically designed to manage acute illness, including infections and injury, while today more than 75% of all health care spending is on chronic diseases such as cardiovascular disease, cancer, chronic obstructive pulmonary disorder, diabetes, and mental illness.

The increased incidence of chronic diseases and conditions presents a huge challenge to health services worldwide. WHO data indicate that 75% of the total population has one chronic condition and 50% has two or more chronic conditions.

As a result of the rising prevalence of chronic conditions, quality and appropriateness of patient care is an area of growing concern in the field of chronic diseases management. Current practices generally focus on acute, episodic care rather than on the effective provision of ongoing treatment and coordination among multiple care providers. Furthermore, current health care largely focuses on treatment rather than on prevention.

Several important aspects emerge. First, it is an acknowledged issue that many health systems dealing with chronic disease are fragmented. Furthermore, their decision making as well as interventions is often hampered by a lack of communication and information as well as standards supporting these.

Second, despite a widespread recognition that lifestyle factors play a determinant role in a person's health, there is however limited use of lifestyle parameters as a significant medical input in clinical diagnosis and treatment. A reversal of this trend, however, could take place with the advent of wearable body monitors that could further assist people at home in managing their lifestyle.

Via sensors in and around the body, these systems provide supplementary parameters in the prevention and treatment of many major conditions confronting health care today.

RISING PATIENT EXPECTATIONS: TOWARD A PATIENT-CENTRIC APPROACH

Patients expect ever more from health services. This can be explained by the fact that demand for health care rises with increasing standard of living, technological progress, and growing public knowledge about health care. In an emerging world of personal wellness, access to basic and vital health care data empowers the patient to further improve his quality of life.

Rising patient expectations ultimately translate into a growing consciousness of consumers to take progressively health matters in their own hands. This in turn reinforces the existing concept of a "patient-centric view" of health care, which emphasizes the patient's experience and journey through a system that provides continuity of care to a proactive patient.

The future points toward a greater reliance on technology enabling fully integrated health-delivery portals. These will be streamlining processes and systems across the different health care domains in and outside the hospital setting with a common access to patient data and correlating in real time with decision-support tools leading to a more effective and cost-efficient disease and health management.

e-HEALTH: TOWARD A PROACTIVE AND CONNECTED HEALTH

A vision shared by many in the industry is one of future health care systems supported by e-health. As defined by WHO, e-health refers to the use of digital data transmitted, stored, and retrieved electronically for clinical, educational, and administrative purposes in the health sector, both at the local site and at a distance.

e-Health is claimed to offer the potential to reduce costs, deliver remote health services, and increase the delivery efficiency in real time. The argumentation can only get stronger as we enter the age of ubiquitous computing.

The migration of patient management and health care delivery activities through the Internet or by means of wireless transmission—sometimes referred to as "connected health" and implying the patient to play a proactive role—becomes almost inevitable. Trends such as rising patient expectations, the emergence of portable health solutions, and the ongoing adoption toward electronic medical records are all converging to the same: e-health.

Whether and how health care could take advantage of achieving the cost-saving potential of information and telecommunication technology as seen in other sectors over the last decade is a very complex matter and remains to be seen. However, an important consideration with respect to the cost-reduction potential is the shift from hospital-centered health care to citizen-centered health care. Performing health care functions at home for the elderly population, for example, rather than requiring patients to attend the primary care or the hospital setting may yield considerable cost savings and waiting times. Patient satisfaction and compliance could be additional beneficial effects as remote-monitoring techniques infer and disrupt to a lesser extent the patient's life.

A profound consequence of adopting new health care system interfaces based on real-time information to and from the patient is the further adoption of self-care and home telecare as it also facilitates the health care service delivery models in changing their focus. In other words, we can anticipate a shift from episodic care to continuity of care, from institutional care to community and home-based care, from individual care to a multidisciplinary team approach, and from disease management to health management.

The era of e-health has only started. An increasing number of initiatives around ubiquitous health care around the world are paving the way to answering the question how it can enhance the health care system efficiency and resolve the associated cost burden.

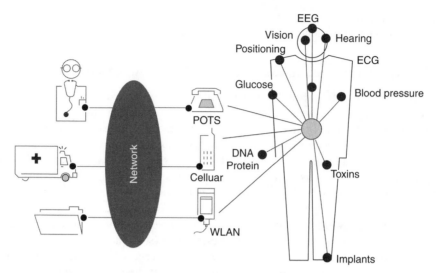

FIGURE 8.1   The technology vision for the year 2010: people will be carrying their personal body area network and be connected with service providers regarding medical, sports, and entertainment functions.

BODY AREA NETWORKS: AN ENABLING e-HEALTH TECHNOLOGY

In this chapter one component of e-health, the body area network (BAN) will be analyzed. This technology will enable people to carry their personal BAN [2] that provides medical, sports, or entertainment functions for the user (see Figure 8.1). This network comprises a series of miniature sensor/actuator nodes each of which has its own energy supply, consisting of storage and energy-scavenging devices. Each node has enough intelligence to carry out its task. Each node is able to communicate with other sensor nodes or with a central node worn on the body. The central node communicates with the outside world using a standard telecommunication infrastructure such as a wireless local area or cellular phone network. The network can deliver services to the person using the BAN. These services can include the management of chronic disease, medical diagnostic, home monitoring, biometrics, and sports and fitness tracking.

The successful realization of this vision requires innovative solutions to remove the critical technological obstacles. First, the overall size should be compatible with the required form factor. This requires new integration and packaging technologies. Second, the energy autonomy of current battery-powered devices is limited and must be extended. Further, interaction between sensors and actuators should be increased to enable new applications such as multiparameter biometrics or closed-loop disease management systems. Intelligence should be added to the device so that it can store, process, and transfer data. The energy consumption of all building blocks needs to be drastically reduced to allow energy autonomy.

The Human++ program [1] is looking into all of these generic BAN challenges. In the following sections a closer look at the technologies under development will be examined. The discussion will start with an overview of the test case, and then the enabling technologies such as wireless communication, micropower generation, and sensors. Finally, how advanced integration technology can bring together all the heterogeneous subcomponents in a compact form factor will be described.

AMBULATORY MULTIPARAMETER MONITORING AS TEST CASE

Ambulatory multiparameter monitoring is selected as a driving application for Human++. The target of such a monitoring system is to acquire process, store, and visualize a number of physiological parameters in an unobtrusive way. In one case, the focus is on the simultaneous acquisition of electroencephalogram (EEG)/electrocardiogram (ECG)/electromyogram (EMG) biopotential signals. Traditionally, such signals are either captured in a clinical setting for immediate interpretation or recorded in an ambulant setting for *post factum* analysis via a Holter monitor. With a wireless ambulatory monitoring system, the real-time features of the clinical system can be combined with the benefits of ambulatory monitoring from a Holter monitor. Figure 8.2 shows a schematic drawing of a typical setup that consists of the following:

- 1 EEG sensor node that can acquire, process, and transmit 1–24 EEG signals
- 1 ECG sensor node that can acquire, process, and transmit 1 ECG signal
- 1 EMG sensor node that can acquire, process, and transmit 1 EMG signal
- 1 basestation that collects the information from the 3 sensors

All the sensors have very similar functionality (see Figure 8.3). First, the incoming signals are amplified and filtered. The resulting signals are sampled at 1024 Hz with a 12 bit resolution. Finally, the digital

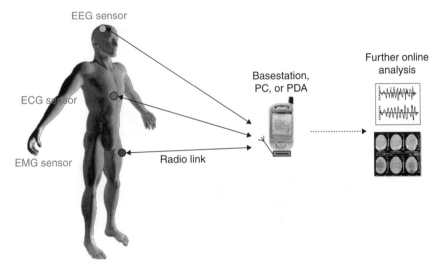

FIGURE 8.2 Schematic overview of the BAN setup, consisting of EEG/ECG/EMG sensors wirelessly connected to a PC or PDA.

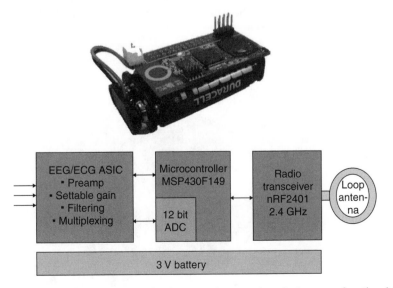

FIGURE 8.3 Close-up of the sensor node. On the top a picture and on the bottom a functional diagram.

signals are transmitted over a wireless link operating in the 2.4 GHz industrial scientific medical (ISM) band. Because the sensors are very similar, they can be realized with the same programmable hardware.

The basestation acts as a data collector. The collected data are passed on to a personal computer (PC) or personal digital assistant (PDA) through a universal serial bus (USB) interface. Further, the basestation also acts as a master for the network, which makes use of time division multiple access (TDMA) scheme, shown in Figure 8.4.

A key design criterion for such system is the power of the sensor nodes because this will directly determine the size and the operational lifetime of the system. Analysis of the operation of the sensors shows that they are alternating between four different modes of operation:

- Listen: During this mode the sensors receive their parameters from the basestation
- Processing: During this mode the biopotential signals are monitored and preprocessed
- Transmit: During this mode the sensors send their data to the basestation
- Sleep: Power save mode in which most of the electronics are switched off to save power

The time spent in each of these modes is very much application dependent. As an example Figure 8.5 shows the relative time spent in each of the modes for a particular EMG measurement. It is clear that

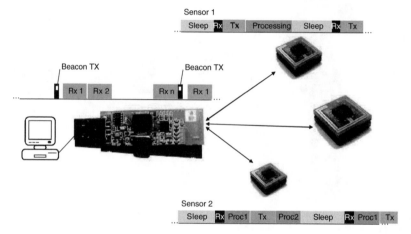

FIGURE 8.4 Basestation with three integrated sensors sharing the radio interface via a TDMA scheme. The basestation sends out a periodic beacon. The sensors synchronize themselves to this beacon and send their data back in the timeslots allocated to them.

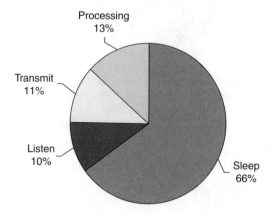

FIGURE 8.5 Relative time spent in different operation modes.

for this sensor the idle time is very important and that the system needs to have a very low standby power consumption.

Each of these modes has its own power consumption. Figure 8.6 shows that the current consumption in listen and transmit modes is much higher than that in processing or sleep mode. This is a direct consequence of the radio that is switched on in these modes and consumes about 90% of the power when it is active.

Bringing all of these data together, will get to the total average power consumption for the sensor node depicted in Figure 8.7. It becomes clear that with the current system consisting mainly of off-the-shelf components a prototype can be designed that consumes less than 1 mW of power if the measurement interval is longer than 1 s.

If two AA batteries are used in series with a capacity of 2500 mAh, the battery lifetime becomes approximately 3 months.

Cost is a very important parameter for sensor systems that have to be applied in large volume. The breakdown of a typical sensor node signifies that it has a bill-of-materials of a few tens of euros (Figure 8.8). Add to this a similar price for the basestation and double the price for all of the software running on these systems and the prices will be in the order of 100€ for low volumes, which lower as volumes pick up.

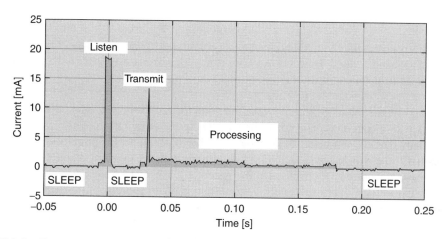

FIGURE 8.6 Current consumption in different operating modes.

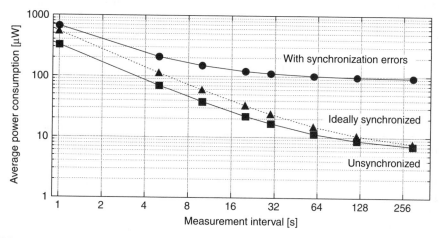

FIGURE 8.7 Average power consumption of a sensor node for different measurement intervals. For long intervals the sensor spends more time in idle mode, and therefore the power consumption decreases.

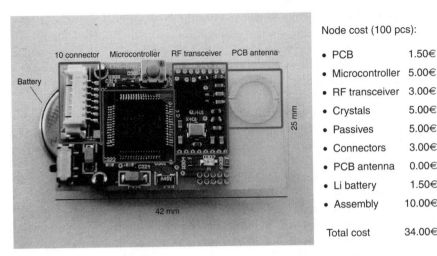

Node cost (100 pcs):

- PCB   1.50€
- Microcontroller 5.00€
- RF transceiver 3.00€
- Crystals   5.00€
- Passives   5.00€
- Connectors   3.00€
- PCB antenna   0.00€
- Li battery   1.50€
- Assembly   10.00€

Total cost   34.00€

FIGURE 8.8 Cost breakdown of a sensor node (prices in euros): PCB 1.50, microcontroller 5.00, RF transceiver 3.00, crystals 5.00, passives 5.00, connectors 3.00, PCB antenna 0.00, Li battery 1.50, and assembly 10.00. Total: 34€.

This clearly shows that with today's technology, first realistic demonstrators with a reasonable lifetime can be manufactured. However, a couple of major challenges still have to be solved to come to a widespread deployment of BANs:

- Run on smaller batteries: AA batteries are good for demonstration, but one would like to work with a coin or planar type of battery. These batteries have roughly 100 times less capacity than the AA cells. To keep the same battery lifetime, the power of the electronics has to be reduced by a factor 100. Alternatively, one can scavenge energy from the environment during the operation of the system. If the average scavenged energy is larger than the average consumed energy, the system can run eternally with the battery or a super-capacitor acting only as a temporary energy buffer.
- Longer maintenance-free operation: The system demonstrated can run for months. However, to come to a truly autonomous system it should be able to operate over its full lifetime without maintenance. The lifetime of today's electronic systems is often a few years; the current system has a battery lifetime of only a few months. Therefore, we need a 10× power reduction in order to make the system run of a single battery charge during its entire lifetime.
- Providing more functionality: Today's sensor system can provide limited autonomy. Most of the devices act as simple gateways, passing on the information to a central hub where the data are converted into actionable information. By adding intelligence to the sensors they can take decisions locally and the signaling overhead in terms of data and latencies can be reduced.
- Become manufacturable at low cost: Today's systems cost anywhere between 10 and 100€. A major reason for this is the low volume of the market so far, but one technical reason is that there are no commercially available packaging technologies that can efficiently integrate such heterogeneous components as batteries, micro-electromechanical systems (MEMS), processors, and radios in a single package.
- Integration of novel sensor and actuator concepts: The quality of the information resulting from a BAN is only as good as what you measure. Today, often only simple physical properties are measured such as biopotentials, temperature, and movement. Sensors that can measure these parameters in an ambulatory setting are much needed. Motion artifacts are often a major source of data corruption. However, in a next step also biochemical measurements will be required to interact with the biochemical reactions that drive our daily life.

In the remainder of this chapter, advances in wireless communication, energy scavenging, sensors, and system integration that can enable such systems in the near future will be demonstrated. The hunt for sensor systems that are 1000× more power efficient, that have ample intelligence to make decisions, that cost less than 1€, and that are unobtrusive is open!

WIRELESS COMMUNICATION

Today's low-power radios such as Bluetooth [4], Zigbee [3], or proprietary radios such as Nordic [17] cannot meet the stringent wireless BAN power requirements that we are looking for. Typical chipsets for these radios consume in the order of 10–100 mW for data rates of 100–1000 kbps. This leads to a power efficiency of roughly 100–1000 mW/Mbps or nJ/bit. As demonstrated in the previous section, we need a radio that is one to two orders more power efficient. None of the traditional radio approaches have proven the ability to reduce the power consumption by the required orders of magnitude while offering the necessary communication performances. Recently, a novel air interface based on communication using wideband signals has attracted a large attention from the wireless community, the so-called ultra-wideband (UWB) communication [6,7]. The Federal Communications Commission (FCC) has authorized UWB communications between 3.1 and 10.6 GHz. Although the regulations on UWB radiation define a power spectral density (PSD) limit of −41 dBm/MHz, there are very few regulations on the definition of the time domain waveform. The latter can then be tailored for low hardware complexity as well as low system power consumption. In pulse-based UWB, the transmitter only needs to operate during the pulse transmission, producing a strong duty cycle on the radio and the expensive baseline power consumption is minimized. Moreover, since most of the complexity of UWB communication is in the receiver, it allows the realization of an ultra-low power, very simple transmitter and shifts the complexity as much as possible to the receiver in the master.

However, the impact of the type of UWB signal chosen on the communication performance and on the complexity of the radio implementation must be carefully analyzed. The minimum bandwidth of a UWB signal is usually 500 MHz. Indeed, various UWB standard proposals [5,8,9] have subdivided the entire UWB spectrum in 500 MHz subbands as a solution to mitigate against strong interferers, to improve the multiple access, and to compose with the different regulations on UWB emissions worldwide. Therefore, to comply with these regulations and standards, the generated pulses of UWB impulse-radio (UWB-IR) approaches must fulfill stringent spectral masks that can feature such low bandwidths. This poses a serious challenge for the pulse generation of UWB-IR transmitters.

UWB PULSE GENERATOR

A possible way to realize short high-frequency UWB signals is by gating an oscillator as proposed in Ref. [10]. The oscillator center frequency is then defined independently from the bandwidth by the gate duration. This class of UWB systems are defined here as "carrier-based UWB" impulse radio due to the presence of a weak carrier inside the pulse. However, since a direct gating of the center frequency features a rectangular shape, the high sidelobe power must be filtered. Since a variable radio frequency (RF) channel select filter is not an option, the shape must be filtered before being translated to an RF signal. In low-cost and low-power applications, this filtering operation should require circuitry with minimal complexity and preferably compatible with full-complimentary metal oxide semiconductor (CMOS) integration. Therefore, the triangular signal pulse appears to be a perfect compromise. The smooth shape of the triangular waveform provides a sidelobe rejection of more than 20 dB with most of the power confined in the useful bandwidth. The triangular signal is relatively easy to generate in standard CMOS circuits by, for instance, charging and discharging a linear capacitor.

The pulse generator architecture is presented in Figure 8.9. A triangular pulse generator and a ring oscillator are activated simultaneously. The triangular signal is multiplied with the carrier

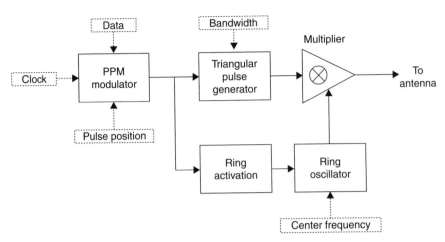

FIGURE 8.9   Architecture of the pulse generator. (From Ryckaert, J., et al., *2005 International Conference on Ultrawideband*, 2005. With permission. © [2005] IEEE.)

created by the oscillator, resulting in an up-converted triangular pulse at the output. The triangular waveform has a duration that can be adapted in accordance with the desired bandwidth. A gating circuit (ring activation circuit in Figure 8.9) activates the ring oscillator when a pulse must be transmitted, avoiding useless power consumption between the pulses. This motivates the choice of a ring type of oscillator since its startup time is very low while the high phase noise problem is obviously not an important issue in the generation of wideband signals.

The pulser circuit has been implemented in a logic 0.18 μm CMOS technology [11]. The system can deliver a pulse rate up to 40 MHz. The pulses are modulated in position and the position modulation can be tuned from 4 to 15 ns. Figure 8.10 shows the chip micrograph.

Figure 8.11a shows the measured spectrum of pulses with a bandwidth of 528 MHz together with one corresponding time domain waveform in the top left corner. Three traces are depicted showing three different center frequency settings. Figure 8.11b shows the 2 GHz spectrum together with its time domain waveform.

The measured energy consumption is 50 pJ per pulse at a 40 MHz pulse repetition rate for a pulse bandwidth of 1 GHz. In other words, if 10 pulses are used to code 1 bit, the pulser provides an average data rate of 10 kbps with 5 μW average power consumption. This is exactly the type of breakthrough low power consumption wanted!

UWB Analog Receiver

Processing wideband analog signals in the digital domain requires an extremely fast sampling analog-to-digital converter (ADC) with a wide input bandwidth [14]. Such solutions [12] have all flexibility in terms of digital signal processing but are often less attractive due to their required power consumption. To minimize the overall sampling rate and total power consumption, analog preprocessing is an interesting alternative. In this section, an analog-based correlation receiver architecture, well suited for low data rate impulse-based UWB applications is proposed. Since the correlation is done in the analog domain, the accuracy in the ADC sampling instants is shifted to a precise timing for the template generation. Usually, to optimally receive an UWB signal, the incoming pulse must be correlated with a template signal that precisely coincides with the incoming waveform. However, as shown in Figure 8.12, the carrier-based UWB signal is first down-converted in quadrature baseband, and the matched template must then correlate with the down-converted envelope of the pulse. Hence, any timing inaccuracy is translated into a phase shift in the complex plane and almost no information is lost. The phase shift can be processed in the digital baseband to track the timing inaccuracy. The matched filtering

FIGURE 8.10   Micrograph of the pulser die. (From Ryckaert, J., et al., *2005 International Conference on Ultrawideband*, 2005. With permission. © [2005] IEEE.)

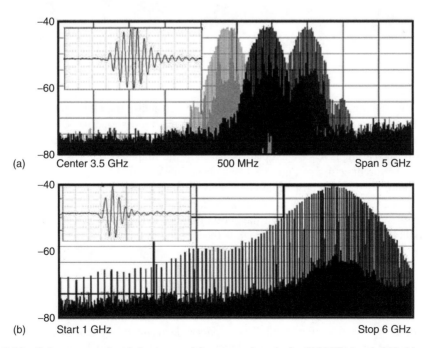

FIGURE 8.11   Pulser output signals in time and frequency domain for 528 MHz bandwidth (a) and 2 GHz bandwidth (b). (From Ryckaert, J., et al., *ISSCC Dig. Tech. Pap.*, 114–115, 2006. With permission. © [2006] IEEE.)

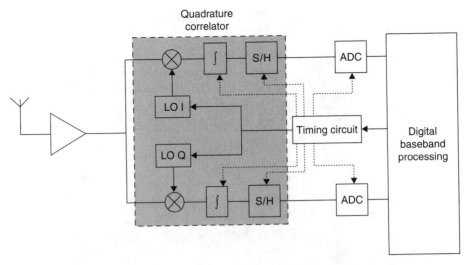

FIGURE 8.12 UWB analog receiver. (From Ryckaert, J., et al., *ISSCC Dig. Tech. Pap.*, 114–115, 2006. With permission. © [2006] IEEE.)

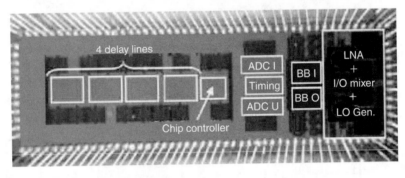

FIGURE 8.13 UWB analog receiver chip micrograph.

is achieved through an analog integration operation over a precise time window. The time window is defined by the duration between the start of the integration process and the sampling instant. This operation corresponds to the correlation of the incoming triangular pulse with a rectangular window. The choice of such a simple receive template is driven by the simplicity of the implementation.

The receiver architecture of Figure 8.12 has been implemented on a 0.18 μm CMOS technology [13] (Figure 8.13). The total current consumption of the chip including the digital baseband is 16 mA measured on a 1.8 V supply at 20 MHz clock rate. This power consumption is substantially higher than that of the pulser. However, the good thing is that for BAN only the transmitter in the sensor should be of extremely low power, while the receiver in the central station has slightly more relaxed power budget.

MICROPOWER GENERATION AND STORAGE

Today, the batteries that are needed to power wireless autonomous transducer systems seriously limit the possibilities of this emerging technology. Modern electronic components become smaller and smaller, while the scaling of electrochemical batteries faces technological restrictions. As a consequence, either large batteries are used that give a longer autonomy but make the system bigger, or small batteries are used that make the system less autonomous. For this reason, a worldwide effort is

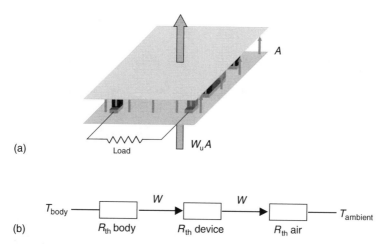

FIGURE 8.14   (a) Schematic of a thermoelectric generator and (b) schematic thermal circuit representing the generator and its environment.

ongoing to replace batteries with more efficient, miniaturized power sources. We aim at generating and storing power on the microscale to improve the autonomy or reduce the size of wireless autonomous transducer systems. The envisaged solution takes its energy—thermal or mechanical—from the human body and converts it into electrical energy, stored in a microbattery or (super) capacitor.

The choice of scavenging principle depends on the application and the environment in which it is used. In this section, thermal scavengers are prevented because they are well suited to convert the thermal body heat into electricity.

Thermal energy scavengers are thermoelectric generators that exploit the Seebeck effect to transform the temperature difference between the environment and the human body into electrical energy. A thermoelectric generator is made of thermopiles sandwiched between a hot and a cold plate. Thermopiles are in turn made of a large number of thermocouples connected thermally in parallel and electrically in series, as shown schematically in Figure 8.14a. The maximum electrical power is generated when the load is matched to the electrical resistance of the generator and when the thermal conductance of the thermocouples equals the one of the air between the plates. (This is exactly true when considering that the heat flow from the body is not influenced by the thermoelectric generator, a condition well approximated in this case.) Under this condition, power increases on increasing the height of the pillars.

In commercially available thermopiles, typically based on Bi_2Te_3, the pillars have a lateral size of 0.3–1 μm and a height of 1–3 μm. A thermo electric generator (TEG) optimized to obtain the maximum power will have a thermal resistance of about 200 cm² K/W per square centimeter of surface. In operating conditions, the generator is inserted in a thermal circuit, which includes the thermal resistance of the body and the equivalent thermal resistance of the air (Figure 8.14b). To have a sizeable temperature drop over the device, these series resistances must not be too large with respect to that of the generator.

The thermal resistance of the body depends on the position of the TEG. It has been measured in two different parts of the wrist. At the location where a watch is normally worn (the outer side in Figure 8.15), the average thermal resistance per unit area is about 300 cm² K/W, while on the radial artery (the inner side in Figure 8.15) it is more than two times smaller.

As for the equivalent thermal resistance of the air, it can be reduced by using appropriate radiators mounted on the cold plate of the thermoelectric generator. For example, assuming that the device thickness is limited to 1 cm, a thermal resistance of about 500 cm² K/W can be obtained for a sitting person in a still air and of about 200 cm² K/W for a walking person. With the values given, about one-third of the total temperature difference will drop across the thermoelectric generator. This is a

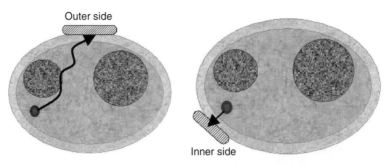

FIGURE 8.15 Schematic position of the measuring device on the wrist. The arrows show a heat path from the artery to the device.

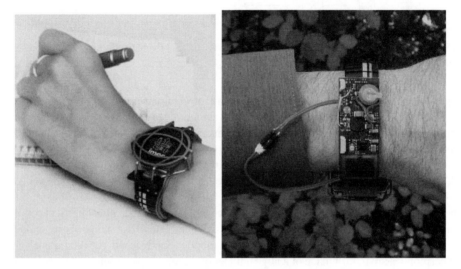

FIGURE 8.16 (Left) A thermoelectric generator fabricated using commercial thermopiles. (Right) The power-conditioning electronics and a transceiver mounted on the wrist strap.

satisfactory result which provides about 10–15 μW/cm^2. Further decreasing the air and body resistance will result in a too large heat flow from the body and in a discomfort for the user (feeling of cold).

The number of thermocouples corresponding to the optimal power for the generator described turns out to be very small (10–20 cm^{-2}), resulting in a very low voltage (20–30 mV/cm^2). To drive a simple power management system for recharging a battery, a minimum voltage of 0.8 V is necessary (with its following up-conversion). To obtain this voltage it would be necessary to increase the number of thermocouples and, at the same time, to decrease their cross section to fulfill the maximum power condition (equivalent heat flow through the thermocouple and the air). If commercial thermopiles are used, their cross section is limited by technology to the values mentioned, and the required voltage is obtained by increasing the number of thermopiles at the expenses of a reduced power. A system based on this compromise has been fabricated and is shown in Figure 8.16. The power-conditioning electronics, together with a low-power radio transmitter, is mounted on a flexible substrate and glued to the wrist strap. The power generated by the device is in excess of 0.1 mW; the output voltage is above 1 V. This power is enough to charge a small battery and to transmit, e.g., the temperature of the body each 1–2 s to a nearby receiving station.

The use of commercial thermopiles has proven that human heat can be used to power a sensor node. Nevertheless, the solution is nonoptimal for two reasons: (i) it does not offer the possibility of optimizing the power and the voltage at the same time, and (ii) it is a very expensive one because thermopiles fabrication techniques cannot be easily automated. A possible solution could be the

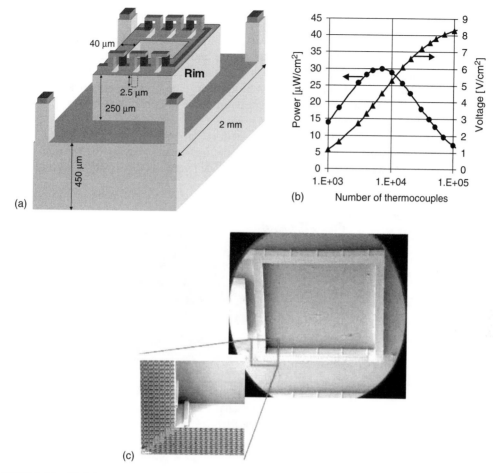

FIGURE 8.17 (a) Schematic of the TEG capable of combining large power and large voltage (rim and thermopiles are not scaled to overall device dimensions). (b) Simulated performance of the TEG shown in (a). (c) Photograph of the fabricated thermoelectric generator. Thermocouples are made of SiGe.

use of micromachined thermopiles. These have already been presented in the scientific literature, and are used in miniaturized commercial thermoelectric coolers. Micromachining has the potential advantage of reducing the lateral size of the thermocouple. This means that a much larger number of thermocouples can be fabricated per unit area thus maintaining the condition of equal thermal conductance of thermocouple and air, needed for power optimization. Such approach would allow combining a large voltage and a large power. Unfortunately, micromachined thermocouples have a height of a few microns only, which drastically reduces the thermal resistance of the generator; consequently, the temperature drop on the device is small and the generated power is negligible.

To overcome this difficulty, a special design of a micromachined thermoelectric generator for application on humans has been developed. The design combines a large thermal resistance of the device with a large number of thermopiles. The schematic is shown in Figure 8.17a. Several thousands of thermocouples are mounted on a silicon rim. The function of this rim is to increase the parasitic plate-to-plate thermal resistance of the generator. If Bi_2Te_3 is used as thermoelectric material, an optimized device fabricated according to this scheme and positioned on the human wrist can generate up to $30\,\mu W/cm^2$ at a voltage exceeding 4V, in indoor application. Figures 8.17b and 8.17c show a realization of such a device based on SiGe thermocouples. Because of the inferior thermoelectric properties of this material with respect to Bi_2Te_3, an optimized device is expected to generate $4.5\,\mu W/cm^2$ at a voltage of 1.5V.

SENSOR AND ACTUATORS

The multiparameter monitoring system needs to acquire biopotential signals in a very power-efficient way. The different biopotentials have different amplitude and frequency characteristics as shown in Figure 8.18. For this purpose, a low-power 25-channel biopotential application specific integrated circuit (ASIC) [18] has been developed. The ASIC allows the preprocessing of typical biopotentials such as ECG and EEG signals. It can be configured in different operational modes, thanks to its variable bandwidth and gain settings.

The mixed-signal ASIC consists of 25 channels (Figure 8.19). In a typical configuration, 24 channels are configured for EEG measurements and 1 channel is configured as ECG channel. Each channel of the ASIC consists of a high common mode rejection ratio (CMRR) instrumentation

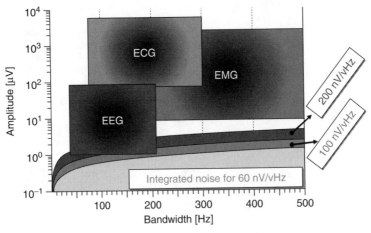

FIGURE 8.18   Amplitude and frequency characteristics of different biopotential signals.

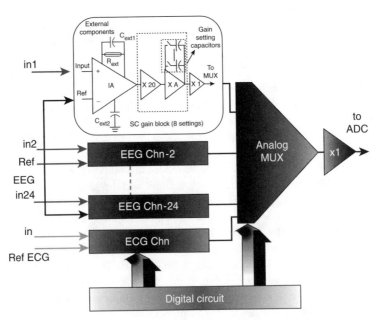

FIGURE 8.19   25-Channel biopotential ASIC. (From Yazicioglu, R. F., Merken, P., and Van Hoof, C., *Electron. Lett.*, 41, 2005. With permission. © [2005] IEEE.)

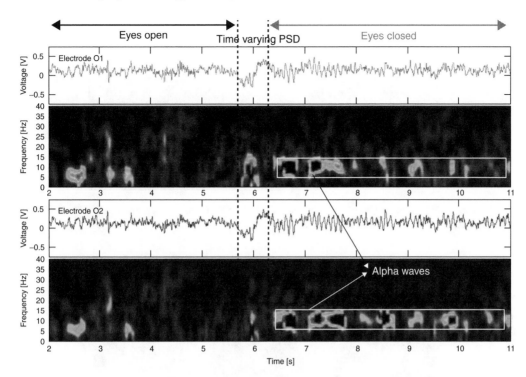

FIGURE 8.20 Alpha activity from the two electrodes at occipital cortex, and their short-time Fourier transform. (From Yazicioglu, R. F., Merken, P., and Van Hoof, C., *Electron. Lett.*, 41, 2005. With permission. © [2005] IEEE.)

amplifier, followed by a variable gain amplifier. There are eight different gain modes ranging from 200 to 10,000 for the EEG channels and from 20 to 1000 for the ECG channel.

The front-end instrumentation amplifier has bandpass filter characteristics, where in-band gain and the cut-off frequencies are settable with external components. With an external capacitor of 1 μF, a bandwidth of 0.5–80 Hz is selected. The CMRR is larger than 90 dB at 50 mV electrode offset. The total input-referred voltage noise of each channel is less than 1 μV rms in the 0.5–80 Hz bandwidth. These features allow suppressing the input common mode voltages coupled to the human body, while amplifying the microvolt-level biopotential signals.

The mixed signal ASIC is designed and fabricated in 0.5 μm CMOS process. The ASIC can operate from a voltage supply ranging from 2.7 to 3.3 V while dissipating less than 10.5 mW. All the channels of the ASIC are multiplexed with a frequency of 1 kHz per channel and buffered at the output. Therefore, a single ADC with a maximum input capacitance of 50 pF can sample all the channels of the ASIC.

In a test setup, two channels of the ASIC are connected to Ag/AgCl electrodes for reading the brain activity at the occipital cortex (backside of the head). A microcontroller with integrated ADC is directly connected to the ASIC. Operation and gain settings of the ASIC are controlled from the microcontroller. When the patient closes his eyes, the typical alpha rhythm becomes clearly visible at the output (Figure 8.20).

More recently an alternative instrumentation amplifier architecture (Figure 8.21) with a power consumption of 20 μW per channel while maintaining a CMRR of 110 dB has also been fabricated [19]. This ASIC provides an additional factor of 20 in power savings per channel compared to the 24-channel version.

INTEGRATION TECHNOLOGY

One form factor suitable for many sensor applications is a small cubic sensor node. To this end, a prototype wireless sensor node has been integrated in a cubic. In this so-called three-dimensional

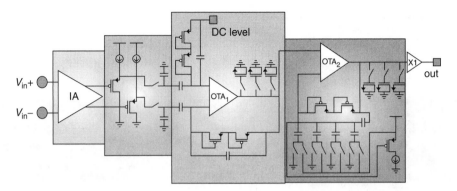

FIGURE 8.21   Architecture of single-channel biopotential ASIC. (From Yazicioglu, R. F., et al., *ISSCC Dig. Tech. Pap.*, 2006. With permission. © [2006] IEEE.)

FIGURE 8.22   3D SiP Wireless autonomous sensor node.

system-in-a-package approach (3D SiP) [5,15], the different functional components are designed on separate boards and afterwards stacked on top of each other through a dual row of 0.7 mm solder balls with a pitch of 1.27 mm (Figure 8.22). This system has the following advantages: (i) modules can be tested separately, (ii) functional layers can be added or exchanged depending on the application, and (iii) each layer can be developed in the most appropriate technology. The first-generation 3D stack offers a complete SiP solution for low-power intelligent wireless communication. The integrated stack includes the following layers: (1) radio, (2) microprocessor, (3) sensor, and (4) power.

The radio layer contains a Nordic 2.4 GHz transceiver [17], crystals, passives, and a matched folded dipole antenna. The antenna is folded around the transceiver to maximize the total perimeter of the antenna, while keeping the consumed area in the module low. The larger outside perimeter of the folded antenna improves the antenna performance in terms of bandwidth and efficiency compared to a standard dipole antenna. Two shielding layers are inserted below the antenna to protect the low-power microcontroller from the radiated output power from the antenna. The antenna design includes the influence of the packaged transceiver and shielding layers on its performance and the effect of antenna placement on the human body. The designed antenna impedance is matched to the output impedance of the integrated amplifier of the transceiver thereby maximizing the radiated output power and thus the range of the sensor node. A careful optimization of the whole assembly using 3D high-frequency modeling tools and the proper placement of grounding planes are required, which is crucial for the correct prediction of the performance of the radiating element with minimal dimensions. In next generation, the radio will be replaced by the ultra-low power UWB radio discussed earlier in this chapter in the section "Wireless Communication."

The microprocessor layer contains a commercial low-power eight-million instructions per second (MIPS) microcontroller [16]. The bare die processor is wirebonded to the printed circuit board (PCB). The back of the board contains the crystal and the passives. The solder ball-to-ball stacking provides the necessary height to integrate the components between the layers, without the need for additional spacers. Figure 8.3 shows the microprocessor and radio two-layer subassembly.

The sensor layer contains a custom programmable ECG/EEG/EMG read-out ASIC [2] discussed earlier. This 0.5 μm CMOS ASIC is mounted as a bare die and wirebonded to the carrier PCB. The passives, needed to tune the gain and bandwidth of this component, are mounted on the backside of the PCB.

The power layer contains a small button cell battery such as a V6HR. Since these NiMH cells offer a voltage of approximately 1.2 V, two cells in series are required. In next generation, a move will be made to a more integrated solution making use of thin film Li-ion microbatteries. Since these will offer a capacity below 1 mAh, the power consumption of the system will have to be radically reduced to keep the current lifetime of the system. This will be achieved through integrating lower power integrated circuits (ICs). As a complementary approach, also introduced will be the power scavengers discussed in the section "Micropower Generation and Storage" that can recharge the battery continuously.

Parallel research was started to implement the same technology on a flexible carrier. The ultimate target is to create a small and smart bandage containing all the necessary technology for sensing and communication with a basestation. It will provide a generic platform for various types of applications (wound healing, EEG, ECG, EMG, etc.). The first prototype (Figure 8.23) is 10 times smaller than a credit card (12×35 mm^2) and about as thin as a compact disc (1–2 mm). The flexible 25 μm polyimide carrier contains a microprocessor and a wireless communication module (2.4 GHz

FIGURE 8.23 Prototype sensor in a flexible bandage.

radio). It enables the antenna to be optimized for its activity on human skin. Current focus lies on adding the necessary sensors and energy equipment (rechargeable battery, energy scavenger, and advanced electronics to keep energy consumption as low as possible). An ultimate device thickness of approximately 100 μm is targeted.

The biggest challenges in developing this kind of modules are the extreme miniaturization and its effects on the functionality of the used components. Some of the many problems to tackle are the use of naked chips, chip scaling, assembly processes such as wire bonding and flip-chip on a flexible substrate, application of thin-film batteries and solar cells, and integration of the entire technology in a biocompatible package.

CONCLUSIONS

This chapter provided overview of the Human++ research program, which is targeted at developing key technologies and components for future wireless BANs for health monitoring applications. Several enabling technologies and integrated modules were discussed.

Over the next years, more BAN technologies will be developed all over the world. Step by step these will bring us closer to the end goal: an unobtrusive portable system that keeps track of our health and fitness level at an affordable cost.

REFERENCES

1. http://www.imec.be/ovinter/static_research/human++.shtml.
2. Schmidt, R., et al., "Body area network BAN, a key infrastructure element for patient-centered medical applications," *Biomed. Tech. (Berl.)*, vol. 47, suppl. 1, part 1, pp. 365–368, 2002.
3. http://www.bluetooth.com, http://www.zigbee.org.
4. http://www.bluetooth.com/Bluetooth/Learn/Technology/Compare/Technical/.
5. IEEE 802.15.4a: http://www.ieee802.org/15/pub/tg4a.html.
6. Win, M. Z. and Scholtz, R. A., "Impulse radio: How it works," *IEEE Commn. Lett.*, pp. 36–38, February 1998.
7. Federal Communications Commission (FCC). Revision of part 15 regarding ultra-wideband transmission systems. First Report and Order, ET Docket, 98–153, FCC 02-48, adopted February 2002, released April 2002, available at http://www.fcc.gov.
8. IEEE 802.15.3a www.ieee802.org/15/pub/TG3a.html.
9. Standard draft proposal. IEEE 802.15.4a available at http://www.ieee802.org/15/pub/TG4a.html.
10. Choi, Y. H., "Gated UWB pulse signal generation," *IEEE Joint International Workshop of UWBST and IWUWBS*, pp. 122–124, May 2004.
11. Ryckaert, J., et al., "Carrier-based UWB impulse radio: Simplicity, flexibility, and pulser implementation in 0.18 μm CMOS," *2005 International Conference on Ultrawideband*, 2005.
12. Blazquez, R., et al., "Direct conversion pulsed UWB transceiver architecture," *Design Automation Test Eur.*, March 2005.
13. Ryckaert, J., et al., "A 16 mA UWB 3-to-5 GHz 20Mpulses/s quadrature analog correlation receiver in 0.18 μm CMOS," *ISSCC Dig. Tech. Pap.*, pp. 114–115, February 2006.
14. Verhelst, M., et al., "Architectures for low power ultra-wideband radio receivers in the 3.1–5 GHz band for data rates <10 Mbps," *ISLPED'04, International Symposium on Low Power Electronics and Design*, pp. 280–285, August 2004.
15. Stoukatch, S., et al., "Miniaturization using 3-D stack structure for SIP application," *SMTA (Surface Mount Technology Association) International Conference*, 21–29 September 2003.
16. TI MSP430F149, www.ti.com.
17. Nordic nRF2401, www.nvlsi.com.
18. Yazicioglu, R. F., Merken, P., and Van Hoof, C., "Integrated low-power 24-channel EEG front-end," *Electron. Lett.*, vol. 41, 2005.
19. Yazicioglu, R. F., et al., "A 60 μW 60 nV/√Hz readout front-end for portable biopotential acquisition systems," *ISSCC Dig. Tech. Pap.*, February 2006.

9 Progress toward a Single-Chip Radio

*Ken K. O, Paul Gorday, Jau-Jr. Lin, Changhua Cao,
Yu Su, Zhenbiao Li, Jesal Mechta, Joe E. Brewer,
and Seon-Ho Hwan*

CONTENTS

Introduction 241
On-Chip Antennas 242
24-GHz CMOS RF Circuits 244
    Integrated 24-GHz Receiver 245
    Integrated 24-GHz Transmitter 247
    Node-to-Node Communication 249
    Node-to-Base-Station Communication 251
Crystal Elimination Approach 251
    Differential Detection 252
    Modulation Format 253
    Noise Analysis 253
    Simulation Results 255
    Digital Implementation Issues and Verification 257
Management of Noise and Unwanted Signal Coupling 258
Conclusions 261
Acknowledgments 261
References 261

INTRODUCTION

The vision for a single-chip radio has been in existence for many years. There also have been numerous claims of single-chip radios. An example of fully featured radio that is close to being a single chip is TI BRF6100, which is ~8 mm by 9 mm in dimensions and includes radio frequency (RF), baseband analog and logic, memory, and power management circuits. Impressive as it is, this device requires about 16 external components. The parts list includes baluns, inductors, and capacitors for off-chip matching, a crystal reference, as well as an antenna.

In this chapter, the term "single-chip radio" is more strictly interpreted. Here, a single-chip radio is an integrated circuit (IC) that can accomplish transmit and receive functions without any off-chip components. A single-chip radio shown in Figure 9.1 integrates antennas [1–3], RF circuits, analog circuits, digital functions, and the crystal reference function. Potentially, sensor functions can also be integrated in the same IC. The only external connections will be those for power and ground. This single-chip radio will be low cost due to reduction of component and assembly cost

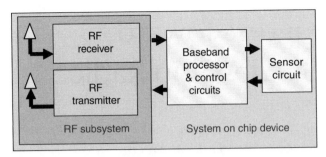

FIGURE 9.1 Block diagram of a single-chip radio. (From O et al., *2006 IEEE Custom Integrated Circuits Conference*, San Jose, CA, pp. 473–480, September 2006. With permission. © 2006 IEEE.)

as well as use of less expensive packaging and printed circuit boards because the RF interface is no longer needed. It will also be compact and reliable due to reduction of the number of solder joints and will be extremely easy to use since the only assembly requirement is to place the chip on a board and make connections to a power source.

Presently, a 24 GHz single-chip radio, which can be utilized to build mm-sized disposable sensor nodes (μNodes), is being used to drive the development of necessary technology. A μNode is similar to a mote as defined in the original Smart Dust [4] concept except that μNodes use RF instead of optical communications. The radio cost target is less than a dollar. The radio is intended to support short-range (1–5 m) communications among nodes and longer range communications to a base station ~100 m away. Four fundamental technology capabilities must be established to provide practical μNode-class devices: on-chip antennas, 24 GHz RF circuits in low-cost complementary metal oxide silicon (CMOS), integration of frequency reference function, and management of unwanted on-chip signal and noise coupling. This chapter describes progress toward development of these technologies. Results to date suggest that the single-chip radio is an item on the immediate horizon.

ON-CHIP ANTENNAS

An on-chip antenna is a key component of a single-chip radio. The size of an on-chip antenna is limited by the maximum allowed dimensions of ICs, which are ~2 cm on a side. To keep the IC cost low, antenna size should be kept small. Operation at high frequencies allows the use of physically small antennas. At 24 GHz (free-space wavelength 12 mm), a quarter-wave antenna is 3 mm long, which is a small fraction of the maximum allowed IC dimension [5,6]. Figure 9.2 shows a 3 mm-long zigzag dipole antenna interfaced to a test circuit. In volume production, such an antenna is expected to cost about $0.015. A simplified cross-section of the antenna is shown in Figure 9.3. The antenna is fabricated using a metallization scheme commonly used in silicon IC processes and is separated from the silicon substrate by a SiO_2 layer. The antenna can also be formed with multiple metal layers shunted together to lower the conduction loss.

Use of on-chip antennas eliminates the need for high-frequency wired input/output (I/O), thus avoiding the losses of RF chip-to-PC-board and transmission line connections [7–9]. Since the antennas are fabricated using normal chip metallization, it is practical to have separate transmit and receive antennas, and thus a transmit and receive (T/R) switch with associated loss can be bypassed. Additionally, baluns needed to interface a single-ended antenna to on-chip differential circuits can be eliminated using dipole antennas that are balanced. This once again lowers loss. These of course need to be weighed against the performance degradation of on-chip antennas due to the losses in the substrate and on-chip metallization used to form the antennas.

To evaluate and to improve the performance of on-chip antennas, their dependences on antenna length, metal thickness, width, substrate resistivity, substrate material, and placement of antennas

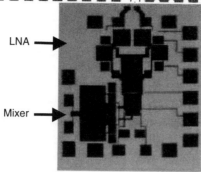

FIGURE 9.2 Zigzag dipole antenna interfaced to a test circuit. (From Yu, S., Lin, J. J., and O, K. K., *2005 IEEE International Solid-State Circuits Conference*, San Francisco, CA, pp. 270–271, February 2005. With permission. © 2005 IEEE.)

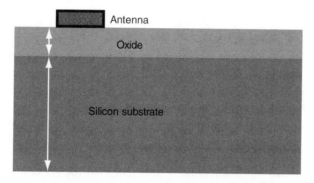

FIGURE 9.3 Simplified cross-section of an antenna. An antenna is fabricated using the metallization process commonly used in silicon integrated-circuit processes and is separated from the silicon substrate by an oxide layer.

within an IC have been evaluated [6]. The performance of antennas has been characterized using antenna-pair gain symbolized as G_a in the following equation:

$$G_a = \frac{|S_{21}|^2}{\left(1-|S_{11}|^2\right)\left(1-|S_{22}|^2\right)} = G_t G_r \left(\frac{\lambda}{4\pi R}\right)^2$$

The antenna pair gain is extracted from scattering parameter (S-parameters) measurements [10] and is a function of the individual antenna gains (G_t transmitter, G_r receiver), the signal wavelength (λ), and the separation between the antennas (R). The operation of any antenna is influenced by its proximity to earth ground. Special mobile RF probe stations that allow control of both elevation (5 mm to 80 cm) and range while making contact with small antenna pads using probes have been constructed [5,11]. The mobile probe stations were constructed with a type of plastic called Delrin with relative permittivity of 3.7.

Figure 9.4 shows plots of G_a versus antenna separation measured at 24 GHz using a 3 mm-long zigzag antenna pair fabricated on 20 Ω-cm silicon substrates. The silicon substrate thickness is varied from 670 to 50 μm. The figure also shows the gain for a pair of high-frequency probes. The antennas are measured at ~50 cm from the ground. The addition of antenna improves G_a by more

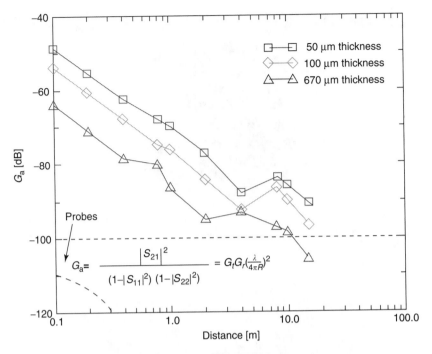

FIGURE 9.4   Dependence of antenna pair gain on substrate thickness.

than 50 dB. G_a decreases with separation following the $1/R^2$ dependence up to 1–2 m. Beyond that, the plots deviate due to the multiple-path effects. As the silicon substrate is thinned to 50 µm, G_a improves by ~15 dB due to the reduction in the substrate loss. Similar G_a improvement can be achieved by increasing the substrate resistivity to higher than 1000 Ω-cm [12,13] or by changing the substrate material (i.e., from silicon to sapphire) [10,14]. The cost benefit of using a standard CMOS logic process makes it desirable to choose an approach that is available as part of a volume process. In the case of currently available technology, in which the substrate resistivity is typically 10 or 20 Ω-cm and the thickness is about 700 µm, the wafer-thinning option is attractive. Substrates are routinely thinned to 100–250 µm before dicing and packaging.

The efficiency of antennas on 50-µm-thick substrate is ~25%, which is excellent for an antenna on a lossy substrate. Comparing this ~6 dB loss to that associated with off-chip and PC board connections and an off-chip antenna as well as that of a balun for the conventional approach, the overall system performance for a single-chip radio should be comparable to that using an off-chip antenna assuming that the sizes of antenna and system are kept the same. This is especially so at frequencies above 10 GHz.

Figure 9.4 also shows a dotted line at the −100 dB level [5] to indicate the minimum G_a a communication link can tolerate for the specific design being considered. The plots indicate that at 50 cm from the ground, the communication range greater than 10 m can be achieved using 3-mm-long zigzag dipole antennas fabricated on 20 Ω-cm substrates. Although not shown, at 4.5 cm from the ground, the communication range is ~5 m. These results have shown that it is indeed possible to communicate over 1–5 m using tiny on-chip antennas fabricated in conventional silicon IC processes.

24-GHz CMOS RF CIRCUITS

Since low cost is a key reason for the drive toward a single-chip radio, it should be manufactured in CMOS technology that has the lowest cost in volume production. For this, the feasibility of fabricating the required circuits in 130-nm CMOS technology has been studied. Figure 9.5 shows the block

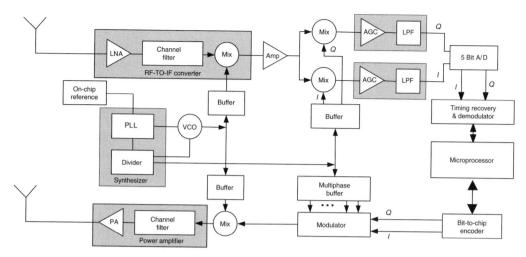

FIGURE 9.5 Block diagram of the 24-GHz single-chip radio. (From Brewer et al., *IEEE Circuits and Devices Magazine,* November 2006. With permission. © 2006 IEEE.)

diagram of a 24-GHz transceiver. It is a half-duplex system in which the transmitter and the receiver are on, one at a time. This eliminates the unwanted signal-coupling problem between the transmitter and receiver. The radio employs a dual-conversion architecture requiring one synthesizer. The intermediate frequency (IF) is ~2.7 GHz. The local oscillator (LO) signals for the IF mixer and modulator are generated by frequency dividing the first LO signal near 21 GHz by 8 [15]. As will be discussed in the section "Management of Noise and Unwanted Signal Coupling," this use of dual-conversion architecture should effectively eliminate the unwanted coupling between the transmitted signal from the antenna, and on-chip voltage-controlled oscillator (VCO) and frequency reference. The radio utilizes the offset quadrature phase shift keying (OQPSK) with half-sine pulse shaping or minimum shift keying (MSK), which is a constant-envelope modulation. This allows the use of a power-efficient nonlinear power amplifier (PA). The noise figure (NF) and gain targets for the receiver are 10 and 80 dB, respectively. The transmitter output power target is 10 dBm.

INTEGRATED 24-GHz RECEIVER

To investigate the feasibility of implementing the 24 GHz receiver in CMOS, a down-converter consisting of a low-noise amplifier (LNA) and a mixer has been fabricated in a 130 nm CMOS technology [16]. Figure 9.6 shows the circuit schematic. Although full differential circuits consume more power, they are utilized to make the connection to the dipole antenna without using a balun as well as to better reject the common-mode noise from the digital circuits that will eventually be co-integrated. The LNA utilizes the commonly used cascode topology. Source degeneration is used to generate resistance for input matching. The output of LNA is capacitively coupled to the mixer. The mixer is a double-balanced Gilbert cell type. The conversion gain of the mixer is ~1 dB. A differential inductor is inserted between the drain nodes of two transconductors (M6 and M7) to tune out the capacitance at these nodes. This increases the mixer gain and lowers flicker noise at mixer output [17]. This also improves the image rejection of down-converter. For testing, each IF output port is matched to 50 Ω using a capacitive transformer. The current sources at common-mode nodes in the LNA and the mixer are replaced by inductors to provide larger voltage headroom. The key design challenge at the 130 nm node and ~20 GHz operating frequency is the low output resistance of transistors and cascode stage (~1–2 kΩ). The die micrograph is shown in Figure 9.2. The antenna trace cannot be seen in photographs since it is built in the lowest four metal layers. The antenna in Figure 9.2 is drawn to illustrate the size and shape. The antenna area is 3 mm × 120 μm, whereas the

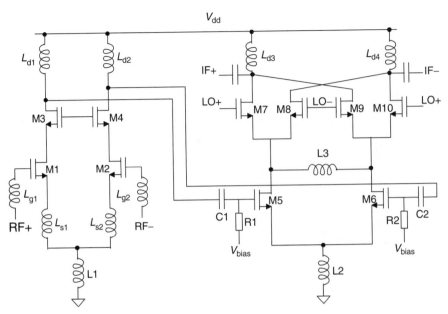

FIGURE 9.6   Receiver functional block diagram. (From Yu, S., Lin, J. J., and O, K. K., *2005 IEEE International Solid-State Circuits Conference*, San Francisco, CA, pp. 270–271, February 2005. With permission. © 2005 IEEE.)

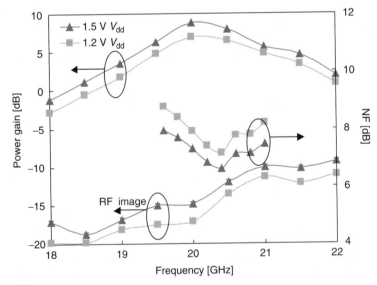

FIGURE 9.7   Power gain and noise figure plots of the RF down-converter. (From Yu, S., Lin, J. J., and O, K. K., *2005 IEEE International Solid-State Circuits Conference*, San Francisco, CA, pp. 270–271, February 2005. With permission. © 2005 IEEE.)

RF front-end area excluding the antenna is $880 \times 830\,\mu m^2$. The total area is $1.1\,mm^2$. To characterize the down-converter, the antenna is cut from the die. Figure 9.7 shows the measured power gain with IF of 3 GHz. It also shows the power gain of corresponding RF image. The image rejection ratio at 20 GHz is ~24 dB. The single side-band (SSB) NF is also shown in Figure 9.7. It is measured with a fixed LO of 17 GHz with 2-dBm power. At $V_{dd} = 1.5$ V and RF of 20.4 GHz, transducer power gain and SSB NF are 9 and 6.6 dB, respectively. The measured third order input intercept point (IIP3) is –10.9 dBm. At $V_{dd} = 1.2$ V, transducer power gain and NF are 7 and 7.2 dB, respectively, IIP3 is –11 dBm, and power consumption is 7.3 mW, which is extremely low for a 24 GHz down-converter. The on-wafer measurement results are summarized in Table 9.1.

INTEGRATED 24-GHz TRANSMITTER

In the transmitter shown in Figure 9.5 [18], the baseband digital signal directly modulates a carrier at IF to generate a modulated signal at IF. The signal is amplified and fed into a double-balanced Gilbert cell up-conversion mixer. The RF signal is amplified by a three-stage driver and fed to a power amplifier. Finally, the PA drives a 3-mm-long on-chip zigzag dipole antenna with a bend angle of 30°, similar to that reported in Refs. 6,10. All the circuits in the signal path are once again fully differential, which should attenuate the common-mode noise injected to the VCO and any noise from the digital circuits that will eventually be integrated on the same chip. Just as in the receiver, this also allows the connection to the dipole antenna to be made without a balun. Figure 9.8 shows the die micrograph. The transmitter including the antenna and bond pads occupies 1.8 mm².

TABLE 9.1
Measured Performance of Down-Converter

Supply voltage	1.5 V		1.2 V
RF		20 GHz	
IF		3 GHz	
Chip area		1.1 mm²	
Total power consumption	12.8 mW		7.3 mW
LNA current	4.7 mA		2.8 mA
Mixer current	3.8 mA		3.3 mA
Power gain	9.0 dB		7.0 dB
Noise figure (SSB)	6.6 dB		7.2 dB
Input IIP3	−10.9 dBm		−11 dBm

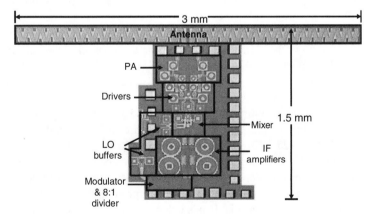

FIGURE 9.8 Micrograph of the transmitter. (From Brewer et al., *IEEE Circuits and Devices Magazine*, November 2006. With permission. © 2006 IEEE.)

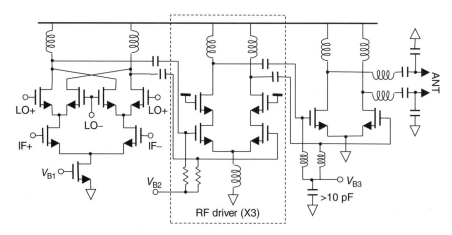

FIGURE 9.9 Schematic of the mixer, RF drivers, and PA. (From Cao et al., *Symposium on VLSI Circuits,* Honolulu, HI, pp. 184–185, June 2006. With permission. © 2006 IEEE.)

The modulator implements pseudo-MSK [19]. For MSK modulation, the envelope remains constant and phase is continuously changed. To simplify the circuit, the possible phases are limited to 16 discrete steps, or four per quadrant. This leads to ~8 dB higher side lobes compared to that for the standard MSK. However, the peak side lobes occur at frequency offset of four times the data rate, which makes the filtering of the side lobes easier. The phases are implemented by combining quadrature signals using a current-summing circuit [19], whereas the total current is unchanged to maintain a constant envelope.

Figure 9.9 shows the schematics of up-conversion mixer, one of the driver stages, and PA. Since a constant-envelope modulation is used, a nonlinear class-E PA is used to improve efficiency. The PA utilizes a single 100 μm transistor. To reduce the AC-coupling capacitor value, the gates of transistors are tuned by shunt inductors. The driver stages are implemented using cascode amplifiers to achieve higher gain and better stability. The sizes of the transistors in the three stages are 14, 20, and 48 μm. The current sources at the common-mode nodes like in the down-converter are replaced by inductors to increase the voltage headroom.

The transmitter is first characterized on-wafer without the antenna. It delivers 8 dBm output power to a 50-Ω load near 24 GHz and exhibits a 3-dB bandwidth of 3 GHz. The transmitter including the divider and LO buffers consumes 100.2 mW (1.2–V V_{dd} for the PA and 1.5-V V_{dd} for the others). The required LO power is −5 dBm. The PA should be able to deliver saturated output power of ~10 dBm at 1.2 V V_{dd} if a larger voltage swing were provided by the driver. Figure 9.10 shows the measured output power spectrum density (PSD) for 100 Mb/s pseudorandom digital input. The measured main-lobe bandwidth is about 1.5 times the data rate and contains nearly all the output power as expected for the MSK spectrum. Outside the 24–24.25 GHz industrial scientific medical (ISM) band, the peak PSD is −36 dBm/MHz, or 5 dB higher than the equivalent isotropic radiation power (EIRP) of −41.25 dBm/MHz (measured with 1 MHz resolution bandwidth) required by the FCC. This can be improved by properly tuning the IF amplifier and also increasing the number of possible phases per quadrant from four to eight. The spectrum measured at 3 m from the transmitter includes the LO spur with EIRP of −27.5 dBm/MHz primarily due to the amplification of the LO signal coupled through the mixer by the driver and PA. To reduce this below-the-Federal Communication Commission (FCC) limit, notch filters should be added in the transmitter. Although, the direct coupling of LO signal to the antenna is believed to be small, presently, it cannot be completely ruled out. More work is needed to confirm this. If needed, approaches to reduce this coupling must be developed. Table 9.2 summarizes the measured performance of the transmitter.

These down-converter and transmitter results clearly indicate that 24 GHz transceivers with adequate performance can be implemented. The transmitter power consumption of ~100 mW

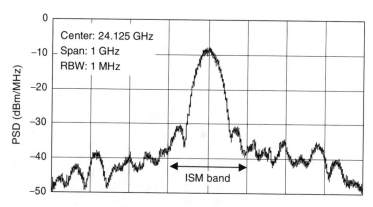

FIGURE 9.10 Measured output spectrum (100 Mb/s data rate). (From Cao et al., *Symposium on VLSI Circuits*, Honolulu, HI, pp. 184–185, June 2006. With permission. © 2006 IEEE.)

TABLE 9.2
Summary of the Transmitter Performance

Technology	0.13 μm CMOS
Chip size	1.8 mm² (TX), 0.7 mm² phase locked loop (PLL)
Output power	8 dBm
Data rate	100 Mb/s
Error vector magnitude (EVM)	7.7% (rms), 16.8% (peak)
Transmitter power consumption	100.2 mW
	PA: 21 mA × 1.2 V;
	others: 50 mA × 1.5 V

is ~2× higher than desired. Using more advanced process technologies should lower the power consumption. The power consumption limits the minimum physical volume of nodes that can be built because increased power consumption increases the battery size, and thus node size. To support the peak current requirement from hearing aid batteries, super capacitors should be utilized. It should be possible to provide 150 mA peak current from three size-10 ZnO batteries in combination with one 1 nF chip capacitor. The batteries and capacitor have a total volume less than that of the volume of an M&M. These batteries should be sufficient to support operation for ~1 month assuming 0.1% duty cycle.

Node-to-Node Communication

As discussed above, the goal for local μNode communication range is 1–5 m at 24 GHz. The communication over 5 m using the down-converter with an on-chip antenna is demonstrated by transmission of an amplitude modulation (AM) signal and down-conversion of this signal to IF. The measurement setup is shown in Figure 9.11. The down-converter is placed on a probe station inside a cage, which is probably a difficult environment for radio operation. The transmitted signal is generated by an Agilent 8254A. The carrier frequency is 20 GHz and modulating signal is 100 kHz with modulation depth of 50%. The signal is converted from single ended to balanced using a balun and fed to a 3 mm on-chip zigzag dipole antenna with a signal–signal probe. The power delivered to the antenna is 10 dBm. The antenna is placed on a mobile probe stand [5]. The down-converter is located 5 m away. The IF output is amplified using an external amplifier with 26-dB gain. The received spectrum with a resolution bandwidth of 300 Hz is shown in Figure 9.12. The power at

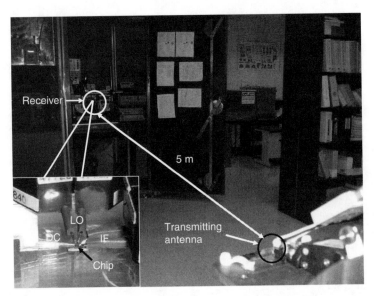

FIGURE 9.11 Measurement setup and environment for communication using on-chip antennas. (From Yu, S., Lin, J. J., and O, K. K., *2005 IEEE International Solid-State Circuits Conference*, San Francisco, CA, pp. 270–271, February 2005. With permission. © 2005 IEEE.)

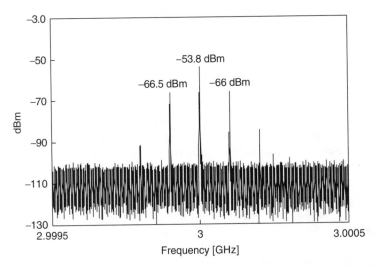

FIGURE 9.12 Received spectrum. The two sidebands 100 kHz away from the IF have power levels of 12.7 and 12.2 dB lower than that for the carrier, which are close to the theoretical values. The extra sidebands in the spectrum are due to the spurs from the signal generator. (From Yu, S., Lin, J. J., and O, K. K., *2005 IEEE International Solid-State Circuits Conference*, San Francisco, CA, pp. 270–271, February 2005. With permission. © 2005 IEEE.)

3 GHz IF is −53.8 dBm. This indicates that the power gain between the transmitting and the receiving antennas is ~−99 dB. The two sidebands 100 kHz away from IF have power levels of 12.7 and 12.2 dB lower than that for the carrier, which are close to the theoretical values. The extra sidebands in the spectrum are due to the spurs from the signal generator. This demonstrates that it is possible to communicate over 5 m using a CMOS radio with an on-chip antenna.

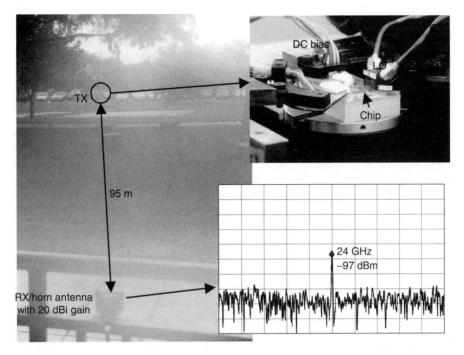

FIGURE 9.13   Composite picture of node-to-base-station experimental arrangement. (From Cao et al., *Symposium on VLSI Circuits*, Honolulu, HI, pp. 184–185, June 2006. With permission. © 2006 IEEE.)

NODE-TO-BASE-STATION COMMUNICATION

A group of μNodes could also be interrogated by a base station to obtain visibility of the local situation. The exact nature of this base station has been beyond the current scope of research, but a variety of arrangements can be envisioned. One rather dramatic scenario would be for a low-flying unmanned aerovehicle (UAV) to communicate with the μNodes and command an uplink of data. Another less exciting, but perhaps more likely, case might be to have a base station unit mounted on a pole at some distance from the monitored region.

The feasibility of this operation scenario is demonstrated by transmitting a 24 GHz single tone (constant digital input) using the 24 GHz transmitter. A horn antenna with 20-dBi gain located at an entrance of a building is used as the receiver, and the transmitter is placed in a parking lot 95 m away on a foggy morning (Figure 9.13). The horn could be the antenna of a base station. The received signal is −97 dBm, and the antenna-pair gain between the transmitting and the receiving antennas is about −105 dB. This suggests that communication between a base station and an IC with an on-chip antenna over a distance of ~100 m is possible at 24 GHz.

CRYSTAL ELIMINATION APPROACH

One of the main challenges in designing a single-chip radio is the integration of a high-stability frequency reference. Most modern communications systems depend on crystal-based frequency references for accurate channel tuning and low phase noise, both of which impact quality of the received signal. Developing modulation and receiver processing methods that will withstand the low stability of a crystal-less frequency reference is a key to realizing a single-chip radio.

Channel-tuning error generally leads to loss of signal energy and pulse shape distortion and can result in higher detector noise levels if the receiver filtering is made wider than necessary to accommodate frequency offsets. The use of frequency acquisition and tracking algorithms is one possible solution to the problem as long as there is sufficient signal time prior to data arrival to allow the algorithms to settle. An alternate strategy is to broaden the desired signal bandwidth using spread-spectrum techniques such that the expected frequency offsets are a small fraction of the signal bandwidth. Direct-sequence spread-spectrum (DS/SS) methods in particular are well suited to digital implementation, which fits well with the goal of a single-chip radio.

Phase noise has long been known as a source of degradation in communications systems, with effects ranging from degraded signal-to-noise ratio (SNR) to irreducible error-rate floors in some digital systems. Comparisons have shown that narrowband modulation is more susceptible than wideband, and coherent detection techniques are more sensitive to its effects than noncoherent techniques [20]. Differential detection in particular is known to be tolerant of phase noise, and when applied at the chip level in a DS/SS system, it leads to relaxation of both phase noise and frequency offset requirements. This section describes the application of differential chip detection (DCD) and its role in relaxing stability requirements for a frequency reference.

DIFFERENTIAL DETECTION

Differential detection is often used as a low-complexity alternative to coherent detection in systems that can withstand a modest sensitivity penalty [21]. Differential phase shift keying (DPSK) and differential quadrature phase shift keying (DQPSK) are common examples in which a delayed version of the previous symbol is used as a phase reference for demodulating the present symbol. These methods can tolerate a small amount of phase drift between adjacent symbols, and frequency offsets are limited to a fraction of the symbol rate. For DS/SS systems, performing differential detection at the chip level instead of symbol level extends the frequency offset tolerance to a fraction of the chip rate [22–25]. For typical DS/SS systems, with processing gains of 10–30 dB, this represents a relaxation of one or more orders of magnitude in the frequency stability requirements of the transmitter and receiver.

Chip-level differential detection has also been shown to improve robustness to oscillator phase noise [20,21]. Intuitively, differential detection can be thought of as having a high-pass response to phase errors. Constant phase errors are removed completely, whereas other slowly varying (i.e., low frequency) phase errors are partially removed. Since oscillator phase noise has a low-pass spectral characteristic, it will tend to be suppressed by differential detection. Increasing the chip rate effectively moves out the corner frequency of the high-pass response and provides higher attenuation to the phase noise. The ability of chip-level differential detection to relax both frequency stability and phase noise makes it an attractive technique.

The basic processing steps used in DCD are illustrated in Figure 9.14. An input chip at time index k includes unwanted phase noise θ_k and frequency offset ω_ε. The differential detector multiplies the present chip by the conjugate of the previous chip, thereby converting the frequency offset to a phase term, $\omega_\varepsilon T_c$, and producing a differential phase noise term $\Delta\theta_k = \theta_k - \theta_{k-1}$. If the pseudo-random noise (PN) sequences representing each symbol are differentially encoded prior to transmission, then the DCD, $c_k c_{k-1}{}^*$, will produce the desired PN sequence values at its output.

Assuming real-valued chips, small frequency offset, and low phase noise, the output of the multiplier will be approximately real. A typical differential detector will therefore ignore the imaginary part of the output and keep only $(c_k c_{k-1}) \cos(\omega_c T_c + \Delta\theta_k)$. As the frequency offset and differential phase noise terms

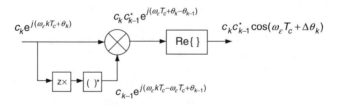

FIGURE 9.14 Differential chip detection block diagram.

increase, the cosine scaling factor becomes less than unity, effectively reducing the power of the desired signal. For example, for a frequency offset of $0.1R_c$ (such that $\omega T_c = \pi/5$), the cosine scaling factor would be 0.8 and the signal power loss would be nearly 2 dB. Increasing the chip rate will tend to move the cosine factor closer to unity, reducing the loss associated with frequency errors and phase noise. However, as shown later, increasing the chip rate also degrades the noise performance of DCD.

MODULATION FORMAT

The modulation format extends the use of DCD to 16-ary orthogonal signaling, offering improved detector performance at the expense of increased demodulator complexity. For general M-ary orthogonal signaling, a group of B information bits is used to select one of $M = 2^B$ orthogonal waveforms for transmission during a symbol period. The $M = 16$ orthogonal waveforms are actually different PN sequences, making it possible to apply DCD during demodulation. The set of sequences $\{s_0, s_1, \ldots, s_{M-1}\}$ comprising the M-ary symbol alphabet consists of M cyclic shifts of an m-sequence. Since m-sequences are known to have good autocorrelation properties, the resulting set of sequences will have good cross-correlation properties (nearly orthogonal), as well as good autocorrelation properties for each symbol. To simplify implementation, the m-sequence is augmented with an extra zero-chip to give it an even length.

Figure 9.15 shows a block diagram for a general M-ary orthogonal modulator. The information sequence is formatted into B-bit data symbols, and data symbols are mapped into L-chip binary PN sequences. The symbol sequences are serially concatenated and passed to the chip-level differential encoder. Although the differential chip encoding is shown here as an explicit operation, in actual implementation it is possible to precode the differentially encoded values of the stored set of symbol sequences $\{s_0, s_1, \ldots, s_{M-1}\}$ to avoid this additional operation. Finally, the encoded chip sequence is modulated onto the carrier using OQPSK with half-sine pulse shaping.

Figure 9.16 shows a block diagram of the M-ary orthogonal demodulator. Functionally, the demodulator consists of a differential chip detector followed by an optimal coherent detector for orthogonal signaling. The differential chip detector removes phase offsets between transmitter and receiver, and it mitigates frequency offsets as well as phase noise. However, the differential chip detector also enhances noise, thereby degrading SNR. For DS/SS systems, whose SNRs are typically low to begin with, the degradation can be larger than the 1–3 dB usually associated with symbol-level differential detection.

NOISE ANALYSIS

Consider the DCD function with chip values c_k and additive noise n_k (Figure 9.17). Here, the frequency offset and phase noise are ignored so that we can focus on the receiver's additive noise, usually modeled as additive white Gaussian noise (AWGN).

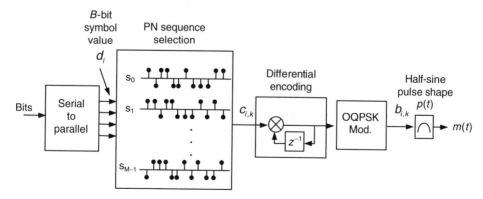

FIGURE 9.15 Block diagram for M-ary quasi-orthogonal modulator.

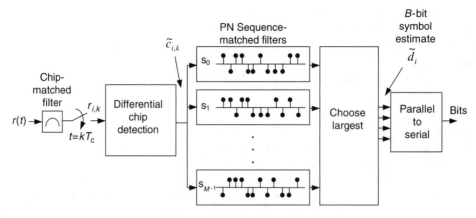

FIGURE 9.16 Block diagram for M-ary quasi-orthogonal demodulator.

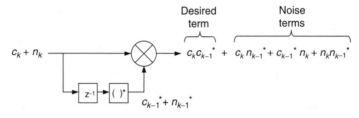

FIGURE 9.17 Noise products in differential detection.

For a simple analysis, assume that the noise samples, n_k and n_{k-1}, and the chip samples, c_k and c_{k-1}, are all uncorrelated. This is only approximately true but leads to a good prediction of the noise performance. Let σ_n^2 represent the noise power and σ_c^2 the desired signal power; then the input and output SNR expressions are

$$\mathrm{SNR_{IN}} = \frac{\overline{c_k^2}}{\overline{n_k^2}} = \frac{\sigma_c^2}{\sigma_n^2}$$

$$\mathrm{SNR_{OUT}} = \frac{\overline{c_k^2}\,\overline{c_{k-1}^2}}{\overline{c_k^2}\,\overline{n_{k-1}^2} + \overline{c_{k-1}^2}\,\overline{n_k^2} + \overline{n_k^2}\,\overline{n_{k-1}^2}} = \frac{\sigma_c^4}{2\sigma_c^2\sigma_n^2 + \sigma_n^4} = \frac{\mathrm{SNR_{IN}^2}}{2\mathrm{SNR_{IN}} + 1} \tag{9.1}$$

For large input SNR, the output SNR is approximately equal to half of the input SNR. This is the 3 dB sensitivity loss commonly associated with differential symbol detection schemes. However, for the low input SNR levels typical of DS/SS receivers (below 0 dB), the output SNR asymptotically approaches the square of the input SNR. In this case, the sensitivity loss associated with differential detection is not constant but is instead a function of the input SNR. Specifically, each halving of the input SNR will be accompanied by 1.5 dB additional sensitivity degradation.

Using a derivation similar to that in Ref. 22, a more complete SNR expression can be obtained for the M-ary system that includes the effects of noise, frequency errors, and phase noise. Written in terms of E_b/N_0 (energy-per-bit to noise power spectral density ratio) the effective SNR for the DCD demodulator is

$$\left(\frac{E_b}{N_0}\right)_{\mathrm{DCD}} = \frac{\varepsilon^2 (E_b/N_0)^2}{(L/B) + 2(E_b/N_0)} \tag{9.2}$$

This effective E_b/N_0 value can be substituted into a theoretical bit error rate (BER) formula for coherent M-ary orthogonal detection to predict the BER performance. The factor ε represents the average energy loss associated with frequency offset ω_ε and differential phase noise $\Delta\theta_k = \theta_k - \theta_{k-1}$.

$$\varepsilon = \overline{\cos(\omega_\varepsilon T_c + \Delta\theta_k)} = \cos(\omega_\varepsilon T_c)\exp(-\sigma_{\Delta\theta}^2/2) \tag{9.3}$$

L and B are the processing gain and number of information bits. As described previously, the variance of the differential phase noise will depend on the chip rate and the noise spectral density of the oscillator. The formal expression of this relationship is given by

$$\sigma_{\Delta\theta}^2 = 2\int_0^\infty P_\theta(f)|H_{\Delta\theta}(f)|^2\,df$$

$$= 4\int_0^\infty P_\theta(f)[1-\cos(2\pi f T_c)]df \tag{9.4}$$

where $P_\theta(f)$ is the power spectral density of the phase noise and $H_{\Delta\theta}(f)$ is the high-pass transfer function created by the differential phase detection. Simulation results are shown later for the particular case of $1/f^2$ phase noise.

SIMULATION RESULTS

Figure 9.18 shows analytical predictions along with MATLAB simulation results for the 16-ary system in AWGN with ideal symbol synchronization. To keep the plot simple, theoretical results are only shown for the lowest and highest processing gain values. For processing gains in between, the results were similar—the theoretical curves were optimistic by 0.2–0.5 dB, with higher processing gains showing closer agreement. For larger processing gains, where the input SNR is low, the results do exhibit the expected 1.5 dB loss of sensitivity for each doubling of the processing gain (i.e., each halving of the input SNR).

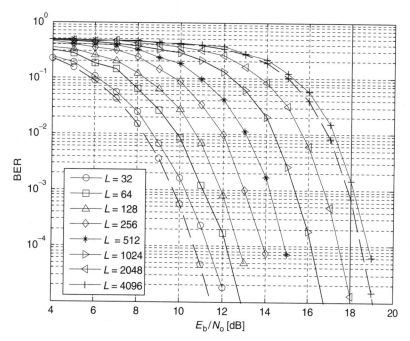

FIGURE 9.18 Simulation (solid) and analytical (dashed) performance of 16-ary system in AWGN.

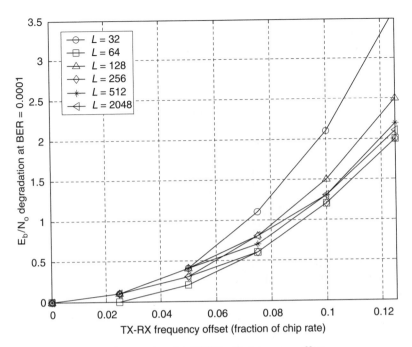

FIGURE 9.19   Performance of 16-ary system in AWGN with frequency offset.

Differences between theoretical and simulated performance are attributed to assumptions that the noise and chip samples are completely uncorrelated. With OQPSK chip demodulation, the matched filter causes interchip interference and noise correlation between adjacent chip periods. In addition, the set of $M = 16$ PN sequences are not purely orthogonal, which leads to minor degradation at lower processing gains. Nevertheless, the theoretical predictions are fairly accurate and provide a useful means to predict performance for systems with high processing gains, where simulation times can be long.

Figure 9.19 shows simulation results for the 16-ary system in AWGN with ideal symbol timing and various frequency offsets. A data rate of 100 kbps is assumed throughout the simulations. Although there is some variation due to the processing gain L and sequence selection, the sensitivity loss is primarily a function of the relative frequency offset as expected. Note that 2 dB loss occurs at frequency offsets of $0.1R_c$ to $0.12R_c$, which is in agreement with the prediction based on the cosine scaling factor.

One of the goals for the integrated frequency reference is to achieve ± 100 ppm stability, which could produce a worst-case frequency offset of ± 200 ppm between the transmitter and the receiver. At 24 GHz operating frequency, this translates into a 4.8 MHz offset. Using 10% of the chip rate as our rule of thumb for acceptable offset, the chip rate must be on the order of 48 Mchips/s. The processing gain L required to achieve this chip rate will depend on the desired data rate. For the 100 kbps assumed here, the 16-ary symbol rate would be 25 kbaud, and a processing gain of $L = 2048$ would give a chip rate of 51.2 Mchips/s. This is slightly larger than the desired 48 Mchips/s, but it allows L to keep a power-of-2 property that simplifies implementation. Table 9.3 summarizes the signaling format.

Figure 9.20 summarizes performance for the 16-ary system in the presence of $1/f^2$ phase noise. The $1/f^2$ characteristic represents a conservative upper bound for phase noise in frequency generation units comprising noisy oscillators controlled by phase-locked loops [26]. These results are similar to previous findings for DS/SS-binary phase shift keying (BPSK), where phase noise tolerance increases by 3 dB for each octave change in chip rate. For the case with $L = 2048$, phase noise levels as high as -65 dBc/Hz at 1 MHz offset can be tolerated with acceptable (0.5 dB) degradation of E_b/N_0.

Results in this section reiterate the effect of chip rate on mitigating both frequency offset and phase noise at the expense of receiver sensitivity. Another cost of using DCD is the increased receiver complexity associated with processing very high chip rates—as high as 51.2 Mchips/s for a 100 kbps

TABLE 9.3
Summary of Signaling Format

Signal format	QPSK with half-sine pulse shape
	Direct-sequence spread-spectrum (DS/SS)
	2048 chips per symbol
	~50 Mchips/s
	Constant envelope
	Meets FCC part-15 mask
Coding	16-ary orthogonal coding (4 bits per symbol)
	~25 k-symbols/s
	~100 kbps
	Improved E_b/N_o performance
Standard	Scaled version of IEEE 802.15.4 PHY
	Compatible with 802.15.4 Zigbee upper open systems interconnection (OSI) layers

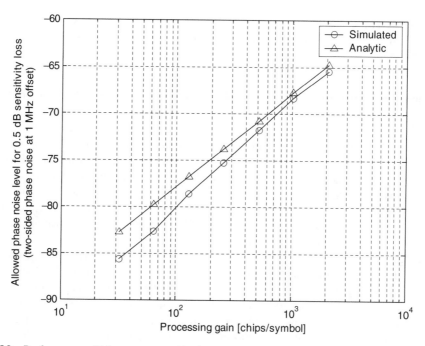

FIGURE 9.20 Performance of 16-ary system with phase noise.

data system. The next section on "Digital Implementaion Issues and Verification" discusses several of the implementation considerations for the demodulator.

DIGITAL IMPLEMENTATION ISSUES AND VERIFICATION

The use of M-ary quasi-orthogonal signaling, together with the high processing gains needed to make crystal-less operation feasible, leads to significant design complexity for the digital section of the receiver. Two key issues are the analog-to-digital (A/D) converter precision and M-ary detector architecture.

The A/D converter will be assumed to follow the chip-matched filter shown in Figure 9.16. For a processing gain of 2048, the simulation results from Figure 9.18 indicate that the SNR (E_c/N_0) at

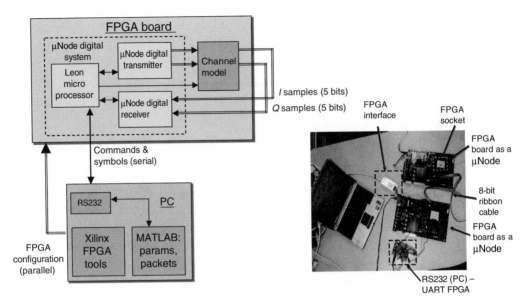

FIGURE 9.21 FPGA assembly used to demonstrate DS/SS differential chip detection technique. (From O, K. K., Brewer, J. E., and Taubenheim, D., *2005 GOMAC*, paper 9.5, Las Vegas, April 2005. With permission. © 2005 IEEE.)

the sampling point will be $-7\,dB$ at threshold sensitivity (BER $= 10^{-4}$). In other words, when the receiver is at its sensitivity threshold, the signal being sampled is predominantly noise. Therefore, with automatic gain control (AGC), it is possible to operate with as little as 4 bits of A/D resolution. At this level, the quantization noise will be a small fraction of the thermal noise and will have very minor impact on receiver sensitivity. Precision at the output of the DCD block grows to 9 bits, reflecting the squaring effect of the differential detection and magnifying the dynamic range needed to represent signal-plus-noise. Processing all resulting 9 bits in the PN sequence–matched filters is not economical, so the DCD output must be quantized to as few bits as possible. Again, recognizing that the DCD output will be dominated by noise at receiver sensitivity, its resolution can be reduced to 4 bits with minimal impact on performance.

The digital baseband circuit implementing DS/SS DSS has been verified in an field programmable gate array (FPGA) including the effects of frequency offsets and white noise [3]. Figure 9.21 shows the functional arrangement of the setup. The system can tolerate frequency stability of $\sim\pm150$ or 300 ppm total. This represents $\sim300\times$ relaxation of frequency tolerance requirements compared to that for radios using conventional modulation/demodulation techniques. Presently, the best on-chip frequency reference has stability of ~1000 ppm over the process and supply variations [27], which is $\sim3\times$ larger than the required one. The oscillator uses an open-loop compensation technique. Research is underway to improve the stability using a closed-loop compensation technique.

MANAGEMENT OF NOISE AND UNWANTED SIGNAL COUPLING

A concern for a highly integrated system such as a single-chip radio is unwanted signal and noise coupling. Integration of antennas is expected to exacerbate this. Large signals from an on-chip power amplifier and a transmit antenna could couple to other sensitive RF circuits, such as LNA and VCO, and disrupt their operation. The noise from digital circuits can couple to RF circuits through the antenna(s). These potentially can make the single-chip radio unrealizable. These must be addressed using all possible means available at different levels of implementation including the system, circuit, and component.

Unwanted signal coupling to a VCO can add noise to a VCO output as well as pull the output frequency. To study the VCO pulling, a 5.8-GHz N-channel MOS (NMOS) VCO along with ground-signal-signal-ground bond pads for injection of pulling signal was fabricated in a 0.18-μm CMOS process. The circuit schematic and die photograph of the structure are shown in Figure 9.22. The pulling signal is injected into the substrate through the parasitic capacitors of bond pads. Applying RF signal with 0.86 V amplitude results in injection of ~0.2 V amplitude signal into the substrate. The measurements show that the pulling signal at frequencies ~40 MHz away or 0.7% of the carrier frequency from the carrier has negligible impact on VCO operation [28]. The measurements also have shown that ground-shielding the inductors and grounding the substrate of transistors using large-area substrate contacts are not effective in reducing the unwanted coupling. Additionally, adding bypass capacitors C_{bp1} and C_{bp2} can actually increase the pulling effect. These counterintuitive experimental results are attributed to 0.3–0.5 nH on-chip common ground inductance, which allows the noise to be spread throughout the VCO via low-impedance connections provided by the ground shields, substrate connections, and bypass capacitors. Use of the dual-conversion radio architecture shown in Figure 9.5 with an IF of ~3 GHz separates the VCO frequency and that for the transmitted signal by 3 GHz. This large separation (14% of carrier frequency) should eliminate any VCO injection-pulling problem. Use of a half-duplex system eliminates the unwanted RF signal coupling to the LNA from a power amplifier.

Another concern is the coupling of noise from digital circuits into RF circuits in particular through antennas. This has been investigated using 4-mm-long linear dipole antennas fabricated on a sapphire substrate and a frequency divide-by-128 circuit fabricated using a 0.8 μm CMOS process mounted onto a 16-pin metal header as shown in Figure 9.23 [29]. The antennas fabricated on sapphire typically have higher efficiency due to lower substrate loss and should more efficiently

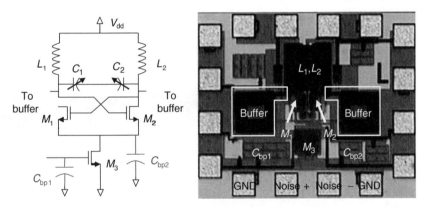

FIGURE 9.22 5.8 GHz voltage-controlled oscillator with ground-signal-signal-ground bond pads for RF-pulling signal injection.

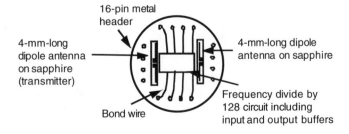

FIGURE 9.23 Top view of the packaged chip using a 16-pin metal header. (From Mehta, J. and O, K. K., *IEEE Transactions on Electro-Magnetic Compatibility*, vol. 44, pp. 282–290, May 2002. With permission. © 2002 IEEE.)

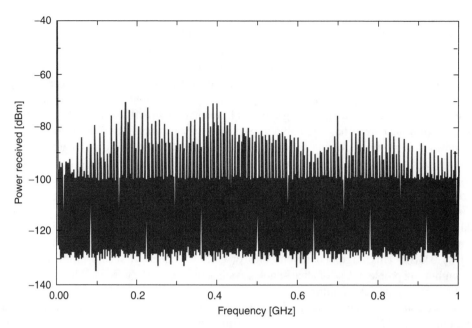

FIGURE 9.24 Single-ended power spectrum when the divider is on with input frequency of 700 MHz. (From Mehta, J. and O, K. K., *IEEE Transactions on Electro-Magnetic Compatibility*, vol. 44, pp. 282–290, May 2002. With permission. © 2002 IEEE.)

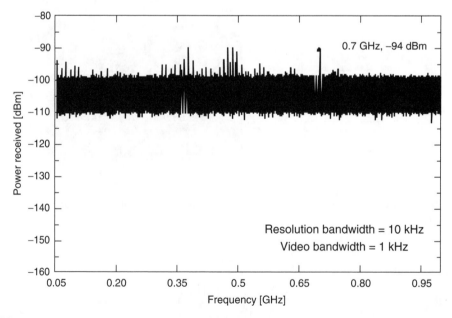

FIGURE 9.25 Power spectrum seen using differential measurement setup. The spectrum is almost clean and harmonics picked up are below –90 dBm, which points out that most of the low-frequency substrate and interconnect crosstalk generate common-mode noise and is rejected by the differential measurement. (From Mehta, J. and O, K. K., *IEEE Transactions on Electro-Magnetic Compatibility*, vol. 44, pp. 282–290, May 2002. With permission. © 2002 IEEE.)

couple noise. The frequency divider input frequency was set to 700 MHz. The noise picked up by the antenna was measured in both single-ended and balanced configurations. In the single-ended configuration, one arm of the dipole antenna is grounded. Figures 9.24 and 9.25 show the spectra of signal picked up by the antenna in single-ended and balanced configurations, respectively. The noise picked up in balanced configuration is less than −90 dBm and around 20 dB lower, indicating that the use of balanced antennas like using differential circuits reduces the coupling of noise from digital circuits, which tend to be common mode in nature. This and existence of several ICs incorporating RF, analog, and digital baseband and control circuits including TI BRF6100 in production suggest that the digital noise coupling can be managed.

CONCLUSIONS

On the basis of the progress in research on on-chip antennas, 24 GHz RF CMOS circuits, relaxation of frequency reference stability requirement using DS/SS DCD, and mitigation of noise and unwanted signal coupling, a single-chip radio integrating antennas, RF transceiver, analog and digital baseband circuits, and a frequency reference seems to be feasible. Presently, there are three pending issues that have not been clearly resolved, although they appear to be manageable: achieving the necessary stability requirement for an on-chip frequency reference (±150 ppm), LO feed-thru to the transmitter output, and noise coupling from digital circuits to RF circuits through on-chip antennas. Lastly, power consumption should be lowered to reduce the minimum size and to increase the lifetime of nodes that can be built using single-chip radios.

ACKNOWLEDGMENTS

This work has been primarily funded by DARPA (N66001-03-1-8901). The authors are also grateful to UMC for chip fabrication.

REFERENCES

1. K. K. O, K. Kim, B. Floyd, J. Mehta, H. Yoon, C.-M. Hung, D. Bravo, T. Dickson, X. Guo, R. Li, N. Trichy, J. Caserta, W. Bomstad, J. Branch, D.-J. Yang, J. Bohorquez, L. Gao, A. Sugavanam, J.-J. Lin, J. Chen, F. Martin, and J. Brewer, "Wireless communications using integrated antennas," *Proceedings of 2003 IEEE International Interconnect Conference*, pp. 111–113, June, 2003, San Francisco, CA.
2. K. K. O, K. Kim, B. A. Floyd, J. Mehta, H. Yoon, C.-M. Hung, D. Bravo, T. Dickson, X. Guo, R. Li, N. Trichy, J. Caserta, W. Bomstad, J. Branch, D.-J. Yang, J. Bohorquez, E. Seok, L. Gao, A. Sugavanam, J.-J. Lin, J. Chen, and J. Brewer, "On-chip antennas in silicon integrated circuits and their applications," *IEEE Transactions on Electron Devices*, vol. 52, no. 7, pp. 1312–1323, July 2005.
3. K. K. O, J. E. Brewer, and D. Taubenheim, "RF Subsystem for μNode SOC," *2005 Government Microcircuit Applications and Critical Technology Conference (GOMAC)*, paper 9.5, Las Vegas, NV, April 2005.
4. B. Warneke, M. Last, B. Liebowitz, and K. S. J. Pister, "Smart dust: Communicating with a cubic-millimeter computer," *IEEE Computer*, vol. 34, no. 1, pp. 44–51, January 2001.
5. J. J. Lin, L. Gao, A. Sugavanam, X. Guo, R. Li, J. E. Brewer, and K. K. O "Integrated antennas on silicon substrates for communication over free space," *IEEE Electron Device Lett.*, vol. 25, no. 4, pp. 196–198, April 2004.
6. J. J. Lin, X. Guo, R. Li, J. Branch, J. E. Brewer, and K. K. O, "10× improvement of power transmission over free space using integrated antennas on silicon substrates," *2004 IEEE Custom Integrated Circuits Conference*, pp. 697–700, October 2004, Orlando FL.
7. B. A. Floyd, K. Kim, K. K. O, "Wireless interconnection in a CMOS IC with integrated antennas," *International Solid-State Circuits Conference Dig. Tech. Papers*, pp. 328–329, February 2000.
8. B. A. Floyd, C.-M. Hung, and K. K. O, "A 15-GHz wireless interconnect implemented in a 0.18-μm CMOS technology using integrated transmitters, receivers, and antennas," *IEEE J. Solid-State Circuits*, vol. 37, no. 5, pp. 543–552, May 2002.

9. F. Touati and M. Pons, "On-chip integration of dipole and VCO using technology for 10 GHz applications," *2003 European Solid-State Circuits Conference*, pp. 494–497, October 2003.

10. K. Kim and K. K. O, "Characteristics of integrated dipole antennas on bulk, SOI, and SOS substrates for wireless communication," *Proceedings of the 1998 IEEE International Interconnect Conference*, pp. 21–23, San Francisco, CA, June 1998.

11. J. J. Lin, A. Sugavanam, L. Gao, and K. K. O, "On-wafer measurement setups for on-chip antennas fabricated on silicon substrates," *64th ARFTG Conference Digest*, pp. 221–225, December 2004, Orlando, FL.

12. J. Buechler, E. Kasper, P. Russer, and K. M. Strohm, "Silicon high-resistivity-substrate millimeter-wave technology," *IEEE Transactions on MTTS*, vol. mtt-34, no. 12, pp. 1516–1521, December 1986.

13. A.B.M.H. Rashid, S. Watanabe, T. Kikkawa, "High transmission gain integrated antenna on extremely high resistivity Si for ULSI wireless interconnect," *IEEE Electron Device Letters*, vol. 23, no.12, pp. 731–733, December 2002.

14. K. Kim, H. Yoon, and K. K. O, "On-chip wireless interconnection with integrated antennas," *Technical Digest of International Electron Device Meeting*, pp. 485–488, San Francisco, 2000.

15. M. Zargari et al., "A 5-GHz CMOS transceiver for IEEE 802.11a wireless LAN systems," *IEEE J. Solid State Circuits*, vol. 37, no. 12, pp. 1688–1694, December 2002.

16. S. Yu, J. J. Lin, and K. K. O, "A 20-GHz CMOS RF down-converter with an on-chip antenna," 2005 *IEEE International Solid-State Circuits Conference*, pp. 270–271, February 2005, San Francisco, CA.

17. A. K Sanjaani, H. Sjoland and A. A Abidi, "A 2 GHz merged CMOS LNA and mixer for WCDMA," *2001 Symposium on VLSI Circuits*, pp. 19–20, June 2001.

18. C. Cao, Y. Ding, X. Yang, J.-J. Lin, A. K. Verma, J. Lin, F. Martin, and K. K. O,"A 24-GHz transmitter with on-chip antenna in 130-nm CMOS," *2006 Symposium on VLSI Circuits*, pp. 184–185, June 2006, Honolulu, HI.

19. X. Yang, C. Cao, J. Lin, K. K. O, and J. E. Brewer, "A 2.5-GHz constant envelope phase shift modulator for low-power wireless applications," *2005 IEEE Radio Frequency Integrated Circuits Symposium*, pp. 667–670, Long Beach, CA, June 2005.

20. A. G. Burr, "Comparison of coherent and noncoherent modulation in the presence of phase noise," *Communications, Speech and Vision, IEE Proceedings-1*, vol. 139, no. 2, pp. 147–155, April 1992.

21. L. W. Couch II, *Digital and Analog Communication Systems,* 4th ed., Macmillan Publishing Company: New York, 1993.

22. P. Gorday, Q. Shi, and F. Martin, "Performance of chip-level differential detection with phase noise," *IEEE Wireless Communications and Networking Conference (WCNC)*, March 21–25, 2004.

23. G. Colavolpe and B. Raheli, "Improved differential detection of chip-level differentially encoded direct sequence spread-spectrum signals," *IEEE Transactions on Wireless Communications*, vol. 1, no. 1, pp. 125–133, January 2002.

24. Q. Shi, R. J. O'Dea, and F. Martin, "A new chip-level differential detection system for DS-CDMA," *IEEE International Conference on Communications (ICC)*, vol. 1, pp. 544–547, April 2002.

25. A. Cavallini, F. Giannetti, M. Luise, and H. T. Nguyen, "Chip-level differential encoding/detection of spread-spectrum signals for CDMA radio transmission over fading channels," *IEEE Transactions on Communications*, vol. 45, no. 4, pp. 456–463, April 1997.

26. A. Hajimiri, "Noise in phase-locked loops," *IEEE 2001 Southwest Symposium on Mixed-Signal Design*, February 2001, pp. 1–6.

27. Y. Wu and A. Aparin, "A temperature stabilized CMOS VCO for zero-IF cellular CDMA receivers," *2005 Symposium on VLSI Circuits*, pp. 399–401, June 2005, Kyoto, Japan.

28. Z.-B. Li, *Radio Frequency Circuits for Tunable Multi-Band CMOS Receivers for Wireless LAN Applications*, University of Florida, December 2004.

29. J. Mehta, and K. K. O, "Switching noise of integrated circuits (IC's) picked up by a planar dipole antenna mounted near the IC's," *IEEE Transactions on Electro-Magnetic Compatibility*, vol. 44, no. 5, pp. 282–290, May 2002.

Part II

Chip Architectures and
Circuit Implementations

10 Digital RF Processor (DRP™)

Robert Bogdan Staszewski

CONTENTS

Introduction 265
Overview of DRP 266
RF Circuits in Scaled CMOS 270
Local Oscillator 271
All-Digital Transmitter 275
    Frequency Response of the ADPLL-Based Transmitter 275
Discrete-Time Receiver 279
    Receiver Architecture 279
    Direct Sampling Mixer 282
    Temporal Moving-Average 282
    High-Rate IIR Filtering 284
    Additional Spatial MA Filtering Zeros 285
    Lower-Rate IIR Filtering 286
    Cascaded MTDSM Filtering 288
    Near-Frequency Interferer Attenuation 289
    Signal-Processing Example 289
    MTDSM Feedback Path 290
Script Processor 293
    Digital Compensation 293
    Built-In Self-Test 294
Energy Management 295
Behavioral Modeling and Simulation in VHDL 295
    VHDL TX Simulations 296
    VHDL RX Simulations 296
Summary 298
References 301

INTRODUCTION

The power and sheer volume of consumer demand has driven cellular handsets to an incredible level of complexity featuring diverse modulation schemes and multiple protocols, as well as coverage of multiple frequency bands. In addition, multiple radios now coexist to support Bluetooth personal area networking, global positioning system (GPS) location technology, wireless local area networks (WLAN) connectivity for high-speed local-area data access, and mobile digital TV reception. Sophisticated applications, such as MP3 audio playback, camera functions, and MPEG video are also sought after. Such application support dictates high level of memory integration together with large digital signal-processing capacity [1]. At the same time users continue to demand handsets that are lighter weight, less expensive, and more power efficient [2]. These demands put handset manufacturers in the unenviable position of meeting customer expectations while maintaining their own profitability.

Delivering increased levels of functionality in a shrinking form-factor while continuing to improve battery life and lower cost is possible only through the aggressive integration of the handset electronics. Analog radio frequency (RF) components are an obvious integration target since they typically occupy 30–40% of the total board real estate, and additional Bluetooth, GPS, and WLAN radios only increase that requirement. However, the RF integration has repeatedly proven to be a complex technological challenge [3]. As any engineer will attest, RF design is probably the trickiest of design challenges [4], even when working with highly specialized, stand-alone components. The conventional design approach has been to make the RF components as linear as they need to be through the customization of process technologies and sophisticated analog circuit linearization techniques. Optimized wafer processes that meet the special demands of analog and RF functions are normally employed and RF circuit designers are trained with such technologies in mind.

While being a favorable implementation for stand-alone transceiver integrated circuits (ICs), this approach poses major challenges for single-chip full radio integration with digital baseband (DBB) and application processors necessarily implemented, due to cost reasons, in the most advanced CMOS process available, which usually does not offer any analog extensions and has very limited voltage headroom. It is a very significant challenge to integrate RF, analog, and CMOS logic to create a production system-on-chip (SoC) device. The circuit design techniques, test approaches, key transistor and other circuit element requirements, and other considerations for analog and RF versus CMOS logic are normally quite different. There is no doubt that highly integrated analog radios can be manufactured [5], but if they do not reduce design complexity, cost, and power consumption, then the value of such integration is questionable.

Around 1999, a few engineers in Texas Instruments (TI) recognized the difficulties of RF integration and took it up as their next big challenge [6] to develop digital alternatives that could leverage the cost and power benefits of high-volume CMOS process manufacturing. The use of CMOS technology (now at the nanometer, i.e., $<0.1\ \mu m$ feature scale) has a potential for unprecedented integration of RF and analog circuits with large-scale digital logic that could not only satisfy, but also further stimulate, the consumer desire for highly sophisticated but inexpensive wireless devices. The nanometer scale CMOS processes are capable of acceptable RF performance [7–9], but their optimization is driven by the lowest cost of digital application specific integrated circuit (ASIC) gates and static random access memory (SRAM) memory. The RF circuits not only lose the benefit of specialized, process-controlled linearized components, but also have to cope with always decreasing supply voltages, increasing level of coupled interference, and area scaling requirements.

Our solution presented in this chapter maps the RF functionality into digital and discrete-time operations. This way, the RF analog circuit design complexity is transformed into the digital domain so that it enjoys the benefits of digital approach, such as process node scaling and design automation. The resulting synergy with digital logic profoundly shifts the paradigm of RF radio design [10]. Today, digital RF processor (DRP) architecture provides an efficient and cost-effective migration path for RF analog integration that promises to have a profound effect on the future of wireless technology.

All-digital phase-locked loop (ADPLL), all-digital control of phase and amplitude of a polar transmitter, and direct RF sampling techniques allow great flexibility in reconfigurable radio design. Digital signal-processing (DSP) concepts are used to help relieve analog design complexity, allowing one to reduce cost and power consumption in a reconfigurable design environment. VHSIC hardware description language (VHDL) hardware description language is universally used throughout this SoC. TI has already developed two generations of the DRP architecture that produced commercial Bluetooth and global system for mobile communication (GSM) SoC radios. We now see a future where RF is easily configurable and modularized so that it can be added to our applications as easily as other wired interfaces are today.

OVERVIEW OF DRP

The key principles presented in this chapter have been used to develop two generations of a DRP: single-chip Bluetooth [11] and GSM/enhanced data for GSM evolution (EDGE) [12,13] radios realized

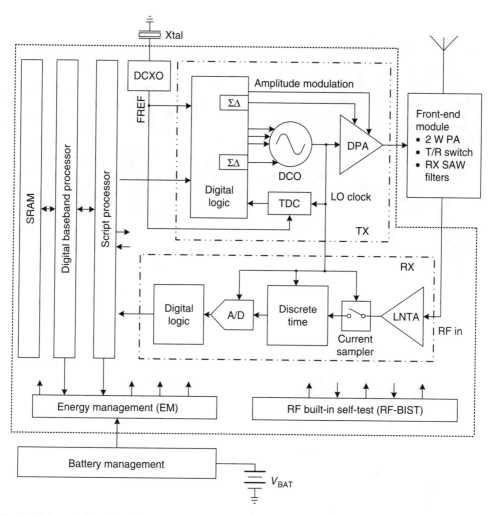

FIGURE 10.1   Single-chip radio based on the second generation of DRP.

in 130 and 90 nm digital CMOS process technologies, respectively. Figure 10.1 highlights the common architecture with added features specific to the cellular radio. The ADPLL-based transmitter employs a polar architecture with all-digital phase/frequency and amplitude modulation paths. The receiver employs a discrete-time architecture in which the RF signal is directly sampled and processed using analog and digital signal-processing techniques. A digitally controlled crystal oscillator (DCXO) [14] generates a high-quality basestation-synchronized frequency reference such that the transmitted carrier frequencies and the received symbol rates are accurate to within 0.1 ppm. An energy management (EM) system consists of a bandgap generator and multiple low dropout regulators to supply voltage to various radio subsystems as well as to provide good noise isolation between them. An RF built-in self-test (RF-BIST) executes an autonomous transceiver performance and compliance testing of the GSM standard. A script processor (SCR) [15] handles various transmit (TX) and receive (RX) calibration, compensation, sequencing, and low-rate datapath tasks, and encapsulates the transceiver complexity to present a much simpler software-programming model. The transceiver is integrated with the DBB, SRAM memory in a complete SoC solution.

The chip micrographs are shown in Figure 10.2. The GSM SoC consists of the DBB with digital logic and SRAM memory on the left part, and the DRP that integrates RF, analog, digital logic, SCR, and memory, on the right part. Closer examination of the IC reveals easily discernable

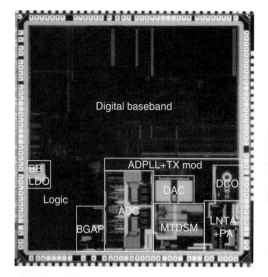

FIGURE 10.2 Die micrographs of the commercial single-chip SoCs employing two generations of DRP: (left) first generation—Bluetooth; (right) second generation—GSM.

TABLE 10.1
Technology Parameters of Texas Instruments' 90 nm CMOS Process Optimized for Wireless Applications

Interconnection Material	Copper
Number of metal layers	5
Minimum metal pitch	0.27 μm
Transistor nominal voltage	1.2 V
Gate oxide thickness	2.6 nm
MOS threshold voltage V_t	0.4 V
f_T of N-channel MOS (NMOS) transistor	100 GHz
ASIC gate density	250 kgates/mm^2
SRAM cell density	1.0 Mb/mm^2

RF inductors and other large analog elements that occupy silicon area equivalent to tens or even hundreds of thousands of digital gates. Consequently, to be cost-effective, the number of classical RF components shall be minimized with architectural and circuit design choices. Table 10.1 summarizes characteristics of the 90 nm CMOS process technology optimized for low-leakage applications. This technology is currently in volume production.

Of a special note is measured RX sensitivity of −82 dBm for Bluetooth and −110 dBm for GSM, versus the respective specifications of −70 dBm and −102 dBm—it is among the best in class. The overall GSM RX noise figure is only 2 dB. TX performance is also excellent. Figure 10.3 shows that the modulated GSM spectrum has 8 dB of margin at the most critical frequency offset of 400 kHz. The margin is even greater with EDGE (8 phase shift keying [8PSK]) modulation scheme, as shown in Figure 10.4. The measured error vector magnitude (EVM) is 1.24%.

Application of the presented digital RF processing techniques results in a high level of phone integration that leads to a substantial reduction in cost, board area, and power consumption. It is now possible to build cellular phones with extremely small dimensions, as demonstrated in Figure 10.5.

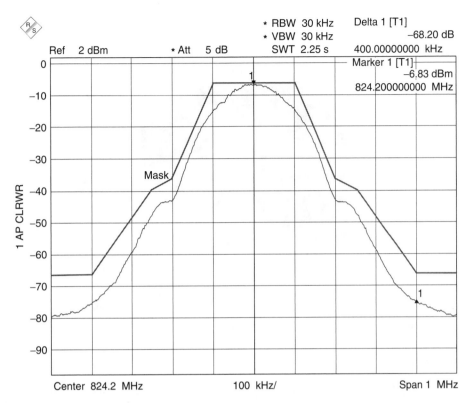

FIGURE 10.3 Measured GSM output spectrum. (From Staszewski, R. B., Wallberg, J., Rezeq, S., Hung, C.-M., Eliezer, O., Vemulapalli, S., Fernando, C., Maggio, K., Staszewski, R., Barton, N., Lee, M.-C., Cruise, P., Entezari, M., Muhammad, K., and Leipold, D., *Proceedings of IEEE Solid-State Circuits Conference*, sec. 17.5, pp. 316–317, 600, 2005, San Francisco, CA. © 2005 IEEE.)

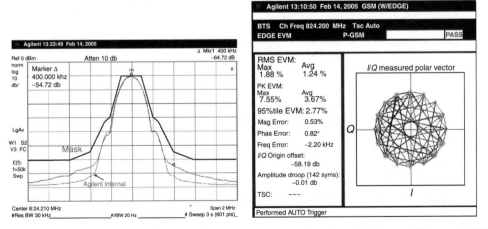

FIGURE 10.4 Measured EDGE TX modulation: (left) output spectrum; (right) constellation and EVM. (From Staszewski, R. B., Wallberg, J., Rezeq, S., Hung, C.-M., Eliezer, O., Vemulapalli, S., Fernando, C., Maggio, K., Staszewski, R., Barton, N., Lee, M.-C., Cruise, P., Entezari, M., Muhammad, K., and Leipold, D., *IEEE J. Solid-State Circuits*, 40, 2469, 2005. © 2005 IEEE.)

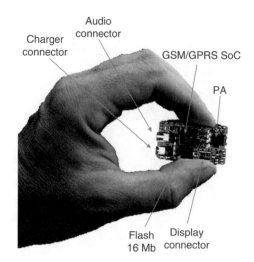

FIGURE 10.5 Prototype phone (without mechanics).

With further advances in battery technology, the long-ago-predicted "Dick Tracey watch"-style cell phone could soon find its ways *onto* consumer hands.

RF CIRCUITS IN SCALED CMOS

An early attempt at designing RF circuits in the nanometer process environment has revealed a new paradigm [17], which forms a foundation of a digital RF processor: In a nanometer-scale CMOS technology, time-domain resolution of a digital signal edge transition is superior to voltage resolution of analog signals. A successful design approach for highly integrated RF circuits in this environment would exploit the paradigm by emphasizing the following:

- Fast switching characteristics or high f_T of MOS transistors (40 ps and 100 MHz in 90 nm CMOS, and 20 ps and 250 GHz in 45 nm CMOS): high-speed clocks and fine control of timing transitions.
- High density of digital logic (250 kgates/mm^2 in 90 nm CMOS and 1 Mgates/mm^2 in 45 nm CMOS) and SRAM memory (1 Mb/mm^2 in 90 nm CMOS and 4 Mb/mm^2 in 45 nm CMOS) makes digital functions and assistant software extremely inexpensive.
- Small device geometries and precise device matching made possible by the fine lithography.

while avoiding the following:

- Biasing currents that are commonly used in analog designs
- Reliance on voltage resolution with ever-decreasing supply voltages and increasing noise and interferer levels
- Nonstandard devices that are not needed for memory and digital circuits, which constitute majority of the silicon die area

The resulting architecture will likely be more robust than a conventional one by producing lower phase noise and spurious degradation of the transmitter chain and lower noise figure of the receiver chain in face of millions of active logic gates on the same silicon die, as proven in Refs. 12 and 13, respectively, for the 90 nm CMOS. Additionally, the new architecture would be highly reconfigurable with analog blocks that are controlled by software to guarantee the best achievable performance and parametric yield. Another benefit of the new architecture would be an easy migration from one process node to the next without significant rework. Comparative study in Refs. 11 and 12 gives testimony to the good design modularity, portability, and scaling.

LOCAL OSCILLATOR

The frequency up-conversion and down-conversion of information carrying signals in a wireless communication channel is performed by a local oscillator (LO), which is implemented as an RF frequency synthesizer, as shown in Figure 10.6. The circuit takes a frequency reference (FREF) clock of frequency f_R (typically 8–40 MHz) and generates a variable frequency f_V (i.e., indirect frequency synthesis) at the multi-GHz RF output according to either an integer or fractional frequency multiplication ratio $N = f_V/f_R$ or frequency command word (FCW), where FCW $\equiv N$. The FREF source is usually built as a tunable crystal oscillator, which features an excellent long-term accuracy and stability, at least when compared to the variable RF oscillator, and it contains the only reference timing information for the frequency synthesizer to which phase and frequency of the RF output are to be synchronized.

In older process technologies, the frequency synthesizer has been traditionally based on a charge-pump phase locked loops (PLL) [18], as shown in Figure 10.7a, but this architecture is not easily amenable to scaled CMOS integration. For example, the loop filter capacitor needs to be large to suppress reference spurs of the charge pump. If realized as a metal-insulator-metal (MIM) capacitor, its size could be prohibitively large. MOS capacitors offer about 10 times area density improvement but the leakage current, which is due to gate electron tunneling, is getting worse with each process node. The leaky capacitor would introduce an equivalent parallel resistance whose value strongly depends on temperature, thus changing the loop characteristics. Efforts have been made to extend the architecture's lifetime by, for example, replacing the loop filter capacitor with a digital integrator or accumulator [19]. Since the capacitor's input and output are analog, the replacing accumulator needs to be preceded by an analog-to-digital converter (ADC) and followed by a digital-to-analog converter (DAC).

Moreover, the charge-pump PLL architecture suffers from high level of reference spurs generated by the correlative phase detection method, which require better filtering and thus slower loop transients that degrade frequency-settling times. To relax this trade-off a fractional-N PLL architecture with $\Sigma\Delta$ dithering of the clock division ratio is often used but at a cost of higher noise.

The new ADPLL [17] frequency synthesizer architecture that is amenable to the scaled CMOS technology and is free of the above problems is presented in Figure 10.7b. It is built from the ground up using digital techniques that exploit the new paradigm described in the section "RF Circuits in Scaled CMOS".

The digitally controlled oscillator (DCO) [20,21], which deliberately avoids any analog tuning controls, splits its tuning capacitance into a large number of tiny capacitors, whose states are selected digitally. As shown in Figure 10.8, this is in contrast with a conventional voltage-controlled oscillator (VCO) method in which the frequency tuning is achieved in analog fashion through a linear capacitance change of a single large varactor. The fully digital control allows for the loop control circuitry to be implemented in a fully digital manner as first proposed in Ref. 22 and then demonstrated as a novel phase-domain all-digital PLL [11]. The advanced lithography allows creation of extremely fine variable capacitors (varactors)—about 40 aF (attofarads) of capacitance per step, which equates to the control of only 250 electrons entering or leaving the resonating LC tank. Despite the small

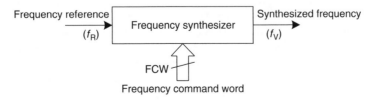

FIGURE 10.6 Frequency synthesizer as a frequency multiplier.

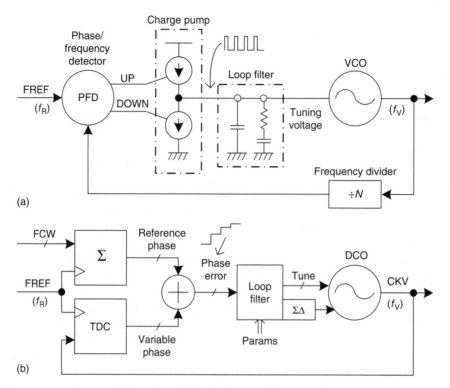

FIGURE 10.7 RF frequency synthesizers: (a) conventional charge-pump PLL; (b) all-digital phase-domain PLL based on TDC and DCO. FCW $\equiv N$ is a fractional frequency division ratio.

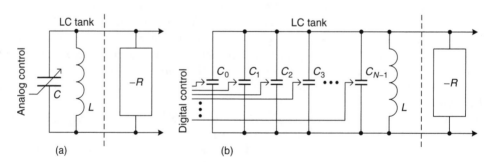

FIGURE 10.8 LC tank based-oscillators: (a) conventional with analog control; (b) with all-digital control. The negative resistance $-R$ perpetuates the lossy LC tank resonance.

capacitance step, the resulting frequency step at the 2 GHz RF output is 10–20 kHz, which is too coarse for wireless applications. Thus, the fast switching capability of the transistors is utilized by performing high-speed 225–900 MHz $\Sigma\Delta$ dithering of the 250 electrons in the finest varactors. The duty cycle of the high/low capacitive states establishes the time-averaged resonating frequency resolution, now better than 1 kHz. The finest varactors are realized as either p-poly/n-well (130 nm CMOS) or n-poly/n-well (90 nm CMOS) MOS capacitor (MOSCAP) devices.

Figure 10.9 shows a simplified schematic of the DCO core that operates in the 3.2–4.0 GHz range. The tuning control is split into several banks of varying degree of frequency step size and range: coarse d^P for process, voltage, and temperature (PVT) calibration; medium d^A for channel acquisition; and fine d^T for tracking of the oscillator drift. The d^P frequency range is the largest since it has to cover all the frequency bands and margin for the oscillator variability. The oscillator phase noise

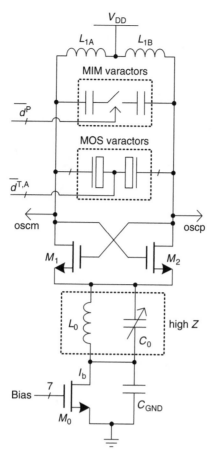

FIGURE 10.9   Oscillator core and the varactor state driver array.

is proportional to the dissipated current, which is established by the 7 bit bias control. The capacitor banks are built using MIM and MOS varactors. In agreement with the principle of avoiding biasing currents, which has been discussed in the section "RF Circuits in Scaled CMOS," the M_0 transistor array operates in linear region instead of in saturation. The current is set through automatic calibration at a minimum value at which the oscillator still produces acceptable RF phase noise.

The time-to-digital converter (TDC), as shown in Figure 10.10, generates the variable phase or timestamps of the FREF edges in the units of the DCO clock period [23]. The variable phase $R_V[k]$, where k is the FREF edge index, is a fixed-point digital word in which the fractional part is measured with a resolution of an inverter delay (less than 20 and 10 ps in 90 and 45 nm CMOS, respectively). Integer part of the variable phase is determined by counting the number of rising clock transitions of the DCO oscillator clock. The 48-bit TDC output forms a pseudothermometer code, which is then converted to binary and normalized to the variable clock (CKV) period, T_0. The number of inverters is set to cover one T_0. To arbitrarily increase the dynamic range, the CKV edge counter with a sufficient wordlength is added. The fixed-point TDC output timestamp consists of the sampled CKV edge count (integer part) and the T_0-normalized delay from CKV to FREF (fractional part). The time difference between the two FREF events is the difference between the two consecutive outputs. In PLL applications, the absolute timestamps (phase) are more useful [11,23] than the time difference (instantaneous frequency). Also, the reference edge locations are quite predictable, so the power is significantly saved by gating off the TDC activity during 90% of the time between the reference edges. To avoid metastability between the counter and the TDC core, a FREF resampling by the

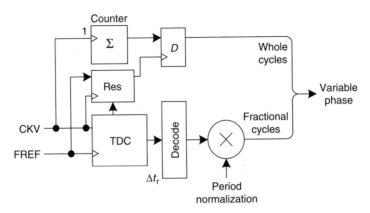

FIGURE 10.10   TDC system with large dynamic range.

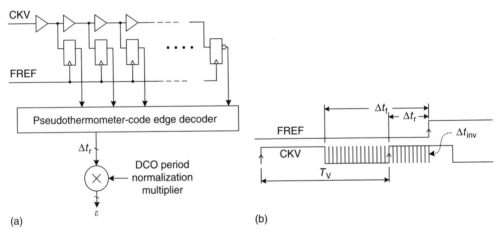

(a)   (b)

FIGURE 10.11   Time-to-digital converter (TDC) core: (a) structure; (b) quantization of the timing difference between the DCO and FREF edges. The integer counter of DCO edges is not shown.

opposite phases of CKV was used [17]. The selection method is similar to that given in Ref. 24 but only one counter is used.

The actual timestamps $R_V[k]$ are compared to the ideal timestamps or the reference phase $R_R[k]$, which is calculated as a summation of FCW: $R_R[k] = \Sigma FCW[k]$. The timing departure or the phase error $\phi_E[k] = R_R[k] - R_V[k]$ is filtered, and adjusts the DCO in a negative feedback manner.

Since the conventional phase/frequency detector and charge pump are replaced by the TDC, the phase-domain operation does not fundamentally generate any reference spurs, thus allowing for the digital loop filter to be set at an optimal performance point between the reference phase noise and the oscillator phase noise.

The TDC core measures and quantizes the time differences between the FREF and DCO edges (see Figure 10.11b), i.e., the fractional part of the variable phase.

Because of the full digital nature of the phase error correction, sophisticated control algorithms through a dynamic change of the loop filter parameters (refer back to Figure 10.7b) could be employed, which would not have been feasible with conventional architectures.

1. Dynamic gear shifting of the ADPLL bandwidth to speed up the frequency settling [25] and to respond to unexpected and expected disturbances in the SoC, such as ramping up the power amplifier and DBB, keyboard, or display activities

2. Adaptable characteristic of the ADPLL loop depending on the communication channel conditions or quality of the DCO and FREF clocks
3. Dynamic change of the ADPLL loop characteristics, such as dynamically switching from type I to type II loop after the settling is complete

ALL-DIGITAL TRANSMITTER

An RF transmitter that is well suited for a deep-submicron CMOS implementation is shown in Figure 10.12 [26]. It performs the quadrature modulation in polar domain [11,12]. The transmitter architecture is fully digital and takes advantage of the wideband frequency modulation capability of the ADPLL by adjusting its digital FCW. The modulation method is an exact digital two-point scheme, with one feed directly modulating the DCO frequency deviation while the other compensating for the developed excess phase error. The DCO gain characteristics are constantly calibrated through digital logic to provide the lowest possible distortion of the transmitted waveform [27,28].

The digitally controlled power amplifier (DPA) circuit, shown in Figure 10.13, which acts as a digital-to-RF-amplitude converter (DRAC) [29], is used for the power ramp as well as amplitude modulation in more advanced modulation schemes, such as the extended data rate (EDR) mode of Bluetooth, EDGE, or wideband CDMA (WCDMA). The DPA operates as a near-class-E RF power amplifier and is driven by the square wave output of the DCO. A large number of core NMOS transistors are used as on/off switches and are followed by a matching network that interfaces with an antenna or an external power amplifier, such as one providing 2 W in GSM. The number of active switches is controlled digitally and establishes the instantaneous amplitude of the output RF envelope. Fine amplitude resolution is achieved through high-speed $\Sigma\Delta$ transistor switch dithering. Despite the high speed of digital logic operation, the overall power consumption of the transmitter architecture is lower than that of architectures to date.

FREQUENCY RESPONSE OF THE ADPLL-BASED TRANSMITTER

Figure 10.14 shows a z-domain model of the ADPLL-based transmitter of Figure 10.12, but it operates here on the excess, that is, deviation from the nominal or expected, rather than absolute phase values. All the signals that are external to the components, except for the multi-GHz DCO output, correspond to fixed-point digital words with information contained in their binary values. The DCO output, however, is a single-bit clock with information contained in its continuously valued transition timestamps. The timestamps are sampled and quantized on rising edges of FREF using linear interpolation by the TDC [22]. The TDC quantization error of about 20 ps is modeled as an internal additive uniform phase noise $\phi_{n,\text{TDC}}$.

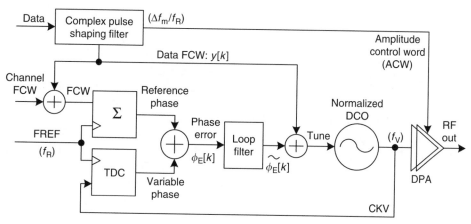

FIGURE 10.12 ADPLL-based polar transmitter.

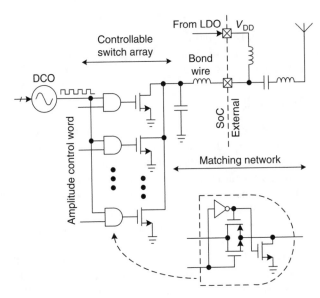

FIGURE 10.13 Digitally controlled power amplifier (DPA).

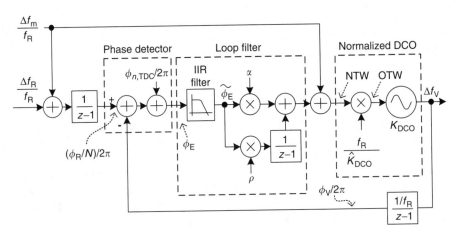

FIGURE 10.14 Discrete-time z-domain model of the ADPLL with wideband frequency modulation. The sampling rate is the reference frequency f_R.

It should be noted that the ADPLL is a multirate system with two clock domains, FREF and CKV, which have vastly different sampling rates, so for clarity and compactness of expression only a single-rate sampled system operating at the lower frequency is considered. As a consequence, it is not possible to thoroughly and accurately express the phase noise of the oscillator since the frequency aliasing will occur. However, for the purposes of the ADPLL phase-domain operation, only the CKV events in the immediate neighborhood of the FREF transitions are considered.

The equidistant sampling is expressed by the sampling interval $1/f_R$. Small period deviations due to the normal noise or jitter are inconsequential for the validity of this approach [45]. The frequency departure Δf_V from the nominal carrier frequency corresponds to a DCO period deviation ΔT_V. Integration of ΔT_V over the sampling interval $1/f_R$ denotes the timing deviation (TDEV) of the variable clock, which is related to the conventional definition of variable phase as $\phi_V = 2\pi \cdot \text{TDEV}/T_V$ [22,45]. The interpolator mentioned above is needed since there can be a noninteger number of variable clock cycles per the reference cycle (e.g., nonzero fractional part of FCW). Similarly, the timing deviation of the reference clock edges, which could be due to the frequency-domain noise that has the same effect as Δf_R, is related to the conventional definition of reference phase as $\phi_R = N \cdot 2\pi \cdot TDEV/T_V$. The reference and variable integrators $1/(z-1)$ have two interpretations. They are both the mathematical integrators of frequency deviations to obtain timing deviations, as well as hardware integrators that sum the deviation in FCW (reference TDEV) and sum the DCO clock edge timestamp deviations (variable TDEV).

Referring back to Figure 10.14, the open-loop z-domain transfer function is

$$H_{ol}(z) = r \cdot L(z) \cdot \frac{1}{z-1} \tag{10.1}$$

where $L(z)$ is a transfer function of the loop filter, which consists of an infinite impulse response (IIR) filter of transfer function $H_{iir}(z)$ and a proportional-integral gain of transfer function $H_{PI}(z)$. The $H_{iir}(z)$ filter is a cascade of four single-stage IIR filters, each with a leakage coefficient λ_i, where $i = 0 \ldots 3$. The $H_{PI}(z)$ filter consists of a proportional loop attenuator α and integral coefficient ϱ. The $r = K_{DCO}/\hat{K}_{DCO}$ factor, where $\hat{K}_{DCO}$ is an estimate of the DCO gain, K_{DCO} is indicative of the calibration error, and $r \approx 1$ is safely assumed:

$$L(z) = H_{iir}(z) \cdot H_{PI}(z) = \prod_{i=0}^{3} \frac{\lambda_i}{z - (1 - \lambda_i)} \cdot \left(\alpha + \frac{\varrho}{z-1} \right) \tag{10.2}$$

Combining Equation 10.1 with 10.2 results in

$$H_{ol}(z) = r \cdot \frac{1}{z-1} \cdot \prod_{i=0}^{3} \frac{\lambda_i}{z - (1 - \lambda_i)} \cdot \left(\frac{\alpha(z-1) + \varrho}{z-1} \right) \tag{10.3}$$

This type II sixth-order loop shows two poles at zero frequency $z_{p1} = z_{p2} = 1$, four poles at $z_{p,3+i} = 1 - \lambda_i$, for $i = 0 \ldots 3$, and one zero at $z_{z1} = 1 - \varrho/\alpha$.

It should be noted that the frequency multiplier $N \equiv FCW$ is not part of the open-loop transfer function and, hence, does not affect the loop bandwidth. Phase deviation of the frequency reference, on the contrary, needs to be multiplied by N since it is measured by the same phase detection mechanism normalized to the DCO clock period. This accounts for the $1/N$ factor in the $(\phi_R/N)/2\pi$ expression in Figure 10.14.

On the basis of the open-loop z-domain transfer function $H_{ol}(z)$, various close-loop transfer functions could be derived. It is convenient to operate with such two general types, frequency and phase, depending on a specific interpretation of the signals of interest. The closed-loop transfer function of the reference phase is low pass with the gain multiplier $N \equiv FCW$:

$$H_{cl,R}(z) = \frac{\Phi_V(z)}{\Phi_R(z)} = N \frac{H_{ol}(z)}{1 + H_{ol}(z)} \tag{10.4}$$

where $\Phi(z)$ is a z-transform of ϕ. The closed-loop phase transfer function for the TDC quantization noise is also low pass but it does not contain the N multiplier:

$$H_{cl,TDC}(z) = \frac{H_{ol}(z)}{1 + H_{ol}(z)} \tag{10.5}$$

The TDC code change of 1.0 corresponds to the DCO timestamp change of 1 UI (unit interval) or variable phase ϕ_V change of 2π. This applies to both the excess and absolute phase.

The closed-loop frequency transfer function from the digital word reference feed $\Delta f_R/f_R$ to the frequency deviation Δf_V at the PLL output is expressed as

$$H_{cl,R,f}(z) = f_R \cdot \frac{H_{ol}(z)}{1 + H_{ol}(z)} \tag{10.6}$$

but is of less practical importance since the preferred method of the ADPLL frequency modulation is through the $\Delta f_m/f_R$ digital input. To see why, let $H_{f_m}(z)$ be the transfer function from the modulating feed $\Delta f_m/f_R$ to the frequency deviation Δf_V at the PLL output. It consists of two parallel paths: one equivalent to the above reference path $H_{cl,R,f}(z)$ and the other is the direct DCO modulation with the transfer function $r \cdot f_R/(1 + H_{ol}(z))$:

$$H_{f_m}(z) = f_R \cdot \frac{r + H_{ol}(z)}{1 + H_{ol}(z)} \tag{10.7}$$

If the DCO gain K_{DCO} is estimated accurately, i.e., $r = 1$, then the modulating transfer function will be ideal with a constant gain of f_R, and there will be no effect on the phase error. This observation is the basis for the proposed RF modulation distortion estimation by the ϕ_E signal.

As an example, if the $\Delta f_m/f_R$ digital word steps from 0 to 1/100, it will immediately cause the DCO frequency deviation of $\Delta f_V = f_R/100$. This will shorten the DCO period by $\Delta T_V = \Delta f_V/f_V^2 = (f_R/100)/f_V^2 = (T_V/100)/N$ for the duration of one FREF cycle. Since there are N CKV clock cycles per FREF cycle, at the next ADPLL sample, the timing deviation detected by the TDC would be $\Delta T_V \cdot N = (T_V/100)/N \cdot (f_V/f_R) = T_V/100$. The TDC and phase detector digital outputs are normalized to the CKV period T_V, so the TDC output will be 1/100. This will be subtracted from 1/100 at the first phase detector input, so the phase detector output will be zero. On the next FREF cycle, both phase detector inputs will be $2 \times 1/100$, on the third $3 \times 1/100$, and so on, so the phase detector output will remain zero.

The closed-loop transfer function for the variable phase perturbations has a high-pass characteristic:

$$H_{cl,V}(z) = \frac{1}{1 + H_{ol}(z)} \tag{10.8}$$

The transfer function from both the reference and variable phase perturbations to the raw ϕ_E or filtered $\tilde{\phi}_E$ phase error is identical for both sources:

$$H_{cl,\phi_E}(z) = \frac{1}{1 + H_{ol}(z)} \tag{10.9}$$

$$H_{cl,\tilde{\phi}_E}(z) = \frac{H_{iir}(z)}{1 + H_{ol}(z)} \tag{10.10}$$

Note that Equation 10.9 is identical to Equation 10.8, which suggests that the raw ϕ_E be used as an estimator of the DCO phase contribution. However, due to the 20 or 30 dB/Hz slope of the DCO phase noise in the close-in region, the amount of integrated phase noise power at high frequencies is very small and $\tilde{\phi}_E$ should be used instead. The ϕ_E filtering provides bandlimiting that allows decimation to lower the calculation complexity.

Figure 10.15 gives a graphical example of various closed-loop transfer functions. The transfer functions of the two primary phase noise sources, reference and variable, to the RF output are shown as solid lines. Depicted as a single dotted line is the transfer function from both the reference and variable phase to the output of the IIR filter or the $\tilde{\phi}_E$ signal. The variable phase noise can be

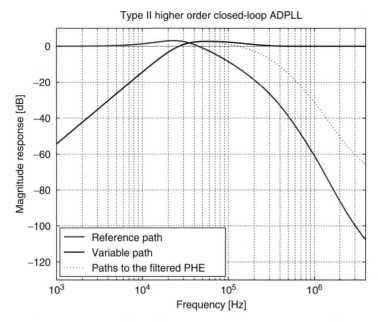

FIGURE 10.15 ADPLL closed-loop transfer function for the reference (normalized by N) and variable paths with the following loop setting: $\alpha = 2^{-7}$, $\varrho = 2^{-15}$, and $\lambda = [2^{-3}, 2^{-3}, 2^{-3}, 2^{-4}]$. The settings establish the closed-loop bandwidth of about 40 kHz and provide 33 dB of attenuation of the FREF phase noise and TDC quantization noise at 400 kHz offset. The type II setting provides 40 dB/dec filtering of the DCO $1/f$ noise.

estimated with great accuracy, especially if the raw ϕ_E is used. The reference phase noise could also be well estimated, albeit with lower precision than the variable noise, since the typical phase noise profile of a crystal oscillator is flat, so the integrated phase noise or phase error (within interval of 1–100 kHz in GSM) is mainly contributed by frequency offsets of tens of kilohertz. Figure 10.15 together with Equations 10.4 and 10.9 suggest that the higher frequency offsets (40 kHz–4 MHz) could be estimated accurately by applying additional filtering to $\tilde{\phi}_E$:

$$H(z) = r \cdot H_{PI} \cdot \frac{1}{z-1} \tag{10.11}$$

However, in this RF environment, the simple statistics on $\tilde{\phi}_E$ are sufficient to detect faults and to determine the GSM specification compliance.

On the basis of typical values of the phase noise sources and transfer functions of Figure 10.15, the composite phase noise at the RF output and that estimated by the phase error signal are depicted in Figure 10.16. It shows that the phase trajectory error (equivalent to the integrated phase noise in this two-point modulation scheme) could be well estimated by $\tilde{\phi}_E$.

DISCRETE-TIME RECEIVER

RECEIVER ARCHITECTURE

The receiver architecture shown in Figure 10.17 [13] uses direct RF sampling [11,31–33] in the receiver front-end path. In the past, only subsampling mixer receiver architectures have been demonstrated: They operate at lower intermediate frequency (IF) frequencies [34,35] and suffer from noise folding and exhibit susceptibility to clock jitter. A recent study [36] uses a high sampling frequency of 480 MHz after the mixer but adds an RC filtering stage. In this architecture, discrete-time analog signal

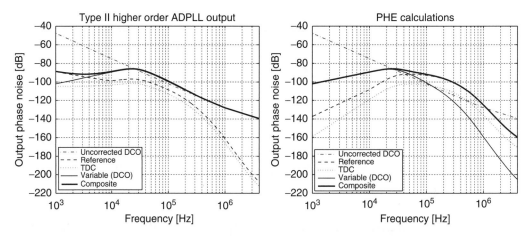

FIGURE 10.16 Spectra of various phase noises at (left) the ADPLL output; and (right) the filtered phase error $\tilde{\phi}_E$. The integrated phase noise within 100 kHz bandwidth is 0.88° and 0.83°, respectively, although the logarithmic axes appear to suggest larger difference.

processing is used to sample the RF input signal at Nyquist rate of the carrier frequency as it is then down-converted, down-sampled, filtered, and converted from analog to digital with a discrete-time $\Sigma\Delta$ ADC. This method achieves great selectivity right at the mixer level. The selectivity is digitally controlled by the LO clock frequency and capacitance ratios, both of which are extremely well controlled and precise in deep-submicron CMOS processes. The discrete-time filtering at each signal-processing stage is followed by successive decimation. The main philosophy in building the receive path is to provide all the filtering required by the standard as early as possible using a structure that is quite amenable to migration to the more advanced deep-submicron processes. This approach significantly relaxes the design requirements for the following baseband amplifiers.

Following the low-noise amplifier (LNA), the signal is converted into current using a trans-conductance amplifier (TA) stage and down-converted into a programmable low-IF frequency by integrating it on a sampling capacitor. After initial decimation through a sinc filter response, a series of IIR filtering follows RF sampling for close-in interferer rejection. These signal-processing operations are performed in the multitap direct sampling mixer (MTDSM) that receives its clocks from the digital control unit (DCU). A $\Sigma\Delta$ ADC containing a front-end gain stage follows. A feedback control unit (FCU) provides a single-bit feedback to the MTDSM to establish the common mode voltage for the MTDSM while canceling out differential offsets. The output of the I/Q ADCs is passed on to digital receive (DRX) chain. The first rate change filter (RCF1) provides antialiasing and decimation filtering to reduce the clock rate by 16. Prefiltering (PREF) is then performed to assist digital resampling (RES) operation. The residual DC offset that could not be corrected by the FCU is corrected by digital offset compensation (DIGOC). The resampler follows and converts the sample rate from LO-dependent clock rate to a fixed output rate of 8.66 MS/s. Next, the sample rate is decimated by a second rate change filter to the following I/Q mismatch block. The IF frequency is then converted from the low-IF to DC by the ZERO IF block. The final filtering is performed using a fully programmable 64-tap channel select finite-impulse response (FIR) filter.

One significance of this work is in demonstrating the feasibility of obtaining low noise figure in a receive chain in the presence of more than a million digital gates. Another significance is the development of very low-area, simple, and highly programmable analog blocks that are controlled by software to guarantee the best achievable performance. A third significance is the architecture of analog structures that are amenable to migration from one process node to the next without significant rework. Signal processing is used to reduce analog area and complexity. The radio solution was targeted to meet quad-band GSM specification in addition to supporting several experimental modes of operation.

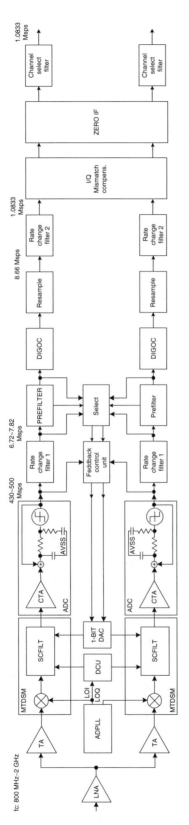

FIGURE 10.17 Block diagram of the receiver. (From Muhammad, K., Murphy, T., and Staszewski, R. B., *Proceedings of 2006 IEEE Radio Frequency Integrated Circuits (RFIC) Symposium*, sec. RTU1C-1, pp. 407–410, 2006, San Francisco, CA. ©2006 IEEE.)

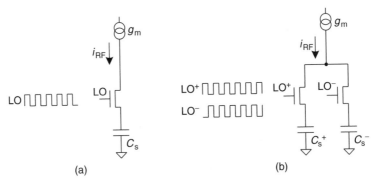

FIGURE 10.18 Temporal MA operation at RF rate: (a) single-ended, (b) pseudodifferential configurations. (From Staszewski, R. B., Muhammad, K., and Leipold, D., *Proceedings of IEEE International Conference on Computer Aided Design (ICCAD)*, pp. 122–129, 2005, San Jose, CA. © 2005 IEEE.)

DIRECT SAMPLING MIXER

The basic idea of the current-mode direct sampling mixer [31,32] is illustrated in Figure 10.18a. The low-noise transconductance amplifier (LNTA) converts the received RF voltage v_{RF} into i_{RF} in current domain through the transconductance gain g_m. The current i_{RF} gets switched by the half-cycle of the LO and integrated into the sampling capacitor C_s. Since it is difficult to switch the current at RF rate, it could be merely redirected to an identical sampler that is operating on the opposite half-cycle of the LO clock, as shown in Figure 10.18b for a pseudodifferential configuration.

If the LO oscillating at f_0 frequency is synchronous and in phase with the sinusoidal RF waveform, the voltage gain of a single RF half-cycle is

$$G_{v,RF} = \frac{1}{\pi} \cdot \frac{1}{f_0} \cdot \frac{g_m}{C_s} \qquad (10.12)$$

and the accumulated charge on the sampling capacitor is

$$G_{q,RF} = \frac{1}{\pi} \cdot \frac{1}{f_0} \cdot g_m \qquad (10.13)$$

In the above equations, the $1/\pi$ factor is contributed by the half-cycle sinusoidal integration. As an example, if $g_m = 30\,\text{mS}$, $C_s = 15.925\,\text{pF}$, and $f_0 = 2.4\,\text{GHz}$, then $G_{v,RF} = 0.25$.

TEMPORAL MOVING-AVERAGE

Continuously accumulating the charge as shown in Figure 10.18 is not very practical if it cannot be read out. In addition, a mechanism to prevent the charge overflow is needed. Both of these operations are accomplished by fixing the integration window length followed by charge readout phase that will also discharge the sampling capacitor such that the next period of integration would start from the same zero condition. The RF sampling and readout operations are cyclically rotated on both C_s capacitors as shown in Figure 10.19. When LO_A rectifies N RF cycles that are being integrated on the first sampling capacitor, LO_B is off and the second sampling capacitor charge is being read out. On the next N RF cycles the operation is reversed. This way, the charge integration and readout occur at the same time and no RF cycles are missed.

The sampling capacitor integrates the half-rectified RF current over N cycles. The charge accumulated on the sampling capacitor and the resulting voltage ($V = Q/C_s$) increase with the integration window, thus giving rise to a discrete signal-processing gain of N.

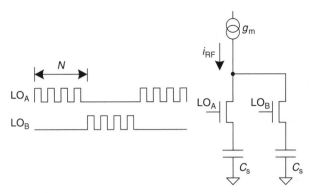

FIGURE 10.19 Temporal MA operation at RF rate with cyclic charge readout. (From Staszewski, R. B., Muhammad, K., and Leipold, D., *Proceedings of IEEE International Conference on Computer Aided Design (ICCAD)*, pp. 122–129, 2005, San Jose, CA. © 2005 IEEE.)

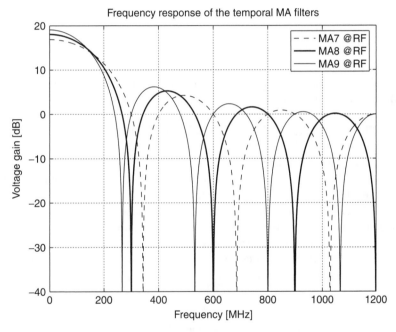

FIGURE 10.20 Transfer function of the temporal MA operation at RF rate. (From Staszewski, R. B., Muhammad, K., and Leipold, D., *Proceedings of IEEE International Conference on Computer Aided Design (ICCAD)*, pp. 122–129, 2005, San Jose, CA. © 2005 IEEE.)

The temporal integration of N half-rectified RF samples performs an FIR operation with N all-one coefficients, also known as moving-average (MA), according to the following equation:

$$w_i = \sum_{l=0}^{N-1} u_{i-l} \tag{10.14}$$

where u_i is the ith RF sample of the input charge sample and w_i the accumulated charge. Since the charge accumulation is performed on the same capacitor, this formula could also be used in the voltage domain. Its frequency response is a sinc function and is shown in Figure 10.20 for $N = 8$ (solid line) and $N = 7,9$ (dashed line and normal line, respectively) with sampling rate $f_0 = 2.4\,\text{GHz}$.

It should be noted that this filtering is performed on the same capacitor in the time domain, resulting in a most faithful reproduction of the transfer function.

Because the MA output is read out at the lower rate of N RF clock cycles, there is an additional aliasing with fold-over frequency at $f_0/2N$ and located halfway to the first notch. Consequently, the frequency response of MA = 7 with decimation of 7 exhibits less aliasing and features wider notches than MA = 8 or MA = 9 with decimation of 8 or 9, respectively.

It should be emphasized that the voltage G_V and charge G_q signal-processing gains of the temporal moving-average (TMA) (followed by decimation) are merely due to the sampling time interval expansion of this discrete-time system (the sampling rate of the input is at the RF frequency): $G_{v,tma} = G_{q,tma} = N$.

In the following analysis, the RF half-cycle integration voltage gain of $g_m/\pi C_s f_0$ is tracked separately. Since this gain depends on the absolute physical parameters of usually low tolerance (g_m value of the preceding LNTA stage and the total integrating capacitance of the sampling mixer), it is advantageous to keep it decoupled from the discrete signal-processing gain of the MTDSM.

HIGH-RATE IIR FILTERING

Figure 10.19 is now modified to include recursive operation that gives rise to the IIR filtering capability, which is generally considered stronger than that of FIR.

A history sampling capacitor C_H is added in Figure 10.21. The integration is continually performed on the history capacitor $C_H = a_1 C_s$ and one of the two rotating charge-and-readout capacitors $C_R = (1 - a_1)C_s$ such that the total RF integrating capacitance, as seen by the LNTA, is always $C_H + C_R = C_s$. When one of the C_R capacitors is being used for readout, the other is being used for RF integration.

The IIR filtering capability comes into play in the following way: The RF current is integrated over N RF cycles, as described before. This time, the charge is shared on both C_H and C_R capacitors proportionately to their capacitance values. At the end of the accumulation cycle, the active C_R capacitor that stores $(1 - a_1)$ of the total charge stops further accumulation in preparation for charge readout. The other rotating capacitor joins the C_H capacitor in the RF sampling process and, at the same time, obtains $(1 - a_1)/[a_1 + (1 - a_1)] = 1 - a_1$ of the total remaining charge in the history capacitor, provided it has no initial charge at the time of commutation. Thus the system retains a_1 portion of the total system charge of the previous cycle.

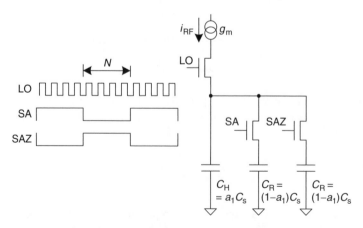

FIGURE 10.21 IIR operation with cyclic charge readout. (From Staszewski, R. B., Muhammad, K., and Leipold, D., *Proceedings of IEEE International Conference on Computer Aided Design (ICCAD)*, pp. 122–129, 2005, San Jose, CA. © 2005 IEEE.)

If the input charge accumulated over the most-recent N RF samples is w_j then the charge s_j stored in the system at sampling time j, where $i = N \cdot j$ (as stated earlier, i is the RF cycle index), could be described as a single-pole recursive IIR equation:

$$s_j = a_1 s_{j-1} + w_j \tag{10.15}$$

$$x_j = (1 - a_1) s_{j-1} \tag{10.16}$$

$$a_1 = \frac{C_H}{C_H + C_R} \tag{10.17}$$

The output charge x_j is $(1 - a_1)$ of the system charge in the most-recent cycle. This discrete-time IIR filter operates at f_0/N sampling rate and introduces a single pole with the frequency attenuation of 20 dB/dec. The equivalent pole location in the continuous-time domain for $f_{c1} \ll f_0/N$ is

$$f_{c1} = \frac{1}{2\pi} \frac{f_0}{N} \cdot (1 - a_1) = \frac{1}{2\pi} \frac{f_0}{N} \cdot \frac{C_R}{C_H + C_R} \tag{10.18}$$

Since there is no sampling time expansion for the IIR operation, the discrete signal-processing charge gain is 1. In other words, because of the charge conservation principle, the input charge per sample interval is on average the same as the output charge. For the voltage gain, however, there is an impedance transformation of $C_{input} = C_s$ and $C_{output} = (1 - a_1)C_s$, thus resulting in a gain:

$$G_{q,iir1} = 1 \tag{10.19}$$

$$G_{v,iir1} = \frac{1}{1 - a_1} = \frac{C_H + C_R}{C_R} \tag{10.20}$$

As an example, the IIR filtering with a single coefficient of $a_1 = 0.9686$, placing the pole at $f_{c1} = 1.5$ MHz ($C_R = 0.5$ pF, $C_H = 15.425$ pF), is performed at $f_0/N = 2.4$ GHz/8 $= 300$ MHz sampling rate and it follows the FIR MA $= 8$ filtering of the input at f_0 RF sampling rate. The voltage gain of the high-rate IIR filter is 31.85 (30.06 dB).

ADDITIONAL SPATIAL MA FILTERING ZEROS

For practical reasons, it is difficult to read out the x_j output charge of Figure 10.21 at $f_0/N = 300$ MHz rate. The output charge readout time is extended $M = 4$ times by adding redundancy of four to each of the two original C_R capacitors, as shown in Figure 10.22. The input charge is cyclically integrated within the group of four C_R capacitors. Adding the redundant capacitors gives rise to an additional antialiasing filtering just before the second decimation of M. This could also be considered as equivalent to adding additional $M - 1$ zeros to the IIR transfer function in Equation 10.15. After the first bank of four capacitors gets charged ($S_{A1} - S_{A4}$ in Figure 10.22), the charge on the first bank of capacitors is summed and read out (R_1), while the second bank ($S_{B1} - S_{B4}$) is in the charging mode. By physically connecting together the four capacitors, an FIR filtering is performed, which is described as the spatial moving-average (SMA) of $M = 4$:

$$y_k = \sum_{l=0}^{M-1} x_{k-l} \tag{10.21}$$

where y_k is the output charge and sampling time index $j = M \cdot k$. R_A and R_B in Figure 10.22 are the readout/reset cycles during which the output charge on the four nonsampling capacitors is transferred out and the remnant charge is reset before the capacitors are put back into the sampling operation.

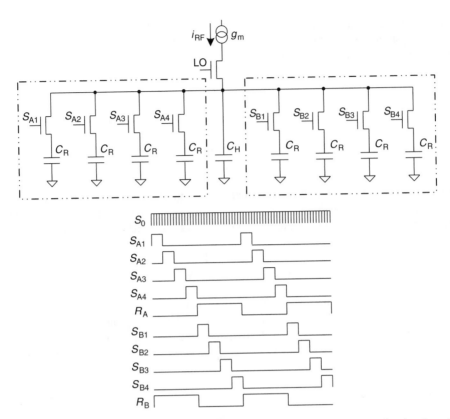

FIGURE 10.22 IIR operation with additional FIR filtering. The readout and reset circuitry is not shown. (From Staszewski, R. B., Muhammad, K., and Leipold, D., *Proceedings of IEEE International Conference on Computer Aided Design (ICCAD)*, pp. 122–129, 2005, San Jose, CA. © 2005 IEEE.)

It should be noted that after the reset phase, but before the sampling phase, the capacitors are unobtrusively precharged [33] to implement a DC-offset cancellation or to accomplish a feedback summation for the $\Sigma\Delta$ loop operation.

Since the charge of four capacitors is added, there is a charge gain of $M = 4$ and a voltage gain of 1. Again, as explained before, the charge gain is due to the sampling interval expansion: $G_{q,sma} = M$ and $G_{v,sma} = 1$.

Figure 10.23 shows frequency response of the TMA with a decimation of 8 ($G_v = 18.06\,dB$), the IIR filter operating at RF/8 rate ($G_v = 30.06\,dB$), and the SMA filter operating at RF/32 rate ($G_v = 0\,dB$) with a decimation of 4. The solid line is the composite transfer function with the DC gain of $G_v = 48.12\,dB$. The first decimation of $N = 8$ reveals itself as aliasing. It should be noted that it is possible to avoid aliasing of a very strong interferer into the critical IF band by simply changing the decimation ratio N. This brings out advantages of integrating RF/analog with digital circuitry by opening new avenues of novel signal-processing solutions not possible before.

LOWER-RATE IIR FILTERING

The voltage stored on the rotating capacitors cannot be readily presented to the MTDSM block output without an active buffer that would isolate the high impedance of the mixer from the required low driving impedance of the output. Figure 10.24 shows the mechanism to realize the second lower-rate IIR filtering through passive charge sharing. The active element, the operational amplifier, does not actually take part in the IIR filtering process. It is merely used to sense voltage of the buffer feedback capacitor C_B and present it to the output with a low driving impedance. Figure 10.24

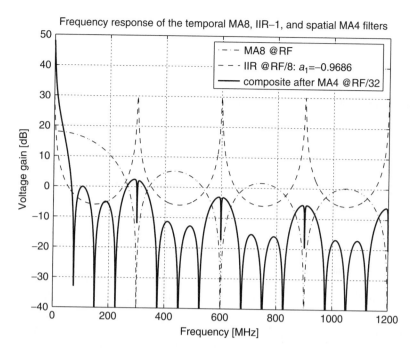

FIGURE 10.23 Transfer function of the temporal MA filter and the IIR filter operating at RF/8 rate. The solid line is the composite transfer at the output of the spatial MA filter. (From Staszewski, R. B., Muhammad, K., and Leipold, D., *Proceedings of IEEE International Conference on Computer Aided Design (ICCAD)*, pp. 122–129, 2005, San Jose, CA. © 2005 IEEE.)

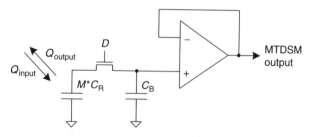

FIGURE 10.24 Second IIR filter. (From Staszewski, R. B., Muhammad, K., and Leipold, D., *Proceedings of IEEE International Conference on Computer Aided Design (ICCAD)*, pp. 122–129, 2005, San Jose, CA. © 2005 IEEE.)

additionally suggests possibility of differentially combining, through the operational amplifier, the opposite (180° apart) processing paths.

The charge y_k accumulated on the $M = 4$ rotating capacitors is shared during the dumping phase with the buffer feedback capacitor C_B. At the end of the dumping phase, the $M \cdot C_R$ capacitors get disconnected from the second IIR filter and their charge reset before they could be re-engaged in the MTDSM operation of Figure 10.22. This charge loss mechanism gives rise to IIR filtering. If the input charge is y_k, then the charge z_k stored in the buffer capacitor C_B at sampling time k is

$$z_k = a_2(z_{k-1} + y_k) = a_2 z_{k-1} + a_2 y_k \tag{10.22}$$

$$a_2 = \frac{C_B}{C_B + M C_R} \tag{10.23}$$

Equation 10.22 describes a single-pole IIR filter with coefficient a_2 and input y_k scaled by a_2, where a_2 corresponds to the storage-to-total capacitance ratio $C_B/(C_B + MC_R)$. Conversely, due to the linearity property, it could also be thought of as an IIR filter with input y_k and output scaled by a_2.

This discrete-time IIR filter operates at f_0/NM sampling rate and introduces a single pole with the frequency transfer function attenuation of 20 dB/dec. The equivalent pole location in the continuous-time domain for $f_{c2} \ll f_0/(NM)$ is

$$f_{c2} = \frac{1}{2\pi} \frac{f_0}{NM} \cdot (1 - a_2) = \frac{1}{2\pi} \frac{f_0}{NM} \cdot \frac{MC_R}{C_B + MC_R} \tag{10.24}$$

The actual MTDSM output is the voltage sensed on the buffer feedback capacitor z_k/C_B. The previously used charge stream model cannot be directly applied here because the output charge z_k is not the one that leaves the system.

The charge lost or reflected back into the $M \cdot C_R$ capacitor for subsequent reset is $(1 - a_2)$ $(z_{k-1} + y_k)$. On the basis of the charge conservation principle, the time-averaged values of charge input, y_k, and charge leaked out, $(1 - a_2)(z_{k-1} + y_k)$, should be equal. As stated before, the leak-out charge is not the output from the signal-processing standpoint. It should be noted that the amplifier does not contribute to the net charge change of the system and, consequently, the only path of the charge loss is through the same $M \cdot C_R$ capacitors that are reset after the dumping phase.

The output charge z_k stops at the IIR-2 stage and does not further propagate; therefore, it is of less importance for signal-processing analysis. The charge discrete signal-processing gain of the second IIR stage is

$$G_{q,iir2} = \frac{a_2}{1 - a_2} = \frac{C_B}{MC_R} \tag{10.25}$$

The input/output impedance transformation is MC_R/C_B. Consequently, the voltage gain of IIR-2 is unity:

$$G_{v,iir2} = 1 \tag{10.26}$$

CASCADED MTDSM FILTERING

The cascaded discrete signal-processing gain equations of the MTDSM mixer are as follows [38]:

$$G_{q,dsp} = G_{q,tma} \cdot G_{q,iir1} \cdot G_{q,sma} \cdot G_{q,iir2} \tag{10.27}$$

$$= N \cdot 1 \cdot M \cdot \frac{C_B}{MC_R} \tag{10.28}$$

$$= \frac{NC_B}{C_R} \tag{10.29}$$

$$G_{v,dsp} = G_{v,tma} \cdot G_{v,iir1} \cdot G_{v,sma} \cdot G_{v,iir2} \tag{10.30}$$

$$= N \cdot \frac{C_H + C_R}{C_R} \cdot 1 \cdot 1 \tag{10.31}$$

$$= \frac{N(C_H + C_R)}{C_R} \tag{10.32}$$

Including the RF half-cycle integration (Equations 10.12 and 10.13) the total single-ended gain becomes

$$G_{q,\text{tot}} = G_{q,\text{RF}} \cdot G_{q,\text{dsp}} \tag{10.33}$$

$$= \frac{1}{\pi} \cdot \frac{1}{f_0/N} \cdot g_m \tag{10.34}$$

$$G_{v,\text{tot}} = G_{v,\text{RF}} \cdot G_{v,\text{dsp}} \tag{10.35}$$

$$= \frac{1}{\pi} \cdot \frac{1}{f_0/N} \cdot \frac{g_m}{C_R} \tag{10.36}$$

Note the similarity between Equations 10.36 and 10.12. In both cases, the term $R_{\text{sc}} = 1/f_s C_s$ is an equivalent resistance of a switched-capacitor C_s sampling at rate f_s. For example, if $f_s = 300\,\text{MHz}$ and $C_R = 0.5\,\text{pF}$, then the equivalent resistance is $R_{\text{sc}} = 6.7\,\text{k}\Omega$. Since the MTDSM output is differential, the gain values in the above equations are actually doubled.

The DC-frequency gain $G_{v,\text{tot}}$ in Equation 10.36 requires further elaboration. The gain depends only on the g_m of the LNTA stage, rotating capacitor value, and the rotation frequency. Amazingly, it does not depend on the other capacitor values, which contribute only to the filtering transfer function at higher frequencies.

NEAR-FREQUENCY INTERFERER ATTENUATION

Most of the lower frequency filtering could be realistically done only with the first and second IIR filters. The two FIR filters do not have appreciable filtering capability at low frequencies and are mainly used for antialiasing.

It should be noted that the best filtering could be accomplished by making 3-dB corner frequency of both IIR filters the same and placing them as close to the higher end of signal band as possible:

$$f_{c1} = f_{c2} \tag{10.37}$$

This gives the following constraint:

$$C_B = C_H - (M-1)C_R \tag{10.38}$$

SIGNAL-PROCESSING EXAMPLE

Figure 10.25 shows the block diagram from the signal-processing standpoint for our specific implementation of $f_0 = 2.4\,\text{GHz}$, $N = 8$, $M = 4$. The following equations describe the time-domain signal

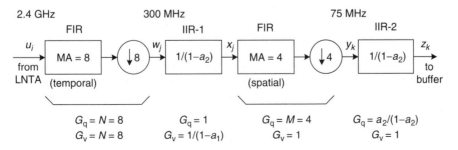

FIGURE 10.25 Discrete signal processing in the MTDSM.

processing: Equation 10.14 for w_j, Equations 10.15 and 10.16 for x_j, Equation 10.21 for y_k, and Equation 10.22 for z_k.

The first aliasing frequency (at $f_0/N = 300\,\text{MHz}$) is partially protected by the first notch of the TMA = 8 filter. However, for higher order aliasing and overall system robustness, it has to be protected with a truly continuous-time filter, such as an antenna filter. A typical low-cost Bluetooth-band duplexer can attenuate up to 40 dB at 300 MHz offset.

For the above system with an aggressive cut-off frequency of $f_{c1} = f_{c2} = 1.5\,\text{MHz}$, using $C_R = 0.5\,\text{pF}$ will result in a DC-frequency voltage gain of 63.66 or 36 dB (Equation 10.36) and the required capacitance is $C_H = 15.425\,\text{pF}$ (Equation 10.18) and $C_B = 13.925\,\text{pF}$ (Equation 10.24). The z-domain coefficients of the IIR filters are $a_1 = 0.9686$ and $a_2 = 0.8744$. The DC-frequency gains are $G_{v,\text{iir1}} = 31.85$ and $G_{v,\text{iir2}} = 1$. The transfer function of these IIR filters is shown in Figure 10.26. The SMA = 4, which follows IIR-1, does not appreciably contribute to filtering at lower frequencies but serves as an antialiasing filter for the low-rate IIR-2. Since the 3-dB point of IIR-2 is slightly corrupted by the discrete-time approximation, the composite attenuation at the cut-off frequencies $f_{c1} = f_{c2} = 1.5\,\text{MHz}$ is about 5.5 dB. The attenuation drops to 13 dB at 3 MHz.

Within the 1 MHz band of interest, there is a 3 dB signal attenuation. For the most optimal detector operation, this in-band filtering should be taken into consideration in the matched-filter design. Figure 10.27 shows the phase response of the above structure versus the ideal constant group delay.

MTDSM Feedback Path

The MTDSM feedback correction could be unobtrusively injected into either group of the four rotating capacitors of Figure 10.28 when they are not in the active sampling state. This way, the main signal path is not perturbed. The feedback correction is accomplished through charge injection/equalization between the "feedback capacitor" C_F and the rotating capacitors C_R in the MTDSM structure by shorting all of them together after the C_R group of capacitors gets reset, but before they are put back to the sampling system. The feedback charge accumulation structure is shown in Figure 10.28. Each feedback capacitor C_F is associated with one of the two rotating capacitors of group "A" and "B". These two groups commutate the charging process.

Voltage on the feedback capacitor can be calculated as follows. Charging the feedback capacitor C_F with the current i_{fbck} for the duration of T will result in incremental accumulation of $\Delta Q_{\text{in}} = i_{fbck} \cdot T$ charge. This charge gets added to the total charge $Q_F(k)$ of the feedback capacitor at the kth time instance:

$$Q_F(k) = Q_F(k-1) + \Delta Q_{\text{in}} = Q_F(k-1) + i_{fbck} \cdot T \qquad (10.39)$$

During the charge distribution moment, the feedback capacitor gets connected with the previously reset group of rotating capacitors $M \cdot C_R$. The charge depleted from C_F is dependent on the relative capacitor values:

$$\Delta Q_{\text{out}}(k) = \frac{MC_R}{C_F + MC_R} Q_F(k) \qquad (10.40)$$

The charge transferred to the rotating capacitors is proportional to the total accumulated charge Q_F or voltage on the feedback capacitor $V_F = Q_F/C_F$. At first, the accumulated charge is small, so the outgoing charge is small. Since the incoming charge is constant, the Q_F charge will continue accumulation until the net charge intake becomes zero. Equilibrium is reached when $\Delta Q_{\text{in}}(k) = \Delta Q_{\text{out}}(k)$:

$$i_{fbck} \cdot T = \frac{MC_R}{C_F + MC_R} Q_F(k) \qquad (10.41)$$

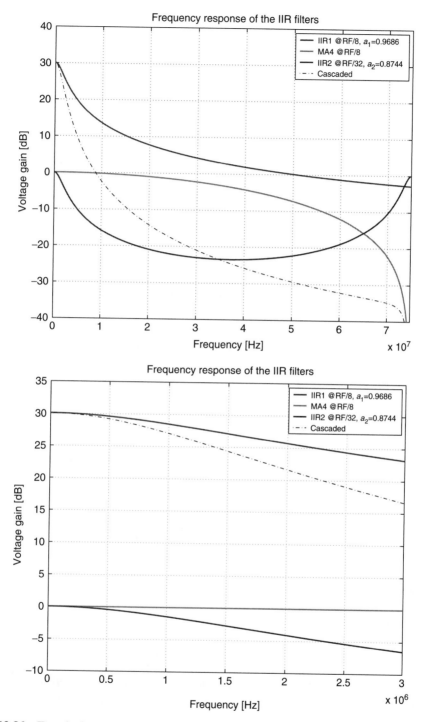

FIGURE 10.26   Transfer function of the IIR filters with two poles at 1.5 MHz (bottom zoomed).

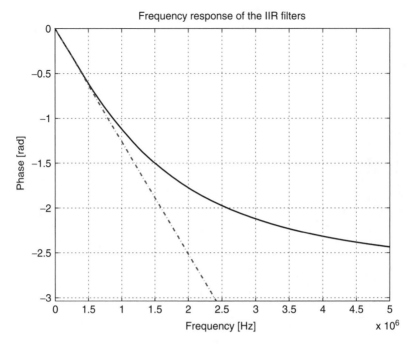

FIGURE 10.27   Phase response the IIR filters with two poles at 1.5 MHz.

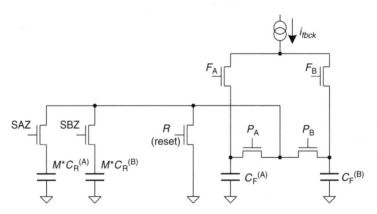

FIGURE 10.28   Feedback into the rotating capacitors.

Transformation of the above gives the equilibrium voltage:

$$V_{F,eq} = i_{fbck} \cdot T \cdot \frac{C_F + MC_R}{C_F \cdot MC_R} \tag{10.42}$$

The $\Delta Q_{out,eq}$ charge transfer into the rotating capacitors at equilibrium will create voltage on the bank of rotating capacitors:

$$V_R = \frac{i_{fbck} \cdot T}{MC_R} \tag{10.43}$$

As shown in the section "Lower-Rate IIR Filtering," the voltage transfer function from the rotating capacitors to the history capacitor is unity. Therefore, the bias voltage developed on C_H is

$$V_H = \frac{i_{fbck} \cdot T}{MC_R} \tag{10.44}$$

SCRIPT PROCESSOR

At the center of the software-driven digital RF processing techniques lies an SCR. The SCR is a dedicated small but efficient reduced instruction set microprocessor inside the DRP. In addition to basic scalar operations, vector instructions for arithmetic, logic, and load/store have been added to compress the program code. The instruction set and feature set have been optimized to minimize silicon area and firmware memory footprint while maximizing processing throughput for the applications handled by this microprocessor. SCR performs the following functions:

1. Provides a traditional RF-transceiver application programming interface (API) view of DRP to the DBB. This abstracts the internal complexity of the highly configurable DRP through a simple API. The simple API commands from DBB trigger execution of more complicated sequences of programming commands and control loops necessary to operate and calibrate the RF radio system.
2. Executes software to calibrate the wafer process strength, and to compensate for the die temperature and supply voltage variations of the analog and RF circuits as part of the overall RF radio system performance optimization.
3. Controls operation of TX and RX datapath processing, assisted by dedicated hardware accelerators, such as an RX channel select FIR filter.
4. Performs processing of lower rate receiver and transmitter datapath functions such as DC offset adjustment in RX circuits or ADPLL loop gain calculations. This saves the silicon area by avoiding dedicated hardware.
5. Configures and performs system-level RF performance RF-BIST, greatly increasing test coverage and decreasing test time. All the internal TX signals and large number of important RX signals are in the digital format to allow for digital signal processing to ascertain the RF performance without an external test equipment.
6. Addresses unexpected process and RF circuit problems or imperfections through software modifications and algorithmic changes. An RF performance/yield issue that would normally require an expensive and time-consuming silicon or metal layer change is now typically fixed in software. As an example to support this claim, the first-reported single-chip Bluetooth radio SoC in a deep-submicron CMOS [11] went into volume production with the original base silicon layers.

Figure 10.29 illustrates details of the SCR and its communication channels within the system. The processor is based on the von Neumann architecture, sharing program and data memory. In addition, the memory is shared with the receiver and transmitter datapaths via a scratch pad memory buffer. The data movement between the receiver/transmitter datapaths and memory buffer channels is handled by micro direct memory access (μ-DMA) engines programmed by SCR. This enables processing of data streams in real time or capturing data based on programmed GSM radio-related events. For example, the GSM mid-amble pattern can be captured and its frequency response calculated using fast fourier transform (FFT) routine acting as low-cost on-chip spectrum/logic analyzer.

DIGITAL COMPENSATION

The RF analog circuit performance imperfections are adaptively calibrated and compensated in the background with the assistance of the SCR. In the transmitter of Figure 10.12, the FFT, rms, and peak of the digital phase error samples are processed in real time to estimate the phase noise, and rms and peak of the phase error trajectory at the RF output [39]. The calculated statistics are then compared against the GSM specifications. This allows to trade off supply voltage and current consumption versus the required circuit performance. For example, the maximum DCO bias current of 18 mA can be reduced to as low as 6 mA if the wafer process is not weak and the die temperature

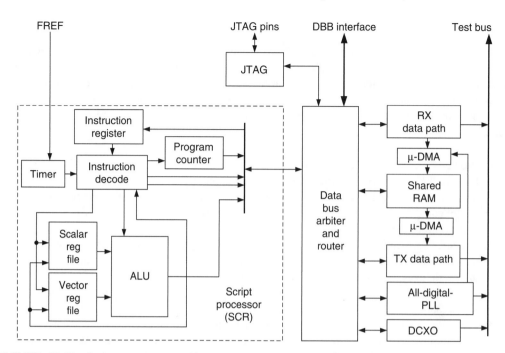

FIGURE 10.29 Script processor architecture. (From Staszewski, R., Jung, T., Staszewski, R. B., Muhammad, K., Leipold, D., Murphy, T., Sabin, S., Wallberg, J., Larson, S., Entezari, M., Fresquez, J., Dondershine, S., and Syed, S., *Proceedings of 2006 IEEE Custom Integrated Circuits Conference*, sec. 6.1, pp. 81–84, 2005, San Jose, CA. © 2006 IEEE.)

is not high [22]. Further reduction is possible at lower transmitted output power. As suggested in Ref. 23, the processed TDC output can be used as an indicator of the CMOS process strength to set biasing of RF/analog circuits.

BUILT-IN SELF-TEST

A considerable portion of the overall IC fabrication cost is in its testing. The testing costs are high in case of a complex mixed-signal SoC for RF wireless applications involving extensive and time-consuming defect, performance, and standard compliance measurements. These factory measurements are traditionally made using expensive and sophisticated test equipment. Furthermore, due to the complexity of the equipment and test settings, these measurements cannot be executed at-probe on wafer, before the IC chip is packaged, nor in the field after the chip leaves the factory environment. Consequently, it is desirable to improve testing costs and coverage during the complete life cycle of an SoC in order to maximize wafer yield, profitability, and customer satisfaction.

Frequency synthesizers and transmitters are conventionally tested for RF performance and wireless standard compliance by measuring their output RF port for the correct carrier frequency, phase noise spectrum, integrated phase noise, spurious content, modulated spectrum, and modulated phase error trajectory (see Ref. 40 for GSM) while stimuli and control signals are applied. In this chapter we propose a new BIST measurement method, which performs signal-processing calculations on a lower frequency internal signal to ascertain the RF performance without external test equipment. This significantly saves test time and costs, and increases coverage. The proposed RF-BIST techniques are now becoming possible.

Several RF-BIST functions are now feasible with the all-digital transmitter shown in Figure 10.12 and digitally intensive discrete-time receiver shown in Figure 10.17. They include digital loop-back, mixed-signal feedback loop (for DC offset cancellation), and TX-RX RF loop-back at the mixer. Coupling at the package can be used to realize an external TX-RX feedback loop that incorporates the entire transceiver. A programmable sine/cosine waveform generates feedback signals that are fed to the mixer through the offset correction loop to establish an additional analog feedback. This loop can be used to perform several calibration and test functions. Because of reuse of the on-die processor, very little hardware overhead is required for RF-BIST.

ENERGY MANAGEMENT

The EM system is a critical component of the SoC and consists of a bandgap (BGAP) voltage/current reference and multiple low dropout (LDO) regulators supplying power to various radio subsystems. One goal of the EM system is to provide good isolation of noise between various subsystems, such that the design of the entire system remains tractable. If isolation is not good, signals in one subsystem can find their way into a completely independent subsystem through the power supply. Consider, as an example, a full duplex system in which a signal is transmitted in the transmit band. All circuits sharing the same power supply will see the frequent perturbation of the power supply at the transmitted frequency as well as their harmonics. The spectrum at the output of these blocks will show spurs at these frequencies with amplitudes representative of the isolation between the source and the destination. In the receiver, for example, an external interferer can mix with one of these spurs and fall on top of the signal of interest due to the nonlinearity of the analog baseband amplification stages, thereby creating a degraded bit error rate (BER) performance. Hence, the EM system design isolates cross-talk of independent subsystems such that the performance of each of these can be predicted and kept under control.

Moreover, the EM system provides load and line regulation over a range of output voltages. The output voltage can be reduced for blocks that exceed the required performance to the extent that extra performance is not worth the extra power consumed by the block. It is then possible to cut back on the power consumption of the entire SoC by adaptively reducing the power supply of such blocks. For other blocks having marginal performance due to inadequate models in the analog simulators, this parameter can be adjusted to improve the block performance, if needed. These functions are provided by voltage regulators that have good programmability to provide sufficient margin in analog performance for a digital CMOS process.

BEHAVIORAL MODELING AND SIMULATION IN VHDL

With an extremely high cost of several million dollars for the complete mask set in the latest CMOS technology, it is imperative to fully validate the SoC radios prior to tape-out. A preferred simulator would allow seamless integration of RF, analog, digital, and software at the top level. A successful use of the standard VHDL has been demonstrated for SoCs containing high-frequency analog and RF [30,41,42]. To accommodate the greater role of software, new event-driven based system simulators, such as SystemC, will find its way into the future of RF designs.

The verification methodology is based on the philosophy that all verifications need to be automated as early as possible with maximum reuse of development efforts. In a complex SoC, the most valuable tests are those that are exercised on a system-wide level. Key specifications are translated to test cases that would be exercised in the final IC. Next, we need stimulus generators and the test equipment. For transmitter, the stimulus is the data sequences that are transmitted and the test equipment consists of analysis routines that can analyze the output generated by the simulation, such as the EVM, phase noise at different offset frequencies, and achievable signal-to-noise ratio (SNR). The RF output is stored in a compressed form as zero-crossing time instants and post processed by analysis routines in MATLAB.

With the first demonstrations of a fully digital frequency synthesizer and transmitter, and a digitally intensive receiver for wireless applications, a need has arisen to model and simulate RF circuits using the same simulation engine as that used for the digital back-end, which nowadays is likely to contain over a million gates. This way, complex interactions and performance of the entire SoC could be validated and verified prior to tape-out. Figure 10.1 offers some examples of these complex interactions:

1. Effect of the TDC resolution and nonlinearity on the close-in PLL phase noise performance and generated spurs
2. Effect of the DCO phase noise on the PLL phase noise performance and generated spurs, especially when the PLL contains a higher order digital loop filter and operates in fractional-N mode
3. Effect of the DCO frequency resolution on the close-in phase noise of the PLL
4. Effect of the $\Sigma\Delta$ DCO dithering on the far-out phase noise
5. Effect of the DCO varactor mismatches on the modulated spectrum
6. Effect of the DPA resolution and nonlinearity on the RF output spectrum
7. Effect of the DCO phase noise on the degradation of the SNR in the direct RF sampling receiver
8. Effect of the mixer capacitor mismatches on the receiver performance
9. Operation of the common and differential mode feedback loops [33] in the receiver

While SPICE-based simulation tools are extremely useful for small RF circuits containing several components (such as an RF oscillator or an LNA), their long simulation times prevent investigation of larger circuits (such as an RF oscillator with a PLL loop and a transmitter or a receiver). In fact, using the presented techniques, we were able to determine that the entire transceiver (with 100s of 1000s of gates) meets the RF GSM and EDGE specifications prior to tape-out. This level of validation seems to be nowadays a requirement given over a million dollar price tag for the reticle set in the 90 nm CMOS process.

The behavioral modeling and simulation environment is based on a standard event-driven single-core simulator, e.g., VHDL. This environment [41] is well suited for digitally intensive SoC solutions with a fair amount of analog/RF circuitry. The main advantage of the single simulation engine at the top level is that it allows seamless integration of all hardware abstraction levels (such as behavioral, register transfer language (RTL), gate level) in a uniform environment. The single most important feature of the standard VHDL hardware description language, which makes it far superior to Verilog for mixed-signal designs, is its support of real or floating-point type signals. Extensive simulation and synthesis support by the standard VHDL language makes it possible for a complex communication system to achieve a "build what we simulate, and simulate what we build" goal. Simulator performance, stability, multivendor support, mature standard, and widespread use are all advantages of this environment.

The RF/analog circuit behavior is modeled in VHDL using real-valued signals. The rest of the RF transceiver is synthesizable from RTL subset of VHDL, autoplaced and autorouted. A portion of the digital logic operates at the multi-GHz clock carrier frequency. The use of VHDL allows for a tight and seamless integration of RF with the digital logic. Simulation of the entire transceiver, including the microcode processor, is carried out to determine the RF performance at the communication packet level.

VHDL TX SIMULATIONS

Figure 10.30 demonstrates capability and usefulness of the event-driven simulation environment to model the RF phase noise behavior of an oscillator. It includes the upconverted flicker and thermal noise. The simulated noise profile of the DCO matches closely with the measured phase noise. DCO modeling in VHDL is described in detail in Ref. 42.

VHDL RX SIMULATIONS

The discrete-time mixer operation is modeled and simulated in VHDL directly in the charge-domain according to the equations described in the section "Direct Sampling Mixer" and Refs. 32 and 33.

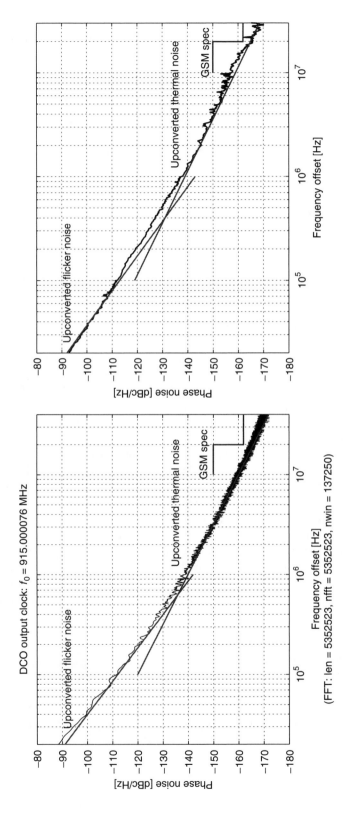

FIGURE 10.30 DCO phase noise of the last channel of the GSM900 band: (left) simulated in VHDL; (right) measured using Aeroflex PN9000. (From Staszewski, R. B., Muhammad, K., and Leipold, D., *Proceedings of IEEE International Conference on Computer Aided Design (ICCAD)*, pp. 122–129, 2005, San Jose, CA. © 2005 IEEE.)

For example, the amount of charge distributed through charge-sharing between two capacitors is tracked on the basis of the relative capacitance ratio. This section demonstrates capability and usefulness of the event-driven simulation environment by specifically showing simulation results of a single-tone frequency input to the first-generation DRP receiver while referring to the already published circuit topology examples and signal flow description in Refs. 32 and 33. The input signal is corrupted by the thermal noise and a large interferer.

The software-programmable mixer is a two-pole IIR filtering system with a certain amount of an FIR antialiasing filtering. Both poles are programmed at 300 kHz (1.5 times the 200 kHz GSM channel spacing with the 100 kHz IF frequency) by controlling the capacitance ratio of the switched capacitor mixer circuit. Consequently, the filtering capability at the IF output is 40 dB/dec with the 6 dB cutoff at 300 kHz. The ADC is a five-level $\Sigma\Delta$ converter [43] operating at 1800 MHz/ $64 = 28.125$ MHz.

The channel of interest lies at 1820 MHz and its power level is at -99 dBm. The data is a continuous wave and is not GMSK modulated, as shown in Figure 10.31 (left). A large interferer 3 MHz away at -23 dB power level is added. The main purpose is to demonstrate the sensitivity and selectivity capabilities along various points of the receive path.

The resolution bandwidth (actually, the FFT bin separation) is related to the number of FFT samples, so the signal power is obtained by integrating it over the windowing bandwidth. In addition, there is a simple noise model for the 50 Ω RF thermal noise input at -174 dBm and the 10 dB noise figure of the receiver is lumped into the RF input. All the power spectral density (PSD) plots use the averaged periodogram method. The power scale is normalized to the highest peak level.

After demodulation to near-zero IF, the channel of interest lies at 100 kHz. Figure 10.31 (right) shows the power spectral density for the IF buffer output after the second passive charge-sharing IIR filter, which is the first practical place one could extract the mixer output in a voltage, not charge, domain. The IF output has the sampling rate of 56.25 MHz. The 76 dB of the interferer-to-signal power ratio is now reduced to 37 dB, so the mixer selectivity realized only through passive devices provides almost 40 dB of the 3 MHz interferer reduction. Note the 40 dB/dec noise filtering profile at higher frequency offsets.

Figure 10.32 (left) shows the spectrum of the five-level second-order $\Sigma\Delta$ ADC output operating at the sampling rate of 28.125 MHz. Since the $\Sigma\Delta$ ADC does not provide any close-in filtering, the 37 dB of the interferer-to-signal power ratio at the mixer is maintained. The large amount of quantization noise gets shaped and pushed into higher frequencies as the 40 dB/dec profile. The task of the following decimating FIR (DFIR) filter is to remove this high-frequency noise, as well as near interferers beyond the channel of interest (not applicable for this particular simulation).

Spectrum at the DFIR output on Figure 10.32 (right) shows some filtering capability of the higher frequency quantization noise. This plot clearly shows the filter bandwidth slightly greater than 200 kHz. At the neighborhood of 100 kHz, the SNR is better than 20 dB, indicating a satisfactory level for baseband detector operation. The large 3 MHz interferer is now completely eliminated.

SUMMARY

In this chapter the authors presented key ideas and techniques used at Texas Instruments to develop two generations of DRP realized in aggressively scaled digital CMOS technology: (1) Bluetooth personal area networking SoC and (2) GSM cellular phone SoC. The DRP technology combines the RF, analog, digital, and software functionality. The main radio blocks, such as the local oscillator, transmitter and receiver, are built from the ground up to be compatible with digital scaled CMOS technology and be readily integrated with digital baseband and application processors. The transmitter performs phase and amplitude modulation employing fully digital techniques. The receiver operates discrete-time using direct RF sampling techniques. Its selectivity is digitally controlled by the LO clock frequency and capacitance ratio of a switched-cap network, both of which here are extremely precise. Imperfections of few remaining RF analog circuits are mitigated by a massive but inexpensive

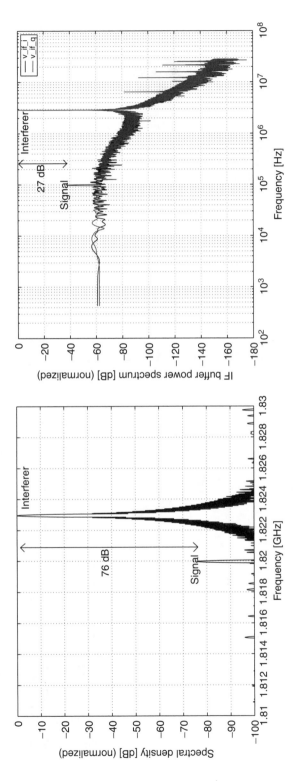

FIGURE 10.31 PSD at (left) the RF input; (right) after the second IIR filter (mixer IF buffer output). (From Staszewski, R. B., Muhammad, K., and Leipold, D., *Proceedings of 2006 IEEE Custom Integrated Circuits Conference*, sec. 27.1, pp. 789–796, 2005, San Jose, CA. © 2006 IEEE.)

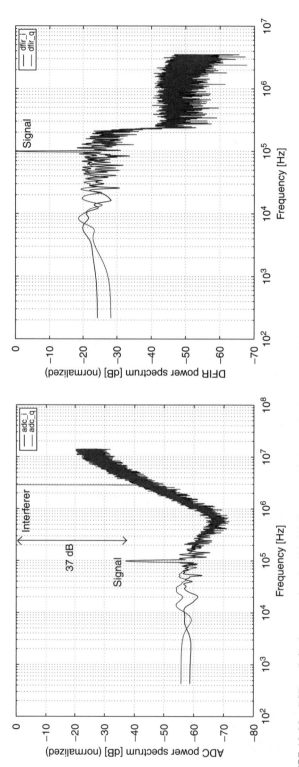

FIGURE 10.32 PSD at (left) the second-order $\Sigma\Delta$ ADC output; (right) the decimating FIR (DFIR) output. (From Staszewski, R. B., Muhammad, K., and Leipold, D., *Proceedings of 2006 IEEE Custom Integrated Circuits Conference*, sec. 27.1, pp. 789–796, 2005, San Jose, CA. © 2006 IEEE.)

digital logic that forms adaptation loops under software control. The modeling and simulation methodology of the entire radio is based on the standard VHDL hardware description language.

REFERENCES

1. W. Krenik, D. Buss, and P. Rickert, "Cellular handset integration—SIP vs. SOC," *Proceedings of 2004 IEEE Custom Integrated Circuits Conference*, pp. 63–70, Oct. 2004.
2. D. Buss, B. L. Evans, J. Bellay, et al., "SOC CMOS technology for personal internet products," *IEEE Transactions on Electron Devices*, vol. 50, no. 3, pp. 546–556, Mar. 2003.
3. A. A. Abidi, "RF CMOS comes of age," *IEEE Journal of Solid-State Circuits*, vol. 39, no. 4, pp. 549–561, Apr. 2004.
4. T. H. Lee, *The Design of CMOS Radio-Frequency Integrated Circuits*, Cambridge, UK: Cambridge University Press, 1998.
5. P.-H. Bonnaud, M. Hammes, A. Hanke, J. Kissing, R. Koch, E. Labarre, and C. Schwoerer, "A fully integrated SoC for GSM/GPRS in 0.13 µm CMOS," *Proceedings of IEEE Solid-State Circuits Conference*, sec. 26.7, pp. 1942–1951, Feb. 2005.
6. P. Mannion, "Seven-year odyssey nets one-chip phone," *EE Times*, Jan. 24, 2005.
7. W. Krenik and J. Yang, "Cellular radio integration directions," *Proceedings of Bipolar/BiCMOS Circuits and Technology Meeting*, pp. 25–30, Sept. 2003.
8. J. Pekarik, D. Greenberg, B. Jaganathan, et al., "RFCMOS technology from 0.25 µm to 65 nm: the state of the art," *Proceedings of 2004 IEEE Custom Integrated Circuits Conference*, sec. 11-5, pp. 217–224, Oct. 2004.
9. H. S. Bennett, R. Brederlow, J. C. Costa, et al., "Device and technology evolution for Si-Based RF integrated circuits," *IEEE Transactions on Electron Devices*, vol. 52, no. 7, pp. 1235–1257, Jul. 2005.
10. K. Muhammad, R. B. Staszewski, and D. Leipold, "Digital RF processing: toward low-cost reconfigurable radios," *IEEE Communications Magazine*, vol. 43, no. 8, pp. 105–113, Aug. 2005.
11. R. B. Staszewski, K. Muhammad, D. Leipold, C.-M. Hung, Y.-C. Ho, J. L. Wallberg, C. Fernando, K. Maggio, R. Staszewski, T. Jung, J. Koh, S. John, I. Y. Deng, V. Sarda, O. Moreira-Tamayo, V. Mayega, R. Katz, O. Friedman, O. E. Eliezer, E. de-Obaldia, and P. T. Balsara, "All-digital TX frequency synthesizer and discrete-time receiver for Bluetooth radio in 130-nm CMOS," *IEEE Journal of Solid-State Circuits*, vol. 39, no. 12, pp. 2278–2291, Dec. 2004.
12. R. B. Staszewski, J. Wallberg, S. Rezeq, C.-M. Hung, O. Eliezer, S. Vemulapalli, C. Fernando, K. Maggio, R. Staszewski, N. Barton, M.-C. Lee, P. Cruise, M. Entezari, K. Muhammad, and D. Leipold, "All-digital PLL and transmitter for mobile phones," *IEEE Journal of Solid-State Circuits*, vol. 40, no. 12, pp. 2469–2482, Dec. 2005.
13. K. Muhammad, Y.-C. Ho, T. Mayhugh, C.-M. Hung, T. Jung, I. Elahi, C. Lin, I. Deng, C. Fernando, J. Wallberg, S. Vemulapalli, S. Larson, T. Murphy, D. Leipold, P. Cruise, J. Jaehnig, M.-C. Lee, R. B. Staszewski, R. Staszewski, and K. Maggio, "The first fully integrated quad-band GSM/GPRS receiver in a 90 nm digital CMOS process," *IEEE Journal of Solid-State Circuits*, vol. 41, no. 8, pp. 1772–1783, Aug. 2006.
14. J. (J.-C.) Lin, "A low-phase-noise 0.004-ppm/step DCXO with guaranteed monotonicity in the 90-nm CMOS process," *IEEE Journal of Solid-State Circuits*, vol. 40, no. 12, pp. 2726–2734, Dec. 2005.
15. R. Staszewski, T. Jung, R. B. Staszewski, K. Muhammad, D. Leipold, T. Murphy, S. Sabin, J. Wallberg, S. Larson, M. Entezari, J. Fresquez, S. Dondershine, and S. Syed, "Software assisted digital RF processor for single-chip GSM radio in 90 nm CMOS," *Proceedings of 2006 IEEE Custom Integrated Circuits Conference*, sec. 6.1, pp. 81–84, Sept. 2005, San Jose, CA.
16. R. B. Staszewski, J. Wallberg, S. Rezeq, C.-M. Hung, O. Eliezer, S. Vemulapalli, C. Fernando, K. Maggio, R. Staszewski, N. Barton, M.-C. Lee, P. Cruise, M. Entezari, K. Muhammad, and D. Leipold, "All-digital PLL and GSM/EDGE transmitter in 90 nm CMOS," *Proceedings of IEEE Solid-State Circuits Conference*, sec. 17.5, pp. 316–317, 600, Feb. 2005, San Francisco, CA.
17. R. B. Staszewski, "Digital deep-submicron CMOS frequency synthesis for RF wireless applications," Ph.D. dissertation, University of Texas at Dallas, Aug. 2002.
18. F. M. Gardner, "Charge-pump phase-locked loops," *IEEE Transactions on Communications*, vol. COMM-28, pp. 1849–1858, Nov. 1980.
19. M. H. Perrott, R. T. Rex, and Y. Huang, "Digitally-synthesized loop filter circuit particularly useful for a phase locked loop," US patent 6,630,868, issued Oct. 7, 2003.

20. R. B. Staszewski, C.-M. Hung, D. Leipold, and P. T. Balsara, "A first multigigahertz digitally controlled oscillator for wireless applications," *IEEE Transactions on Microwave Theory and Techniques*, vol. 51, no. 11, pp. 2154–2164, Nov. 2003.

21. C.-M. Hung, R. B. Staszewski, N. Barton, M.-C. Lee, and D. Leipold, "A digitally controlled oscillator system for SAW-less transmitters in cellular handsets," *IEEE Journal of Solid-State Circuits*, vol. 41, no. 5, pp. 1160–1170, May 2006.

22. R. B. Staszewski, D. Leipold, K Muhammad, and P. T. Balsara, "Digitally controlled oscillator (DCO)-based architecture for RF frequency synthesis in a deep-submicrometer CMOS process," *IEEE Transactions on Circuits and Systems II*, vol. 50, no. 11, pp. 815–828, Nov. 2003.

23. R. B. Staszewski, S. Vemulapalli, P. Vallur, J. Wallberg, and P. T. Balsara, "1.3 V 20 ps time-to-digital converter for frequency synthesis in 90-nm CMOS," *IEEE Transactions on Circuits and Systems II*, vol. 53, no. 3, pp. 220–224, Mar. 2006.

24. M. Mota and J. Christiansen, "A high-resolution time interpolator based on a delay locked loop and an RC delay line," *IEEE Journal of Solid-State Circuits*, vol. 34, no. 10, pp. 1360–1366, Oct. 1999.

25. R. B. Staszewski, G. Shriki, and P. T. Balsara, "All-digital PLL with ultra fast acquisition," *Proceedings of IEEE Asian Solid-State Circuits Conference*, sec. 11-7, pp. 289–292, Nov. 2005, Taipei, Taiwan.

26. R. B. Staszewski, K. Muhammad, and D. Leipold, "Digital signal processing for RF at 45-nm CMOS and beyond," *Proceedings of 2006 IEEE Custom Integrated Circuits Conference*, sec. 13.1, pp. 517–522, Sept. 2005, San Jose, CA.

27. R. B. Staszewski, D. Leipold, and P. T. Balsara, "Just-in-time gain estimation of an RF digitally-controlled oscillator for digital direct frequency modulation," *IEEE Transactions on Circuits and Systems II*, vol. 50, no. 11, pp. 887–892, Nov. 2003.

28. R. B. Staszewski, J. Wallberg, C.-M. Hung, G. Feygin, M. Entezari, and D. Leipold, "LMS-based calibration of an RF digitally-controlled oscillator for mobile phones," *IEEE Transactions on Circuits and Systems II*, vol. 53, no. 3, pp. 225–229, Mar. 2006.

29. P. Cruise, C.-M. Hung, R. B. Staszewski, O. Eliezer, S. Rezeq, D. Leipold, and K. Maggio, "A digital-to-RF-amplitude converter for GSM/GPRS/EDGE in 90-nm digital CMOS," *Proceedings of 2005 IEEE Radio Frequency Integrated Circuits (RFIC) Symposium*, pp. 21–24, June 2005.

30. K. Muhammad, T. Murphy, and R. B. Staszewski, "Verification of RF SoCs: RF, analog, baseband and software," *Proceedings of 2006 IEEE Radio Frequency Integrated Circuits (RFIC) Symposium*, sec. RTU1C-1, pp. 407–410, June 2006, San Francisco, CA.

31. K. Muhammad, D. Leipold, B. Staszewski, Y.-C. Ho, C.-M. Hung, K. Maggio, C. Fernando, T. Jung, J. Wallberg, J.-S. Koh, S. John, I. Deng, O. Moreira, R. Staszewski, R. Katz, and O. Friedman, "A discrete-time Bluetooth receiver in a 0.13 µm digital CMOS process," *Proceedings of IEEE Solid-State Circuits Conference*, pp. 268–269, 527, Feb. 2004.

32. K. Muhammad and R. B. Staszewski, "Direct RF sampling mixer with recursive filtering in charge domain," *Proceedings of 2004 IEEE International Symposium on Circuits and Systems*, sec. ASP-L29.5, pp. I-577–I-580, May 2004.

33. K. Muhammad, R. B. Staszewski, and C.-M. Hung, "Joint common mode voltage and differential offset voltage control scheme in a low-IF receiver," *Proceedings of 2004 IEEE Radio Frequency Integrated Circuits (RFIC) Symposium*, sec. TU3C-2, pp. 405–408, June 2004.

34. S. Karvonen, T. Riley, and J. Kostamovaara, "A low noise quadrature subsampling mixer," *Proceedings of IEEE International Symposium on Circuits and Systems*, pp. 790–793, 2001.

35. S. Lindfors, A. Parssinen, and K. A. Halonen, "A 3-V 230-MHz CMOS decimation subsampler," *IEEE Transactions on Circuits and Systems II*, vol. 50, no. 3, pp. 105–117, Mar. 2003.

36. R. Bagheri, A. Mirzaei, S. Chehrazi, M. Heidari, M. Lee, M. Mikhemar, W. Tang, and A. Abidi, "An 800 MHz to 5 GHz software-defined radio receiver in 90 nm CMOS," *Proceedings of IEEE Solid-State Circuits Conference*, sec. 26.6, pp. 480–481, 667, Feb. 2006.

37. R. B. Staszewski, K. Muhammad, and D. Leipold, "Digital RF Processor (DRP™) for cellular phones," *Proceedings of IEEE International Conference on Computer Aided Design (ICCAD)*, pp. 122–129, Nov. 2005, San Jose, CA.

38. Y.-C. Ho, R. B. Staszewski, K. Muhammad, C.-M. Hung, D. Leipold, and K. Maggio, "Charge-domain signal processing of direct RF sampling mixer with discrete-time filters in Bluetooth and GSM receivers," *EURASIP Journal on Wireless Communications and Networking*, vol. 2006, pp. Article ID 62905, 14 pages, 2006.

39. I. Bashir, R. B. Staszewski, O. Eliezer, and E. de-Obaldia, "Built-in self testing (BIST) of RF performance in a system-on-chip (SoC)," *Proceedings of 2005 IEEE Dallas/CAS Workshop: Architectures, Circuits and Implementation of SoC (DCAS-05)*, pp. 215–218, Oct. 2005, Dallas, TX.

40. "GSM Specification: Radio Transmission and Reception, GSM 05.05 version 8.5.1," ETSI EN 300 910, http://www.etsi.org, Nov. 2000.

41. R. B. Staszewski and S. Kiriaki, "Top-down simulation methodology of a 500 MHz mixed-signal magnetic recording read channel using standard VHDL," *Proceedings of Behavioral Modeling and Simulation Conference*, sec. 3.2, Oct. 1999.

42. R. B. Staszewski, C. Fernando, and P. T. Balsara, "Event-driven simulation and modeling of phase noise of an RF oscillator," *IEEE Transactions on Circuits and Systems I*, vol. 52, no. 4, pp. 723–733, Apr. 2005.

43. J. Koh, K. Muhammad, B. Staszewski, G. Gomez, and B. Horoun, "A sigma-delta ADC with a built-in anti-aliasing filter for Bluetooth receiver in 130 nm digital process," *Proceedings of 2004 IEEE Custom Integrated Circuits Conference*, sec. 25-6, pp. 535–538, Oct. 2004.

44. R. B. Staszewski, K. Muhammad, and D. Leipold, "Digital RF processor techniques for single-chip radios (invited)," *Proceedings of 2006 IEEE Custom Integrated Circuits Conference*, sec. 27.1, pp. 789–796, Sept. 2005, San Jose, CA.

45. R. B. Staszewski and P. T. Balsara, "Phase-domain all-digital phase-locked loop," *IEEE Transactions on Circuits and Systems II*, vol. 52, no. 3, pp. 159–163, Mar. 2005.

11 Low Noise Amplifiers

Leonid Belostotski and James Haslett

CONTENTS

Introduction..305
High-Frequency Noise Mechanisms in CMOS MOSFETs...306
    Classical LNA Optimization 306
Power-Matched LNA Topologies..308
    Narrowband LNAs...309
        Narrowband LNA Gain 314
        Effect of the Signal Source Impedance on LNA Noise Figure 315
    Wideband LNAs 315
        Wideband LNA Gain..320
LNA Linearity 321
Wideband LNA Design Examples for Radio Astronomy...322
    CMOS and Radio Astronomy...322
    Examples of Wideband LNAs 323
    LNA Optimization Summary..325
Conclusions and Future Trends...325
References..326

INTRODUCTION

The growing consumer market has pushed CMOS technology into sub-0.1 μm feature sizes that allow a large number of very fast transistors to be integrated on to the same die. The advance in transistor speed also benefits the wireless sector, but to a lesser degree since the performance is limited by the physical properties of the material. This, however, has not stopped designers from systematically improving the performance of the wireless circuitry by overcoming the fundamental physical limitations with new and clever circuit topologies and design techniques.

This chapter discusses the current state-of-the-art in radio frequency (RF) low noise amplifier (LNA) design and presents the trends that the reader should expect as the CMOS technology scales below the 90 nm technology. The chapter starts with a brief introduction to MOSFET noise modeling to give the reader background information necessary to understand how various LNA topologies and circuits are affected by the device physics. The chapter focuses on the review of narrowband and wideband LNA topologies and discusses the advantages and disadvantages of the topologies as related to the noise figure. The LNA is the only active wireless component where high-frequency noise is crucial to its operation. Therefore, most of the chapter is dedicated to the review of LNA designs from the noise figure point of view. The other parameters that are not unique to the LNA, such as gain and linearity, are also briefly discussed in this chapter. The chapter concludes with two examples of wideband LNAs in 90 nm CMOS and 180 nm CMOS followed by a summary of future trends in LNA designs.

HIGH-FREQUENCY NOISE MECHANISMS IN CMOS MOSFETs

An LNA cannot be designed without at least some understanding of the origin of noise in a transistor and the ways in which the noise is modeled. This section gives only a high-level insight into the noise origin but hopefully this would be enough for the reader to follow the discussion of various LNA topologies presented.

The thermal noise in a MOSFET originates in the channel of the transistor due to charge fluctuations commonly modeled with a channel noise current [1,2]. These fluctuations capacitively couple through the gate-channel capacitance and result in induced gate noise current. Because the coupled gate noise originates from the channel noise these two appear partially correlated. There are other noise sources such as flicker noise or $1/f$ noise and shot noise. Flicker noise is not important for LNA designers operating above 100 MHz as is the case in wireless communication systems. Shot noise may become more and more important as the gate oxide becomes thinner with decreasing feature sizes of CMOS technologies. The thin gate oxide allows leakage current to flow between the gate terminal and the channel. The noise due to this leakage current can be modeled as an additional shot noise component [3].

The simplest noise model of a CMOS transistor is shown in Figure 11.1. The noise powers due to the drain noise and gate noise are modeled with noise generators $i_{nd}^2 = 4kTB\gamma g_{d0}$ and $i_{ng}^2 = 4kTB\delta g_{gg}$, respectively, where g_{d0} is the conductance of the transistor when $V_{ds} = 0$, $g_{gg} = \omega_0^2 C_{gs}^2/5g_{d0}$, B is the noise bandwidth, k is Boltzmann's constant, δ and γ are the coefficients of drain- and gate-induced noise, respectively, and T is are the absolute temperature [1,4]. For long-channel transistors $\gamma = 2/3$, $\delta = 4/3$, and the correlation coefficient between the two noise sources $c = \overline{i_{ng}^* i_{nd}} / \sqrt{\overline{i_{ng}^2 i_{nd}^2}} = j0.395$ [1]. For short-channel transistors, the values for γ, δ, and c are not well known and are under intensive research [3,5,6]. Measurements reported in Ref. 6 show $\gamma = 1.3$ and $\delta = 3.8$ for 0.13 μm technology. The increase in the noise for short-channel transistors is attributed to channel length modulation and carrier heating [7] and the noise parameters γ and δ should increase with the smaller transistor gate lengths.

The other important noise contribution in a CMOS LNA is that of the MOS transistor gate finger resistance, which can be reduced by contacting the gate on both sides and using multiple gate fingers [8]. Of course good layout techniques are important to minimize the overall noise figure. Triple-well transistors are recommended as well as good shielding of transmission lines and bond pads from noisy epi-currents in the substrate.

CLASSICAL LNA OPTIMIZATION

Before proceeding with the discussion of LNA topologies, it is important to review the classical LNA optimization, which shows that there is an optimum signal source admittance:

$$Y_{opt} = G_{opt} + jB_{opt} \tag{11.1}$$

$$= \alpha\omega C_{gs}\sqrt{\frac{\delta}{5\gamma}\left(1 - |c|^2\right)} - j\omega C_{gs}\left(1 - \alpha|c|\sqrt{\frac{\delta}{5\gamma}}\right) \tag{11.2}$$

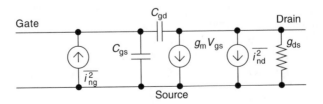

FIGURE 11.1   Two-port model of a CMOS transistor.

and this for a given MOS transistor results in the lowest noise figure [9,10]. In Equation 11.2, $\alpha = g_m/g_{d0}$. The minimum noise factor of the common-source transistor with the small signal model shown in Figure 11.1, ignoring C_{gd} and g_{ds}, and the signal source admittance equal to Y_{opt} is

$$F_{min} = 1 + 2R_nG_{opt} \tag{11.3}$$

$$= 1 + 2\frac{\omega}{\omega_T}\sqrt{\frac{\delta\gamma}{5}\left(1 - |c|^2\right)} \tag{11.4}$$

where $\omega_T = g_m/C_{gs}$ and

$$R_n = \frac{\gamma\alpha}{g_m} \tag{11.5}$$

is the equivalent noise resistance [10]. This optimization also shows that the optimum susceptance B_{opt} corresponds to a negative capacitor, which at a single frequency is commonly approximated by an inductor. This narrowband approximation limits the bandwidth where the LNA noise factor can approach the minimum noise factor. As the signal source admittance $Y_s = G_s + jB_s$ deviates from Y_{opt} the noise factor degrades according to

$$F = F_{min} + \frac{R_n}{G_s}\left[(G_s - G_{opt})^2 + (B_s - B_{opt})^2\right] \tag{11.6}$$

With the transistor feature size scaled, the unity short-circuit current gain frequency ω_T increases, which results in a decrease of the minimum noise figure as shown in Figure 11.2. Interestingly, scaling of the transistor gate length does not affect an optimum current density of approximately 0.15 mA/µm (Figure 11.3b) at which the minimum noise figure reaches its lowest values [11,12]. A representative plot of the minimum noise figure as a function of the transistor DC current density is shown in Figure 11.3. The behavior shown in this figure is only strictly true when the transistor finger

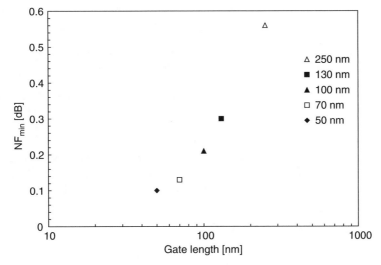

FIGURE 11.2 Dependence of the transistor minimum noise figure on the nominal transistor gate length. (Data are from Woerlee, P. H. et al., *IEEE Trans. Electron Devices*, 48, 1776, 2001. © [2001] IEEE.)

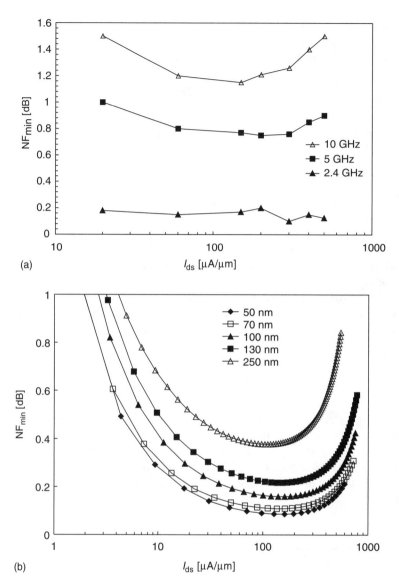

FIGURE 11.3   Representative dependence of the minimum noise figure on the transistor drain current density. (a) Frequency dependence of F_{min} versus current density for 0.18 µm technology. Note that F_{min} increases with operating frequency. (Data are taken from Chern, J.G.J., *IEEE RFIC Symposium Workshop*, June 2004. © [2004] IEEE.) (b) Technology dependence of F_{min} on the current density. The total device width is 192 µm and frequency is 2 GHz. (From Woerlee, P.H. et al., *IEEE Trans. Electron Devices*, 48, 1776, 2001. With permission. © [2001] IEEE.)

width is kept constant. The reason for the increase in the minimum noise figure at high current densities is partially attributed to the saturation of the transistor's transconductance, the presence of the parasitic resistance at the transistor gate, and the nearing of the triode region.

POWER-MATCHED LNA TOPOLOGIES

LNA topologies can be classified according to their bandwidth. A narrowband LNA exhibits acceptable input power match over bandwidths that are less than 25% of the center frequency. This is a somewhat arbitrary number but it encompasses most LNAs that are used in commercial applications. The rest are classified as wideband LNAs.

The classical optimization is not often used as it does not guarantee power match at the LNA input. In the next section, other optimization techniques are discussed that do allow for the power match. Generally, it is desirable to achieve the power-matching condition without employing any matching components at the LNA input as the losses in the matching circuitry worsen the LNA noise figure. Usually, this is not possible and LNA optimization needs to optimize noise contributions of both the matching circuitry and the LNA transistor [14].

Input power match for wideband LNAs is more demanding. The section "Wideband LNAs" addresses wideband LNA topologies and discusses state-of-the-art approaches to minimize the noise contribution of losses in the matching circuitry.

Narrowband LNAs

This section discusses recent narrowband LNA implementations and their advantages and disadvantages. The most common narrowband LNA utilizes the source-degenerated topology (Figure 11.4a). Without inductive source degeneration, the input impedance of the common-source MOSFET approximately equals the gate-source capacitance C_{gs} of the transistor. This approximation

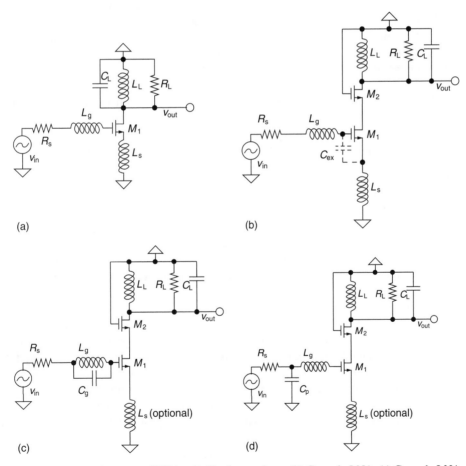

FIGURE 11.4 Source-degenerated LNAs. (a) Single transistor. (b) Cascode LNA. (c) Cascode LNA with a parallel tank at the gate. (d) Cascode LNA with a capacitor in shunt with R_s.

is valid when gate-drain capacitance C_{gd} is significantly smaller than C_{gs}. The inductive source degeneration produces a real part of the input impedance by resonating with C_{gs}. The gate inductor is then added to resonate with the remaining portion of C_{gs} to produce the power match at the desired operating frequency. The input impedance of this LNA is well known and is given by

$$Z_{gs} = \frac{1}{sC_{gs}} + sL_g + sL_s + \frac{g_m L_s}{C_{gs}} \tag{11.7}$$

when C_{gd}, gate parasitic resistance, and channel charging resistance are ignored. An improved cas-coded version of this circuit (Figure 11.4b) can be found in many designs. The cascode transistor is used to reduce the effect of the input transistor gate-drain capacitor and to increase the reverse isolation of the LNA, which helps with stability and makes the LNA input impedance less sensitive to the LNA load impedance.

It has been shown that this source-degenerated topology nearly achieves the minimum noise figure of the transistor as prescribed by the classical optimization and therefore in theory is very close to achieving the goal of providing the input match and best noise performance simultaneously [14–16]. In practice, however, the situation is more complicated. The gate inductor, required to resonate with the gate-source capacitance, can add significant noise to the LNA due to very low quality factors of on-chip inductors in CMOS. This creates a problem for designers because the lossless impedance match can no longer be called lossless and the LNA design must somehow account for the extra loss. By analyzing Figure 11.4b, it can be observed that since the inductor L_g is added to resonate with C_{gs}, to minimize the detrimental effect of the inductor on the noise figure, the size of C_{gs} can be increased, thus making L_g smaller. The size of L_g can be decreased either by making the transistor wider, which perhaps can result in higher power consumption, or by adding an extra capacitor C_{ex} (Figure 11.4b) between the gate and source terminals, which does not affect the power consumption directly but does affect the unity short-circuit current gain frequency ω_T of the transistor [17,18]. Remember that the classical noise figure optimization suggests that there is a signal source impedance (or admittance) that would result in the lowest noise figure for a given transistor. Analysis of Figure 11.1 shows that if the signal source impedance is approximately zero, the gate current noise will appear shorted and only the drain noise will contribute to the noise of the LNA. On the contrary, if the generator impedance is high, the quality factor of the input network will amplify the gate current noise that will become the dominant source of the LNA noise. Intuitively, somewhere in between these two extreme cases there is an optimum generator impedance. Adding the noise contribution of the gate inductor complicates the derivation of the optimum signal source impedance but does not affect the conclusion made about the existence of the optimum signal source impedance, as the noise due to the losses in the inductor can be incorporated into the transistor drain noise current and gate noise current by performing a series two-port transformation [19]. The extra capacitor C_{ex} has a very high quality factor compared with the inductor and adds an extra degree of design freedom in the LNA optimization. The addition of this capacitor may be counterintuitive to some LNA designers who welcome diminishing CMOS feature size due to rapid increase in ω_T, which reduces the drain noise contribution to the noise figure. Unfortunately, as the drain noise contribution decreases, the noise due to the matching circuitry at the gate terminal, in particular the gate inductor, becomes dominant. The capacitor C_{ex} however allows a compromise to be reached between the noise contribution from the gate circuitry and from the transistor itself. The value of C_{ex} must be kept relatively small so that the gain of the LNA is not affected. C_{ex} also allows the LNA to be designed with both a power consumption constraint and a gain constraint, which was not possible prior to the introduction of the capacitor [14].

The noise factor of the LNA in Figure 11.4b, ignoring g_{ds} and C_{gd} and representing the losses of L_g with a resistor, is [14]

$$F = \frac{R}{R_s}\left(1 + \frac{R}{R_s}\frac{\omega_0^2 R_s g_m \gamma}{\omega_T^2 \alpha}\chi\right) \tag{11.8}$$

and

$$\chi = \frac{\delta\alpha^2}{5\gamma}\left[1 + Q_s^2\right]\frac{C_{gs}^2}{C_t^2} + 1 - 2|c|\frac{C_{gs}}{C_t}\sqrt{\frac{\delta\alpha^2}{5\gamma}} \tag{11.9}$$

where $R = R_s + R_g$, $R_g = R_{ind} + R_{fing}$ is the total parasitic resistance at the gate of the transistor consisting of the parasitic resistances of the gate inductor and transistor gate fingers, $C_t = C_{gs} + C_{ex}$, $\alpha = g_m/g_{d0}$, $\omega_T = g_m/C_t$, and $Q_s = 1/\omega_0 C_t R$ [14]. The LNA input impedance is

$$Z_{in} = R_g + \frac{1}{sC_t} + sL_g + sL_s + \frac{g_m L_s}{C_t} \tag{11.10}$$

The first term in Equation 11.9 is due to the gate noise. This term becomes insignificant when C_{ex} is large, as the gate noise source sees low impedance and is virtually shorted. The optimization of the LNA in Figure 11.4b requires that all terms in Equation 11.8 are expressed as functions of either the gate-source overdrive voltage V_{od} or C_{ex}. The second-order DC model for the MOS transistor, expressed as [20]

$$I_D = WC_{ox}v_{sat}\frac{V_{od}^2}{V_{od} + LE_{sat}} \tag{11.11}$$

where C_{ox} is the oxide capacitance, v_{sat} the saturation velocity, E_{sat} the velocity saturation field, and L and W the transistor effective length and width, accomplishes this under a constant power consumption constraint [14,15]. Once the expressions are found the optimization finds numerically a pair of $(V_{od,opt}, C_{ex,opt})$ such that $\min[F(V_{od}, C_{ex})] = F(V_{od,opt}, C_{ex,opt})$. A typical set of curves for 0.18 μm technology is shown in Figure 11.5. Each combination of C_{ex} and V_{od} results in a particular noise figure, transconductance gain of the LNA $G_m = Q_s g_m$, and the gate inductor parasitic resistance and gate finger parasitic resistance shown as a combined parasitic resistance R_g. Figures 11.5a and 11.5b show that a nonzero C_{ex}, even though reducing transistor ω_T, helps to lower the noise figure with a proper choice of V_{od}. Interestingly, but not surprisingly, the region where the lowest noise figures are found in Figure 11.5a is not coincident with the region of the lowest total parasitic resistance R_g in Figure 11.5c as a balance is achieved between the noise contribution of the parasitic resistances at the LNA input and the LNA transistor itself. The aforementioned second-order DC model allows all parameters in Equation 11.8 to be expressed in terms of $(V_{od,opt}, C_{ex,opt})$. Therefore, having determined the optimum $(V_{od,opt}, C_{ex,opt})$, the optimum LNA design can be readily obtained. The surface plot of the noise figure as a function of $(V_{od,opt}, C_{ex,opt})$ and the power dissipation P_D is shown in Figure 11.5d, which clearly indicates that higher power consumption lowers and broadens the noise figure minimum. The noise figure dependence on the DC power dissipation, quality factor of the gate inductor, and the noise parameters γ and δ are shown in Figure 11.6.

The optimization technique can be made more accurate by incorporating the C_{gd} capacitor into Equations 11.8 and 11.9 and by recognizing that its effect on the noise figure and the input

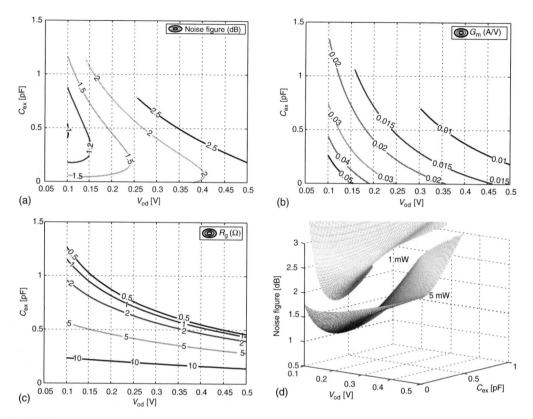

FIGURE 11.5 Typical source-degenerated LNA optimization curves for 0.18 μm CMOS technology. In (a), (b), and (c) the power dissipation is fixed to 5 mW. The long-channel noise model is used in all cases. In the simulation $V_{od} \geq 0.1$ V. Quality factor of the gate inductor $Q_{ind} = 10$; a 1 nH input bond wire is assumed with a quality factor of 50. (a) Contours of constant noise figure as a function of C_{ex} and V_{od}. (b) Contours of constant gain G_m as a function of C_{ex} and V_{od}. (c) Parasitic gate resistance R_g as a function of C_{ex} and V_{od}. (d) Noise figure as a function of C_{ex} and V_{od} for $P_D = 1$ mW and $P_D = 5$ mW.

impedance is similar to that of C_{ex}. Consider the input impedance of the LNA when C_{gd} is included in the calculation:

$$Z_{in} \approx R_g + sL_g + sL_s \frac{C_t' - C_{gd}}{C_t'} + \frac{1}{sC_t'} + \frac{g_m L_s}{C_t'} \tag{11.12}$$

where $C_t' = C_{gs} + C_{ex}'$ and $C_{ex}' = C_{ex} + 2C_{gd}$. The similarity between Equations 11.12 and 11.10 is apparent. The only term that experiences a slight modification is the sL_s term. However this slight reduction in source inductor impedance can be compensated by a fractional increase in L_g. The optimization described in Ref. 14 can then be performed if C_{gd} is approximated as a fraction of C_{gs}, i.e., $C_{gd} = C_{gs}/k$, where $k \approx 3$.

As shown in Figure 11.3 there is an optimum current density at which the minimum noise figure reaches its lowest point. The power-constrained optimizations in Refs. 14, 15 do not address this point directly. The power-constrained optimization in Ref. 15 is only partially aware of the optimum current density because the second-order DC model only incorporates the g_m saturation function α. When the gate parasitic resistance is not modeled, the optimum current density appears higher than that in Figure 11.3. The overdrive voltage that allows the transistor to be biased at the optimum current density is found with the second-order model in Equation 11.11 to be ~0.5 V

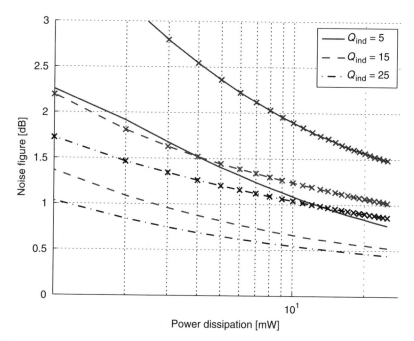

FIGURE 11.6 Noise figure dependence on DC power dissipation, quality factor of the gate inductor, and the noise parameters γ and δ for 0.18 μm CMOS technology with 1 nH bond wire inductor at the LNA input. Solid lines correspond to the long-channel transistor noise model and "X" marks result with a pessimistic short-channel transistor noise model with $\gamma = 2$ and $\delta = 4$. (From Lee, T.H., *The Design of CMOS Radio-Frequency Integrated Circuits*, Cambridge University Press, New York, 2004.)

using parameters from 0.18 μm technology. As described in Ref. 15, the power-constrained optimization tends to increase the overdrive voltage (and decrease the transistor width due to the power constraint) because it follows the F_{min} versus current density curve, such as in Figure 11.3. Nevertheless, biasing a transistor at the optimum current density requires very large power consumption, so the power-constrained optimization cannot bias the LNA at the point of optimum current density. The power-constrained optimization in Ref. 14 models both the g_m saturation and the parasitic gate resistance and thus the optimization is aware of the point of the optimum current density. When searching through the space of solutions of Equation 11.8, the minimum LNA noise figure is selected rather than the transistor's F_{min}. As a result, the power-constrained optimization in Ref. 14 biases the transistor at a current density that is lower than the optimum to minimize the size of the gate inductor and produce a balance between the noise due to the inductor loss and the transistor noise.

Another approach to deal with the noise introduced by the gate inductor is discussed in Ref. 21 in which the authors introduced a capacitor C_g in parallel with L_g (Figure 11.4c). When designed properly, this capacitor increases the apparent inductance of L_g, thus requiring a smaller gate inductor to resonate with C_{gs}. The parasitic resistance of L_g also appears increased, which is exploited to produce the real part of input impedance for power match, and this allows an LNA designer to remove the source-degeneration inductor altogether. The noise contribution of the gate inductor is minimized by using a smaller sized inductor to achieve the resonance. The self-resonance of $L_g \| C_g$ increases the noise figure of the LNA and therefore the designers must make sure that the frequency of LNA operation is significantly lower than the self-resonant frequency. The noise figure still follows Equation 11.8, and optimization strategies similar to those described in Ref. 14 can be used to find the optimum V_{od} from which the rest of the LNA parameters follow.

Matching to parasitic resistance has also been applied in Ref. 22 where a capacitor C_p in shunt with the signal source was introduced (Figure 11.4d). This capacitor and L_g are sized such that they

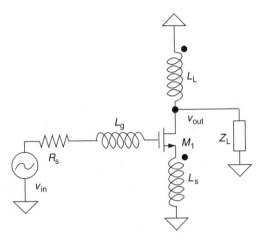

FIGURE 11.7   Transformer feedback narrowband LNA.

transform the parasitic resistance at the LNA gate to equal the real part of the signal source impedance while achieving resonance in the middle of the band. Again the optimization in Ref. 14 can
be applied to this topology to optimize the noise contribution of the gate inductor versus the noise
contribution of the transistor.

Adding a cascode transistor is not the only possible approach to reducing the effect of the input
transistor gate-drain capacitor. In Ref. 23 an inductive feedback is introduced that cancels the current
through C_{gd} (Figure 11.7). This approach allows the supply voltage to be reduced as the LNA consists of
only one transistor. The increased gain of the LNA can be thought of as equivalent to g_m boosting that
results in improvement to the LNA noise figure similar to that discussed in Ref. 24.

Narrowband LNA Gain

The gain in a source-degenerated LNA is the cascade of the gain due to the passive input network,
$Q_s = (\omega C_t R_s)^{-1}$, and the transconductance gain of the input transistor stage driving load R_L is given
by the following equation:

$$\text{Gain}_{LNA} = G_m R_L = g_m Q_s R_L \tag{11.13}$$

$$= \frac{\omega_T R_L}{2\omega_0 R_s} \tag{11.14}$$

From Equation 11.13, having large passive gain in terms of large Q_s may be beneficial if large g_m
cannot be achieved due to power consumption constraints. However, the amplification of the various
noise signals by the increased Q_s sets the maximum practical value for the passive gain. The LNA
gain is greatly benefited by the scaling of the CMOS transistors due to increase of the transistor
unity gain frequency (see Equation 11.14). Interestingly, when the LNA is power matched to the signal
source resistance with $R_s = g_m L_s / C_{gs}$, the gain of the source-degenerated LNA gain becomes

$$\text{Gain}_{LNA} = \frac{R_L}{2\omega_0 L_s} \tag{11.15}$$

Equation 11.15 shows that the amount of feedback introduced by the source inductor L_s controls the LNA
gain regardless of what the transconductance of the LNA input transistor is. For this reason the
design shown in Figure 11.4d produces higher gain as it allows the designer to completely remove
the source inductor. The circuit topology in Figures 11.4c does not attain higher gain, however,

as the gain is affected by the apparent increase of the L_g's parasitic resistance due to the parallel combination of L_g and C_g. Of course the load resistance enters the three expressions, and it is obvious that the higher the load resistance the higher the gain. Recently in Ref. 25 the authors added a Q-enhancement circuit to the LNA load inductor to increase the total load resistance and improve the LNA gain at the expense of stability.

As the LNA operating frequencies increase, the gate-drain capacitor may result in significant reduction of the LNA gain. To combat this, transformer feedback can be employed [23,25]. When designed properly the feedback can cancel the effect of the gate-drain capacitor and increase the overall gain of the LNA.

Effect of the Signal Source Impedance on LNA Noise Figure

The LNA noise factor in Equation 11.8 is inversely proportional to the signal source resistance. If the signal source resistance becomes very large, expect a reduction in the noise factor. Of course, changing the signal source resistance from the conventional $50\,\Omega$ is not an easy decision as all the equipment in most laboratories are based on the standard $50\,\Omega$ impedance level. But when very low noise figures are required, for example, in the field of radio astronomy, the signal source resistance can in fact be optimized [26]. It turns out that higher than $50\,\Omega$ impedance is required to reduce the noise figure. If the off-chip antenna is closely integrated with the LNA, the antenna design may in fact be simplified since antenna designs with high impedance are generally easier to achieve. In a more conventional system, a filter (often a ceramic filter) is placed between the LNA and the antenna. But if a non-standard impedance is selected, the off-the-shelf filter can no longer be used. On the bright side, with sufficiently high manufacturing volumes, such as a few tens of thousands of units per year, a custom filter can be tailored to the application. It turns out that filters consisting of high-impedance ceramic resonators are easier to design than their $50\,\Omega$ counterparts, which may in fact translate not only to lower noise figures due to better-optimized LNAs but also to lower overall product costs. The disadvantages of the higher signal source resistance are the decrease in the LNA gain (see Equation 11.14), increase of noise contribution from the biasing network of the main transistor, and increase of the noise coupled from the substrate. Therefore, there is an optimum signal source resistance in the order of 100–$150\,\Omega$. Good layout techniques, such as substrate shielding, and lower noise design of the biasing network must be employed to minimize the adverse effects associated with increasing R_s.

WIDEBAND LNAS

Wideband LNAs dominate current LNA research largely due to spectrum allocation for the ultra-wideband (UWB) services and due to consumer desire to have one wireless device providing many different functions. From the noise figure point of view, achieving minimum noise figure across the whole band of interest is not possible since the optimum signal source susceptance is emulated with an inductor at a single frequency. The minimum noise factor increases with the operating frequency (see Equation 11.4); therefore, a common wideband design strategy is to optimize a narrowband LNA at the high-frequency band edge and to allow the LNA's noise figure to deviate from F_{min} at low frequencies, where F_{min} is low, and, as shown in Figure 11.8, deviation from F_{min} does not produce noise figures higher than that at the high-frequency end [27].

The conceptually simplest wideband LNA is designed by adding a shunt resistor at the input to a common-source LNA (Figure 11.9a). This provides the input match but degrades the noise figure and the gain. This circuit will not attain noise figures lower than 6 dB; however, this topology is still valuable for very wideband applications [12]. A better approach is to add a shunt-feedback resistor between the gate and the drain terminals of a common-source transistor (Figure 11.9b). The feedback resistor is larger than the shunt resistor in Figure 11.9a and therefore contributes less noise. This approach, however, still does not guarantee that the classical condition of the optimum noise impedance is achieved and the feedback resistor adds additional thermal noise that pushes the total noise figure even further from the minimum noise figure of the transistor. Nevertheless, for many

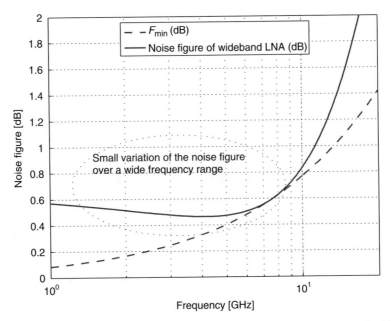

FIGURE 11.8 Common wideband design strategy of optimizing LNA noise figure at the high-frequency band edge.

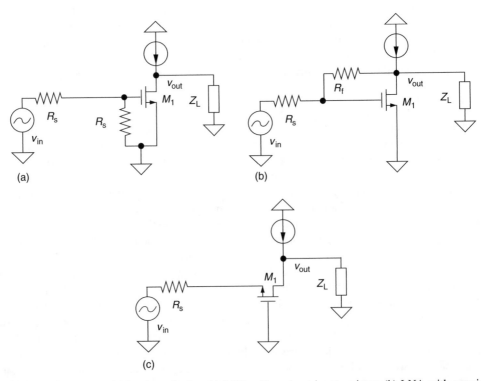

FIGURE 11.9 Simple wideband topologies. (a) LNA with a shunt input resistor. (b) LNA with a resistive shunt feedback. (c) Common-gate LNA.

wireless systems, noise figures as high as 5 dB are quite tolerable. The optimization techniques for both LNA topologies are discussed in detail in Ref. 12 in which the authors derive the expression for the optimum transistor width that minimizes the LNA noise figure and then bias the LNA at the optimum current density (see Figure 11.3b). If power-constrained design was required, the LNA transistor would have to be biased at lower than optimum current density.

Another resistive wideband matching can be accomplished by configuring the input LNA as a common-gate amplifier as shown in Figure 11.9c. The input impedance is controlled by the transistor transconductance, $Z_{in} \approx 1/g_m$, and by sizing the transistor properly the power match can be achieved. As in the case of the resistive shunt-feedback topology, the common-gate LNA provides wideband response and is straightforward to design. The noise factor of a common-gate LNA is no less than $F \geq 1 + \gamma/\alpha$, which gives a reasonable 2.2 dB noise figure with a long-channel transistor.

Although the aforementioned topologies suffer from poor noise figure, their broadband match makes them useful nonetheless. Numerous works have been dedicated to improving the noise performance of these topologies. In Ref. 28, the authors applied a capacitive g_m-boosting technique that was shown to reduce the noise figure while preserving the wideband response. A similar idea was later used in Ref. 24, where a common-gate wideband LNA was modified by adding a transformer feedback that boosted the LNA transconductance gain. In theory the transformer can reduce the noise contribution of the drain noise to arbitrarily low values; however, when the gate noise and the transformer implementation are considered an optimum value for the transformer coupling and turns ratio must be determined [24].

In Ref. 29, the authors decided to systematically search for the lowest noise figure broadband topology that consisted of two transistors. The search resulted in a new topology shown in Figure 11.10a that has a common-gate LNA with a common-drain transistor connecting the input and output of the common-gate transistor. The input impedance of this configuration is still broadband

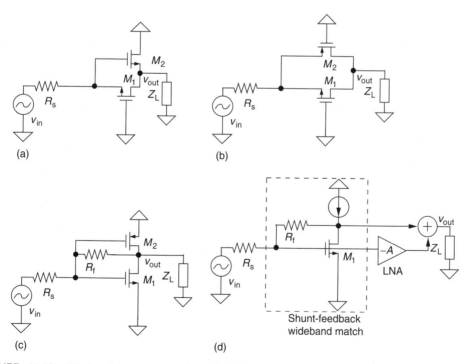

FIGURE 11.10 Wideband LNA topologies. (a) LNA with noise cancellation. (From Bruccoleri, F., Klumperink, E.A.M., and Nauta, B., *IEEE J. Solid-State Circuits*, 36, 1032, 2001.) (b) A common-gate LNA with PMOS g_m boosting. (c) A shunt-feedback LNA with PMOS g_m boosting. (d) LNA with noise cancellation. (From Bruccoleri, F., Klumperink, E.A.M., and Nauta, B., *IEEE J. Solid-State Circuits*, 39, 275, 2004.)

$1/g_m$ as in the standard common-gate amplifier. However, this topology has an interesting benefit of providing a negative feed forward path of the common-gate transistor's drain noise current through the common-drain transistor, which results in noise cancellation. Although two transistors are used in this topology, the noise of the common-drain transistor is the one that determines the overall noise figure. At the power-matched condition $g_{m1} = 1/R_s$, the LNA voltage gain becomes

$$\text{Gain} = \frac{1}{2}\left(\frac{g_{m1} + g_{m2}}{g_{m2} + 1/R_L}\right) \tag{11.16}$$

and the output noise voltage due to the drain noise current of the common-gate transistor is expressed as

$$v_{out} = i_{nd} \frac{1 - g_{m2}R_s}{1 + R_s(g_{m1} + g_{m2}) + R_s R_L g_{m1} g_{m2}} \tag{11.17}$$

which shows that the condition for the total noise cancellation is $g_{m2} = 1/R_s$; however, with $g_{m1} = g_{m2} = 1/R_s$ the LNA voltage gain is less than unity. As the g_{m2} deviates from $1/R_s$ the LNA gain increases at the expense of the noise figure. When the drain current noise contribution of the two transistors is the same, the noise cancellation still occurs and the LNA gain can be made larger than unity (see Ref. 29).

The common-gate and shunt-feedback LNAs can be improved further with the addition of a P-channel MOS (PMOS) device in the common-gate LNA topology (see Figure 11.10b) and in the shunt-feedback LNA topology to form what looks like an inverter circuit (see Figure 11.10c), which boosts the g_m of the LNA without sacrificing the power consumption, due to current reuse. This improves the noise figure slightly but decreases the bandwidth as the extra PMOS adds additional capacitance. Nevertheless, this topology could be used for applications with data rates as high as 40 Gb/s [12]. In Ref. 31, the authors have taken a differential version of the current-reuse shunt-feedback LNA to form a shunt-feedback/common-Gate (SFBCG) topology by connecting the sources of an NMOS transistor and a PMOS transistor to the other polarity input signal as shown in Figure 11.11a. By using the current-reuse technique and by topology reuse a power advantage of four is achieved compared to the conventional differential feedback LNA or common-gate LNA. The authors went even further by using triple-well transistors and connecting the bulk terminals to the source terminals of the corresponding transistors in the other branch (not shown in Figure 11.11a), slightly enhancing the effective g_m of the transistors and decreasing the noise figure to some extent.

The shunt-feedback topology provides a relatively easy method of designing broadband LNAs as the input impedance is controlled by resistors. However, the topology does not allow designers to achieve the minimum noise figure. In Ref. 30, the authors reused their previously published idea [29] of providing a negative feed-forward path for the drain noise so that the two noise signals could destructively add (Figure 11.10a), but this time the feed-forward was accomplished by noting that the noise signals due to the transistor drain noise at the drain of the shunt-feedback LNA and at the gate of the shunt-feedback stage appear out of phase (Figure 11.10d). Therefore, when these two signals are combined the noise contribution of the input transistor is reduced. The destructive addition of the noise signals can be accomplished in a negative gain LNA ("LNA" in Figure 11.10d) when its gain is set to

$$|A| = 1 + \frac{R_f}{R_s} \tag{11.18}$$

This LNA does not need to provide broadband power match, as the shunt-feedback stage accomplishes this, and therefore an extra design freedom is available for the designer to achieve low noise figures. This topology still does not allow desired noiseless matching circuitry to be achieved. The overall

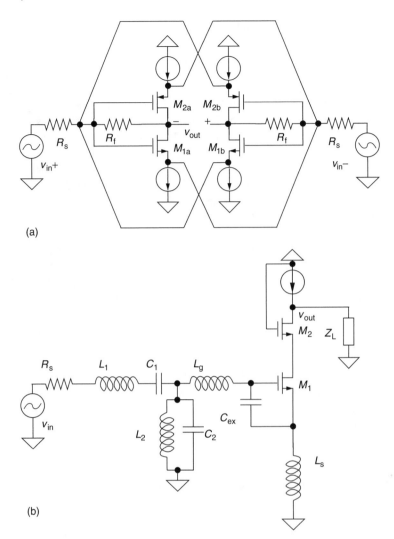

FIGURE 11.11 Improved wideband topologies. (a) Shunt-feedback, common-gate LNA. (b) Chebyshev broadband match.

noise figure of this LNA is limited by the feedback resistor, the size of which is restricted by the input power-matching conditions, the LNA gain, and the desire of having the negative replica of the noise for the proper cancellations. The gain of this LNA is

$$\text{Gain}_{\text{n-c,LNA}} = \frac{1}{2}(1 - g_m R_f - A) = -\frac{1}{2} R_f \left(g_m + \frac{1}{R_s} \right) \tag{11.19}$$

which is higher than what is achievable with the noise-canceling LNA in Figure 11.10a. This topology has been recently modified in Refs. 32, 33.

The implementations described attempt to use a resistive power match but do not achieve the minimum noise factor F_{min}. The common-source topology, usually with source degeneration, is however the best at coming close to achieving F_{min}. Authors in Ref. 34 have decided to use the inductively degenerated common-source CMOS LNA and provide the broadband input match by using C_{gs}, L_s, and the real part of input impedance due to L_s as a part of a Chebyshev broadband-matching network (Figure 11.11b). Again as was the case with the narrowband designs, the wideband matching circuit is not noise free and requires two series inductors and one shunt inductor that can dominate the LNA noise figure.

All of the aforementioned techniques have employed topologies that do not achieve minimum noise figure or use lossy (noisy) matching components at the LNA input. Although they all can be used to achieve wideband power match, they cannot achieve F_{min}. Therefore, another way of achieving the broadband match is desired. Although seemingly an impossible task, it can actually be achieved by introducing a second feedback element. In the narrowband design a single feedback with L_s is used to provide a match at one frequency. If the feedback through gate-drain capacitor C_{gd} is exploited then it is possible to produce a second resonance, thus creating a broadband power match. In Refs. 35–37, a similar idea has been used where a capacitive feedback at the output of the source-degenerated LNA is used to create a second resonance in the LNA input match. This approach allows the design of the narrowband source-degenerated LNA by using optimization techniques described in Ref. 14 that come very close to achieving the minimum noise figure and then modifying its input impedance by introduction of the second feedback loop that does not affect the noise figure [38]. A UWB LNA based on this idea is presented in Ref. 39. In this LNA, the feedback path through C_{gd} of the input transistor is enhanced with an extra capacitor between the drain and gate terminals as suggested in Ref. 37, and the capacitive-resistive feedback elements are generated with a differential amplifier circuit stacked on top of the input transistor.

The distributed amplifier, first described in Ref. 40, has been used to create a wideband CMOS LNA. Only recently, however, has the advance in CMOS technology allowed the design of the transmission lines with high enough Q to implement wideband LNAs using this topology at frequencies approaching $\omega_T/2$ [41]. For UWB applications, the transmission lines can be simulated with integrated lumped circuits. In Ref. 42, the authors have implemented such an LNA by coupling common-source cascode LNAs with inductors and have derived closed-form noise figure expressions. They have also successfully used the expressions to determine the optimum number of stages required for achieving an impressive noise figure of 2.1 dB across the UWB band. The main drawback of this topology is of course the large power consumption.

Wideband LNA Gain

The voltage gain of the shunt-feedback LNA in Figure 11.9b under the input power match constraint is related to the amount of the feedback, the transistor transconductance, and the load resistance:

$$\text{Gain}_{s\text{-f,LNA}} = \frac{1}{2}\left(\frac{1-g_m R_f}{1-R_f/R_L}\right) \tag{11.20}$$

However, because of the input power match requirement, one is not free to select any gain by changing the value of the feedback and the g_m. Both of these are linked to the value of the signal source resistance R_s:

$$R_s = \frac{R_f - R_L}{1 + g_m R_L} \tag{11.21}$$

As the frequencies increase, capacitors C_{gs} and C_{gd} begin to reduce the gain and spoil the input impedance. Equation 11.20 also represents the voltage gain of the LNA in Figure 11.10c when g_m represents the sum of the transconductances of PMOS and N-channel MOS (NMOS) transistors.

The transistor transconductance of the input power-matched common-gate LNA in Figure 11.9c is inversely proportional to the signal source resistance, $g_m = 1/R_s$, while the voltage gain can be found from a very simple relationship:

$$\text{Gain}_{c\text{-g,LNA}} = \frac{1}{2}g_m R_L \tag{11.22}$$

Similarly to the shunt-feedback LNA, C_{gs} and C_{gd} begin to reduce the gain and detune the input impedance at high frequencies. Adding the transconductances of the PMOS and NMOS transistors to form a total g_m allows Equation 11.22 to express the gain of the improved common-gate LNA in Figure 11.10b. The voltage gain of the SFBCG LNA in Figure 11.11a is also related to the gain summation of the shunt-feedback and common-gate stages in Equations 11.20 and 11.22. The voltage gains of the wideband LNA based on the source-degenerated LNA topology have been covered in the section "Narrowband LNA Gain."

LNA LINEARITY

Linearity of an LNA is often an important design consideration. In a system the input to the LNA can range from the thermal noise level to levels around 0 dBm. From a system-level point of view, any close-in distortions generated in the LNA will propagate through the whole system chain desensitizing the receiver. The third-order distortions usually fall into the band of interest. The design challenge is to design an LNA that produces low noise figures but at the same time does not generate the distortions even when either a strong interferer is present or the input signal has large peak-to-average ratio, for example, in CDMA systems. The two measures commonly used to describe the linearity of an LNA are the 1 dB compression point (P_{1dB}) and the third-order intercept point (IP3). These are related to each other with IP3 approximately 10 dB higher than P_{1dB}.

Although CMOS scaling has benefited the gain and noise figure of LNAs by the significant increase of the transistor ω_T, the decreasing power supply voltages reduce the maximum amplitude that the received signals are allowed to achieve before the LNA circuitry saturates, which results in nonlinear transfer characteristics and a high level of distortion. In addition to the decreasing supply voltages, the scaled transistors also experience a shift from the third-order voltage intercept point to higher current densities requiring large power consumptions to achieve high linearity. A plot of $V_{IP3} = \sqrt{24 g_m / g_{m3}}$, where V_{IP3} is the input voltage amplitude at which the first and third output voltage amplitudes are equal and g_{m3} is the second derivative of g_m, is shown in Figure 11.12. The plot shows that there is a strong singularity in the moderate inversion region. The singularity, at which the transistor linearity is significantly enhanced, shifts to higher current densities with technology scaling. These optimum current densities increase from 5 to 50 µA/µm as the technology scales.

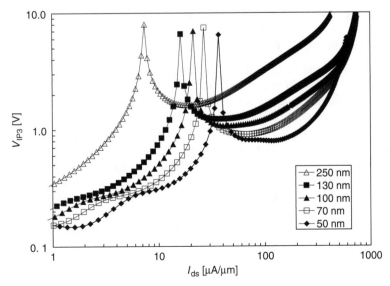

FIGURE 11.12 Simulated third-order voltage intercept point of nominal gate length NMOS transistors. (From Woerlee, P.H. et al., *IEEE Trans. Electron Devices*, 48, 1776, 2001. With permission. © [2001] IEEE.)

The decreasing supply voltages and increasing optimum current densities have resulted in the necessity for linearization techniques. One of the successful linearization methods for a source-degenerated CMOS LNA is known as the derivative superposition technique. The most successful implementation of this idea was demonstrated by Aparin and Larson [43]. In the derivative super-position approach, a parallel transistor operating near the weak-inversion region is introduced. The sign of the third-order derivative of its DC transfer characteristics is positive, which allows the cancellation of the negative third-order derivative of the main transistor DC transfer characteristic. This approach was demonstrated to achieve 20 dB of improvement in IP3 performance of an LNA [43]. This approach increases the noise figure only by a fraction of a decibel due to a noise contribution of the transistor operating in the weak-inversion region.

The noise penalty and additional parasitic capacitance of the transistor in the weak-inversion region are drawbacks of the derivative superposition approach. Active postdistortion (APD) techniques provide an improved linearization method in which the main transistor of an auxiliary common-source cascode is driven by the drain of the main transistor in a source-degenerated LNA (designed for the minimum noise figure using LNA optimization techniques described in the section "Narrow-band LNAs") and the outputs of the two cascodes are connected together [44]. The additional common-source cascode has a low noise contribution, and by sizing the transistors in both cascodes properly, cancellation of the third-order terms is achieved. An IP3 improvement of 6 dB was demonstrated in Ref. 44.

Recently, the concept of adding an IMD sinking circuit to a source-degenerated cascode LNA was introduced [45]. In this postlinearization technique, the common-source LNA is optimized for low noise operation as discussed in Ref. 14. The IMD sinker consisting of a PMOS transistor is added at the drain of the LNA common-source transistor to absorb the IMD3 current generated by the common-source transistor. One should expect that some of the useful signal would flow into the IMD sinker and reduce the LNA gain. The noise current of the IMD sinker adds directly to the drain noise current of the LNA input transistor impacting the noise figure; however, this impact is somewhat minimized by employing low current draw in the PMOS IMD sinker. To reduce the effect of the second-order distortions that contribute to IMD due to the feedback of the LNA as reported in Ref. 43, an extra gate-source capacitor is employed in this technique. An IP3 improvement of 8 dB has been demonstrated in Ref. 45.

WIDEBAND LNA DESIGN EXAMPLES FOR RADIO ASTRONOMY

CMOS AND RADIO ASTRONOMY

CMOS LNAs are now common in many commercial applications from global positioning system (GPS) to UWB systems that do not require extraordinarily low noise figures. The CMOS implementations are very common in these applications and allow for high levels of integration and low product costs. Radio astronomy, on the contrary, can only tolerate very little of added noise due to extremely low signal levels received by the radio telescopes. All radio telescope designers implement LNAs made of exotic and expensive materials, such as GaAs and InP high electron mobility transistors (HEMTs), that add very little noise. For a telescope that requires only a handful of LNAs, the price advantage of MOSFET is overshadowed by the performance advantage of HEMT.

With the development of the square kilometer array (SKA) radio telescope [46], scaled CMOS is emerging as a valuable alternative to GaAs and InP, as the number of receivers is estimated to be as high as a few millions, resulting in price-sensitive receiver designs. At this large volume, the cost and integration of the LNA with the rest of the receiver chain become very important. Because of this, CMOS LNA implementations may now overtake the LNAs designed with GaAs and InP technologies, provided that sufficiently low noise figures are achievable. Two examples of such LNA implementations targeted for the SKA telescope are discussed in the rest of this section.

Examples of Wideband LNAs

As discussed in the section "Wideband LNAs" in conventional wideband designs, telescope designers were mostly concerned with LNA noise figures and therefore chose to optimize the noise figure at the upper frequency band edge such that it approached the transistor's F_{min}. Below that point the noise figure is relatively flat as shown in Figure 11.8. The LNA input impedance was allowed to deviate from the power-matched condition.

A way of achieving low noise figures and power match simultaneously over a wide bandwidth is with the addition of a second feedback path through the C_{gd} capacitor as described in the section "Wideband LNAs" and Ref. 38. A schematic of this LNA topology and its impedance representation is shown in Figure 11.13. A representation of the network due to feedback through C_{gd} is denoted by Y_{gd} in Figure 11.13b. The network Y_{gd} is used to control the input impedance by adjusting the size of the capacitor C_L until the input impedance appears resonant at the band center. Since the value of C_L has very little effect on the noise figure, this topology allows the independent tuning of the frequency at which the LNA noise figure approaches the transistor's F_{min} and the frequency at which the input impedance is resonant. Therefore, this topology allows to achieve both the power match and the low noise figures.

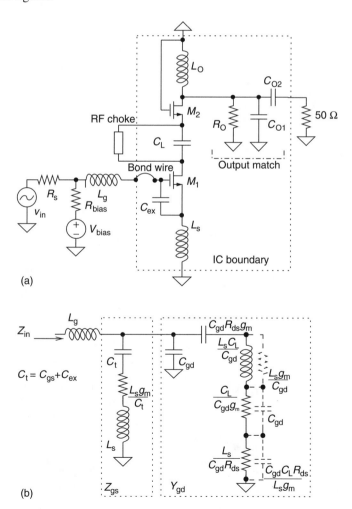

FIGURE 11.13 (a) Circuit diagram of the wideband LNA. (b) Input impedance representation of the LNA. Components shown with broken lines appear when the cascode transistor is represented by g_m during derivation of the Y_{gd} component expressions.

In contrast with consumer electronic designs, in radio astronomy the bandwidth requirement supersedes the power consumption requirement. In the LNA examples discussed in this section, the LNA input impedance band of 700–1400 MHz is met by selection of the total gate-source capacitance, $C_t = C_{ex} + C_{gs}$, such that the quality factor of the input network meets the bandwidth. A noise figure optimization with the fixed Q_s based on the bandwidth requirement is desired. This bandwidth-constrained noise figure optimization is a reverse procedure to the power-constrained optimization described in the section "Narrowband LNAs." To optimize the transistor size and bias, in contrast with the power-constrained LNA noise figure optimization resulting in an optimum

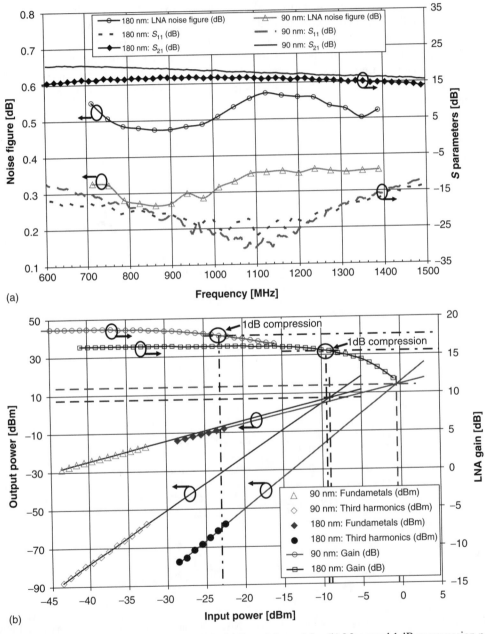

FIGURE 11.14 (a) Measured noise figure of the LNA and S_{11} and S_{21}. (b) Measured 1 dB compression point and IP3 at 1 GHz with the two tones were separated by 10 MHz.

TABLE 11.1
Performance Summary of the Two CMOS LNAs

Specification	90 nm CMOS	180 nm CMOS
Frequency	0.7–1.4 GHz	0.7–1.4 GHz
Noise temperature	25 K max	42 K max
Input return loss	15 dB min	16 dB min
Power gain	16 dB min	15 dB min
Input IP3	−9 dBm typ.	−0.5 dBm typ.
Input P_{1dB}	−23 dBm typ.	−9.4 dBm typ.
DC power	45 mW typ.	50 mW typ.

$Q_s(P_D,Q_g)$ for minimum noise figure [14], in this reverse optimization the noise figure is minimized by searching for an optimum power consumption $P_D(Q_s,Q_g)$ [49]. The optimum P_D allows for determining all remaining LNA parameters [14].

The LNA design procedure starts by setting capacitor C_L to a large value, to appear as an AC short in the frequency range of interest. The resultant cascode LNA is optimized for the best noise figure at the high band edge given the design constraints: the bandwidth and the quality factor of the gate inductor. Once the optimization is completed, C_L is lowered until the input impedance appears resonant in the middle of the band.

An LNA designed according to this procedure was implemented in both 180 nm RF CMOS and 90 nm bulk CMOS. The source and choke inductors were integrated on the die, and triple-well transistors with gate fingers contacted on both sides were used. Substrate shielding was employed for all inductors, bond pads, capacitors, and transmission lines. Figure 11.14a shows measured noise figure and the S parameters. The LNA was designed to drive a 50 Ω load by lowering the quality factor of the output matching network with $R_O = 100\,\Omega$, which resulted in a modest reduction in the gain and an increase in the noise figure. In the ultimate application R_O will not be required as it will come from the input impedance of the following receiver chain, and the noise figure of the circuit without the resistor is the most relevant for the application. The 1 dB compression points and IP3 points of the LNAs are shown in Figure 11.14b. The 180 and 90 nm LNA power consumptions are approximately 50 mW from a 1.5 V supply and 45 mW from a 1 V supply, respectively [49].

LNA OPTIMIZATION SUMMARY

The LNAs were optimized using narrowband LNA optimization technique discussed in the section "Narrowband LNAs." The measured results in Table 11.1 show that the 90 nm design confirms that scaled CMOS technology is applicable for low-frequency radio astronomy. In its intended application, the 90 nm LNA achieves noise figures that are less than 0.35 dB (25 K) across the band of interest at room temperature and reach levels as low as 0.27 dB (19 K) at 850 MHz.

CONCLUSIONS AND FUTURE TRENDS

Discussions in the section "High-Frequency Noise Mechanisms in CMOS MOSFETs" showed that as the CMOS technology moves beyond 45 nm processing node the transistor minimum noise figure is expected to improve due to increase in the cut-off frequency ω_T, which is expected to outpace the accompanying increase in transistor noise parameters. The transistor minimum noise figure falling below 0.1 dB when operated at a few gigahertz is remarkable but cannot be achieved unless the losses in passive components such as inductors, interconnects, and bond pads are reduced. Fortunately, CMOS scaling is accompanied by the introduction of low-loss copper metal layers replacing the aluminum, high-k dielectrics, thick metal layers, and large separation between the substrate and top metal layers: changes that are required to mitigate the losses in the passive components.

The equivalent noise resistance of a MOSFET, represented by its small signal model in Figure 11.1, as expressed in Equation 11.5, has decreased, as CMOS technology scaling has resulted in increasing g_m's [11]. The decrease in R_n implies that LNAs designed in the advanced processes will be less sensitive to their operating conditions and will achieve very low minimum noise figures since R_n is a measure of the LNA sensitivity to deviation of the LNA signal source impedance from the optimum according to the classical optimization. The low noise figures of scaled CMOS can be traded off against the power consumption. The trend of increasing g_m with the decrease in the gate length may not continue forever as the velocity saturated region of the transistor channel will eventually become nearly equal to the length of the channel itself, thus limiting the improvement in g_m with scaling [47]. Although more work is required to assess exactly the RF performance of CMOS transistors scaled beyond 65 nm, it is reasonable to assume that the CMOS scaling will continue much further beyond 65 nm and the possible limit on the transistor transconductance will lead to innovative devices and materials [48].

In addition, increase in ω_T and the quality factor of inductors will have a positive effect on the LNA gain as seen from Equation 11.14. Unfortunately, the linearity of the LNA will suffer as discussed in the section "LNA Linearity." Sophisticated linearization techniques will be required to compensate for the deficiency of downscaled transistors. Some of the emerging linearization techniques have been addressed in the section "LNA Linearity." The design examples in the section "Wideband LNA Design Examples for Radio Astronomy" highlight the expected trends in performance of scaled CMOS. In the two examples, the noise figure improvement and the LNA linearity degradation have been verified experimentally (see Figure 11.14a,b and Table 11.1).

REFERENCES

1. van der Ziel, A. *Noise in Solid State Devices and Circuits*. New York: John Wiley and Sons, 1986.
2. Trofimenko, F. N., Haslett, J. W., and Smallwood, R. E. "Hot electron thermal noise models for FETs," *International Journal of Electronics*, vol. 44, pp. 257–272, March 1978.
3. Scholten, A. J., Tiemeijer, L. F., van Langevelde, R., Havens, R. J., Zegers-van Duijnhoven, A. T. A., and Venezia, V. C. "Noise modeling for RF CMOS circuit simulation," *IEEE Transactions on Electron Devices*, vol. 50, pp. 618–632, March 2003.
4. Haslett, J.W. and Trofimenko, F.N. "Thermal noise in field-effect devices," *Proceedings of the Institution of Electrical Engineers*, vol. 116, pp. 1863–1868, November 1969.
5. Scholten, A. J., Tromp, H. J., Tiemeijer, L. F., van Langevelde, R., Havens, R. J., de Vreede, P. W. H., Roes, R. F. M., Woerlee, P. H., Montree, A. H., and Klaassen, D. B. M. "Accurate thermal noise model for deep-submicron CMOS," in *International Electron Devices Meeting*, Washington: IEEE, pp. 155–158, December 5–8, 1999.
6. Scholten, A. J., Tiemeijer, L. F., van Langevelde, R., Havens, R. J., Van Duijnhoven, A. T. A. Z., de Kort, R., and Klaassen, D. B. M. "Compact modelling of noise for RF CMOS circuit design," *IEE Proceedings-Circuits, Devices and Systems*, vol. 151, pp. 167–174, April 2004.
7. Roy, A.S. and Enz, C.C. "Compact modeling of thermal noise in the MOS transistor," *IEEE Transactions on Electron Devices*, vol. 52, pp. 611–614, April 2005.
8. Razavi, B., Yan, R.-H., and Lee, K. F. "Impact of distributed gate resistance on the performance of MOS devices," *IEEE Transactions on Circuits and Systems I: Fundamental Theory and Applications*, vol. 41, pp. 750–754, November 1994.
9. Haus, H. A., Atkinson, W. R., Fonger, W. H., Mcleod, W. W., Branch, G. M., Harris, W. A., Stodola, E. K., Davenport, Jr., W. B., Harrison, S. W., and Talpey, T. E. "Representation of noise in linear twoports," *Proceedings of the Institution of Radio Engineers*, vol. 48, pp. 66–74, January 1960.
10. Lee, T. H. *The Design of CMOS Radio-Frequency Integrated Circuits*. New York: Cambridge University Press, 2004.
11. Woerlee, P. H., Knitel, M. J., van Langevelde, R., Klaassen, D. B. M., Tiemeijer, L. F., Scholten, A. J., and Zegers-van Duijnhoven, A. T. A. "RF-CMOS performance trends," *IEEE Transactions on Electron Devices*, vol. 48, pp. 1776–1782, August 2001.

12. Voinigescu, S. P., Dickson, T. O., Chalvatzis, T., Hazneci, A., Laskin, E., Beerkens, R., and Khalid, I. "Algorithmic design methodologies and design porting of wireline transceiver IC building blocks between technology nodes," in *IEEE Custom Integrated Circuits Conference*, San Jose: IEEE, pp. 111–118, September 18–21, 2005.

13. Chern, J. G. J. "CMOS technology for RF/mixed-signal applications–progresses and challenges," in *IEEE RFIC Symposium, Workshop: RFIC Technology Evolution and Reality*, Fort Worth: IEEE, June 6–8, 2004.

14. Belostotski, L. and Haslett, J. W. "Noise figure optimization of inductively-degenerated CMOS LNA's with integrated gate inductors," *IEEE Transactions on Circuits and Systems I*, vol. 53, pp. 1409–1422, July 2006.

15. Shaeffer, D. K. and Lee, T. H. "A 1.5-V, 1.5-GHz CMOS low noise amplifier," *IEEE Journal of Solid-State Circuits*, vol. 32, no. 5, pp. 745–759, May 1997.

16. Goo, J.-S. et al. "A noise optimization technique for integrated low-noise amplifiers," *IEEE Journal of Solid-State Circuits*, vol. 37, pp. 994–1002, August 2002.

17. Girlando, G. and Palmisano, G. "Noise figure and impedance matching in RF cascode amplifiers," *IEEE Transactions on Circuits and Systems II*, vol. 46, pp. 1388–1396, November 1999.

18. Andreani, P. and Sjöland, H. "Noise optimization of an inductively degenerated CMOS low noise amplifier," *IEEE Transactions on Circuits and Systems II: Analog and Digital Signal Processing*, vol. 48, pp. 835–841, September 2001.

19. Hillbrand, H. and Russer, P. H. "An efficient method for computer aided noise analysis of linear amplifier networks," *IEEE Transactions on Circuits and Systems*, vol. 23, pp. 235–238, April 1976.

20. Ko, P.K. *Advanced MOS Device Physics*, vol. 18 of *VLSI Electronics: Microstructure Science*, chap. 1, pp. 1–37. New York: Academic Press, 1989.

21. Shouxian, M. et al. "A modified architecture used for input matching in CMOS low-noise amplifiers," *IEEE Transactions on Circuits and Systems II*, vol. 52, pp. 784–788, November 2005.

22. Asgaran, S. et al. "A 4-mW monolithic CMOS LNA at 5.7 GHz with the gate resistance used for input matching," *IEEE Microwave and Wireless Components Letters*, vol. 16, pp. 188–190, April 2006.

23. Cassan, D. J. and Long, J. R. "A 1-V transformer-feedback low-noise amplifier for 5-GHz wireless LAN in 0.18-μm CMOS," *IEEE Journal of Solid-State Circuits*, vol. 38, pp. 427–435, March 2003.

24. Li, X., Shekhar, S., and Allstot, D. J. "g_m-boosted common-gate LNA and differential Colpitts VCO/QVCO in 0.18-μm CMOS," *IEEE Journal of Solid-State Circuits*, vol. 40, pp. 2609–2619, December 2005.

25. Vitzilaios, G. et al. "A low-voltage CMOS LNA with multiple magnetic feedback for WLAN applications," in *IEEE International Symposium on Circuits and Systems*, Kos, Greece: IEEE, pp. 4503–4506, May 21–24, 2006.

26. Belostotski, L. and Haslett, J. W. "On selection of optimum signal source impedance for inductively-degenerated CMOS LNAs," in *IEEE Canadian Conference on Electrical and Computer Engineering*, Ottawa: IEEE, pp. 1435–1440, May 7–10, 2006.

27. Maas, S. *Noise in Linear and Nonlinear Circuits*. Boston: Artech House, 2005.

28. Zhuo, W., Embabi, S., de Gyvez, J. P., and Sanchez-Sinencio, E. "Using capacitive cross-coupling technique in RF low noise amplifiers and down-conversion mixer design," in *Proceedings of the 26th European Solid-State Circuits Conference*, Stockholm, Sweden: IEEE, pp. 77–80, September 19–21, 2000.

29. Bruccoleri, F., Klumperink, E. A. M., and Nauta, B. "Generating *all* two-MOS-transistor amplifiers leads to new wide-band LNAs," *IEEE Journal of Solid-State Circuits*, vol. 36, pp. 1032–1040, July 2001.

30. Bruccoleri, F., Klumperink, E.A.M., and Nauta, B. "Wide-band CMOS low-noise amplifier exploiting thermal noise canceling," *IEEE Journal of Solid-State Circuits*, vol. 39, pp. 275–282, February 2004.

31. Wang, S. B. T., Niknejad, A. M., and Brodersen, R. W. "A sub-mW 960-MHz ultra-wideband CMOS LNA," in *IEEE Radio Frequency Integrated Circuits Symposium*, Long Beach: IEEE, pp. 35–38, June 12–14, 2005.

32. Jackson, S. A. "RF design of a wideband CMOS integrated receiver for phased array applications," *Experimental Astronomy*, vol. 17, pp. 201–210, June 2004.

33. Bagheri, R., Mirzaei, A., Chehrazi, S., Heidari, M., Lee, M., Mikhemar, M., Tang, W., and Abidi, A. "An 800 MHz to 5 GHz software-defined radio receiver in 90nm CMOS," in *IEEE International Solid-State Circuits Conference*, San Francisco: IEEE, pp. 480–481, February 5–9, 2006.

34. Bevilacqua, A. and Niknejad, A. M. "An ultrawideband CMOS low-noise amplifier for 3.1–10.6-GHz wireless receivers," *IEEE Journal of Solid-State Circuits*, vol. 39, pp. 2259–2268, December 2004.

35. Hu, R. "An 8-20-GHz wide-band LNA design and the analysis of its input matching mechanism," *IEEE Microwave and Wireless Components Letters*, vol. 14, pp. 528–530, November 2004.

36. Hu, R. and Yang, M. S. C. "Investigation of different input-matching mechanisms used in wide-band LNA design," *International Journal of Infrared and Millimeter Waves*, vol. 26, pp. 221–245, February 2005.

37. Hu, R. "Wide-band matched LNA design using transistor's intrinsic gate-drain capacitor," *IEEE Transactions on Microwave Theory and Techniques*, vol. 54, pp. 1277–1286, March 2006.

38. Belostotski, L., Haslett, J. W., and Veidt, B. "Wide-band CMOS low noise amplifier for applications in radio astronomy," in *IEEE International Symposium on Circuits and Systems*, Kos, Greece: IEEE, pp. 1347–1350, May 21—24, 2006.

39. Townsend, K. A., Belostotski, L., Haslett, J. W., and Nielsen, J. "Ultra-wideband front-end with tunable notch filter," in *IEEE Northeast Workshop on Circuits and Systems*, Gatineau, Canada: IEEE, pp. 177–180, June 18–21, 2006.

40. Ginzton, E. L., Hewlett, W. R., Jasber, J. H., and Noe, J. D. "Distributed amplification," *Proceedings of IRE*, vol. 36, no. 8, pp. 956–969, August 1948.

41. Kleveland, B., Diaz, C. H., Vook, D., Madden, L., Lee, T. H., and Wong, S. S. "Exploiting CMOS reverse interconnect scaling in multigigahertz amplifier and oscillator design," *IEEE Journal of Solid-State Circuits*, vol. 36, pp. 1480–1488, October 2001.

42. Heydari, P., Lin, D., Shameli, A., and Yazdi, A. "Design of CMOS distributed circuits for multiband UWB wireless receivers [LNA and mixer]," in *Radio Frequency Integrated Circuits Symposium*, Long Beach: IEEE, pp. 695–698, June 12–14, 2005.

43. Aparin, V. and Larson, L. E. "Modified derivative superposition method for linearizing FET low-noise amplifiers," *IEEE Transactions on Microwave Theory and Techniques*, vol. 53, pp. 571–581, February 2005.

44. Kim, N., Aparin, V., Barnett, K., and Persico, C. "A cellular-band CDMA 0.25-μm CMOS LNA linearized using active post-distortion," *IEEE Journal of Solid-State Circuits*, vol. 41, pp. 1530–1534, July 2006.

45. Kim, T.-S. and Kim, B.-S. "Post-linearization of cascode CMOS low noise amplifier using folded PMOS IMD sinker," *IEEE Microwave and Wireless Components Letters*, vol. 16, pp. 182–184, April 2006.

46. Hall, P.J. "The square kilometre array: An international engineering perspective," *Experimental Astronomy*, vol. 17, pp. 5–16, June 2004.

47. Asgaran, S., Deen, M., and Chen, C.-H. "Analytical modeling of MOSFETs channel noise and noise parameters," *IEEE Transactions on Electron Devices*, vol. 51, pp. 2109–2114, December 2004.

48. Goo, J.-S., Xiang, Q., Takamura, Y., Wang, H., Pan, J., Arasnia, F., Paton, E. N., Besser, P., Sidorov, M. V., Adem, E., Lochtefeld, A., Braithwaite, G., Currie, M. T., Hammond, R., Bulsara, M.T,. and Lin, M.-R. "Scalability of strained-Si nMOSFETs down to 25 nm gate length," *IEEE Electron Device Letters*, vol. 24, pp. 351–353, May 2003.

49. Belostotski, L. and Haslett, J. W. "Wide band room temperature 0.35-dB noise figure LNA in 90-nm bulk CMOS," *Radio and Wireless Symposium*, Long Beach: IEEE, pp. 221–224, , January 9–11, 2007.

12 Design of Silicon Integrated Circuit W-Band Low-Noise Amplifiers

Sean T. Nicolson, Keith W. Tang, T. O. Dickson,
P. Chevalier, B. Sautreuil, and Sorin P. Voinigescu

CONTENTS

Introduction...329
    What Is a Low-Noise Amplifier?...329
    LNA Design in the V-Band and W-Band 330
    A Brief History of LNA Design Philosophy 330
Original Methodology for Simultaneous Noise and Input Impedance Matching 331
    Step 1: Optimum Biasing..331
    Step 2: Device Size 331
    Step 3: Input Impedance Matching...333
    Step 4: Gain Optimization 334
    Step 5: Extension to Multistage Designs 334
Matching Methodology for W-Band LNAs...335
Accounting for Bond-Wire Inductance..336
Intrastage Matching for Cascode Amplifiers...338
Layout Techniques for Millimeter-Wave CMOS 339
    Gate Contact Arrangements 340
    Finger Width..340
    Optimizing Finger Width and Gate Contacts...341
    Drain and Source Metallization and Gate Pitch...343
CMOS W-Band LNA Designs...343
Measurement Results...345
Conclusion..345
Acknowledgments..347
References...347

INTRODUCTION

WHAT IS A LOW-NOISE AMPLIFIER?

Digital signal processing (DSP) systems require interfaces to the analog world. The first amplification stage in a radio receiver is typically the tuned low-noise amplifier (LNA), which provides enough amplification to the input signal to make further signal processing insensitive to noise. Although the primary goal in LNA design is to minimize the system noise figure, the LNA must also meet other equally important specifications in gain, linearity, power consumption, input and output impedance matching, and bandwidth. This chapter provides a systematic method for achieving these design goals in V-band and W-band LNAs, based on an existing gigahertz-range simultaneous noise and impedance matching technique.

LNA Design in the V-Band and W-Band

With the measured f_{MAX} of the latest SiGe heterojunction bipolar transistor (HBT) and 65 nm CMOS technologies at 300 GHz and with noise figure values as low as 1 dB at 40 GHz [1], as illustrated in Figure 12.1, V-band (50–75 GHz) and W-band (75–110 GHz) radios are now possible in silicon. Several SiGe HBT V-band and W-band LNAs have been reported [2–7]. Although 60 GHz CMOS LNAs have been developed almost at the same time [8,9], it is only at the 90 nm node that their power dissipation and noise figure [10,11] have become competitive and that W-band operation has been demonstrated [12]. LNA design for the W-band is particularly challenging for several reasons. First, the performance of silicon devices in the W-band is poor, and in the case of MOSFETs, highly layout dependent. Second, passive devices are lossy, and obtaining a low-impedance analog ground is difficult, particularly with the strict metal density and slotting rules found in state-of-the-art silicon processes. Perhaps most importantly, existing gigahertz-range techniques for simultaneous noise and input impedance matching need adjustments in the W-band.

A Brief History of LNA Design Philosophy

Traditional LNA design employed lossless reactive components to transform the signal source impedance to the optimum noise impedance of a transistor biased at optimum minimum noise figure (NF_{MIN}) current density [13]. Because the real part of the optimum noise impedance and the real part of the input impedance of transistors are in general different, this approach compromises the input reflection coefficient (and hence gain) to improve noise performance. When the designer has no control over the relationship between the real parts of the input impedance and optimal noise impedance, as in LNA designs involving discrete off-the-shelf components, this simple design methodology yields a unique and optimal solution that minimizes the amplifier noise factor.

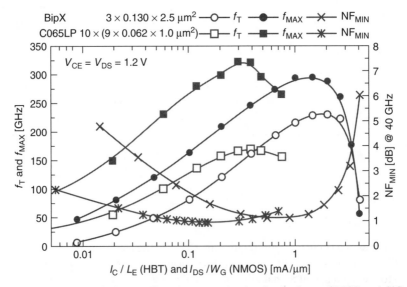

FIGURE 12.1 f_T/f_{MAX} versus drain or collector current density for the latest CMOS and SiGe processes. (From Chevalier, P., et al., *IEEE CSICS*, San Antonio, 2006. With permission. © IEEE 2006.)

In integrated circuit LNAs, however, the designer can control the optimum noise impedance and bias current of the input transistor by adjusting its emitter length (or gate width). Thus, by choosing the correct transistor size, it is possible to obtain simultaneous noise and impedance match at the input of a SiGe or CMOS integrated circuit LNA. The method employed to achieve simultaneous noise and input impedance match was first presented in Refs. 14, 15, and will be reviewed in the next section. Unfortunately, like all other LNA design methodologies to date, it suffers from several limitations that arise at millimeter-waves.

The first limitation is caused by the pad capacitance at the LNA input, which appears in parallel with the signal source impedance. The pad capacitance, which might be as large as 50 fF, is a negligible 530 Ω at 6 GHz, but only 53 Ω at 60 GHz. A second limitation arises because bond wires are often used to connect an LNA to an off-chip antenna, and the original technique does not describe how the bond-wire inductance can be incorporated into the low-noise design methodology. Finally, the methodology does not explain how to optimize the transistor layout to minimize transistor noise, other than ensuring that the real parts of the optimal noise and input impedances are equal.

In this chapter, the original simultaneous noise and impedance matching methodology of Voinigescu and Maliepaard [14] is reviewed, and a new methodology that addresses its limitations through design and layout techniques is presented. Finally, experimental results of 77 and 94 GHz LNAs fabricated in 90 nm [12] and 65 nm [26] CMOS are presented.

ORIGINAL METHODOLOGY FOR SIMULTANEOUS NOISE AND INPUT IMPEDANCE MATCHING

The original methodology for achieving simultaneous noise and input impedance match in integrated LNAs can be summarized neatly in five steps. References 14, 15 provide many details omitted here for brevity. For the interested reader, there are two alternate design approaches for sub–10 GHz CMOS LNAs, which focus on meeting power dissipation constraints [16,17]. To save power, they sacrifice the noise figure [16,17], the gain, and noise resistance of the LNA [17]. However, the power dissipation is quite low at millimeter-wave frequencies because the device sizes and currents required for simultaneous noise matching are smaller than 40 μm and 6 mA, respectively [10–12].

STEP 1: OPTIMUM BIASING

At a given frequency, SiGe HBTs and MOSFETs feature at optimum current density J_{OPT} that minimizes the transistor noise figure [13–15]. The LNA transistors should be biased at this current density. In n-MOS LNAs, J_{OPT} is 0.15 mA/μm, and has been independent of technology since the application of constant-field scaling rules begin at the 0.5 μm node [18]. Figures 12.2a and 12.3 show the trend from the 0.18 μm node to the 65 nm node. To maximize gain with minimal increase in noise figure, the amplifier can be biased instead at peak f_{MAX} current density J_{pfMAX}, which is 0.2 mA/μm in CMOS, as illustrated in Figure 12.2b. In addition to being independent of CMOS technology, J_{OPT} is also independent of frequency, as illustrated in Figure 12.3 [19].

In contrast, the optimum noise figure and peak f_{MAX} current densities of SiGe technologies depend on the technology and the design frequency, as shown in Figure 12.4. Furthermore, in SiGe, J_{OPT} is higher for cascode amplifiers than for single-transistor amplifiers. Regardless, the optimal noise figure current density can be found by performing noise figure simulations at a variety of bias current densities for the particular LNA topology.

STEP 2: DEVICE SIZE

In CMOS LNAs, the total gate width (W_G) is controlled by connecting a number of fingers N_f in parallel, each with a fixed finger width W_f. In this case, the noise parameters of a MOSFET can be

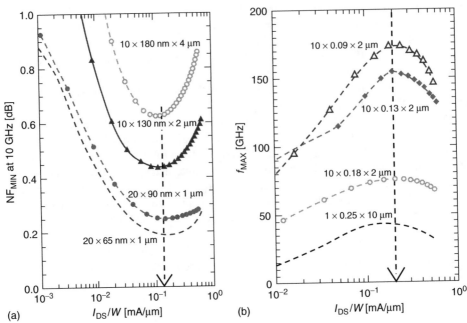

FIGURE 12.2 The optimal noise figure current density (a) and the peak f_{MAX} current density (b) are constant for CMOS technology nodes from 0.18 μm to 65 nm. (From Dickson, T. O., et al, *IEEE J. Solid State Circuits.* 41, 1830, 2006, With permission, © IEEE 2006.)

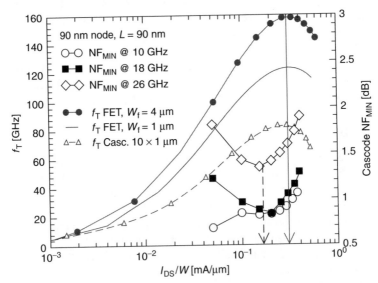

FIGURE 12.3 NF_{MIN} versus design frequency and f_T versus finger width in 90 nm n-channel MOSFETs. (From Yao, T., et al., *IEEE RFIC Symposium*, 147, 2006. With permission. © IEEE 2006.)

expressed in terms of the total gate width $W_G = N_f W_f$, as shown in Equations 12.1 through 12.4 where parameters R_{NMOS}, $G_{u,NMOS}$, $G_{cor,NMOS}$, and $B_{cor,NMOS}$ are bias-dependent technology parameters:

$$R_n = \frac{R_{NMOS}}{N_f W_f} \tag{12.1}$$

$$G_u = G_{u,NMOS}\, \omega^2 N_f W_f \tag{12.2}$$

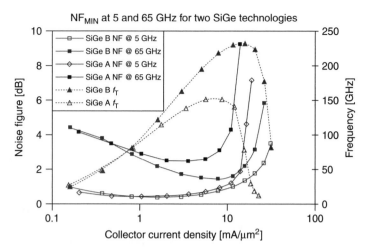

FIGURE 12.4 In SiGe, J_{OPT} depends on the technology and frequency.

$$G_{cor} = G_{cor,NMOS}\, \omega^2 N_f W_f \tag{12.3}$$

$$B_{cor} = B_{cor,NMOS}\, \omega^2 N_f W_f \tag{12.4}$$

The total gate width is chosen such that the real part of the optimum noise impedance (Equation 12.5) is equal to the source impedance at the design frequency [14]. Note that this design methodology places no constraints on N_f or W_f individually, but only on W_G. However, as illustrated by the solid f_T curves in Figure 12.3, W_f plays an important role in the MOSFET f_T and hence is critical to LNA performance. Also note that because G_{cor}, B_{cor}, and G_u all increase with frequency, lower frequency LNAs require larger devices and counterintuitively have higher power dissipation [15]. In HBT LNAs, the emitter length (l_E) is substituted for W_G in Equations 12.1 through 12.4, and exactly the same design procedure applies:

$$\frac{1}{Z_{sopt}} = \sqrt{G_{cor}^2 + \frac{G_u}{R_n}} - jB_{cor} \tag{12.5}$$

STEP 3: INPUT IMPEDANCE MATCHING

The input impedance (Z_{IN}) of the source-degenerated amplifier, given by Equation 12.6, is now tuned to the source impedance (Z_0) using two inductors L_S and L_G, as illustrated in Figure 12.5 [14,20]. Decomposing Equation 12.6 into its real and imaginary parts yields the required values of L_S (Equation 12.7) and L_G (Equation 12.8). Note that L_S changes the real part of the input impedance but not that of the optimum noise impedance [14,15,21]. If a pad is now added at the LNA input, the amplifier is no longer matched correctly:

$$Z_{IN} = \omega_T L_S + j\left(\omega L_S + \omega L_G - \frac{\omega_T}{\omega g_m}\right) \tag{12.6}$$

$$L_S = \frac{Z_0}{\omega_T} \tag{12.7}$$

$$L_G = \frac{\omega_T}{\omega^2 g_m} - L_S \tag{12.8}$$

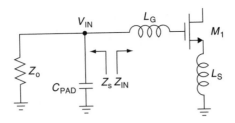

FIGURE 12.5   Schematic of LNA input.

STEP 4: GAIN OPTIMIZATION

Finally, an inductive load is employed to maximize amplifier gain and linearity. When the LNA is matched at input and output, the power gain is given by Equation 12.9, where R_P is the output impedance of an inductively loaded LNA at resonance. Note that although the source degeneration inductance (L_S) is independent of the design frequency, the impedance of L_S increases with the design frequency. Therefore, in W-band SiGe LNAs, the large degeneration impedance and the large transconductance of the HBT can cause the series feedback to be quite strong. In these cases the gain can be simplified as shown:

$$|G| = \frac{1}{2} R_P Z_0 \left| \frac{g_m}{1 + j(\omega/\omega_T) Z_0 g_m} \right|^2 \approx \frac{1}{2} \left(\frac{\omega_T}{\omega} \right)^2 \frac{R_P}{Z_0} \qquad (12.9)$$

Given that g_m was fixed in steps 1 and 2 when we fixed bias current density and total gate width, and Z_0 is normally 50 Ω, the only means of increasing the amplifier gain is to increase R_P using a higher Q inductive load, in turn reducing the amplifier bandwidth. Ultimately, the gain is limited by the device output resistance.

STEP 5: EXTENSION TO MULTISTAGE DESIGNS

In a single-stage amplifier, all design parameters are either beyond the designer's control (Z_0, ω) or optimized to minimize the amplifier noise figure (W_G, I_{DS}, ω_T) or gain (R_P). In a multistage amplifier, the noise contribution of the second stage to the overall amplifier noise factor is reduced by the gain of the first stage, as given in Equation 12.10, where A_1 is the first-stage gain and F_n is the nth-stage noise factor:

$$F_{overall} = F_1 + \frac{F_2 - 1}{A_1} + \cdots + \frac{F_n - 1}{A_1 A_2 \ldots A_n} \qquad (12.10)$$

Therefore, the second stage can be optimized for bandwidth, gain, and linearity with reduced impact on the overall amplifier noise figure. For example, the second stage need not be biased at the optimum noise figure current density but instead at a higher current density to improve gain and linearity. Alternatively, to increase the amplifier bandwidth, the second stage can be tuned to a slightly different frequency to produce a wider, flatter gain response, as shown in Figure 12.6. Finally, the output impedance of the first stage and input impedance of the second stage are no longer part of the 50 Ω environment, and can be adjusted to meet gain and linearity specifications, albeit at the expense of increased noise if the stages are not noise matched.

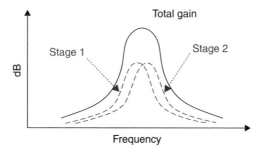

FIGURE 12.6 Stagger-tuned stages increase amplifier bandwidth.

MATCHING METHODOLOGY FOR W-BAND LNAS

In the W-band, the pad capacitance C_{PAD} (see Figure 12.5) cannot be neglected in the matching process. The optimum noise impedance of the transistor must now be matched to the real part of Z_S, and the amplifier input impedance Z_{IN} must be conjugately matched to Z_S. Accounting for C_{PAD}, Z_S is now given by Equation 12.11, and has a real part which is a factor of k, given by Equation 12.12, smaller than Z_0. Intuitively, Z_S decreases because the amplifier now sees admittance through C_{PAD}, in addition to admittance through the original $50\,\Omega$ source.

$$Z_S = \frac{Z_0}{1+\omega^2 C_{PAD}^2 Z_0^2} - j\frac{\omega C_{PAD} Z_0^2}{1+\omega^2 C_{PAD}^2 Z_0^2} \tag{12.11}$$

$$k = 1 + \omega^2 C_{PAD}^2 Z_0^2 \tag{12.12}$$

The new emitter length ($l_{E(new)}$) or gate width ($W_{G(new)}$) required for optimum noise match increases by the same factor of k, as shown in Equation 12.13. The larger device size leads to increased transconductance (Equation 12.14) and bias current (Equation 12.15), again by a factor of k, as well as a larger output 1 dB compression point:

$$W_{G(new)} \text{ or } l_{E(new)} = kW_G \text{ or } kl_E \tag{12.13}$$

$$g_{m(new)} = kg_m \tag{12.14}$$

$$I_{DS(new)} = kI_{DS} \tag{12.15}$$

To find the new values of L_S and L_G, consider that Equation 12.16 must be satisfied to provide conjugate matching between the amplifier and source impedance. Substituting Equation 12.11 into Equation 12.16 and solving, the new values for L_S and L_G are given by Equations 12.17 and 12.18, respectively:

$$Z_S^* = \omega_T L_{S(new)} + j\left(\omega L_{S(new)} + \omega L_{G(new)} - \frac{\omega_T}{\omega g_{m(new)}}\right) \tag{12.16}$$

$$L_{S(new)} = \frac{Z_0}{k\omega_T} \tag{12.17}$$

TABLE 12.1
Numerical Design Examples of LNAs from 60 to 94 GHz

Parameter	65 GHz Cascode, 0.18 μm SiGe HBT [6]	60 GHz Cascode, 90 nm CMOS [10]	77 GHz Cascode, 90 nm CMOS [12]	94 GHz Cascode, 90 nm CMOS [12]
f_T device/cascode	125 GHz/110 GHz	128 GHz/80 GHz	128 GHz/80 GHz	128 GHz/80 GHz
Device size	9 μm by 0.18 μm	34 μm	30 μm	26 μm
C_{PAD}	12 fF	20 fF	20 fF	20 fF
K	1.166	1.142	1.234	1.349
J_{DS} (J_{CE})	2.4 mA/μm²	0.2 mA/μm	0.3 mA/μm	0.3 mA/μm
I_{DS} (I_{CE}) and g_m	3.9 mA and 30 mS	6.9 mA and 34 mS	8.7 mA and 30 mS	7.8 mA and 26 mS
$L_{S/E}$, $L_{G/B}$	60 pH, 90 pH	55 pH, 155 pH	50 pH, 110 pH	45 pH, 80 pH

Source: Nicolson, S. T. and Voinigescu, S. P., *IEEE CSICS*, San Antonio, 279–282, 2006. With permission. © IEEE 2006.

$$L_{G(new)} = \frac{Z_0^2 C_{PAD}}{k} - L_{S(new)} + \frac{\omega_T}{\omega^2 g_{m(new)}} \tag{12.18}$$

With this new methodology, L_S is decreased by a factor of k, resulting in weaker feedback, and correspondingly higher gain (helped further by increased g_m). To determine the conditions under which L_G is also reduced by the addition of C_{PAD}, consider solving Equation 12.19, the inequality $L_{G(new)} < L_G$, using Equations 12.8 and 12.18. The inequality can be simplified as shown in Equation 12.20, by substitution of Equation 12.7 for L_S, Equation 12.17 for $L_{S(new)}$, and finally Equation 12.12 for k. The resulting constraint on C_{PAD}, which ensures $L_{G(new)} < L_G$, is given by Equation 12.21. Because C_{PAD} can be augmented using on-chip capacitors, L_G can be easily reduced by the appropriate selection of C_{PAD}:

$$\frac{Z_0^2 C_{PAD}}{k} - L_{S(new)} + \frac{\omega_T}{\omega^2 g_{m(new)}} < \frac{\omega_T}{\omega^2 g_m} - L_S \tag{12.19}$$

$$\frac{Z_0^2 C_{PAD}}{k} - \frac{Z_0}{k\omega_T} + \frac{\omega_T}{\omega^2 k g_m} < \frac{\omega_T}{\omega^2 g_m} - \frac{Z_0}{\omega_T} \tag{12.20}$$

$$C_{PAD} > \left(\frac{\omega_T}{g_m} - Z_0 \frac{\omega^2}{\omega_T} \right)^{-1} \tag{12.21}$$

The reduction in L_S and L_G results in lower series resistance and higher self-resonance frequency for these passives, and correspondingly improved noise figure. Table 12.1 provides numerical examples of the new design technique, applied to SiGe and CMOS millimeter-wave LNAs at 60, 77, and 94 GHz. The CMOS examples are based on simulations that show $W_G = 30$ μm is required for simultaneous noise and impedance match to 50 Ω at 60 GHz, without C_{PAD}.

ACCOUNTING FOR BOND-WIRE INDUCTANCE

Often, a bond wire is used to connect the LNA input to an off-chip antenna. At millimeter-waves, the bond-wire impedance must be accounted for to achieve simultaneous noise and impedance matching. The bond wire and bond pad are represented by the T-network shown in Figure 12.7, which can

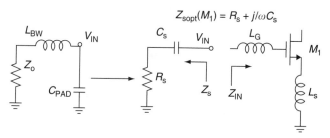

FIGURE 12.7 Noise matching with bond wire and bond pad.

TABLE 12.2
LNA Parameters When Noise and Impedance Matched to $n \times Z_0$

Parameter	Transformation
Device size	$W_{G(new)} = W_G/n$
g_m	$g_{m(new)} = g_m/n$
I_{DS}	$I_{DS(new)} = I_{DS}/n$
L_S	$L_{S(new)} = nL_S$
L_G	$L_{G(new)} = nL_G$

be thought of as an impedance transformer. Repeating the analysis of the section "Matching Methodology for W-Band LNAs," this time including L_W yields Equation 12.22 for k and Equation 12.23 for Z_S. The values of L_S and L_G can now be recalculated using Equations 12.16 and 12.23, where k is given by Equation 12.22 in place of Equation 12.12:

$$k = \left(1 - \omega^2 L_W C_{PAD}\right)^2 + \omega^2 Z_0^2 C_{PAD}^2 \tag{12.22}$$

$$Z_S = \frac{Z_0}{k} + j\omega \frac{\left[L_W \left(1 - \omega^2 L_W C_{PAD}\right) - Z_0^2 C_{PAD} \right]}{k} \tag{12.23}$$

To minimize the sensitivity of the match to variations in L_W that may arise in the packaging process, components L_W and C_{PAD} should be chosen to resonate at the design frequency. This causes k to become independent of L_W, which also makes the emitter or source degeneration inductance (L_E/L_S), g_m, and therefore, the transistor size and bias current, independent of L_W.

When C_{PAD} and L_W resonate, they transform $\text{Re}(Z_S)$ to another real impedance $n \times Z_0$ at the design frequency, where n is given by Equation 12.24. The input transistor size and bias current must thus be modified to match $n \times Z_0$. The new LNA parameters are given in Table 12.2, in terms of the original parameters, derived in the section "Original Methodology for Simultaneous Noise and Input Impedance Matching." The gain of the new LNA is given by Equation 12.25. Note that n is just a simplification of k under the condition that L_W and C_{PAD} resonate:

$$n = \left(\frac{1}{\omega^2 Z_0^2 C_{PAD}^2} \right) \tag{12.24}$$

$$|G|_{(new)} = \frac{1}{2n} R_P Z_0 \left| \frac{g_m}{1 + j(\omega/\omega_T) Z_0 g_m} \right|^2 \approx \frac{1}{2n} \left(\frac{\omega_T}{\omega} \right)^2 \frac{R_P}{Z_0} \tag{12.25}$$

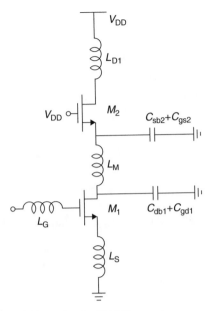

FIGURE 12.8 Adding L_M to the middle node of a CMOS cascode.

For greater gain, we would like to select $n<1$. However, selecting $n<1$ places the constraint given by Equation 12.26 on L_W, if C_{PAD} and L_W are assumed to resonate. For W-band LNAs in a $50\,\Omega$ environment, this places L_W in the range of 70–105 pH, maximum. To perform matching with larger, more realistic bond-wire inductances requires $n>1$ and a correspondingly decreased gain:

$$L_W \leq \frac{Z_0}{\omega}$$

$$(12.26)$$

Alternatively, if greater dependence on manufacturing variation in L_W can be tolerated, then the constraint that L_W and C_{PAD} resonate can be relaxed. In this case, $k<1$ can be obtained for larger values of L_W. For example, for a 77 GHz automotive radar LNA in a $50\,\Omega$ environment with 30 fF pad capacitance, $k=3$ can be obtained for $L_W = 370$ pH.

INTRASTAGE MATCHING FOR CASCODE AMPLIFIERS

In sub–100 nm n-channel MOSFETs, because C_{GD} is 50% of C_{GS}, the f_T of the cascode is at least 33% smaller than the f_T of a single transistor. Because the middle pole degrades the cascode frequency response, CMOS cascode amplifiers require bandwith extension techniques [22–24] to achieve acceptable gain at millimetre-waves. One approach is to create a parallel resonance at the middle node between the common-source (CS) and the common-gate (CG) transistors [6]. However, this resonance is usually quite narrowband. Another technique is to form an atificial transmission line by inserting a series inductor between the drain of the common-source FET, and the source of the common-gate FET, as illustrated in Figure 12.8. As shown in Figure 12.9a, the required inductance is a strong function of the size of the transistors; an optimal inductance maximizes the cascode f_T as well as reducing the noise figure. The optimal inductance is also a strong function of the layout parasitics of the transistor. In Figure 12.9b, we repeat the same sweep of L_M after transistor layout parasitic extraction. Relative, to Figure 12.9a, layout parasitic have caused the optimal value of L_M and

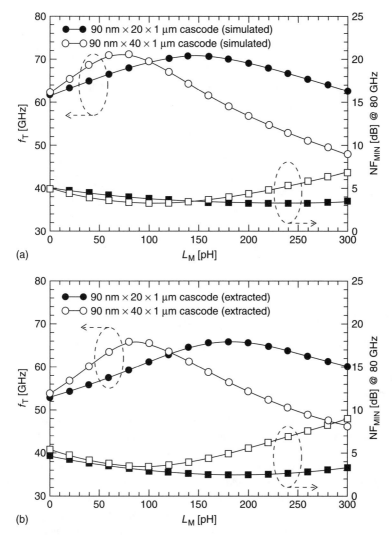

FIGURE 12.9 The simulated f_T and NF_{MIN} of a 90 nm CMOS cascode amplifier when L_M is swept from 0 to 250 pH (a) before and (b) after transistor layout parasitics extraction.

the cascode f_T to decrease by approximately 10%. Thus, the LNA design methodology described earlier must be refined to include the extraction of tansistor layout parasitic, and the maximization of the cascode f_T (through adding L_M), just before step 2.

LAYOUT TECHNIQUES FOR MILLIMETER-WAVE CMOS

Unlike SiGe HBTs, MOSFET performance is extremely sensitive to device layout. The f_T and f_{MAX} of MOSFETs depend on finger width, drain and source metallization, gate finger spacing, and gate and substrate contact arrangement [21]. In W-band CMOS circuits, finding the optimal layout for the MOSFET is critical to achieving the best possible circuit performance. Here the strategies for obtaining the highest f_T and f_{MAX} possible in a given technology are discussed. In general, the layout geometry with the highest f_{MAX} coincides with that for the minimum noise figure. Also note that in the expressions for f_{MAX}, NF_{MIN}, and noise resistance R_n, R_{gate} and R_S always appear as a

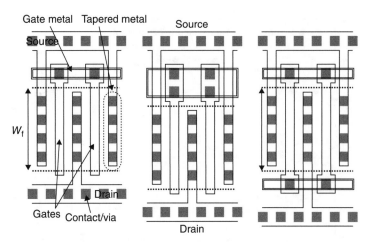

FIGURE 12.10 MOSFETs with different gate connections. (From Nicolson, S. T. and Voinigescu, S. P., *IEEE CSICS*, San Antonio, 279–282, 2006. With permission. © IEEE 2006.)

sum $R_S + R_{gate}$. Unlike R_{gate}, R_S only depends on the total gate width of the device and not on contact arrangements or finger width. It is typically $200–300\,\Omega\,\mu m$ in 130, 90, and 65 nm CMOS. Therefore, once R_{gate} becomes much smaller than R_S, no further improvement in f_{MAX} or NF_{MIN} can be achieved.

GATE CONTACT ARRANGEMENTS

Shown in Figure 12.10 are three MOSFET layouts, with one single-sided contact per gate (a), two single-sided contacts per gate (b), and double-sided gate contacts (c). Double-sided gate contacts have the advantage that the gate finger is treated as a transmission line driven from both ends; however, because the drain and gate are overlapped, the layout suffers from increased C_{GD}. Generally speaking, the more the gate contacts are added, the more reliable the layout and the lower the gate resistance. These benefits come at the expense of increased parasitic capacitance. The gate resistances for single-sided and double-sided contacts are given by Equations 12.27 and 12.28, respectively, where R_{CON} is the contact resistance, N_{CON} the number of contacts per gate finger, R_{sq} the gate poly sheet resistance per square, W_{ext} the gate extension beyond the active region, W_f the finger width, N_f the number of gate fingers connected in parallel, and l_{phys} the physical gate length.

$$R_G = \frac{\dfrac{R_{CON}}{N_{CON}} + \dfrac{R_{sq}}{L} \Big/ \left[W_{ext} + \dfrac{W_f}{3} \right]}{N_f} \tag{12.27}$$

$$R_G = \frac{\dfrac{R_{CON}}{N_{CON}} + \dfrac{R_{sq}}{L} \left[W_{ext} + \dfrac{W_f}{6} \right]}{2N_f} \tag{12.28}$$

FINGER WIDTH

Finger width also affects MOSFET performance. Longer fingers result in higher f_T and shorter fingers initially result in higher f_{MAX} until the degradation of f_T offsets the reduction in gate resistance, as illustrated for 90 nm CMOS in Figures 12.11a and 12.11b, respectively. For finger widths of 1–2 μm

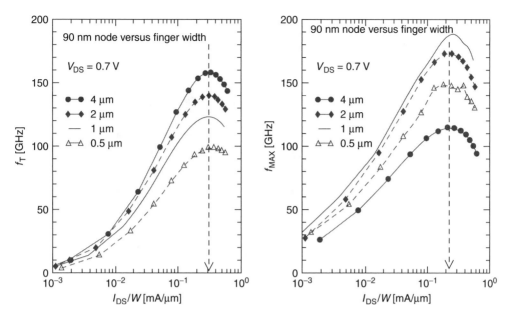

FIGURE 12.11 f_T and f_{MAX} versus finger width for 90 nm CMOS.

in 90 nm CMOS, f_{MAX} and f_T are reasonably large, demonstrating that there is probably an optimum finger width between 1 and 2 µm for the best LNA performance.

OPTIMIZING FINGER WIDTH AND GATE CONTACTS

Shown in Figure 12.12 are measured f_T and f_{MAX} for three different MOSFET layouts in 130 nm CMOS. The layout with 2 µm fingers and single gate contacts has f_T and f_{MAX} of about 85 GHz. The other two layouts suffer degradation in either f_T or f_{MAX} for an improvement in the other. Such degradation is unacceptable in a millimeter-wave LNA tuned near f_T/f_{MAX}. In a technology with higher f_T/f_{MAX}, a greater number of gate contacts could be used for improved yield and reliability; however, 130 nm CMOS does not have enough performance at millimeter-wave frequencies for such considerations.

If W_f is held constant as the CMOS gate length is scaled by $\sqrt{2}$ to the next technology node, the capacitance of the gate finger is scaled down by $\sqrt{2}$ and the resistance of the gate finger is scaled up by $\sqrt{2}$. However, the contact resistance is scaled by a factor of 2 because the contact size is scaled in both dimensions. The longer finger width relative to the new gate length, combined with greater contact resistance per finger, will cause f_T to increase relative to f_{MAX}. Therefore, the starting point for finding the optimum MOSFET layout in a new technology is to scale W_f by $\sqrt{2}$ along with the length. This strategy should yield optimum finger widths of 1.4 µm in 90 nm CMOS and 1 µm in 65 nm CMOS (Table 12.3) based on the 2 µm finger width in 130 nm CMOS. In subsequent sections we demonstrate that this scaling strategy is reasonably accurate. Regardless, it is important to note that the contact resistance becomes an ever-larger proportion of the gate resistance as CMOS technology is scaled.

Example 1: 10 µm × 90 nm device contacted on one side:
$R_{sq} = 10\,\Omega$, $l_{phys} = 65\,nm$, $N_{CON} = 1$, $R_{CON} = 20\,\Omega$, $W_{ext} = 200\,nm$; $R_S = R_D = (1/W_G) \times 300\,\Omega\,\mu m = 30\,\Omega$

a. $W_f = 1\,\mu m$; $N_f = 10$, $R_{gate} = 10.2\,\Omega$, $R_{gate} + R_S = 40.2\,\Omega$
b. $W_f = 2\,\mu m$; $N_f = 5$, $R_{gate} = 30.66\,\Omega$, $R_{gate} + R_S = 60.66\,\Omega$

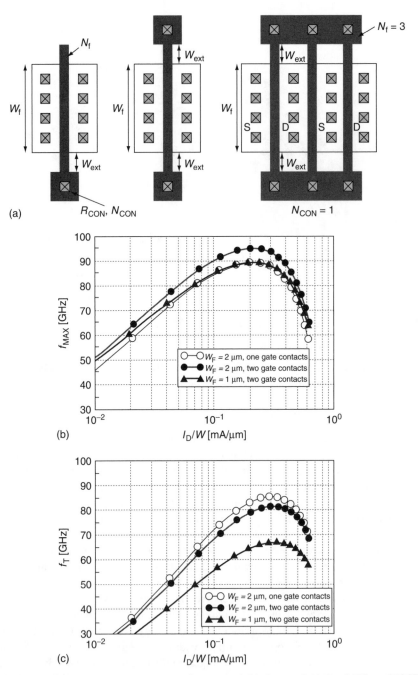

FIGURE 12.12 (a) Gate contact arrangement and measured (b) f_{MAX} and (c) f_T of 130 nm MOSFETs with different layout styles. (From Dickson, T. O., PhD Thesis, ECE Department, University of Toronto, 2006.)

Example 2: $10 \mu m \times 90$ nm device contacted on both sides:
$R_{sq} = 10 \Omega$, $l_{phys} = 65$ nm, $N_{CON} = 1$, $R_{CON} = 20 \Omega$, $W_{ext} = 200$ nm; $R_S = R_D = (1/W_G) \times 300 \Omega \mu m = 30 \Omega$

a. $W_f = 1 \mu m$; $N_f = 10$, $R_{gate} = 3.82 \Omega$, $R_{gate} + R_S = 33.82 \Omega$
b. $W_f = 2 \mu m$; $N_f = 5$, $R_{gate} = 10.2 \Omega$, $R_{gate} + R_S = 40.2 \Omega$

TABLE 12.3
Typical Parameters for 65 nm n-MOS General-Purposed (GP) and Low-Power (LP) Devices

Parameter	GP	LP
Physical gate length (nm)	45	57
Effective oxide thickness (EOT) (nm)	1.3	1.8
W_f (μm)	1	1
N_{CON}	1	1
Contact on both sides	N	N
R_{CON} (Ω)	40	40
R_{sq} (Ω/sq)	20	20
W_{ext} (nm)	200	200
N_f	1	1
R_{gate} (Ω)	198	160
R_S (Ω)	300	300

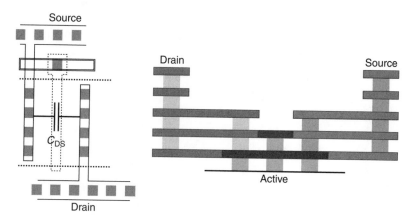

FIGURE 12.13 Tapered metallization on the drain and source minimize parasitic C_{DS} while meeting electromigration rules.

Drain and Source Metallization and Gate Pitch

To meet electromigration rules at 100°C, and to minimize parasitic drain-source capacitance (C_{DS}) caused by the closely spaced, vertically stacked metals on the drain and source fingers, the drain and source metallization should be tapered as shown in Figure 12.13.

CMOS W-BAND LNA DESIGNS

To verify the new noise and impedance matching methodology, and to evaluate CMOS for W-band applications, 77 and 94 GHz LNAs were designed and fabricated in STMicroelectronics (STM's) 90 nm CMOS technology. The schematics of the LNAs are shown in Figure 12.14, and consist of three topologies: a one-stage cascode, a two-stage cascode, and a two-stage transformer-coupled cascode. The designs employ a combination of lumped inductors over substrate, allowing compact layout, and transmission lines over metal, allowing easy routing of power and ground planes.

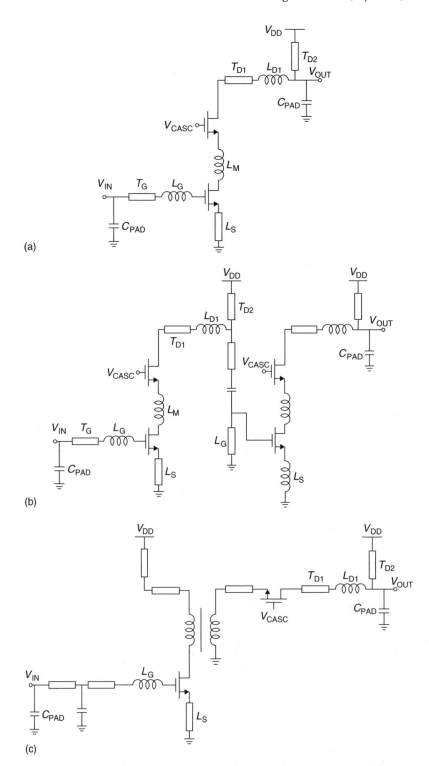

FIGURE 12.14 (a) One-stage, (b) two-stage cascode, and (c) transformer-coupled LNA schematics.

TABLE 12.4
Description of MOSFET Layouts

	W_f (µm)	W_G (µm)	Contacts
A	1	36	One, single-sided
B	2	36	One, double-sided
C	1.5	36	One, single-sided

TABLE 12.5
LNA Performance Summary (Simulations)

	S21 (dB)	NF (dB)
A	3.63	4.78
B	4.36	4.88
C	5.04	4.63

To investigate the effects of the layout issues, discussed in the section "Layout Techniques for Millimeter-Wave CMOS," upon LNA performance, three single-stage cascode LNAs were designed, each with a different device layout but identical gate width. Based on simulation results with extracted RC parasitics, the layout that yielded the best performing single-stage cascode was chosen to design the remaining two LNAs. Note that although the variation of f_{MAX} with finger width is captured in simulation, the variation of f_T seen in measurements of 90 nm MOSFETs is not captured in simulation [19,21]. The three MOSFET layouts are described in Table 12.4. Because only digital MOSFET models were available, the gate resistances of the MOSFET layouts summarized in Table 12.4 were calculated and manually added to postlayout extracted netlists of the MOSFETs. The postlayout simulation results for the three LNAs are summarized in Table 12.5, and indicate that layout C, with single-sided gate contacts and 1.5 µm fingers, is superior. This result offers some supporting evidence to our theory of CMOS layout scaling presented in the section "Layout Techniques for Millimeter-Wave CMOS."

MEASUREMENT RESULTS

Shown in Figure 12.15 are the S-parameter measurements for the fabricated LNAs [12]. The measured S-parameters and DC performance of the LNAs are summarized in Table 12.6. Because of lack of equipment, we cannot measure the LNA noise figure above 65 GHz. However, good agreement between simulations and receiver noise figure measurements was found in a 60 GHz radio receiver fabricated in the same process [11]. Die photo micrographs of the LNAs are shown in Figure 12.16.

CONCLUSION

An algorithmic design methodology for simultaneous noise and input impedance matching in millimeter-wave LNAs has been presented and verified using design examples and measurement results. It directly accounts for the pad capacitance and bond-wire inductance without requiring iteration. Finally, the first CMOS W-band LNAs at 77 and 94 GHz have been experimentally demonstrated.

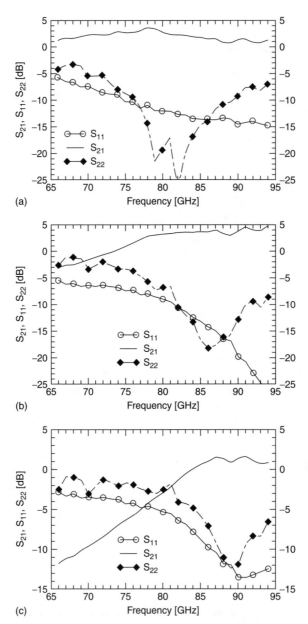

FIGURE 12.15 LNA S-parameter measurements for (a) single-stage cascode, (b) two-stage cascode, and (c) transformer coupled cascade. (From Nicolson, S. T. and Voinigescu, S. P., *IEEE CSICS*, San Antonio, 279–282, 2006. With permission. © IEEE 2006.)

TABLE 12.6
LNA Performance Summary (Measurements)

LNA	S21 (dB)	Frequency (GHz)	Power Supply (V)	Current (mA)
One-stage	3.8	78	1.8	8
Two-stage	4.8	94	1.8	16
XFMR	1.65	92	1.5	23

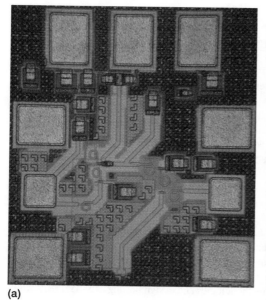
(a)

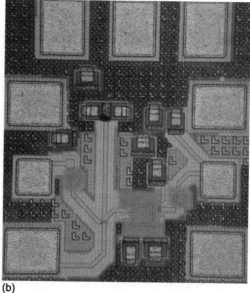
(b)

FIGURE 12.16 Die photos of (a) two-stage cascode and (b) transformer-coupled common-source, common-gate (CS–CG) LNA. (From Nicolson, S. T. and Voinigescu, S. P., *IEEE CSICS*, San Antonio, 279–282, 2006. With permission. © IEEE 2006.)

ACKNOWLEDGMENTS

The authors thank CMC and the University of Toronto for CAD support, CMC for fabrication, and CITO and STMicroelectronics for funding.

REFERENCES

1. Chevalier, P., Gloria, D., Scheer, P., Pruvost, S., Gianesello, F., Pourchon, F., Garcia, P., Vildeuil, J.-C., Chantre, A., Garnier, C., Noblanc, O., Voinigescu, S. P., Dickson, T. O., Laskin, E., Nicolson, S. T., Chalvatzis, T., and Yau, K. H. K., Advanced SiGe BiCMOS and CMOS platforms for optical and millimeter-wave integrated circuits, *IEEE CSICS*, San Antonio, pp. 12–15, November 2006.
2. Floyd, B. A., 60 GHz transceiver circuits in SiGe bipolar technology, *IEEE ISSCC*, vol. 1, pp. 442–538, 2004.
3. Floyd, B. A., V-band and W-band SiGe bipolar low-noise amplifiers and voltage-controlled oscillators, *RFIC Symposium*, pp. 295–298, 2004.
4. Gordon, M., and Voinigescu, S. P., An Inductor-based 52-GHz, 0.18 mm SiGe BiCMOS Cascode LNA with 22 dB Gain, *IEEE ESSCIRC*, Leuven, Belgium, pp. 287–291, September 2004.
5. Dehlink, B., Wohlmuth, H.-D., Aufinger, K., Meister, T. F., Bock, J., and Scholtz, A. L., Low-noise amplifier at 77 GHz in SiGe:C bipolar technology, *Compound Semiconductor Integrated Circuit Symposium*, p. 287, 2005.
6. Gordon, B., Yao, T., and Voinigescu, S. P., 65-GHz receiver in SiGe BiCMOS using monolithic inductors and transformers, *IEEE SiRF*, p. 265, 2006.
7. Reuter, R. and Yin, Y., A 77 GHz (W-band) SiGe LNA with a 6.2 dB noise figure and gain adjustable to 33 dB, *IEEE BCTM*, pp. 1–4, 2006.
8. Doan, C. H., Emami, S., Niknejad, A. M., and Brodersen, R. W., Design of CMOS for 60 GHz applications, *IEEE ISSCC*, vol. 1, pp. 440–538, 2004.
9. Razavi, B., A 60-GHz CMOS receiver front-end, *IEEE ISSCC*, 2005; *IEEE J. Solid-State Circuits*, vol. 41, no. 1, p. 17, 2006.
10. Yao, T., Gordon., M., Yau, K., Yang, M. T., and Voinigescu, S. P., 60-GHz PA and LNA in 90-nm RF-CMOS, *IEEE RFIC Symposium*, pp. 147–150, June 2006.

11. Alldred, D., Cousins, B., and Voinigeseu, S. P., A 1.2 V, 60 GHz radio receiver with on-chip transformers and inductors in 90 nm CMOS, *IEEE CSICS*, San Antonio, pp. 51–54, November 2006.
12. Nicolson, S. T. and Voinigescu, S. P., Methodology for simultaneous noise and impedance matching in W-band LNAs, *IEEE CSICS*, San Antonio, pp. 279–282, November 2006.
13. Gonzales, G., *Microwave Transistor Amplifiers*, Prentice Hall, New Jersey, 1997.
14. Voinigescu, S. P. and Maliepaard, M. C., U.S. Patent No: 5789799, High frequency noise and impedance matched integrated circuits.
15. Voinigescu, S. P., Maliepaard, M. C., Showell, J. L., Babcock, G., Marchesan, D., Schroter, M., Schvan, P., and Harame, D. L., A scalable high-frequency noise model for bipolar transistors with application to optimal transistor sizing for low-noise amplifier design, *IEEE BCTM*, vol. 32, p. 1430, 1996.
16. Shaeffer, D. K. and Lee, T. H., A 1.5-V, 1.5-GHz CMOS low noise amplifier, *IEEE J. Solid-State Circuts*, vol. 32, pp. 745–759, May 1997.
17. Nguyen, T. K., Kim, C.-H, Ihm, G.-J., Yang, M.-S., and Lee, S.-G., CMOS low-noise amplifier design and optimization techniques, *IEEE Trans. Microwave Theory Tech.*, vol. 52, no. 5, pp. 1433–1442, May 2004.
18. Voinigescu, S. P., Tarasewicz, S. W., MacElwee, T., and Ilowski, J., An assessment of the state-of-the-art 0.5 μm bulk CMOS technology for RF applications, *IEDM Tech.*, p. 721, 1995.
19. Dickson, T. O., Yau, K. H. K., Chalvatzis, T., Mangan, A., Beerkens, R., Westergaard, P., Tazlauanu, M., Yang, M. T., and Voinigescu, S. P., The invariance of characteristic current densities in nanoscale MOSFETs and its impact on algorithmic design methodologies and design porting of Si(Ge) (Bi)CMOS high-speed building blocks, *IEEE J. Solid-State Circuits*, vol. 41, pp. 1830–1845, 2006.
20. Lee, T. H., *The Design of CMOS Radio-Frequency Integrated Circuits*, 2nd ed., Cambridge University Press, Cambridge, 2004.
21. Voinigescu, S. P., Dickson, T. O., Gordon, M., Lee, C., Yao, T., Mangan, A., Tang, K., and Yau, K., RF and millimeter-wave IC design in the nano-(Bi)CMOS era, in *Si-based Semiconductor Components for RF Integrated Circuits*, pp. 33–62, Transworld Research Network, 2006.
22. Kim, W. S., Li, X., and Ismail, M., A 2.4 GHz CMOS low noise amplifier using an inter-stage matching inductor, *Midwest Symposium on Circuits and Systems*, vol. 2, pp. 1040–1043, 1999.
23. Suzuki, T., Nakasha, Y., Takahashi, T., Makiyama, K., Imanishi, K., Hirose, T., Watanabe, Y., A 90 Gb/s 2:1 multiplexer IC in InP-based HEMT technology, *IEEE Int. Solid-State Circuits Conf. Tech. Dig.*, vol. 1, pp. 192–193, February 2002.
24. Zhang, C., Huang, D., and Lou, D., Optimization of cascode CMOS low noise amplifier using inter-stage matching network, *IEEE Elec. Dev. SS Cir.*, pp. 465–468, 2003.
25. Dickson, T. O., PhD Thesis, ECE Department, University of Toronto, 2006.
26. Voinigescu, S. P., Nicolson, S. T., Khanpour, M., Tang, K. K. W., Yau, K. H. K., Seyedfathi, N., Timonov, A., Nachman, A., Eleftheriades, G., Schvan, P., and Yang, M. T., CMOS SOCs at 100 GHz: System architectures, device characterization, and IC design examples. *IEEE ISCAS*, May 2007.

13 Power Amplifier Principles and Modern Design Techniques

Vladimir Prodanov and Mihai Banu

CONTENTS

Introduction...350
Review of Prerequisite Knowledge...350
    Relative Signal Bandwidth for Most Modern PAs...350
    What Is a Power Amplifier?...351
    PA Efficiency...352
    Matching for Maximum Output Power...353
    The Meaning of Linear PA..353
The Classical Approach to PA Design...354
    Types of PAs and the Concept of Conduction Angle..354
    Class A, B, and C Operations...354
    Class AB Operation...358
    Switching PAs..359
A Unified General Approach to PA Analysis and Design...362
    The Mathematics of Efficient DC-to-RF Conversion..362
    Zero Harmonic Power..363
    Fundamental-to-DC Ratios..363
    Pairing the Voltage and Current Signals Appropriately.......................................365
    Finite Bandwidth Signals Internal to PA..365
    Efficiency in the Presence of Finite Bandwidth...366
PA Techniques for Power Back-Off Applications..367
    Reasons for Back-Off Requirements and Efficiency Penalties.............................367
    PA Subranging...368
    Envelope Tracking...369
    Envelope Following...369
    Envelope Elimination and Reconstruction...369
    The Out-Phasing PA..370
    The Doherty PA...371
Additional PA Considerations...372
    Power Control..372
    Linearity..372
Current PA Technology and Recent Developments..372
    Wireless System PAs in the Real World..372
    PAs for AMPS, GSM, and GPRS...373
    PAs for EDGE...374

PAs for CDMA and WCDMA 374
PAs for IEEE 802.11a/b/g 375
Class AB g_m Ratio Biasing 376
A Historical Perspective and Conclusions 377
References 379

INTRODUCTION

Enabled by Lee de Forest's invention of the vacuum tube triode in 1906, power amplification of electrical signals has played a key function in electronic systems ever since. Mundane devices we take for granted such as the telephone, the radio, or the television would not exist without this capability. Given such a wide application space, it is not surprising that early on electrical engineers have worked out the details of designing good power amplifiers (PAs), first with vacuum tubes, and then with discrete transistors [1]. They did such a fine job that by the second part of the twentieth century, the art of designing PAs became a mature electrical engineering (EE) specialty, which seemed to have little room left for breakthroughs or major innovations. However, the late-century market explosion of mobile digital communication systems and devices, such as cellular phones and wireless local area networks (LANs), and the massive introduction of integrated circuit (IC) technology in everyday life have changed the electronic landscape dramatically, opening new challenges and opportunities for PAs.

In this chapter, the issues and appropriate techniques for modern PAs are discussed, focusing on IC implementations for wireless communication systems. To familiarize the reader with the general PA design approach, which is rather different from the regular analog circuit approach, a few important points are clarified, as a prerequisite for the following material. Then, the classical theory of PA design in the case of constant magnitude signals is reviewed and the trade-offs for different classes of transistor operation are pointed out. The important class AB case is discussed in more details. Next, the PA design problem from a unified, general point of view based on the internal PA signal harmonic content is revisited. This will give the reader a further insight into the PA design problem and high-level solution possibilities. The following section concerns the important topic of efficiency in the presence of back-off and briefly mentions other important design considerations. Finally, recent PA results are reviewed and conclusions drawn.

REVIEW OF PREREQUISITE KNOWLEDGE

RELATIVE SIGNAL BANDWIDTH FOR MOST MODERN PAS

The main motivation for the renewed interest in PA technology comes from the technical challenges and the economics of modern digital communication systems. The very high production volumes of consumer wireless mobile devices have created a large market for high-quality, low-cost PAs operating in the medium output power range (0–30 dBm). The allocated radio frequency (RF) bands for such typical applications are shown in Table 13.1. A simple calculation of the relative bandwidth compared to the average RF frequency for each system clearly shows that the signals at the RF front-end are narrow band-pass signals on the absolute frequency scale. This fact is not in conflict with the usual categorization of some of these systems as wideband because the latter refers to the baseband signal bandwidth and not to the RF relative bandwidth. More precisely, wideband signals carry a substantially larger amount of information than traditional voice-band signals, but when placed at a high RF carrier frequency, they become relatively narrow, as shown in Figure 13.1.

The relevance of the previous discussion is the realization that on a relatively short time span, that is, over a small number of carrier frequency cycles, the PA signals are practically sinusoidal. At the RF timescale, the magnitude and phase of this sinusoidal signal slowly change only over many carrier cycles. This justifies the common practice in the PA literature to analyze the circuit under

TABLE 13.1
Frequency Bands and Available Bandwidths for Common Wireless Systems

	Licensed Bands				
	US Cellular	R-GSM	DCS	PCS	IMT2000
Uplink (MHz)	824–849	876–915	1710–1785	1850–1910	1920–1980
Downlink (MHz)	869–894	921–960	1805–1880	1930–1990	2110–2170
Total BW (MHz)	25	39	75	60	60
Relative BW (%)	~3.0	~4.4	~4.3	~3.2	~3.1

	Unlicensed Bands		
	ISM-2.4 2400–2483.5	UNII-5.2 5150–5350	UNII-5.8 5725–5825
Total BW (MHz)	83.5	200	100
Relative BW (%)	~3.4	~3.8	~1.7

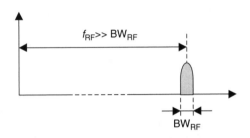

FIGURE 13.1 A typical frequency diagram illustrating that even wideband wireless systems (e.g., wideband code division multiple access [WCDMA]) have small signal bandwidths compared to the carrier frequency f_{RF}.

sinusoidal signal conditions. This adequately represents the PA behavior over short time durations, which is a necessary but not a sufficient criterion for a valid design. Later, in the section "PA Techniques for Power Back-Off Applications," the PA performance over long time spans is discussed in detail, but until then, assuming sinusoidal signals for the PA input and output will be sufficient to explain many important PA properties.

WHAT IS A POWER AMPLIFIER?

Despite the deep-rooted terminology, PAs do not amplify power! Power is energy per unit of time, and as the first law of thermodynamics states, energy cannot be created. Then, what are PAs? And why are they given this name?

A defining property of a PA is that its output signal power delivered to a load is larger than the input signal power it absorbs from a driver. In this respect and outside any energy balance considerations, the PA produces the effect of a nonphysical power amplification device, hence the name. The way the PA accomplishes this effect is by converting the DC power supplied through the DC biasing lines into output signal power. Therefore, a PA is an energy conversion circuit very much like a DC-to-DC converter or an RF oscillator, which converts DC power into constant wave (CW) power. However, unlike DC-to-DC converters or oscillators, an ideal PA converts the DC power into output signal power under the linear control of an RF input. A wireless system PA is simply a DC-to-modulated-RF converter.

The simple observation regarding power conversion in PAs is crucial to understanding the design and operation of this type of circuits, as will be explained later. Here, it suffices to notice

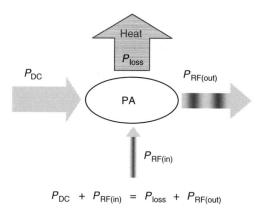

$$P_{DC} + P_{RF(in)} = P_{loss} + P_{RF(out)}$$

FIGURE 13.2 Power flow and balance diagram in a typical PA.

that the very PA concept implies a nonlinear operation since linear networks cannot shift power from one frequency to another.

As described so far, the PA concept is still nondistinguishable from a regular voltage or current amplifier since the latter may (or may not) generate a power-amplified output with respect to its input. What sets the PA apart is the matter of power conversion efficiency. Although the design of a regular voltage or current amplifier is not concerned with efficiency, this performance aspect is paramount in PA designs. In addition, very often the PAs are required to deliver much higher levels of power into the load than regular amplifiers do and may need the capability for power control.

Figure 13.2 illustrates the PA functionality in terms of a power flow diagram. The input power at DC is shown on the horizontal axis to emphasize the key role it plays in this circuit. The very purpose of the PA is to transfer most of this power to the modulated-RF output. The portion that is not transferred is lost through heat. The output modulation information is provided through a low-power RF input in a similar way as with regular analog amplifiers.

Related to the artificial power amplifier terminology are the concepts of power gains. Several output-power-to-input-power ratios are commonly defined under various operating and power-accounting conditions [2]. The PA power gains lack any deep physical meaning but are useful in practice for the purpose of specifying the driving requirements of the circuit in relation to matching and stability conditions.

PA EFFICIENCY

Figure 13.3 shows the simplest classical nonswitched single-transistor PA configuration. DC bias is provided through a large inductor (choke) and the PA load is connected via an ideally lossless matching network. Two most important figures of merit of any PA are the following power efficiency ratios using the notation from Figure 13.2:

$$PE = \frac{P_{out}}{P_{DC}} \tag{13.1a}$$

$$PAE = \frac{P_{out} - P_{in}}{P_{DC}} \tag{13.1b}$$

PE is the power conversion efficiency reflecting the percentage of the DC power drawn from the power supply, which has been converted into output signal power. This figure of merit is also called drain/collector efficiency. Power-added efficiency (PAE) is calculated by subtracting the input power from

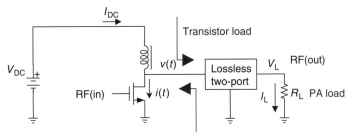

FIGURE 13.3   A classical single-transistor PA, often called linear PA or current PA.

the output power to include the effect of the PA driver in the efficiency metric. Obviously, for large power gains, PAE approaches power efficiency (PE).

The various power quantities can be calculated in the circuit from Figure 13.3 as follows:

$$P_{DC} = V_{DC}I_{DC} \qquad (13.2a)$$

$$P_{out} = \frac{1}{2}V_L I_L \qquad (13.2b)$$

$$P_{loss} = \frac{1}{T}\int_0^T i(t)v(t)\,dt \qquad (13.2c)$$

These relations can be used in Equations 13.1a and 13.1b to calculate the PA efficiencies.

MATCHING FOR MAXIMUM OUTPUT POWER

On the basis of linear system theory hastily applied to the circuit shown in Figure 13.3, one would tend to believe that conjugate matching between the transistor output impedance and the transistor-load impedance ("seen" into the matching network input port) would transfer the maximum possible power to the PA load. This is not true because the transistor nonlinear behavior limits the voltage swing at the drain, shifting the maximum-power conditions far from the theoretical linear case. Laboratory experiments and theoretical investigations [2] show that constant-power closed curves exist on the transistor-load-impedance plane, usually shown as a Smith chart. These oval curves nest within each other like the classical constant-gain circles shrinking to a point of maximum power delivery under strong nonlinear operating conditions. The tuning of the transistor-load impedance performed with special equipment to identify the maximum-power case for various operating conditions is called load pulling. RF PA designers regularly use load-pulling laboratory data to guide their work since modeling is rarely accurate enough.

THE MEANING OF LINEAR PA

It was mentioned earlier that the PA is a nonlinear circuit by necessity. Nevertheless, though power conversion is a nonlinear process, it is possible to design an approximately linear modulation transfer characteristic from the RF input to the PA output. This is the second important design criterion in addition to getting high efficiency.

The usual implications of the previous requirements are illustrated in Figure 13.4. The RF input and the PA output are clean band-pass signals carrying the same modulation information. The internal PA voltage is a rather dirty wideband signal with rich and large harmonic content. It will

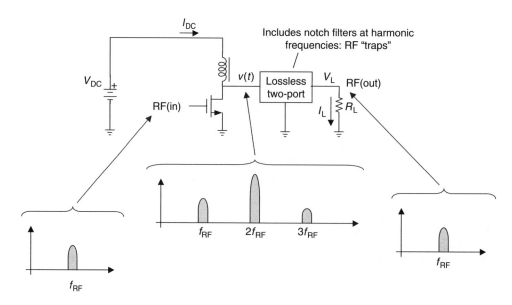

FIGURE 13.4 A highly nonlinear PA with linear RF-in/RF-out characteristic.

become clear later that it is precisely this internal harmonic content that is responsible for obtaining good efficiency. This indicates that the design strategy for a PA is quite different from that for a regular linear amplifier. In the latter, since there are no efficiency concerns, it is not necessary and undesirable to introduce high internal nonlinear behavior, which would have to be transparent to the output. Efficient PAs must be highly nonlinear internally and still be input/output linear in terms of the RF modulation transfer.

THE CLASSICAL APPROACH TO PA DESIGN

TYPES OF PAS AND THE CONCEPT OF CONDUCTION ANGLE

There are two main branches in the PA family tree shown in Figure 13.5. If the main PA transistor operates as a transconductance element converting the RF input signal into a current, the circuit is called a linear or current PA. If the main PA transistor is just a switch, the circuit is called a switching PA. This PA family branch will be discussed in the subsection "Switching PAs."

Current PAs, whose general structure is similar to that shown in Figure 13.3, are further divided into classes of operation on the basis of conduction angle [2]. Figure 13.6 illustrates this concept for the case of an ideal transistor with piecewise linear I/V characteristics. The conduction angle is a measure of the drain current generation process for a given biasing point and a given RF input signal magnitude. If the biasing point and the RF input signal magnitude are such that all input signal excursion is linearly converted into a drain current, the PA operates in class A with 2π conduction angle. Class B operation is defined for π conduction angle when exactly only one side of the RF input sinusoidal signal is converted into current. Lowering the conduction angle bellow π defines class C and increasing it toward 2π defines class AB, not shown in Figure 13.6. Next, the merits of these possibilities are discussed.

CLASS A, B, AND C OPERATIONS

The maximum drain voltage and current waveforms for classes A, B, and C are shown in Figure 13.7. Notice that in all cases, the drain voltage is the same and consists of a full sinusoidal (see the subsection "Relative Signal Bandwidth for Most Modern PAs"). The difference comes from

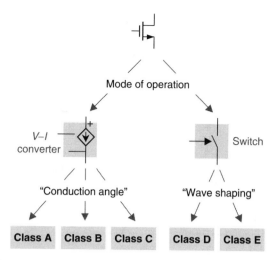

FIGURE 13.5 PA family tree.

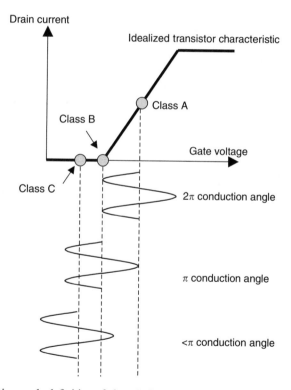

FIGURE 13.6 Conduction angle definition of class A, B, and C operations.

the transistor current, which varies from a full sinusoidal in class A to portions of a sinusoidal for classes B and C. This determines major variations in PA efficiency calculated with Equations 13.1a through 13.2c and in other important performance parameters. Figures 13.8 through 13.10 illustrate these effects.

Theoretically, though class A is limited to 50% maximum efficiency, class B attains 78.5% efficiency and class C tends toward 100% efficiency. A crucial aspect is the loss of efficiency as the PA

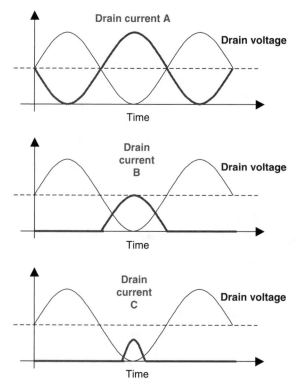

FIGURE 13.7   Drain voltage and current waveforms for class A, B, and C operations.

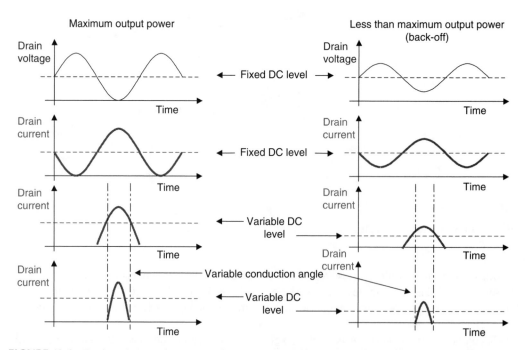

FIGURE 13.8   Drain voltage and current under power back-off conditions for class A, B, and C operations.

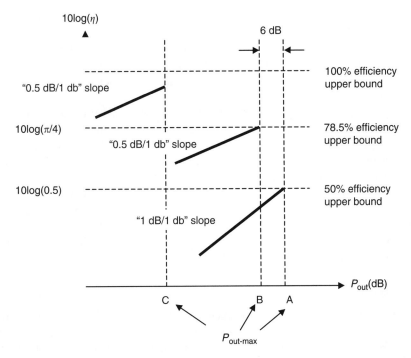

FIGURE 13.9 Power efficiency in decibels under power back-off conditions for class A, B, and C operations; same maximum input RF power is assumed.

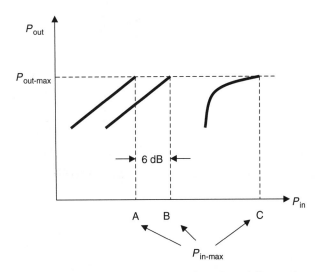

FIGURE 13.10 Output power versus RF input power for class A, B, and C operations.

output power is lowered or backed off from the peak value. Ideally, there should not be any reduction in efficiency but this is not the case. Figure 13.9 shows that class C is best in this respect followed by class B. This can be explained with the help of Figure 13.8. For class A operation, the DC value of the current signal does not change with the output power level, thus wasting efficiency in back-off, which drops a full decibel for every decibel reduction in output power. Classes B and C feature a fundamentally different and valuable behavior: the DC current components decrease with the output power. As a result, class B has only 0.5 dB loss in efficiency for every 1 dB reduction in output power, and so does class C.

The efficiency benefits of classes B and C compared to class A are not without penalties. An important property that only class A has is that its linearity performance is monotonic, that is, in back-off the linearity always improves [2]. This is not true in any other operation classes, including class AB, which will be discussed later.

Class C pays a particularly high price for excellent efficiency. Because of low conduction angle and low device utilization, the output power level is much reduced compared to that for classes A or B for the same input drive, as shown in Figures 13.9 and 13.10. In addition, the input/output signal characteristic changes rapidly with the output power level creating severe nonlinear effects in the output signal. For these reasons, class C operation is rarely used.

On the contrary, an ideal class B seems to be quite a good compromise between increased efficiency and a small 6 dB gain reduction compared to class A, with no loss in linearity. A simple analysis would convince the reader of this theoretical fact. Unfortunately, the ideal class B case is a poor approximation in practice since real transistors have smooth turn-on characteristics. A more appropriate model is class AB operation.

CLASS AB OPERATION

Class AB is the workhorse of linear high-efficiency RF PA applications; yet the reasons for its success cannot be explained from the idealized model shown in Figure 13.11a. According to this model, class AB is very similar to class C, as shown in Figure 13.12: good back-off efficiency due

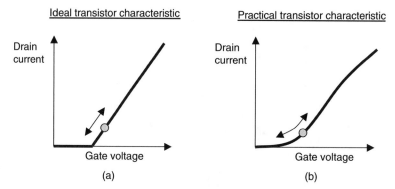

FIGURE 13.11 Class AB operation on (a) ideal transistor *I/V* characteristics and (b) real transistor *I/V* characteristics.

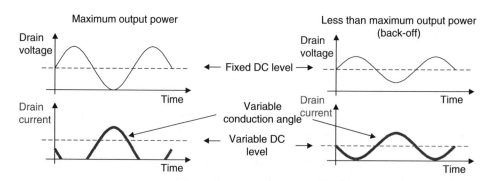

FIGURE 13.12 Drain voltage and current under peak power and back-off conditions for ideal class AB PA.

to variable current DC component but unacceptable nonlinear behavior in the output due to signal-dependent conduction angle.

The main reason class AB works well in practice is the fact that real transistor characteristics are smooth, as illustrated in Figure 13.11b. As a result, the intermodulation components of the drain current in the real transistor are quite different from those generated by the ideal curve shown in Figure 13.11a [3]. More important, these components vary with the biasing point in such a way that an optimum biasing condition exists in terms of odd-order intermodulation distortion (IMD). Care must be taken here to stress that the optimum biasing is very sensitive and difficult to find and maintain over fabrication process and temperature variations. In addition, unlike class A, the linearity performance of class AB is not monotonic and the odd IMD may get worse in back-off [2,3].

A circumstantial proof for the existence of the optimum biasing can be given with the help of Figures 13.13 and 13.14. Figure 13.13 shows the decomposition of a typical LDMOS RF-power FET I/V characteristics into even and odd components at the operating point. The g_m characteristic is decomposed as well and the focus is on the even g_m component, which is directly responsible for setting the IMD values. Figure 13.14 clearly shows how the even g_m component changes shape quite dramatically as a function of the biasing point. By inspection, notice that biasing at half the peak g_m value yields the minimum error ripple. In practice, the situation is complicated by many other practical factors such as transistor-parasitic capacitors, dynamic effects, etc., but the class AB nonlinear behavior remains qualitatively as described.

Starting from peak power level downwards, the back-off efficiency in class AB is practically identical to that of ideal class B, that is, the PE drops 0.5 dB for each decibel of output power reduction. Eventually, however, as the input signal gets small enough, the amplifier approaches a class A behavior due to the smooth transistor I/V characteristic (Figure 13.11b). As a result, the efficiency degradation gradually shifts to "1 dB per dB" roll-off. This is detrimental in applications with large back-off requirements such as code division multiple access (CDMA) PAs, which will be discussed later. A common method for mitigating this effect is to decrease the transistor bias gate voltage dynamically for low input signals and thus maintain class AB behavior. Naturally, this must be done without introducing PA linearity problems.

SWITCHING PAs

The natural way in which a current PA becomes a switching PA is by overdriving the circuit shown in Figure 13.3 to the point of operating the transistor as a switch [2]. Figure 13.15a shows this possibility and Figure 13.15b expands this concept to a two-switch/transistor configuration by eliminating the biasing inductor. Now, the true nature of the PA as an energy converter comes in full view. The amplitude modulation can no longer be transmitted through the input port. Phase modulation is still transferred into the PA through the variable zero crossings defining the switching instances. Therefore, a first important observation about switching PAs is that they can process correctly only input signals that are phase/frequency modulated and have no amplitude modulation. However, it is still possible to pass amplitude modulation information into the PA through the power supply voltage since the output power is proportional to its value.

The main motivation for using a switching PA is the theoretical possibility of obtaining outstanding efficiency. To this end, the traditional approach is to satisfy two conditions: (a) arranging the circuit such that the transistor voltage and current overlap as little as possible, thus minimizing the loss through heat and (b) designing the lossless two-port networks shown in Figure 13.15 such that only the fundamental frequencies are allowed to pass into the output, thus avoiding harmonic power loss. As the two conditions must be met simultaneously and all signals inside the PA are strongly interrelated, a high degree of design skill and knowledge is necessary to obtain a valid solution. Three possibilities, which have been proposed, are known as classes D, E, and F PAs [1,2].

Figure 13.16 shows typical voltage and current waveforms for traditional switching PAs [1]. The class D PA uses the two-transistor architecture and relies on very fast switching of the lossless two-port

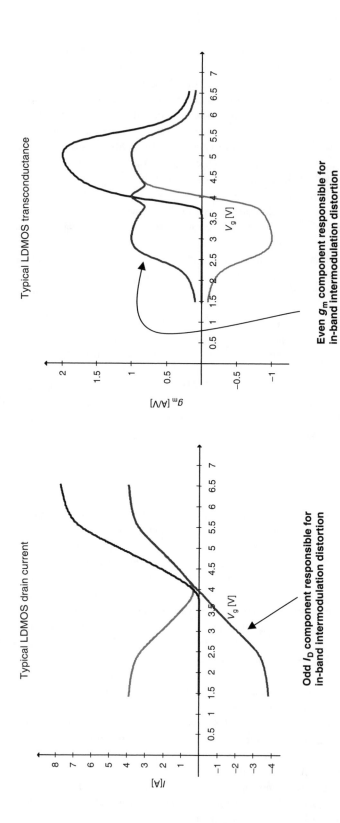

FIGURE 13.13 Decomposition of the *I/V* and g_m characteristics of a typical LDMOS RF power transistor into odd and even parts. (From Banu, M., Prodanov, V., and Smith, K., *Asia Pacific Microwave Conference*, 2004. With permission. © IEEE 2004.)

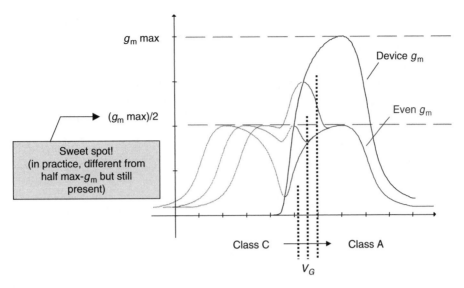

FIGURE 13.14 Simple demonstration of sweet spot biasing in real class AB operation. (From Banu, M., Prodanov, V., and Smith, K., *Asia Pacific Microwave Conference*, 2004. With permission. © IEEE 2004.)

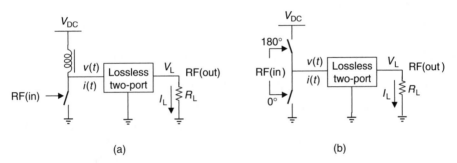

FIGURE 13.15 Switching PA architectures: (a) single-transistor architecture and (b) double-transistor architecture.

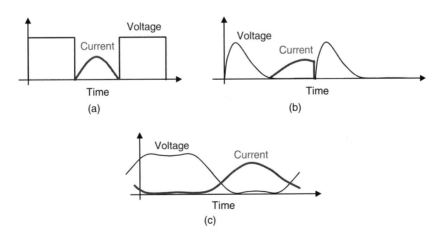

FIGURE 13.16 Typical voltage and current signals in (a) class D PA, (b) class E PA, and (c) class F PA.

input between the power supply line and ground. Figure 13.16a shows the resulting square drain voltage waveforms. For ideal switches, there is no power loss through heat, and assuming the loss-less two-port is a band-pass filter rejecting all harmonics, 100% efficiency is obtained theoretically. In practice, the class D PA technology has been applied successfully at audio frequencies where switching can be realized fast enough compared to the signal bandwidth. Any reasonable application of this technique at RF has not been demonstrated yet and is plagued by unrealistically high demands on the transistor switching speeds. In addition, the losses due to drain-parasitic capacitance charging/discharging are difficult to avoid.

An RF switching PA approach that has been demonstrated in practice in the gigahertz range with better than 70% efficiency is based on class E operation [2–5]. This single-transistor PA switches the current ideally only when the voltage and its derivative are zero, thus avoiding heat losses. This is a promising approach, but it produces inherently large voltage swings, requiring transistors capable of handling such conditions [2,6]. Nevertheless, the voltage waveform shown in Figure 13.16b is substantially less abrupt than in the case of class D, hence the suitability of class E for RF applications.

The class F operation [7–10] employs a single-transistor architecture and voltage shaping improving the efficiency and the transistor utilization. Starting with class AB transistor biasing, the lossless two-port is designed to greatly enhance the third voltage harmonic to obtain an effective squaring of the voltage signal, as shown in Figure 13.16c. This increases the efficiency beyond class AB operation while maintaining the voltage swing within reasonable levels. Theoretically, fifth, seventh, and higher odd harmonics could be also enhanced for the benefit of even higher efficiency. Unfortunately, the design of the lossless two-port is quite challenging, and maintaining class F operation in back-off is problematic.

A UNIFIED GENERAL APPROACH TO PA ANALYSIS AND DESIGN

THE MATHEMATICS OF EFFICIENT DC-TO-RF CONVERSION

An empirical observation clearly stands out from the discussion in the previous section: it seems that the only way to boost the efficiency from one PA scheme to another is by making its internal nonlinear behavior more pronounced. Mathematically this is indeed the case shown in Figure 13.3 by calculating the power flows at various frequencies in the PA and interpreting the results. The lossless two-port network is assumed AC coupled. Therefore, by construction, the DC voltage and current of the power supply are identical to those at the drain of the transistor.

In steady state, under a sinusoidal excitation of angular frequency ω_{RF} applied on the transistor gate, the drain voltage and current are periodic functions represented by the following Fourier series:

$$v(t) = V_{DC} + V_1 \cos(\omega_{RF}t + \varphi_{V_1}) + \sum_{k=2}^{\infty} V_k \cos(k\omega_{RF}t + \varphi_{V_k}) \tag{13.3a}$$

$$i(t) = I_{DC} + I_1 \cos(\omega_{RF}t + \varphi_{I_1}) + \sum_{k=2}^{\infty} I_k \cos(k\omega_{RF}t + \varphi_{I_k}) \tag{13.3b}$$

where V_k and I_k are the amplitudes and ϕ_{V_k} and ϕ_{I_k} the phases of respective harmonics. The total loss at the drain is calculated by multiplying Equations 13.3a and 13.3b and integrating over a period according to Equation 13.2c. Since all orthogonal products (i.e., voltage harmonic different from current harmonic) integrate to zero, we have

$$P_{loss} = V_{DC}I_{DC} + \frac{1}{2}V_1I_1 \cos(\varphi_{V_1} - \varphi_{I_1}) + \frac{1}{2}\sum_{k=2}^{\infty} V_kI_k \cos(\varphi_{V_k} - \varphi_{I_k}) \tag{13.4}$$

Equation 13.4 gives important insights on how the PA converts energy from DC to RF. The total loss P_{loss} must be a positive quantity since the transistor considered as operating with full voltages and currents is a passive device (transistors do not generate power), unlike its customary model used for small-signal analysis. Furthermore, the transistor physics forces the DC drain current as defined in Figure 13.3 to be always positive, which makes the first term in the right-hand side of Equation 13.4 positive. This term is clearly identified as the power delivered into the PA by the DC power supply. The energy conservation law tells us that P_{loss} must be smaller than the DC power flowing into the PA; therefore, the second and third terms in the right-hand side of Equation 13.4 must add to a negative number. The right-hand side of Equation 13.4 can be interpreted as the superposition of the DC power flowing into the transistor from the DC power supply and a portion of it flowing out of the transistor at RF fundamental and harmonics. Since the biasing choke blocks the RF fundamental and harmonics, the only place the outgoing power can go is the PA load resistor through the lossless two-port. Thus, the PA accomplishes energy conversion: it extracts power at DC from the power supply and delivers a portion of it to the load at RF fundamental and harmonics. Next, the possibilities are analyzed to make this process power efficient, i.e., with as small P_{loss} as possible.

The second term in the right-hand side of Equation 13.4 is essential to the very function of the PA since it represents the fundamental RF power to be delivered to the PA load. This term should be negative with magnitude as large as possible. A necessary condition for this objective is to create fundamental voltage and current signals swinging in opposite directions (180° phase shift) to make the cosine factor equal to -1. This is automatically accomplished if the transistor pushes current into a real impedance. Therefore, the two-port lossless network terminated by the PA load resistor must be designed to have a real input impedance at the fundamental frequency. An equivalent way to state this is that the two-port lossless network terminated in the PA load resistor is a filter with a pass-band at the fundamental RF frequency. Naturally, the transistor-parasitic capacitances must be included in the network.

The generation of harmonic power represented by the last summing term in the right-hand side of Equation 13.4 must be eliminated for the following reasons. As discussed above in this section, any negative components in the sum would represent respective harmonic power flowing out of the transistor only to be dissipated in the PA load. This is not allowed by the PA linearity requirements (see Figure 13.4). On the contrary, any positive components in the sum would be dissipated in the transistor to the detriment of power efficiency. The only alternative left is to make the harmonic power summation zero.

ZERO HARMONIC POWER

Ignoring the theoretical but exotic possibility of shifting power between harmonics for a zero net game, four ways of making the last summation term in Equation 13.4 null are illustrated in Figure 13.17. The two signals shown for each case in the frequency domain can be voltage or current signals interchangeably. They are members of a set of four generic signals, each containing DC and fundamental terms. In addition, the first generic signal contains no harmonics, the second generic signal contains only odd harmonics, the third generic signal contains only even harmonics, and the fourth generic signal contains all harmonics. These generic signals will be called, no-harmonic, odd-harmonic, even-harmonic, and all-harmonic, respectively. The four methods shown in Figure 13.17 combine pairs of generic signals such that the products in the summation of Equation 13.4 are only of orthogonal signals, integrating to zero over the RF input signal period.

FUNDAMENTAL-TO-DC RATIOS

Assuming zero harmonic power as per the methods discussed in the previous subsection, the power efficiency from Equation 13.4 is calculated with the last term eliminated:

$$PE = \frac{1}{2} \frac{V_1}{V_{DC}} \frac{I_1}{I_{DC}} \tag{13.5}$$

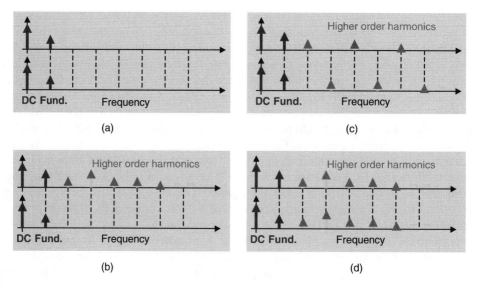

FIGURE 13.17   Pairs of internal PA signals with orthogonal harmonics: (a) two no-harmonic signals, (b) no-harmonic and all-harmonic signals, (c) odd-harmonic and even-harmonic signals, and (d) two quadrature all-harmonic signals.

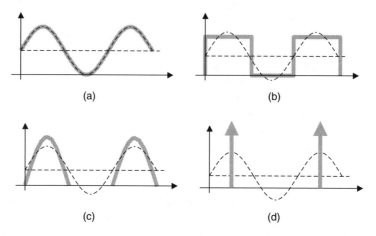

FIGURE 13.18   Signals with optimum FDC ratios: (a) no-harmonic shifted sinusoidal with FDC = 1, (b) odd-harmonic square wave with FDC = $4/\pi$, (c) even-harmonic half-wave rectified sinusoidal with FDC = $\pi/2$, and (d) all harmonic impulse with FDC = 2.

Notice that the overall PA efficiency is the product of two fundamental-component-to-DC-component (FDC) signal ratios. This gives a very important clue of what needs to be done for maximum efficiency, namely, maximizing the FDC ratios for the voltage and the current signals inside the PA. This is explicit evidence that the PA efficiency is directly linked to its internal signal harmonics, whose presence in proper amount and phasing can increase the FDC ratios. Next, this possibility under the condition of zero harmonic power is analyzed.

The transistor drain voltage and current signals as defined in Figure 13.3 must be positive on the basis of proper operation of the device. The question is, which positive functions have the maximum FDC ratio and are of the form of the generic functions discussed in the subsection "Zero Harmonic Power?" The answer is given in Figure 13.18. The positive no-harmonic function is unique and

TABLE 13.2
Efficiencies and Resulting Classes

(a) Maximum theoretical PA efficiency for different waveform pairing.

harmonics	voltage none	voltage odd	voltage even	voltage all
none	50.0 %	63.6 %	78.5 %	100 %
odd	63.6 %			100 %
even	78.5 %	100 %		
all	100 %			100 %

(current)

(b) PA operating classes corresponding to waveform pairing in (a).

harmonics	voltage none	voltage odd	voltage even	voltage all
none	A		"Inverse" B,C,D,F	
odd				
even	B	D/F		
all	C			E

(current)

has an FDC ratio of 1. The odd-harmonic function with maximum FDC ratio of $4/\pi$ is a square wave, the even-harmonic function with a maximum FDC ratio of $\pi/2$ is a half-wave rectified sinusoidal, and the all-harmonic function with a maximum FDC ratio of 2 is an impulse train function. The iterative way in which these functions are constructed in the subsection " Finite Bandwidth Signals Internal to PA" ensures that they are optimum in terms of best FDC ratios for their respective class.

PAIRING THE VOLTAGE AND CURRENT SIGNALS APPROPRIATELY

Equipped with Equation 13.5 and the functions of Figure 13.18, the PA schemes discussed in the section "The Classical Approach to PA Design" can be analyzed from a unified and general point of view. For example, a class A PA uses only nonharmonic internal voltage and current functions. Equation 13.5 gives 50% efficiency, which of course is as calculated before. For ideal class B PA, the internal voltage is a no-harmonic signal, but the internal current is a half-wave rectified sinusoidal, the best even-harmonic signal. The efficiency increases to 78.5% in response to adding current harmonics. If proper odd harmonics are added to the voltage signal, e.g., use a square wave, the best even-harmonic signal, the efficiency reaches 100% and an ideal class D or F PA has been constructed. The same 100% efficiency may be obtained by using a no-harmonic signal for the internal voltage and an impulse train, the best all-harmonic signal for the internal current. This describes an ideal class C PA with infinitely small conduction angle (and infinitely large input signal or infinitely large transistor g_m).

Table 13.2a shows the efficiencies of all possible pairs of best FDC ratio signals, according to the schemes shown in Figure 13.17. Table 13.2b shows the resulting classes of PA operation. Notice that not all possible pairs have known PA configurations. Also, notice that class E operation requires voltage and current harmonics in quadrature to ensure orthogonal conditions.

FINITE BANDWIDTH SIGNALS INTERNAL TO PA

The previous analysis of ideal PAs assumed that internal signals with infinite bandwidths could be used. In reality, of course, this is not the case. For this reason, it is important to determine the effect of limited bandwidths inside the PA on efficiency. To be able to do this, the generic signals discussed

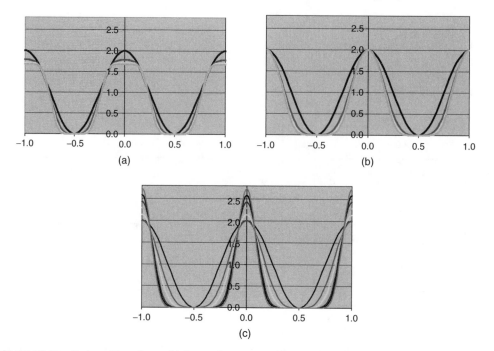

FIGURE 13.19 Series of functions with increasing number of harmonics and FDC ratios: (a) odd-harmonic functions converging to a square wave, (b) even-harmonic functions converging to a half-wave rectified sinusoidal, and (c) all-harmonic functions converging to an impulse.

in the subsection "Zero Harmonic Power" are first constructed not as infinite but rather as limited bandwidth signals, i.e., allowing only a limited number of harmonics [6,8–11].

Figure 13.19a shows graphically a series of odd-harmonic signals with increasingly larger number of harmonics. For each signal, the harmonic content is calculated so as to create zero derivatives at the midpoint in the fundamental cycle up to $(N-1)$th order derivative, where N is the number of harmonics. In this way, it is ensured that the function reaches a minimum at that point and it is as flat as possible. The fundamental component is increased to place the minimum point at zero value. As the number of harmonics increases, these functions resemble more and more a square wave, and in the limit (infinite number of harmonics), they become a square wave.

The same construction can be done for even-harmonic functions; the result is shown in Figure 13.19b. Here, in the limit, the half-wave rectified sinusoidal function is recovered. Finally, the all-harmonic functions shown in Figure 13.19c synthesized in a similar manner converge toward an impulse function. In all three series, the FDC ratio increases with the number of harmonics.

EFFICIENCY IN THE PRESENCE OF FINITE BANDWIDTH

On the basis of the functions from Figure 13.19 and the same signal pairing as in Figure 13.17, PA efficiencies can be calculated for various internal PA bandwidths. Table 13.3a summarizes the results for the pairing case in Figure 13.17c up to the seventh harmonic. The good news is that the efficiency increases rapidly, reaching respectable numbers without an excessive number of harmonics. For example, a class B PA (class AB practically the same) with up to sixth-order harmonics in the current has already 73.1% efficiency. Similarly, Table 13.3b representing the pairing from Figure 13.17b shows that a class C PA with only fourth-order harmonics reaches 80% efficiency. On the contrary, it is also clear that trying to push efficiency to even higher levels would be very challenging for RF PAs due to very high bandwidth demands.

TABLE 13.3
Efficiency Tables

(a) Harmonics in voltage and current waveforms.

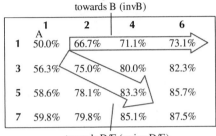

towards B (invB)

	1	**2**	**4**	**6**
1	A 50.0%	66.7%	71.1%	73.1%
3	56.3%	75.0%	80.0%	82.3%
5	58.6%	78.1%	83.3%	85.7%
7	59.8%	79.8%	85.1%	87.5%

towards D/F (or inv D/F)

(b) Harmonics only in voltage or current waveform.

Max. Harmonic	1	2	3	4	5	6	7
Efficiency	50.0%	66.7%	75.0%	80.0%	83.3%	85.7%	87.5%

A $\xrightarrow{\text{toward C (invC)}}$

A final observation is made by comparing the entries in Table 13.3a on a diagonal from top left to bottom right and those of Table 13.3b. The efficiency numbers are identical for identical number of total harmonics irrespective of which signals contain these harmonics. In other words, at this high-level explanation, the efficiency is independent of the actual PA configuration and depends only on the number of harmonics used internally. Given N internal PA harmonics, the following simple relation can be used to estimate efficiency:

$$PE \leq \frac{N}{1+N} \tag{13.6}$$

PA TECHNIQUES FOR POWER BACK-OFF APPLICATIONS

REASONS FOR BACK-OFF REQUIREMENTS AND EFFICIENCY PENALTIES

The efficiency of PAs in back-off operation was considered previously. This aspect is crucial for RF applications using amplitude modulation. In the subsection "Relative Signal Bandwidth for Most Modern PAs," it is mentioned that the PA input RF signal looks sinusoidal for short durations. However, if amplitude modulation is present, the magnitude of this sinusoidal signal varies over long time, as shown in Figure 13.20. A traditional way to describe this magnitude variation is as the ratio between the peak power and the average power of the RF signal, known as peak-to-average ratio (PAR), usually expressed in decibels [1,2]. The typical statistics of real communication signals are such that peak power actually occurs infrequently.

The PAs must be designed and operated to handle the input signal correctly at all times without ever entering compression. The simplest way to accomplish this is by designing the PA for proper operation at expected peak power. Naturally, most of the time, the PA will be underutilized delivering only average power and thus be effectively backed off by the PAR value. The net result is that the PA average efficiency is not as given at peak power value but rather at some effective back-off value, depending on the signal statistics. The larger the signal PAR, the more backed-off the PA will be and the more severe the penalty in average efficiency.

The recent introduction of wideband digital wireless communication systems such as those based on CDMA or 802.11a/g standards has placed to center stage the PA efficiency problem in back-off operation. Nevertheless, this is not a new problem. The commercial amplitude modulation (AM)

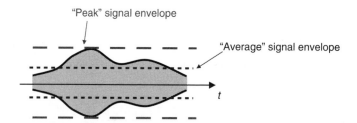

- PAs must operate properly for worst-case signal conditions, i.e., peak power conditions.
- PA must be backed off from rated power by signal PAR (e.g., CDMA-2000: 4–9 dB, W-CDMA 3.5–7 dB, 802.11a/g: 6–17 dB).
- Efficiency suffers under back-off conditions.

FIGURE 13.20 Peak and average power levels in an amplitude-modulated RF signal.

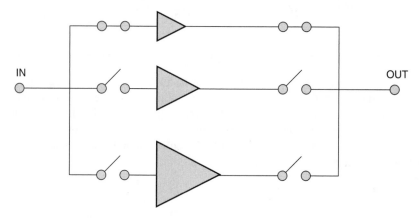

FIGURE 13.21 Conceptual PA subranging architecture.

broadcast industry has encountered and solved this issue for high-power PAs with system-level techniques, albeit using conventional RF technology, which is too expensive and bulky for modern portable devices [12–15]. Recently, there has been a considerable renewed interest in these system techniques [3,16,17] with a focus on trying to apply them to modern low- and medium-power PAs using integrated circuit (IC) technology. The most important system concepts for increased efficiency PAs in back-off is reviewed next.

PA SUBRANGING

A brute-force solution to the back-off efficiency problem is shown conceptually in Figure 13.21. Several PA segments of increasing output power are placed in parallel and switched on and off appropriately by the transmitter system such that the RF output signal is always processed by a PA segment operating close to its peak power and efficiency. This is possible in theory because the transmitter system knows in advance the information to be transmitted and can bring on line the appropriate PA segment at the right time. This strategy increases substantially the average efficiency of the overall PA.

The challenge in implementing the scheme shown in Figure 13.21 comes from the input and output interfacing networks, which must provide "smooth" RF switching without major impedance changes and with low loss. A less demanding application of this architecture is power control, to be discussed in the section "Additional PA Considerations."

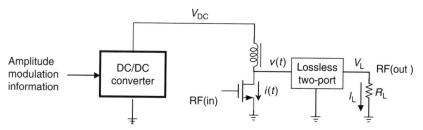

FIGURE 13.22 Envelope tracking and following PA concept.

ENVELOPE TRACKING

A more sophisticated technique is illustrated in Figure 13.22 and is based on the observation that the back-off efficiency of a class AB PA can be boosted by lowering the power supply voltage dynamically when the signal magnitude decreases. The core PA consists of the transistor, the inductor, and the lossless matching network. The DC power supply is an agile DC-to-DC converter capable of delivering the necessary PA current for a discrete set of output voltages under the control of an input terminal. The purpose of this converter is to change the supply voltage dynamically according to the RF amplitude modulation so as to operate the class AB PA at or close to its peak efficiency for all input signal levels between average and peak power. In effect, the transistor drain voltage always swings close to the full power supply voltage, which is dynamically changed. The voltage FDC ratio remains near unity independent of signal magnitude for close to peak efficiency in back-off.

Two conditions must be met for the proper operation of this scheme. First, the agility of the DC-to-DC converter must match or be better than the baseband signal bandwidth, which equals the amplitude modulation bandwidth. Second, the efficiency of the DC-to-DC converter must be good enough to make the overall system more efficient than a classical class AB PA. In the case of power control back-off, the bandwidth condition is relaxed.

ENVELOPE FOLLOWING

The envelope tracking concept requires that the power supply voltage generated by the DC-to-DC converter follows only roughly the signal magnitude for the sole purpose of increasing the average PA efficiency. Theoretically, one can imagine the power supply voltage following exactly the signal envelope, in which case the method is called envelope following [17].

When the circuit uses a current PA, the additional improvement in efficiency envelope following brings is minimal when compared to envelope tracking and does not justify the extra precision requirements for the DC-to-DC converter. Envelope following becomes an attractive option if instead of the current PA we use a switching design, such as a class E PA. In this case, the DC-to-DC converter provides the amplitude modulation information through the power supply line, and the switching PA generates the output power extremely efficiently [18].

ENVELOPE ELIMINATION AND RECONSTRUCTION

A particular version of the envelope following concept, which was historically first proposed by Kahn, is envelope elimination and restoration (EER) [14]. Figure 13.23 illustrates this design, which predates baseband digital signal processors (DSPs). Here, the RF input is first processed by analog blocks, and the amplitude modulation information is separated from the RF signal and converted into a baseband signal. The remaining constant-envelope RF signal drives a switching RF PA with excellent efficiency and the amplitude modulation is reintroduced through the power supply voltage. The latter is driven by an efficient baseband PA. A critical and challenging issue in this technique is the correct synchronization between amplitude and phase.

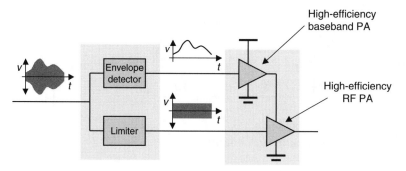

FIGURE 13.23   Envelope elimination and restoration PA concept.

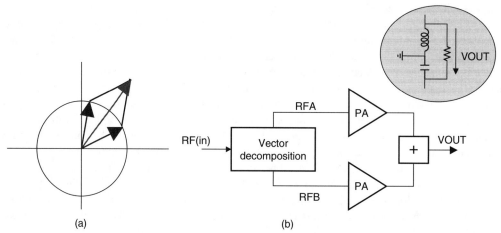

FIGURE 13.24   Out-phasing PA concept.

Kahn's EER scheme is less attractive when digital baseband processing is available. In this case, it makes more sense to generate the amplitude and phase signals directly from the DSP rather than decomposing an analog RF signal, whose phase and amplitude components were originally created by the DSP in the first place. The EER structure without the signal decomposition part is known as a polar transmitter [19].

THE OUT-PHASING PA

This concept is explained in Figure 13.24a where the RF signal is represented as a rotating vector with time-variable magnitude and angular velocity. From simple geometrical considerations, clearly it is possible to decompose this vector into two new rotating vectors with constant magnitudes, as shown in the figure. Therefore, all information contained in a modulated RF signal can be also represented in a pair of constant-envelope signals. This is a convenient representation for efficient power conversion of the two components based on switching PAs.

Figure 13.24b shows the Chireix first implementation of this concept [12], before the availability of DSPs. First, two constant-envelope components are derived from decomposition of the RF signal. Today, this would be done by the DSP [20] like in the polar transmitter case [19]. Then, these components are passed through two efficient switching PAs. Finally, an output power-combining network recreates the original RF vector. The design of the combining network is quite critical and is the potential Achilles' heel of this technique. This is discussed next.

A simple analysis would convince the reader that the process of combining two constant-envelope vectors through a simple vector addition, i.e., by conventional power combining, is fundamentally inefficient if the two signals are not in phase. Any vector components canceling each other dissipate power. For example, in the worst case when the two constant-envelope vectors are in opposite phases, all their power gets dissipated and no power is produced at the output for 0% efficiency. Chireix recognized that traditional power combining is not an acceptable solution and proposed a fully differential load connection as shown in the Figure 13.24 inset. This configuration solves the efficiency problem but introduces a major PA-loading problem. The effective loads seen by each PA are not purely resistive as assumed in the standard design of the PAs but rather contain large reactive components. Even more troublesome is the fact that these reactive components depend on the angle separating the two vectors, which is constantly changing with the modulation. A compensation of these reactive components is possible, as shown in the figure inset, with the addition of a capacitor and an inductor, but this compensation is valid only around a unique separation angle. Chireix made this technique work with substantially better back-off efficiency than the class AB case for the AM broadcasting application.

THE DOHERTY PA

The Doherty concept [13] is shown in Figure 13.25 and in some respects may be viewed as a very ingenious analog version of the PA subranging idea. It contains a main amplifier, which is always on, and a secondary or peaking amplifier, which turns on only when the input signal power exceeds a predetermined threshold, e.g., 6 dB below maximum PA power if the two transistors are identical. The classical implementation uses a class AB main amplifier and a class C peaking amplifier with identical transistors [2,3]. A single inductor is sufficient to bias the drain of both transistors at the power supply voltage.

The key Doherty innovation is combining the two transistor drain currents via a quarter-wave transformer, as shown in Figure 13.25. The quarter-wave transformer, which in practice is a piece of transmission line, converts the input current of one port into a voltage output at the other port. The same action is achieved using a symmetric LC π network. For input power levels, when only the main transistor is on, its drain current is converted linearly into a voltage applied to the load, just as in a regular PA. When the peaking transistor turns on and pushes RF current into the quarter-wave transformer, a differential RF voltage is generated at the drain of the main transistor. Phasing the drain RF current of the peaking transistor correctly, e.g., shifting the input RF signal by a quarter-wavelength before applying it to the peaking transistor gate, has the effect of lowering the RF voltage swing at the main transistor drain. This creates voltage headroom for the main transistor, which now can push higher RF currents before reaching current saturation. The system is adjusted such that the drain RF voltage swing of the main transistor remains constant after the input power increases beyond the triggering level of the peaking transistor. In this way, the main transistor keeps

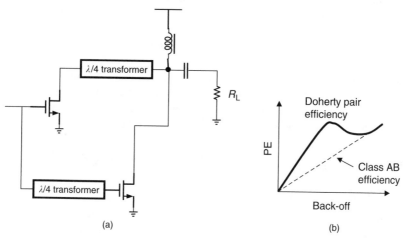

(a)   (b)

FIGURE 13.25   Doherty PA concept.

pumping higher power into the load but at lower rate since only its RF current increases. The peaking transistor naturally supplies exactly the amount of additional power necessary to have a linear power-in/power-out overall characteristic.

The advantage of the Doherty configuration is in terms of the PA efficiency. For low power levels, one quarter of the maximum power or less, when only the main transistor is on, the PA performance equals that of a class AB case. The maximum efficiency is reached when the drain voltage swings approximately as much as the power supply voltage for half of the total current the main transistor can deliver. This happens just before the peaking transistor turns on at a quarter of the PA total power capability, i.e., at 6 dB back-off. At this point, the drain of the peaking transistor, which is still off, swings approximately half the power supply voltage, driven by the main transistor. By increasing the input signal further, the peaking transistor turns on and starts delivering power with good efficiency from the start due to the presence of a substantial RF voltage at its drain. The peaking transistor efficiency increases with the input signal reaching the maximum value when the PA delivers maximum power. The overall PA efficiency is shown in Figure 13.23, a major improvement compared to conventional class AB.

ADDITIONAL PA CONSIDERATIONS

Power Control

An important practical issue related to the PA efficiency is power control. Typical wireless standards prescribe various transmitter output power levels for various operation modes and some require dynamic power changes. RF PAs are usually designed for specific applications and support appropriate power control capabilities. Naturally, having good efficiency in all power modes is highly desirable [16].

Many of the efficiency enhancement techniques discussed in the previous section inherently support power control, since the latter is a form of back-off operation from maximum possible PA power. Nevertheless, power control may require orders of magnitude higher range than in the usual signal-induced back-off operation, but the system response agility to power control changes is not particularly demanding. Some practical details of this PA aspect will be addressed via examples in the section "Current PA Technology and Recent Developments."

Linearity

The important topic of PA linearity and methods for improving it is a vast area of knowledge beyond the scope of this treatment (see, for example, Refs. 1–3). The discussion here is limited to the observation that two general wireless transmitter specifications address the PA linearity performance: (a) the output spectral mask and (b) the error vector magnitude (EVM) of the output signal.

The spectral mask specification is concerned with protecting the communication channels in the system from excessive noise generated by transmitting devices not using the respective channels. The EVM specification is concerned with the communication link signal-to-noise ratio (SNR) budget and allocates a maximum SNR loss in the transmitter including the PA. Depending on the wireless standard, the actual PA design is limited either by the spectral mask specifications as in the global system for mobile communication (GSM) or by the EVM specifications as in 802.11b/g. As a general rule, the higher the data rate supported by a communication system per unit of bandwidth, the more important EVM is since deeper modulation formats must be used with less room for errors.

CURRENT PA TECHNOLOGY AND RECENT DEVELOPMENTS

Wireless System PAs in the Real World

The theory presented thus far can be used to understand the commercially deployed RF PAs and the recent PA research results. Each real PA targets a specific wireless application among many wireless

TABLE 13.4
PAR, Bandwidth, and Power Control Specifications

	PAR (dB)	Signal Bandwidth (MHz)	Power Control (dB)
AMPS, GSM, GPRS, EDGE	Low (~0, ~3.2)	Small (≤ 0.2)	Moderate (≤ 30)
CDMA, CDMA2000, WCDMA	Moderate (3–5)	Large (1.23, 3.84)	Very large (70–80)
IEEE 802.11a, IEEE 802.11g	Large (> 7)	Very large (~17)	N/A

systems and standards. Nevertheless, some systems have enough commonality in their specifications to prompt very similar PA solutions. The key specifications determining the PA design approach are the RF signal PAR and bandwidth, and the requirements for power control. Table 13.4 shows how popular wireless systems compare in terms of these specifications. The PAs we discuss next can be grouped into similar design and performance categories.

PAs for AMPS, GSM, and GPRS

Advanced mobile phone system (AMPS) uses old-fashioned frequency modulation, whereas GSM and general packet radio service (GPRS) employ Gaussian minimum shift keying (GMSK) [18]. Since in all three cases, the RF signal PAR is near 0 dB, the PAs are operated in high-efficiency switching/saturation mode at high power levels. The power control strategies include lowering the drain DC voltage [21] or reducing the input RF signal allowing the PA to enter current mode operation at lower power levels.

Most present 824–849 MHz AMPS PAs are designed as dual-mode CDMA/AMPS solutions implemented with gallium arsenide (GaAs) or gallium indium phosphide (GaInP) transistors. These circuits support CDMA in optimized preferred mode as current PAs and are used for AMPS in legacy mode as switching PAs. The AMPS operation delivers respectable performance such as 31 dBm output power with 50% PAE from a 3.4 V power supply [22]. The 25 dB power control required by AMPS is a subset of the much tougher 73 dB of CDMA [18] to be discussed later.

The typical GSM/GPRS PAs are quad-band multichip modules (MCMs) containing two circuits, each covering adjacent bands: 824–849/880–915 MHz and 1710–1785/1850–1910 MHz. GaAs or InGaP transistors are used and the RF input/output are prematched to 50 Ω. At nominal 3.5 V battery voltage and 25°C, these amplifiers deliver 35 dBm power in the U.S. Cellular/enhanced GSM (EGSM) bands and 33 dBm power in the digital cellular communication system (DCS)/personal communication services (PCS) bands with average PAE (over different products) better than 55 and 52%, respectively [21].

The GSM/GPRS MCMs include a CMOS power control circuit accepting an analog voltage and producing PA internal control signals. The typical power control range is in excess of 50 dB [21], the system requirement being 30 dB [18].

In the last 10 years a substantial research effort has targeted the demonstration of class E CMOS PAs for GSM with 40% or better PAE, motivated by lowering the cost of present solutions [23–25]. Ref. 23 reports a 0.35 μm CMOS differential two-stage design delivering 1 W at 1.9 GHz with 48% PAE from a 2 V supply. A board microstrip balun is used for differential to single-ended conversion. The PAE including the balun is 41%.

The two-stage 0.25 μm CMOS PA reported in Ref. 24 has an output cascode transistor used to avoid the device voltage overstress. Powered from a 1.8 V DC supply, the circuit delivers 0.9 W of power at 900 MHz with 41% PAE. The output drain efficiency is larger than 45% and remains above 40% for supply voltages as low as 0.6 V, demonstrating the excellent power control capability of class E PAs.

Ref. 25 demonstrates a 0.13 μm CMOS class E PA for 1.4–2.0 GHz operation. Similar to the one used in Ref. 24, a cascode device is used in the output stage. The circuit operates from 1.3 V supply and delivers 23 dBm with 67% PAE at 1.7 GHz. The PAE is better than 60% over the entire band.

PAs for EDGE

Enhanced data for GSM evolution (EDGE), a GSM upgrade, uses the same 200 kHz channelization but introduces 8 phase shift keying (8PSK) modulation for a 3× increase in raw data rate. This comes at the expense of 3.2 dB signal PAR, which dictates a different approach to the PA design. To meet the spectral mask and EVM requirements, it is customary to use class AB current/linear PAs operated 2.5–3 dB below 1 dB compression. The resulting PAE penalty compared to GSM PAs is significant. Typically, an EDGE PA with 3.5 V supply providing 29 dBm power in the U.S. Cellular/EGSM bands and 28 dBm power in the DCS/PCS bands has only 25% PAE [26].

Because of moderate PAR and narrow signal bandwidth, EDGE is an excellent system candidate for polar PA application [18,19,27]. Research efforts in this direction [28,29] have focused on using supply-modulated class E PAs.

The 0.18 μm CMOS circuit discussed in Ref. 28 is a three-stage design, the last stage powered by a linear regulator for amplitude modulation insertion. In addition, the last stage and the linear regulator use tick-oxide transistors operating from 3.3 V supply unlike the rest of the circuit using 1.8 V supply. A peak CW output power of 27 dBm was measured at 34% PAE. The design met the EDGE EVM and spectral mask requirements in the DCS band at 23.8 dBm power with 22% PAE.

The class E PA discussed in Ref. 29 is integrated in 0.18 μm BiCMOS SiGe technology, has a single stage, and operates at 881 MHz. All necessary passive components except the choke coil are included on-chip. At the peak 22.5 dBm output power, the CW PE and PAE are 72.5 and 65.6%, respectively, operating from a 3.3 V supply. This PA does not contain the amplitude modulation driver, which is an external, discrete switching converter with 5 MHz bandwidth and 82.6% efficiency. The overall configuration meets the EVM and the spectral mask requirements for EDGE at 20.4 dBm power with better than 44% PAE.

PAs for CDMA and WCDMA

CDMA and WCDMA use MHz signal bandwidths and a coding scheme generating RF signals with high PAR. For example, the downlink signal composed of many superimposed CDMA channels addressing all active mobile units regularly exceeds 10 dB PAR [30]. This makes the design of efficient base station CDMA PAs extremely challenging. The PAR of the uplink signal containing a single channel is smaller and the handset PA design seems easier by comparison, but is no small feat in absolute terms. It is not surprising that the CDMA PA design problem has generated a large amount of activities and ideas [33–40]. The most successful CDMA PA approach to date is class AB with efficiency enhancements, but other techniques are also considered.

Table 13.5 shows typical specifications of commercial CDMA and WCDMA handset PAs. Somewhat surprisingly to a reader unfamiliar to the CDMA systems, the PA vendors quote PAE at 28 dBm and 16 dBm power. The 28 dBm figure is good but almost irrelevant since handsets rarely transmit at this level, the most likely transmit power being 5–6 dBm. The CDMA system's proper operation relies on very wide mobile unit power control, as shown in Table 13.4. This complicates the efficient handset PA design by a large degree.

TABLE 13.5
Typical Power/Gain/PAE Performance for Commercial CDMA and WCDMA PAs

	824–849 MHz	1850–1910 MHz	1920–1910 MHz
CDMA	28 dBm/28 dB/37% at 3.4 V	28 dBm/27 dB/39% at 3.4 V	
	16 dBm/25 dB/8% at 3.4 V	16 dBm/21 dB/8% at 3.4 V	
WCDMA	28 dBm/27 dB/43% at 3.4 V	28 dBm/27 dB/37% at 3.4 V	28 dBm/27 dB/42% at 3.4 V
	16 dBm/16 dB/19% at 3.4 V	16 dBm/25 dB/21% at 3.4 V	16 dBm/21 dB/15% at 3.4 V
	7 dBm/15 dB/14% at 1.5 V		7 dBm/24 dB/20 % at 1.5 V

A common method used in CDMA PAs to mitigate the efficiency problem due to large power control is quiescent operation adaptation [16] as discussed in the subsection "Class AB Operation." Another method is the power supply adaptation according to the envelope tracking technique [16,17]. Agile DC-to-DC converters with over 90% efficiency capable of adjusting the PA output power by 1 dB every 1.2 ms as required in CDMA are readily available [31,32]. Further activities are reported in stand-alone agile DC-to-DC converters [33–35] as well as DC-to-DC converters cointegrated with class AB PAs [36].

PA subranging is also effective for power control without excessive efficiency degradation as adopted in Refs. 37 and 38 using two PA segments. A three-segment PA is described in Ref. 39 and transformer-based subranging in Ref. 40. We also mention a 3-bit binary subranging PA reported in Ref. 41.

Finally, the CDMA efficiency problem has motivated a serious reconsideration of the classical Doherty and Chireix concepts [20,42–44]. In Ref. 42, a 0.5 W extended Doherty (peak efficiency at 12 dB back-off) has been implemented using discrete indium gallium phosphide (InGaP)/GaAs HBTs and microstrip quarter-wave transformers. The circuit operates at 950 MHz, delivering 27.5 dBm power at 1 dB compression with 46% PAE. PAE of better than 39% is maintained over the entire 0–12 dB back-off region and 15% is measured at 20 dB back-off. Design considerations and board-level implementations of three-stage WCDMA (1920–1980 MHz) Doherty amplifiers using 10 V GaAs field effect transistors (FETs) can be found in Ref. 43. The amplifier meets WCDMA linearity requirements up to 33 dBm with 48.5% PAE. The measured PAE at 27 dBm (6 dB back-off) and 21 dBm (12 dB back-off) are 42 and 27%, respectively. The 3 dB bandwidth of these amplifiers is broad enough to accommodate the WCDMA uplink reliably.

Similarly, a Chireix out-phasing PA for WCDMA 2110–2170 MHz downlink is discussed in Ref. 20. The circuit uses a pair of saturated class B amplifiers implemented with two 0.25 μm p-channel high electron mobility transistors (pHEMPTs), which are bare-die bonded on a printed circuit board (PCB). These circuits deliver 34.5 dBm from 5 V supply with PE and PAE of 75 and 54%, respectively. The Chireix combiner and matching circuits are implemented using on-board microstrip lines.

PAs for IEEE 802.11a/b/g

The typical performance of commercial IEEE 802.11 PAs operating from a 3.3 V power supply [45,46] is summarized in Table 13.6. The 6.5–8 dB difference between the 1 dB compression point P_{1dB} and the orthogonal frequency division multiplexing (OFDM) signal power $P_{OFDM(max)}$ is consistent with the large PAR of 64-QAM OFDM. Despite using identical signaling, the 802.11a PA efficiency is approximately 2× smaller than that of the 802.11g PA. This is a consequence of operation at much higher frequencies. The 802.11g parts support IEEE 802.11b CCK signaling with lower PAR. In CCK mode, the PAs can be operated with less back-off and much improved PAE. Typically, an 802.11b/g PA with 26.5 dBm P_{1dB} delivers approximately 23 dBm CCK power with 30% PAE [45].

Current research efforts target the implementation of 802.11 PAs with acceptable PAE in Si technologies [47], the development of methods for reducing EVM and improving efficiency by reduction of AM–PM distortion, and the introduction of previously discussed power-efficient PA schemes [48,49].

TABLE 13.6
Typical Performance for Commercial 802.11 PAs

	Gain (dB)	P_{1dB} (dBm)	$P_{OFDM(max)}$ (64 QAM with EVM ~3%) (dBm)	PAE@ $P_{OFDM(max)}$ (%)
IEEE 802.1g	25.5	26.5	19	25
IEEE 802.11a	21	26	18	3

Class AB g_m Ratio Biasing

This section is concluded with a description of a recent contribution on a promising new biasing technique for the class AB PA [50,51]. The class AB stage is one of the most important PA building blocks either as a stand-alone linear stage or as part of a more sophisticated scheme such as the Doherty PA. The motivation for this work is the fact that maintaining proper class AB biasing under all fabrication and temperature conditions is a challenging circuit design task due to the high sensitivity of the PA linearity to the biasing conditions. During the investigation on this matter, a new circuit design concept was discovered, which not only seems to solve the current PA biasing problem very efficiently but also gives a new insight into the transistor class AB operation.

The standard biasing technique for current/linear PAs is known as constant I_{DQ} biasing, where I_{DQ} is the transistor quiescent drain current. As the name suggests, the main objective of the biasing circuits is to maintain a constant I_{DQ} over all operating conditions. The types of practical circuits trying to accomplish this objective are either open-loop, developing the right quiescent gate voltage through an independent circuit, or closed-loop via an analog or digital control system measuring I_{DQ} and keeping it constant through negative feedback. Both types of techniques have important shortcomings. The open-loop methods are not precise enough for the high PA sensitivity and the closed-loop methods can guarantee the right I_{DQ} only when the RF signal is not present and are not able to correct for biasing drift during the PA operation. Regarding the open-loop methods, it is stressed that the use of a conventional current mirror is quite challenging due to RF coupling from the PA transistor into the mirror transistor, which can shift the quiescent gate voltage enough to create biasing errors.

The reasons why PA designers use constant I_{DQ} biasing are mostly pragmatic rather than based on any solid theoretical justification. Laboratory tests simply show that a good linearity compromise over temperature variations is obtained for constant I_{DQ}. It is also known that fabrication process variations require slightly different I_{DQ} values for different process corners for best linearity performance. For manufacturing cost reasons, I_{DQ} is rarely tuned for individual PA during production, so the actual shipped PA is usually not operating at its best.

The new principle called constant-g_m-ratio biasing is shown in Figure 13.26 and effectively implements a differential current mirror [50]. Instead of copying a current as in conventional current

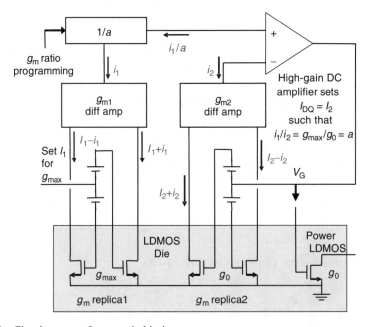

FIGURE 13.26   Circuit concept for g_m-ratio biasing.

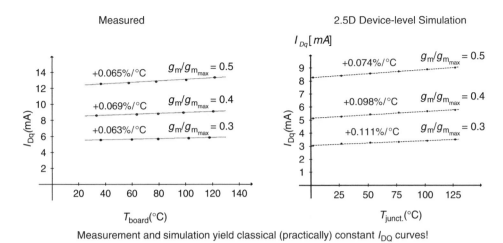

FIGURE 13.27 Simulated and measured drain current curves versus temperature under constant g_m-ratio conditions, demonstrating approximately constant I_{DQ} behavior. (From Banu, M., Prodanov, V., and Smith, K., *Asia Pacific Microwave Conference*, 2004. With permission. © IEEE 2004.)

mirrors, the difference between two slightly dissimilar currents is copied, which is a measure of the transistor transconductance g_m. The master g_m is set and tracks the peak transistor g_m value as shown in Figure 13.14. This is not difficult to accomplish even in open-loop fashion due to zero sensitivity of g_m as a function of the drain current at that point. The slave g_m is automatically set through high-gain negative feedback such that under all temperature and fabrication variations, the ratio between the peak g_m and the slave g_m is kept constant. The resulting gate voltage on the slave g_m transistors is used as the quiescent gate voltage of the PA main RF transistor. Naturally, the transistors realizing the two g_m values are assumed to be matched to the main PA transistor.

Constant-g_m-ratio biasing is justified mathematically by the fact that the maximum error in the transistor even g_m component compared to an ideally flat characteristic is determined by the g_m ratio to a high-order approximation [50]. In other words, the ripple magnitudes in the curves of Figure 13.14 are practically determined only by the g_m ratio at the operating point. Then, it is reasonable to expect that the constant-g_m-ratio strategy should maintain a consistent linearity performance over temperature and process variations.

Figure 13.27 shows simulated and measured I_{DQ} curves of an lateral double-diffused MOS (LDMOS) transistor under constant-g_m-ratio conditions. The resulting practically constant I_{DQ} curves agree with and for the first time explain the traditional PA biasing strategy. Figure 13.28 shows that a constant g_m ratio is substantially better than constant I_{DQ} biasing under fabrication process variations. In addition, this method is fully compatible with IC implementation requirements and the differential nature of the circuits makes them insensitive to RF coupling effects, as discussed for single-ended current mirrors.

A HISTORICAL PERSPECTIVE AND CONCLUSIONS

The PA has been an essential component since the dawn of electronics, and its history has been closely entangled with that of wireless information transmission technology, from the traditional analog radio to the sophisticated digital systems of today. Lee De Forest invented the triode tube in 1906 [52], the first electrical power-amplifying device, whose gain was boosted by Edwin H. Armstrong of Columbia University in 1915 through positive-feedback circuit techniques [53]. These advancements enabled the start of AM broadcasting in 1920. In 1926, Bernhard D. H. Tellegen of Philips Research Labs invented the pentode tube [54], the first voltage-controlled-current-source device, which opened the door for the development of amplifier circuit techniques known today as

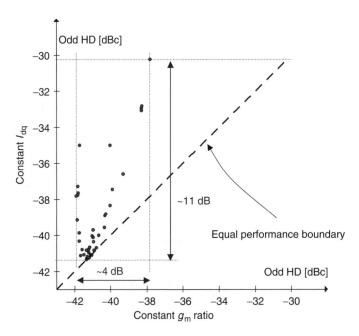

FIGURE 13.28   Simulated scatter plot showing the advantage of constant g_m-ratio biasing compared to conventional constant-I_{DQ} biasing for class AB PAs. (From Banu, M., Prodanov, V., and Smith, K., *Asia Pacific Microwave Conference*, 2004. With permission. © IEEE 2004.)

classes A, B, AB, and C. The effort to resolve linearity issues led to the discovery of feed-forward and negative-feedback principles in 1924 and 1927, respectively, both by Harold Black [55,56] of Bell Telephone Laboratories. All these early inventions consolidated the AM broadcasting industry, and even though the FM modulation was invented in 1933 by E. Armstrong and essentially modern FM transmitters were installed as early as 1940, AM remained the dominant method for RF broadcasting until 1978 when finally the FM exceeded AM in number of listeners.

During the golden years of AM radio broadcasting (1920–1960), much emphasis was put on power-efficient amplifiers, and in the mid-1930s, two important techniques were invented. In 1935, Henry Chireix described the out-phasing PA topology [12] and in 1936, William Doherty of Bell Telephone Laboratories invented a load modulation technique [13]. Today, these techniques bear the name of their inventors, the Doherty amplifier being sometimes called the crown jewel of RF power amplification. Both the Chireix and the Doherty architectures were successfully commercialized. Western Electric deployed Doherty-based AM transmitters in 1938, and RCA used the Chireix architecture in 1956 under the name ampliphase. Other notable accomplishments in efficient RF PAs are the concept of EER developed by Leonard Kahn in 1952 [14], the concept of odd-harmonic enhancement (class F) developed by V. J. Taylor in 1958 [7] and the zero-voltage-switching amplifier (class E) developed by Nathan Sokal and Alan Sokal in 1972 [4,5].

Since FM has fundamental advantages over AM in terms of higher quality audio signals and constant-envelope RF signals (0 dB PAR) for easy and efficient PA implementation, it was selected as the preferred modulation format for the original cellular systems, AMPS in 1983 and GSM in 1992. Other wireless systems followed suit. The great consumer market opportunity opened by these applications focused most EE talent on designing PAs for constant-envelope modulation formats. The new circuit design technology available and appropriate for low-cost implementations was now based on ICs. There was no need for realizing efficient PAs under back-off conditions, so the respective generation of circuit designers could just ignore and forget most of the wealth of PA knowledge developed earlier. Class E switching PA is a perfect example of an appropriate architecture for constant-envelope applications. The powerful Doherty, Chireix, and EER techniques, among others, had all but fallen into obscurity, but not for long.

In the last 10 years, the constant push for higher data rates has reintroduced AM modulation with a vengeance, as discussed in the subsection "Reasons for Back-Off Requirements and Efficiency Penalties." Reviewing their limited present options, PA designers have quickly realized that by looking into the past, they can see the future. We are witnessing a true renaissance in the PA field, and as was the case in the sixteenth-century art, we expect the rebirth of classical PA techniques such as class AB, Doherty, Chireix, etc., to surpass the originals in mastery.

REFERENCES

1. F. H. Raab, P. Asbeck, S. Cripps, P. B. Kenington, Z. B. Popovic, N. Pothecary, J. F. Sevic, and N.O. Sokal, "Power amplifiers and transmitters for RF and microwave," *IEEE Trans. Microwave Theory Tech.*, vol. 50, pp. 814–825, March 2002.
2. S. C. Cripps, *RF Power Amplifiers for Wireless Communication*, Artech House, 1999.
3. S. C. Cripps, *Advanced Techniques in RF Power Amplifier Design*, Artech House, 2002.
4. N. O. Sokal and A. D. Sokal, *"High-Efficiency Tuned Switching Power Amplifier,"* U.S. Patent 3919656, November 11, 1975.
5. N. O. Sokal and A. D. Sokal, "Class E—A new class of high-efficiency tuned single-ended switching power amplifiers," *IEEE J. Solid-State Circuits*, vol. SC-10, no. 3, pp. 168–176, June 1975.
6. S. D. Kee, I. Aoki, A. Hajimiri, and D. Rutledge, "The class-E/F family of ZVS switching amplifiers," *IEEE Trans. Microwave Theory Tech.*, vol. 51, no. 6, pp. 1677–1690, June 2003.
7. V. J. Tayler, "A new high-efficiency high power amplifier," *Marconi Rev.*, vol. 21, pp. 96–109, 1958.
8. F. H. Raab, "Class-F power amplifiers with maximally flat waveforms," *IEEE Trans. Microwave Theory Tech.*, vol. 45, no. 11, pp. 2007–2012, November 1997.
9. F. H. Raab, "Class-F power amplifiers with reduced conduction angles," *IEEE Trans. Broadcasting*, vol. 44, no. 4, pp. 455–459, December 1998.
10. F. H. Raab, "Maximum efficiency and output of class-F power amplifiers," *IEEE Trans. Microwave Theory Tech.*, vol. 49, no. 6, pp. 1162–1166, June 2001.
11. F. H. Raab, "Class-E, class-C, and class-F power amplifiers based upon a finite number of harmonics," *IEEE Trans. Microwave Theory Tech.*, vol. 49, pp. 1462–1468, August 2001.
12. H. Chireix, "High power outphasing modulation," *Proc. IRE*, vol. 23, pp. 1370–1392, November 1935.
13. W. H. Doherty, "A new high efficiency power amplifier for modulated waves," *Proc. IRE*, vol. 24, pp. 1163–1182, September 1936.
14. L. Kahn, "Single-sided transmission by envelope elimination and restoration," *Proc. IRE*, pp. 803–806, July 1952.
15. D. C. Cox, "Linear amplification with non-linear components," *IEEE Trans. Commn.*, vol. COM-22, pp. 1942–1945, December 1974.
16. T. Fowler, K. Burger, N. Cheng, A. Samelis, E. Enobakhare, and S. Rohlfing, "Efficiency improvement techniques at low power levels for linear CDMA and WCDMA power amplifiers," *IEEE Radio Frequency Integrated Circuits Symposium*, pp. 41–44, 2002.
17. J. Staudinger, "An overview of efficiency enhancement with application to linear handset power amplifiers," *IEEE Radio Frequency Integrated Circuits Symposium*, pp. 45–48, 2002.
18. E. McCune, "High-efficiency, multi-mode, multi-band terminal power amplifiers," *IEEE Microwave Magazine*, March 2005.
19. E. McCune, "Multi-mode and multi-band polar transmitter for GSM, NADC, and EDGE," *Proc. IEEE Wireless Commn. Networking*, vol. 2, pp. 812–815, March 2003.
20. I. Hakala, D. K. Choi, L. Gharavi, N. Kajakine, J. Koskela, and R. Kaunisto, "A 2.14-GHz Chireix outphasing transmitter," *IEEE Trans. Microwave Theory Tech.*, vol. 53, pp. 2129–2138, June 2005.
21. *"RF3140: Quad-band GSM850/GSM900/GCS/PCS,"* RF Micro Devices.
22. *RMPA0951AT*, datasheet, Fairchild Semiconductor.
23. K. Tsai and P. Gray, "A 1.9-GHz, 1-W CMOS class-E power amplifier for wireless communications," *IEEE J. Solid-State Circuits*, vol. 34, no. 7, pp. 962–970, July 1999.
24. C. Yoo and Q. Huang, "A common-gate switched 0.9-W class-E power amplifier with 41% PAE in 0.25 μm CMOS," *IEEE J. Solid-State Circuits*, vol. 36, pp. 823–830, May 2001.

25. A. Mazzanti, L. Larcher, R. Brama, and F. Svelto, "Analysis of reliability and power efficiency in cascaded class-E PAs," *IEEE J. Solid-State Circuits*, vol. 41, pp. 1222–1229, May 2006.

26. *RF3145*, datasheet, RF Micro Devices.

27. J. Johnson, "Power amplifier design for open loop EDGE large signal polar modulation systems," *RF Design*, pp. 42–50, June 2006.

28. P. Reynaert and M. Steyaert, "A 1.75-GHz polar modulated CMOS RF power amplifier for GSMEDGE," *IEEE J. Solid-State Circuits*, vol. 40, no. 12, pp. 2598–2608, December 2005.

29. J. Popp, D. Y. C. Lie, F. Wang, D. Kimball, and L. Larson, "A fully-integrated highly-efficient RF class E SiGe power amplifier with an envelope-tracking technique for EDGE applications," *IEEE Radio and Wireless Symposium*, pp. 231–234, January 2006.

30. V. Lau, "Average of peak-to-average ratio (PAR) of IS95 and CDMA2000 systems—Single carrier," *IEEE Commn. Lett.*, vol. 5, pp. 160–162, April 2001.

31. *"LM3202: 650mA Miniature, Adjustable, Step-Down DC–DC Converter for RF Power Amplifiers,"* datasheet, National Semiconductor.

32. *"MAX8506/MAX8507/MAX8508: PWL Step-Down Converters with 75mΩ Bypass FET for WCDMA and cdmaONE Handsets,"* datasheet, Maxim.

33. B. Sahu and G. Rincon-Mora, "A high-efficiency linear RF power amplifier with a power-tracking dynamically adaptive buck-boost supply," *IEEE Trans. Microwave Theory Tech.*, vol. 52, pp. 112–120, January 2004.

34. M. Hoyerby and M. Andersen, "High-bandwidth, high-efficiency envelope tracking power supply for 40 W RF power amplifier using parallel bandpass current sources," *Proceedings of the IEEE Conference on Power Electronics Specialists*, pp. 2804–2809, September 2005.

35. V. Yousefzadeh, E. Alarcon, and D. Maksimovic, "Three-level buck converter for envelope tracking applications," *IEEE Trans. Power Electron.*, vol. 21, no. 2, pp. 549–552, March 2006.

36. S. Abeinpour, K. Deligoz, J. Desai, M. Figiel, S. Kiaei, "Monolithic supply modulated RF power amplifier and DC–DC power converter IC," IEEE MTT-S Digest, pp. A89–A92, 2003.

37. *"MAX2266 Power Amplifier for the CDMA Cellular Band with PIN Diode Switch"*, Application note 275, Dallas Semiconductor.

38. Joon Hyung Kim, Ji Hoon Kim, Y. S. Noh, and Chul Soon Park, "An InGaP-GaAs HBT MMIC smart power amplifier for W-CDMA mobile handsets," *IEEE J. Solid-State Circuits*, vol. 38, no. 6, pp. 905–910, June 2003.

39. Junxiong Deng, P. S. Gudem, L. E. Larson, D. F. Kimball, and P. M. Asbeck, "A SiGe PA with dual dynamic bias control and memoryless digital predistortion for WCDMA handset applications," IEEE J. Solid-State Circuits, vol. 41, pp. 1210–1221, May 2006.

40. G. Liu, T. Liu, and A. Niknejad, "A 1.2V, 2.4GHz fully integrated linear CMOS power amplifier with efficiency enhancement," *Proceedings of the IEEE Custom Integrated Circuit Conference*, pp. 141–144, 2006.

41. A. Shirvani, D. Su, and B. Wooley, "A CMOS RF power amplifier with parallel amplification for efficient power control," *IEEE J. Solid-State Circuits*, vol. 37, pp. 684–693, June 2002.

42. M. Iwamoto, A. Williams, Pin-Fan Chen; A. G. Metzger, L. E. Larson, and P. M. Asbeck, "An extended Doherty amplifier with high efficiency over a wide power range," *IEEE Trans. Microwave Theory Tech.*, vol. 49, pp. 2472–2479.

43. N. Srirattana, A. Raghavan, D. Heo, P. E. Allen, and J. Laskar, "Analysis and design of a high-efficiency multistage Doherty power amplifier for wireless communications," *IEEE Trans. Microwave Theory Tech.*, vol. 53, pp. 852–860.

44. Seongjun Bae, Junghyun Kim, Inho Nam, and Youngwoo Kwon, "Bias-switched quasi-Doherty-type amplifier for CDMA handset applications," *Proceedings of the IEEE Radio Frequency Integrated Circuits Symposium*, pp. 137–140, 2003.

45. *MGA-412P8*, datasheet, Avago.

46. *LX5503E*, datasheet, Microsemi.

47. A. Scuderi, D. Cristaudo, F. Carrara, and G. Palmisano, "A high performance silicon bipolar monolithic RF linear power amplifier for W-LAN IEEE802.11g applications," *IEEE Radio Frequency Integrated Circuits Symposium*, pp. 79–82, 2004.

48. Y. Palaskas, S. S. Taylor, S. Pellerano, I. Rippke, R. Bishop, A. Ravi, H. Lakdawala, and K. Soumyanath, "A 5 GHz class-AB power amplifier in 90 nm CMOS with digitally-assisted AM-PM correction," *IEEE Custom Integrated Circuits Conference*, pp. 813–816, 2005.

49. M. Elmala, J. Paramesh, and K. Soumyanath, "A 90-nm CMOS Doherty power amplifier with minimum AM-PM distortion," *IEEE J. Solid-State Circuits*, vol. 41, pp. 1323–1332, June 2006.

50. M. Banu, V. Prodanov, and K. Smith, "A differential scheme for LDMOS power transistor class-AB biasing using on-chip transconductance replicas," *Asia Pacific Microwave Conference*, 2004.

51. V. I. Prodanov, "*Automatic Biasing of a Power Device for Linear Operation*," U.S. Patent No. 7084705, August 1, 2006.

52. P. Delogne, "Lee de Forest, the inventor of electronics: A tribute to be paid," *Proc. IEEE*, vol. 86, no. 9, pp. 1878–1880, Sept. 1998.

53. J. E. Brittain, "Scanning our past: Electrical engineering hall of fame—Edwin H. Armstrong," *Proc. IEEE*, vol. 92, no. 3, pp. 575–578, March 2004.

54. I. J. Blanken, "Scanning our past from the Netherlands: Bernard Tellegen and the pentode valve," *Proc. IEEE*, vol. 91, no. 1, pp. 238–239, January 2003.

55. C. McNeilage, E. N. Ivanov, P. R.Stockwell, and J. H. Searls, "Review of feedback and feedforward noise reduction techniques," *Proceedings of the IEEE Frequency Control Symposium*, pp. 146–155, May 1998.

56. R. Kline, "Harold Black and the negative-feedback amplifier," *IEEE Control Systems Magazine*, pp. 82–85, August 1993.

14 Phase-Locked Loop–Based Integer-*N* RF Synthesizer

Vikas Choudhary and Krzysztof (Kris) Iniewski

CONTENTS

Introduction 384
PLL System Specification 386
    Synthesis Range (Tuning Range) 387
    Reference Frequency and Feedback Division Ratio 387
    Frequency Accuracy 387
    Settling Time 388
    Spectral Purity 389
        Reference Spurs 389
        Random Phase Noise 390
Linear Modeling of PLL 392
    Continuous Time Modeling 393
        PFD/QP 393
        Loop Filter 395
        Voltage-Controlled Oscillator 395
        Frequency Divider 397
        Loop Delay 397
    A Control-Centric Discussion of PLL Dynamics 397
        Type and Order 398
        Phase-Error Response and Origin of Higher Order Systems 398
        Loop Stability for a Type II, Order III PLL 400
    Noise Modeling 402
    Discrete Time Modeling 403
Architectures 405
    Design Trade-Off 406
        Settling Time 406
        Phase Noise 406
        Reference Feed-Through 406
    Integer-*N* Architecture 407
        Single Loop with External Frequency Divider 408
        Single Loop with Frequency Offset 408
        Dual-Loop Architecture 409
        Adaptive Loops for Fast Settling 409
        Low-Jitter Loops 410
Circuit Implementation for Building Blocks 411
    Phase-Frequency Detector 411
    Charge Pump 412

Loop Filter 414
Voltage-Controlled Oscillators 417
Prescalers (Dividers)... 418
Conclusion.. 424
References... 424

INTRODUCTION

The revolution in the space of consumer electronics has spurred research for electronically controlled digital tuning systems used to generate (synthesize) accurate and stable frequencies. This search has led the evolution of frequency synthesis methods from a manual tuning system to that of the modern sophisticated phase-locked loop (PLL) techniques. The manual tuning system consisted of many stable resonators or oscillators for each tuning frequency. Quartz crystals offered good stability and were often used for this purpose. However, this brute force technique quickly became impractical as the number of supported frequency channels and demands on frequency stability increased.

Modern synthesis landscape can be broadly classified as direct, indirect, and hybrid methods as shown in Figure 14.1. Direct analog synthesis (DAS) approach is a feed-forward method that can be implemented through mixers, band-pass filters, and multipliers in analog or digital domain. Indirect synthesis method consists of a feedback loop approach, wherein the phase and frequency of a reference source are locked with a feedback clock source as shown in Figure 14.2. Hybrid method consists of combination of direct and indirect methods. Several references for each class of PLL have been provided in Figure 14.1 for readers who wish to further explore the topic.

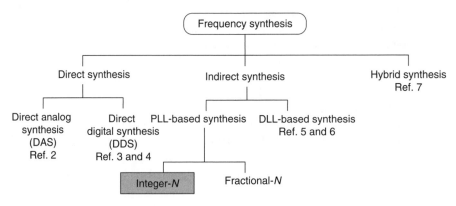

FIGURE 14.1 Classification of frequency synthesis techniques.

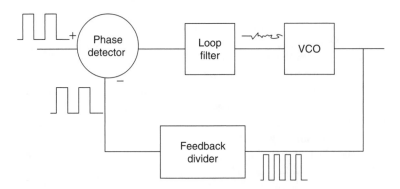

FIGURE 14.2 Conceptual block diagram for PLL-based frequency synthesis.

The choice of approach depends on several factors, such as cost, complexity, frequency step size, switching rate, purity of synthesized frequency, etc. However, over a period of time, PLL-based synthesis technique has come out as the most preferred method, enabled mostly through the advancements in integrated circuit (IC) process technology and CAD tools that facilitate simulation of complex behaviors. PLL-based techniques are classified into two broad categories as shown in Figure 14.1. Integer-*N* PLL uses feedback divider whose modulus can change only in integer ratio, whereas in fractional-*N* counterpart the division ratio could be a fraction. In this chapter only integer-*N* PLL system, architecture, specifications, and circuits are discussed. However, the topics discussed remain largely applicable to all classes of PLL including fractional-*N*, which has been discussed in a different chapter.

The use of PLL-based frequency synthesis has its roots in the evolution of coherent communication system. The first wireless receiver, called superheterodyne receiver invented by Edwin Armstrong, suffered from multiplicity of tuned stages forging research into better methods. In 1924, a team of British engineers led by F. M. Colebrook came up with homodyne (later renamed as synchrodyne) method, which consisted of a local oscillator, a mixer, and an audio amplifier (very close to the direct-conversion receivers prevalent today). However, the method suffered from the fact that the reception was valid only when the local oscillator's phase and frequency were very close to the incoming signal. Any slight drift in the phase and the frequency of this local oscillator would cause loss in signal strength. This spurred the research in techniques for automatic frequency control (AFC). In 1931, French engineer H. de Bellescize applied for a U.K. patent for an improved homodyne radio tuning circuit. This was the first AFC system and the first circuit to incorporate the basic features of a PLL. In 1932, Bellescize described his design in a paper in the French journal *Onde Electrique* [1].

In spite of the fact that homodyne receiver was superior to the classical superheterodyne, the requirement of a PLL outweighed its advantages. The first widespread use happened to be in television receivers for the synchronization of horizontal and vertical scans around 1943, and later on for synchronization of colors with the advent of color television in 1954. The first IC PLL arrived around 1965 spurring an explosion in application of PLL. The idea slowly spread to other fields including frequency synthesis. The term frequency synthesis was first coined in 1940 in a paper by Wendt and Frendall [7]. One of the earliest patents describing frequency synthesis with a PLL utilizing feedback divider appeared in 1970 [8].

Over a period of time, the PLL has been so widely used that it is easy to get confused between the application and the analysis technique applied to different classes of PLL, although the embracing principles are the same. Each application area requires its own set of analysis methods and expertise domain. However, a broad classification can be done on the basis of the signal-to-noise-ratio (SNR) of the signal being phase-locked, as shown in Figure 14.3. It can be seen that the analysis

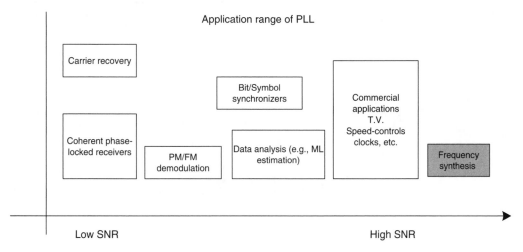

FIGURE 14.3   Application range of PLL.

technique employed in the case of PLL used in Costas carrier recovery (at the low end of the SNR) requires knowledge from the domain of the estimation and communication theory. The PLL employed for frequency synthesis application (at the high end of SNR) will require knowledge from the domain of radio frequency (RF) and digital design and continuous and discrete control system. In this chapter the focus is on methods and analysis for frequency synthesis only.

After the foregoing discussion from a historical perspective as to how PLL started being used for frequency synthesis, the rest of the chapter is organized as follows. The section "PLL System Specification" outlines the system-level specifications. These specifications determine the complexity of the PLL system to be designed. The issue of linear modeling of PLL in both continuous time and discrete time domain is discussed in the section "Linear Modeling of PLL" and will serve as the fundamental basis for the rest of the chapter. The section "Architectures" will discuss the myriad architectural choices that can be made for different classes of RF applications. Frequent visits to concepts established from the section "Linear Modeling of PLL" will be made while discussing the architectures. The section "Circuit Implementation for Building Blocks" will finally discuss the circuit-level implementation challenge associated with the building blocks of a PLL. The final section will briefly summarize the chapter.

Throughout the text the emphasis has been on developing intuitive understanding of the fundamentals employed in the design of PLL-based synthesizer. While there are several tutorials and books available today, covering PLL in great detail in their own way, for this chapter the approach has been to focus on the aspects that lead the path for extremely low-power and highly integrated solutions. The chapter builds enough mathematical basis upfront so that the system and circuit solutions can be discussed with both intuitive and quantitative ease. A list of references categorized at the end of the chapter is provided for readers who wish to further explore this interesting and ever-fascinating topic.

PLL SYSTEM SPECIFICATION

Although integer-N PLL as frequency synthesis has found application in both wireline and wireless applications, this section will focus on PLL specification from a wireless system standpoint. Figure 14.4 shows how frequency synthesis is applied in any modern wireless transceiver. The transceiver architecture shown in the figure is a simplified version of a time domain duplexed (TDD) superheterodyne system with the spectrum of the modulated signals annotated. The RF PLL synthesizer generates

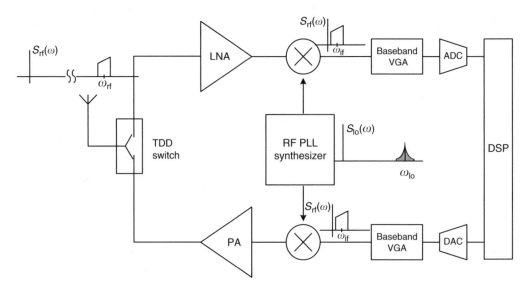

FIGURE 14.4   Modern RF transceiver block diagram.

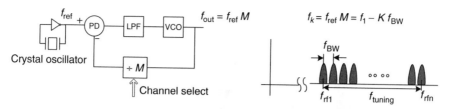

FIGURE 14.5 Illustration of channel selection and reference frequency requirement.

frequencies, which down-convert or up-convert the information from RF down to baseband or vice versa. The next few subsections will discuss the demands on RF synthesizer from this perspective.

SYNTHESIS RANGE (TUNING RANGE)

Synthesis range (also referred to as the tuning range) refers to the frequency range of the application band. The synthesized output frequency f_{out} for any standard ranges from f_1 for the first channel up to f_k for the kth channel (Figure 14.5). Each channel spacing is approximated as f_{BW}, the channel bandwidth. The synthesis range f_{tuning} has direct implications on the voltage-controlled oscillator (VCO) complexity. In general, VCO design becomes more challenging with increase in absolute frequency and the tuning range required.

REFERENCE FREQUENCY AND FEEDBACK DIVISION RATIO

For an integer-N PLL, the choice of reference frequency and the modulus division ratio is a critical parameter. From a behavioral integer-N system shown in Figure 14.5, the synthesized frequency f_{out} is an integer multiple M of the reference frequency. Now f_k, the center frequency for kth channel, is equal to $f_1 + kf_{BW}$. The frequency separation of f_{BW} between two consecutive channels can be achieved by integer variation of modulus of the feedback divider only if the reference frequency f_{ref} is equal to the channel separation (approximated as channel bandwidth) f_{BW}. This imposition on the reference frequency is very restrictive for the integer-N PLL design from loop dynamics perspective, since it results in a limitation on the settling time and noise performance that can be achieved. Fractional-N PLL have an edge over the integer-N PLL as the feedback modulus can be a fraction and hence reference frequency can be much larger than the channel spacing. However, fractional-N suffer from their own set of implementation and analysis challenge; this has been explored in a different chapter in the book.

The modulus of the feedback divider can vary only in integer multiples as stated above. For example, for a digital enhanced cordless telecommunications (DECT) standard for the receiver chain, $f_{ref} = 1.728\,\text{MHz}$ and $N = 128\text{–}137$ for down-converting the entire channel down to baseband (dc). The design of such variable modulus is dealt with later in the chapter.

FREQUENCY ACCURACY

Achieving accurate frequency is one of the major motivations for frequency synthesis using PLL. Frequency accuracy is typically classified as long-term frequency instability (as compared to short term that will be covered in the subsection "Phase Noise") and is specified as parts per million (ppm) deviation from the intended frequency value. The motivation to achieve such accurate frequencies is to account for the spectral shift in information and hence loss in SNR due to fixed frequency filtering as shown in Figure 14.6. Typically the baseband filter that selects the desired channel from the down-converted signal has a fixed cut-off frequency and any shift in information could result in excessive loss of information. For example, a 10.0 ppm offset at 900.0 MHz will result in 4.5 KHz of relative frequency shift. Also in many modern orthogonal frequency division multiplexing (OFDM)-based direct down-conversion receivers, this relative frequency offset between receiver and transmitter

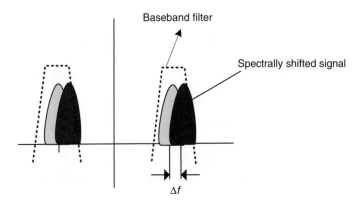

FIGURE 14.6   Effect of frequency inaccuracy on receiver selectivity.

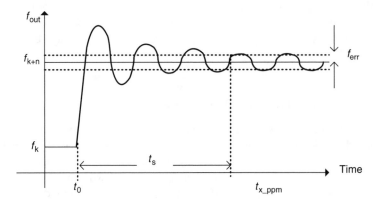

FIGURE 14.7   Settling time definition for integer-N PLL.

results in shift of lowest frequency bin. The dc-offset correction circuitry (which has a high-pass frequency response) ends up chopping some of the information content from low-frequency bins.

Integer-N PLL, however, can synthesize such frequencies accurately since the synthesized clock is always in phase and frequency lock with the reference clock. The reference clock can be accurately generated using a crystal oscillator (sometimes within 1.0 ppm). Hence, the frequency accuracy parameter becomes more of a specification on choice of the crystal oscillator. However, there is cost associated with tight frequency tolerance on crystals and hence this specification should be carefully chosen. A potential solution to relaxing the frequency tolerance is to use an on-chip AFC.

SETTLING TIME

Settling time is defined as the time it takes for a PLL to move from one synthesized frequency to another within a given accuracy as shown in Figure 14.7. Settling time is part of the linear dynamics of the system and is independent of the initial nonlinear acquisition process. Hence, the PLL design would typically maintain near-lock conditions all throughout the process of acquiring new frequency. Managing settling time in integer-N PLL is probably one of the most challenging aspects of the design. The section on architecture will cover the fundamental limitation on settling time and ways to get around it.

The settling time specification is dictated by some key system requirements. For example, in a TDD system, the transmission and reception is performed on the same carrier frequency and hence the synthesizer need not hop. However, in a frequency domain duplexed (FDD) system the transmitter and receiver duplex carrier frequency, and hence synthesizer, has to quickly hop from one channel

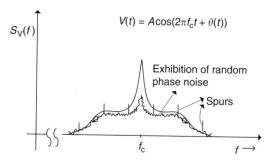

FIGURE 14.8 Spectral purity of synthesized signal.

during receive mode to another channel in transmit mode. Also in many frequency-hopped spread spectrum (FHSS) systems, both transmitter and receiver have to keep jumping between channels to mitigate the channel fading effect in a mobile environment. For such a system, the standards dictate the time period and settling accuracy for a synthesizer.

SPECTRAL PURITY

An ideal spectrum of the synthesized frequency instead of being an impulse looks more like that shown in Figure 14.8. The origin of such spectrally shaped signal is the phase fluctuation due to electronics or extrinsic noise in the building blocks of the PLL. This fluctuation leads to short-term frequency instability and through narrowband phase modulation results in the signal (voltage) fluctuation with spectrum like that in Figure 14.8. Mathematically stated, the synthesized frequency $V(t) = A \cos(2\pi f_c t + \theta(t))$, where $\theta(t)$ is the phase fluctuation or modulation. Also, $\theta(t)$ can be of two types: random and periodic, resulting in either continuous skirts or spurs (spikes) around the carrier. Both these impurities have different implication on the system, and the next subsection will derive a limit on the specification for each kind.

Reference Spurs

The reference spurs happen due to periodic modulation of the phase, meaning phase perturbation $\theta(t)$ is periodic. In an integer-N architecture, this modulation is typically at the reference frequency rate, which in turn is the same as RF channel spacing, resulting in spurs spaced apart by approximately the channel bandwidth. On the receive side, this causes the mixing down of the adjacent channels on the desired channel, thereby degrading the SNR of the system (SNR_{system}) as shown in Figure 14.9. Typically for a good system design, this degradation in system SNR due to adjacent channel mixing down should be negligible. Hence, the SNR with just this impairment (SNR_{spur}) should be at least 6–9 dB above the SNR_{system}.

Once SNR_{spur} has been determined on the basis of SNR_{system} and P_{adj} (dBc) (Figure 14.9), the spur levels P_{spur} (dBc) can be determined. Further, a limit on the peak phase deviation can be derived assuming narrowband phase modulation. If $V(t) = A \cos(2\pi f_c t + \theta(t))$, and $\theta(t) = A_m \sin(2\pi f_{ref}t)$, then assuming $A_m \leq 1$ rad (a narrowband phase-modulation assumption) results in two sidebands around the carrier frequency f_c, as shown in Figure 14.9. Thus, the modulated signal can be rewritten as:

$$V(t) = A\cos(2\pi f_c t) - \frac{AA_m}{2}\cos(2\pi(f_c \pm f_{ref})t) \tag{14.1}$$

From this it can be calculated that:

$$P_{spur}(dBc) = 10\log\left(\frac{A_m^2}{2}\right) \tag{14.2}$$

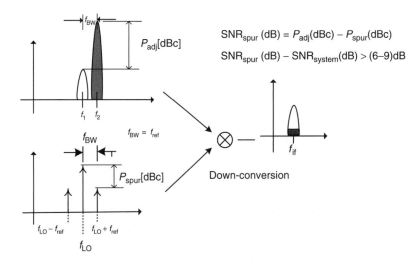

FIGURE 14.9 Illustration of reference spurs and adjacent channel rejection.

On the transmit side, spurs result in spurious emissions on adjacent channels. Typically the transmitter has to meet a spectral mask defined for every RF standard. The amount of spur attenuation can be derived on the basis of similar considerations as above.

Random Phase Noise

Mathematics of Phase Noise
If $\theta(t)$ is random, created due to the random electronic noise, then the synthesized signal has a frequency spectrum with skirts around the carrier frequency. Although the origin and shape of this phase modulation will be discussed later, it is important to understand the mathematical basis for transference of the phase modulation into voltage fluctuations. Once again this can be easily understood by revoking the equation $V(t) = A\cos(2\pi f_c t + \theta(t))$ and the narrowband phase modulation theory. If $S_\theta(f)$ is the power spectral density (PSD) of the random phase fluctuation, then from Wiener–Khinchin–Einstein theorem

$$S_V(f) = \Im^{-1}(E[V(t)V(t+\tau)]) \tag{14.3}$$

where $S_V(f)$ is the PSD of the synthesized clock signal and operator E stands for the expectation (or mean) of a stochastic process. Expanding Equation 14.1 with appropriate terms will result in

$$S_V(f) = \frac{A^2}{2}\delta(f - f_c) + \frac{A^2}{2}\left[S_\theta(f - (f_c \pm f_m))\right] \tag{14.4}$$

which is essentially the PSD of phase fluctuation being transferred exactly as voltage sidebands around the carrier frequency as shown in the one-sided PSD in Figure 14.10.

This voltage (or current) fluctuation can now be measured on a spectrum analyzer and a figure of merit for the random phase fluctuation can hence be stated as relative power in this noise compared to the power in signal itself (often approximated as power in carrier for practical reasons), all quantities being normalized to a bandwidth of 1 Hz. One such figure of merit is the single-sideband (SSB) phase noise defined as:

$$L(f_m) = 10\log\left(\frac{S_\theta(f_m)}{2}\right) \tag{14.5}$$

where the factor of 2 accounts for the noise power in SSB only.

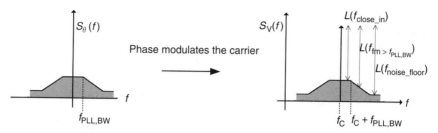

FIGURE 14.10   Transference of baseband phase modulation to signal-level fluctuations at RF.

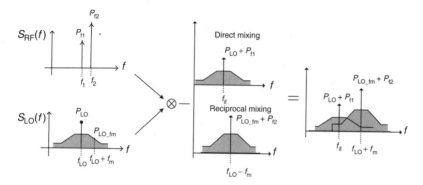

FIGURE 14.11   Direct and reciprocal mixing effects due to random phase noise.

This derivation for phase noise assumes that all noise sources are wide sense stationary (WSS) stochastic processes and that the amplitude fluctuation is negligible. As will be discussed in the next section, the shape of the phase noise is determined by the loop dynamics, to be more specific by the bandwidth of the loop $f_{PLL,BW}$. This typically results in three sets of specifications for phase noise, namely the close-in phase noise ($f_m \leq f_{PLL,BW}$), phase noise at some reasonable offset from carrier ($f_m \geq f_{PLL,BW}$), and the phase noise floor. Considerations for each phase-noise specification are discussed next.

Direct and Reciprocal Mixing
Phase noise impacts the receiver operation by influencing its sensitivity and selectivity, exhibited through direct and reciprocal mixing as shown in Figure 14.11, wherein for simplicity, the power in the RF signal is shown as impulses. The close-in phase noise will result in the blurring of the signal itself causing sensitivity degradation (sometimes also called residual phase deviation in literature). Integrating the phase-noise plot in Figure 14.10, over the channel bandwidth and assuming a $1/f^2$ shape for phase noise beyond the bandwidth of the PLL, the close-in phase noise can be approximated from Equation 14.6 as:

$$L(f_{\text{close\_in}}) \leq \frac{3}{8} \frac{\theta^2 f_{BW}^3}{f_{BW}(f_{BW}^3 - 2f_{PLL,BW}^3)} \tag{14.6}$$

where f_{BW} is the channel bandwidth, $f_{PLL,BW}$ the PLL open-loop bandwidth, and θ the phase error in radians. The extent of θ allowed is determined from SNR consideration for the system. For single-carrier modulation schemes,

$$\text{SNR} = \sqrt{\int_{f_{BW}} S_\theta(f) df} \tag{14.7}$$

Again this SNR should be 6–9 dB (SNR_{margin}) above the system SNR to provide minimal degradation. The far-off phase noise essentially results in degrading the selectivity of the receiver, by down-converting the adjacent channel onto the desired ones. This once again results in degraded system SNR. The phase-noise requirement can be derived simply as:

$$L(f_m) \leq P_{adj}(dBc) - [SNR_{system}(dB) + SNR_{margin}(dB)] - 10\log(f_{BW}) \tag{14.8}$$

For example, for DCS-1800 system, for reference sensitivity level of -99 dBm, a blocker of -43 dBm is allowed at an offset frequency of 600.0 KHz from the desired channel center frequency. Detecting Gaussian minimum shift keying (GMSK)-modulated signal with a 0.1% bit error rate (BER) requires an SNR of about 8 dB for channel bandwidth of 200.0 KHz. Assuming no margin for system SNR, the phase noise required would be -117 dBc/Hz at 600.0 KHz offset.

The direct and reciprocal mixing can be considered to cause similar effect on the transmit side and will result in an out-of-band (OOB) emission as well as degraded SNR for the system. The derivations for the specifications are very similar to the ones derived for the receiver. Many systems use error vector magnitude (EVM) as figure of merit for SNR. From phase-noise perspective, EVM for single-carrier modulation schemes is given as:

$$EVM = 100\% \sqrt{\int_{f_{BW}} S_\theta(f)df} \tag{14.9}$$

Limit on EVM is a system specification and can be used to derive phase-noise requirement for different frequency region for the synthesizer. For multicarrier modulation schemes, a system simulation should be performed to derive the phase noise and EVM relationship.

I/Q Accuracy

Many modern fully integrated transceivers down-convert the same RF signal twice, however offset in phase by exactly 90° and have the same amplitude. This is done to reject the image signal, which if down-converted would reduce the SNR for the system. A similar reverse approach is applied on the transmit side. Details of this kind of transceivers are discussed in Ref. 9. This requirement has resulted in RF synthesizer to produce quadrature signals, called I and Q. For a receiver, the mismatch in I and Q results in degradation in the image rejection, given as image rejection ratio (IRR) in Equation 14.10 [9]. For the transmit side, the same equation captures the loss in SNR or EVM.

$$IRR = \left| \frac{1 - 2(1+\varepsilon)\cos\phi + (1+\varepsilon)^2}{1 + 2(1+\varepsilon)\cos\phi + (1+\varepsilon)^2} \right| \tag{14.10}$$

Power and Area

Every system and application will dictate its own set of demands on the power and area for the PLL synthesizer. In general, power and area tend to make the frequency synthesizer design extremely challenging and in some cases limit the system performance, which can be achieved for a given application. In general, there has been a trend to fully integrate these subsystems with power consumption in the range of few tens of milliwatt. Literature has been replete with several circuit and architectural approaches to address this growing demand and some of it will be covered in the sections "Architectures" and "Circuit Implementation for Building Blocks."

LINEAR MODELING OF PLL

Before getting into the details of the PLL system architecture, it is important to introduce linear modeling of a PLL to understand the dynamics of the system. In this section, a simple linear model for each of the building blocks is presented and then eventually the complete PLL subsystem is discussed. The approach will be to grasp the parameters that control the various dynamics of the PLL and then manipulate those to achieve specifications like the settling time, phase noise, etc.

The charge pump–based PLL (hereafter referred to as QP-PLL) has become the *de facto* topology for a fully integrated PLL as discussed in the previous sections. This chapter will assume this topology and discuss linear modeling of building bocks of such a PLL system. Following this, the PLL dynamics will be discussed from a control-centric standpoint. This viewpoint will then be extended to discuss the noise modeling in a PLL. The very early PLLs were in fact continuous time systems. Even with the advent of modern QP-PLL, this attitude has continued and engineers have come up with several boundary conditions to make continuous time modeling work. The nexus between the continuous time modeling and the limitations set on the performance of PLL, however, is too restrictive for modern systems. Hence, the discrete domain modeling technique will be introduced briefly.

CONTINUOUS TIME MODELING

As shown in Figure 14.12, the QP-PLL is a feedback loop, wherein a phase detector (PD) compares the reference phase (θ_{ref}) with the fed-back phase (θ_{fb}), and the error phase (θ_e) in turn drives the charge pump (QP). The QP delivers a current (I_{qp}) for a sliver of time proportional to the phase error θ_e. The charge pump drives a loop filter (LPF) with a certain impedance transfer function. The filter produces a filtered voltage signal $V_c(t)$ forming the control voltage for the VCO. VCO is essentially a phase integrator (and hence an infinite integrator, since the phase could grow without bounds) and produces an output phase $\theta_o(t)$. This output phase is divided with a frequency divider by a factor N before being fed-back to the PD. The servo system makes sure that the error phase θ_e is driven to zero, resulting in equality of the reference and the fed-back phase. This in turn ensures that the output phase θ_o is N times the reference phase θ_{ref}. Since phase and frequency are linearly related, this relationship also holds true for frequency, i.e., the output frequency is N times the reference frequency.

In a PLL, it is the phase that is being manipulated throughout the loop and hence a PLL system is typically studied in phase domain. It should be pointed out that a frequency and phase domain model would be exactly the same (in Laplace variable) since phase and frequency are linearly related. The follow-up text will elaborate on linear time invariant (LTI), Laplace-domain phase-transfer function for each of the building blocks. In this context it is worthwhile noting that while the overall input–output for the system will be modeled in phase domain, the individual blocks will have input–output transfer function in different domains. For example, the combination of phase-frequency detector (PFD)/QP has input as phase while the output is in current. Similarly for the LPF, the input–output is current–voltage and for VCO it is voltage–phase.

PFD/QP

Traditionally PD has been implemented as a multiplier (mixer) and hence can be viewed as a nonlinear time-varying component, albeit continuous time. Linearization of such a PD is made under

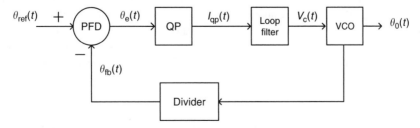

FIGURE 14.12 Block-level signal flow for PLL as a feedback system.

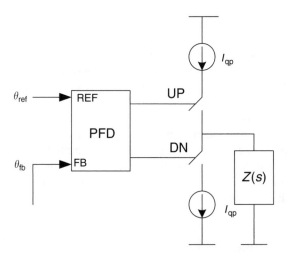

FIGURE 14.13 Implementation method for a PFD/QP.

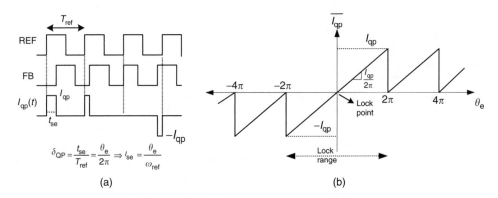

FIGURE 14.14 (a) Phase to current signaling and (b) transfer function for the PFD/QP.

assumptions of near-lock condition when θ_e is very small.* However, this element has undergone a change from being a nonlinear continuous time to the modern implementation methodology of linear and discrete time systems as shown in Figure 14.13. Although this technique renders the QP-PLL a discrete time nature, continuous time approximations can be made under restrictive assumption of loop bandwidth being at least 10 times lower [10] than the reference frequency. In other words, per-cycle change is insignificant compared to the overall dynamics of the loop. To understand the model of PFD/QP, it is imperative to understand their operation in slight detail.

While there are several implementation techniques for a PFD, the tri-state PFD is the most used and understanding of their behavior can be extended to any other type. A tri-state PFD has three output signal states depending on the sign of the error phase, i.e., θ_{ref} leading, lagging, or equal to θ_{fb}. Pursuant to this sign, the charge pump either asserts the up (UP) current source, injecting charge into the LPF, asserts down (DN) current source extracting charge from the filter node, or simply stays put if $\theta_{ref} = \theta_{fb}$, or is tri-stated. An exhibition of this behavior is represented in Figure 14.14a. Charge transferred to or from the filter node per cycle (of the reference clock) is simply $I_{qp}t_{\theta e}$, where $t_{\theta e}$ is the time for which the UP or DN signal is ON.

---

*The PFD as implemented in Figure 14.13 is in fact both a nonlinear and time-variant element. This aspect has been elaborated in Ref. 16. However, for most practical purposes they can be replaced by an LTI model as discussed in this section.

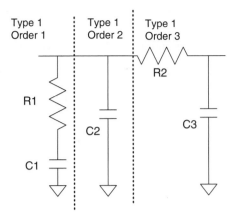

FIGURE 14.15   Passive implementation for loop filter and catalog of transfer functions (Table 14.1).

Eventually the relationship between the average output current $\overline{I_{qp}}$ (which is in fact the average charge transferred) and input phase error is given as:

$$\overline{I_{qp}} = \left(\frac{I_{qp}}{2\pi}\right)\theta_e \tag{14.11}$$

This transfer function is illustrated in Figure 14.14b for wide phase differences. The gain of the PFD/QP combination (K_{PD}) is given by the slope of the curve within phase difference range of $\pm 2\pi$ (the lock range) and is:

$$K_{PD} = \frac{I_{qp}}{2\pi}[\text{A/rad}] \tag{14.12}$$

Loop Filter

LPF determines most of the specifications for a PLL and is a critical component of the system that determines its stability and jitter characteristics. There are several implementation techniques for the LPF; however, the passive technique shown in Figure 14.15 has found its way through in most low cost–integrated applications. The type and order* for the filter refer to the number of poles located at dc and total number of poles for the impedance transfer functions $Z(s)$, respectively. As will be elaborated later in this section, the type I filter is the most widely used. The filter impedance transfer function for type I and various other orders of filter in modern use is tabulated in Table 14.1.

Voltage-Controlled Oscillator

Heart of the PLL operation, a VCO is a highly nonlinear and time-varying circuit and harnessing its exact phase behavior is an active area of research. However, the VCO can be very simply modeled as an LTI element if the transfer function is output phase to input voltage (or control voltage). The transfer characteristic of the behavior is shown in Figure 14.16.

Mathematically:

$$\omega_{VCO} = \omega_{fr} + K_{VCO}V_{ctrl} \tag{14.13}$$

---

*For more details on these two terms, refer to the section "Control-Centric Discussion of PLL Dynamics."

TABLE 14.1
Passive Filter Transfer Function and Pole-Zero Location

Filter Order	Transfer Function $Z(s)$	DC Gain	Pole-Zero Location
1	$K_F \dfrac{1+s\tau_z}{sC_1}$	$K_F = 1$	$\tau_z = R_1 C_1$
2	$K_F \dfrac{1+s\tau_z}{s\left(1+s\tau_p\right)}$	$K_F = \dfrac{1}{\left(C_1+C_2\right)}$	$\tau_p = R_1 C_{eq};\ C_{eq} = \dfrac{C_1 C_2}{C_1+C_2}$
3	$K_F \dfrac{1+s\tau_z}{s\left(1+s\tau_{p1}\right)}\dfrac{1}{\left(1+\tau_{p2}\right)}$	$K_F = \dfrac{1}{\left(C_1+C_2\right)}$	$\tau_{p1} = R_1 C_{eq}$ $\tau_{p2} = R_2 C_3$

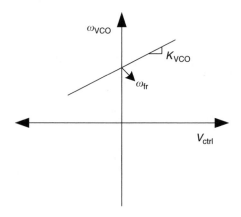

FIGURE 14.16 Transfer function for a VCO.

where ω_{VCO} is the output frequency of the oscillator, ω_{fr} the free running oscillator frequency, K_{VCO} the VCO gain in units of (rad/s)/V, and V_{ctrl} the input control voltage. Now

$$\theta_{VCO} = \int \omega_{VCO} dt = \omega_{fr} dt + K_{VCO} \int_0^t V_{ctrl}(t) dt \tag{14.14}$$

For the excess (or incremental) VCO phase

$$\delta\theta_{VCO} = K_{VCO} \int_0^t V_{ctrl}(t) dt \tag{14.15}$$

The Laplace domain transfer function is hence given as:

$$\delta\theta_{VCO}(s) = \frac{K_{VCO} V_{ctrl}(s)}{s}$$

$$\Rightarrow \frac{\delta\theta_{VCO}(s)}{V_{ctrl}(s)} = \frac{K_{VCO}}{s} \tag{14.16}$$

It is important to understand that the transfer function derived above is an idealized version of what might occur in practice. For example, the transfer curve could be nonlinear and K_{VCO} would be different at different locking frequency. Such behavior has to be accounted for when dealing with real systems.

Frequency Divider

Frequency dividers form the component which in fact renders the frequency synthesis property to an ordinary PLL. Frequency at the output is $1/N$th of the frequency at the input. The linear relationship between the phase and frequency essentially means that phase at the output is $1/N$th of the input phase as well. This relationship is captured mathematically for a very general case in set of Equation 14.17, where $\theta_m(t)$ stands for a random phase modulation, $\theta_{VCO,m}(t)$ the modulated phase from VCO, and $\theta_{DIV,m}(t)$ the modulated divider output phase. Note that the subscript m depicts the modulation in phase.

$$\theta_{VCO,m} = \frac{1}{N}\theta_{DIV,m}$$

$$\theta_{VCO,m} = \omega_{VCO}\,t + \theta_m(t)$$

$$\theta_{VCO,m} = \frac{d\theta_{VCO,m}}{dt}\,\omega_{VCO} + \frac{d\theta_m(t)}{dt}$$

$$\theta_{DIV,m} = \frac{\omega_{VCO,m}}{N}\frac{\omega_{VCO}}{N} + \frac{1}{N}\frac{d\theta_m(t)}{dt} \tag{14.17}$$

$$\theta_{DIV,m} = \omega_{DIV,m}\,t\frac{\omega_{VCO}t}{N} + \frac{\theta_m(t)}{N}$$

$$\Rightarrow \theta_{DIV,m} = \frac{\theta_{VCO,m}}{N}$$

$$\Rightarrow \frac{\theta_{DIV,m}}{\theta_{VCO,m}} = \frac{1}{N}$$

Loop Delay

Loop delay is the practical aspect of the circuit implementation for each PLL subcomponent. In Laplace domain, pure time delay is modeled as e^{-st_d}, where t_d is the loop delay. Loop delay has two aspects to it, the first being the pure signal propagation delay due to processing time in each circuit. Second, the presumed sampled nature of the PLL will introduce a delay in the loop. However, this part can be ignored as long as the dynamics of the loop are constrained to be too slow as compared to the phase-comparison rate at the input of the PLL.* The effect of the delay is phase shift only and acts as pseudoparasitic pole for the system. In this chapter, the loop delay aspect will be ignored; however, their significance should be cautiously evaluated on a per system basis.

A CONTROL-CENTRIC DISCUSSION OF PLL DYNAMICS

The study of individual component modeling can be put together now to finally study the complete loop behavior. This is where the knowledge of control system can be invoked to understand some key behaviors. A classical feedback loop is generally classified by a feed-forward gain $G(s)$ and a feedback factor $H(s)$ as shown in Figure 14.17.† For a PLL, $G(s)$, $H(s)$, and thereby the open-loop

---

*A discrete time model discussed later in the chapter will cover this in more detail.

†In a signal flow diagram, one always assumes unidirectional flow. This is a fair assumption for comprehending the important top-level behavior; however, for practical implementation, one should be careful of such assumption.

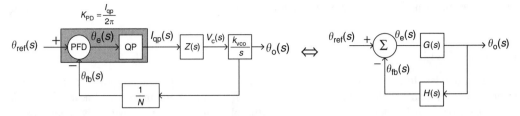

FIGURE 14.17 System transfer function for a PLL.

gain (or loop transmission) $GH(s)$ and closed-loop gain $G_{CL}(s)$ can be derived from Figure 14.17 and is given by set of Equation 14.18.

$$G(s) = \frac{K_{PD}K_{VCO}Z(s)}{s}$$

$$H(s) = \frac{1}{N}$$ (14.18)

$$G_{CL}(s) = \frac{\theta_{out}(s)}{\theta_{ref}(s)} = \frac{G(s)}{1 + GH(s)} = \frac{K_{PD}K_{VCO}NZ(s)}{sN + K_{PD}K_{VCO}Z(s)}$$

Type and Order

Although the two terms have already been introduced while discussing LPF, it is imperative to discuss them in slight detail. There is no established standard for defining these terms, and to some extent, they have been used indiscriminately in the literature. The most common definition for "type" of system refers to the number of poles at origin in the open-loop transfer function, in this case $GH(s)$. The order of the system refers to the highest degree of polynomial in the characteristic equation $1 + GH(s) = 0$. For example, if the loop impedance transfer function is set to unity, i.e., $Z(s) = 1$ (the simplest PLL one can come up with), then

$$GH(s) = \frac{K_{PD}K_{VCO}}{s}\frac{1}{N}$$ (14.19)

$$1 + GH(s) = sN + K_{PD}K_{VCO}$$

This is then a type I and order I system. It should be obvious by now that LPF ends up determining the order and type of the system. Also given that VCO due to its integrative nature has a pole at dc, all PLL are at least type I system. This is also the reason why the type and order for filters as defined in Table 14.1 is one less than that for a complete PLL.

Phase-Error Response and Origin of Higher Order Systems

In examining the behavior of a PLL system, the phase error as a function of time $\theta_e(t)$ is an important function to be monitored. PLL is a negative feedback system and a proper loop should generally drive this error toward zero. However, depending on the input and the loop order and type, this may or may not be true. The phase error as a function of time $\theta_e(t)$ with respect to the various input types $\theta_i(t)$ is depicted in Figure 14.18. The steady-state value in time domain can be easily calculated by invoking the final value theorem stated as:

$$\underset{t \to \infty}{lt}\ \theta(t) = \underset{s \to 0}{lt}\ s\theta(s)$$ (14.20)

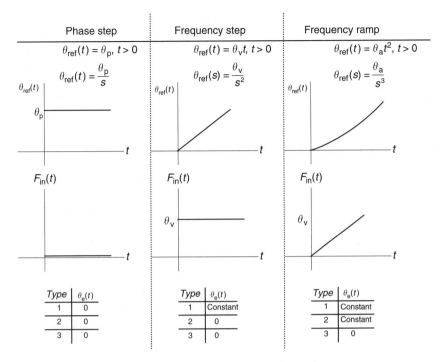

FIGURE 14.18   Error response of a PLL for various phase variations at input.

The equation can be recast for a generalized PLL as:

$$\lim_{t \to \infty} \theta_e(t) = \lim_{s \to 0} \left[\frac{s\theta_i(s)}{1 + GH(s)} \right] = \lim_{s \to 0} \left[\frac{s^2 N \theta_i(s)}{sN + K_{PD}K_{VCO}Z(s)} \right] \tag{14.21}$$

Equation 14.21 was evaluated for different order of filters and has been tabulated in Figure 14.18. This will now lead to some important conclusions regarding evolution of the LPF. It is obvious that as long as the phase disturbance at the input of a PLL is a simple phase step (i.e., frequency change); the phase error will eventually be tracked out by the loop. However, if there were a frequency step at the input, as would be the case for a RF frequency synthesizer, the only way for the loop to track and reduce this error to zero would be by introducing another pole at dc (or another integration) in the open-loop transfer function. Intuitively this can be understood by the fact that a finite phase error will persist in a type I system to drive the VCO to correct frequency. However, if this error has to be tracked out then another integration factor or pole at dc should be introduced. This leads to a type II system.

A simple way to create another integrator is by pumping a current (or charge) onto a capacitor, resulting in genesis of QP-PLL. The integration of charge onto the capacitor results in LPF voltage.* As will be discussed later in more detail, two integrators (another one due to the VCO) cause a full 180° phase shift resulting in stability issues. To counter this effect, a zero is inserted in the transfer function by adding a resistor in series with the capacitor. However, the impulsive current at every phase correction instance (at f_{ref}) causes a ripple on the control voltage of magnitude $I_{qp}R$. This ripple voltage could have twofold deleterious effect. First, it ends up modulating the VCO at every reference

---

*The discrete nature of this charge integration on the loop filter capacitor is worth noting. At every reference cycle, a pulse of charge is either pumped on or pumped out during the locking process. Once lock is achieved, the loop filter gets tristated (or disconnected). In continuous time modeling, it is assumed that the rate of charge transfer is too fast compared to the overall loop dynamics and hence average behavior can be still continuous charge integration. This assumption will be revisited when discussing the discrete time modeling of PLL.

clock cycle, causing spurious tones spaced f_{ref} Hz apart in frequency domain. As discussed earlier, for an integer-N synthesizer this results in adjacent channel being folded down on desired channel for a receiver and in spurious emission for a transmitter. Second, if the magnitude of the ripple is too big, it could result in overloading of the VCO [10], either resulting in PLL getting out of lock or causing a large recovery time from this state. To mitigate this, another capacitor is added to the LPF node, finally arriving at the topology shown in Figure 14.15. This results in an increase in the order of PLL to 3, although the loop type remains the same. This is one of the most popular filter structures prevalent today, and it is worthwhile spending a small section studying the loop stability for this kind of PLL. Another subtle point underscoring the pervasiveness of type II system stems from the dynamic (transient) behavior of the PLL. Even if the application does not require tolerating frequency steps at input, they are inevitably present in many phase disturbances (which is rarely a step). Hence, a type II system has a larger lock-in range than type I and subsequently ends up being universally used.

Finally the type III PLL could even track frequency ramps at its input. In general, such systems are extremely rare for couple of reasons. First, the requirement for PLL to handle frequency ramp is required only in few application areas such as deep space where Doppler shifts cause frequency ramps. And second, the stability issue for a type III system is too difficult and overloading for most engineers and hence has been avoided.

The next subsection will now examine the loop stability of a type II, order III PLL. This could very well be the heart of mathematics one needs to know for a PLL and will be frequently invoked in rest of the chapter.

Loop Stability for a Type II, Order III PLL

As briefly mentioned while discussing the evolution of type II system, the stability of the loop plays a vital role and forms the core for any PLL design. The open-loop transfer function for such a PLL is given by Equation 14.18 and has been plotted in Figure 14.19.

$$GH(s) = \frac{I_{CP}K_{VCO}}{2\pi N(C_1 + C_2)} \frac{(1 + s\tau_z)}{(1 + s\tau_p)} \tag{14.22}$$

Figure 14.19 clearly marks the location of the two poles at dc, the zero due to R and C in series and the third pole due to additional C (see the sub-section "Loop Filter" under section "Linear Modeling of PLL"). Such a gain-phase plot called Bode plot* can now be used to study the stability of the system. To quantify stability, the parameter of interest is phase margin, as defined in Figure 14.19.[†] Two most critical parameters now, the crossover frequency ω_c (or open-loop bandwidth) and the phase margin Φ_m, can be found by the following equation:

$$|G(j\omega_c)| = \frac{K|1 + \omega_c\tau_z|}{\omega^2|1 + \omega_c\tau_{p1}|} = 1 \tag{14.23}$$

$$\Phi_m(\omega_c) = -\pi + \angle(1 + j\omega_c\tau_z) - \angle(1 + j\omega_c\tau_{p1})$$

where K is given by:

$$K = \frac{I_{CP}K_{VCO}}{2\pi N(C_1 + C_2)} \tag{14.24}$$

---

*Named after the famous Dutch scientist Hendrik Wade Bode, who among his several contributions is most well known for his discovery of the technique to analyze stability in frequency domain. Remarkably the stability of the complete system can be assessed by looking at gain and phase behaviors at just two frequency points. Nyquist plots, however, are still the most comprehensive method for stability and analyze the system at all frequencies.

†Another quantity of interest is Gain Margin; however, for QP-PLL it is rarely a problem. Still it should be evaluated on a case-by-case basis.

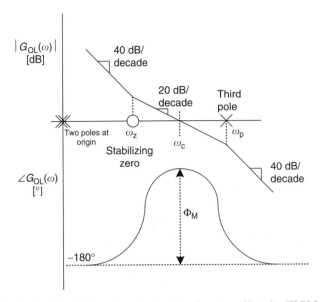

FIGURE 14.19 Open-loop gain and phase transfer function for a type II, order III PLL.

The crossover frequency ω_c is a classical designers' dilemma and is primarily determined by concerns of valid continuous time modeling, settling time, spurious tones rejection, and rejection of noise from various sources. On the contrary, phase margin Φ_m determines the stability of the system and controls the time domain behavior of the system. There is no closed-form solution to Equation 14.23; however, significant strides can be made by realizing the fact that from design robustness point of view, it is always beneficial to place the top of the phase curve at ω_c for two reasons: first, it gives the maximum phase margin for the system. Second, the design is least sensitive to parametric variations in LPF component values. It is not very difficult to obtain the resulting solution by differentiating

$$\Phi(\omega) = -\pi + \angle(1+j\omega\tau_z) - \angle(1+j\omega\tau_p) \tag{14.25}$$

resulting in the crossover frequency $\omega_{c,max}$ given by:

$$\omega_{c,max} = \sqrt{\frac{1}{\tau_z\tau_p}} = \sqrt{\omega_z\omega_p} \tag{14.26}$$

Worthwhile noting is the fact that this maxima is obtained if the crossover frequency is crafted at the geometric mean of the zero and third pole frequency, namely ω_z and ω_p. Ratio of these two frequencies, ω_z/ω_p, hence is an important design criteria and deserves a name of its own and will be henceforth called b. When expanded, b is also equal to $1 + C_1/C_2$, or almost equal to the ratio of two capacitors in the LPF. Now, b can be further derived as a function of the maximum phase margin $\Phi_{m,max}$ desired as:

$$b = \frac{1+\sin\Phi_{m,max}}{1-\sin\Phi_{m,max}} \tag{14.27}$$

The equation essentially states that a large phase margin is obtained by a large b, which means a high C_1/C_2 ratio. Obviously a right bound has to be established on the maximum attainable phase

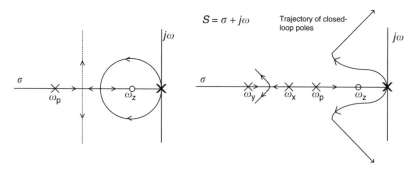

FIGURE 14.20   Root locus plot for type II, order II and multiorder systems.

margin for reasons of integrability of the filter components and settling issues. Also, a lower bound is obtained by assessing stability over all design corners, jitter peaking, and excessive settling time due to underdamped system.

At this point, it is important to state that although stability was discussed with respect to Bode plots for an intended third-order system, in practice a real system is a multiorder system with lots of parasitic elements and delay in the loop creating deleterious effect mostly by shifting the peak of the phase curve from loop crossover at ω_c. This effect can be further underscored by invoking the root-locus diagram. The root locus plots the variation of the closed-loop poles with variation in loop gain.* Figure 14.20 clearly shows the variation of poles for a type II, order III system and then a multiorder system. As depicted, an ideal order III system is always stable, while a multiorder system could easily get unstable (i.e., the poles could end up in right-half s-plane at some high enough loop gain). This aspect should be carefully evaluated while practicing real design.

Noise Modeling

Apart from stability issues, which are more of a design necessity than a design specification, one aspect which preoccupies RF synthesizer designers most is the aspect of low noise design. As will be explored in this section, the classical dilemma of balancing noise from various sources and thereby engineering the system's poles and zero is paramount. The noise source from each component can be modeled as an additive noise source [11,12] in the system as shown in Figure 14.21. The root mean square (rms) power in each noise source at a modulating frequency f_m is given as:

$$S_\phi(f_m) = \phi^2(f_m) \tag{14.28}$$

---

*It is not possible to cover the root-locus technique in detail here and readers should refer to plethora of texts on theory of control systems to comprehend this topic. However, there is one point that is worth remembering: the pole of a closed-loop system starts from the pole of open-loop system and ends at the zero of open-loop system as the loop-gain changes from low to high. Hence, the behavior of the closed loop can be simply assessed by looking at the open-loop poles and zeros and their variation with the loop gain. This can be further understood by representing the loop gain as a ratio of polynomial $Z(s)$ and $P(s)$, i.e., $G(s) = KZ(s)/P(s)$, where K is the loop gain. Assuming the feedback factor to be unity, the closed-loop gain is

$$G_{CL}(s) = \frac{KZ(s)}{P(s) + KZ(s)}$$

The poles of the closed-loop system can now be analyzed by evaluating the characteristic equation (CE) $P(s) + KZ(s) = 0$. At low frequency, when K is small, solution to the CE is simply $P(s) = 0$. Or simply stated the pole of the closed loop is given by pole of the open loop. Similarly when K is very big, the solution to CE is given by $KZ(s) = 0$, i.e., the pole of closed loop ends up at zero of the open loop. For stability the poles of the closed loop should never end in right-half plane of the s-plane.

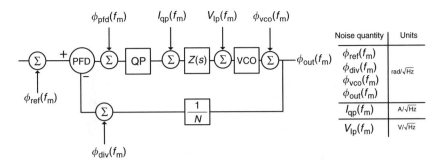

FIGURE 14.21   Additive noise sources in a PLL.

Total output noise from each source can then be carried forward through respective transfer function and assembled together as:

$$S_{\phi_{out}}(f_m) = \left|T_{LP}(f_m)\right|^2 S_{\phi_{in\_pfd}}(f_m) + \left|T_{HP}(f)\right|^2 S_{\phi_{out\_vco}}(f) \tag{14.29}$$

where each term is further elaborated in the set of Equation 14.30 as:

$$S_{\phi_{in\_pfd}}(f_m) = \phi_{ref}^2(f_m) + \phi_{div}^2(f_m) + \phi_{PFD}^2(f_m) + \frac{i_{qp}^2(f_m)}{K_{PD}^2}$$

$$S_{\phi_{out\_vco}}(f_m) = \phi_{VCO}^2(f_m) + \left(\frac{K_{VCO}}{2\pi f_m}\right)^2 v_{lp}^2(f_m)$$

$$T_{HP}(f_m) = \frac{1}{1 + GH(j2\pi f_m)}$$

$$T_{LP}(f_m) = \frac{G(j2\pi f_m)}{1 + GH(j2\pi f_m)} = N\frac{1}{\dfrac{1}{G(j2\pi f_m)} + 1} \tag{14.30}$$

T_{HP} and T_{LP} are the high-pass and low-pass transfer functions, respectively, and $S_{\phi_{in\_pfd}}$ and $S_{\phi_{out\_vco}}$ are the PSD of noise sources referred to the input of PFD and output of VCO, respectively. Figure 14.22 plots the two transfer functions. Note that both the transfer functions have the same transition frequency defined as f_o. Evidently the noise from reference, frequency divider, and the PFD/QP is low-pass filtered, while the noise from VCO and LPF is high-pass filtered. Also, the input-referred noise gets multiplied by N^2, as $G(j2\pi f)$ tends to infinity at low frequencies. Finally a balancing act has to be performed for choosing the bandwidth of the loop depending on the intensity of different noise sources. The section "Architectures" will further explore the choices to be made for loop bandwidth for optimizing the noise from different sources.

DISCRETE TIME MODELING

So far the focus has been continuous time modeling in spite of the fact that a QP-PLL is a discrete time system for at least two reasons. First, the PFD/QP combination results in a discrete time event, since unlike analog PDs, the phase corrections are made for every reference cycle. Second, the digital frequency dividers are in essence discrete time systems since phase information at the output is present for every *N*th VCO cycle. Having convinced that modern QP-PLL is indeed a discrete time system, one has to arrive at the motivation to build discrete time model foregoing the luxury of ever so convenient Laplacian domain.

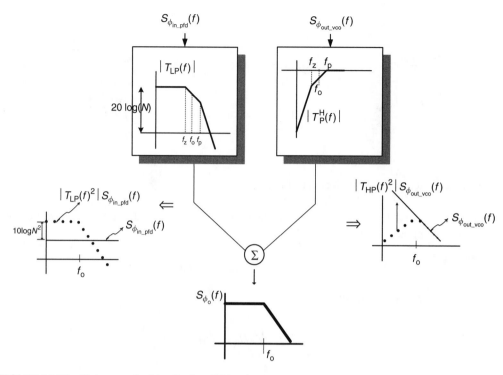

FIGURE 14.22   Noise transfer function in a PLL.

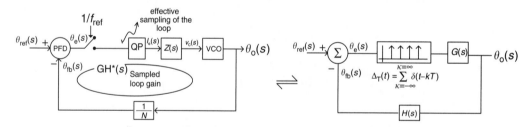

FIGURE 14.23   Concept of sampling in a QP-PLL.

The sampling nature of the system can be embodied as in Figure 14.23. Thus, an ideal impulse sampler can follow the PFD,* with the impulses spaced $1/f_{ref}$ apart. A hold function should follow the error samples for rest of the electronics to process it. In Figure 14.23, this hold function will be assumed folded in $G(s)$. Essentially this results in system's loop gain being sampled at every $1/f_{ref}$. The effect of this sampling on the loop gain is represented in Figure 14.24, thereby exemplifying the aliasing that happens if the sampling rate is too low compared to the bandwidth of the system. Thus, the loop bandwidth should be sufficiently low compared to the sampling frequency (or $1/f_{ref}$) if aliasing is to be reduced to insignificant level. Continuous time model remains valid if the aforesaid

---

*The impulse sampler in mathematics is also referred to as Dirac comb function, which is a periodic Schwartz distribution (or generalized functions) constructed from Dirac delta functions.

$$\Delta_T(t) = \sum_{k=-\infty}^{k=\infty} \delta(t - kT)$$

This function is also called Shah function by few authors.

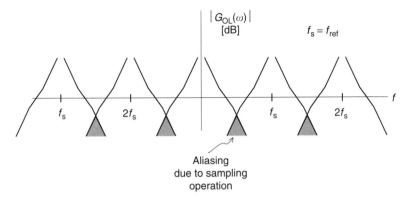

FIGURE 14.24   Aliasing due to sampling in a PLL.

statement is valid qualitatively. Quantitatively the limit for a type II, order II system has been derived by Gardener [10]. This is also the genesis of the often-employed rule of thumb for keeping the open-loop bandwidth at least 10 times less than the reference frequency.

However, this constraint is too restrictive for an integer-*N* design from either settling time requirement or say choice of bandwidth for rejecting the noise from VCO. Comprehension of discrete time model, hence, is paramount and is an active area of research.

There are two popular methods to derive the discrete time model. The first method involved accurately capturing the behavior of the system by writing difference equation for each of the independent state variable in the system [10,14,16]. This method captures both the locking acquisition and steady-state tracking behavior accurately. The thoroughness and complexity, however, is too overpowering, and a simpler method of impulse invariant transform was proposed by Scott and Hein [13]. The concept of impulse invariant transformation is extremely simple. For arriving at an equivalent discrete time system from its continuous time counterpart, the idea is to match their impulse responses. This is particularly applicable in a QP-PLL example since the phase corrections are followed by sliver of current from the charge pump, which can be approximated as impulses. This can be thought of as if the phase error is being sampled by a Dirac comb function. Although the results derived in Ref. 13 did not match the results derived in Ref. 10, the error can be patched by correct initial condition as shown in Ref. 15. Finally the loop gain for a type II, order II system in *z*-domain is given as:

$$GH(z) = \frac{K}{C}\left[\frac{RCz^{-1}}{1-z^{-1}} + \frac{Tz^{-1}}{(1-z^{-1})^2}\right] \tag{14.31}$$

where $K = K_{PD}K_{VCO}$ and T is the sampling interval $1/f_{ref}$. The equation for a type II, order III system has been derived by both methods in Refs. 14 and 15.

The comprehension of linear behavior will now be employed to study various integer-*N* architectures of a QP-PLL.

ARCHITECTURES

Before delving into architecture of integer-*N* PLL, it is important to discuss the fundamental design trade-off. Essentially the loop parameters have to be engineered to meet the key specifications, namely the settling time, phase noise, and spurious response, for any RF application. The section will first highlight the conflicting requirements for a modern high-performance RF synthesizer and then discuss the various architectures to counter such conflicts.

DESIGN TRADE-OFF

Settling Time

As discussed in the section "PLL System Specification," settling time is a key requirement for RF synthesizers. This new frequency is synthesized by changing the modulus of the feedback divider, say from N to $N + 1$. For conceptual and mathematical ease, this change in modulus of the feedback divider can be considered the same as changing the frequency of the reference by f_{ref}/N while keeping the modulus the same, i.e., N. In other words, the reference frequency has a scaled step change. Now the continuous-time LTI phase transfer function model developed in previous section can be invoked to calculate the output response for frequency in time domain. Noteworthy is the fact that since frequency and phase are linearly related, the transfer functions are exactly the same. And finally from the time domain response, the settling time, t_s, can be calculated for a given settling error, ε, in ppm ($1\,\text{ppm} = 1\,e^{-6}$). For a first-order system, the settling time can be evaluated as:

$$t_s = \frac{-\ln(\varepsilon)}{\omega_c} \tag{14.32}$$

Thus, t_s is inversely proportional to the crossover frequency ω_c (or f_c). Settling time for higher order system is best calculated through simulations of the corresponding linear model. Vaucher [17] and Rategh and Lee [44] have provided empirical formulae for settling time of higher order systems under the constraint that ω_c is optimally placed to get the maximum phase margin achievable. Two important conclusions can be drawn from their studies. First, the settling time is still inversely proportional to ω_c. Second, the least settling time for a given bandwidth is achieved when the phase margin (Φ_m) is about 50°. The intuitive explanation for the first conclusion is obvious from a control theoretic point of view, i.e., for any negative feedback system, the system state variables are updated at inverse bandwidth rate. The best settling time condition can be intuitively understood by pole location of the closed-loop transfer function in a complex s-plane. For example, for a third-order system, the three poles should be equidistant from the origin of the s-plane for all poles to equally contribute to the settling. Any departure from this equilibrium will result in either excessive damping from real pole versus complex or vice versa. This pole placement also corresponds to the case of phase margin of about 50°. The first hints of design conflicts are underway now, once it is realized that integer-N is constrained in its choice of f_c. The choice is hard dictated by channel-spacing specification and soft dictated by engineering choice to stick with a linear model and hence chose f_c at least 1/10 of the reference frequency f_{ref}. Also, one cannot be conservative in stabilizing the loop through excessive phase margin, since there is an optimality. Here we underscored the conflicting requirements for the choice of f_c from settling time perspective.

Phase Noise

Now follows a discussion on how noise properties of a PLL would dictate that the optimal choice f_c should be the frequency where the scaled-noised sources $N^2 S_{in\_pfd}$ equals S_{out_vco}. This is illustrated in Figure 14.25. Clearly, deviation from this optimality will lead to excessive noise either from contribution from PFD, feedback divider, and reference or from VCO and LPF. Depending on the noise performance of each component, this optimality will rarely be the same as that for settling time or 1/10 of f_{ref}. Also, a phase margin of 50° will lead to jitter peaking, thereby resulting in extra noise in the system. Thus, system noise consideration will guide yet another choice for f_c.

Reference Feed-Through

Suppression of spurious tones at reference frequency (reference feed-through) demands low impedance of the LPF at f_{ref}. Magnitude of the ripple on VCO control voltage at nth harmonic, nf_{ref} due to reference feed-through is simply $i_{qp}(nf_{ref})|Z(nf_{ref})|$, where $i_{qp}(nf_{ref})$ is the ac component of charge pump current

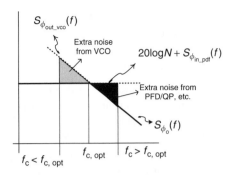

FIGURE 14.25 Noise shaping in an integer-N PLL.

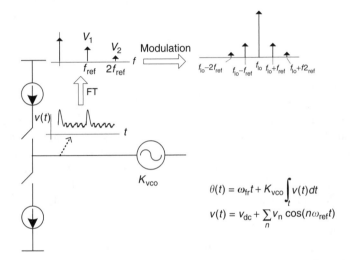

FIGURE 14.26 Reference noise modulation in a PLL.

and $|Z(nf_{\text{ref}})|$ is the magnitude of impedance of LPF. An illustration of VCO modulation through relative reference tones is shown in Figure 14.26. Important conclusion from aforementioned discussion is that suppression of reference tones is independent of choice of f_c; however, it is dependent on LPF impedance, VCO sensitivity K_{VCO}, and relative separation of f_c and f_{ref}. Thus, additional RC sections adding poles in the transfer function above f_c (thereby increasing the order of PLL to higher than 3) would further diminish the reference spurs.

So far the design considerations have been dictated by either application standards or design need. However, the end application imposes a practical set of consideration for low-power and low-area design for RF synthesizers. Several trade-offs are involved at a system level for balancing the need for power and area. A low-power PLL will typically entail higher noise of the components, especially the VCO. For such a PLL, the best approach would be to design a wideband PLL to suppress VCO noise as much as possible. Similarly, low-area RF-PLL will entail lower values of LPF capacitors, resulting in either low suppression of reference spurs, settling time, or even higher noise from LPF resistor (assuming no compromise on stability of the loop).

Having discussed briefly the design trade-offs, a discussion on possible integer-N RF-PLL is in order.

INTEGER-N ARCHITECTURE

A formal classification of integer-N PLL architectures is virtually impossible due to richness of literature in this field. In this section, the architecture classification has been made by addressing some of the fundamental design issues that were discussed in the previous section. The list is in no way exhaustive, but will serve as the basis for more advanced and novel architectures.

Single Loop with External Frequency Divider

As mentioned earlier, the classical single-loop architecture with channel select as the modulus control suffers from the drawback that minimum resolution in output frequency is determined by the reference frequency. This in turn constraints the loop bandwidth to 1/10 of channel spacing causing an upper limit on settling time, phase noise, etc. The simplest solution to the problem could be as shown in Figure 14.27, wherein an external frequency divider is used with the same modulo as the multiplier for f_{ref}. Thus, the bandwidth of the loop can be now N times larger than before. Also, the phase noise of the output frequency is divided by a factor of N and hence is better by $20 \log N$ dB. On the flip side, the VCO has to generate a frequency N times larger now. This, in general, is a problem for extended tuning range and also results in higher phase noise. Hence, this approach has severe limitation in GHz range.

Single Loop with Frequency Offset

Another approach to generating fine frequency steps is by realizing that one could get very close to the carrier frequency through the phase-lock process and then add fine frequency steps either inside or outside the loop. The concept is illustrated in Figure 14.28, wherein an offset frequency f_{offset} is introduced and added to the frequency Mf_{ref}. This addition of frequency can be done through a non-linear system like a mixer. A mixer produces both the sum and difference of its input frequencies. To select one out of two possible outputs, one could use either a real filter like band-pass filter or a complex filter like SSB mixer. Thus, f_{offset} could be either the channel bandwidth/spacing or simply the frequency modulation information. This last concept has gained popularity on the transmit side for constant envelop modulation schemes like frequency modulation (FM) and GMSK. However, as an RF synthesizer (especially for the receive side), the generation of precise frequency remains a problem. Also noteworthy is the concept that f_{offset} could be added at any location, either inside or outside the loop. Armed with these observations, the dual-loop architecture will be discussed next.

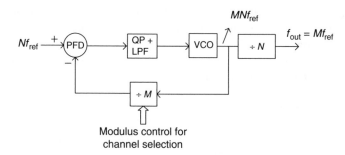

FIGURE 14.27   Single-loop PLL with external frequency divider.

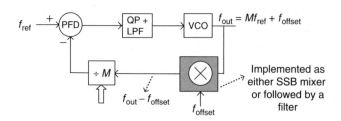

FIGURE 14.28   Single-loop PLL with frequency offset.

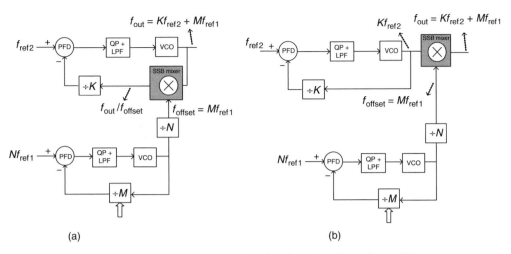

(a)   (b)

FIGURE 14.29   (a) Dual-loop PLL with two loop in series (b) and two loops in parallel.

Dual-Loop Architecture

The problem of generating finite frequency steps can be tackled through addition of another loop, either in series (Figure 14.29a) or in parallel (Figure 14.29b). The second PLL could further utilize an external divisor as discussed at the beginning of this section. Given the narrow spacing of the two sidebands produced by a mixer, the offset frequency addition is mostly achieved through an SSB mixing approach and hence both the VCO need to generate in-phase (I) and quadrature (Q) phases.

The dual-loop approach has at least two severe limitations. First, the accuracy of SSB generation results in residual sidebands (spurs) at the center of alternate channels. Modern digital signal processing (DSP) algorithms could be used to calibrate out some of the I–Q gain-phase mismatch. Alternatively, the spurs could be placed at the edge of the alternate channels instead of center and thereby mitigate the effect of alternate down-conversion to some extent, as discussed in Ref. 19. The second drawback of dual PLL approach is the use of two oscillators in the system. The two oscillators could couple to each other through substrate or parasitic elements and could modulate each other (also referred to as injection locking). Hence, isolation of the two oscillators becomes a major challenge in the design of dual PLL systems. Other than these two limitations, dual PLL also need two references (and hence two crystal oscillators) unless one reference is derived from the other through some careful frequency planning. Also, the variation of modulus in the second PLL results in wide variation of the loop gain, and stability of such loops in general is a challenge.

Adaptive Loops for Fast Settling

So far the single loop posed the problem of conflict between high bandwidth (as close as possible to $f_{ref}/10$) for fast locking and a much lower bandwidth than $f_{ref}/10$ for low phase noise and suppression of spurs in the system. Obviously it is not possible to satisfy both the requirements at the same time, but the fact that they might not be needed at the same time paves the way for a potentially adaptive architecture. From a system's perspective, whenever the feedback modulus is changed (say when the RF system is scanning for a new channel), the system provisions some time for settling to happen. However, during this time, while the loop still needs to be stable, it need not be optimal for noise or spurs. Hence if a means were to be devised wherein the bandwidth (or crossover) frequency of the loop can be adapted, while maintaining its stability, such a system would be an adaptive loop PLL. The core requirement is further illustrated in Figure 14.30.

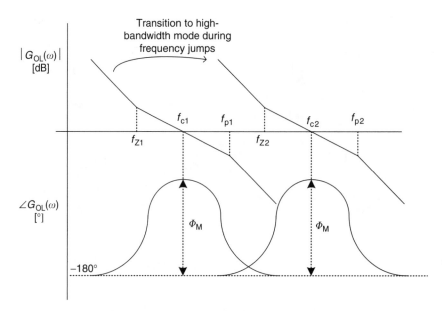

FIGURE 14.30   Adaptive shifting of gain and phase during new channel acquisition.

Concept for implementation of such system can be explained lucidly if one were to make the assumption that $f_c \gg f_z$ and $f_p \gg f_c$. Thus, Equation 14.23 for f_c can be approximated as:

$$f_c = \frac{I_{QP} K_{VCO}}{2\pi N} R_1 \frac{C_1}{C_1 + C_2} \tag{14.33}$$

From the knobs available for tweaking f_c, one could potentially choose any; however, the one which can be easily programmed is the charge-pump current I_{QP}. However, this mere change in f_c would lead to instability if phase margin were not compensated for simultaneously. Observation of Equation 14.23 reveals that phase margin can be compensated for by halving both the zero and third pole time constants by switching in another resistor R_1 in parallel with LPF resistor, where the LPF topology is as shown in Figure 14.15. Since R_1 also affects f_c, I_{qp} should be scaled by a factor of 4 for doubling the bandwidth. Although conceptually very simple, practical implementation of such systems should make sure that the LPF voltage is minimally disturbed during such shifts. A frequency lock circuit most conveniently does adaptation of such loops from one mode to the other. Thus, when output frequency has settled within desired accuracy, the frequency lock circuit would command the PLL to shift to low-bandwidth mode. Refs. 18 and 19 discuss other potential techniques to adapt to the loop. Alternative technique to hasten settling time by controlling the LPF charge is discussed in Ref. 25. Finally Refs. 26 and 27 discuss an approach where the limitation on $f_c < f_{ref}/10$ is violated through careful discrete-time modeling of the system.

Low-Jitter Loops

Designing low-jitter loops essentially entails a balance between the input-referred jitter sources (such as PFD, QP, reference, etc.) and output-referred jitter sources such as VCO, LPF resistor, etc. As discussed earlier, the crossover frequency for minimal jitter in the loop happens when the two scaled jitter sources contribute equally. If the dominant source of noise is the input sources, then the loop to be designed is called a narrowband loop (since the loop should act as a low-pass filter as much as possible). However, when the dominant source is VCO, then the loop should be wideband. In a wideband loop, to further reduce the noise from the feedback divider, the output could be latched by VCO output, suppressing the noise even further [29]. Ref. 30 discusses a novel concept to reduce the accumulated jitter from ring-oscillator VCO by realigning the output to a pure reference clock. Ref. 24 presents a very good discussion on jitter optimization based on loop parameters, and Ref. 23 discusses an adaptive architecture for minimizing jitter over process and parametric spreads. To reduce the sensitivity of the VCO (K_{VCO}), Refs. 21 and 21 discuss a novel approach through fine and coarse setting for tuning of VCO and through gain-shaping of the loop, respectively.

CIRCUIT IMPLEMENTATION FOR BUILDING BLOCKS

Finally the circuit implementation detail for individual building blocks is studied. This section will focus on CMOS circuit implementation (although the concepts discussed herein can be extended to any other technology), and nonidealities associated to these circuits, and then review some of the emerging circuit topologies, which address these impairments. A numerical basis, though not thoroughly discussed, has been brought out with appropriate references, wherever applicable.

PHASE-FREQUENCY DETECTOR

As mentioned earlier, PFD has evolved from being continuous time systems to modern implementation of discrete time and in fact render that discrete nature to PLL. PFD has emerged from being implemented as multipliers to modern D flop-flop–based sequential circuits. An ideal PFD and QP combination circuit example was discussed briefly in the section "Linear Modeling of PLL." This section will elaborate on PFD methods while still assuming an ideal QP.

Historically the requirement for PLL was phase detection and hence topologies addressing this need emerged. However, PLL employing only PD had limited pull-in and hold-in range,* apart from other renderings such as nonlinear gain, limited lock-in range, and quadrate locking [30]. This set of limitation can be encountered by employing the method depicted in Figure 14.31a. By studying the ideal transfer characteristics in Figure 14.31b, it is important to identify the two unique characteristics of this PFD. First, it can be seen that this transfer curve has a net average current when phase error is larger than $\pm 2\pi$, the lock-in range. Second, the average current has a sign corresponding to the sign of phase difference, i.e., the transfer curve has an asymmetry. These two characteristics make this method capable of frequency-difference detection. Thus, the PLL can be pulled into lock even with large initial frequency offset. Other than that it can be seen that ideally it has a lock-in range of $\pm 2\pi$ and is highly linear.

This paragraph will discuss the practical impairments the technique suffers from. One is the problem of dead zone. Because of finite switching time of following charge-pump current sources

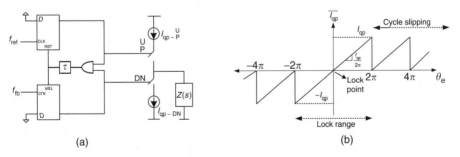

(a)   (b)

FIGURE 14.31 (a) Implementation scheme for a PFD and (b) associated transfer function.

---

*Pull-in range is defined as frequency range over which the PLL will eventually pull the VCO into lock even with many cycle slips. This is also called the "acquisition range" or "capture range." Hold-in range is defined as the frequency range over which the PLL operation is statistically stable. For QP/PFD-based PLL, this is limited only by the tuning range of VCO. Lock-in range is the frequency range over which the PLL acquires lock without cycle slips, in one beat note for linear-PLL. This is also the range over which the PLL operation can be well approximated to be linear.

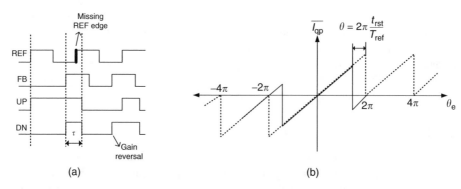

FIGURE 14.32   (a) Missing edge due to finite delay in a classical PFD and (b) transfer function.

when in the vicinity of lock, the UP and DN pulses are very narrow. A very common practice is to insert additional delay, τ, in the feedback reset path (Figure 14.32a). This will result in UP and DN pulses being wide enough for current sources to respond. This reset delay, however, results in finite frequency operation of the PFD expressed as $f_{ref} < (1/2t_{rst})$, where t_{rst} is the total reset delay including the explicit delay of τ and implicit delay through the flip-flops and logic gates [31]. This limitation arises essentially from the fact that PFD is immune to rising edges when both the UP and DN pulses are high or when reset is high. One such occasion of missing reference (REF) pulse is depicted in Figure 14.32a. Pursuant to this missing edge, the next feedback (FB) edge will lead the PLL in wrong direction. The impaired transfer function is shown in Figure 14.32b. It shows that when δ is equal to π, the two unique properties of this PFD (as discussed earlier) are lost and PLL would not lock unconditionally anymore. One technique to circumvent this issue is to simply move out the additional delay from the feedback path and put it in feed-forward path. Ref. 32 discusses two methods to overcome this problem resulting in enhanced pull-in time and pull-in range.

Also, noise in PFD could be deleterious, especially given that the low-frequency content is enhanced by $20 \log N$ dB in an integer-N architecture. The phase noise in a PFD is dependent on rise time (for positive edge-triggered implementation) and is given as:

$$\phi_{PFD} = \frac{2\pi}{T_{ref}} \frac{v_n}{V_{sr}} \tag{14.34}$$

where v_n is the noise voltage and V_{sr} the slew rage in volts per second. Hence, the edge rates should be sufficiently high to mitigate intrinsic device noise. Dynamic logic could be used for such fast implementations.

CHARGE PUMP

Charge-pump circuit, simplistically represented as switched current source, has several practical hurdles to be scaled. Broadly the impairments can be categorized as current mismatch related and switch related (Figure 14.33). Ideally speaking, there should be no residual current in the LPF once PLL has attained phase lock. However, due to timing mismatch between UP and DN pulses, mismatch of UP and DN current itself, or leakage in LPF, a residual charge is integrated on LPF node for every f_{ref} cycle. The PLL can be considered to be virtually open loop at this frequency due to extremely low gain. Hence, the ripples on control voltage are not corrected by the servo action and end up modulating the VCO creating spurs, which are f_{ref} away from the carrier frequency f_c. Magnitude of spurs can be approximated as:

$$P_{spur}[dBc] = 20 \log \left(\frac{1}{\sqrt{2}} \frac{f_c}{f_{ref}} N\phi_e \right) - 20 \log \left(\frac{f_{ref}}{f_p} \right) \tag{14.35}$$

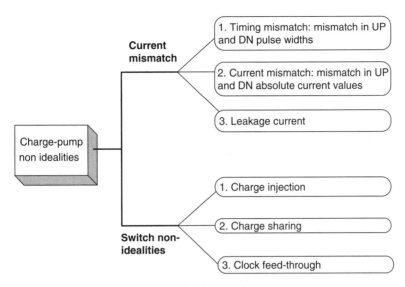

FIGURE 14.33   Nonidealities in a charge-pump implementation.

where ϕ_e is the static phase error and f_p the third pole in an order III PLL. Noteworthy is the fact that the Equation 14.34 is very general and can be extended to quantify all nonidealities in terms of static phase error at the input of the PFD [33].

Switch-related errors also result in periodic modulation of the VCO creating reference break-throughs at output of VCO. The error is associated with the nonideal behavior of the switches such as finite switch-on or switch-off time, charge injection, finite switch capacitance and resistance, etc. Several circuit methods have emerged to address these issues as shown in Figures 14.34a through 14.34d. The simplest, however quite robust and low-power, topology is the one shown in Figure 14.34a. The switches are placed on either the drain or the gate of the current source transistor. However, such topologies suffer either from current spikes when the switches are toggled or from finite switch on–off time [33]. The switch on the source side ensures that the transistor M1 and M2 are always in saturation, thereby minimizing any spike current and also charge injection from switches, since they tend to flow into source instead of drain. To further improve the switching time, Figure 14.34b uses current steering instead of completely switching on or off the current sources. This speed is obviously gained at the expense of redundant power dissipation through the dummy steering branch. The UP and DN signals could be arranged to either hard or soft switch the top and bottom current sources. To alleviate the effect of charge sharing, Figure 14.34c shows the scheme to connect the output node and the voltage reference (VREF) node through a unity gain buffer. Since the two nodes are always at the same potential, the charge redistribution is avoided. All schemes in Figures 14.34a through 14.34c still suffer from finite matching of top and down current sources. Figure 14.34d shows one possible scheme, wherein a replica bias of the transistors on output side is used to create a V_{replica} node, which is then used in a feedback to set the top and bottom currents precisely. A simple illustration of this concept is presented in Ref. 44.

All the schemes in Figure 14.34 can be further improved by using dummy switches or comple-mentary switches to mitigate charge-injection effect [34]. Pass transistor logic can be used to balance the timing mismatch of the UP and DN pulses. Also, Ref. 35 presents a scheme to balance top and bottom current through use of all n-channel MOS (nMOS) current sources only. Finally it should be pointed out that differential version of charge pump is far more robust; however, it suffers from huge area and power penalty. Hence, unless the application permits, they are generally avoided.

Finally regarding noise contribution from the charge pump, had the situation been ideal, no residual charge would have deposited on LPF and hence noise would not have mattered. However,

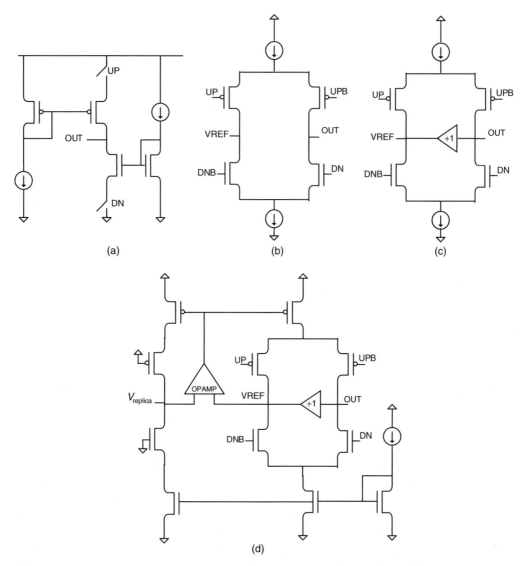

FIGURE 14.34 (a) Voltage-switched charge pump with nMOS switches on the source side; (b) current-steering charge pump; (c) current steering with buffer to mitigate charge sharing; and (d) current-steered charge pump with UP/DN current balancing.

due to finite reset delay, τ, both the top and bottom current sources are periodically on and end up depositing noise charge on LPF in that period. The magnitude of noise is inversely proportional to the absolute magnitude of charge-pump current I_{qp} and directly proportional to the duty cycle τ/T_{ref}. Hence from design perspective, the charge-pump current should be maximized while the reset delay τ be minimized.

LOOP FILTER

LPF has been one of the biggest stumbling blocks in the path of complete integration of PLL system. Generally the zero time constant τ_z is very high resulting in either high value of the series resistor R_1 or the loop capacitor C_1. However, R_1 cannot be very high due to proportionally high noise. Hence, synthesis of zero has received special attention from the research community. The subtle point to

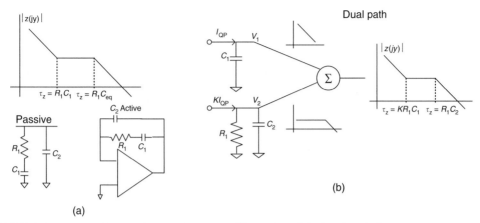

FIGURE 14.35 (a) Implementation and (b) concept illustration of dual-path technique.

be observed in this context is that while the poles are natural time constant of the system, zero is a forced response and can be synthesized by manipulation of external stimuli. The exploration toward this end can be started with the classical second-order passive and active filter topology shown in Figure 14.35a. Active methods have the advantage of higher voltage compliance range at the output of the charge pump at the expense of more complexity, noise, and power and do not address the issue of integration. Hence, most modern implementations are passive in nature, although active methods have their place too.

As discussed earlier, an example of synthesizing zero with manipulation of external stimuli is shown in Figure 14.35b. The core idea is that the final impedance transfer can be synthesized by the addition of two separate responses, one being an integrator and the other being that of a low-pass filter with dc value set by the scaled value of charge-pump current. The zero location can thus be manipulated by changing the scaling of the charge-pump current [37,38]. This architecture is called dual-path technique. In spite of its advantages it suffers from the issue of increased noise and power due to active devices and from the integration path working with a very low current. Besides, the mismatch between the two paths and voltage decay on low-pass path could cause undesirable ripples on VCO control voltage.

Another technique to synthesize zero was proposed in Ref. 39 and is shown in Figure 14.36a. Here the zero has been synthesized using delay. The concept can be simple stated by expanding the Laplace domain transfer function of delay, i.e., $e^{-s\Delta T} \approx 1 - s\Delta T$. The delay can be implemented through switch capacitors. The zero location is precisely defined by the ratio of currents and the ratio of capacitors in two paths, as shown in Figure 14.36a. Another example of discrete time method of mimicking a resistor and hence creating a zero is shown in Figure 14.36b and elaborated in Ref. 40.

A sample-reset LPF with a current-controlled oscillator was proposed in Ref. 41. A conceptual block diagram of the scheme is shown in Figure 14.37a. The idea is to generate a proportional current equal to the average of the current that a normal charge pump would have generated and this current is constant over the entire update period as shown in Figure 14.37b. This results in the same location for stabilizing zero as for a normal charge pump; however, there are no more ripples. In implementing this, one pole at dc is altogether avoided and hence results in a highly stable PLL system, albeit at the expense of less low-frequency noise rejection.

All the ideas discussed so far suffer from the limitations of charge sharing and clock feed-through, a classical feedback with any sampled system. Hence, efforts have been made in the direction of active multiplication of capacitor (a subset concept of gyrators). Two methods have been shown in Figures 14.38a and 14.38b from Refs. 42 and 43, respectively. Both methods have the advantage of highly scaled capacitor and no clock feed-through and charge-sharing effect; however, they suffer from the noise and power dissipation of the amplifiers. Also, these implementations are heavily process dependent.

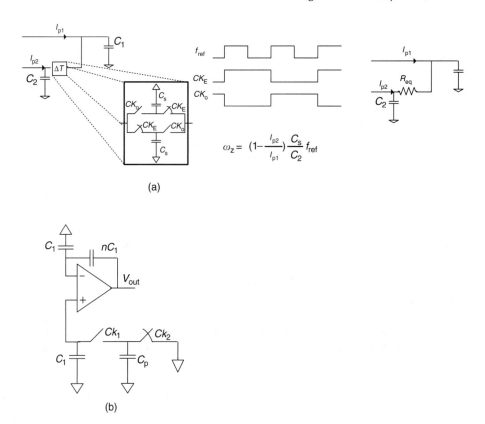

FIGURE 14.36   (a) Zero synthesis by precise delay and (b) discrete time sampling.

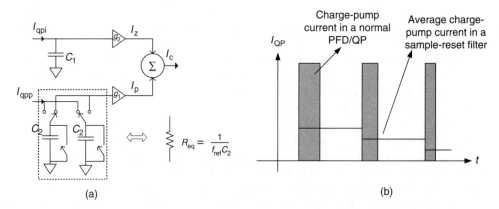

FIGURE 14.37   (a) Sample-reset loop filter and (b) average current generation.

In conclusion, in spite of the hectic research in the area of LPF implementation, this area holds the key to future advancements in low-power and low-area PLL-based frequency synthesizers. Miniaturization of technology has made it possible to implement several advanced DSP algorithms on chip. Such DSP-based filtering methods can significantly enhance the performance of future PLL-based frequency synthesizers.

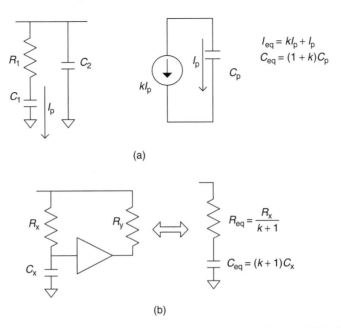

$$I_{eq} = kI_p + I_p$$
$$C_{eq} = (1 + k)C_p$$

(a)

$$R_{eq} = \frac{R_x}{k+1}$$

$$C_{eq} = (k+1)C_x$$

(b)

FIGURE 14.38 Capacitor multiplication technique as described in (a) Ref. 42 and (b) Ref. 43.

VOLTAGE-CONTROLLED OSCILLATORS

Apart from LPF, purity of on-chip VCO has been another stumbling block in complete integration of PLL system. A low-noise VCO design has engaged generations of researchers. Table 14.2 reviews some important phase-noise theories, which have evolved specifically for circuit designers. The first model presented by Leeson [45] is a phenomenological model, yet quite insightful for a tuned LC resonator–based design. Hence, this model for the first time predicted the classical 20 dB/decade roll-off of phase noise with respect to frequency. On the basis of LTI considerations, this model could not predict either the exact value of phase noise in $1/f^2$ region (Figure 14.39a) (patched through excess noise factor F) or close-in phase noise behavior (patched as shown in Ref. 45). Although it did indicate that the quality of the tank (Q) dictates the phase noise of a resonator-based oscillator, provided there is no guideline for the active part of the circuit design. Figure 14.39b presents a simplified version of a very popular VCO structure. For oscillations to occur, the negative resistance provided by the cross-coupled nMOS differential pair compensates the tank loss. Rael and Abidi [46] derived an equation for F (Table 14.2) for the circuit configuration in Figure 14.39b. On the basis of the insights gained, it was found that the bottom current source transistor is a major source of noise. An implementation to this effect was presented in Ref. 47, wherein the noise from current sources was LC filtered. Razavi [48] presented a Q-based phase noise model for ring oscillators, extended to other LC-based structures in Ref. 49. Finally Hajimiri and Lee present a simulation-based phase-noise model [50]. They propose an impulse sensitivity function (ISF), which is essentially a charge perturbation response of an oscillator. This model does provide sufficient guidelines for circuit design and predicts all regions of experimentally observed phase noise. This concept was extended to design a noise-shifting differential Colpitts oscillator in Ref. 53. Additional design guidelines for LC-based oscillators have also been presented by Ham and Hajimiri in Ref. 52. The key idea presented in the work is that LC oscillators for optimal noise should be operated at the juncture of current-limited regime (a regime where the swing of oscillator depends on bias current in contrast to a voltage-limited regime, where the swing saturates irrespective of the increase in bias current) and a voltage-limited regime.

After best efforts to minimize the intrinsic noise sources, modern integrated VCO are highly susceptible to extrinsic noise sources such as those originating from supply or substrate. The frequency

TABLE 14.2
Catalog of VCO Phase-Noise Theory for Circuit Design

Reference	Phase Noise Equation	Comments
[45]	$10.0 \log \left[\dfrac{2kTF}{P_c} \dfrac{1}{4Q_L{}^2} \left(\dfrac{f_c}{\Delta f} \right)^2 \right]$	P_c F f_c Δf Q_L
[48]	$10.0 \log \left[\dfrac{1}{4Q^2} \left(\dfrac{f_c}{\Delta f} \right)^2 \right]$	$Q = \dfrac{\omega_c}{2} \sqrt{\left(\dfrac{dA}{d\omega} \right)^2 + \left(\dfrac{d\phi}{d\omega} \right)^2}$ A ϕ ω
[46]	$10.0 \log \left[\dfrac{2kTF}{P_c} \dfrac{1}{4Q_L{}^2} \left(\dfrac{f_c}{\Delta f} \right)^2 \right]$	$F = 1 + \dfrac{4\gamma IR}{pV_0} + g \dfrac{4}{9} g_{\text{mbias}} R$ g_{mbias} γ V_0 I R
[50]	$10.0 \log \left[\dfrac{\Gamma^2{}_{\text{rms}}}{q^2{}_{\text{max}}} \dfrac{\overline{i_n^2} \big/ \Delta f}{4\Delta\omega^2} \right]$	Γ_{rms} q_{max}

P_c, power in the carrier frequency; F, effective noise figure; f_c, carrier frequency or oscillation frequency; Δf, offset frequency from the carrier frequency; Q_L, operating or loaded quality factor; A, gain of the delay-stage in a ring oscillator; ϕ, phase of the delay stage, ω frequency, g_{mbias} transconductance of the cross-coupled pair; γ, excess noise factor in devices; V_0, peak amplitude of oscillation; I, tail current; R, effective load resistance at frequency of oscillation; Γ_{rms}, value for the impulse sensitivity function; q_{max}, maximum charge displacement on an oscillating node.

change at the output of VCO in hertz for every millivolt of change in supply is called supply pushing. Hence, the VCO should be designed to minimize the supply-pushing figure. Differential structures are highly immune to these perturbations. However, most VCO have to be tuned over an extended frequency range and they do so using varactors (variable capacitors). These varactors are highly nonlinear and respond to common-mode voltage variations. Hence, care has to be taken to minimize such effects even in a differential structure. To alleviate the effect of substrate noise, VCO should be well shielded by using deep N-well technologies.

Finally, fully integrated VCO have to integrate very high-quality passives, such as inductors and varactors. A study of on-chip passives will not be covered in this chapter; however, literature has been replete with such studies and readers are encouraged to explore those papers.

PRESCALERS (DIVIDERS)

An integer-N prescaler (or divider) is what renders the PLL the characteristics of frequency synthesis. This section examines first the evolution of modern-day prescaler architecture and then delves into implementation details for dividers.

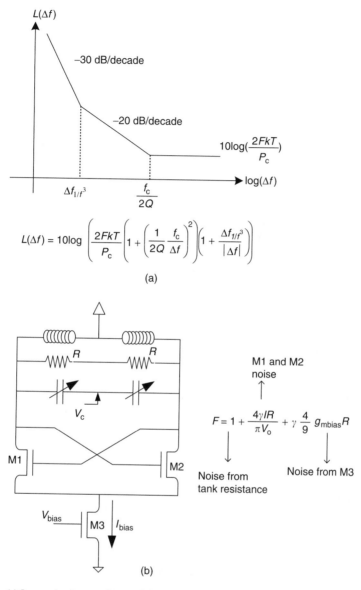

FIGURE 14.39 (a) Leeson's phase noise model. (b) CMOS implementation of a VCO.

Although dividers do not pose much challenge for integration of the PLL system, they pose a substantial one for speed, power, and design reuse. The challenge has been to define the delineation between extent of programmability and power. Figure 14.40 shows the progression of divider architecture, finally showing the most popular one. Conceptually a divider is nothing but a digital counter; however, such an implementation is very power hungry and sometimes impossible to implement due to speed limitations. This motivated to split the dividers into two parts, the first part being a fixed one (and hence could be optimized for power and speed) and the second part being the programmable one (and hence could not be optimized for power; however, it runs at a lower speed and is a low-power block). Resolution of such dividers is limited to the modulus of the first divider. Finally the dual-modulus dividers (DMDs) emerged to overcome the aforediscussed limitations. The first block is the DMD with division ratio as either P or $P + 1$. This is followed by a set of two counters (essentially a down counter) N and M. Both N and M are programmable; however, N is always greater than M. The idea

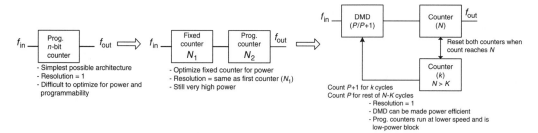

FIGURE 14.40   Evolution of a power-efficient prescaler topology.

is that DMD divides by $P + 1$ for M cycles and then by P for the rest of $N - M$ cycles. The division ratio achieved thereby is $NP + M$. Thus, the resolution is essentially determined by the resolution of M counter. DMD thereby acts as a pulse swallower or inserter when it divides by $P/P + 1$ or $P + 1/P$, respectively. In spite of various advantages, such divider architecture still suffers from finite speed limitation due to loading at output of the DMD and reset time periods.

This paragraph will discuss the implementation technique for both fixed divider and DMD. The divider topology can be broadly categorized into analog and digital. Figure 14.41 shows various kinds of analog implementation techniques. Figure 14.41a shows what is called regenerative or Miller method for division. The technique first proposed by Miller in 1939 [54] is based on mixing the output with input and passing the result through a selective filter, which selects the desired frequency at the output. An example implementation for divide-by-2 is also shown in Figure 14.41a. Such regenerative dividers can operate at really high frequencies, even close to $f_T/2$. A 40 GHz regenerative divider in 0.18 μm has been reported in Ref. 55. Although regenerative dividers are capable of operating at very high frequency, they are not best suited for low-power applications. Parametric divider is another popular structure in microwave applications. Parametric division is a process in which subharmonic oscillation is generated from a nonlinear reactive element, for example, the nonlinear $C–V$ characteristics of a varactor. An example implementation method is shown in Figure 14.41b [56]. Such dividers operate over a wide frequency range and are very low-power solutions; however, they are not amenable to integration due to high Q components required. Another analog technique is the method of injection-locked frequency division (ILFD). The principle of forced oscillations (or injection locking) in nonlinear circuits was first studied formally by van der pol [57]. Injection-locked oscillators (ILO) can lock themselves to the fundamental mode (called first-harmonic oscillators). However, they can be made to lock themselves to either subharmonic of or superharmonic of the injected frequency. Thus, an ILO could be operated as frequency divider when the incident frequency is a harmonic of the oscillation of frequency. This class of ILO has been shown in Figure 14.41c and has been extensively studied in Ref. 44.

Analog dividers have been successfully used in practice, for applications requiring high-speed operation albeit fixed division ratios (like 2). Hence, they have found wide usage as front of the complete prescaler topology to lower the speed of operation for following circuits. However, analog dividers are yet to achieve the kind of programmability that digital dividers can. This sets the stage for a brief review of digital divider topologies. A very popular approach is the Johnson counter shown in Figure 14.42a, configured as div2 (i.e., divide-by-2). The flip-flops themselves can be implemented using CMOS logic, dynamic logic, or current mode logic (CML) (Figure 14.42c). CML implementation typically achieves very high speed of operation, although consumes more power than their CMOS and dynamic logic counterparts. Other methods to implement div2 are true single phase clock (TSPC) register [58] or the ones described in Refs. 59 and 60.

DMD is also implemented using digital flop-flops (FF); however, the dual modulus is obtained by logical insertion between the two flops, which serves to swallow an extra pulse on command of the control signal. An implementation of 2/3 DMD is shown in Figure 14.42b. Cascading fixed

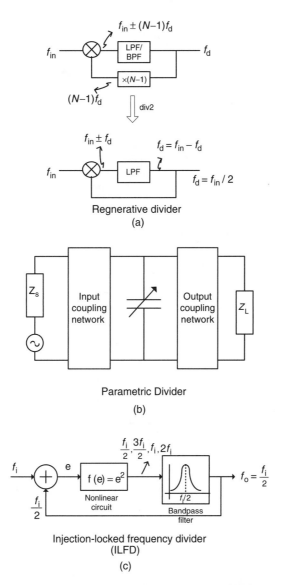

FIGURE 14.41 (a) Regenerative divider; (b) parametric divider; and (c) ILFD.

dividers with 2/3 DMD can obtain higher dual modulus. This cascading of divide values can be done either synchronously or asynchronously. Synchronous methods have the advantage of retiming the jitter at the output at the expense of higher power (as each flop works at same frequency), while in asynchronous methods each subsequent stage works at a lower frequency and hence lower power; however, they suffer from jitter accumulation. DMD implemented as in Figure 14.43 suffers from sever speed limitations, mostly due to the delay and loading associated with the critical paths. Phase-switching topology [62] overcomes this speed limitation. An example method is shown in Figure 14.43. Essentially the pulse swallow or insertion is achieved by shifting from one phase to the next. Implementation of such topologies should make sure that glitches do not appear while switching from one phase to the next.

To summarize this section, a generic divider architectural approach is illustrated in Figure 14.44. The approach shows the myriad possibilities and best practices associated while portioning a complete divider.

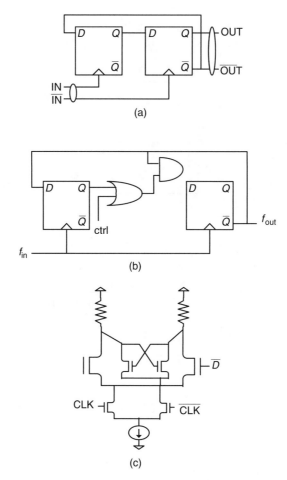

FIGURE 14.42 (a) Johnson counter; (b) 2/3 DMD prescaler; and (c) CML style transistor-level implementation of a DFF.

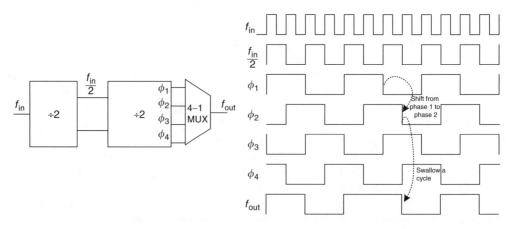

FIGURE 14.43 Phase-switching topology for $P/P+1$ modulus control.

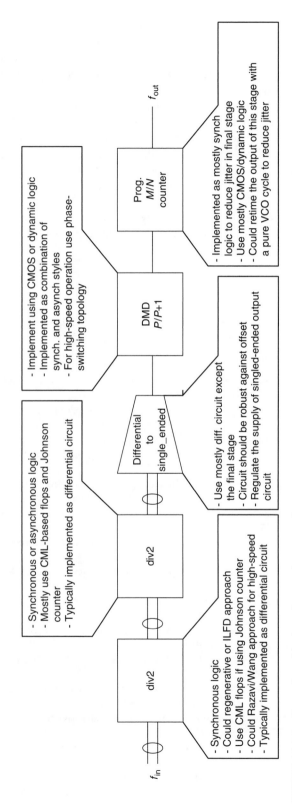

FIGURE 14.44 General partitioning of a complete prescaler.

CONCLUSION

The chapter started with a historical perspective on the advent of PLL-based frequency synthesis. The "Introduction" section also emphasizes the kind of background required while studying such PLL. The section "PLL System Specification" explores the system-level specifications for the design of an integer-N frequency synthesizer. The section also explores quantitative and qualitative basis for the origin of these specifications. The section "Linear Modeling of PLL" examines the linear modeling aspect for PLL-based synthesizer. This section forms the mathematical base for the rest of the chapter. The first part of the section "Architectures" discusses the design trade-offs for balancing various system specifications. This section emphasizes the conflicting requirements that a PLL designer is confronted with. The next part of the section then offers some architectural solutions to counter such conflicts. The section focuses only on key frequency synthesizer requirements, such as settling time, phase noise, spurious response, and continuous time-modeling assumptions. Finally the section "Circuit Implementation for Building Blocks" discusses some of the circuit implementation challenge associated with the building blocks. The section emphasizes the nonidealities for each block and then poses possible solutions with appropriate references.

REFERENCES

1. H. de Bellescize, "La Reception Synchrone," *Onde Electr.*, vol. 11, pp. 230–240, 1932.
2. J. Yong, S. Kim, S. Kim, and B. Jeon, "Fast switching frequency synthesizer using direct analog technique for phase-array radar," *Radar 97 (Conf. Pub. No. 449)*, pp. 386–390, October 1997.
3. A. Rokita, "Direct analog synthesis modules for an X-band frequency source," *International Conference on Microwaves and Radar*, vol. 1, pp. 63–68, May 1998.
4. V. Reinhardt, "Spur reduction technique in direct digital synthesizers," in *Proceedings of the International Frequency Control Symposium*, pp. 230–241, June 1993.
5. G. Chien and P. Gray, "A 900 MHz local oscillator using a DLL based frequency multiplier for PCS applications," *IEEE J. Solid-State Circuits*, vol. 35, pp. 1996–1999, December 2000.
6. A. A. Abidi, "Radio-frequency integrated circuits for portable communications," in *Proceedings of the IEEE 1994 Custom Integrated Circuits Conference (CICC)*, San Diego, USA, pp. 151–158, May 1994.
7. K. R. Wendt and G. L. Frendall, "Automatic frequency and phase control of synchronization in Television Receivers," *Proceedings of the IRE*, vol. 42, pp. 106–132, January 1954.
8. R. B. Sepe and R. I. Johnston, "Frequency multiplier and frequency waveform generator," *US patent No. 3,553,826*, December 29, 1970.
9. V. Choudhary, "Spectral transformation in front end of rf transceivers," Tutorial presented at *IWSOC 2005*, Banff.
10. F. M. Gardener, "Charge-pump phase-lock loop," *IEEE Trans. Commn.*, vol. COM-28, pp. 1849–1858, November 1980.
11. V. F. Kroupa, "Noise properties of PLL systems," *IEEE Trans. Commn.*, vol. COM-30, pp. 2244–2252, October 1988.
12. A. Hajimiri, "Noise in phase-locked loops," A Mixed-Signal Design, 2001 SSMDS, 2001 *South-West Symposium*, pp. 1–6, February 25–27, 2001.
13. J. W. Scott and J. P. Hein, "z-Domain model for discrete-time PLL's," *IEEE Trans. Circuits Sys.*, vol. 35, pp. 1393–1400, November 1988.
14. P. K. Hanumolu, M. Brownlee, K. Mayaram, and Un-Ku Moon, "Analysis of charge-pump phase-locked loops," *IEEE Trans. Circuits Sys. I*, vol. 51, no. 9, pp. 1665–1674, September 2004.
15. J. Lu, B. Grung, S. Anderson, and S. Rokhsaz, "Discrete Z-domain analysis of high order phase locked loops," *Circuits Sys., 2001 ISCAS 2001, The 2001 IEEE International Symposium*, vol. 4, no. 6–9, May 2001.
16. Z. Wang, "An analysis of charge-pump phase-locked loop," *IEEE Trans. Circuits Sys. I*, vol. 52, no. 10, pp. 2128–2138, October 2005.
17. C. Vaucher, "An adaptive PLL tuning system architectures combining high spectral purity and fast settling time," *IEEE J. Solid-State Circuits*, vol. 35, no. 4, pp. 490–502, April 2000.

18. C. Hur, YoungShig Choi, HyekHwan Choi, and TaeHa Kwon, "A low jitter phase-lock loop based on a new adaptive bandwidth controller," *Proceedings of the 2004 IEEE Asia-Pacific Conference on Circuits and Systems*, vol. 1, no. 6–9, pp. 421–424, December 6–9, 2004.

19. C. Lam and B. Razavi, "A 2.6-GHz/5.2-GHz frequency synthesizer in 0.4-μm CMOS technology," *IEEE J. Solid-State Circuits*, vol. 35, no. 5, pp. 788–794, May 2000.

20. B. Razavi, "Challenges in the design of frequency synthesizers for wireless applications," in *Proceedings of 1997 IEEE Custom Integrated Circuit Conference (CICC)*, pp. 395–402.

21. R. Nonis, N. Da Dalt, P. Palestri, and L. Selmi, "Modeling, design and characterization of a new low-jitter analog dual tuning LC-VCO PLL architecture," *IEEE J. Solid-State Circuits*, vol. 40, no. 6, pp. 1303–1309, June 2005.

22. K. Iniewski, S. Magierowski, and M. Syrzycki, "Phase locked loop gain shaping for gigahertz operation," *2004 Proceedings of the 2004 International Symposium on Circuits and Systems (ISCAS '04)*, vol. 4, pp. 157–160, May 23–26, 2004.

23. S. D. Vamvakos, C. Werner, and B. Nikolic, "Phase-locked loop architecture for adaptive jitter optimization," *Proceedings of the 2004 International Symposium on Circuits and Systems (ISCAS '04)*, vol. 4, pp. 161–164, May 23–26, 2004.

24. M. Mansuri, "Jitter optimization based on phase-locked loop design parameters," *IEEE J. Solid-State Circuits*, vol. 37, no. 11, pp. 1375–1382, November 2002.

25. J. Kakkinen, "Speeding up an integer-N PLL by controlling the loop filter charge," *IEEE J. Solid-State Circuits*, vol. 50, no. 7, pp. 343–354, July 2003.

26. S . Levantino, "Fast-switching analog PLL with finite-impulse response," *IEEE Trans Circuits Sys. I*, vol. 51, no. 9, pp. 1697–1701, September 2004.

27. I. Hwang, "A digitally controlled phase-locked loop with fast locking scheme for clock synthesis application," *Proc. ISSCC Dig. Tech.*, pp. 204–205, February 2000.

28. S. Ye, L. Jansson, and I. Galton, "A multiple-crystal interface PLL with VCO realignment to reduce phase noise," *IEEE J. Solid-State Circuits*, vol. 37, no. 12, pp. 1795–1803, December 2002.

29. L. Lin and P. Gray, "A 1.4 GHz differential low-noise CMOS frequency synthesizer using a wideband PLL architecture," *Proc. ISSCC Dig. Tech.*, pp. 204–205, February 2000.

30. R. Best, *Phase-Locked Loop*, 3rd ed., McGraw Hill, 1996.

31. M. Soyuer, "Frequency limitations of a conventional phase-frequency detector," *IEEE J. Solid-State Circuits*, vol. 25, no. 4, pp. 1019–1022, August 1990.

32. M. Mansuri, "Fast frequency acquisition phase frequency detectors for G-samples/s phase-locked loops," *IEEE J. Solid-State Circuits*, vol. 37, no. 10, pp. 1331–1334, October 2002.

33. W. Rhee, "Design of high-performance CMOS charge-pumps in phase-locked loops," *Proceedings of International Symposium on Circuits and Systems (ISCAS)*, vol. 2, pp. 545–548, Orlando, FL, July 1999.

34. V. Kaenel, "A 320 MHz, .5 mW @1.35 V CMOS PLL for microprocessor clock generation," *IEEE J. Solid-State Circuits*, vol. 31, no. 11, pp. 1715–1722, November 1996.

35. J. G. Maneatis, "Low-jitter process-independent DLL and PLL based on self-biased techniques," *IEEE J. Solid-State Circuits*, vol. 31, no. 11, pp. 1723–1732, November 1996.

36. K. Shu, E. Sanchez-Sinencio, and J. Silva-Martinez, "A 2.1-GHz monolithic frequency synthesizer with robust phase switching prescaler and loop capacitance scaling," *Proceedings of International Symposium on Circuits and Systems (ISCAS 2002)*, vol. 4, pp. 791–794, Phoenix, AZ, May 2002.

37. D. Mijuskovic, "Cell-based fully integrated CMOS frequency synthesizers," *IEEE J. Solid-State Circuits*, vol. 29, pp. 271–279, March 1994.

38. J. Craninckx and M. Steyaert, "A fully integrated CMOS DCS-1800 frequency synthesizer," *IEEE J. Solid-State Circuits*, vol. 33, pp. 2054–2065, December 1998.

39. T. Lee and B. Razavi, "A stabilization technique for phase-locked frequency synthesizer," in *Proceedings of 2001 Symposium VLSI Circuits*, Kyoto, Japan, pp. 39–42, June 2001.

40. M. Perrot, T. Tewksbury, and C. Sodini, "A 27-mW CMOS fractional-N frequency synthesizer using digital compensation for 2.5-Mb/s GFSK modulation," *IEEE J. Solid-State Circuits*, vol. 32, pp. 2048–2060, December 1997.

41. A. Maxim, B. Scott, E. M. Schneider, M. L. Hagge, S. Chacko, and D. Stiurca, "A low-jitter 125-1250-MHz process-independent and ripple-poleless 0.18-μm CMOS PLL based on a sample-reset loop filter," *IEEE J. Solid-State Circuits*, vol. 36, no. 11, pp. 1673–1683, November 2001.

42. J. Craninckx and M. S. J. Steyaert, "A fully integrated CMOS DCS-1800 frequency synthesizer," *IEEE J. Solid-State Circuits*, vol. 33, no. 12, pp. 2054–2065, December 1998 (DOI 10.1109/4.735547).

43. P. Larsson, "An offset cancelled CMOS clock-recovery/demux with a half-rate linear phase detector for 2.5 Gb/s optical communication," in *Int. Solid-State Circuits Conf. (ISSCC) Dig. Tech. Pap.*, pp. 74–75, February 2001.

44. H. Rategh and T. Lee, *"Multi-GHz Frequency Synthesis and Division,"* Kluwer Academic Publishers, 2001.

45. D. B. Leeson, "A simple model of feedback oscillator noises spectrum," *Proceedings of IEEE*, vol. 54, pp. 329–330, February 1966.

46. J. J. Rael and A. A. Abidi, "Physical processes of phase noise in differential LC oscillators," in *Proceedings of IEEE Custom Integrated Circuits Conference*, pp. 569–572, Orlando, FL, 2000.

47. E. Hegazi, H. Sjoland, and A. A. Abidi, "A filtering technique to lower LC oscillator phase noise," *IEEE J. Solid-State Circuits*, vol. 36, no. 12, pp. 1921–1930, December 2001.

48. B. Razavi, "A study of phase noise in CMOS oscillators," *IEEE J. Solid-State Circuits*, vol. 31, no. 3, pp. 331–343, March 1996.

49. J. van der Tang and D. Kasperkovitz, *High-Frequency Oscillator Design For Integrated Transceivers*, Kluwer Academic Publishers, 2003.

50. A. Hajimiri and T. Lee, "A general theory of phase noise in electrical oscillators," *IEEE J. Solid-State Circuits*, vol. 33, no. 2, pp. 179–194, February 1998.

51. A. Hajimiri, S. Limotyrakis, and T. H. Lee, "Jitter and phase noise in ring oscillators," *IEEE J. Solid-State Circuits*, vol. 34, no. 6, pp. 790–804, June 1999.

52. D. Ham and A. Hajimiri, "Concepts and methods in optimization of integrated LC VCOs," *IEEE J. Solid-State Circuits*, vol. 36, no. 6, pp. 896–909, June 2001.

53. R. Aparicio and A. Hajimiri, "A CMOS differential noise-shifting Colpitts VCO," *Int. Solid-State Circuits Conf. (ISSCC) Dig. Tech. Pap.*, pp. 288–289, February 2002.

54. R. L. Miller, "Fractional frequency generation utilizing regenerative modulation," *Proceedings of the IRE*, pp. 446–457.

55. J. Lee and B. Razavi, "A 40-GHz frequency divider in 0.18 μm CMOS technology," *IEEE J. Solid-State Circuits*, vol. 39, no. 4, pp. 594–601, April 2004.

56. I. Bahl and P. Bhartia, *Microwave Solid State Circuit Design*, Wiley, New York, 1988.

57. B. van der pol, "Forced oscillations in a circuit with non-linear resistance (reception with reactive triode)," *Philos. Mag.*, vol. 3, pp. 361–376, January 1927.

58. J. Yuan and C. Svensson, "High-speed CMOS circuit technique," *IEEE J. Solid-State Circuits*, vol. 24, pp. 62–70, February 1989.

59. B. Razavi, "Design of high speed, low power frequency dividers and phase-locked loops in deep submicron CMOS," *IEEE J. Solid-State Circuits*, pp. 101–109, February 1995.

60. H. Wang, "A 1.8 V 3 mW 16.8 GHz frequency divider in 0.25 μm CMOS," *Intl. Solid-State Circuits Conf. (ISSCC) Dig. Tech. Pap.*, pp. 196–197, February 2000.

61. C. S. Vaucher, I. Ferencic, M. Locher, S. Sedvallson, U. Voegeli, and Z. Wang, "A family of low-power truly modular dividers in standard 0.35 μm CMOS technology," *IEEE J. Solid-State Circuits*, vol. 35, no. 7, pp. 1039–1045, July 2000.

62. J. Craninckx and M. S. Steyaert, *Wireless CMOS Frequency Synthesizer Design*, Kluwer Academic Publishers, 1998.

15 Frequency Synthesis for Multiband Wireless Networks

John W. M. Rogers, Foster F. Dai, and Calvin Plett

CONTENTS

Introduction...427
Basic Phase Noise Concepts 428
Phase-Locked Loop Synthesizer Architectures 431
    Integer-*N* PLL Synthesizers...431
    Fractional-*N* PLL Frequency Synthesizers..431
Building Block Phase Noise Models for PLL Synthesizer 434
    VCO Noise...434
    Crystal Reference Noise 434
    Frequency Divider Noise 435
    Phase Detector Noise 435
    Charge Pump Noise 435
    Loop Filter Noise..435
    Phase Noise Due to $\Sigma\Delta$ Converters 436
In-Band and Out-of-Band Phase Noise in PLL Synthesis 438
Implementation of a Multiband Fractional-*N* $\Sigma\Delta$ Synthesizer 440
    Programmable $\Sigma\Delta$ Modulator with Precalculated Seeds 440
    Programmable Multimodulus Divider Phase Frequency
        Detector and Charge Pump 442
    Multiband VCO Designs..445
Complete Phase Noise Analysis and Comparison with Measurements 447
Conclusions 452
References...452

INTRODUCTION

High-speed frequency synthesis is one of the most challenging areas in radio frequency integrated circuit (RFIC) design. It requires a diverse knowledge of both high-speed analog and digital circuits as well as a deep knowledge of system-level issues. The performance requirements on circuits used for frequency synthesis are often extremely demanding making the design of these blocks even more challenging. However, a high-performance frequency synthesizer is a key component in many wired (fiber or cable) and wireless communication systems.

For modern multistandard applications, it is often difficult to cover multiple frequency bands using classical integer-*N* frequency synthesizers whose step size is limited by the reference frequency. To achieve fine step size to cover the multiband channel frequencies, one has to lower the reference frequency in an integer-N synthesizer design, which results in high division ratio of the phase-locked loop (PLL) and thus high in-band phase noise. In contrast, a fractional-N synthesizer allows the PLL to operate with a high reference frequency while achieving fine step size by constantly swapping the loop division ratio between integer numbers; thus, the average division ratio is a fractional number [1–4]. However, fine step size and low in-band phase noise are achieved with the penalty of fractional spurious tones, which come from the periodical division ratio variation. To remove the fractional spurious components for a synthesizer with fine step size, the best solution is to employ a $\Sigma\Delta$ noise shaper to control a programmable divider. A $\Sigma\Delta$ noise shaper will help to move large spurs to higher frequencies where they can be easily filtered. Although spurs are often one of the most important design considerations for a frequency synthesizer, they will not be treated in detail in this chapter. Since these techniques are becoming more and more common in modern synthesizer design, noise in this type of synthesizer will be the focus of this chapter.

Here, a theoretical analysis of phase noise in modern frequency synthesizers will be presented. Phase noise is often the most challenging and crucial performance specification that must be met by a synthesizer. It is also the specification that often proves the most difficult to model and simulate. In this chapter, a review of basic phase noise concepts will be presented, followed by a model that will allow the designer to take noise data from individual circuit simulations and predict the overall phase noise performance of an entire PLL frequency synthesizer.

The proposed analytical model will then be used to predict and optimize the phase noise performance of a $\Sigma\Delta$ fractional-N frequency synthesizer designed for multiband wireless local area network (WLAN) applications. The comparison between the simulated and the measured phase noise demonstrates that the analytical model can accurately predict the performance of the complete synthesizer and provide the designer with a quick and reliable means to predict the phase noise performance of a synthesizer RFIC prior to its fabrication.

BASIC PHASE NOISE CONCEPTS

Noise in synthesizers is contributed from all the building block circuits and components that make up the PLL. Synthesizer noise performance is usually expressed as phase noise, which is a measure of how much the output differs from an ideal impulse function in the frequency domain. The primary concern is with noise that causes fluctuations in the phase of the output rather than noise that causes fluctuations in the amplitude, since the output typically has a fixed and limited amplitude. The output of a synthesizer can be described as

$$v_{out}(t) = V_0 \sin(\omega_{LO}t + \varphi_n(t)) \tag{15.1}$$

where $\omega_{LO}t$ is the desired phase of the output and $\varphi_n(t)$ the time-variant random phase fluctuation of the output signal due to any noise sources in the PLL. Phase noise is often quoted in units of dBc/Hz or rad^2/Hz.

The phase fluctuation term $\varphi_n(t)$ in Equation 15.1 may be random phase noise or discrete spurious tones, also called spurs, as shown in Figure 15.1. The discrete spurs at a synthesizer output are most probably due to the fractional-N mechanism, whereas the phase noise in an oscillator is mainly due to thermal, flicker or $1/f$ noise and the finite Q of the oscillator tank. Assume the phase fluctuation is of a sinusoidal form as

$$\varphi(t) = \varphi_p \sin(\omega_m t) \tag{15.2}$$

where φ_p is the peak phase fluctuation and ω_m the offset frequency from the carrier. Substituting Equation 15.2 into Equation 15.1 gives:

$$v_{out}(t) = V_0 \cos[\omega_c t + \varphi_p \sin(\omega_m t)]$$

$$= V_0[\cos(\omega_c t)\cos(\varphi_p \sin(\omega_m t)) - \sin(\omega_c t)\sin(\varphi_p \sin(\omega_m t))] \tag{15.3}$$

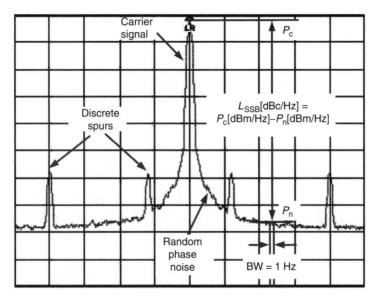

FIGURE 15.1   An example of phase noise and spurs at the synthesizer output observed using a spectrum analyzer. (From Rogers, J. W. M., Plett, C., and Dai, F. F., *Integrated Circuit Design for High-Speed Frequency Synthesis*, Artech House, 2006. With permission. © Artech House 2006.)

For a small phase fluctuation, the Equation 15.3 can be simplified as

$$v_0(t) = V_0[\cos(\omega_c t) - \varphi_p \sin(\omega_m t)\sin(\omega_c t)]$$

$$= V_0[\cos(\omega_c t) - \frac{\varphi_p}{2}[\cos(\omega_c + \omega_m)t - \cos(\omega_c - \omega_m)t]] \tag{15.4}$$

It is now evident that the phase-modulated signal includes the carrier signal tone and two symmetric sidebands at any offset frequency. A spectrum analyzer measures the phase-noise power in dBm/Hz, but often phase noise is reported relative to the carrier power as

$$PN(\Delta\omega) = \frac{\text{Noise}(\omega_{LO} + \Delta\omega)}{P_{\text{carrier}}(\omega_{LO})} \tag{15.5}$$

where *Noise* is the noise power in a 1-Hz bandwidth and P_{carrier} is the power of the carrier or local oscillator (LO) tone at the frequency at which the synthesizer is operating. In this form, phase noise has the units of rad^2/Hz. Often this is quoted as so many decibels down from the carrier in units of dBc/Hz. To further complicate this, both single-sideband (SSB) and double-sideband phase noise can be defined. SSB phase noise is defined as the ratio of power in one phase modulation sideband per hertz bandwidth, at an offset $\Delta\omega$ away from the carrier, to the total signal power. The ratio of SSB phase noise power spectral density (PSD) to carrier power, in dBc/Hz, is defined as

$$PN_{SSB}(\Delta\omega) = 10\log\left[\frac{N(\omega_{LO} + \Delta\omega)}{P_{\text{carrier}}(\omega_{LO})}\right] \tag{15.6}$$

Combining Equations 15.4 and 15.6 gives

$$PN_{SSB}(\Delta\omega) = 10 \log \left[\frac{\frac{1}{2}\left(\frac{V_0\varphi_p}{2}\right)^2}{\frac{1}{2}V_0^2} \right]$$

$$= 10 \log \left[\frac{\varphi_p^2}{4} \right] = 10 \log \left[\frac{\varphi_{rms}^2}{2} \right] \tag{15.7}$$

where φ_{rms}^2 is the rms phase-noise power density in rad^2/Hz. Note that SSB phase noise is by far the most common type reported and often it is not specified as SSB, but rather simply reported as phase noise. However, alternatively double sideband phase noise can be expressed by

$$PN_{DSB}(\Delta\omega) = 10 \log \left[\frac{N(\omega_{LO} + \Delta\omega) + N(\omega_{LO} - \Delta\omega)}{P_{carrier}(\omega_{LO})} \right]$$

$$= 10 \log \left[\varphi_{rms}^2 \right] \tag{15.8}$$

From either the SSB or double-sideband phase noise, the rms phase noise can be obtained in the linear domain as

$$\varphi_{rms}(\Delta\omega) = \frac{180}{\pi} \sqrt{10^{PN_{DSB}(\Delta\omega)/10}}$$

$$= \frac{180\sqrt{2}}{\pi} \sqrt{10^{PN_{SSB}(\Delta\omega)/10}} \left[\deg/\sqrt{Hz} \right] \tag{15.9}$$

It is also quite common to quote integrated phase noise over a certain bandwidth. The rms integrated phase noise of a synthesizer is given by

$$IntPN_{rms} = \sqrt{\int_{\Delta\omega_1}^{\Delta\omega_2} \varphi_{rms}^2(\omega)\, d\omega} \tag{15.10}$$

The limits of integration are usually the offsets corresponding to the lower and upper frequencies of the bandwidth of the information being transmitted.

In addition, it should be noted that dividing or multiplying a signal in the time domain also divides or multiplies the phase noise. Similarly, if a signal is translated into frequency by a factor of N, then the phase noise power is increased by a factor of N^2 as

$$\varphi_{rms}^2(N\omega_{LO} + \Delta\omega) = N^2 \cdot \varphi_{rms}^2(\omega_{LO} + \Delta\omega)$$

$$\varphi_{rms}^2\left(\frac{\omega_{LO}}{N} + \Delta\omega\right) = \frac{\varphi_{rms}^2(\omega_{LO} + \Delta\omega)}{N^2} \tag{15.11}$$

Note that this assumes that the circuit that did the frequency translation is noiseless. Otherwise, additional phase noise will be added. Also, note that the phase noise is scaled by N^2 rather than N because we are dealing with noise in units of power rather than units of voltage.

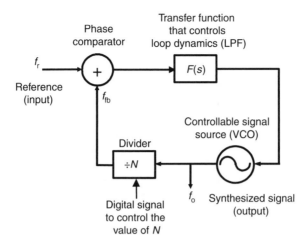

FIGURE 15.2 A simple integer-N PLL frequency synthesizer. (From Rogers, J. W. M., Plett, C., and Dai, F. F., *Integrated Circuit Design for High-Speed Frequency Synthesis*, Artech House, 2006. With permission. © Artech House 2006.)

PHASE-LOCKED LOOP SYNTHESIZER ARCHITECTURES

In this section, an overview of system-level configurations associated with phase-locked loop synthesizers will be given. Additional information on synthesizer architectures can be found in Ref. 6.

INTEGER-N PLL SYNTHESIZERS

An integer-N PLL is the simplest type of phase-locked loop synthesizer and is shown in Figure 15.2. Note that N refers to the divide-by-N block in the feedback of the PLL, and the two choices are to divide by an integer, or integer N, or to divide by a fraction, fractional N, essentially by switching between two or more integer values such that the effective divider ratio is a fraction. PLL-based synthesizers are among the most common ways to implement a synthesizer. The PLL-based synthesizer is a feedback system that compares the phase of a reference f_r to the phase of a divided down output of a controllable signal source f_{fb} (also known as a voltage-controlled oscillator [VCO]). The summing block in the feedback is commonly called a phase detector. Through feedback, the loop forces the phase of the signal source to track the phase of the feedback signal, and therefore their frequencies must be equal. Thus, the output frequency, which is a multiple of the feedback signal, is given by

$$f_o = N \cdot f_{ref} \qquad (15.12)$$

Since N is an integer, the minimum step size of this synthesizer is equal to the reference frequency f_r. Therefore to get a smaller step size, the reference frequency must be smaller. This is often undesirable, so instead a fractional-N design is often used.

FRACTIONAL-N PLL FREQUENCY SYNTHESIZERS

In contrast to an integer-N synthesizer, a fractional-N synthesizer allows the PLL to operate with high reference frequency while achieving a fine step size by constantly swapping the loop division ratio between integer numbers. As a result, the average division ratio is a fractional number. As will be shown, a higher reference frequency leads to lower in-band phase noise and faster PLL transient responses. In addition, for multiband applications, the channel spacing of the different bands is often skewed, requiring an even lower reference frequency if the synthesizer is to cover both bands.

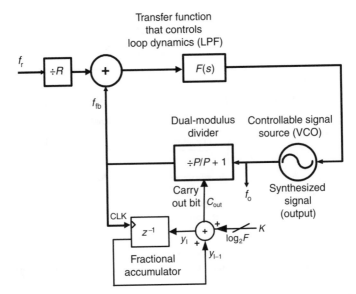

FIGURE 15.3   A fractional-N frequency synthesizer with a dual-modulus prescaler. (From Rogers, J. W. M., Plett, C., and Dai, F. F., *Integrated Circuit Design for High-Speed Frequency Synthesis*, Artech House, 2006. With permission. © Artech House 2006.)

Figure 15.3 illustrates one way to implement a simple fractional-N frequency synthesizer with a dual-modulus prescaler $P/P + 1$. Note that it is called a dual-modulus prescaler because it can be programmed to two division ratios. As discussed in the previous section, the fractionality can be achieved by toggling the divisor value between two values P and $P + 1$. The modulus control signal (C_{out}) is generated using an accumulator (also called an integrator or adder with feedback, or a counter) with size of F (or $\log_2 F$ bits). That is, an overflow occurs whenever the adder output becomes equal to or larger than F. At the ith rising clock edge, the accumulator's output y_i can be mathematically expressed as

$$y_i = (y_{i-1} + K_i) \bmod F \tag{15.13}$$

where y_{i-1} is the output on the previous rising clock edge and K_i is a user-defined input, and its value will determine the fractional divider value. Note that the modular operation ($A \bmod B$) returns the remainder of ($A \div B$) and is needed for modeling the accumulator overflow.

If the dual-modulus prescaler divides by P when C_{out} is low and divides by $P + 1$ when C_{out} is high, the average VCO output frequency is

$$f_o = \frac{f_r}{R} \left[\frac{(P+1)K + P(F-K)}{F} \right] = \frac{f_r}{R} \left[P + \frac{K}{F} \right] \tag{15.14}$$

Because fractionality is achieved by using this accumulator, it is often called a fractional accumulator. It has a fixed size F due to a fixed number of accumulator bits built into the hardware. The dual-modulus prescaler ratio P is normally fixed as well. The only programmable parameter for the architecture shown in Figure 15.3 is the accumulator input K, which can be programmed from one to a maximum of F. Thus, since K is an integer, the step size of this PLL architecture is given by

$$\text{Step size} = \frac{f_r}{RF} \tag{15.15}$$

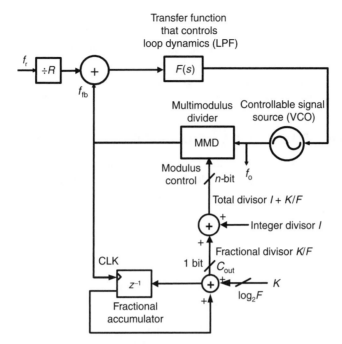

FIGURE 15.4 A fractional-N frequency synthesizer with a multimodulus divider. (From Rogers, J. W. M., Plett, C., and Dai, F. F., *Integrated Circuit Design for High-Speed Frequency Synthesis*, Artech House, 2006. With permission. © Artech House 2006.)

where R is normally fixed to avoid changing the comparison frequency at the input. Note that R is normally as small as possible to minimize the in-band phase noise contribution from the crystal. Thus, step size is inversely proportional to the number of bits ($\log_2 F$); as a result, the accumulator is normally used to reduce synthesizer step size without increasing R and degrading the in-band phase noise.

Replacing the dual-modulus divider with a multimodulus divider (MMD), the fractional-N synthesizer architecture shown in Figure 15.3 can be modified as illustrated in Figure 15.4. Using an MMD has the advantage that the range of frequencies over which the synthesizer can be tuned is expanded, compared to the architecture using dual-modulus divider. The synthesizer output frequency is given by

$$f_o = \frac{f_r}{R}\left[I + \frac{K}{F}\right] \tag{15.16}$$

where I is integer portion of the loop divisor, and, depending on the complexity of the design, I could have many possible integer values. For instance, if loop division ratio 100.25 is needed, we can program $I = 100$, $K = 1$, and $F = 4$. The MMD division ratio is toggled between 100 and 101.

A fractional-N synthesizer achieves fine step size and low in-band phase noise with the penalty of fractional spurious tones, which comes from the periodic division ratio variation. Fractional spurs may be removable by using a high-order loop transfer function if the closest spur is outside of the PLL bandwidth. Note that the spacing of the closest spur to the carrier is determined by the synthesizer step size. For a synthesizer with fine step size smaller than the transfer function bandwidth, it is thus practically impossible to remove fractional spurs by using a loop low-pass filter (LPF). Reducing the loop bandwidth to combat the fractional spurs results in the penalty of longer lock time and increased out-of-band phase noise due to the VCO. Even if the closest spur is outside of the loop filter bandwidth, removing those spurs normally requires a high-order loop filter with sharp roll-off, which increases the complexity and cost of the synthesizer. To remove the fractional spurious components for a synthesizer with fine step size, the best solution is to employ a $\Sigma\Delta$ noise shaper in the fractional accumulator. Its function is to break up the repeated patterns of the loop divisor time sequence without affecting its average division ratio. This will result in the reduction or elimination of the spurs in the spectrum.

BUILDING BLOCK PHASE NOISE MODELS FOR PLL SYNTHESIZER

Next, we will present the phase noise models for all PLL synthesizer building blocks such as the crystal oscillator, divider, phase-frequency detector (PFD), charge pump (CP), loop LPF, and VCO. Although the circuit or block-level simulation of a typical synthesizer design will not be discussed in detail in this chapter, some basic theory will be presented to show how the noise in each block can affect the loop performance.

VCO Noise

The phase noise from a VCO can be described as [5,6]

$$\varphi_{VCO}^2(\Delta\omega) = \left(\frac{\omega_o}{(2Q\Delta\omega)}\right)^2 \left(\frac{GkT}{2P_S}\right)\left(1+\frac{\omega_c}{\Delta\omega}\right) \tag{15.17}$$

where P_S is the signal power of the carrier, T the temperature, Q the quality factor of the oscillator's resonator, k the Boltzmann's constant, ω_o the frequency of oscillation, ω_c the flicker noise corner frequency, and G a constant of proportionality, which takes into account the excess noise from the VCO transistors, and nonlinearity. Note that many additional refinements have been made to this formula; however, as given here, it is sufficient to capture the shape of most integrated VCO's phase noise. Thus, at most frequencies of interest, the phase noise produced by the VCO will decrease at the rate of 20 dB/decade for an increasing offset frequency away from the carrier. This will not continue indefinitely because thermal noise will put a lower limit on this phase noise, which is somewhere between -120 and -150 dBc/Hz for most integrated VCOs. VCO phase noise is usually dominant outside the loop bandwidth and of less importance at low offset frequencies.

Crystal Reference Noise

Crystal resonators are widely used in frequency control applications because of their unequaled combination of high Q, stability, and small size. The resonators are classified according to cut, which is the orientation of the crystal wafer (usually made from quartz) with respect to the crystal-lographic axes of the material. The total noise PSD of a crystal oscillator can also be found from Leeson's formula and by making use of a typical empirical multiplier [7]:

$$\varphi_{XTAL}^2(\Delta\omega) = 10^{-16\pm1} \cdot \left[1+\left(\frac{\omega_o}{2\Delta\omega \cdot Q_L}\right)^2\right]\left[1+\frac{\omega_c}{\Delta\omega}\right] \tag{15.18}$$

where ω_o is the oscillator output frequency, ω_c the corner frequency between $1/f$ and thermal noise regions, which is normally in the range 1–10 kHz, and Q_L the loaded quality factor of the resonator. Since Q_L for crystal resonator is very large (normally in the order of 10^4–10^6), the reference noise contributes only to the very close-in noise, and it quickly reaches thermal noise floor at offset frequency around ω_c.

FREQUENCY DIVIDER NOISE

Frequency dividers consist of switching logic circuits, which are sensitive to the clock timing jitter. The jitter in the time domain can be converted to phase noise in the frequency domain. Time jitter or phase noise occurs when rising and falling edges of digital dividers are superimposed with spurious signals such as Johnson noise and flicker noise in semiconductor materials. Ambient effects result in variation of the triggering level due to temperature and humidity. Frequency dividers generate spurious noise especially for high-frequency operation. Dividers do not generate signals, but rather simply change their frequency. Kroupa provided an empirical formula, which estimates the amount of phase noise that frequency dividers add to a signal [8,9]:

$$\varphi_{\text{Div\_Added}}^2(\Delta\omega) \approx \frac{10^{-14\pm1} + 10^{-27\pm1}\omega_{\text{do}}^2}{2\pi\cdot\Delta\omega} + 10^{-16\pm1} + \frac{10^{-22\pm1}\omega_{\text{do}}}{2\pi} \tag{15.19}$$

where ω_{do} is the divider output frequency and $\Delta\omega$ is the offset frequency. Notice that the first term in Equation 15.19 represents the flicker noise and the second term gives the white thermal noise floor. The third term is caused by timing jitter due to coupling, ambient effects, and supply variations.

PHASE DETECTOR NOISE

Phase detectors experience both flicker and thermal noise. At large offsets, phase detectors generate a white phase noise floor typically about -160 dBc/Hz, which is thermal noise dominant. The noise PSD of phase detectors is estimated empirically by [9]

$$\varphi_{\text{PD}}^2(\Delta\omega) \approx \frac{2\pi\cdot10^{-14\pm1}}{\Delta\omega} + 10^{-16\pm1} \tag{15.20}$$

CHARGE PUMP NOISE

The noise of the CP can be characterized as an output noise current and is usually given in pA/$\sqrt{\text{Hz}}$. Note that this point in the loop current represents the phase. The CP output current noise can be a strong function of the reference frequency and width of the current pulses. Therefore, for low-noise operation, it is desirable to keep the CP sink and source currents well matched. This is because current sources only produce noise when they are on. When an ideal loop is locked, the sink and source current sources in a CP are turned off, resulting in zero net current charge or discharge of the holding capacitor. However, nonidealities result in finite pulses that will turn on the source and sink currents for about the same amount of time. The closer reality matches the ideal case, the less noise will be produced. Also, note that as the offset frequency is decreased, $1/f$ noise will become more important, causing the noise to increase. This noise can often be the dominant noise source at low-frequency offsets. CP noise can be simulated with proper tools such as Cadence periodic steady state (PSS) Pnoise analysis. The results depend on the design in question, so no simple general analytical formula will be given here; however, an example will be given later.

LOOP FILTER NOISE

Loop filters can be analyzed for noise in the frequency domain in a linear manner. The most common loop filters that will be examined in this chapter will now be analyzed. A loop filter consists of two capacitors and one resistor. For off-chip filters, the loss experienced by capacitors is negligible. Thus, the loop filter contains only one noise source, the thermal noise associated with the resistor R. The loop filter with its associated noise source can be drawn as shown in Figure 15.5. Now the

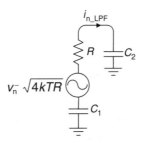

FIGURE 15.5 Loop filter with thermal noise added. (From Rogers, J. W. M. et al., *EURASIP J. Wireless Commn. Networking*, vol. 2006, Article ID 48489, pp. 1–11, May 2006. With permission. © Hindawi 2006.)

noise voltage develops a current flowing through the series combination of C_1, C_2, and R (assuming that the CP and VCO are both open circuits), which is given by

$$i_{n\_LPF} = \frac{1}{R} \cdot \frac{v_n s}{s + \dfrac{C_1 + C_2}{C_1 C_2 R}} \approx \frac{1}{R} \cdot \frac{v_n s}{s + \dfrac{1}{C_2 R}} \tag{15.21}$$

Thus, this noise current will have a high-pass characteristic; therefore, the loop will not produce any noise at DC (zero frequency), and this noise will increase until the high-pass corner is reached, after which it will be flat. Other filters can be analyzed in a similar manner.

PHASE NOISE DUE TO ΣΔ CONVERTERS

Fractional-N synthesizers often include ΣΔ modulators to shift the spurious components to a higher frequency band, where the loop filter can filter randomized spurs. In a ΣΔ fractional-N synthesizer, the average loop divisor value corresponds to the desired output frequency, and the instantaneous divisor value is dithered around the correct value by the ΣΔ modulator. The ΣΔ noise shaping can be modeled as a linear gain stage with an additive quantization noise source, which is shaped by a high-pass transfer function. Hence, the quantization error component at the synthesizer output is composed of mostly high-frequency noise that can be filtered by the PLL [21]. A block diagram of a typical ΣΔ modulator that is widely used in synthesizer applications is shown in Figure 15.6 [3]. This three-loop ΣΔ topology is called a multi-stage noise shaping (MASH) 1-1-1 structure because it is a cascaded ΣΔ structure with three first-order loops. Each of the three loops is identical. The input of the second loop is taken from the quantized error E_{q1} of the first loop, whereas the input of the third loop is taken from the quantized error E_{q2} of the second loop. Thus, only the first loop has a constant input, which is the fractional portion of the desired rational divide number $\cdot F(z)$, i.e., the fine tune word. The integer part of the frequency word $I(z)$, the coarse tune word, is added at the output of the three-loop ΣΔ modulator. Thus, $N_{div}(z) = I(z) + F(z)$ is the time sequence used to control the integer-restricted divider ratios. The modulator is clocked at the divider output frequency, reflecting the sampled nature of the circuit.

The first loop generates the fractional divisor value $F(z)$ with the by-product of quantization error E_{q1}, which is further fed to the input of the second loop for further processing. The second loop cancels the previous loop's quantization error E_{q1} by the additional filter block $(1 - z^{-1})$ in its output path. The only quantization noise term left after summing the first and second loop outputs is the quantization error E_{q2}, which is second-order noise-shaped. When this noise term is further fed to the input of the third loop, the loop generates a negative noise term to cancel the previous loop's quantization error E_{q2} by the additional filter block $(1 - z^{-1})^2$ in its output path. Summing the outputs of the three loops, we obtain the modulated divisor value as

$$N(z) = I(z) + N_1(z) + N_2(z) + N_3(z)$$

$$= I(z) + F(z) + (1 - z^{-1})^3 E_{q3}(z) \tag{15.22}$$

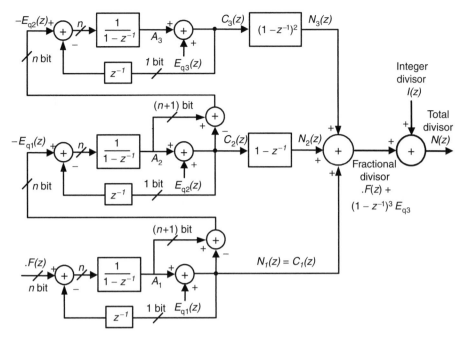

FIGURE 15.6 A three-loop MASH 1-1-1 $\Sigma\Delta$ modulator for fractional-N synthesis. (From Rogers, J. W. M., Plett, C., and Dai, F. F., *Integrated Circuit Design for High-Speed Frequency Synthesis*, Artech House, 2006. With permission. © Artech House 2006.)

where $I(z)$ and $\cdot F(z)$ are the integer portion and the fractional portion of the division ratio, respectively. As desired, the fractional divisor value $\cdot F(z)$ is not affected by the modulator, whereas the quantization error generated in the last loop E_{q3} is noise shaped by a third-order high-pass function of $(1 - z^{-1})^3$. The quantization errors generated in the first and second loops are totally canceled, and as a result, the total quantization noise is equal to that of a single loop, although three loops are used. Therefore, the multiloop $\Sigma\Delta$ architecture provides high-order noise shaping without additional quantization noise.

Discrete fractional spurs are generated by this circuit at multiples of the reference frequency, but these spurs become more like random noise after $\Sigma\Delta$ noise shaping. The SSB phase noise of the noise-shaped fractional spurs can be analyzed as follows. The 1-bit quantization error power is $\Delta^2/12$, where Δ is the quantization step size. For $\Delta = 1$, which is the case for a truncated binary word, the quantization error power is $1/12$. This error power is spread over the sampling bandwidth, or equivalently the reference bandwidth of $f_r = 1/T_s$. Thus, the error PSD becomes $1/(12f_r)$. Considering the noise shaping with an mth-order MASH $\Sigma\Delta$ modulator as expressed in Equation 15.22, the frequency noise PSD is obtained as

$$S_\Omega(z) = \frac{\left|(1 - z^{-1})^m f_r\right|^2}{12 f_r} = \frac{1}{12}(1 - z^{-1})^{2m} f_r \tag{15.23}$$

where the subscript Ω denotes the frequency fluctuations referred to the *input* of the divider. To obtain the phase fluctuations, consider the following relationship between frequency and phase:

$$\omega(t) = \frac{d\phi(t)}{dt} \approx \frac{\phi(t) - \phi(t - T_s)}{T_s} \tag{15.24}$$

and its z-domain representation of

$$2\pi \cdot \Omega(z) = \frac{\Phi(z)(1-z^{-1})}{T_s} \tag{15.25}$$

where $T_s = 1/f_r$ is the sample period and where multiplication by z^{-1} represents a delay of T_s. Rearranging this expression yields

$$\Phi(z) = \frac{2\pi \cdot \Omega(z)}{f_r(1-z^{-1})} \tag{15.26}$$

Noting that $S_\Omega(z)$ is given in terms of power, the double-sideband phase noise PSD is obtained as

$$S_\Phi(z) = S_\Omega(z)\frac{(2\pi)^2}{\left|1-z^{-1}\right|^2 f_r^2}$$

$$= \frac{(2\pi)^2}{\left|1-z^{-1}\right|^2 f_r^2} \cdot \frac{1}{12}(1-z^{-1})^{2m} f_r$$

$$= \frac{(2\pi)^2}{12 f_r} \cdot (1-z^{-1})^{2m-2} \tag{15.27}$$

where the subscript Φ denotes phase fluctuations. Noting that

$$(1-z^{-1}) = \left|1-e^{-j\omega T}\right| = 2\sin\left(\frac{\omega T}{2}\right) = 2\sin\left(\frac{\pi f}{f_r}\right) \tag{15.28}$$

the SSB phase noise PSD in the frequency domain is given by

$$\frac{\varphi^2_{\Sigma\Delta}(f)}{2} [\text{rad}^2/\text{Hz}] = \frac{(2\pi)^2}{24 f_r} \cdot \left[2\sin\left(\frac{\pi f}{f_r}\right)\right]^{2(m-1)}$$

$$\text{PN}(f) [\text{dBc/Hz}] = 10\log\left\{\frac{(2\pi)^2}{24 f_r} \cdot \left[2\sin\left(\frac{\pi f}{f_r}\right)\right]^{2(m-1)}\right\} \tag{15.29}$$

IN-BAND AND OUT-OF-BAND PHASE NOISE IN PLL SYNTHESIS

A system-level diagram of a typical PLL-based synthesizer that will be analyzed in this chapter is shown in Figure 15.7. It consists of a PFD, a CP, a loop filter, a VCO, a programmable divider, a reference oscillator (typically a crystal reference source), and a fractional accumulator with $\Sigma\Delta$ modulation circuit to achieve the fine synthesizer step size without impacting the phase noise performance.

The noise transfer functions (NTFs) for the various noise sources in the loop can be derived using conventional control theory [9,10]. There are three additive NTFs: one for the VCO noise that is the contributor of the synthesizer out-of-band noise, one for the $\Sigma\Delta$ modulator noise that could contribute to both in-band and out-of-band noise, and one for all other noise sources such as the PFD, CP, divider, and loop filter that are the contributors of the in-band noise. All in-band noise sources are referred back to the input of the PLL and shown as $\varphi_{\text{noise I}}$ in Figure 15.7. The noise from

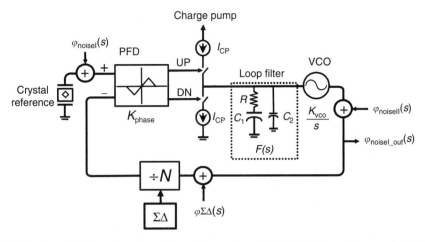

FIGURE 15.7 A synthesizer showing places where noise is injected. (From Rogers, J. W. M. et al., *EURASIP J. Wireless Commn. Networking*, vol. 2006, Article ID 48489, pp. 1–11, May 2006. With permission. © Hindawi 2006.)

the VCO is referred to the output and represented by $\varphi_{\text{noise II}}$ in Figure 15.7, whereas the noise from the $\Sigma\Delta$ modulator is shown as $\varphi_{\Sigma\Delta}$. The NTF for in-band noise $\varphi_{\text{noise I}}(s)$ is given by

$$\frac{\varphi_{\text{noise\_out}}(s)}{\varphi_{\text{noise I}}(s)} = \frac{\dfrac{IK_{\text{VCO}}}{2\pi \cdot C_1}(1 + RC_1 s)}{s^2 + \dfrac{IK_{\text{VCO}}}{2\pi \cdot N}Rs + \dfrac{IK_{\text{VCO}}}{2\pi \cdot NC_1}} \tag{15.30}$$

As shown, the in-band NTF has a low-pass characteristic. Note that for low frequencies inside the loop bandwidth, the loop will track the input phase including the input phase noise. Therefore, this noise will be transferred to the PLL output. At higher offset frequencies, this noise is suppressed by the loop's LPF. Thus, the noise coming from the PFD, CP, divider, and loop filter contributes to the in-band noise at the PLL output. Also, note that the division ratio plays a very important role in this transfer function. Within the loop bandwidth, the in-band phase noise is magnified N times by the loop. Therefore, choosing smaller divisor value N will benefit the in-band noise reduction.

The VCO NTF is slightly different. In this case, setting the input reference and input noise source to zero, the VCO NTF is given by

$$\frac{\varphi_{\text{noise\_out}}(s)}{\varphi_{\text{noise II}}(s)} = \frac{s^2}{s^2 + \dfrac{IK_{\text{VCO}}}{2\pi \cdot N}Rs + \dfrac{IK_{\text{VCO}}}{2\pi \cdot NC_1}} \tag{15.31}$$

As shown, the VCO NTF has a high-pass characteristic. Thus, at low offsets inside the loop bandwidth, the VCO noise is suppressed by the feedback loop; yet outside the loop bandwidth, the VCO is essentially free running without noise attenuation. Thus, the out-of-band PLL noise approaches the VCO noise.

The NTF of the $\Sigma\Delta$ modulator is very similar to the in-band NTF, except that an extra $1/N$ term in the numerator as the $\Sigma\Delta$ is not input referred. Note that due to the high-pass nature of the $\Sigma\Delta$ NTF, the order of the loop roll-off is very important. The noise-shaping slope of an mth-order MASH $\Sigma\Delta$ modulation is $20(m-1)$ dB/decade according to Equation 15.29, whereas an nth-order low-pass loop filter has a roll-off slope of $20n$ dB/decade. Therefore, the order of loop filter must be higher than or equal to that of the $\Sigma\Delta$ modulator to attenuate the out-of-band noise due to $\Sigma\Delta$ modulation.

Thus, for instance, when calculating the effect of the $\Sigma\Delta$ modulator on out-of-band noise on the typical loop, it is necessary to include additional capacitor C_2 in the loop filter because this will provide extra attenuation out-of-band. In this case, the $\Sigma\Delta$ NTF to the output would be

$$\frac{\varphi_{\text{noise\_out}}(s)}{\varphi_{\Sigma\Delta}(s)} = \frac{K_{\text{VCO}}K_{\text{phase}}(1+sC_1R)}{s^2N(C_1+C_2)(1+sC_sR)+K_{\text{VCO}}K_{\text{phase}}(1+sC_1R)} \tag{15.32}$$

where $C_s = C_1C_2/(C_1 + C_2)$.

IMPLEMENTATION OF A MULTIBAND FRACTIONAL-N $\Sigma\Delta$ SYNTHESIZER

With WLAN standards operating in very different frequency bands, market-leading WLAN solutions have to offer multimode interoperability with transparent worldwide usage. Moreover, the frequency allocation of the WLAN standards in the "unlicensed" 5 GHz band is constantly evolving. In particular, the Japanese government recently proposed four additional radio frequency (RF) channels in the 4.9–5.0 GHz band and further three channels in the 5.03–5.09 GHz band for this standard. This change could significantly increase the available channels for 5 GHz WLAN in Japan and create yet another difficulty for WLAN chipmakers by requiring them to enable access to this lower frequency band. It is thus desirable to design a WLAN transceiver with a future-proofed multiband frequency synthesis scheme against evolutions and changes in allocated spectrum worldwide.

This section presents a synthesizer designed for multiband multiple input multiple output (MIMO) WLAN applications. It uses a $\Sigma\Delta$ fractional-N architecture as described in Ref. 11. The synthesizer allows a high reference frequency, fine step size, and low divide ratio to achieve low in-band phase noise. The use of a resettable divide-by-4 block in the synthesizer enables the use of a walking intermediate frequency (IF) radio architecture and alleviates the need for an IF synthesizer in the radio. Next, we will present the details on the synthesizer implementation.

PROGRAMMABLE $\Sigma\Delta$ MODULATOR WITH PRECALCULATED SEEDS

The synthesizer includes $\Sigma\Delta$ accumulators to shift the spurious components associated with fractional-N synthesis to a higher frequency band, where randomized spurs can be filtered by the loop filter. The presented $\Sigma\Delta$ accumulator is a multiloop MASH structure, as illustrated in Figure 15.8. The MASH $\Sigma\Delta$ gains the advantage of superior close-in noise-shaping capability [2,3]. It is unconditionally stable and can be easily implemented for high-speed applications. As shown in Figure 15.8, the $\Sigma\Delta$ modulator consists of three cascaded accumulators with a single-bit carry from each accumulator. The carries from three accumulators are delayed properly and are then used to select coefficient banks derived from the MASH transfer function $(1 - z^{-1})^n$. Using precalculated coefficients speeds up the MASH calculation by eliminating a few adders. Moreover, every accumulator output is buffered, allowing pipelined operation. Thus, the speed constraint is reduced to only one accumulator calculation per $\Sigma\Delta$ clock cycle. The first accumulator in the MASH architecture provides fractional programmability with programmable accumulator size of F and accumulator input K. The three accumulators can be selected and powered down separately to allow programmable $\Sigma\Delta$ architectures in the following format:

1. An integer-N synthesizer, when only the coarse tune is selected and all three accumulators are powered down
2. A fractional-N synthesizer without noise shaping, when coarse tune and the first accumulator is selected and the second and the third accumulators are powered down
3. A fractional-N synthesizer with second-order $\Sigma\Delta$ noise shaping, when coarse tune and the first and the second accumulators are selected and the third accumulator is powered down
4. A fractional-N synthesizer with third-order $\Sigma\Delta$ noise shaping, when coarse tune and all three accumulators are selected

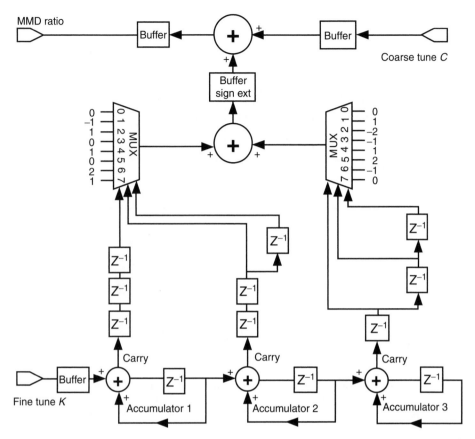

FIGURE 15.8 The second-order and third-order multiloop MASH $\Sigma\Delta$ fractional accumulators. (From Rogers, J. W. M., Dai, F. F., Cavin, M. S., and Rahn, D. G., *J. Solid State Circuits*, vol. 40, pp. 678–689, March 2005. With permission. © IEEE 2005.)

The programmable $\Sigma\Delta$ synthesizer architecture is desired for multiband applications, where different fractionality and noise-shaping effects are required. A lower order of $\Sigma\Delta$ should be chosen as long as it has enough spur rejection. For a multistage MASH structure, a higher order $\Sigma\Delta$ modulator corresponds to more output bits, which dither the loop divisor ratio in a wider range around its desired value (the average value of the instantaneous divisor values). The CP would be turned off only when the divisor value is equal to the desired value. Dithering the divisor value in a wider range will upset the CP with larger source or sink currents, which needs a longer time to be corrected by the loop. Thus, a higher order $\Sigma\Delta$ increases the modulated turn-on time of the CP in the locked condition, which makes the synthesizer more sensitive to substrate noise coupling and reference spur feed-through.

The presented $\Sigma\Delta$ accumulator has a special feature of loading precalculated start values to three accumulators to avoid artificial spurs at the synthesizer output. It is known that any repetition of time sequence will cause artificial spurs in the spectrum with a frequency inversely proportional to the repetition's period. Loading different start values (seeds) to different accumulators can break the repetition in the time domain. Thus, artificial spurs can be reduced using a proper start value for each accumulator. In addition, a different order of $\Sigma\Delta$ noise shaper can be selected simultaneously. Resetting the first accumulator with a loaded seed is more important than for the second and the third accumulators, since only the first accumulator has a constant input, which may result in a repeated accumulator value, i.e., artificial spurs. The second and the third accumulators take the outputs from previous accumulators as their inputs, which are more like random numbers. Thus, the values in the

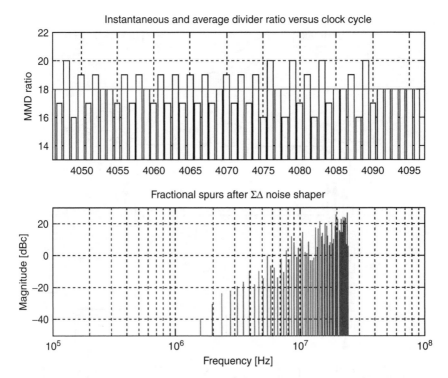

FIGURE 15.9 Simulated division ratio and its spectrum shaped by a third-order MASH $\Sigma\Delta$ fractional accumulator for the worst spur case with $F = 64$. (From Rogers, J. W. M., Dai, F. F., Cavin, M. S., and Rahn, D. G., *J. Solid State Circuits*, vol. 40, pp. 678–689, March 2005. With permission. © IEEE 2005.)

second and the third accumulators are not likely to be repeated. All required WLAN channels have been carefully simulated, and the desired order as well as seed values for the $\Sigma\Delta$ modulators has been determined for each channel to meet the spur requirement of less than $-50\,\text{dBc}$. The proposed spur reduction scheme with precalculated accumulator seed values can be simply implemented with the existing accumulator reset and hence does not require additional hardware, whereas other spur randomization schemes such as dithering the least significant bit (LSB) of the $\Sigma\Delta$ input not only require additional hardware but also cause large frequency variation for small input words.

The $\Sigma\Delta$ fractional-N block is designed in $0.5\,\mu\text{m}$ complementary metal oxide semiconductor (CMOS) technology using a Verilog-driven digital design flow. Figure 15.9 gives Verilog simulation results of the MMD ratio and its spectrum for the worst spur case with $N = 81 + 63/64$, where coarse tune $C = 81 - 64 = 17$ (MMD ratio $= 64 + C$), fractionality $F = 64$, and fine tune $K = 63$. All three accumulator outputs are selected, and the fractional spurs are thus shaped by a third-order MASH $\Sigma\Delta$ fractional accumulator. Figure 15.9 shows how the third-order $\Sigma\Delta$ dithers the division ratio around its average value ($17 + 63/64$, as shown by the straight line in the upper plot) in the time domain. As a result, it clearly demonstrates the third-order noise-shaping effect of $60\,\text{dB/decade}$ (bottom plot) in the frequency domain.

Programmable Multimodulus Divider Phase Frequency Detector and Charge Pump

A 6-bit programmable MMD is designed on the basis of the topology of cascaded two-third dual-modulus prescalers as shown in Figure 15.10. The divide-by-2/3 cells can be implemented in various topologies such as those described in Ref. 8. An set reset (SR) latch-based digital tristate-type PFD is chosen to provide phase and frequency comparison between the reference frequency and the VCO divided frequency as illustrated in Figure 15.10. It contains an exclusive-OR gate that can be used as

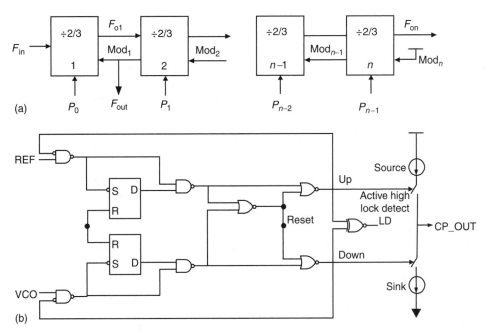

(a)

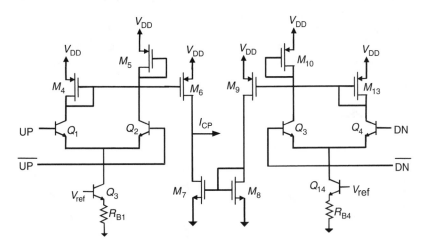

(b)

FIGURE 15.10 (a) MMD using cascaded divide-by-2/3 cells and (b) PFD and CP. (From Rogers, J. W. M., Dai, F. F., Cavin, M. S., and Rahn, D. G., *J. Solid State Circuits*, vol. 40, pp. 678–689, March 2005. With permission. © IEEE 2005.)

FIGURE 15.11 Differential charge pump circuitry with CMOS current mirrors. (From Rogers, J. W. M., Dai, F. F., Cavin, M. S., and Rahn, D. G., *J. Solid State Circuits*, vol. 40, pp. 678–689, March 2005. With permission. © IEEE 2005.)

a lock indicator. This PFD has only one zero-crossing over its phase detection range, which ensures that the device does not lock to the second and the third harmonics.

The CP following the PFD is differential with single-ended output as shown in Figure 15.11. The PFD input phase error is represented by the PFD up (UP) and down (DN) output pulses, which control the CP source and sink current sources, resulting in current flow in or out of the CP.

The source/sink current I_{CP} can be programmed for eight different settings from 125 to 1875 μA. These currents can be used to adjust the loop gain and therefore to compensate for VCO gain variations. A 3-bit select signal controls the current mirrors whose transistors size is binary weighted. Thus, I_{CP} can be adjusted in eight steps.

Nonlinearity associated with the phase detector and CP has a great impact on the synthesizer performance. Nonlinearity is even a bigger concern when the spur compensation is done in the analog domain where a correction signal is produced by a digital-to-analog converter (DAC) and the CP needs to linearly transfer the spur content to an analog signal for cancellation [20]. LO phase accuracy is desired in MIMO applications where multiple synthesizers are used and their output phases need to be synchronized. There are many sources that can cause slow LO phase drift including PFD/CP, dead zone, and unbalanced propagation delay variation in separate synthesizer paths. Dead zone occurs when the PFD and CP gain drops to zero and thus the loop loses the sensitivity for any phase difference smaller than this range. PFD/CP gain deviating from the ideal linear curve is not necessarily a dead zone problem as long as the slope of the gain curve is not zero and the loop can still respond to the input phase difference. For a tristate PFD, its output at lock contains narrow pulses and it is very difficult for a CP with a capacitive load to linearly convert those narrow pulses into corresponding currents. To minimize the dead zone problem, various antibacklash circuitries can be added to the PFD, and transistors in PFD and CP should also be optimized and balanced. As discussed earlier, the optional LO porting block eliminates many of these LO drift issues associated with separate PLLs locked to a common reference.

Figure 15.12 compares the transfer functions of the PFD and PFD/CP designed with field-effect transistors (FETs) at different temperatures and VCO tuning voltages. In the simulation, the input contains only one pulse with a width of 10 ns (50 MHz) and the output is averaged over one reference period (20 ns) right after the propagation delay. It can be seen that the PFD itself contributes very little nonlinearity under temperature variations. The CP dominates the nonlinearity. At the center of the VCO tuning voltage (~1.35 V) and around room temperature (20–40°C), the temperature coefficient is about 0.037°/°C. Note that the flat part around the lock point corresponds to the dead zone of the PFD and CP transfer curve, where the loop loses the gain to correct any phase error smaller than the dead zone. As a result, the synthesizer output signal may jitter within the dead zone, causing phase error and phase noise. The speed limit of the CP transistors for responding to the narrow pulses at the PFD outputs and the imbalance between the CP source and sink mirrors are the major source of the CP dead zone. For this reason, n-p-n transistors at the CP input differential pairs are used, although FETs are used in the CP mirrors. In addition to the dead zone problem,

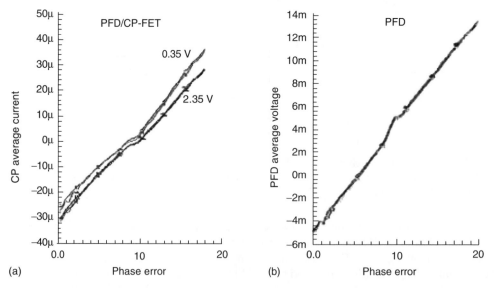

FIGURE 15.12 Transfer functions of (a) PFD/CP-FET and (b) PFD at different temperatures (20, 30, and 40°C) and VCO tuning voltages (0.35 and 2.35 V). Charge pump current setting I_{cp} = 1.875 mA. (From Rogers, J. W. M., Dai, F. F., Cavin, M. S., and Rahn, D. G., *J. Solid State Circuits*, vol. 40, pp. 678–689, March 2005. With permission. © IEEE 2005.)

the leakage of the loop filter capacitor increases the CP average turn-on time in the lock condition, which degrades the spurs and phase noise due to increased substrate noise coupling. The nonlinearity of the PFD/CP is caused by the inability of the PFD/CP to respond to extremely narrow pulses due to bandwidth limitations. To improve the PFD/CP linearity, a delay unit can be inserted in the PFD reset path to increase the pulse width at the PFD outputs in the locked state. Thus, the switching of the CP transistors does not need to operate at high speed. Introducing a small amount of leakage current at the CP output can also widen the narrow PFD pulses. Both schemes increase the CP turn-on time in the locked state and thus degrade the loop noise and spur performance. Moreover, the delay unit degrades the frequency performance of the PFD as well.

Multiband VCO Designs

The core of the oscillator is formed using two p-doped metal-oxide semiconductor (PMOS) transistors M_1 and M_2 connected in a negative resistance configuration and attached to the LC tank of the VCO as shown in Figure 15.13 [15,16]. The tank itself consists of two inductors L and a pair of p-n junction varactors C_{var}. p-n junction varactors are preferred in this technology because of their high Q at the frequencies of interest. However, they have one critical drawback. They have a parasitic substrate diode associated with the n side of the junction that has a low Q and parasitic capacitance. Unless the n side of the diode is connected to alternating current, or small-signal (AC) ground, the structure will include this lossy substrate diode impacting VCO performance [6,17]. Specifically, these diodes will cause a reduction in tank Q, and it was also observed that their presence can lead to significant power supply noise rejection issues. For this reason, the PMOS VCO core shown in Figure 15.13 is preferred. In this case, the tank can be connected to ground rather than to V_{CC} so that the diodes can be connected in the proper polarity. PMOS transistors also offer the advantage of higher output swing than bipolar transistors, provided high phase noise at offset frequencies below 100kHz (due to high flicker noise) can be tolerated. This is because the PMOS transistors can be operated into the triode region without affecting the VCO noise performance.

The tank inductors are made as large as possible to maximize the tank equivalent parallel resistance while still allowing the varactors to be large enough so that parasitic capacitance does not reduce the available frequency tuning range to an unacceptable level. The PMOS transistors themselves are sized so that they have a large DC V_{GS} voltage leaving enough headroom to accommodate the current source M_3. The inductor L_{Tail} and the capacitor C_{Tail} form a filter that is designed to filter out noise from the bias circuitry so that it does not affect the phase noise of the VCO [18].

The current through the VCO must be set large enough to maximize the voltage swing at the tank to minimize the phase noise of the circuit. The swing will be maximized when the transistors are made to alternate between saturation and cut-off at the top and bottom of the VCO voltage swing. Once the transistors reach this voltage swing level, raising the current will not cause the swing to grow any more, will increase the phase noise, and will waste current.

The automatic amplitude control (AAC) loop used in this design is also shown in Figure 15.13. Here transistors Q_1 and Q_2 are set nominally in cut-off and behave as class C amplifiers. They do nothing until the VCO amplitude reaches the desired level (set by choosing an appropriate value for R_E). Once this level is reached, Q_1 and Q_2 turn on briefly at the top of the swing and steal current away from Q_4. This in turn reduces the current flowing in the oscillator until it is just enough to provide that level of output swing. Diodes D_2 and D_3 are used to provide a low impedance to short out the noise generated by the band gap. C_1 is included to form a dominant and controllable pole in the feedback loop so that stability of the feedback loop is assured for all operating conditions. Note also that Q_1 and Q_2 are connected in such a manner so that they do not load the tank and provide additional loss when they turn on, impacting phase noise. This is a key advantage over previous designs that loaded the tank with the dynamic emitter resistance (r_e) of the limiting transistors [19]. In this case, these transistors do not act as clamping diodes. An output buffer is also included that isolates the tank from all the circuitry in the transceiver that it must eventually drive. This is a bipolar buffer with a current output that is combined with other cores so that they can be conveniently switched on or off without loading each other.

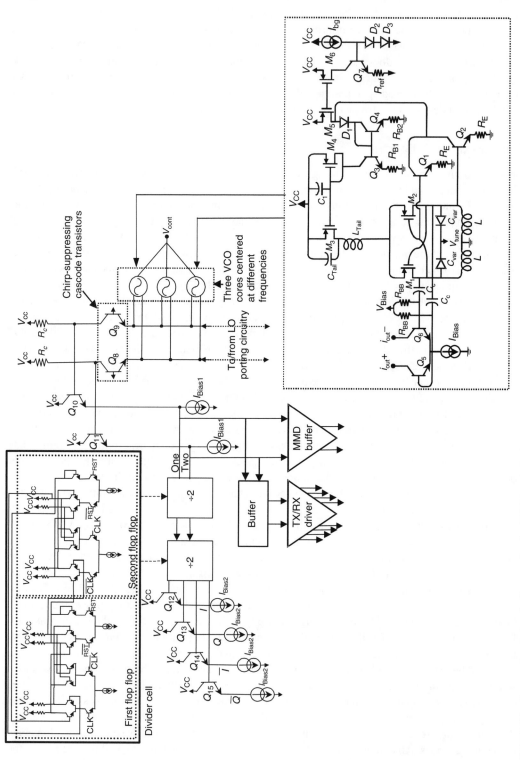

FIGURE 15.13 System-level circuit diagram for the VCOs and clock tree.

The complete VCO subsystem is also shown in Figure 15.13. All three VCO cores' open collector buffers are tied together and brought to load resistors R_C through cascode transistors, which are included to increase isolation to the RF buffers and therefore reduce chirp. Next, the signal is driven into two buffer transistors Q_{10} and Q_{11} that are driven with enough current to drive the 5 GHz divider and three additional output buffers. The three buffers are fed to the transceiver RF mixers and to the synthesizer buffer. The outputs of the second divider are used to produce quadrature differential signals that drive both transmit and receive IF stage mixers.

COMPLETE PHASE NOISE ANALYSIS AND COMPARISON WITH MEASUREMENTS

The methods for dealing with phase noise will now be considered with application to an actual synthesizer RFIC design case [22]. The results of the analysis can then be verified against measurement data. The synthesizer to be considered was designed using a 47 GHz 0.5 μm combined bipolar CMOS (BiCMOS) process using primarily the CMOS part of the technology. The only exceptions were some high-speed bipolar current-mode logic (CML) in the divider and the output buffer circuits. The rest of the synthesizer including the VCO cores was all CMOS. It was designed for multiband WLAN applications and had a reference frequency of 40 MHz, a fairly standard CP and PFD configuration with gain K_{phase} of 750 μA/2π, an MMD programmable between 64 and 127, and an LC-based VCO with a K_{VCO} of approximately 120 MHz/V. The synthesizer was designed to generate carrier frequencies in the range from 3.2 to 3.3 GHz and from 4.1 to 4.3 GHz. The MMD gave a total division ratio of 86–88 and 102–108 under normal operating conditions and was controlled by a third-order ΣΔ modulator to provide the needed step size and noise shaping. The crystal oscillator used as a reference for this design had a Q_L of 8×10^4 and a noise floor of −150 dBc/Hz. The details of the actual circuit implementation will not be discussed in this chapter but are similar to those given in Ref. 11. The raw VCO phase noise can be either predicted from a calculation [6] or simulated with the aid of Spectra or some other simulator. Output current noise from the CP/PFD combination can also be simulated or predicted from transistor-level noise calculations. This simulation must be done using driving signals in the locked state to simulate accurately the amount of time the CP spends in the on state. This simulation can be used to predict how much noise current is on average produced by the circuit. Likewise simulations on the divider can be performed. The crystal oscillator is normally a commercially available part, and data on its phase noise performance are often available from the manufacturer. The ΣΔ phase noise can be estimated from Equation 15.29. Note that the maximum fractionality used in this design was 1/32. Although this had an impact on the spurs of the system in different channels, the third-order ΣΔ has kept all the spurs below −50 dBc level such that the fractional spurs did not affect the phase noise of the system. Such simulations and calculations were performed for the sample design. The results of all raw phase noise due to circuit components are plotted in Figure 15.14. All phase noise is referred to the VCO output frequency for easy comparison of the relative importance of the phase noise sources.

Next the optimal loop bandwidth for best phase noise performance must be determined. This task can be accomplished using the following steps:

a. Plot all phase noise components.
b. Determine the intercept point of ΣΔ and VCO noise.
c. Compare it to the intercept between VCO noise and in-band noise (normally dominated by CP noise).
d. If the ΣΔ intercepts the VCO noise at a lower frequency than the in-band noise does, a higher order ΣΔ is needed to prevent in-band noise degradation. Then make sure that the higher order ΣΔ noise curve intercepts the VCO noise curve at a higher frequency than the in-band noise does.
e. Choose the intercept between the out-band noise (VCO noise) and the in-band noise (CP noise, reference noise, divider noise, etc.) as the loop optimal bandwidth.

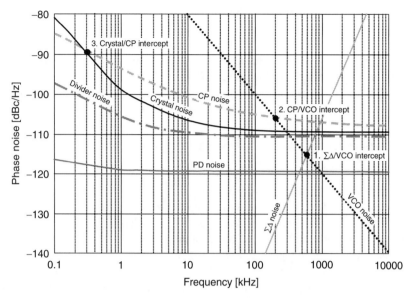

FIGURE 15.14 A plot of all raw phase noise components for the design referred to the VCO output frequency. (From Rogers, J. W. M. et al., *EURASIP J. Wireless Commn. Networking*, vol. 2006, Article ID 48489, pp. 1–11, May 2006. With permission. © Hindawi 2006.)

As an example, consider the plot shown in Figure 15.14. First the $\Sigma\Delta$ modulator used in the design must be considered. Since this noise increases with offset frequency, the loop bandwidth must be set low enough to properly attenuate this noise and prevent it from growing to dominate the phase noise of the design. Thus, the loop bandwidth must be set lower than the intercept of the VCO noise and the $\Sigma\Delta$ noise (see point 1 in Figure 15.14 at 600 kHz offset). For this design at frequencies between 300 Hz and 200 kHz, the in-band noise is dominated by CP, which is a fairly typical occurrence. This noise must also be weighed against the VCO noise and the intercept of these two noise sources (see point 2 in Figure 15.14 at 200 kHz offset). Note that this point is lower than the $\Sigma\Delta$ intercept with the VCO noise, and therefore it is the crucial point in this case that sets the loop bandwidth. Thus, the loop bandwidth should be set at the point where these two noise sources are equal. Setting the loop bandwidth wider would result in the loop phase noise being dominated by the CP when it could instead follow the lower VCO noise, and setting the loop bandwidth lower than this will result in the loop phase noise being dominated by the VCO, when it could instead follow the lower CP/PFD noise. Thus, in this design, the optimum loop bandwidth can be determined from the plot as the crossover point between these two curves at an offset frequency of 200 kHz. Therefore, the best possible out-of-band phase noise is the raw phase noise of the VCO, and the in-band phase noise will be dominated by the CP above a frequency of 300 Hz. Below this frequency the crystal oscillator noise will dominate the in-band noise (see point 3 in Figure 15.14 at 300 Hz offset).

Having determined the optimum loop bandwidth for best phase noise performance, the overall loop phase noise performance can be predicted with the aid of the theory developed in the section "Building Block Phase Noise Models for PLL Synthesizer." The loop filter components were chosen as shown in Table 15.1. A ratio of only 5:1 was chosen for C_1 and C_2 to help attenuate high-frequency $\Sigma\Delta$ phase noise and also to provide additional spur rejection. This can cause slight additional peaking in the phase noise at the loop corner frequency, but had a negligible impact on the integrated phase noise. Note that

TABLE 15.1
Loop Filter Components

Parameter	Value
C_1	3 nF
C_2	600 pF
R	600 Ω

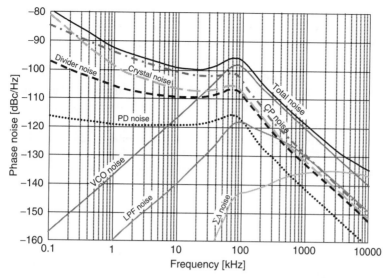

FIGURE 15.15 A plot of all phase noise including the effect of the loop. (From Rogers, J. W. M. et al., *EURASIP J. Wireless Commn. Networking*, vol. 2006, Article ID 48489, pp. 1–11, May 2006. With permission. © Hindawi 2006.)

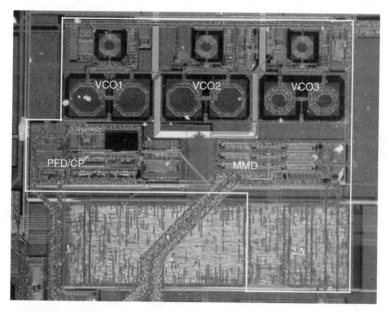

FIGURE 15.16 Die photograph of the synthesizer. (From Rogers, J. W. M. et al., *EURASIP J. Wireless Commn. Networking*, vol. 2006, Article ID 48489, pp. 1–11, May 2006. With permission. © Hindawi 2006.)

additional poles in the loop filter could lead to improved out-of-band performance, but since the loop filter was external in this experiment, this would have required additional package pins.

The overall phase noise as well as all noise components is plotted in Figure 15.15 for a divider ratio of 87. The phase noise for this design integrated from 100 Hz to 10 MHz was predicted to be 0.44° rms.

The synthesizer was fabricated and was embedded with the rest of the circuitry that formed the WLAN transceiver. The back end of the process featured thick aluminum metallization designed to provide high-quality inductors. A die photo of the synthesizer is shown in Figure 15.16. This particular design implemented three VCO cores; however, only two were required to cover all required WLAN frequencies. Each VCO had a tuning range of approximately 600 MHz. The synthesizer occupies an area of 2.3 mm × 1.4 mm. The synthesizer drew a current of 36 mA from a 2.75 V supply.

The measured and simulated phase noise is compared in Figure 15.17 for a division ratio of 87 and in Figure 15.18 for a division ratio of 105. The comparison demonstrates that the overall PLL noise performance is predicted very closely by simulation and calculation. Thus, the proposed analytic model provides a rigorous model for analyzing PLL synthesizer phase noise performance. The model can serve as a design guide for synthesizer designers to optimize their circuits and meet their design goals prior to the expensive fabrication. The measured integrated phase noise of the WLAN synthesizer was 0.5° rms for the lower band and 0.535° rms for the upper band, which are close to the predicted phase noise values. These results are summarized in Table 15.2.

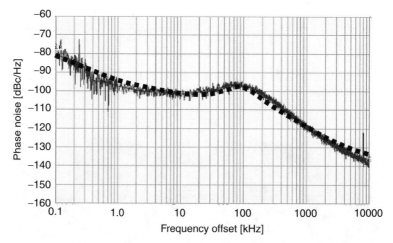

FIGURE 15.17 Comparison of measured and simulated phase noise for the 3.2–3.3 GHz band. The square dots are the simulated data. (From Rogers, J. W. M. et al., *EURASIP J. Wireless Commn. Networking*, vol. 2006, Article ID 48489, pp. 1–11, May 2006. With permission. © Hindawi 2006.)

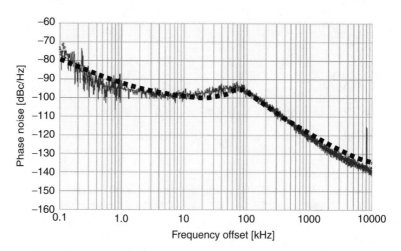

FIGURE 15.18 Comparison of measured and simulated phase noise for the 4.1–4.3 GHz band. The square dots are the simulated data. (From Rogers, J. W. M. et al., *EURASIP J. Wireless Commn. Networking*, vol. 2006, Article ID 48489, pp. 1–11, May 2006. With permission © Hindawi 2006.)

Owing to the accuracy of the proposed phase noise model, we were able to optimize the synthesizer circuits for improved noise performance prior to fabrication. The overall measured and simulated phase noise performance of the synthesizer RFIC is summarized in Table 15.3. Note that in this work, the synthesizer was integrated with a superheterodyne front-end with an IF of approximately 1 GHz, and thus the LO frequencies are offset from the WLAN frequency bands. Translating the frequency of the LO up or down will improve or degrade the phase noise by the ratio the center frequency is scaled. The achieved phase noise is also compared to the most recently published WLAN synthesizer designs in Table 15.4. As shown, this design achieved one

TABLE 15.2
Comparison of Measured and Simulated Phase Noise

Frequency Band (GHz)	Simulated Phase Noise (° rms)	Measured Phase Noise (° rms)
3.2–3.3	0.44	0.50
4.1–4.3	0.50	0.535

TABLE 15.3
Summary of Synthesizer Performance

Parameter	Performance
Technology	0.5 μm BiCMOS
VCO phase noise	−120 dBc/Hz @1 MHz
In-band phase noise	−100 dBc/Hz @10 kHz
Loop corner frequency	200 kHz
Reference frequency	40 MHz
No of accumulators/MMD bits	6
Order of ΣΔ accumulator	Third
Synthesizer step size	468.75 kHz
Spurious	< -50 dBc
Power supply	2.75 V
Current consumption	36 mA
Synthesizer die area	3.22 mm^2

TABLE 15.4
Comparison of Synthesizer Performance

Ref.	Freq. Band (GHz)	Technology	Phase Noise (dBc/Hz @1 MHz)	Phase Noise (dBc/Hz @10 kHz)	Integrated Phase Noise of the System
[12]	2.4, 5.1–5.8	0.25 μm CMOS	−115	−105	0.7° rms, 5.3 GHz, 1 kHz to 10 MHz
[13]	5.1–5.8	0.18 μm CMOS	−115	−92	0.8° rms, 1 kHz to 10 MHz
[14]	5.1–5.3	0.18 μm CMOS	−110	−92	1.5–2° rms, 10 kHz to 10 MHz
This work	2.4, 5.1–5.3	0.5 μm BiCMOS	−120	−98	0.4° rms, 2.4 GHz; 0.7° rms, 5.3 GHz; 100 Hz to 10 MHz

of the best phase noise performances for integrated WLAN transceiver RFICs. Note that in this table, the phase noise quoted was for the transceiver system and not simply of the synthesizers themselves.

CONCLUSIONS

In this chapter, a rigorous analytical model for determining the phase noise performance of PLL-based fractional-N $\Sigma\Delta$ synthesizers has been presented. Noise due to VCOs, CPs, crystal oscillators, PFDs, CPs, loop filters, and $\Sigma\Delta$ modulator has been analyzed. Analyzing an example synthesizer RFIC designed for multiband MIMO WLAN applications has validated the theory. The analytical model achieved good agreements with measured synthesizer phase noise performance. The predicted phase noise values of 0.44 and 0.50° rms at 3 and 4 GHz bands, respectively, agreed closely with the respective measured values of 0.5 and 0.535° rms.

REFERENCES

1. T. A. Riley, M. Copeland, and T. Kwasniewski, "Delta–sigma modulation in fractional-N frequency synthesis," *IEEE J. Solid-State Circuits*, vol. 28, pp. 553–559, May 1993.
2. J. N. Wells, "*Frequency Synthesizers*," U.S. Patent, No. 4609881, September 1986.
3. B. Miller and R. Conley, "A multiple modulator fractional divider," *Proc. 44th Annual Symposium on Frequency Control Baltimore*, pp. 559–568, May 1990.
4. B. Muer and M. Steyaert, "A CMOS monolithic $\Sigma\Delta$-controlled fractional-N frequency synthesizer for DCS-1800," *IEEE J. Solid-State Circuits*, vol. 37, pp. 835–844, July 2002.
5. D. B. Leeson, "A simple model of feedback oscillator noise spectrum," *Proc. IEEE*, vol. 54, no. 2, pp. 329–330, February 1966.
6. J. W. M. Rogers, C. Plett, and F. F. Dai, *Integrated Circuit Design for High-Speed Frequency Synthesis*, Artech House, Boston, London, 2006.
7. Y. Watanabe, T. Okabayashi, S. Goka, and H. Sekimoto, "Phase noise measurements in dual-mode SC-cut crystal oscillators," *IEEE Trans. Ultrason. Ferroelectr. Freq. Control*, vol. 47, pp. 374–378, March 2000.
8. V. F. Kroupa, "Jitter and phase noise in frequency dividers," *IEEE Trans. Instrum. Meas.*, vol. 50, pp. 1241–1243, October 2001.
9. V. F. Kroupa, "Noise properties of PLL systems," *IEEE Trans. Commn.*, vol. 30, no. 10, pp. 2244–2252, October 1982.
10. W. F. Egan, *Frequency Synthesis by Phase Lock*, John Wiley & Sons Inc., New York, 2000.
11. J. W. M. Rogers, F. F. Dai, M. S. Cavin, and D. G. Rahn, "A multi-band $\Sigma\Delta$ fractional-N frequency synthesizer for a MIMO WLAN transceiver RFIC," *J. Solid State Circuits,* vol. 40, pp. 678–689, March 2005.
12. M. Zargari, M. Terrovitis, S. H.-M Jen, B. J. Kaczynski, M. Lee, M. P. Mack, S. S. Mehta, S. Mendis, K. Onodera, H. Samavati, W. W. Si, K. Singh, A. Tabatabaei, D. Weber, D. K. Su, and B. A. Wooley, "A single-chip dual-band tri-mode CMOS transceiver for IEEE 802.11a/b/g WLAN," *IEEE J. Solid-State Circuits,* vol. 39, pp. 2239–2249, December 2004.
13. J. Bouras, S. Bouras, T. Georgantas, N. Haralabidis, G. Kamoulakos, C. Kapnistis, S. Kavadias, Y. Kokolakis, P. Merakos, J. Rudell, S. Plevridis, I. Vassiliou, K. Vavelidis, and A. Yamanaka, "A digitally calibrated 5.15–5.825 GHz transceiver for 802.11a wireless LANs in 0.18 μm CMOS," *IEEE International Solid-State Circuits Conference (ISSCC)*, pp. 352–498, San Francisco, February 2003.
14. P. Zhang, T. Nguyen, C. Lam, D. Gambetta, C. Soorapanth, B. Cheng, S. Hart, I. Sever, T. Bourdi, A. Tham, and B. Razavi, "A direct conversion CMOS transceiver for IEEE 802.11a WLANs," *IEEE International Solid-State Circuits Conference (ISSCC)*, p. 354, San Francisco, February 2003.
15. B. De Muer and M. Steyaert, "A CMOS monolithic $\Sigma\Delta$-controlled fractional-N frequency synthesizer for DCS-1800," *IEEE J. Solid-State Circuits*, vol. 37, pp. 835–844, July 2002.
16. J. W. M. Rogers and C. Plett, "*Radio Frequency Integrated Circuit Design*," Artech House, Boston, London, 2003.

17. J. W. M. Rogers, J. A. Macedo, and C. Plett, "The effect of varactor non-linearity on the phase noise of completely integrated VCOs," *IEEE J. Solid-State Circuits*, vol. 35, pp. 1360–1367, September 2000.

18. E. Hegazi, H. Sjoland, and A. Abidi, "A filtering technique to lower LC oscillator phase noise," *IEEE J. Solid-State Circuits*, vol. 36, pp. 1921–1930, December 2001.

19. J. W. M. Rogers, D. Rahn, and C. Plett, "A study of digital and analog automatic-amplitude control circuitry for voltage-controlled oscillators," *IEEE J. Solid-State Circuits*, vol. 38, pp. 352–356, February 2003.

20. I. Bietti, E. Temporitil, G. Albasini, and R. Castello, "An UMTS $\Sigma\Delta$ fractional synthesizer with 200 kHz bandwidth and -128 dBc/Hz @ 1 MHz using spurs compensation and linearization techniques," *Custom Integrated Circuits Conference*, pp. 463–466, September 2003.

21. I. T. W. Rhee, B. Song, and A. Ali, "A 1.1-GHz CMOS fractional-N frequency synthesizer with a 3-b Third-Order $\Sigma\Delta$ Modulator," *IEEE J. Solid-State Circuits*, vol. 35, pp. 1453–1460, October 2000.

22. J. W. M. Rogers, F. F. Dai, C. Plett, and M. S. Cavin, "Design and characterization of a 5.2 GHz/2.4 GHz $\Sigma\Delta$ fractional-N frequency synthesizer for low-phase noise performance," *EURASIP J. Wireless Commn. Networking*, vol. 2006, Article ID 48489, pp. 1–11, Hindawi Publishing Corporation, May 2006.

16 Design of a Delta-Sigma Synthesizer for a Bluetooth® Transmitter

Jan-Wim Eikenbroek

CONTENTS

Introduction...456
The Zero-IF Transmitter Architecture 456
Transmitter Imperfections in Relation to the Bluetooth
    "Modulation Characteristics" Test...458
    Spurious Modulation of the VCO 460
    Finite Image Rejection..461
    Third-Order Distortion in the Quadrature Low-Frequency Paths
        and Carrier Feed-Through 462
    Evaluation of the Influence of Transmitter Imperfections on the
        Ripple in the Momentary Frequency of the RF Output Signal 463
Synthesizer Considerations...465
    Why a Synthesizer with a ΔΣ-Modulator-Controlled Fractional Divider?......465
    Synthesizer Requirements 465
    Synthesizer Transfer Derivations..466
    Important Integrals for Residual FM Calculations 471
Residual FM Contribution of the Synthesizer 472
    Residual FM Calculations 472
    Residual FM Contribution of the Noise Sources at the Input of the Synthesizer 473
    Contribution of the VCO Phase Noise to the Residual FM.................474
    Combination of the Dominant Residual FM Contributions 474
    Contribution of a Dither Signal in the ΔΣ Modulator 475
Characterization of the Synthesizer Blocks...476
    The ΔΣ Modulator...476
    VCO Design Considerations..476
    The Divider Chain 478
    PFD Requirements...479
    The Charge Pump 480
    The Loop Filter...482
Simulation Results 482
Discussion...483
Summary and Conclusions 484
References..484

INTRODUCTION

Bluetooth is a low-power wireless technology for short-range personal connectivity [1]. Its core specifications are steadily evolving (see www.bluetooth.org). The design of a single-chip Bluetooth system, a radio and a baseband processing unit together, is a prerequisite for a broad acceptance of Bluetooth; therefore, CMOS is the technology to use. But the design of a CMOS Bluetooth transceiver system is a challenging task and the first step to take is to translate the Bluetooth radio specifications into block specifications. In addition, the power consumption should (usually) be as low as possible, which makes the design of the radio not a trivial task.

In this chapter we will focus on the derivation of the synthesizer requirements and demonstrate that its phase noise requirements are closely related to the performance of the transmitter. Especially the combination of a zero intermediate frequency (zero-IF) up-conversion transmitter architecture and a synthesizer whose voltage controlled oscillator (VCO) frequency is equal to the double radio frequency (RF) frequency impose stringent requirements on the phase noise performance of the synthesizer.

The chapter is organized as follows: The section "The Zero-IF Transmitter Architecture" briefly describes the zero-IF architecture and its associated frequency-generation system. In the section "Transmitter Imperfections in Relation to the Bluetooth 'Modulation Characteristics' Test," an important test specification with respect to the synthesizer requirements is discussed. In addition, the ripple in the momentary frequency, induced by imperfections in the transmitter chain, is derived. A fast-settling, fourth-order, $\Delta\Sigma$-modulator-controlled fractional-N synthesizer is characterized in the section "Synthesizer Considerations." After that the residual frequency modulation (FM) contribution due to the dominant phase noise sources of the synthesizer are treated in the section "Residual FM Contribution of the Synthesizer." In the section "Characterization of the Synthesizer Blocks," various blocks of the synthesizer are described and specified. Attention is paid to the relation between the linearity of the charge-pump transfer and noise folding of the high-pass-shaped $\Delta\Sigma$-modulator-induced noise. Finally, in the sections "Simulation Results" and "Discussion," some simulation results are shown and discussed.

THE ZERO-IF TRANSMITTER ARCHITECTURE

A common method of generating a band-limited RF spectrum is by means of a zero-IF up-conversion system. The signal spectrum at a zero-carrier frequency is first generated in the digital domain and after analog-to-digital (AD) conversion and low-pass filtering (to suppress the spectral replicas at multiples of the clock frequency of the digital-to-analog (DA) converters) the zero-IF signal is up-converted to the desired RF frequency by means of two mixers, driven by quadrature local oscillator signals. Such a transmitter architecture is a generic system for many modulation types: amplitude modulation (AM), phase modulation (PM), or frequency modulation (FM) is possible. For Bluetooth, which employs frequency modulation, the envelope of the output signal is constant; all information is embedded in the momentary phase of the signal.

A block diagram of a typical zero-IF up-conversion transmitter architecture is shown in Figure 16.1.

In Figure 16.1, $m(t)$ represents the input bit stream. The advantage of the zero-IF up-conversion architecture is the fact that it is a proven concept, but perhaps the most important drawback in practice is its relatively high power consumption due to the presence of the quadrature mixers and the local oscillator (LO)-buffer circuitry that has to drive these mixers.

Any modulated band-pass signal can be described by

$$u_{\mathrm{rf}}(t) = a(t)(\cos(\theta(t))\cos(\omega_{\mathrm{o}}t) - \sin(\theta(t))\sin(\omega_{\mathrm{o}}t))$$

$$= x(t)\cos(\omega_{\mathrm{o}}t) - y(t)\sin(\omega_{\mathrm{o}}t) \tag{16.1}$$

where ω_{o} is the frequency of the quadrature local oscillator signals. The signals $x(t)$ and $y(t)$ together represent the modulated signal at a zero-carrier frequency ("zero-IF") and are low-pass signals

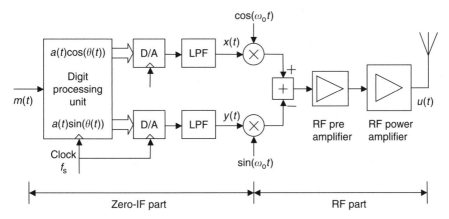

FIGURE 16.1   Block diagram of a zero-IF up-conversion transmitter.

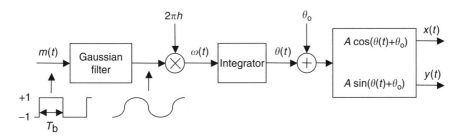

FIGURE 16.2   Construction of the momentary phase $\theta(t)$ and signals $x(t)$ and $y(t)$ for Bluetooth.

with a relatively small bandwidth compared to the oscillator frequency. The phase signal $\theta(t)$ is the momentary phase of the modulated signal.

In fact, the system in Figure 16.1 is a direct implementation of the mathematical description in Equation 16.1: the low-pass signals $a(t)\cos(\theta(t))$ and $a(t)\sin(\theta(t))$ are first generated in the digital domain and are subsequently translated into frequency by the quadrature LO signals.

The momentary phase $\theta(t)$ is the actual information-bearing quantity. The low-frequency signals $x(t)$ and $y(t)$ can be constructed from the incoming bit stream $m(t)$ in a manner as shown in Figure 16.2 [2]. Note that, in practice, these mathematical manipulations are performed in the digital domain.

The mathematical representation of the momentary phase $\theta(t)$ is (see also Figure 16.2)

$$\theta(t) = 2\pi h \int_0^t m(t) \otimes h(t)\,\mathrm{d}t + \theta_\mathrm{o} \tag{16.2}$$

where h is the peak modulation index, $m(t)$ represents the (bipolar) input bit stream, $h(t)$ the impulse response of the Gaussian low-pass filter (LPF), $\otimes$ the convolution operator, and θ_o an initial phase value.

By definition, the momentary frequency $\omega(t)$ is the time derivative of the momentary phase

$$\omega(t) = \frac{\mathrm{d}(\theta(t))}{\mathrm{d}t} \tag{16.3}$$

Some characteristics of the basic-rate modulation of Bluetooth, together with some other characteristics, are listed in Table 16.1.

TABLE 16.1
Bluetooth Characteristics of the Physical Layer

Modulation Type[a]	GFSK[b]
Symbol shaping	Gaussian, BT[c] = 0.5
Bitrate[a]	1 Mb/s
Modulation index (h)	$0.28 < h < 0.35$
Number of channels	79
Channel spacing	1 MHz
Frequency band	2402–2480 MHz
Slot time	625 µs
Max. switching time synthesizer	250 µs
Frequency drift in single slot	< 25 kHz

[a] This modulation and bitrate only hold for the basic rate of the Bluetooth 2.0 standard.
[b] GFSK = Gaussian frequency shift keying.
[c] BT = bandwidth symbol-time product of the filter.

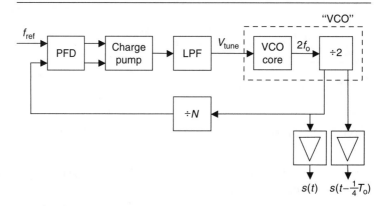

FIGURE 16.3 Frequency generation system.

The up-conversion transmitter architecture requires quadrature oscillator signals, which are provided by the frequency generation system of the transceiver. A typical frequency generation system is shown in Figure 16.3. It consists of a synthesizer system and buffer circuitry that drive the quadrature mixers.

The quadrature LO signals are derived by means of a combination of a VCO, running at the double RF frequency, and a divide-by-2 circuit. When the output signal of the VCO exhibits an accurate 50% duty cycle, a single divide-by-2 circuit can be used to generate accurate quadrature signals.

The advantage of this concept of quadrature signal generation is that the highest Q of an integrated inductor can be exploited in combination with a relatively small chip area. Also, an important feature is that the VCO frequency is separated from the desired RF frequency, which should prevent pulling of the VCO frequency by the RF output signal of the transmitter.

In the following sections, the focus is on the requirements of a zero-IF transmitter in combination with a synthesizer in relation to one of the Bluetooth radio specifications.

TRANSMITTER IMPERFECTIONS IN RELATION TO THE BLUETOOTH "MODULATION CHARACTERISTICS" TEST

Which synthesizer parameters affect the radio requirements [3,4]? Table 16.2 shows these relations (only for basic-rate Bluetooth).

TABLE 16.2
Synthesizer Requirements and Their Origin in the Radio Specifications

	Radio Requirement	Synthesizer Parameter	Other Related Blocks
Transmitter	Modulation characteristics	Close-by phase noise	LF[a], LO, and RF path
	Spurious emission	Ref. osc. spurs	LF, LO, and RF path
	Radio frequency tolerance	Settling time	
Receiver	Interference performance	Phase noise	Rx[b] front-end
	Out-of-band blocking	Phase noise	Rx front-end
Application imposed		Power consumption	Power consumption

[a] LF = low frequency part of the transmitter.
[b] Rx = receiver.

From Table 16.2 it is clear that the phase noise characteristics of the synthesizer play an important role. The imperfections in the signal path of the transmitter and the LO path also play a dominant role in the performance of the transmitter (Tx) "Modulation Characteristics." Hereafter, it will be demonstrated how closely related the signal-path imperfections of the transmitter and the synthesizer imperfections are in meeting this particular radio requirement.

The radio test we will focus on is the transmitter "Modulation Characteristics" test [4]. In this test, the modulation index of the basic-rate FM modulation is verified. The test equipment is fed with the Tx output signal. The desired channel is selected by means of a measurement filter. The shape of this band-pass filter is defined in the test specifications. The output signal of the filter is demodulated and evaluated, according to the requirements of the test specifications.

According to the specifications, the ripple in the momentary frequency of the demodulated transmitter signal must fulfill the following requirements:

- When a periodic sequence of "11110000" is transmitted as a payload, the average frequency deviation of all frequencies as measured shall be between 140 and 175 kHz.
- When a periodic sequence of "10101010" is transmitted as a payload, then at least 99.9% of all frequency deviations as measured shall be greater than 115 kHz.

Details of the measurement method and its definitions can be found in the Bluetooth test specifications [4].

An important observation is that the first test imposes limits on the average frequency deviation, but the second test specifies limits on the peak deviation. It is expected that the second test imposes the most stringent requirements on the system.

Suppose that the peak frequency deviation of the frequency modulation in the transmitter is set to 160 kHz. This corresponds to a modulation index of 0.32, well within the allowable range and it leaves some headroom for an average frequency ripple when a periodic sequence of "11110000" is transmitted. Because of the Gaussian pulse shaping of the baseband pulses (a Gaussian LPF with a bandwidth symbol–time product (BT product) of 0.5), the peak frequency deviation of a periodic sequence of "01" is less than the peak value of 160 kHz and turns out to be 141 kHz. This implies that the peak value of the ripple in the momentary frequency must not exceed $141 - 115$ kHz $= 26$ kHz to fulfill the "Modulation Characteristics" requirement.

In the Bluetooth radio requirements the bandwidth of the measurement filter is specified. The -3 dB bandwidth of this band-pass filter is 1.3 MHz. An obvious contributor to the ripple in the frequency deviation is the phase noise–induced ripple, caused by the noise in the various blocks of the synthesizer. The contribution of these stochastic quantities can be expressed by the spread σ_{FM} of the noise in the demodulated signal and is called residual FM.

But it is important to be aware that all undesired signals within the measurement bandwidth will contribute to the ripple in the frequency deviation. For example, the presence of a spurious signal will influence the momentary phase of the desired signal and consequently will also affect the frequency deviation.

Therefore, we can identify two major contributors to the overall ripple in the peak frequency deviation of the system: the phase noise–induced ripple and a spurious signal-induced ripple.

Several imperfections of the transmitter and synthesizer determine the ripple in the peak frequency deviation. The following error sources can be identified:

Transmitter:

- Spurious modulation of the VCO by the second harmonic of Tx signal
- Image rejection property of the up-conversion system
- Distortion in the low-frequency path of the system + LO feed-through

Synthesizer:

- Phase noise (will be dealt with in the next section)
- Spurious signals (like multiples of the reference frequency)

Note that only those unwanted signals are important for the "Modulation Characteristics" test that are located within the passband of the measurement bandwidth.

In the following subsections, the focus is on the spurious signal sources of the transmitter and after that the contribution of the different phase noise sources of the synthesizer will be considered.

SPURIOUS MODULATION OF THE VCO

Although the synthesizer with a VCO core running at the double RF output frequency has the advantage that the desired output signal of the transmitter will not pull the frequency of the VCO, there is still a potential source of pulling present when such a synthesizer is used in combination with a zero-IF up-conversion transmitter. The cause of this potential source of pulling is the frequency contents of the signals in the supply and ground lines of the RF part of the transmitter and the signals injected into the substrate by these RF circuits. The spectrum of these signals contains signal components at even harmonics of the desired output spectrum because these signals contain (full-wave) rectified versions of the actual RF signal. This implies that when the fundamental tone of the desired signal has a frequency ($\omega_c t + d\theta(t)/dt$), the frequency spectrum of the signals in the supply lines of the RF part of the transmitter, such as the power amplifier (PA), will contain a signal component with frequency $2\omega_c t + 2d\theta(t)/dt$, where $\theta(t)$ represents the momentary phase of the desired signal. The carrier frequency of this signal component is exactly the same as the frequency of the VCO core in case of a zero-IF up-conversion system and the VCO is very sensitive to spurious signals with frequency contents equal or close to the resonance frequency of the VCO. This reasoning also holds for the signals injected into the substrate by the RF part.

This undesired coupling is schematically shown in Figure 16.4.

To avoid (magnetic) coupling between the VCO and the signals of the PA, very careful layout and routing of the supply lines of the PA circuitry is required. In addition, a carefully chosen substrate contacting and circuit-isolation strategy must be adopted, together with a low-impedance grounding solution with an appropriate package.

Suppose that a weak coupling exists between the RF signals in the supply lines and the signals injected into the substrate by the PA and the VCO core. What is the influence of this coupling on the spectrum of the RF output signal?

Note that the VCO is sensitive to spurious RF signals with a frequency very close to its oscillation frequency due to the band-pass nature of the positive feedback loop of the VCO core. So, only the second harmonic of the RF signal will be considered here.

Assuming that the level of this interference signal is small compared to the wanted oscillator signal, the interfering signal can easily be decomposed into an amplitude modulation (AM) component

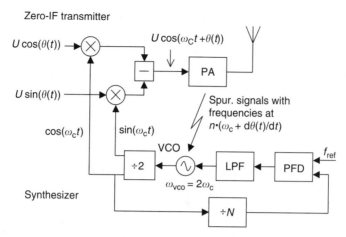

FIGURE 16.4 Possible coupling between PA harmonics and the VCO.

and a phase modulation (PM) component. The amplitude-limiting mechanism of the VCO will remove the AM component and we will focus on the phase modulation that is associated with this interferer. Let us say that the perturbed VCO signal phase is denoted by

$$\phi_{VCO}(t) = \omega_{VCO}t + \beta_{spur}\cos(2\theta(t) + \varphi_{spur}) \tag{16.4}$$

where β_{spur} is much smaller than 1 (only a very weak coupling is present) and φ_{spur} is an initial phase. Given this signal phase, the fundamental tones of the quadrature signals behind the divide-by-2 circuit can be described by

$$s_i(t) = \hat{U}\cos\left(\omega_c t + \frac{\beta_{spur}}{2}\cos(2\theta(t) + \varphi_{spur})\right)$$

$$s_q(t) = \hat{U}\sin\left(\omega_c t + \frac{\beta_{spur}}{2}\cos(2\theta(t) + \varphi_{spur})\right) \tag{16.5}$$

where $\omega_c = \omega_{VCO}/2$ and $\hat{U}$ is the amplitude of the fundamental tone of the signals at the output of the divide-by-2 circuit.

Because of these quadrature LO signals, the RF output signal will be of the form

$$u_{RF}(t) = \hat{U}_{RF}\cos\left(\omega_c t + \theta(t) + \frac{\beta_{spur}}{2}\cos(2\theta(t) + \varphi_{spur})\right) \tag{16.6}$$

So, the momentary frequency of the output signal, excluding the carrier, is denoted by

$$\omega_{tot}(t) = \omega(t)\left(1 - \beta_{spur}\sin(2\theta(t) + \varphi_{spur})\right) \tag{16.7}$$

where $\omega(t)$ represents the desired momentary frequency: $\omega(t) = d\theta(t)/dt$.

Because of the spurious injection of the second harmonic of the RF output signal into the VCO, a spurious FM modulation will be present, whose peak value (β_{spur}) is directly proportional to the injection level.

The period time of the unwanted part is the same as that of the desired signal $\omega(t)$.

FINITE IMAGE REJECTION

When a gain mismatch is present between the quadrature low-frequency paths of the transmitter and a phase error exists between any of the quadrature signals, an image signal will be present in the RF output signal. Because of the zero-IF architecture, this image signal will be located on top of the wanted signal.

Suppose that the amplitude difference between the quadrature low-frequency signals at the input of the mixers is denoted by ΔA and the sum of the quadrature phase errors of the low-frequency signals and the local oscillator signals is denoted by $\Delta \varphi$, then the RF output signal $u_{rf}(t)$ can be expressed as

$$u_{rf}(t) = \text{Re}\left\{ Ae^{j\theta(t)} + \left(\frac{\Delta A - jA\Delta \varphi}{2} \right) e^{-j\theta(t)} \right\} \tag{16.8}$$

where A is the nominal amplitude of the low-frequency signals at the input of the mixers and $\theta(t)$ represents the momentary phase of the FM signal. The first part between brackets represents the desired signal and the second part the image signal. It is assumed that the relative gain error and the total phase error are small.

If the image rejection ratio (IRR) is defined by the ratio of the magnitude of the image signal and the wanted signal,

$$\text{IRR} = \frac{1}{2} \sqrt{ \left(\frac{\Delta A}{A} \right)^2 + (\Delta \varphi)^2 } \tag{16.9}$$

then the momentary phase of the RF signal, excluding the carrier phase, is denoted by

$$\theta_{tot}(t) = \theta(t) - \text{IRR} \sin(2\theta(t) - \varphi_{IRR}) \tag{16.10}$$

where φ_{IRR} is the phase of the complex quantity $\Delta A/A - j\Delta\phi$.

The momentary frequency associated with Equation 16.10 is the time derivative of the momentary phase

$$\omega_{tot}(t) = \omega(t)(1 - 2\text{IRR} \cos(2\theta(t) - \varphi_{IRR})) \tag{16.11}$$

This signal is in fact similar to the signal due to spurious injection into the VCO. If both imperfections are present, the overall ripple in the momentary frequency is equal to the (vectorial) sum of the two ripple signals.

THIRD-ORDER DISTORTION IN THE QUADRATURE LOW-FREQUENCY PATHS AND CARRIER FEED-THROUGH

The last two types of imperfections in the signal path of the transmitter which will be discussed here are the third-order distortion (HD3) in the two low-frequency branches of the transmitter and the presence of any carrier feed-through, due to DC offsets in the low-frequency branches and LO leakage at the mixers.

Suppose that both imperfections can be described by the following expressions:

$$x(t) = x_o + A\cos(\theta(t)) - \gamma A^3 \cos^3(\theta(t))$$
$$y(t) = y_o + A\sin(\theta(t)) - \gamma A^3 \sin^3(\theta(t)) \tag{16.12}$$

where γ denotes the coefficient of the third-order distortion whose sign is positive, while x_o and y_o represent the presence of any DC offsets and LO leakage of the mixers, respectively, which can be represented by equivalent DC offsets.

The complex equivalent baseband signal $s(t) = x(t) + jy(t)$ is denoted by

$$s(t) = s_o e^{j\varphi_o} + A\left(1 - \frac{3}{4}\gamma A^2\right)e^{j\theta(t)} - \frac{1}{4}\gamma A^3 e^{-j3\theta(t)} \tag{16.13}$$

with

$$s_o e^{j\varphi_o} = x_o + jy_o \tag{16.14}$$

Assuming that the third-order distortion is small and that $s_o/A \ll 1$, the expression can be approximated by

$$s(t) \approx A e^{j\theta(t)}\left[1 + \frac{s_o}{A}e^{-j(\theta(t)-\varphi_o)} - \frac{\gamma A^2}{4}e^{-j4\theta(t)}\right] \tag{16.15}$$

The momentary frequency of this signal is

$$\omega_{tot}(t) = \omega(t)\left[1 - \frac{s_o}{A}\cos(\theta(t) - \varphi_o) + 4\text{HD3}\cos(4\theta(t))\right] \tag{16.16}$$

where

$$\text{HD3} = \frac{\frac{1}{4}\gamma A^2}{1 - \frac{3}{4}\gamma A^2} \approx \frac{1}{4}\gamma A^2$$

So, the presence of a carrier in the output signal and the third-order distortion in the low-frequency paths produce sinusoidal ripples in the FM signal. The period time of all ripple signals is equal to the period time of the momentary frequency $\omega(t)$. When an alternating "10" pattern is transmitted, the fundamental frequency of $\omega(t)$ is 500 kHz.

EVALUATION OF THE INFLUENCE OF TRANSMITTER IMPERFECTIONS ON THE RIPPLE IN THE MOMENTARY FREQUENCY OF THE RF OUTPUT SIGNAL

A summary of the various contributors to the ripple in the momentary frequency is shown in Table 16.3.

(The 3rd harmonic distortion (HD3)-induced ripple is given an initial phase (φ_{HD3}) to take any frequency-dependent distortion into account.) Note that the period time of all ripple signals is equal

---

TABLE 16.3
Summary of the Various Ripple Contributions in $\omega_{out}(t)$

Momentary phase wanted signal	$\theta(t)$
Momentary frequency wanted signal	$\omega(t) = d\theta(t)/dt$
Imperfection	**Ripple Signal in $\omega_{out}(t)$ (rad/s)**
Spurious injection into VCO	$\omega(t) \cdot \beta_{spur} \cdot \sin(2\theta(t) + \phi_{spur})$
Image signal	$\omega(t) \cdot 2\text{IRR} \cdot \sin(2\theta(t) + \phi_{IRR})$
HD3 in LF path transmitter	$\omega(t) \cdot 4\text{HD3} \cdot \sin(4\theta(t) + \phi_{HD3})$
DC offsets + LO leakage	$\omega(t) \cdot s_o/A \cdot \sin(\theta(t) - \phi_o)$

to the period time of the momentary frequency $\omega(t)$ and, consequently, the ripple signals will be hardly attenuated by the measurement filter.

Because it is expected that the alternating bit sequence test of the "Modulation Characteristics" test is the most difficult one to fulfill, we will now focus on this test.

The momentary phase $\theta(t)$ of an alternating zero-one sequence can be described by

$$\theta(t) \cong \beta_{max} \sin(\omega_{in}t) \tag{16.17}$$

where $\beta_{max} = 0.28\,\mathrm{rad}$. The associated momentary frequency is

$$\omega(t) = \omega_{in}\beta_{max} \cos(\omega_{in}t)$$

$$= \omega_{max} \cos(\omega_{in}t) \tag{16.18}$$

where $f_{max} = 141\,\mathrm{kHz}$.

In the radio test, it is stated that the peak deviation per symbol must be determined. The overall influence of all distorting mechanisms together is difficult to predict in practice because either the magnitude of an imperfection is difficult to predict (notably the ripple due to spurious injection into the VCO) or the relative phases are difficult to determine. Therefore, only an assumption regarding the overall possible peak ripple can be made.

Example:

Suppose the transmitter has the following typical values (no initial calibration assumed):
Spurious injection into VCO ($=\beta_{spur}$): $-20\,\mathrm{dBc}$
Image signal: $-26\,\mathrm{dBc}$
HD3 at the LF inputs of the mixers: $-40\,\mathrm{dB}$
DC offsets in the LF chain and LO leakage: $-20\,\mathrm{dBc}$

Then their contributions to the peak frequency ripple become (peak values):

Peak ripple due to spurious injection into the VCO: $14\,\mathrm{kHz}$
Image signal: $14\,\mathrm{kHz}$
HD3-induced peak ripple: $5.6\,\mathrm{kHz}$
DC offsets and carrier feed-through contribution: $14\,\mathrm{kHz}$

As mentioned earlier, all these values do not necessarily add up in practice. It is even possible that some (partially) cancel each other at the time instances where the peak deviation is measured of each symbol. However, it should be clear by now that the overall influence of all these contributions can impose very stringent requirements on the remaining contribution of the synthesizer phase noise, which is the other contributor to the frequency ripple.

Although the imperfections of the transmitter can increase the peak of the momentary frequency, one must be prepared for a situation wherein the peaks will be reduced. Because it is not really possible to predict all the contributions by means of (circuit) simulations, an initial assumption must be made on the partitioning between the transmitter signal-path-induced contribution and the synthesizer contribution. Therefore, the following partitioning is chosen (remember that we can only allow for a reduction in the peak value of the momentary frequency of at most $26\,\mathrm{kHz}$):

Contribution due to Tx imperfections: $16\,\mathrm{kHz}$
Residual FM contribution due to phase noise of the synthesizer: $10\,\mathrm{kHz}$

Because the test requirement states that 99.9% of the measured values must be higher than $115\,\mathrm{kHz}$, the residual FM due to the phase noise of the synthesizer, which can be considered to be a

Gaussian-distributed random signal, must fulfill the following requirement: $\sigma_{FM} \leq 3.3\,kHz$. $(3\sigma \equiv 10\,kHz$ corresponds approximately to a 99.9% likelihood that the ripple will be less than $10\,kHz$.)

Next, we can start with the (system-level) characterization of the synthesizer and its phase noise performance.

SYNTHESIZER CONSIDERATIONS

Bluetooth is designed to coexist with other (wireless) applications. An example is the inclusion of a Bluetooth transceiver in mobile terminal, like a global system for mobile (GSM) handheld in Europe. In such host systems, a stable reference oscillator is already present and it is cost-effective when the Bluetooth transceiver can use the already present crystal oscillator of the host system as the reference oscillator for the synthesizer.

This implies that a universal applicable Bluetooth transceiver must be able to handle different crystal frequencies.

WHY A SYNTHESIZER WITH A ΔΣ-MODULATOR-CONTROLLED FRACTIONAL DIVIDER?

The channel spacing for Bluetooth is 1 MHz. So, an obvious choice for a reference frequency for the synthesizer would be 1 MHz, implying that an integer-N divider can be used. But a 1 MHz reference frequency is a relatively low frequency compared to the closed-loop bandwidth of the synthesizer. Common synthesizer bandwidths are in the range 20–50 kHz. This implies that it is likely that any spurious pulses at the output of the charge pump with a fundamental frequency equal to the reference frequency will be insufficiently suppressed by the loop filter and that the transmitter spurious requirements are violated. In addition, deriving a stable and spectrally clean 1 MHz signal from a variety of crystal frequencies is not a straightforward task too.

Therefore, another strategy is chosen: use the (already present) crystal oscillator of the host system as the reference for the synthesizer and realize the 1 MHz step size by means of a fractional divider in the feedback path. Although a fractional division factor can be realized by switching periodically between several integer division values and simultaneously adding a compensation signal to the output of the charge pump, to compensate for the unwanted periodic phase error (otherwise spurious signals will be present in the output spectrum of the VCO), a much more elegant solution is to randomize the switching sequence between several integer-divider values by means of a ΔΣ modulator [5,6]. In such a system, the energy of the original spurious tones is randomly spread in frequency and will appear as noise, which can be suppressed by the low-pass transfer of the synthesizer. This concept will be described in this section.

SYNTHESIZER REQUIREMENTS

The basic requirements for the Bluetooth synthesizer are set by the radio requirements imposed by regulatory bodies and the Bluetooth radio requirements that should guarantee good interoperability between different units. But, usually other requirements are imposed by the application as well. One of those requirements is to minimize the power consumption of the Bluetooth unit. Such a requirement affects the allowable switching time of the synthesizer: because the output frequency of the synthesizer must be (almost) settled to its final value before the actual transmission or reception can start, the synthesizer must be turned on in advance. But the power consumption of the synthesizer constitutes a major part of the overall power consumption during transmission or reception. Therefore, the faster the synthesizer is able to settle, the later it can be switched on and the less (average) power it will consume. This implies that on top of the maximum switching time imposed by the Bluetooth protocol [1,3] a more stringent settling-time requirement is imposed by the power consumption requirements.

On the basis of the power-consumption consideration, the Bluetooth radio requirements, and given the residual FM requirement imposed on the synthesizer, synthesizer requirements can be compiled as shown in Table 16.4.

TABLE 16.4
Requirements for the Synthesizer

	Set By	Remark	Requirement
Reference frequency	Application	GSM mobile app.	13 MHz
Settling time (80 MHz freq. step)	Tx/Rx switching time	User choice (power consumption)	$\leq 100\,\mu s$
Δf_{res} in a bandwidth of 1.3 MHz	Tx modulation char.		$\leq 3.3\,kHz$
$L(\Delta f)$ @ 10 kHz	Tx modulation char.	No $1/f$ noise assumed	$\leq -80\,dBc/Hz$
$L(\Delta f)$ @ 2 MHz	Rx interference req.		$\leq -114\,dBc/Hz$
$L(\Delta f)$ @ 3 MHz	Rx interference req.		$\leq -124\,dBc/Hz$
$L(\Delta f)$ noise floor	Rx interference req.		$\leq -140\,dBc/Hz$

As derived in the section "Evaluation of the Influence of Transmitter Imperfections on the Ripple in the Momentary Frequency of the RF Output Signal," the amount of residual FM (Δf_{res}), which the synthesizer is allowed to contribute to the overall ripple in the peak frequency deviation, is determined by the residual FM requirement imposed by the Tx "Modulation Characteristics" test, which in our case allows for a peak dip of at most 26 kHz and the contribution induced by the imperfections in the signal path of the transmitter, to which a maximum contribution of 16 kHz is allocated.

The phase noise levels at a distance of 2 MHz and more from the carrier are predominantly determined by the receiver interference performance requirements [1,3]. Although the phase noise levels at 2 and 3 MHz distance are not important for the Tx "Modulation Characteristics" test, they do however determine the phase noise level outside the passband of the synthesizer but still within the passband of the measurement filter.

The spurious emission requirements by the transmitter are not considered here, because these spurs are usually located at multiples of the reference frequency with respect to the carrier frequency and will be suppressed by the measurement filter.

SYNTHESIZER TRANSFER DERIVATIONS

The synthesizer has to fulfill several conflicting requirements: the settling time must be as short as possible and a minimum of residual FM must be generated. In addition, it has to be a low-power design as well.

Because the divider is controlled by a $\Delta\Sigma$ modulator, the spectrally shaped noise at the output of the quantizer is converted to a spectrally shaped phase-error signal. Because of the high-pass shape of the power spectral density (PSD) of this noise, the closed-loop transfer of the synthesizer must be able to suppress the out-of-band noise to fulfill the residual FM requirements and the other phase noise requirements. This might impose an upper limit to the closed-loop bandwidth of the synthesizer.

A block diagram of the synthesizer is shown in Figure 16.5.

When the closed-loop bandwidth is much smaller than the reference frequency (reference frequency greater than 20× closed-loop bandwidth), the synthesizer can be considered to be a continuous-time system [7,8]. Assuming that the synthesizer transfer is linear and time-continuous, the loop behavior can be described by its phase transfer in the frequency domain. This transfer represents small excursions around the bias points of the blocks.

The VCO consists of the actual oscillator circuit and a divide-by-2 circuit. When the frequency-selective network of the VCO is formed by a combination of an inductor and a capacitor (LC resonator), the output signal of the VCO core is a sinusoidal signal.

In Refs. 7 and 8, the third-order, type 2 synthesizer loop is analyzed. When a continuous-time approach can be adopted, the closed-loop system exhibits the fastest settling time (for a given

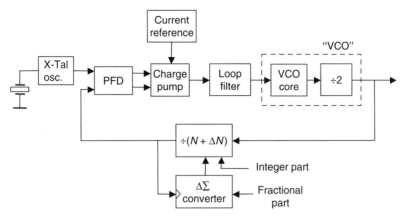

FIGURE 16.5 Block diagram of a synthesizer with a $\Delta\Sigma$-controlled fractional divider.

closed-loop bandwidth) when the system has three equal closed-loop poles, located at $s = -\omega_N$. The associated closed-loop bandwidth is $f_c = 1.53 \cdot \omega_N$ and the phase margin is 53° [8].

But in many practical cases, the out-of-band attenuation must be increased compared to the attenuation of such a third-order loop. For instance, reference spurs are too high, the high-pass-shaped quantization noise of a $\Delta\Sigma$-modulator-controlled fractional divider must be suppressed even more outside the passband because it has become the dominant phase noise source, and so on. Usually, an addition low-pass section is added to the loop filter of the third-order system. But how should the closed-loop poles be placed?

Suppose we want to use a fourth-order loop, what is the best closed-loop transfer to fulfill all our requirements in this application? A similar strategy can be adopted as for the fast-settling third-order, type 2 loop: it can be expected that the fastest settling time is obtained when all dominant poles are real (implying no ringing in the step response) and simultaneously, for a given closed-loop bandwidth, the maximum out-of-band attenuation can be obtained when these closed-loop poles are located as close as possible.

A block diagram of the system is depicted in Figure 16.6.

To fulfill the above-stated goals for the fourth-order loop, we have to solve the following equation:

$$\text{Denominator closed-loop transfer} \equiv \left(1 + \frac{s}{\omega_N}\right)^3 \left(1 + a\frac{s}{\omega_N}\right)$$

where a is a constant that must be determined. The denominator of the closed-loop transfer must be expressed as a function of all the loop components. So, the goal is to find a solution where we have a closed-loop transfer with three equal poles at $-\omega_N$ (similar to the third-order loop) and a fourth pole as close as possible to the other three poles at a location $-\omega_N/a$.

Equating both fourth-order polynomials gives four equations with four unknowns, which can be solved and the following solution for the closed-loop transfer is obtained, where ω_N is the design parameter of the system:

$$H(s) = \frac{\theta_o(s)}{\theta_r(s)} = N \frac{(1 + 3.15s/\omega_N)}{(1 + s/\omega_N)^3 (1 + 0.155s/\omega_N)} \qquad (16.19)$$

So, the constant a turns out to be 0.155.

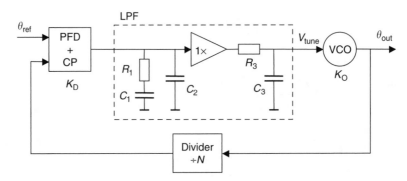

FIGURE 16.6 The fourth-order, type 2 phase locked loop (PLL).

For a given ω_N, all component values can be expressed as functions of the loop parameters:

$$C_2 = \frac{K_D K_O}{4.309\omega_N^2 N} \qquad R_1 C_1 = \frac{3.15}{\omega_N}$$

$$C_1 = 13.927 C_2 \qquad R_3 C_3 = \frac{0.211}{\omega_N} \tag{16.20}$$

where K_D = phase detector transfer (A/rad), K_O = VCO transfer (rad/s/V), and N = division factor.

Either R_3 or C_3 can be chosen arbitrarily, but in practice, C_3 must be larger than the parasitic capacitance at the input of the VCO.

The LPF transfer is

$$H_F(s) = \frac{U_o(s)}{I_i(s)} = \frac{0.289 N\omega_N^2}{K_D K_O} \frac{(1+3.15s/\omega_N)}{s(1+0.211s/\omega_N)^2} \tag{16.21}$$

The normalized step response for a phase-step at the input (φ_{eo}) and the normalized step response for a desired frequency-step at the output (Δf_{out}) are depicted in Figure 16.7. Note that $t_N = \omega_N t$.

The most important peak values that are required to determine the dynamic range of the tune voltage and the phase error range before the onset of cycle slips are:

a. Phase step: $\omega_{o\_max} = 0.81$ at $t_N = 0.88$ and $\varphi_{e\_min} = -0.29$ at $t_N = 3.08$
b. Frequency step: $\omega_{o\_max} = 1.29$ at $t_N = 3.08$ and $\varphi_{e\_max} = 0.96$ at $t_N = 1.69$

The root locus of the system and the location of the closed-loop poles are shown in Figure 16.8.

Because the closed-loop transfer has three dominant poles at ω_N, it is expected that the behavior of this system will resemble the behavior of the third-order system with three equal real poles at ω_N.

The relation between the normalized frequency error at the output ($f_{error}/\Delta f_{out}$) for a given frequency step at the input ($\Delta f_{out}/N$) and the normalized time (t_N) required to settle to within this error is given by Equation 16.22 and is graphically shown in Figure 16.9 (note that this only holds for the tail of the step response).

$$\frac{f_{error}}{\Delta f_{out}} \cong 10^{(1-t_N/3)}$$

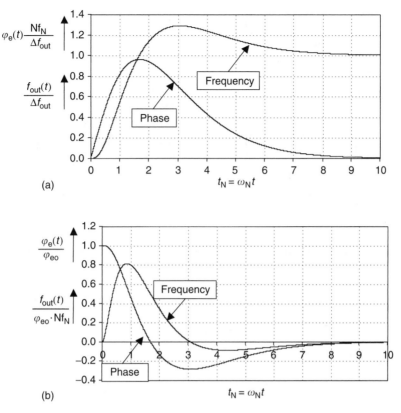

(a)

(b)

FIGURE 16.7 Step responses of the fourth-order, type 2 synthesizer, with four real poles. (a) Normalised phase error and output frequency after a frequency step $\Delta f_{out}/N$ at the input (initial phase error = 0). (b) Normalised phase error and output frequency after a phase step φ_{eo} at the input.

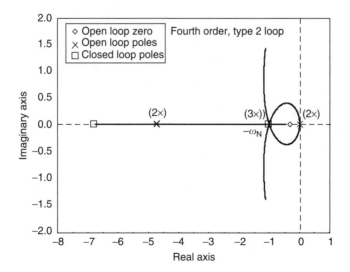

FIGURE 16.8 Root locus of the fourth-order, type 2 system.

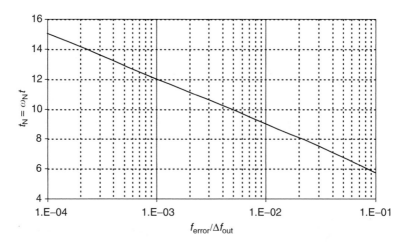

FIGURE 16.9   Normalized settling time t_N versus the ratio of frequency error and frequency step size ($f_{error}/\Delta f_{out}$).

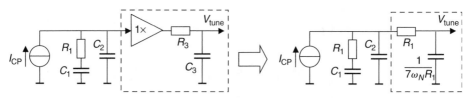

FIGURE 16.10   Alternative loop-filter configuration without a voltage buffer.

or alternatively

$$t_N \cong 3\left\{1 - 10\log\left(\frac{f_{error}}{\Delta f_{out}}\right)\right\} \tag{16.22}$$

Example:

When an 80 MHz step is required, the VCO frequency will settle to within 25 kHz of its final value at $t_N = 13.6$ ($f_{error}/\Delta f = 25\text{e}3/8\text{e}7 = 3.125\text{e}{-}4$). While the time required to settle within 25 kHz error for a minimum step of 1 MHz is $t_N = 7.8$. This is only a factor 1.75 faster, while the step size is 80 times smaller.

Although the results are based on an LPF where the impedance of the additional low-pass section (R_3, C_3) does not load the other part of the filter by means of a voltage follower, it turns out that a good approximation of the performance can also be obtained without the voltage follower. The alternative LPF configuration is shown in Figure 16.10.

The new values for R_3 and C_3 are:

$$R_3 = R_1$$

$$C_3 = \frac{1}{7R_1\omega_N} \tag{16.23}$$

The Bode plot of the original fourth-order loop and the one without the voltage buffer is shown in Figure 16.11. The closed-loop -3 dB bandwidths of the two systems are

$$f_{c\_ideal} = 1.66 f_N$$

$$f_{c\_alt} = 1.54 f_N \tag{16.24}$$

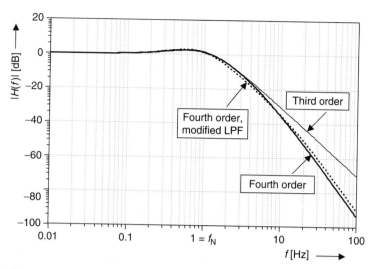

FIGURE 16.11 Magnitude versus frequency of the original fourth-order system and of the system without the LPF buffer.

Although the bandwidths slightly differ, both systems have virtually the same time-domain performance, while for the version without the voltage buffer, the attenuation in the stopband slightly decreased by 3 dB, compared with the original fourth-order loop (see Figure 16.11).

IMPORTANT INTEGRALS FOR RESIDUAL FM CALCULATIONS

Given the closed-loop transfer $H(f)$ of the fourth-order, type 2 synthesizer (Equation 16.19), the numerical evaluation of the following integrals is required for the residual FM calculations later on. Note that $H_{meas}(f)$ is the (single-sided) transfer of the Bluetooth measurement filter, which must be used during the transmitter "Modulation Characteristics" test [4].

a.

$$\int_0^\infty \left|\frac{H(f)}{N}\right|^2 df = c_{dith} f_N \tag{16.25}$$

where $c_{dith} = 2.5$

b.

$$\int_0^\infty f^2 \left|\frac{H(f)}{N}\right|^2 df = c_o f_N^3 \tag{16.26}$$

where $c_o = 4.2$

c.

$$\int_0^\infty f^4 \left|\frac{H(f)}{N}\right|^2 df = c_{\Delta\Sigma\infty} f_N^5 \tag{16.27}$$

where $c_{\Delta\Sigma\infty} = 78$

d.

$$\int_0^\infty f^4 \left| \frac{H(f)}{N} \right|^2 |H_{\mathrm{meas}}(f)|^2\, \mathrm{d}f = c_{\Delta\Sigma} f_N^5 \qquad (16.28)$$

where $c_{\Delta\Sigma} = 55$

The result of this section will be used to determine the filter-component values and to calculate the residual FM contribution of the various noise sources of the synthesizer.

RESIDUAL FM CONTRIBUTION OF THE SYNTHESIZER

Now, we will focus on the noise performance of the synthesizer system. The phase noise at the output of the synthesizer is predominantly determined by the noise of the VCO and the noise from the reference signal source, charge pump, and $\Delta\Sigma$-modulator-driven fractional divider. The noise contribution of the LPF is ignored for the moment so as not to complicate the derivations.

Figure 16.12 shows the location of the dominant noise sources in the synthesizer.

In the following subsections, the contribution to the residual FM of all these phase noise sources will be determined.

RESIDUAL FM CALCULATIONS

The variance σ_{FM}^2 (or $\Delta f_{\mathrm{res}}^2$) of the residual FM at the output of the synthesizer is

$$\sigma_{\mathrm{FM}}^2 = \Delta f_{\mathrm{res}}^2 = \int_{f_a}^{f_b} f^2 S_{\varphi_0}(f)\, \mathrm{d}f$$

$$= 2 \int_{f_a}^{f_b} f^2 L_{\mathrm{o}}(f)\, \mathrm{d}f \qquad (16.29)$$

where $S_{\varphi_0}(f)$ represents the PSD of the phase noise of the momentary phase $\theta_{\mathrm{o}}(f)$ of the synthesizer output signal. The single-sided phase noise of the output signal is denoted by $L_{\mathrm{o}}(f)$. Note that f in $S(f)$ and $L(f)$ represents the frequency distance from the carrier frequency.

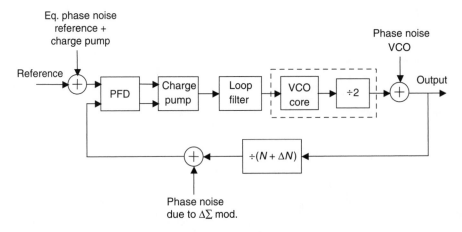

FIGURE 16.12   Location of the dominant noise sources in the synthesizer.

The integration interval $(f_b - f_a)$ is usually imposed by the application. For Bluetooth, the bandwidth is set by the measurement filter bandwidth, which is 650 kHz (single-sided, -3 dB corner frequency).

More precisely, when we denote the transfer of the measurement filter by $H_{meas}(f)$, the residual FM at the output of this filter is described by

$$\sigma^2_{FM} = 2 \int\limits_0^\infty f^2 L_o(f) |H_{meas}(f)|^2 \, df \tag{16.30}$$

The residual FM of the phase noise at the output of the synthesizer that has passed the measurement filter is

$$\sigma^2_{FM} \approx \int\limits_0^\infty f^2 \left(S_{eq\_ref}(f)|H(f)|^2 + S_{VCO}(f) \left| 1 - \frac{H(f)}{\langle N \rangle} \right|^2 \right) |H_{meas}(f)|^2 \, df \tag{16.31}$$

where $H(f)$ is the closed-loop phase transfer of the synthesizer and $\langle N \rangle$ is the average division factor. $S(f)$ represents the PSD of a noise source. $S_{eq\_ref}(f)$ consists of the noise of the reference signal source, the $\Delta\Sigma$-modulator-induced phase noise at the output of the divider chain, and the noise contribution of the charge pump which can be converted to an equivalent noise source at the input of the synthesizer. $S_{VCO}(f)$ represents the phase noise at the output of the divide-by-2 circuit.

RESIDUAL FM CONTRIBUTION OF THE NOISE SOURCES AT THE INPUT OF THE SYNTHESIZER

First, we will deal with the noise due to the $\Delta\Sigma$-modulator-driven divider chain.

A general expression of the PSD of the phase noise of a single-loop, mth-order $\Delta\Sigma$-modulator-driven divider at the output of the divider chain is $(f < f_{ref}/2)$ [9,10]:

$$S_{\Delta\Sigma}(f) = \frac{2\sigma^2_q}{\langle N \rangle^2 f_{ref}} 4\pi^2 \left(2\sin\left(\frac{\pi f}{f_{ref}} \right) \right)^{2(m-1)} \tag{16.32}$$

where σ^2_q denotes the unshaped quantization noise power of the quantizer, $\langle N \rangle$ is the average division factor, m the order of the modulator, and f_{ref} denotes the clock frequency of the modulator, which is equal to the reference frequency of the synthesizer, once the synthesizer is in lock.

For this application, a single-loop, second-order $\Delta\Sigma$ modulator turns out to be adequate. So, $m = 2$.

In the frequency band of interest (which is set by the measurement filter), the combined PSD of the phase noise of the reference frequency source and the charge-pump noise-current can be described by a white noise source. Usually, the in-band single-sided phase noise level at the *output* of the synthesizer is specified. If we denote this output phase noise level by $L_o(0)$, the residual FM, expressed as a function of the noise of the reference source, charge-pump current, and the second-order $\Delta\Sigma$ modulator becomes:

$$\sigma^2_{FM} = 2 \int\limits_0^\infty f^2 \left[L_o(0) + \frac{4\pi^2 \sigma^2_q}{f_{ref}} \left(2\sin\left(\frac{\pi f}{f_{ref}} \right) \right)^2 \right] \frac{|H(f)|^2}{\langle N \rangle^2} |H_{meas}(f)|^2 \, df$$

$$\approx 2 \int\limits_0^\infty f^2 \left[L_o(0) + \frac{16\pi^4 \sigma^2_q}{f_{ref}^3} f^2 \right] \frac{|H(f)|^2}{\langle N \rangle^2} |H_{meas}(f)|^2 \, df \tag{16.33}$$

The last approximation holds as long as the measurement bandwidth is much smaller than half the reference frequency, which is valid here.

Note that the charge-pump noise can be converted to an equivalent phase noise at the input of the synthesizer by $S_{\varphi\_eq} = S_I/K_D^2$.

CONTRIBUTION OF THE VCO PHASE NOISE TO THE RESIDUAL FM

The other major contributor is the phase noise of the VCO. Its contribution to the residual FM at the output of the synthesizer is

$$\sigma_{FM}^2 = 2\int_0^\infty f^2 \frac{L_{VCO}(\Delta f)\Delta f^2}{f^2} \left|1 - \frac{H(f)}{\langle N \rangle}\right|^2 |H_{meas}(f)|^2 \, df$$

$$= 2L_{VCO}(\Delta f)\Delta f^2 \int_0^\infty \left|1 - \frac{H(f)}{\langle N \rangle}\right|^2 |H_{meas}(f)|^2 \, df \tag{16.34}$$

where $L_{VCO}(\Delta f)$ denotes the single-sided phase noise number of the VCO at a frequency distance Δf from the carrier and is imposed by the receiver interference performance requirements [1,3]. In Table 16.4, the most important values of L_{VCO} are listed. As long as the closed-loop bandwidth of the synthesizer is much smaller than the measurement bandwidth, the integral can be approximated by f_{BWn} which is the equivalent noise bandwidth of the measurement filter and is approximately equal to its $-3\,dB$ bandwidth ($\approx 650\,kHz$). So,

$$\sigma_{FM}^2 \approx 2L_{VCO}(\Delta f)\Delta f^2 f_{BWn} \tag{16.35}$$

COMBINATION OF THE DOMINANT RESIDUAL FM CONTRIBUTIONS

The overall residual FM can now be determined. Combining Equations 16.33 and 16.35 gives the expression of the overall residual FM:

$$\sigma_{FM}^2 \approx 2\int_0^\infty \left\{ f^2\left[L_0(0) + \frac{16\pi^4\sigma_q^2}{f_{ref}^3}f^2\right] \frac{|H(f)|^2}{\langle N \rangle^2} |H_{meas}(f)|^2 \right\} df + 2L_{VCO}(\Delta f)\Delta f^2 f_{BWn} \tag{16.36}$$

For the fourth-order synthesizer transfer (see Equation 16.19) and with the help of the evaluation results of the integrals from the section "Important Integrals for Residual FM Calculations," the integral can be replaced by a function of the design parameter f_N of the synthesizer.

The overall expression for the residual FM as a function of f_N becomes:

$$\sigma_{FM}^2 \approx 2L_0(0)c_o f_N^3 + \frac{32\pi^4\sigma_q^2}{f_{ref}^3}c_{\Delta\Sigma}f_N^5 + 2L_{VCO}(\Delta f)\Delta f^2 f_{BWn} \tag{16.37}$$

where $c_o = 4.2$ and $c_{\Delta\Sigma} = 55$.

It is clear from Equation 16.37 that only the contributions of the equivalent reference noise source and the $\Delta\Sigma$-modulator-induced noise can be influenced by the closed-loop bandwidth. Equation 16.37 can be used to determine the maximum value of f_N to meet a given residual FM number.

According to the phase noise requirements of Table 16.4, the most demanding out-of-band VCO phase noise requirement is set by the phase noise level at $3\,MHz$ distance from the carrier ($L_{VCO}(f) = -124\,dBc/Hz$ at $3\,MHz$ distance).

Substituting all data in Equation 16.37, the following relation between f_N and the residual FM is obtained ($f_{ref} = 13\,MHz$ and the variance of the unshaped quantization noise $[\sigma_q^2]$ is 1/12 [quantization step size = 1]):

$$\sigma_{FM}^2 \approx 9.1 \cdot 10^{-8} \cdot f_N^3 + 6.5 \cdot 10^{-18} \cdot f_N^5 + 4.7 \cdot 10^6 \tag{16.38}$$

For an imposed upper limit of 3.3 kHz for the overall residual FM, it turns out that the allowable maximum value for f_N is 40 kHz. It is interesting to observe that the minimum residual FM is set by the VCO and the bandwidth of the measurement filter. This minimum value is $\sigma_{FM} = 2.2\,kHz$, which is already close to the maximum value of 3.3 kHz! A single-sided phase noise level of $-120\,dBc/Hz$ @3 MHz of the VCO would already give a residual FM of 3.3 kHz.

When f_N is 40 kHz, the settling time of the step response, for a frequency step of 80 MHz at the output, is $13.6/(2\pi40e3) = 54\,\mu s$ (see Equation 16.22), which is well below the required value of $100\,\mu s$. (It is assumed that the loop remains linear during the step-response dynamics. So, no cycle slips and tune-voltage clipping occurs.)

CONTRIBUTION OF A DITHER SIGNAL IN THE $\Delta\Sigma$ MODULATOR

Because the input signal of the modulator is a static signal, a dither signal must be added to avoid idle patterns in the output signal. Idle patterns give rise to spurious sidebands in the synthesizer output signal. These idle patterns can be avoided in two manners: either a dither signal can be applied to the input of the quantizer, or a dither signal can be applied to the input of the modulator.

Applying a dither signal at the quantizer input is the usual solution, because the PSD of the dither signal will exhibit a high-pass transfer which will be sufficiently suppressed (ideally) by the low-pass transfer of the synthesizer. The backside of this solution (assuming a single-loop modulator) is that the overall high-pass-shaped PSD of the noise at the output of the modulator is differently shaped compared with the theoretical-shaped PSD of the quantization noise (Equation 16.32). It turns out that the variance is increased and the noise peak is already reached before $f = f_{ref}/2$. All this can have an undesirable influence on the receiver interference performance and the residual FM performance.

Therefore, the other option is preferred here because this mainly generates a white noise floor in addition to the already present high-pass-shaped quantization noise.

When a dither signal is applied at the input of the modulator, an additional white noise source must be added to the input of the modulator. This additional noise source will give rise to an additional residual FM contribution of the synthesizer which is roughly equal to

$$\sigma_{FM}^2 \approx \int_0^\infty \left\{ f^2 \left[\frac{2\sigma_{dith}^2}{f_{ref}} \frac{1}{f^2} \right] \frac{|H(f)|^2}{\langle N \rangle^2} |H_{meas}(f)|^2 \right\} df \tag{16.39}$$

where σ_{dith}^2 is the variance of the frequency dither signal at the input and whose PSD is assumed to be white. Also, it is assumed that the dither signal is clocked at the same rate as the $\Delta\Sigma$ modulator (f_{ref}).

Evaluation of the integral, when $H(f)$ is the transfer of the fourth-order synthesizer, gives for the contribution to the residual FM:

$$\sigma_{FM}^2 \approx \frac{2\sigma_{dith}^2}{f_{ref}} c_{dith} f_N \tag{16.40}$$

where $c_{dith} = 2.5$ (see Equation 16.25).

Usually, σ_{dith}^2 can be chosen small enough so that its contribution to the overall residual FM can be neglected. A calculation of this contribution will be carried out, once the $\Delta\Sigma$ modulator is characterized.

CHARACTERIZATION OF THE SYNTHESIZER BLOCKS

Now the (upper limit of the) bandwidth of the synthesizer is determined ($f_N \approx 40\,\text{kHz}$); hence, all the synthesizer blocks can be further characterized. It was already decided that the closed-loop synthesizer transfer would be a fourth-order, type 2 transfer as described by Equation 16.19.

THE $\Delta\Sigma$ MODULATOR

The $\Delta\Sigma$ modulator sets the fractional part of the overall division factor of the synthesizer. An example of a single-loop, second-order modulator is shown in Figure 16.13. Note that the quantizer has three levels "−1," "0," and "1." These three levels correspond to division factors $N-1$, N, and $N+1$, respectively. Note that the modulator is a fully digital circuit.

The overall division factor of the synthesizer consists of an integer part and a fractional part. The relation between the reference frequency and the output frequency of the synthesizer is [5]

$$f_{\text{out}} = f_{\text{ref}}\left(N + \frac{x}{X_{\text{max}}}\right) \tag{16.41}$$

where N is the integer part, x the digital input word of the modulator, and X_{max} an integer, which sets the resolution of the frequency steps. The minimum step size is ($x = 1$):

$$\Delta f_{\text{min}} = \frac{f_{\text{ref}}}{X_{\text{max}}} \tag{16.42}$$

For a reference frequency of 13 MHz, a suitable value for X_{max} is 4096, which gives a minimum step size of about 3 kHz. The associated maximum frequency error of the carrier frequency is 1.5 kHz, which is acceptable in practice.

The input range of x that is required to set the fractional part is $-0.5X_{\text{max}} \leq x \leq 0.5X_{\text{max}}$. Beyond this range the integer part N can simply be decreased or increased by 1 to bring x back within this range. This range of x is well within the stability range of the modulator.

When $X_{\text{max}} = 4096$, the dither signal at the input can be chosen to be a randomly varying least significant bit of the digital input word of the $\Delta\Sigma$ modulator. When the following probabilities of occurrences are implemented: $P(-1/X_{\text{max}}) = P(1/X_{\text{max}}) = 1/4$ and $P(0) = 1/2$, σ_{dith}^2 is equal to $0.5f_{\text{ref}}/X_{\text{max}}^2$. Substituting this variance in Equation 16.40 gives $\sigma_{\text{FM}} = 300\,\text{Hz}$ for the contribution of such a dither signal to the residual FM, which is minor compared to the overall residual FM.

VCO DESIGN CONSIDERATIONS

The synthesizer must fulfill the phase noise requirements according to the numbers in Table 16.4. The VCO determines the phase noise outside the closed-loop bandwidth of the synthesizer. Actually, the VCO requirements outside the passband of the synthesizer should be slightly more stringent,

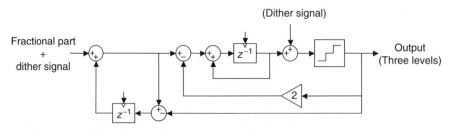

FIGURE 16.13   Block diagram of a second-order $\Delta\Sigma$ modulator.

because in the transition region where the loop gain of the synthesizer has become less than 1, the VCO phase noise will be enhanced slightly.

Because quadrature oscillator signals are generated by a VCO running at the double RF frequency together with a divide-by-2 circuit, the phase noise level of the quadrature output signals of the divider will be reduced by 6 dB compared to the level at its input. But to take into account a small degradation induced by divider circuit imperfections, an improvement of 5 dB is used here.

The phase noise requirement at 3 MHz distance from the carrier is the most stringent out-of-band phase noise requirement. Together with a small margin, let us say 2 dB, the required phase noise level at 3 MHz distance from the carrier of the VCO core should be: $L(\Delta f)_{\Delta f = 3\,\text{MHz}} = -124 + 5 - 2\,\text{dBc/Hz} = -121\,\text{dBc/Hz}$, whereas the associated noise level at the output of the divide-by-2 circuit is $-126\,\text{dBc/Hz}$.

Suppose a CMOS LC cross-coupled differential pair is used as the VCO core. Can we meet the phase noise requirements of $-121\,\text{dBc/Hz}$ @3 MHz distance? In Refs. 11 and 12, a relation between the parameters of the VCO core and the single-sided phase noise is derived. Such a relation is [11]

$$L(\Delta f) = 10 \cdot \log\left(\frac{kT\left(1 + \frac{1}{2}\gamma \cdot A\beta_o\left[1 + \sqrt{\eta_o}\right]\right)R_p}{V_p^2} \cdot \left(\frac{f_o}{Q_{tot}\Delta f}\right)^2 \right)$$ (16.43)

where γ is the excess noise factor of the CMOS transistors [13], $A\beta_o$ the small-signal loop gain at start-up, R_p the overall parallel loss resistance across the tank circuit, η_o the W/L ratio between the tail-current transistor and the oscillator-core transistors, and V_p the peak voltage of the differential signal across the tuned circuit.

Example:

Suppose we have an integrated inductor with a self inductance $L = 1.8\,\text{nH}$, while the overall quality factor of the loaded resonator is $Q_{tot} = 8$ @5 GHz. Together with a small-signal loop gain at start-up $A\beta_o = 2.0$, $V_p = 0.7\,\text{V}$ and a W/L ratio between the tail-current transistor and the oscillator-core transistors $\eta_o = 1$; then the phase noise at 3 MHz distance from the 5 GHz carrier becomes (assuming a noise excess factor $\gamma = 1.33$ for the transistors):

$$L(\Delta f)_{\Delta f = 3\,\text{MHz}} = -121.6\,\text{dBc/Hz} @ f_{VCO} \approx 5.0\,\text{GHz}$$

At the outputs of the divide-by-2 circuit, this will be $-126.6\,\text{dBc/Hz}$. So it seems feasible to meet the phase noise requirements of the VCO. The associated tail current of this oscillator is 2.5 mA.

Note that for a low-power design, the Q of an integrated inductor must be as high as possible. In a CMOS process with a lightly doped substrate, a Q of 10 for the inductor is feasible. But the loaded Q will be less due to other losses across the tank circuit (here, a 20% reduction of the unloaded Q value is used).

In the section "Synthesizer Considerations," it has been demonstrated that the capacitor values C_1 and C_2 of the LPF are proportional to the modulation transfer of the VCO (K_O). So, smaller the K_O, the smaller the occupied chip area will be. Another important reason to go for a small K_O is to minimize the risk of spurious modulation of the VCO due to any spurious signal pickup at the modulation input. Also, the common-mode part of all the noise sources in the VCO core modulate the varactor capacitance and can give rise to an increased $1/f^2$ portion of the phase noise when K_O is large.

But it must be still possible to tune the VCO to the desired channel for all process corners, the desired temperature range, and supply voltage range ("PVT corners"). Both requirements can be satisfied by using a combination of a switchable capacitor bank and a varactor, which only has to cover a small tuning range [12].

During operation, the selection strategy is to select the appropriate capacitors of the capacitor bank, once the desired Bluetooth transmit or receive channel is known at the onset of a transmission or reception and before the synthesizer is switched on. This will bring the remaining frequency error within the tuning range of the varactor.

With such an arrangement, a K_O of about 50 MHz/V is very well feasible in practice. Note that this transfer represents the overall transfer of the VCO and the divide-by-2 circuit.

THE DIVIDER CHAIN

One of the output signals of the divide-by-2 circuit must be further divided in frequency by means of a divider with a variable division factor. The integer part of the division factor must be set before the onset of a transmission or reception, while the fractional part is realized by randomly varying the division factor between three integer values induced by the $\Delta\Sigma$ modulator, during the operation of the synthesizer.

The frequency range of the input signal that the variable divider should be able to handle must cover at least the 2.4 GHz industrial, scientific and medical (ISM) radio band plus some margin. To minimize the power consumption, a multimodulus concept is usually adopted [14,15]. In such a concept, only a small part of the total circuitry operates at the high input frequency.

Because of the high input frequency of the divider system, standard logic cannot be used and custom-made source-coupled logic must be designed.

The architecture of a typical divider system, based on a dual modulus divider, is shown in Figure 16.14.

The operation of the divider is as follows [16]:

a. Dual modulus divider: start with a division factor $P + 1$, corresponding to "division select" = 1.
b. Presetable down counters A and B, with $A > B$: when B becomes 0, "division select" = 0, so the dual modulus divider divides by P. When A becomes 0: preset both A and B to their start values and set "division select" = 1 again.

The overall division factor N of the divider is

$$N = B(P+1) + (A-B)P$$

$$= AP + B \tag{16.44}$$

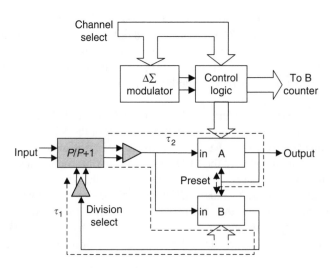

FIGURE 16.14 Dual modulus divider architecture (gray shaded blocks: differential circuits, all other blocks: standard logic cells).

If no holes in the division range are allowed, the minimum division factor is set by (P^2-P). This imposes a maximum value for P, which in this application is 2400/13. But it is good practice to take a margin into account, let us say $f_{rf\text{-min}} = 2200\,MHz$. This implies that $P^2 - P < 169$, so, $P < 13$. A commonly used dual-modulus divider is the 8/9 divider. But a 10/11 divider is also possible. The trade-off here is the difference in power consumption and the risk of spurious generation via supply lines and the substrate due to the transitions of the standard CMOS logic of the A and B counters. When an 8/9 divider is used, the highest input frequency for the A and B counters is roughly 300 MHz. While for the 10/11 divider this frequency is about 250 MHz. The choice depends on the operating range of the standards logic cells under all required operating conditions (PVT corners) of the chosen CMOS technology.

Because the complete divider consists of the combination of differentially driven source coupled logic and single-ended-driven standard cell logic, a differential-to-single-ended circuit and a single-ended-to-differential circuit are required (Figure 16.14).

There are several loop delays involved in the divider system; besides the internal loop delays within the dual modulus divider itself [15,17], there are also the loop delays as shown in Figure 16.14. The most critical one is the delay τ_1. An upper limit for this delay is [8]:

$$\tau_1 < \frac{P}{f_{in}} \tag{16.45}$$

where f_{in} is the frequency of the input signal of the dual-modulus divider.

Suppose we choose $P = 10$, what are the requirements for the A and B counters? The minimum frequency range that must be covered is the 2.4 GHz industrial scientific medical (ISM) band plus some margin for those situations where the VCO has a start-up frequency either much higher or lower than the ISM band. The following counter values fulfill our requirements $A = 18$ and $0 \leq B \leq 17$.

With these values for A and B, the frequency range that can be covered is $2340\,MHz \leq f_{rf} \leq 2561\,MHz$. But it is good practice to increase the frequency range a bit by making the A counter programmable, say $A = 17$, 18, or 19. The associated frequency range becomes: $2210\,MHz \leq f_{rf} \leq 2704\,MHz$. This range should be adequate in practice.

PFD Requirements

The next block to characterize is the phase frequency detector (PFD). To arrive at the fastest settling time, no cycle slips are allowed when the synthesizer switches from one channel to another channel. The largest phase error between the inputs of the PFD occurs when the maximum frequency step is applied (see the section "Synthesizer Considerations"). For Bluetooth, the maximum distance between two channels is 79 MHz. Given a zero-IF transmitter architecture, the actual maximum step size of the synthesizer also depends on the IF frequency of the receiver. A low-IF receiver topology is very attractive for Bluetooth [18]. Let us say that the IF frequency is 1 MHz. So, a maximum frequency step size of 80 MHz is possible. According to the derivations in the section "Synthesizer Transfer Derivations," the normalized peak value of the phase error, when a frequency step is applied, is 0.96. For a synthesizer with $f_N = 40\,kHz$, the denormalized maximum phase error becomes:

$$(\varphi_e)_{max} = 0.96 \frac{(\Delta f_{out})_{max}}{N_{min} f_N} \tag{16.46}$$

When we substitute the following quantities in Equation 16.46: $(\Delta f_{out})_{max} = 80\,MHz$, $N_{min} = 185$ $(f_{ref} = 13\,MHz)$ and $f_N = 40\,kHz$, then $(\varphi_e)_{max} = 10.5\,rad \approx 3.4\pi$ radians. This is far too large for a conventional PFD, which has a peak range of 2π radians. However, the number of states of a conventional PFD can easily be extended from three (up, down, and idle) to five or more (up, up-more, down, down-more, and idle) [19].

The transfer of a five-state PFD–charge pump combination will look as shown in Figure 16.15.

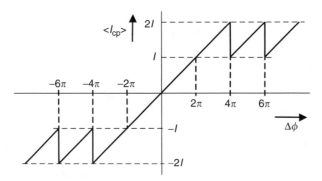

FIGURE 16.15 Combined transfer of a five-state PFD and charge pump.

The more states of the phase detector control additional current sources in the charge pump with the same peak values as the original current sources of a conventional three-state detector. The conversion gain is still $K_D = I/(2\pi)$. Cycle slips will occur for phase errors larger than 4π rad, which is sufficiently larger than the 3.4π peak error we expect.

THE CHARGE PUMP

The up and down pulses, generated by the PFD, are converted to current pulses by the charge pump. Usually, two current sources are used in parallel: one that sinks current pulses and one that sources current pulses.

The synthesizer in-band single-sided phase noise requirement imposed on the charge pump and the reference source of the synthesizer together is -80 dBc/Hz at the output of the synthesizer (Table 16.4). In practice, the charge pump is usually the main contributor. To avoid a dead zone in the PFD + charge-pump transfer, a delay is introduced in the reset line of the PFD. The result is that even when the phase error is zero, the up and down current sources are switched on for a time equal to (at least) twice the switching time of these current sources. Even when both up and down currents cancel (ideal case), their uncorrelated noise will not cancel and inject a noise current into the LPF.

As mentioned earlier, the peak values of both current pulses should be equal, but in practice they will be slightly different. One cause for this is the voltage dependency of the output current of the current sources in practice. In a synthesizer with a passive loop filter, the average voltage at the output of the charge pump depends on the selected channel and, consequently, the peak value of the up and down current source may differ slightly.

What is the effect of unequal current peak values on the noise performance of the loop?

Suppose that the peak value of the down current is smaller than the peak value of the up current by an amount ΔI. This nonlinear transfer can be decomposed into a wanted linear (odd) transfer and an unwanted even transfer (see Figure 16.16) [20]. The even transfer is the transfer of a full-wave rectifier.

Because of the presence of the full-wave rectifier, the feedback loop will regulate to a nonzero phase error to arrive at a zero charge transfer at the output of the charge pump.

To determine the spectral density of the noise at the output of the full-wave rectifier ("b" in Figure 16.17), some assumptions about the statistics of the phase noise are made: if we assume the noise part of the phase error to be Gaussian-distributed with a zero mean value and with spread σ_i, then the average value and the variance of the signal at the output of the rectifier block are [21]

$$\mu_o = \sigma_i \frac{\Delta I}{2I} \sqrt{\frac{2}{\pi}}$$

$$\sigma_o^2 = \sigma_i^2 \left(\frac{\Delta I}{2I}\right)^2 \left(1 - \frac{2}{\pi}\right)$$

(16.47)

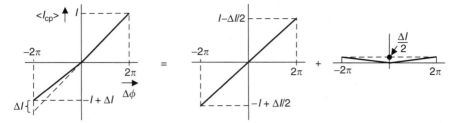

FIGURE 16.16   Decomposition of a nonlinear charge-pump transfer in a linear and a nonlinear transfer.

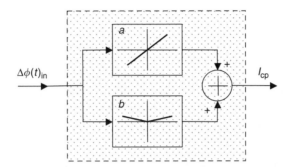

FIGURE 16.17   Block diagram of a nonlinear charge-pump transfer, block a is the desired linear transfer and block b is the undesired nonlinear transfer.

A combination of a PFD and a charge pump that exhibits such a nonlinearity can be described by the block in Figure 16.17. The nonzero mean value of the output noise signal will introduce a small phase error to arrive at a zero average charge transfer at the output of the charge pump when the loop is in lock.

The transfer of the full-wave rectifier can be approximated by a power series with only *even* powers of its input signal. This implies that noise folding takes place and the consequence for the output spectrum is that the spectrum will be whitened.

Suppose that the PSD at the output is approximately white, even when the PSD of the input signal is spectrally shaped. Although this is a crude approximation, it helps us to estimate the PSD of the output signal. The spectral density is simply determined by the ratio of the variance of the output noise and half the reference frequency:

$$S_{\sigma_o}(f) \approx \frac{2}{f_{\text{ref}}} \sigma_i^2 \left(\frac{\Delta I}{2I} \right)^2 \left(1 - \frac{2}{\pi} \right)$$ (16.48)

Example:

Suppose we want to determine the allowable difference in peak currents to keep the single-sided phase noise $L(f)$ due to the full-wave rectification, below $-90\,\text{dBc/Hz}$. When the synthesizer is in lock, the major contribution of the noise power at the input of the phase detector in the frequency band $0 - f_{\text{ref}}/2$ is generated by the $\Delta\Sigma$-modulator-induced phase noise. For a second-order modulator, the PSD of this source at the input of the phase detector is given by Equation 16.32. The noise power in the frequency band $0 - f_{\text{ref}}/2$ turns out to be:

$$\sigma_i^2 = \sigma_\Delta^2 \frac{8\pi^2}{\langle N \rangle^2}$$ (16.49)

TABLE 16.5
Loop Filter Component Values

R_1	C_1	C_2	R_3	C_3
$92\,k\Omega$	$136\,pF$	$9.8\,pF$	$92\,k\Omega$	$6.2\,pF$

where σ_Δ^2 = the unshaped quantization noise power of the $\Delta\Sigma$ modulator which is approximately equal to 1/12. Substituting this noise power in Equation 16.48 and solving for $\Delta I/I$ gives:

$$\frac{\Delta I}{I} < 0.14$$

So, a 14% difference between the source and sink currents is allowed to keep the in-band phase noise PSD below $-90\,dBc/Hz$. Such a requirement is relatively easy to meet in practice. But one must be aware that the calculations are based on crude assumptions and time-domain simulations should be carried out to determine the exact consequence of the nonlinearity.

THE LOOP FILTER

Up to this point, most parameters of the synthesizer are determined, except for the loop filter components. For this, the peak value of the charge-pump current must be chosen. Because chip area is expensive, small capacitor values are desired. For a given synthesizer parameter ω_N, all time constants of the loop filter are set (see the section "Synthesizer Transfer Derivations"). This implies that the smaller the capacitors are chosen, the higher the resistor values will be.

Suppose we choose $10\,\mu A$ for the peak value of the charge-pump current sources. Then we are able to calculate the component values of the LPF. Note that the filter topology without the voltage buffer is used here.

For $f_N = 40\,kHz$, $K_D = 10\,\mu A/(2\pi)$, and $K_O = 2\pi \cdot 50\,Mrad/s/V$, the LPF component values are shown in Table 16.5.

Although the charge-pump current could be chosen even smaller, this will increase the impedance at the filter nodes and these nodes become more vulnerable to spurious pickup. Also, the switching speed of the current sources could become too large, and as a consequence, too much noise might be generated during a zero phase error when both current sources are turned on.

In addition, if the output resistance of the charge pump is not high enough, the settling time will increase substantially. This can be avoided when

$$\frac{1}{T_{ref}} \int_0^{T_{ref}} r_{out}(t)\,dt \geq \frac{50}{\omega_N(C_1 + C_2)} \tag{16.50}$$

where the integral represents the average value of the output resistance of the charge pump when the synthesizer is in lock.

SIMULATION RESULTS

The synthesizer is modeled as a linear phase-transfer system. All noise sources were incorporated, with the levels according to the requirements of Table 16.4, except for the VCO. The output signal of the VCO was passed on to the Bluetooth measurement filter and the output signal of this filter was used to determine the residual FM.

The division factor $\langle N \rangle$ was set to 188, which corresponds to a midband ISM frequency.

TABLE 16.6
Summary of All Residual FM Contributions of the Synthesizer

Source	σ_{FM} [kHz]
Reference + charge pump	2.06
VCO	2.11
$\Delta\Sigma$-modulator driven frac. divider	0.78
R1	0.71
R3	1.00
Total	**3.30**

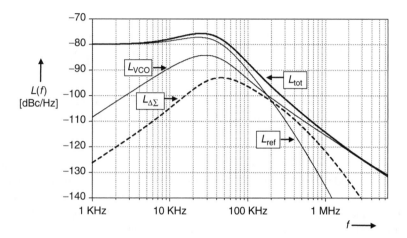

FIGURE 16.18 Single-sided phase noise contributions of the synthesizer noise sources.

The single-sided phase noise of the VCO (behind the divide-by-2) was set to a value of -125 dBc/Hz at 3 MHz distance. The LPF of the synthesizer is the version without the voltage buffer.

The following simulation results were obtained:

- The settling time (= time when the frequency error has become less than 25 kHz) for an 80 MHz step at the output turns out to be 50 µs, which is well below our maximum value of 100 µs.
- The simulated overall residual FM is $\sigma_{FM} = 3.3$ kHz. This is just on the edge. The different contributions are tabulated in Table 16.6.

Note that the residual FM of the system without the contribution of the LPF resistors is 3.1 kHz. So, it was allowed to neglect their contributions during the derivation of the maximum value of f_N in the previous section.

The various phase noise contributions at the output of the VCO (@2.4 GHz) are shown in Figure 16.18.

Note that the fourth pole of the closed-loop transfer provides additional out-of-band suppression of the $\Delta\Sigma$-modulator-induced phase noise.

DISCUSSION

The residual FM and phase noise simulation results indicate that with the choices made, it seems feasible to meet the synthesizer requirements. However, the design steps taken here must only be considered as the first step toward a robust system design.

First of all, all circuits must be designed on a transistor level and should fulfill the noise, speed, and power consumption requirements. In addition, the design must be made robust to be able to fulfill the requirements under all conditions (PVT corners). This will most likely imply that the nominal bandwidth of the synthesizer must be reduced somewhat. Suppose that f_N is reduced from 40 to 30 kHz; then the nominal value of the residual FM reduces from 3.3 to 2.7 kHz, whereas the settling time for an 80 MHz step at the output becomes 67 µs. So, these numbers still leave some margin for variation in loop parameters in practice.

Another design step to take is to carefully evaluate the noise performance of the $\Delta\Sigma$ modulator in the time domain. Usually, such a modulator shows a worse noise performance compared with the theoretical predictions; the PSD is differently shaped with a more pronounced peaking before $f_{ref}/2$ and with a steeper slope. This might influence the overall noise performance of the system in practice.

SUMMARY AND CONCLUSIONS

In this chapter the design of a fourth-order, type 2 synthesizer with a $\Delta\Sigma$-modulator-controlled, fractional divider for Bluetooth is described on a system level. It was demonstrated that the requirements for the synthesizer, regarding its residual FM performance, are highly dependent on the performance of the transmitter when a commonly used zero-IF up-conversion Tx architecture is used. In addition, when a synthesizer with a VCO running at the double RF frequency is employed, spurious injection of harmonics of the RF output signal can be picked up by the VCO and generate additional spurious signal components in the RF output signal, which can degrade the residual FM performance significantly.

Because the sum of all imperfections in the signal path of the transmitter can contribute too much to the overall residual FM and because the level of spurious injection into the VCO cannot be predicted in advance, it is advisable to calibrate the low-frequency signal path of the transmitter to arrive at a reasonable residual FM requirement for the synthesizer. The imperfections in the signal path of the transmitter that usually can be calibrated are the overall quadrature phase error and DC offsets in the LF path.

Besides the residual FM requirements for the synthesizer, all the blocks of the synthesizer are briefly described regarding their basic requirements. Notably, the PFD should have an extended input range to avoid cycle slips for the largest frequency step size and the allowable nonlinearity of the charge-pump transfer is limited to avoid noise folding of the $\Delta\Sigma$-modulator-induced phase noise to in-band frequencies of the synthesizer transfer.

REFERENCES

1. Haartsen, J. C. and Mattisson, S., "Bluetooth—A new low-power radio interface providing short-range connectivity," *Proceedings of the IEEE*, vol. 88, pp. 1651–1661, October 2000.
2. Benedetto, S. and Biglieri, E., *Principles of Digital Transmission, With Wireless Applications*, Kluwer Academic Publishers, New York, 1999.
3. Bluetooth ver. 2.0 + EDR, vol. 3: Radio Specification, November 4, 2004.
4. Bluetooth Test Specification ver. 1.2/2.0/2.0 + EDR, Radio Frequency, October 14, 2005.
5. Riley, T. A. D., Copeland, M. A., and Kwasniewski, T. A., "Delta-sigma modulation in fractional-N frequency synthesis," *IEEE J. Solid-State Circuits*, vol. 28, pp. 553–559, May 1994.
6. Zarkeshvari, F., Noel, P., and Kwasniewski, T., "On $\Delta\Sigma$ fractional-N frequency synthesizers," *International Symposium on Signals, Circuits and Systems, ISSCS*, vol. 2, pp. 509–512, July 14–15, 2005.
7. Gardner, F. M., "Charge-pump phase-lock loops," *IEEE Trans. Commn.*, vol. COM-28, pp. 1849–1858, November 1980.
8. Vaucher, C. S., *Architectures for RF Frequency Synthesizers,* Kluwer Academic Publishers, The Netherlands, 2002.
9. Perrott, M. H. and Troll, M. D., "A modeling approach for $\Sigma-\Delta$ fractional-N frequency synthesizers allowing straightforward noise analysis," *IEEE J. Solid-State Circuits*, vol. 37, pp. 1028–1038, August 2002.

10. Norsworthy, S. R., Candy, J. C., and Temes, G. C. (Eds.), *Delta-Sigma Data Converters, Theory Design, and Simulation*, IEEE Press, Piscataway, NJ, 1997.
11. Soltanian, B. and Kinget, P. R., "Tail current-shaping to improve phase noise in LC voltage-controlled oscillators," *IEEE J. Solid-State Circuits*, vol. 41, pp. 1792–1802, August 2006.
12. Hegazi, E., Sjöland, H., and Abidi, A. A., "A filtering technique to lower LC oscillator phase noise," *IEEE J. Solid-State Circuits*, vol. 36, no. 12, pp. 1921–1930, December 2001.
13. Lee, T. H., *The Design of CMOS Radio-Frequency Integrated Circuits*, 2nd ed., Cambridge University Press, Cambridge, UK, 2004.
14. Rohde, U. L., *Digital PLL Frequency Synthesizers, Theory and Design,* Prentice-Hall, Inc., Englewood Cliffs, NJ, 1983.
15. Foroudi, N. and Kwasniewski, T., "CMOS high-speed dual-modulus frequency divider for RF frequency synthesis," *IEEE J. Solid-State Circuits*, vol. 30, pp. 93–100, February 1995.
16. "Design compromises in single loop frequency synthesisers," in Personal Communications IC Handbook, Plessey Semiconductors, 1990.
17. Romanò, L., Levantino, S., Pellerano, S., Samori, C., and Lacaito, A., "Low jitter design of a 0.35 μm-CMOS frequency divider operating up to 3 GHz," *Proceedings of the 28th European Solid-State Circuits Conference*, ESSIRC 2002, pp. 611–614.
18. Zeijl, P. v., Eikenbroek, J. W., Vervoort, P. P., Setty, S., Tangenberg, J., Shipton, G., Kooistra, E., Keekstra, I. C., Belot, D., Visser, K., Bosma, E., and Blaakmeer, S. C., "A Bluetooth radio in 0.18-μm CMOS," *IEEE J. Solid-State Circuits*, vol. 37, no. 12, pp. 1679–1687, December 2002.
19. Eijselendoorn, J. and den Dulk, R. C., "Improved phase-locked loop performance with adaptive phase comparators," *IEEE Trans. Aerosp. Electron. Syst.*, vol. AES-18, pp. 323–332, May 1982.
20. Eikenbroek, J. W. and Mattisson, S., "Frequency synthesis for integrated transceivers," in *High-Speed Analog-to-Digital Converters: Mixed-Signal Design; PLLs and Synthesizers*, Kluwer Academic Publishers, Dordrecht, The Netherlands, pp. 339–355, 2000.
21. Papoulis, A., *Probability, Random Variables, and Stochastic Processes*, McGraw-Hill Book Company, International Student Edition, 1981.

17 RFIC Parametric Converters: Device Modification, Circuit Design, Control Techniques

Sebastian Magierowski, Howard Chan,
Krzysztof (Kris) Iniewski, and Takis Zourntos

CONTENTS

Introduction...487
      Resistive and Transresistive Amplification...488
      Historical Background...488
Principle of Operation..490
      The Degenerate Amplifier..490
      The Oversampling Parametric Amplifier...492
Configurations of Parametric Amplifiers 493
      Upper Sideband Up-Converter...495
      Lower Sideband Down-Converter..497
      Parametric Transmit and Receive Chains..497
Varactor Structures 500
      Elastance Model 501
      Figures of Merit..502
Parametric Circuit Simulations..508
      Upper Sideband Up-Converter Simulations 508
      Lower Sideband Down-Converter Simulations 514
Conclusions...516
References..517

INTRODUCTION

Parametric converters are circuits that rely on reactive components to affect signal amplification. As their name implies, these circuits can also realize signal frequency conversion. This chapter discusses the theory and operation of parametric circuits in an integrated system setting. As part of this discussion, integrated circuit (IC) parametric converters from three levels are considered: device level (i.e., the selection and optimization of circuit components), circuit level (i.e., the expected performance of parametric converter circuits), and system level (i.e., the potential for parametric circuits in radio frequency (RF) front-ends).

Resistive and Transresistive Amplification

Before focusing completely on parametric circuits, a more customary means of amplification—conductivity modulation—is considered and reviewed with the hope of establishing a familiar counterpoint to the parametric principles. Amplification by conductivity modulation can be divided into two broad classes: negative-resistance amplification and transistor amplification. To emphasize the similarity between these approaches, think of the latter as a transresistive technique.

An idealized schematic of a one-port negative-resistance amplifying circuit is illustrated in Figure 17.1; here $-R_A$ effectively reduces the generator resistance, hence boosting the power available from the source. A variety of diode structures can be configured for negative-resistance amplification. In this case, an external DC power bias can alter the internal device structure in such a way that a higher signal current can flow through the element for a lower energy penalty. A number of physical principles can be exploited to achieve this behavior. For example, in tunnel diodes the negative resistance is achieved on an essentially quantum mechanical level (the variation of resonant energy states based on changes to energy-band distribution). Although capable of high-speed operation, this approach is largely limited to specialized applications for a variety of reasons including difficulty of integration, cost, poor efficiency, and cumbersome biasing requirements.

The limitations of one-port resistive amplification are especially glaring when compared to transistor amplifiers. Rather than directly boosting the energy of a signal, transistors use the input to mediate the connection between a high-energy supply and a load. The three-terminal makeup of a transistor is obviously essential to such operation. The fundamental advantages of a transistor element are twofold; first, the fact that a very small input signal at one port is used to draw a large current through an output port and second, that the device is largely unidirectional. These characteristics enable efficient amplification and the possibility of synthesizing complex functions. For example, a cross-coupled transistor connection succeeds in implementing a negative-resistance element, relegating the latter to a subset of transistor-based amplification.

Common to both of these techniques is the fact that a dissipative property (conductivity modulation) is used in the amplification process. It follows that a certain thermal noise penalty is exacted on signals processed by such systems. This is relevant in comparison to parametric circuits, which employ reactive elements and thus are not intrinsically subject to thermal fluctuations. Thus, low-noise behavior is one obvious advantage of parametric circuits. Other advantages follow, particularly for applications that require low-cost, low-power radio systems that operate above the frequency limits of production-level CMOS technologies. These advantages are discussed in more depth in the following section.

Historical Background

Adhering to its namesake, parametric conversion promotes amplification by using a local power source to modulate an oscillatory system's physical parameters. To clarify this by way of a mechanical

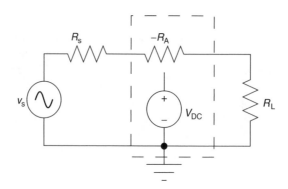

FIGURE 17.1   A one-port negative-resistance amplifier.

analogy, rather than exerting direct force to compress a spring, a parametric system may achieve the same goal by changing the spring constant itself. For example, the oscillation frequency of the damped second-order parametric system described by

$$\mathrm{d}x^2/\mathrm{d}^2t + \gamma \mathrm{d}x/\mathrm{d}t + \omega_0^2[1 + p(t)]x = 0 \tag{17.1}$$

depends on a parametric signal, often referred to as the pump signal, $p(t)$. Given a pumping disturbance at twice the oscillator's natural frequency,

$$p(t) = A_\mathrm{p} \sin(2\omega_0 t) \tag{17.2}$$

a subharmonic (relative to the pump) response

$$x(t) = Ae^{\alpha t} \sin(\omega_0 t) \tag{17.3}$$

can be solicited where α ranges between positive and negative values depending on the oscillator damping and the strength of the pumping signal. Lacking a forcing function, Equations 17.1 to 17.3 are an example of parametric excitation than parametric amplification, *per se*. Nonetheless, the core characteristics are present: a resonant system and variable parameters. Besides parameter variability, the former plays a key role in continuous-time parametric systems (in contrast with discrete-time work [1]) as a means of swapping energy between the circuit and the pump. No such mechanism is necessary in transistor amplification, which uses the signal to directly control the power drawn from a local supply. In summary, one can think of parametric amplification as an implicit scheme compared to the explicit relation between input and output present in transistor amplification.

It is commonly acknowledged that sustained parametric oscillations as described by Equation 17.3 were first observed by Michael Faraday. In 1831, he reported the rise of oscillations at precisely half the forcing frequency as part of experiments on the distribution of granular and liquid matter on acoustically pumped glass surfaces (Chladni plates) [2]. In 1957, following suggestions by Suhl [3], Weiss [4] reported the realization of an experimental solid-state microwave amplifier exploiting the parametric principle. Notable among a large number of intermediate contributions is Hartley's work on electromechanical parametric amplifiers [5,6]. This area has been a subject of interest in the microelectro mechanical systems (MEMS) arena for low-frequency, precision-sensing applications such as atomic force microscopy [7,8], where the low-noise properties of parametric systems are of particular benefit. However, the use of parametric amplification in terrestrial communication systems is extremely rare. One notable exception is a discrete-time parametric amplifier implemented in a 0.25 μm CMOS technology and reported in Ref. 1. The circuit took particular advantage of the three-terminal inversion mode MOS varactor but focused on low-frequency applications in the 100 kHz range.

The absence of parametric converters from the communications mainstream is due mainly to the superior utility of transistor-based circuits for low-frequency commercial applications. A key advantage of the former approach is its ability to operate at higher frequencies early in the technology life cycle. This is because the performance of parametric circuits is less dependent on the lateral dimensions, and hence delay, of their components. However, historically, improvements in transistor technology steadily encroached on the high-frequency reserve of parametric circuits, thus marginalizing this advantage. Since more remote regions of the spectrum were better accommodated by maser and laser amplifiers, a loss of interest in the parametric circuit approach followed.

Today, as personal commercial communications applications migrate to more exotic frequency domains, a niche for the parametric circuit may resurface. This may especially be the case for low-profile millimeter-wave electronics intended for dense sensor or distributed network applications. Size and power constraints exclude many of today's molecular amplifiers from consideration (although integration progress has been substantial [9]), whereas performance, power, and approaching physical limits have relegated millimeter-wave applications to the domain of expensive IC technologies.

PRINCIPLE OF OPERATION

First impressions may indicate that parametric circuits convert energy in a manner similar to that present in the transistor amplification mechanism. That is, both schemes seem to rely on a small signal to modulate the coupling between a power source and a load. The obvious differences include the fact that parametric circuits employ an AC power source and the amplified signal can emerge at a frequency different from that of the input. Of course, with an adequate circuit arrangement, the power delivered by a parametric circuit can be made to emerge at the same frequency as the input signal. However, the fact that we are dealing with a reactive two-terminal device in the parametric topology introduces several key differences that become apparent when we consider the parametric amplification process in greater detail.

THE DEGENERATE AMPLIFIER

The fundamental principles at work in a parametric amplifier are perhaps best illustrated by considering a toy parametric configuration operating in discrete time. A simplified schematic of this circuit and its clocking scheme is shown in Figure 17.2, which is composed of a source voltage, v_s, a variable capacitor, C_A, a pump voltage, v_p, and a load, R_L. When the source switch φ_S is closed (step 1 of the clocking scheme illustrated in Figure 17.2), the capacitor C_A is charged, storing an energy of E_s (energy deposited by signal):

$$E_s = \frac{1}{2} C_A \cdot v_s^2 \tag{17.4}$$

Next, the source voltage is detached and a pump voltage source connected to the capacitor via φ_P. We assume that the pump voltage decreases the capacitance by ΔC but does not alter the net amount of stored signal charge. Mechanically, this may be accomplished by simply increasing the plate separation (assuming a parallel-plate capacitor). Electrically, ΔC can be realized by changing the polarizability of the dielectric material.

At the end of the pumping phase (end of step 2), the excess voltage on the capacitor increases to

$$v_{s,p} = \frac{C_A}{C_A - \Delta C} v_s \tag{17.5}$$

and thus the net energy stored by the capacitor increases to

$$E_{s,p} = \frac{1}{2} \frac{C_A^2}{C_A - \Delta C} v_s^2 \tag{17.6}$$

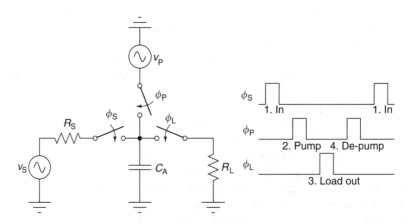

FIGURE 17.2   An idealized parametric amplifier in discrete time.

Thus, the net energy deposited by the pump source in this phase is

$$E_p = E_{s,p} - E_s = \frac{\Delta C}{C_A - \Delta C} E_s \qquad (17.7)$$

Now, the stored energy can be extracted in a variety of ways. The simplest way is for φ_L to close (step 3) until all the stored charge, $E_{s,p}$ is extracted, that is

$$E_{out} = E_{s,p} \qquad (17.8)$$

With all the energy extracted from the capacitor and into the load, the pump is activated again (step 4) to restore the original capacitance. Since the capacitor has been discharged, this does not result in any energy return to the pump.

The energy gain of this toy configuration versus $\Delta C/C_A$ ratio is shown in Figure 17.3. Immediately, we see that a rather large capacitance variation, ΔC, is needed to achieve substantial energy gain (for example, the gain does not exceed 10 dB until the available capacitance variation reaches 90% of nominal). Practical issues aside, this sums up the most fundamental properties of the parametric amplifier: an indirect (implicit) and staged amplification of signal energy.

The first property refers to the use of a pump to operate on the capacitance of the mediating element C_A rather than on the signal charges themselves. In this sense its operation is unlike the processes at work in resistance-based amplifiers. For instance, a typical (classical) negative-resistance element relies on the injection of minority carriers to regulate the conductivity of a material. In this case, the signal (manifest as minority carriers) plays a direct role in establishing the negative-resistance property of the device. Of course, this characteristic is also partly enabled by the presence of a constant energy supply from a bias source. Similarly, in transistor amplifiers the input signal modulates the conductivity of a material, although here the intention is to control the amount of power drained from the bias source.

The second property, that of staged amplification, addresses the three basic phases (φ_S, φ_P, and φ_L) involved in the parametric amplification process. It is essential that these be implemented

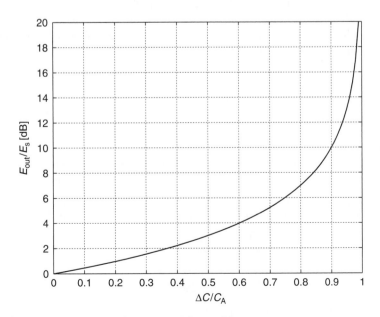

FIGURE 17.3 Energy gain behavior of toy parametric amplifier.

in the proper order because, enabled out of sequence, the circuit will divert energy to the pump rather than the load; in a discrete-time circuit it is straightforward to arrange for this. However, a continuous-time implementation (the preoccupation of this chapter) requires careful phase alignment between the signal and the pump—an arrangement that is referred to as a "degenerate implementation." The resistance-based amplifiers discussed earlier are not subject to such a restriction.

The vast majority of continuous-time parametric designs overcome the degenerate topology's constraints by increasing the pump frequency and the filtering complexity. We will examine the influence of higher pumping rates from the discrete-time degenerate perspective in the following section.

THE OVERSAMPLING PARAMETRIC AMPLIFIER

Figure 17.4 shows the clocking scheme for a modified discrete-time parametric amplifier. Structurally, the main difference between this circuit and the one illustrated in Figure 17.2 is the addition of an energy storage element referred to, for historical reasons, as the idler. The clock regulating the connection between the variable capacitor and the idler is denoted with φ_I. Besides adding the idler, this topology has increased the pump (and de-pump) rate by a factor of N_{up}. Increasing the pump frequency and adding the idler is an essential step away from a degenerate topology and toward a phase-insensitive continuous-time parametric design. This change also facilitates an increase in the gain of a parametric circuit, a property that can be appreciated from the discrete-time (and hence degenerate) perspective.

As observed in the previous section, a rather large capacitance variation is necessary to obtain substantial gain. This issue can be mediated by simply pumping and extracting energy faster than the deposition rate of the signal source. In short, we can effectively oversample the input by some factor N_{up}. In this case, the maximum energy that can be extracted during a single signal interval of the clock φ_S is

$$E_{out} = N_{up} \cdot \frac{\Delta C}{C_A - \Delta C} E_s + E_s \tag{17.9}$$

where it is assumed that the total energy available to a load in one signal-sampling period is the original Equation 17.9 signal energy E_S plus the energy contributed by a pump applied N_{up} times during the sampling period. Without taking the details of operation into account, Equation 17.9 is an idealization. If we consider an amplifying configuration that pumps C_A and extracts E_{samp} of energy N_{up} times in a sample period until completely exhausting the total stored energy (in the variable reactance and the idler) on the last extraction, then a slightly different relation than that expressed in Equation 17.9 results. Specifically, it can be shown that the amount of energy dissipated in the load per pump cycle (relative to the signal energy) is

$$\frac{E_{samp}}{E_s} = \frac{X-1}{1-X^{-N_{up}}} \tag{17.10}$$

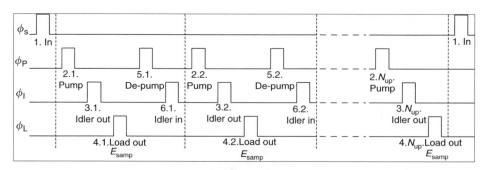

FIGURE 17.4   Clocking scheme for oversampled amplifier.

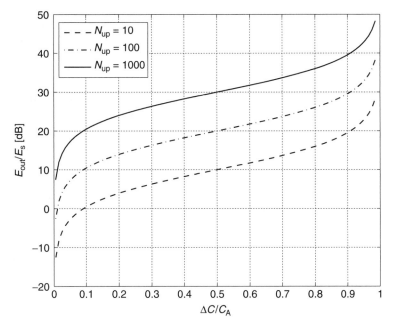

FIGURE 17.5   Energy gain behavior of an oversampled toy parametric amplifier.

where

$$X = \frac{C_A}{C_A - \Delta C} \tag{17.11}$$

This naturally leads to the total energy gain expression of

$$\frac{E_{out}}{E_s} = \frac{N^{up}(X-1)}{1 - X^{-N_{up}}} \tag{17.12}$$

The energy gain of the oversampled system described by Equation 17.12 is shown in Figure 17.5. Although the available capacitance variation, ΔC, remains an important contributor to overall gain, its influence is augmented by the introduction of an oversampling pumping apparatus. For instance, now a 10 dB gain can be obtained with a variation of only 50% around nominal if a 10 times oversampling pump is used. This improvement comes with the price of increased clocking and component counts; however, these costs are rather naturally absorbed in a continuous-time implementation. The idler and the load, for example, can easily be merged into a resonant circuit (e.g., series RLC). In the process of storing energy in the magnetic field of the inductor, there is energy delivered to the resistive load. After this the variable capacitor is de-pumped and energy returned to the capacitor's electric field from the inductor's magnetic field. At this time the charge capacitor can be pumped again and so on until all energy is dissipated and the next signal sample is introduced for amplification.

CONFIGURATIONS OF PARAMETRIC AMPLIFIERS

The discussion in the section "Principle of Operation" presents a time domain description of the parametric converter applicable to discrete-time implementations. For continuous-time

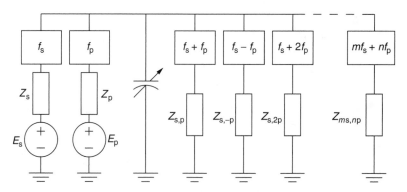

FIGURE 17.6   The ideal parametric converter used to derive the Manley–Rowe relations. Ideal impedance filters allow only a single tone to flow through any one branch.

parametric converters, a succinct frequency domain description is provided by the Manley–Rowe relations [10]

$$\sum_{m=0}^{\infty}\sum_{n=-\infty}^{\infty}\frac{mP_{m,n}}{mf_s+nf_p}=0 \tag{17.13}$$

$$\sum_{n=0}^{\infty}\sum_{m=-\infty}^{\infty}\frac{nP_{m,n}}{mf_s+nf_p}=0 \tag{17.14}$$

The ideal circuit used to derive these equations is shown in Figure 17.6. It consists of some unspecified nonlinear capacitor (although it is assumed that the capacitance is a single-valued function of the applied voltage) driven by two independent harmonic sources operating at frequencies f_s (the signal source) and f_p (the pump source). The capacitor is also connected to an array of loads via a variety of band-pass filters (pass frequencies indicated on filter blocks). These filters determine the signal frequencies we allow to excite the nonlinear component—in theory any subset of filters with pass frequencies mf_s+nf_p (where m and n are integers) can be used.

The Manley–Rowe relations show the constraints on the power, $P_{m,n}$, absorbed by the nonlinear capacitor at frequencies mf_s+nf_p. That is, the power delivered to the capacitor at mf_s+nf_p is denoted by

$$P_{m,n}=\frac{1}{2}\text{Re}\left\{V_{m,n}I_{m,n}^*\right\}$$

where the voltage and current represent peak harmonic amplitudes. It is natural to expect that as a result of the nonlinear capacitance, power components at different frequencies will be present throughout the circuit. Since we are working with a lossless capacitance, we also expect the net power delivered to it to be zero, that is

$$\sum_{m,n}P_{m,n}=0$$

The utility of the Manley–Rowe relations lies primarily in the fact that they express this fundamental characteristic as a function of the frequencies allowed to excite the circuit. As will be seen in the following sections, the selection of band-pass frequencies used in the circuit has a decisive influence on the properties of the parametric converter. This choice constrains the raw circuit performance encapsulated by the Manley–Rowe relations to describe such basic functions as up-conversion (signal at f_s amplified and mixed to higher frequencies), down-conversion (signal at f_s amplified and mixed to lower frequencies), or simply straight amplification (signal at f_s amplified at f_s).

As a final note, it is important to keep in mind that the Manley–Rowe relations cannot be used for the purposes of practical design. This comes as a result of the fact that they do not account for losses in the nonlinear reactive element, a property with a substantial impact on practical circuit behavior.

UPPER SIDEBAND UP-CONVERTER

Imagine a nonlinear capacitor locked in a resonant band-pass filter configuration that allows power to flow only at frequencies $f_s, f_p,$ and $f_u = f_s + f_p$. This configuration is referred to as the upper sideband up-converter (USBUC) because $f_s + f_p$ (the upper sideband) is greater (up-conversion) than f_s. In this scenario, Equations 17.13 and 17.14 simplify to

$$\frac{P_{1,0}}{f_s} + \frac{P_{1,1}}{f_s + f_p} = \frac{P_s}{f_s} + \frac{P_u}{f_u} = 0 \tag{17.15}$$

$$\frac{P_{0,1}}{f_p} + \frac{P_{1,1}}{f_s + f_p} = \frac{P_p}{f_p} + \frac{P_u}{f_u} = 0 \tag{17.16}$$

As stated earlier, the pump is the energy source in this circuit; it is responsible for a positive power flow, P_p into the capacitor. According to Equation 17.16, this means that the power, P_u, at f_u must be negative, implying that it flows out of the capacitor and to a load. Combining Equations 17.15 and 17.16, we can derive the operating power gain:

$$G_{up} = \frac{P_u}{P_s} = \frac{f_u}{f_s} \tag{17.17}$$

From the relations established so far, it is clear that the higher, the pump frequency the greater the circuit gain. For instance, under ideal conditions, converting a 6 MHz signal to a 60 GHz carrier promises a 40 dB operating power gain. In the ideal case described by Equations 17.15 and 17.16, the gain is independent of the pumping power. This immediately begs the question of what the output power limits are. Although losses are not accounted for until the section "Conclusions," thereby leaving questions of efficiency out of place for the moment, it is still instructive to consider the finer points of power transfer in the USBUC as summarized by Equations 17.15 and 17.16. As illustrated in Figure 17.7, the maximum available output power is $P_a f_u/f_p$, where P_a is the power available from the pump. This level will only be reached when the power delivered by the source, P_s, is $P_a f_s/f_p = P_{s,max}$. It is only at this input level that all the available gain power will be utilized. Thus, to make most efficient use of the pump, one should note the maximum expected input signal and set $P_a = P_{s,max} f_p/f_s$. Of course, this recipe may hamper the linearity of the converter because a nonideal pump is bound to exhibit saturation effects near maximum output.

The bandwidth of the lossless parametric amplifier was considered by Rowe [11] in the basic setting shown in Figure 17.8. Arranging for a conjugate match at the input and output ports and considering a capacitance pumped at f_p

$$C_{pump}(t) = \sum_{n=-\infty}^{\infty} C_n e^{j2\pi nf_p t} \tag{17.18}$$

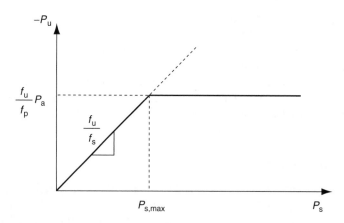

FIGURE 17.7   Input/output power profile of an ideal USBUC.

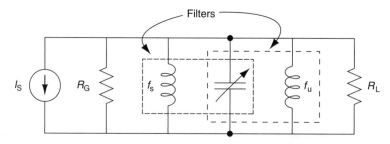

FIGURE 17.8   Basic USBUC structure including signal and up-conversion filter.

Rowe derived the maximum gain-bandwidth product

$$BW_{up} = \frac{C_1}{C_0}\sqrt{2f_uf_s} \qquad (17.19)$$

Returning to the previous example, to accommodate a signal of 10 MHz bandwidth converted from a 6 MHz to a 60 GHz center frequency requires a C_1/C_0 ratio of less than 1/80. This bodes extremely well for the pump. For a well-designed MOS varactor, it is not unreasonable to anticipate a 50% variation around C_0 as indicated by the following approximation valid for a small pump voltage, v_p:

$$C_{MOS} = C_0 + \frac{C_0}{2}\tanh\left(\frac{3}{2}v_p\right) \approx C_0 + \frac{3C_0}{4}v_p \qquad (17.20)$$

This is confirmed by the small C_1/C_0 requirements of our example. From Equations 17.19 and 17.20, a peak-to-peak pump voltage of only 16 mV is needed to sufficiently perturb the varactor so that a 10 MHz bandwidth is established. This must be tempered by the fact that in this case the pump needs to operate at 60 GHz. This is not out of the question for production-level technologies exploiting distributed operation [12], frequency doubling [13], or second-harmonic generation [14]. Further, it is possible to redesign the pumping scheme of the parametric converter to continue meeting the Manley–Rowe predictions while operating with a pump at lower frequencies.

The benefits available to the USBUC become serious impairments when considering this topology for a receiver's down-conversion block. The substantial gain available to the up-converter (Equation 17.17) becomes a tremendous loss as, naturally, $f_u \ll f_s$ for down-converters. Fortunately, a large variety of parametric conversion topologies exist, some of which do allow for gain in the down-conversion arrangement. One such topology is discussed in the following section.

LOWER SIDEBAND DOWN-CONVERTER

The lower sideband down-converter (LSBDC) is one parametric topology capable of amplifying the radio frequency input. In this case, the filtering is aligned such that power only at f_s (the RF signal), f_p, and $f_d = f_p - f_s$ (the down-converted signal) can flow through the circuit. For proper operation, an LSBDC's pump frequency must lie between the bounds $f_s < f_p < 2f_s$. Staying above the lower bound ensures gain, whereas staying below the upper bound avoids up-conversion. Returning to the Manley–Rowe relations, we get

$$\frac{P_{1,0}}{f_s} + \frac{P_{1,-1}}{f_s - f_p} = \frac{P_s}{f_s} - \frac{P_d}{f_d} = 0 \tag{17.21}$$

$$\frac{P_{0,1}}{f_p} + \frac{P_{-1,1}}{-f_s + f_p} = \frac{P_p}{f_p} + \frac{P_d}{f_d} = 0 \tag{17.22}$$

Adding Equations 17.21 and 17.22 results in

$$\frac{P_s}{f_s} + \frac{P_p}{f_p} = 0 \tag{17.23}$$

which, given that the pump power flows into the circuit, implies that power actually emerges from the converter's signal port. This indicates that the impedance looking into the LSBDC's signal port is negative. Thus, the LSBDC doubles as a reflection amplifier. Similarly, Equation 17.22 states that the pump energy causes power to emerge from the down-conversion port as well. The operation of this circuit can be summarized as follows:

The pump generates the highest frequency signal in the circuit. Thus, unlike the USBUC, on average it can couple power from the signal port (RF) to the down-conversion port intermediate frequency (IF) and vice versa. Since the pump switches only slightly faster than the signal, it transfers a relatively small amount of the input power into the down-conversion port's IF frequency. However, being much higher than IF, the pump taps, amplifies, and converts a great deal of the IF back to RF (as predicted by Equation 17.21). Part of the larger input signal is then tapped once again by the pump and fed into the IF port. A positive feedback is established and the circuit functions as a regenerative amplifier. The pump power emerges as RF and IF from the two ports, which results in their having negative effective input impedance (the more power that is pumped into the circuit, the higher the quality factor, Q, of the RF and IF modes). Thinking of the input as a forcing signal on these modes, we can imagine that the higher the Q, the higher the signal gain but the lower the bandwidth. Nonetheless, for signals centered around millimeter-wave carriers, the LSBDC topology has a good deal of relative bandwidth performance available to sacrifice. A drawback to this circuit, however, is that an excess of pump power leads to instabilities (overcompensation of loss) and, simultaneously, a greater sensitivity to component variations (thus increasing the likelihood of instability). The advantage remains the potential low-noise behavior about which the Manley–Rowe relations say nothing. We will return to this feature later in this chapter.

PARAMETRIC TRANSMIT AND RECEIVE CHAINS

How can parametric converters be arranged to realize the transceiver chain? Since they combine oscillator, mixer, and amplifier functions under essentially one circuit, they hold the potential to form the basis for a diverse set of radio systems. Perhaps the most straightforward application is the

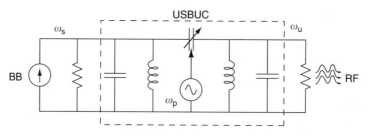

FIGURE 17.9   A simple transmit chain employing the USBUC.

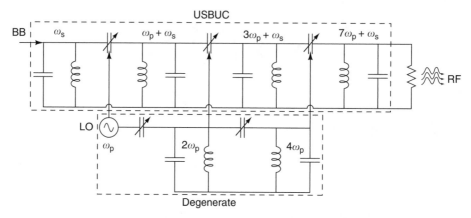

FIGURE 17.10   Multistage USBUC transmitter with degenerate pump.

use of a USBUC as a low-voltage up-converter of IF signals to millimeter-RF. For minimal complexity, the design shown in Figure 17.9 can be used. This diagram suggests interfacing the USBUC directly to the antenna, which, if the antenna is sufficiently narrowband, can serve as the f_u bandpass filter. Employing a standard two-terminal varactor structure in this topology will impose extra gain limiting—significant up-converted signal amplitudes can induce lower sideband signals to flow (i.e., $\omega_p - \omega_s$), thus returning power back to the input source. A simple alternative is to use the USBUC as an up-converting mixer and preamplifier and leave the final millimeter-wave amplification to a dedicated high-frequency (and high-cost) power amplifier. Alternatively, anticipating the encroachment of digital processing toward the antenna, the USBUC can be used as a small-signal low-noise amplifier (LNA) driving a subsampling mixer. This situation will be explored later.

Another transmitter topology, shown in Figure 17.10, incorporates a degenerate local oscillator (LO) in a heterodyne USBUC architecture. In this case, the gain of the USBUC is distributed over several stages. The benefit of such a partition is reaped by the pump, which can also be potentially generated in a staged fashion as well. In Figure 17.10, the staged pump is built out of degenerate parametric converters. In degenerate converters the signal, LO, acts simultaneously as the input and the pump. A self-mixing occurs, which naturally results in a signal at twice the input frequency. As shown, two such stages attached back–to back can produce a signal at four times the driving pump frequency (with the need of a high-power output at ω_p) and can be combined with a multistage USBUC to gradually up-convert a signal from ω_s to $7\omega_p + \omega_s$. Since parametric circuits couple power from low to high frequencies, the receiver's down-conversion function obviously poses a problem. As already described, the LSBDC gets around this issue by employing positive feedback. This can grant substantial gain at the expense of sensitivity.

A possible receiver topology employing an LSBDC is shown in Figure 17.11. Since the circuit functions as a reflection amplifier for both RF and IF frequencies, a circulator is included to

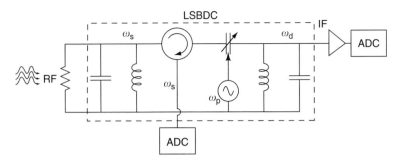

FIGURE 17.11   Receive chain using an LSBDC as a mixer and amplifier.

prevent reradiation and help maintain stability. A number of options are available even within the basic LSBDC receiver. Most simply, it can be treated as an LNA with the amplified RF signal tapped out of the circulator to the remainder of the radio. In this case, we benefit simply from the large gain and low-noise performance of the parametric converter. Any standard down-conversion architecture or subsampling techniques can be employed afterward. Compared to integrated transistor LNAs operating in the microwave region, this benefit is marginal at best; however, at millimeter-wave frequencies, the improvements for amplification, noise, and power consumption become marked (at least compared to production-level CMOS technology). Using the down-conversion port is another possibility, in this case taking advantage of the LSBDC's conversion properties alongside its low-noise performance. The difficulty in this case is gain; as the down-conversion gain is increased, the regenerative design becomes difficult to stabilize under practical conditions.

An obvious drawback to parametric converters is the high pump frequency needed to transfer power. As a result, a number of high-frequency pump generation and conversion techniques have already been mentioned. Another approach is to reconfigure the varactor structure for subharmonic pumping. "Subharmonic pumping" refers to an arrangement in which a certain pumping frequency transfers energy at the same rate as a higher pumping frequency would [15]. One means of realizing this is to utilize one of the higher pumped capacitance harmonics [15]. Herein, the more abrupt MOS C–V characteristics (compared to the junction varactor) can be of substantial benefit. For example, imagine a varactor pumped such that part of its Fourier series expansion from Equation 17.18 is

$$C_A(t) = \cdots + C_{-2}e^{-j2\omega_p t} + C_{-1}e^{-j\omega_p t} + C_0 + C_1 e^{j\omega_p t} + C_2 e^{j2\omega_p t} + \cdots \tag{17.24}$$

Another varactor, C_B, pumped 180° out of phase relative to C_A can be described with

$$C_B(t) = \cdots + C_{-2}e^{-j2(\omega_p t + \pi)} + C_{-1}e^{-j(\omega_p t + \pi)} + C_0 + C_1 e^{j(\omega_p t + \pi)} + C_2 e^{j2(\omega_p t + \pi)} + \cdots \tag{17.25}$$

Combining C_A and C_B,

$$C_A(t) + C_B(t) = \cdots + 2C_{-2}e^{-j2\omega_p t} + 2C_0 + 2C_2 e^{j2\omega_p t} + \cdots \tag{17.26}$$

leads to a net capacitance variation that occurs at twice the actual pump rate. The schematic of a differentially driven subharmonic scheme based on this approach is shown in Figure 17.12. For subharmonic pumping to work properly, the varactors, C_A and C_B, both must have the same terminal (either gate or source) connected to the circuit proper. Aside from exciting the second harmonic, the differential pumping scheme allows the circuit to operate without a dedicated pump filter (despite the use of two-terminal varactors). Alternatively, if the orientation of one varactor is flipped (i.e., the terminal connections are reversed or a complementary structure is used), the subharmonic pumping effect is removed [16]. The benefit of this connection lies in the common-mode isolation it establishes

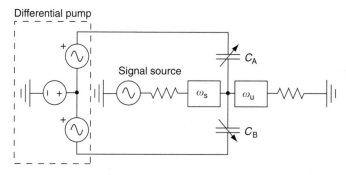

FIGURE 17.12   A differential subharmonic pumping scheme.

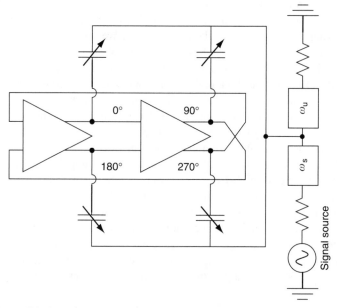

FIGURE 17.13   A possible four-phase subharmonic pumping scheme.

between the pump and the signal frequencies (input and output), allowing the filtering at these terminals to be significantly relaxed. A more extreme attempt at subharmonic pumping, employing a four-phase excitation scheme, is sketched in Figure 17.13. In this case, a ring oscillator (an injection-locked oscillator can be used for better purity) generates differential in-phase and quadrature-phase signals. Altogether four pump signals, offset by 90°, are available. Each pumping signal is sent to a separate varactor with C–V characteristics identical to the other three. Given sufficiently nonlinear (i.e., abrupt) C–V characteristics, the net capacitance seen between the signal (ω_s) and the up-conversion (ω_u) terminals of the varactor will vary at four times the injected pumping frequency. Of course, at this harmonic, a large degradation in capacitance compromising the benefit of low pumping frequencies can be expected.

VARACTOR STRUCTURES

Since the late 1950s, the junction diode has served as the *de facto* standard for all-electronic parametric amplifiers. However, for the applications considered here, this element is generally inferior

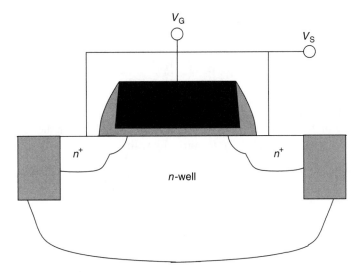

FIGURE 17.14 A sketch of an *n*-type (referring to the body doping) accumulation-mode varactor's cross-section.

to MOS varactor structures. Since the most vigorous research on electronic parametric circuits predates the rise of MOS technology, these devices have only sporadically been considered in the context of modern electronic technologies. In this section, the key varactor characteristics and design options for parametric conversion are looked at more closely.

ELASTANCE MODEL

An important advance in customized MOS varactor technology for RF applications was made when CMOS processes began to accommodate the accumulation-mode varactor [17,18] (Figure 17.14). This simplified the device bias scheme as compared to the more common inversion-mode varactor and simultaneously lowered its resistive losses and parasitic contributions. As a frequency-tuning element, the advantages of the accumulation-mode varactor compared to the junction diode were clear: a large C_{max}/C_{min} ratio, an abrupt capacitive transition implying only the need for low tuning voltages, an isolated bias scheme, and acceptable Q. Optimization of these characteristics for LC-voltage-controlled oscillators (VCOs) is straightforward: increase the C_{max}/C_{min} ratio and reduce losses. For parametric circuits, a more detailed assessment is necessary. First, unlike Manley–Rowe, a more accurate analysis of parametric circuit behavior must account for losses in the varactor. To this end, a rough but physically realistic pumped varactor model employs a nonlinear capacitance in series with a resistance R_s. As emphasized by Penfield and Rafuse [19], this varactor model sidesteps the difficulties and inaccuracies that emerge when a parallel RC equivalent is used or when the series resistance is incorporated into source and load impedances. The terminal characteristics of this physically motivated model are best described with the relation

$$v(t) = \int S(t)i(t)\,dt + R_s(t)i(t) \tag{17.27}$$

This equation directly catalogues the influence of the pump voltage on the varactor as a whole. However, it contains a relatively obscure varactor measure, the incremental elastance, $S(t)$. For the remainder of the chapter, we refer to $S(t)$ as simply "the elastance," with the incremental properties of this value remaining implicit. As with the capacitance of nonlinear devices, practical measurement techniques allow the extraction of only incremental properties. A rough approximation of a MOS varactor elastance per unit area is given by

$$S(V_{GS}) = \frac{Q_{sd}(V_{GS})}{e\varepsilon_s N_d} + \frac{1}{C_{ox}} \tag{17.28}$$

where C_{ox} is the oxide capacitance, e the electronic charge, ε_s the permittivity of the semiconductor, N_d the donor doping in the semiconductor, and Q_{sd} the depletion charge in the semiconductor body. The depletion charge itself is modeled semiempirically with

$$Q_{sd} = \frac{e\varepsilon_s}{C_{ox}N_d}\left(\sqrt{1 + \frac{4V_{MOS}}{\gamma^2}} - 1\right) \tag{17.29}$$

where γ is the device body factor and

$$V_{MOS} = \frac{1}{2}\left[\sqrt{(V_{GS} - V_{FB})^2 + \delta} - (V_{GS} - V_{FB})\right] \tag{17.30}$$

Equation 17.30 follows a modeling technique reported in Ref. 20 and incorporates a small smoothing factor of δ. This correction is used since the transition from full accumulation to flat band is not rigorously accounted for here. With such factors present, it is best to consider this model as a rough design guide (a detailed account of the varactor device physics in compact model form is described in Ref. 21, for example). The value of the simple model described here lies in its direct exposure of the relations between performance and device characteristics. The next section discusses this point, along with device losses, in more detail. A comparison of this approximation to the normalized C–V and S–V characteristics extracted from a full charge-based analysis [22], as well as a simple tanh curve fit, is presented in Figure 17.15. As shown, the tanh curve, a popular approach in empirical compact C–V models, underestimates the elastance in the depletion region. We will return to this point in the next section.

FIGURES OF MERIT

The elastance characteristics must be considered along with device losses in estimating the impact of integrated MOS technology on parametric performance. Penfield and Rafuse [19] highlighted two figures of merit, the cutoff frequency

$$f_c = \frac{S_{max} - S_{min}}{2\pi R_s} \tag{17.31}$$

and the modulation ratio

$$m_n = \frac{|S_n|}{S_{max} - S_{min}} \tag{17.32}$$

The cutoff frequency, which can also be expressed as

$$f_c = \frac{C_{max} - C_{min}}{2\pi R_s C_{max} C_{min}} \tag{17.33}$$

reflects only the influence that device properties bear on the circuit. Ideally, f_c marks the maximum frequency at which it is worth pumping the capacitor. Conversely, the modulation ratio encompasses several contributions. The numerator, $|S_n|$, indicates the size of the elastance harmonic at the pumping

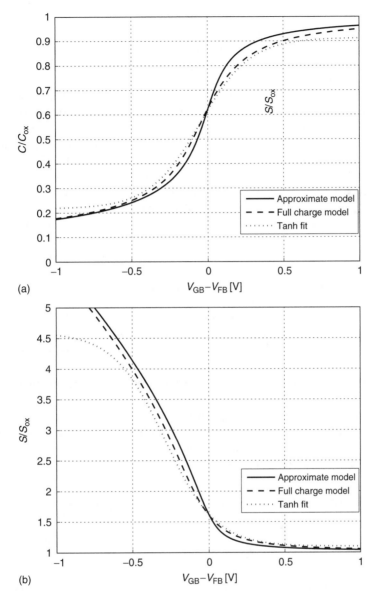

FIGURE 17.15 Comparing the rough semiempirical model to a complete charge-based description and a tanh fit: (a) capacitance–voltage characteristics (b) elastance–voltage characteristics.

frequency nf_p. That is, assuming small-signal conditions, we can treat the elastance as a linear time-varying component controlled by the pump

$$S(t) = \sum_{n=-\infty}^{\infty} S_n e^{j2\pi n f_p t} \tag{17.34}$$

This is the elastance analog to Equation 17.18. The elastance harmonics are influenced by three things: the bias of the pumping signal, the amplitude of the pumping signal, and the steepness of the varactor's elastance characteristics (the steeper the S–V curve, the more efficient the pump in relaying its energy to the varactor). As shown in Equations 17.28 and 17.29, the abruptness of the elastance can be increased by reducing the channel doping. This necessarily increases the series losses, but at a rate proportional to N_d, whereas the S–V slope increases with N_d^2. Similarly, we can see from Equation 17.29 that a decrease in the gate capacitance per unit area, C_{ox}, also contributes to an improvement in the S–V slope. This also comes with the benefit of allowing larger pumping signals to be applied across the gate oxide.

These relationships run counter to the changes employed in scaling MOS devices. Nonetheless the variety present in most modern MOS technologies presents some room for optimization. For instance, many CMOS processes offer devices of various oxide thickness and channel doping. A plot of the S–V characteristics extracted from S-parameter measurements on accumulation-mode devices in a 0.13 μm CMOS technology with varying-channel doping and oxide thickness is shown in Figure 17.16. In this case, only devices with a marginal difference in oxide thickness were examined. As expected, a lower channel doping results in a steeper S–V characteristic. The measured, 4:1 ratio between S_{max} and S_{min} is about 2.5 times greater than that available from a junction diode. The two-channel doping levels (nominal and high) are obtained by employing threshold adjust implants intended for the variety of N-channel MOS (NMOS) and P-channel MOS (PMOS) devices offered in the technology. This feature, correctly combined with the thick-oxide option available in most CMOS technologies, constitutes the most direct approach to device customization for parametric circuit applications. Of note in the measurement results is the manner in which the elastance characteristic saturates in the depletion region. This is a characteristic encompassed by the tanh fit example included in Figure 17.15 but not the basic model of Equation 17.28. The disparity between the predicted and the measured elastance characteristics at large depletion bias can be traced to the fact that the varactor measurements were done with a small-signal, high-frequency (5 GHz) perturbation atop a slowly stepped bias—a common high-frequency C–V extraction technique. Such a setup allows the minority charge to respond to the bias settings, thus preventing the onset of deep depletion as naturally included by the basic model. However, in parametric circuit applications, we can expect a large-signal, high-frequency pump voltage to continuously excite the MOS varactor. Thus, the equilibrium bias conditions present during measurement hardly apply for pumped varactors. This supports the elastance predictions of the basic varactor model, but a convincing answer requires an analysis beyond the scope of this chapter.

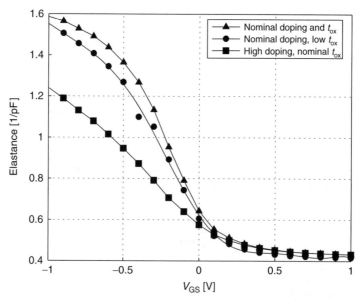

FIGURE 17.16 Elastance measurements for accumulation-mode varactors with varying degrees of channel doping in a 0.13 μm CMOS technology.

As highlighted by the merit relations, S–V performance alone is not a sufficient device selection criterion. Careful consideration must be given to the reduction of series losses as attested by Equation 17.33. For designers, with little control over the varactor's physical characteristics, layout becomes paramount. Without considering special layout techniques (such as differential excitation [23]), four controls are available: gate length (L_g), gate width (W_g), finger number (N_f), and number of stripes/segments (N_s). These combine to give an active varactor area of $L_g W_g N_f N_s$. To clarify, a varactor consists of N_s stripes in parallel, each containing N_f fingers. In turn, each finger has dimensions W_g and L_g. We must consider what arrangement of these terms maximizes f_c. This requires finding the right balance between layout influence on series resistance and capacitance properties.

One finger of an accumulation-mode varactor can be modeled with the schematic shown in Figure 17.17. The contact and via resistances to the polysilicon gate are modeled by R_{cg}, whereas the contact and via resistances to the n^+ diffusion pickups are modeled by R_{csd}. R_g models the resistance of the polysilicon gate. Underneath the gate, the channel resistance is denoted by R_{ch}, whereas R_w is the resistance of the n^+ diffusion bulk pickups and the well. C_{var} is the equivalent series capacitance of each finger. The model of the varactor with multiple fingers is shown in Figure 17.18, where R_{sfg} and R_{sfsd} are the series resistances between two fingers.

For the gate resistance, if the gate poly of each finger is joined from both sides of source/drain, the equivalent poly resistance of one finger is

$$R_g = \frac{1}{12} \frac{W_g}{L_g} R_{g-sh} \tag{17.35}$$

where R_{g-sh} is the gate's sheet resistance. On the contrary, for the channel and well resistance we have

$$R_{ch}, R_w \propto \frac{L_g}{W_g} R_{ch,w-sh} \tag{17.36}$$

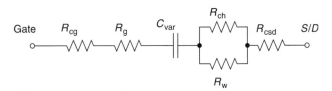

FIGURE 17.17   Model of a single-finger varactor.

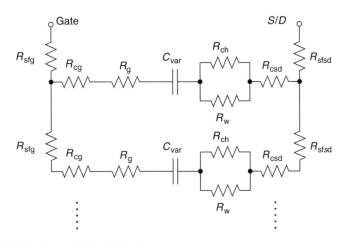

FIGURE 17.18   Model of a parallel multiple-finger varactor.

Being lower doped and unsilicided, the sheet resistance of the well and bulk, $R_{ch,w-sh}$, is greater than that of the polysilicon. This suggests that one use the minimum channel length to reduce the body contribution to the series resistance. However, due to their inverse dependence on finger dimensions some trade-off between the influences of Equations 17.35 and 17.36 on the series resistance is present. This trade-off affects the setting for W_g and L_g, but it is not the only consideration. As shown in Equation 17.33, we want to maximize C_{max}, minimize C_{min}, and minimize R_s. Somewhat arbitrarily choosing a minimum practical value of $C_{min} = 100\,fF$ (in anticipation of parasitic effects and process variations), we are left to consider how L_g, W_g, N_f, and N_s influence the remaining two characteristics. Obviously this complicates selection based purely on an R_g–R_{ch} trade-off. For instance, minimizing $L_g W_g$ maximizes the $N_f N_s$ product and thereby reduces R_s. This is done at the cost of increasing the relative parasitic capacitance contribution and hence reducing $C_{max} - C_{min}$.

Another important consideration is the contact and interconnect resistance introduced between fingers (R_{sfg} and R_{sfsd} in Figure 17.18) and stripes. This is often ignored when assessing device resistance, but can certainly be influential. With R_{sfg} and R_{sfsd} the equivalent resistance will not be reduced simply as a function of $1/N_f$. Instead, as N_f is increased the series resistance will eventually saturate due to the contributions of the interfinger connections, R_{sfg} and R_{sfsd}. Attempts to get a sense of how the characteristics L_g, W_g, N_f, and N_s influence f_c are greatly aided by the availability of Verilog-A-based compact models, such as the one described in Ref. 21. Since these models account for both physical and layout characteristics, a broad comparison between designs can be made. Employing empirically based compact models, the f_c for a variety of accumulation-mode n-type varactors (excited in a single-ended manner) is shown in Tables 17.1 through 17.3. The total active area ($L_g W_g N_f N_s = 43.2\,\mu m^2$) is the only value that all designs have in common. It is chosen such that C_{min} remains above $100\,fF$ over the relevant region of operation (V_{GS} ranges from -1 to $1\,V$).

TABLE 17.1
Cutoff Frequencies for Varactor with $W_g = 1\,\mu m$, $L_g = 0.24\,\mu m$, and Area $= 43.2\,\mu m^2$

N_f	N_s	C_{max} (Ff)	C_{min} (fF)	R_s (Ω)	f_c (GHz)
180	1	477.8	187.3	13.77	32.57
60	3	481.5	188.1	2.429	212.2
30	6	481.8	188.3	0.8025	641.5
15	12	482.3	188.7	0.508	1011
5	36	483.5	190.0	1.275	398.8

TABLE 17.2
Cutoff Frequencies for Varactor with $W_g = 1.41\,\mu m$, $L_g = 0.34\,\mu m$, and Area $= 43.2\,\mu m^2$

N_f	N_s	C_{max} (Ff)	C_{min} (fF)	R_s (Ω)	f_c (GHz)
90	1	475.8	147.5	7.281	102.2
45	2	476.3	147.8	2.777	267.5
30	3	476.5	147.8	1.538	482.8
15	6	476.8	148.1	0.7094	1045
5	18	477.5	149.0	0.9085	808.9

Table 17.1 summarizes the results for varactors consisting of minimum unit area (i.e., W_gL_g) elements, Table 17.2 shows the results for devices composed of twice the minimum unit area, and Table 17.3 summarizes the characteristics of varactors composed of four times the minimum unit area. Note that all R_s values have been calculated for 5-GHz excitations. A layout-dependent self-resonance frequency could not be extracted as the model did not account for inductive parasitics, although it should be noted that self-resonant frequencies do not necessarily pose a problem for parametric circuits. The self-resonant frequency can be exploited as one of the modes of interest in the parametric circuit. The f_c values shown are certainly optimistic, as the compact models do not account for the effects that would limit device performance at such frequencies. Nevertheless, they are useful as a relative measure of the best device type. Judging by the f_c results, it is best to use an intermediate unit area that ably juggles two conflicting characteristics: parasitic capacitance and series resistance. For a given total area, as the unit area shrinks, more devices in parallel imply a smaller total resistance. As can be seen in all cases, this is best achieved by keeping N_f and N_s on the same order. Unfortunately, the capacitance of small unit areas contains a higher relative proportion of parasitic capacitance. This lowers $C_{max} - C_{min}$, which ends up hurting the f_c. Attempts to get around this by increasing the unit area are frustrated by an increase in series resistance (due simply to a decrease in the parallel connection count). The simulated cutoff frequencies associated with these varactors are plotted in Figure 17.19 as a function of stripe count. As can be seen, f_c is relatively forgiving of unit size, but quite sensitive to N_f and N_s distributions.

TABLE 17.3
Cutoff Frequencies for Varactor with $W_g = 2\,\mu m$,
$L_g = 0.48\,\mu m$, and Area $= 43.2\,\mu m^2$

N_f	N_s	C_{max} (fF)	C_{min}(fF)	R_s (Ω)	f_c (GHz)
45	1	475.8	147.5	4.359	222.8
15	3	477.5	123.5	1.266	754.4
5	9	476.3	122.1	0.9787	982.9

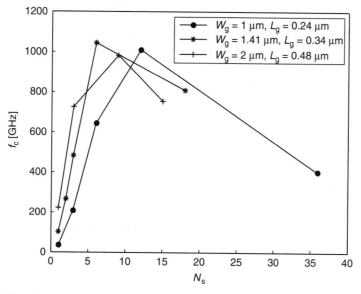

FIGURE 17.19   Plot of varactor cutoff frequencies versus number of stripes for a $43.2\,\mu m^2$ (total active area) varactor.

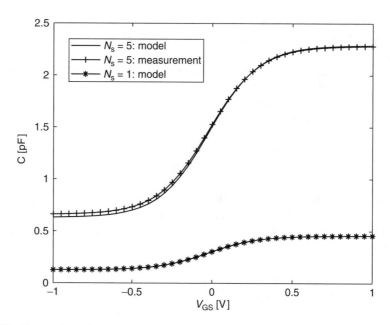

FIGURE 17.20   Comparison of measurement results to compact model predictions.

Measured results are available to double-check the *C–V* characteristics of the scalable varactor model. The experimental varactor design has unit widths and lengths of 5 μm and 0.42 μm, respectively, which are arranged into N_s = 5 parallel stripes of N_f = 20 gate fingers each. In Figure 17.20, the *C–V* curve obtained from the model is plotted alongside the *C–V* data obtained from a high-frequency (5 GHz) *S*-parameter characterization of the varactor. It is observed that the *C–V* characteristic of the fabricated device matches very closely with the scalable model (at the frequency of extraction).

We can attempt to tailor this varactor design for parametric circuits by changing the number of stripes from five to one. The implications of this change on the device characteristics and USBUC and LSBDC are explored in detail in the next section. Even though reducing N_s shifts the *C–V* curve down and decreases the $C_{max} - C_{min}$ (as shown in Figure 17.20), C_{min} is also reduced, thus increasing f_c. The large change in capacitance characteristics from N_s = 5 to N_s = 1 affects the performance quality of the parametric converter but not the substance of its operation. This is unlike, for example, the effect on the operations of a VCO, whose center frequency and tuning range would be severely impacted. The only requirement imposed in this work is that C_{min} exceed 100 fF, which is satisfied.

PARAMETRIC CIRCUIT SIMULATIONS

UPPER SIDEBAND UP-CONVERTER SIMULATIONS

Although useful as a general guide, the Manley–Rowe relations apply only to the case of a lossless nonlinear reactance. In contrast, Penfield and Rafuse [19] emphasized the need to account for series varactor losses when evaluating parametric circuit performance. Following this work, the available gain of a USBUC treated as a two-port (see Figure 17.21) is given by

$$\frac{1}{\text{Ga}} = \left(\frac{\omega_s}{m_1 \omega_c}\right)^2 \frac{|Z_G + R_s - jS_0/\omega_s|^2}{R_G R_s} + \frac{\omega_s}{\omega_u} \frac{R_G + R_s}{R_G} \tag{17.37}$$

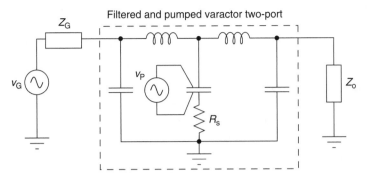

FIGURE 17.21 The upper sideband up-converter represented as a two-port with ideal filters and series loss, R_s only in the varactor.

assuming that signals only at ω_s, ω_p, and $\omega_u = \omega_s + \omega_p$ are allowed to excite the capacitor. The influence of the cutoff frequency, $f_c = 2\pi\omega_c$, and the modulation index, m_1, on the gain is clearly seen in Equation 17.37. From Equation 17.37, we find that maximum available gain is

$$G_{a,max} = \frac{\left(m_1\omega_c/\omega_s\right)^2}{\left(1 + \sqrt{1 + \left(m_1\omega_c/\omega_s\omega_u\right)^2}\right)^2} \tag{17.38}$$

when the jS_0/ω_s term is resonated out and the generator resistance is set to

$$R_G = R_s\sqrt{1 + \frac{\left(m_1\omega_c\right)^2}{\omega_s\omega_u}} \tag{17.39}$$

The varactor structures described earlier (Figure 17.20) are now used to get a sense of the performance attainable by the USBUC. A catalog of their properties along with the simulated composite figure of merit ($m_1\omega_c/2\pi$) is included in Table 17.4. A DC point ($-200\,\text{mV}$) is chosen for the pump signal such that the device is biased evenly between S_{max} and S_{min}. Also, the figures are extracted for a sinusoidal pump signal of 300 mV peak amplitude applied directly across the varactor. This prevents the generation of higher order harmonics, which, under the circuit conditions considered here, do not contribute any useful power gain and may actually decrease it. The simulation results for a USBUC that amplifies and converts a 1 GHz signal into a 60 GHz frequency are shown in Table 17.5. A number of conditions are examined and, for easy reference, compared to theoretical calculations of available gain. The results are determined for no input match (Equation 17.37 with $Z_G = 50\,\Omega$), and for an input conjugate match (Equation 17.38), to extract the maximum available gain. For the first two varactor simulations ($N_s = 5$ measured and modeled), we see that the gain is well below that predicted by the Manley–Rowe relations. We can attribute this to the fact that our upper sideband frequency is greater than the composite figure of merit for the device with five stripes. A simple modification, the one-stripe varactor with a composite merit figure well in excess of the 60 GHz up-conversion frequency, comes within 15% of the Manley–Rowe prediction. Although their performance potential is compelling, a number of complications blur the role that a parametric converter can play in modern consumer wireless applications. One awkward point is the abundance of filtering required around the pumped reactance. Fortunately, there are situations where the presence of a parametric converter can be made rather streamlined; a hypothetical case – in point is shown in Figure 17.22. Pictured is a USBUC acting as a low-noise preamplifier in a subsampling receiver (a standard RF down-converter can be used in place of the subsampler). Thus, although the received

TABLE 17.4
Relevant Varactor Measurements with $V_{DC} = -200\,mV$ and $V_p = 300\,mV$ (Peak)

Design	$S_0\ (\times 10^9/F)$	$S_1\ (\times 10^9/F)$	$R_s\ (\Omega)$	$m_1\omega_c/2\pi$ (GHz)
$N_s = 5$ (measured)	957.2	211.9	1.1	30.67
$N_s = 5$ (model)	987.7	228.0	1.032	35.16
$N_s = 1$ (model)	4938	1140	2.359	76.92

TABLE 17.5
Simulated USBUC Power Gain

Design	Input Condition	$Z_G\ (\Omega)$	G_a (theory)	G_a (sim.)
$N_s = 5$ (Measured)	No match	50	1.94	1.91
	Conjugate match	$4.49 + j152.3$	36.93	36.26
$N_s = 5$ (Model)	No match	50	2.25	2.21
	Conjugate match	$4.8 + j157.2$	38.75	38.61
$N_s = 1$ (Model)	No match	50	1.1	1.09
	Conjugate match	$23.5 + j785.9$	49.07	48.94

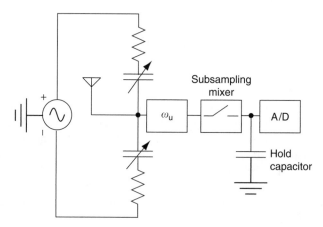

FIGURE 17.22   Example of USBUC in a receiver front-end.

signal power is amplified to a higher frequency, it remains accessible to a subsampling mixer [24]. Of course, the point of using a high up-conversion frequency is to improve the gain. The potential benefits as compared to a standard transistor-based LNA are improved noise performance and built-in filtering, which is needed in any case by the subsampler [24]. A differential pumping scheme is assumed with the USBUC interfaced directly to the antenna. Immediately we have sidestepped two of three filters with only the up-converted frequency filter remaining (although subharmonic differential pumping can be employed, it is not considered in this example). The circuit is designed around a model of the $N_f = 5$, $N_s = 9$ device listed in Table 17.3. Coming from the family of devices with the highest unit size $(2\,\mu m \times 0.48\,\mu m)$, this varactor has a respectable cutoff frequency without attempting to push dimensional tolerances. Compared to the measured structures listed in Table 17.4,

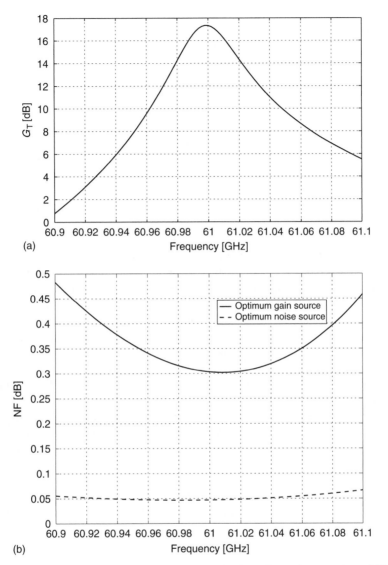

FIGURE 17.23 Gain and noise performance with generator load set for optimum gain at 1 GHz input: (a) transducer gain, (b) noise figure.

this varactor has a much lower average capacitance. This eases somewhat the burden of optimally matching the generator (i.e., antenna) to the circuit. Modeling the subsampling mixer as a 10 Ω switch in series with a 500 fF hold capacitor, the simulated transducer gain and noise figure of the USBUC can be represented as shown in Figure 17.23. These results are obtained with a 300 mV differential oscillator output used as the pump (further specifics of the pumping arrangement are not considered in detail here).

The results employ a source with a generator impedance conjugately matched to the amplifier input at 1 GHz. The output filter is a third-order Chebyshev designed for a 2 GHz bandwidth around the 61 GHz output frequency; thus, our limit to a 40 MHz signal bandwidth in Figure 17.25a is due primarily to a mismatch at the USBUC's input port. The maximum simulated gain is less than a decibel below the Manley–Rowe predictions, an indication that pumped varactors can be run close to optimum performance. To obtain this performance, a net power of 0.841 mW is supplied by the pump. Employing a less dissipative varactor (e.g., see Table 17.1) and running at a lower cutoff frequency (i.e., reducing S_1) can further reduce this value. This result has been facilitated by designing the up-conversion filter to provide a conjugate match to the subsampler at 61 GHz; thus at the center frequency, the circuit is running close to its maximum available power gain. Naturally, maximizing the power delivered to the load maximizes the average energy placed on the hold capacitor. The simulated noise behavior shown in Figure 17.25b is excellent despite the fact that no effort has been made to incorporate an optimum noise match. As shown in Figure 17.25b, driving the circuit with a source optimized for noise performance results in phenomenal noise predictions for an integrated circuit running at room temperature. As the vagaries of interconnect and other circuit components are included, the performance is expected to degrade. As it stands, a 40 MHz gain bandwidth is commensurate with a number of current cellular and WLAN standards and can be expanded if a broadband match is employed. As cited above, that is not the case for the present results where a generator impedance of $6.91 + j353.22\,\Omega$ is employed. At 1 GHz, this implies a generator with a 56 nH series inductance. Had a larger varactor with a larger average capacitance been employed in the amplifier, it would have required an even larger inductive contribution (for conjugate match). At this point, the designer eyeing an efficient solution may attempt to implement the requisite generator directly on a substrate rather than following the more common route of matching between separate blocks designed to a common impedance (e.g., the ubiquitous $50\,\Omega$). It is another advantage to integrated designs that they allow such a freedom. In this case, the benefits are small form factor and direct implementation of front-end filters, which are needed at any rate to block interferers, limit noise, and control the frequencies exciting the varactor. The required generator impedance characteristics ($6.91 + j353.22\,\Omega$) at 1 GHz for the design in question are adequately met by a slot antenna structure. The simultaneous requirements of low-radiation resistance in series with an inductive component are met by an electrically small implementation of this structure. An illustration of the antenna along with the manner in which it can be implemented alongside the remaining filtering and electronics is shown in Figure 17.24. In this conceptual design, the antenna is etched out of the ground plane of a (lossless) silicon substrate; the remaining components can be implemented on top of this substrate as shown in the figure. At 1 GHz, the antenna footprint is bound by dimensions of $3.1\,\text{cm} \times 2.3\,\text{cm}$, which, being electrically small, sharpens the gain characteristics even more. The antenna-driven USBUC performance is shown in Figure 17.25. Most obvious is the narrowbanding of the response (gain bandwidth drops to 10 MHz), though still a high center frequency performance is maintained. Coming from an electrically small antenna, this is not unexpected; with some effort directed at broadbanding the source, it is entirely possible to extend the bandwidth. Similarly, the

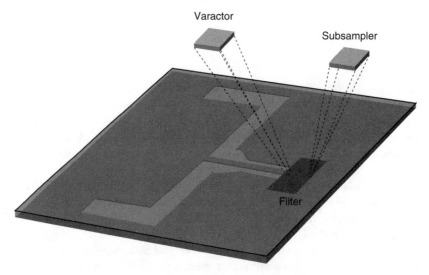

FIGURE 17.24 Possible physical arrangement for USBUC front-end receiver.

unoptimized noise performance remains at less than 0.5 dB in the bandwidth of interest, although the performance drops off rather sharply out of this band. Once again, this effect is due to the sharp impedance characteristics of this electrically small antenna.

A substantial challenge to this design centers once again around the filtering requirements. The output filter whose transmission-line equivalent is shown in Figure 17.26 requires unrealistic component values to implement the narrowband (2 GHz around 61 GHz center frequency) specification. One can tackle this issue by designing a broadband filter centered at a higher frequency offset whose components are much easier to realize (e.g., in a filter with a 70 GHz center and a 20 GHz bandwidth, the transmission line element characteristic impedances range between 80 and 2 Ω). This comes at the price of increased overall circuit sensitivity as the lower sideband signals close to the filter

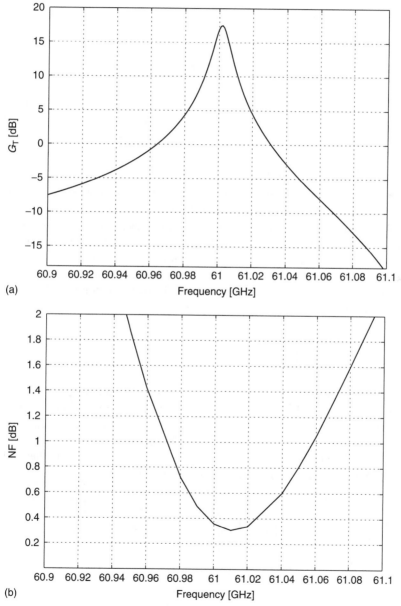

(a)

(b)

FIGURE 17.25 Gain and noise performance with slot antenna generator impedance: (a) transducer gain, (b) noise figure.

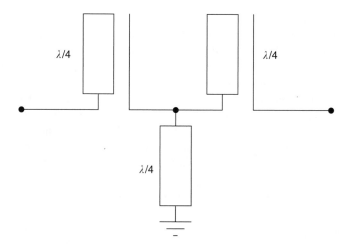

FIGURE 17.26 Sketch of third-order Chebyshev up-conversion filter structure.

transition are less attenuated by the stopband. As already discussed and as explored in more detail in the following section, lower sideband excitations introduce a regenerative effect in the circuit. This greatly increases gain at the cost of stability.

LOWER SIDEBAND DOWN-CONVERTER SIMULATIONS

A lower sideband parametric converter is capable of significant gain in the down-conversion process. On the basis of this alone, the LSBDC would be a far better choice for the receive chain than its upper sideband (up- or down-converter) counterpart. However, since it essentially constitutes a regenerative reflection amplifier, the lower sideband topology suffers from a number of complications including load-dependent gain and instability. In this section, we examine the potential of this structure from a modern context and in more detail than available with the Manley–Rowe description. Once again, following the analysis in Ref. 19 on LSBDC structures with intrinsic loss, the expression for the available gain becomes

$$\frac{1}{\text{Ga}} = \left(\frac{\omega_s}{m_1\omega_c}\right)^2 \frac{\left|Z_G^* + R_s + jS_0/\omega_s\right|^2}{R_G R_s} - \frac{\omega_s}{\omega_d} \frac{R_G + R_s}{R_G} \tag{17.40}$$

From this equation, we can see that it is possible for the available gain to be negative. Negative values of available gain indicate that the real part of the amplifier's output impedance is negative, and for such circuits the expression "exchangeable gain" is preferred. In these cases, the power emerging from the amplifier is dependent on the load. For the exchangeable gain to be negative, it is necessary that $\omega_s\omega_d < (m_1\omega_c)^2$. The noise behavior of the LSBDC topology is identical to that of the USBUC. The single-sideband noise factor is given by the following equation:

$$F = 1 + \left[\frac{R_s}{R_G} + \left(\frac{\omega_s}{m_1\omega_c}\right)^2 \frac{\left|Z_G^* + R_s + jS_0/\omega_s\right|^2}{R_G R_s}\right] \tag{17.41}$$

with an optimal noise impedance setting given by the following equation:

$$Z_G = R_s\sqrt{1 + \left(\frac{m_1\omega_c}{\omega_s}\right)^2} + j\frac{S_0}{\omega_s} \tag{17.42}$$

and a minimum noise factor given by the following equation:

$$F_{\min} = 1 + \left[\frac{R_s}{R_G} + \left(\frac{\omega_s}{m_1 \omega_c} \right)^2 \frac{(R_G + R_s)^2}{R_G R_s} \right] \qquad (17.43)$$

Under optimal noise performance conditions, the exchangeable gain of the amplifier is

$$\mathrm{Ga} = \sqrt{1 + \left(\frac{m_1 \omega_c}{\omega_s} \right)} \cdot \left[\frac{\omega_s}{m_1 \omega_c} + \sqrt{1 + \left(\frac{m_1 \omega_c}{\omega_s} \right)^2} \right]^{-1} \cdot \left[\frac{\omega_s}{m_1 \omega_c} + \sqrt{1 + \left(\frac{m_1 \omega_c}{\omega_s} \right)^2} - \frac{m_1 \omega_c}{\omega_d} \right]^{-1} \qquad (17.44)$$

The exchangeable gain can still be negative under optimum noise matching conditions, implying that the LSBDC still has a negative equivalent output resistance and hence the ability to deliver power to the signal and down-conversion ports. Note that in this case (optimum noise match), it is possible to increase the gain by changing the pumping frequency and hence ω_d. That is, the higher the frequency of the pump, the higher the maximum gain. Of course, given a constant signal frequency, increasing the pump increases the frequency of the down-converted signal. This option is somewhat limited when the LSBDC serves as a down-converter (as opposed to a straight LNA), since the IF frequency is not commonly a design variable on the circuit level. This limit is eased somewhat if subsampling is employed. We examine the (noise) performance of the LSBDC using the same varactor structures and pumping conditions (summarized in Table 17.4) and operating conditions (60 GHz input signal, 1 GHz down-converted signal) as those considered in the USBUC case in the previous sub-section. In Table 17.6, the theoretical and simulated output impedances along with the expected noise figure of each of the LSBDC circuits are given. Note that we present the output impedances as opposed to the available gain from the LSBDC. This is motivated by the fact that the circuit is a reflection amplifier. Thus if the output impedance of the LSBDC is known, the designer can allot an appropriate load to realize a desired power gain. Admittedly, this is difficult to do in board-level designs, which typically adhere to a strict termination standard (i.e., 50 Ω). However, the need to tailor a load impedance is certainly not out of the question in high-frequency integrated circuits where designers have much more control over the interface between circuits. In the case of the five-stripe varactor (measured and modeled), when the source resistance is 50 Ω, the real part of the output impedance will actually be positive, and thus the LSBDC will operate with a power loss. As shown in Table 17.6, the single-strip device has a promising noise figure assuming that the source has been tuned for noise optimization. The optimum source noise resistance (3.8 Ω) is quite low for the particular version of the LSBDC studied here. Fortunately, the design is not impractically sensitive to source impedance variations. For example, a 600% increase in source resistance to 25 Ω results in a noise figure of 6.95 dB, a 0.72 dB increase. This compares favorably with the performance achieved using standard IC radio topologies. For instance, one report [25] describes a 60 GHz transceiver implemented in a 200 GHz f_T 0.12 μm SiGe technology with a measured 11 mW LNA noise figure of 4.5 dB and a 147 mW mixer (driven by a 30 mW VCO) with a noise figure of 14.8 dB. Importantly, it must be emphasized that the simulated parametric behavior reported here is managed with a simple circuit consisting, aside from an oscillator, entirely of passives. Further, the parametric converter is not heavily reliant on the quality of its IC technology. As an example of this, the noise performance estimated in Table 17.6 is built around a CMOS varactor with $L_u = 0.42$ μm. Granted, this analysis considers only the varactor's noise contribution, but it should be noted that the device under question is inferior to a number of other available device designs, which are cataloged in Tables 17.1 through 17.3. Optimizing around one of the better varactors listed (specifically, the minimum unit area varactor with $N_f = 15$ and $N_s = 12$ in Table 17.1) predicts very encouraging results for the operational circumstances under consideration. An optimum noise figure of 2.8 dB is

TABLE 17.6
Simulated LSBDC Performance

Design	Input Condition	Z_G (Ω)	Z_{out} (theory) (Ω)	G_a (sim.) (Ω)	NF (dB)
$N_s = 5$ (Measured)	No match	50	$0.73 - j152.4$	$0.73 - j144.4$	22.63
	Noise Opt.	$1.24 + j2.54$	$-7.02 - j152.3$	$-6.84 - j143.8$	12.37
$N_s = 5$ (Model)	No match	50	$0.603 - j157.2$	$0.609 - j148.3$	21.71
	Noise Opt.	$1.20 + j2.62$	$-8.82 - j157.2$	$-8.63 - j148.3$	11.33
$N_s = 1$ (Model)	No match	50	$-7.5 - j788.4$	$-17.3 - j737.5$	12.06
	Noise Opt.	$3.8 + j13.1$	$-86.22 - j785.9$	$-84.8 - j740$	6.23
$N_f = 15$, $N_s = 12$	No match	50	$-2.41 - j582$	$-2.4 - j563$	10.6
	Generator 1	$25 + j9.7$	$-5.49 - j581$	$-5.45 - j563$	5.6
	Generator 2	$20 + j9.7$	$-6.95 - j581$	$-6.90 - j563$	4.3
	Generator 3	$15 + j9.7$	$-9.36 - j581$	$-9.29 - j563$	3.0
	Noise Opt.	$1.7 + j9.7$	$-69.6 - j581$	$-68.3 - j563$	2.8

predicted under an optimum R_G of 20, and 1.7 Ω. This value increases to 3, 4.3, and 5.6 dB at more reasonable source resistances of 15, 20, and 25 Ω, respectively.

CONCLUSIONS

This chapter has reviewed the parametric conversion/amplification principle and explored its potential in a CMOS context. In summary, parametric converters use a variable reactance to transfer power from an oscillating source to an input signal. Since this transfer is mediated by a reactive element in theory, two fundamental noise sources, thermal and shot noise, are absent. Traditionally, the primary advantage of this amplification technique has been the promise of ultra-low-noise operation. However, this circuit approach is always an option when operational frequencies beyond the reach of transistor technology are the objective.

The Manley–Rowe relations were employed in this chapter to summarily illustrate the behavior of two fundamental parametric topologies: the USBUC and the LSBDC. Besides these, a large array of configurations is possible. In this report, the former was considered for its relative operational simplicity and stability, whereas the latter was considered for the high gain performance it is capable of achieving. For realistic design insights, the analysis employed by Penfield and Rafuse was utilized in, and compared to, integrated designs.

One goal of this chapter is to convince the reader of the promising potential of parametric circuits in a CMOS setting. As indicated, the parametric principle has recently been studied in discrete-time CMOS applications. However, as indirectly discussed in the body of the chapter, so far only the simplest topology (the degenerate configuration) has been considered. For instance, an idler-based discrete-time amplification scheme (common to continuous-time parametric circuits) may be applied to ease the requirements on capacitance variation. Perhaps the most attractive aspect of CMOS technology for parametric amplifiers, aside from structural particulars, is the ready availability of the MOS varactor. The MOS varactor assists the efficacy of this technique with its rich nonlinearity (compared to the junction varactor), broad capacitive range, complementary structure (i.e., n-type and p-type varactor modes), unobtrusive biasing, and three-terminal operation (which is not discussed herein). On the device level, attention was brought to several issues involving accumulation-mode varactors in parametric converters. The importance of elastance-based design was mentioned (for instance, this influences the optimum pump biasing) and a simple elastance model discussed. An outstanding point is the impact of nonequilibrium, deep-depletion effects. RF MOS characterization typically does not elicit this behavior; however, it is expected to play a role in parametric circuit performance. An investigation into optimum device layout based on compact models highlighted the need to balance finger and stripe count of the varactor. Ignored were the possible influences of capacitive well parasitics on the frequency response of parametric ICs. Perhaps the most nagging issues with parametric circuits are their need for high pumping frequencies and copious filtering. A number of proposals were made regarding these, from high-frequency reference generation, to subharmonic pumping, to differential excitation. These approaches are aided by IC implementations (better matching) and the custom device structures (complementary varactors) that can be fabricated with only a layout rearrangement (i.e., without specific need for thermal budget or doping adjustments during processing). Suggestions were also made for customization of physical MOS varactor characteristics to better suit parametric needs. Given the availability of thick oxide analog devices in mixed-signal CMOS technologies and the presence of multiple threshold implants, even this adjustment can be implemented without significant process demands. Further, an efficient USBUC hybrid structure serving as an RF receiver front-end was described, which utilized the natural filtering properties of the driving antenna to simplify the overall converter circuit design.

REFERENCES

1. Ranganathan, S. and Tsividis, Y., "Discrete-time parametric amplification based on a three-terminal MOS varactor," *IEEE J. Solid-State Circuits*, vol. 38, pp. 2087–2093, December 2003.
2. Faraday, M., "On a peculiar class of acoustical figures; and on certain forms assumed by groups of particles upon vibrating elastic surfaces," *Phil. Trans. Roy. Soc., London*, vol. 121, pp. 299–340, March/July 1831.
3. Suhl, H., "Proposal for a ferromagnetic amplifier in the microwave range," *Phys. Rev.*, vol. 106, pp. 384–385, April 1957.
4. Weiss, M. T., "A solid-state microwave amplifier and oscillator using ferrites," *Phys. Rev.*, vol. 107, p. 317, July 1957.
5. Hartley, R. V., "A wave mechanism of quantum phenomena," *Phys. Rev.*, vol. 33, p. 289, February 1929.
6. Hartley, R. V., "Oscillations in systems with non-linear reactance," *Bell Sys. Tech. Jour.*, vol. 15, pp. 424–440, July 1936.
7. Raskin, J. P., Brown, A. R., Khuri-Yakub, B., and Rebeiz, G. M., "A novel parametric-effect MEMS amplifier," *J. Microelectromech. Syst.*, vol. 9, pp. 528–537, December 2000.
8. Olkhovets, A., Carr, D. W., Parpia, J. M., and Criaghead, H. G., "Non-degenerate nanomechanical parametric amplifier," *International Conference on MEMS*, pp. 298–300, IEEE, Interlaken, Switzerland, January 2001.
9. Knappe, S., Shah, V., Schwindt, P. D., Hollberg, L., and Kitching, J., "A microfabricated atomic clock," *Appl. Phys. Lett.*, vol. 85, no. 9, pp. 1460–1462, 2004.
10. Manley, J. M. and Rowe, H. E., "Some general properties of nonlinear elements—Part I. General energy relations," *Proc. IRE*, vol. 44, pp. 904–913, July 1956.
11. Rowe, H. E., "Some general properties of nonlinear elements. II. Small signal theory," *Proc. IRE*, vol. 46, pp. 850–860, May 1958.
12. Kleveland, B., Diaz, C. H., Vock, D., Madden, L., Lee, T. H., and Wong, S. S., "Monolithic CMOS distributed amplifier and oscillator," *ISSCC Digest of Technical Papers*, pp. 70–71, IEEE, San Francisco, February 1999.
13. Hackl, B. and Bock, J., "42 GHz active frequency doubler in SiGe bipolar technology," *International Conference on Microwave and Millimeterwave Technology*, pp. 54–57, IEEE, Beijing, China, August 2002.
14. Lee, C., Yao, T., Mangan, A., Yao, K., Copeland, M. A., and Voinigescu, S. P., "SiGe BiCMOS 65-GHz BPSK transmitter and 30 to 122 GHz LC-varactor VCOs with up to 21% tuning range," *IEEE Compound Semiconductor Integrated Circuit Symposium*, pp. 179–182, October 2004.
15. Chan, H., *Sub-harmonic Pumping in Parametric Amplifiers*, MSc dissertation, University of Calgary, 2007.
16. Engelbrecht, R., "Nonlinear-reactance (parametric) travelling-wave amplifiers for uhf," *IEEE International Solid-State Circuits Conference*, pp. 8–9, IEEE, Philadelphia, February 1959.

17. Soorapanth, T., Yue, C. P., Shaeffer, D., Lee, T., and Wong, S., "Analysis and optimization of accumulation-mode varactor for RF ICs," *Symposium on VLSI Circuits Digest of Technical Papers*, pp. 32–33, IEEE, Honolulu, 1998.

18. Castello, R., Erratico, P., Manzini, S., and Svelto, F., "A ±30% tuning range varactor compatible with future scaled technologies," *Symposium on VLSI Circuits Digest of Technical Papers*, pp. 34–35, IEEE, Honolulu, 1998.

19. Penfield, P. and Rafuse, R. P., *Varactor Applications*. MIT Press: Cambridge, MA, 1962.

20. Maget, J., Kraus, R., and Tiebout, M., "A physical model of a CMOS varactor with high capacitance tuning range and its application to simulate a voltage controlled oscillator," *International Semiconductor Device Research Symposium*, pp. 609–612, IEEE, Washington, D.C., December 2001.

21. Victory, J., Yan, Z., Gildenblat, G., McAndrew, C., and Zheng, J., "A physically based, scalable MOS varactor model and extraction methodology for RF applications," *IEEE Trans. Electron. Devices*, vol. 52, p. 1343, July 2005.

22. Tsividis, Y. P., *Operation and Modeling of the MOS Transistor*. McGraw-Hill: New York, pp. 48–49, 1987.

23. Porret, A. S., Melly, T., Enz, C. C., and Vittoz, E. A., "Design of high-Q varactors for low-power applications using a standard CMOS process," *IEEE J. Solid-State Circuits*, vol. 35, pp. 337–345, March 2000.

24. Pekau, H. and Haslett, J., "A 2.4 GHz CMOS sub-sampling mixer with integrated filtering," *IEEE J. Solid-State Circuits*, vol. 40, pp. 2159–2166, November 2005.

25. Floyd, B. A., Reynolds, S. K., Pfeiffer, U. R., Zwick, T., Beukema, T., and Gaucher, B., "SiGe bipolar transceiver circuits operating at 60 GHz," *IEEE J. Solid-State Circuits*, vol. 40, pp. 156–167, January 2005.

Part III

Device and Process Technology
for Wireless Chips

CMOS Technology for Wireless Applications

John J. Pekarik

CONTENTS

Motivation for CMOS 521
Basic CMOS Process Flow 524
    Isolation 525
    Well Implants 526
    Gate 526
    Active Area Implants 527
    Stress Engineering 528
    Contacts and Wiring 528
    Device Menu 529
    Optional Thick Oxide FETs 529
    Resistors 530
    Varactors 530
    Inductors 531
    Metal Capacitors 532
FET Performance and Scaling 532
    Self-Gain 533
    f_T, f_{MAX} 534
    Noise Figure 535
    Flicker Noise 536
Device Models 537
    High-Frequency Models 537
    Modeling Process Variation 538
    Variation of Flicker Noise 539
Conclusions 541
References 541

MOTIVATION FOR CMOS

Semiconductor technologies are usually referred to by the lithography node, designated by its minimum feature length, e.g., 90 nm CMOS. As lithographic dimensions shrink in advanced CMOS nodes, increasing performance, illustrated by increasing f_T, presents great opportunities for radio frequency (RF) designers. Figure 18.1 shows published values of f_T for silicon n-channel FETs (NFETs) along with the 2003 International Technology Roadmap for Semiconductors (ITRS) projection [1,2]. These data show performance comparable with III–V devices and inverse scaling with gate lengths to 27 nm. A similar plot of f_{MAX} data in Figure 18.2 also shows increasing performance but with considerable

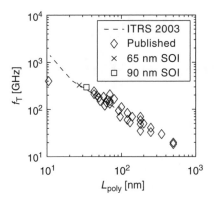

FIGURE 18.1 Unity current gain frequency f_T versus gate length for the 2003 ITRS with reported data. (From Lee, S. et al., *Technical Digest of IEEE International Electron Devices Meeting*, IEEE, 2005, p. 241. With permission. © IEEE 2005.)

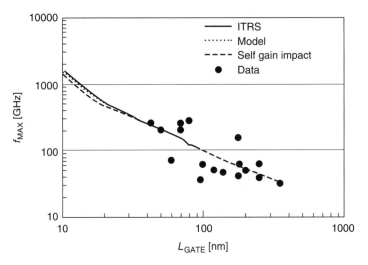

FIGURE 18.2 Unity power gain frequency f_{MAX} versus gate length for the 2003 ITRS and reported data. (From Bennett, H. S. et al., *IEEE Trans. Electron Devices*, 52, 1235, 2005. With permission. © IEEE 2005.)

variability in published results due to the sensitivity of power gain extraction to parasitic impedances and inconsistencies in de-embedding techniques applied to the measured data. These figure of merits (FOMs) imply that usable gain at conventional wireless application frequencies (~800 MHz–5 GHz) is achievable in CMOS technology nodes beginning at 0.25 μm. In fact, a survey of product announcements or conference publications [3,4] will provide many examples of active work.

Bipolar transistors, particularly SiGe heterostructure bipolar transistors (HBTs) implemented in bipolar/CMOS (BiCMOS) technologies, have dominated the market in chip designs for wireless applications. Many factors favor bipolar transistors when compared to FETs including higher transconductance (g_m), higher voltage tolerance, lower flicker noise, and current–voltage relationships that are more easily modeled with analytic expressions. These all contribute to making analog and RF circuit design easier using bipolar transistors. However, FETs have one advantage that is often overlooked. The turn-on, or threshold voltage of bipolar transistors, is determined by the semiconductor band gap whereas for FETs the threshold voltage (V_T) is adjustable with doping levels. Figure 18.3 compares f_T of the SiGe HBT from 130 nm BiCMOS with the NFET from 90 nm CMOS. It shows that for an application where $f_T = 100$ GHz is required to assure adequate small-signal gain,

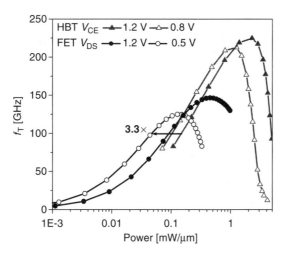

FIGURE 18.3 Unity current gain frequency versus DC power dissipation comparing an NFET from 90 nm CMOS and an SiGe NPN from 130 nm BiCMOS. (From Jagannathan, B. et al., in *2004 Topical Meeting on Silicon Monolithic Integrated Circuits in RF Systems*, pp. 115–118.)

such gain could be had at more than three times decrease in DC power dissipation. This benefit is even more dramatic with aggressively scaled CMOS [1].

Although low production cost is often cited as a motivating factor, care must be taken in assessing all contributions to the product cost. Early in the lifetime of a technology node, limited manufacturing capacity and uncertainty in the reliability of supply translate into high prices for the so-called "bleeding edge" technologies. Demand for high-volume manufacturing is driven by high-performance digital applications especially with the proliferation of multifunction personal digital appliances and electronic gaming. With time, as more manufacturing capacity comes on line, a compelling downward pressure on pricing permeates the market. Mature technologies enjoy stable production control and commodity pricing. Individual suppliers may be differentiated by added value in the form of niche features or excellence in design tools.

Let us consider potential cost reductions when migrating applications from BiCMOS to CMOS. Integrating a high-performance bipolar transistor with CMOS adds masks and processing steps and necessarily raises the cost of producing the wafer. If application specifications can be met with CMOS, market forces will clearly favor those implementations over BiCMOS. Figure 18.4 compares CMOS and BiCMOS approximate price per wafer and f_T of the NFET and SiGe NPN for various technology nodes. Assuming that f_T is an adequate performance metric, the NFET of the 0.18 μm node is at least comparable to the NPN of the 0.35 μm node. The price per wafer of these two technologies is also comparable. If we compare 130 nm BiCMOS and 65 nm CMOS, the SiGe NPN and the 65 nm NFET have essentially the same f_T. However, the wafer price for 65 nm CMOS is considerably higher. The worldwide capacity being developed for 65 nm is staggering [5] and, as demand shifts to higher performance technology nodes, there will be considerable downward pressure on price. This argument applies to any CMOS node.

The single most compelling argument for using CMOS to implement wireless applications is integration density. Returning to our comparison of 0.35 μm BiCMOS with 0.18 μm CMOS, we note comparable pricing but also that the CMOS offers a clear advantage of integration density. We can use the square of the second metal wiring level (M2) pitch to illustrate the integration density of the digital CMOS available at each technology node as shown in Figure 18.5. Any digital content comes at that node's CMOS density (and performance) giving another reason to consider our example of 0.18 μm CMOS over 0.35 μm BiCMOS. For a hypothetical chip that is 50% digital CMOS and 50% RF implemented in 0.35 μm BiCMOS and is being migrated to 0.18 μm CMOS, if we assume that the RF portion, dominated by passive devices, does not scale, we see that the overall chip

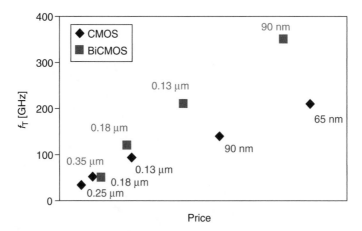

FIGURE 18.4   A qualitative illustration of price-performance comparison between BiCMOS (HBTs) and CMOS (NFETs). Advanced nodes are projected. (From Pekarik, J. et al., *IEEE Custom Integrated Circuits Conference*, IEEE, 2004, p. 217. With permission. © IEEE 2004.)

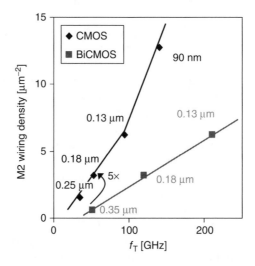

FIGURE 18.5   Integration density (illustrated by second-level metal pitch) versus unity current gain frequency for CMOS and BiCMOS technology nodes. (From Pekarik, J. et al., *IEEE Custom Integrated Circuits Conference*, IEEE, 2004, p. 217. With permission. © IEEE 2004.)

area shrinks by ~25% due to the dramatic increase in the density of the digital circuitry. By fitting proportionally more chips on a wafer that costs about the same, the economic benefit is obvious. Such analyses need to consider all aspects of product cost such as masks, packaging, and volume discounts, and need to use current pricing information as competitive pressures usually force prices to lower with time. Furthermore, many other considerations will factor into a decision on which technology to use for a given product integrated circuit (IC). Designer expertise or the existence of verified circuit designs could drastically affect the time to market. Package, board, and system assembly costs could well dominate the contribution of the IC die to the total product cost.

BASIC CMOS PROCESS FLOW

A summary of the process steps used to fabricate a CMOS IC is shown in Table 18.1. In the subsections below we will describe these process steps and provide examples of physical phenomena or

TABLE 18.1
Process Steps and Their Impact on Design

Process Step	Physical Mechanism	Impact on Design
Isolation	Stress-induced lattice strain	Layout-dependent device properties
Well implant	Resist-proximity effects	Systematic V_T variation
Gate	Flicker noise	Model limitations requiring conservative design
Active area implants	Output conductance	Low self-gain
Stress engineering	Lattice strain-enhanced mobility	Layout-dependent device properties
Contacts	Contact resistance	Resistance–capacitance trade-off
Wiring	Parasitic capacitance, resistance	Metal fill and slotting, reliability-performance trade-off

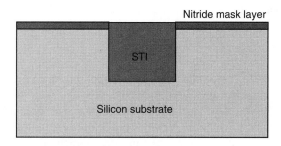

FIGURE 18.6   A shallow trench isolation structure.

mechanisms that have an impact on analog or RF designers through the properties of the devices. The intent is not to provide a detailed review of the state-of-the-art CMOS process—there are many excellent resources, for example Ref. 6—but to illustrate that every process step can contribute to anomalous device behavior. The astute designer will study these and understand how to mitigate their effects in physical design.

ISOLATION

Early CMOS processes employed thick thermal oxides grown in areas opened through a silicon nitride mask layer—the local oxidation of silicon (LOCOS) process—to provide isolation between devices. Because of the high temperatures involved, LOCOS caused relatively little long-range distortion of the surrounding silicon crystal lattice. The use of shallow trench isolation (STI) was introduced in dynamic random access memory (DRAM) processes to reduce memory cell area in the 0.5–0.35 µm generations and found almost universal application in CMOS logic processes by the 0.18 µm node.

A typical process flow to create STI is as follows: open a window through a silicon nitride mask layer, etch a shallow trench in the underlying silicon using an anisotropic reactive ion etch (RIE), clean the surface, grow a thermal oxide to passivate the surface, deposit a near-conformal oxide film using chemical vapor deposition (CVD), planarize the oxide film to be nearly coplanar with the silicon surface using a combination of RIE and chemical–mechanical polish (CMP). The final structure is illustrated in Figure 18.6.

The vertical depth of the STI is determined by requirements for isolation between devices and latch-up immunity when complementary FETs are in close proximity. As CMOS nodes scale to smaller lateral (lithographic) dimensions, the aspect ratio between the depth and width of the STI increases. Any residual stresses in the films comprising the STI impart strain to the nearby silicon lattice. The lattice is pinned by the bulk wafer below and so the amount of strain decreases with distance from the STI-active area boundary. Lattice strain affects carrier mobility and the diffusion

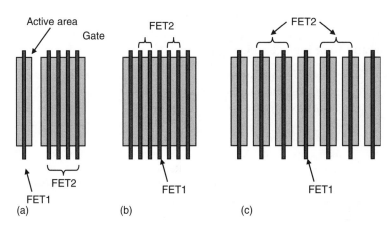

FIGURE 18.7 FET layout examples showing (a) a case with potential for a high degree of stress-induced mismatch, (b) and (c) cases with low potential for stress-induced mismatch.

of impurities. Therefore, device properties dependent on mobility (such as current and g_m) and those dependent on doping distribution (such as V_T and body effect) also vary with distance from STI.

Compact model developers have introduced modules that modify mobility and V_T to account for STI-related stress effects [7,8]. These modules require information about the physical design of the FET and are valid only for simulations run after extraction of layout parameters or with schematic-based simulation of device-library elements with specified layouts and coordinated model calls. Either approach would capture the systematic effects of stress that could introduce layout-induced mismatch between devices. However, depending on and designing for stress effects is not recommended. The variation of stress itself introduces yet another component of mismatch between nonidentical layouts.

Consider the FET structures shown in Figure 18.7. If the goal is to achieve 4:1 current scaling, option (a) is certainly not the way to achieve it. Although options (b) and (c) both consume more area, they both utilize FETs with consistent stress environments.

WELL IMPLANTS

High-energy ion implants, necessary to provide a low-resistance path to body contacts, introduce another source of systematic layout-dependent device mismatch. Typically, the same masking layer is used to define openings in photoresist into which both these high-energy ion implants and low-energy implants, which establish the surface doping concentrations that determine the V_T of the FET, are performed. As shown in Figure 18.8, the high-energy ions can scatter out of the photoresist thereby contributing to the surface doping in the region near the edge of the photo resist. The V_T of the FETs near the boundary of the well region will be higher than those farther away. The deep N-well (or triple-well) isolation scheme for NFETs would have the opposite effect where n-type doping scattered out of the resists will lower the V_T of affected NFETs.

Compact model modules to predict this effect are available but, like those for STI-induced stress, they would only be valid in postextraction simulations. Mixed-mode designs, such as those comprising custom analog circuits and standard-cell library elements, are susceptible to error if the blocks are simulated separately and interactions between well regions of adjoining circuit blocks are not captured.

GATE

Because the active area of the MOSFET is the silicon oxide interface, the capture and release of carriers by interface-related point defects, giving rise to flicker noise, is much more significant in

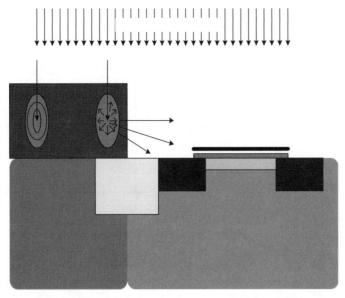

FIGURE 18.8 Implant scattering during a high-energy well implant is illustrated. V_T of the FETs formed near the well boundary will be higher. (From Hook, T.B. et al., *IEEE Trans. Electron Devices*, 50, 1946, 2003. With permission. © IEEE 2003.)

MOSFETs than in bipolar transistors. The increase in flicker noise with technology scaling has, so far, followed expected trends [3,9] and the influence of high-K materials is being studied [10] but is not expected to limit the usability of FETs. However, flicker noise is often an important limiting constraint in optimizing designs of voltage controlled oscillators (VCOs) and mixers where trade-offs in bias point and device size to optimize noise, linearity, or gain are limited by the level of flicker noise [11].

Theory and modeling of flicker noise in MOSFETS are widely discussed in the literature [12,13]. To a rough approximation, flicker noise power is proportional to the number of carriers in the channel. It scales inversely with the active area of the FET, increases directly with gate bias, and is relatively weakly dependent on drain bias. The dependencies are well represented by compact models.

However, flicker noise is directly dependent on the number of interface traps (Nit) and Nit varies statistically with process variation [14]. We will discuss a compact model for the simulation of the process variation of flicker noise in the section "Variation of Flicker Noise."

ACTIVE AREA IMPLANTS

As channel length scales to shorter dimensions, the lateral electric field associated with the drain-to-body junction begins to have significant influence on the energy barrier seen by the carriers at the source side of the FET competing with the influence of the vertical electric field associated with the gate electrode. This is further aggravated with applied drain bias and results in a dramatic reduction in V_T at shorter gate lengths—known as the short-channel effect (SCE). Halo or pocket implants were introduced at the 0.35–0.25 µm nodes to reduce the impact of the SCE. To form the halo, dopant ions of the same type as the well are implanted at an angle of approximately 30–45° so that they extend under the gate. This implant is typically repeated four times at 90° rotations of the wafer forming the halo under gate-edges at all orientations. A vertical implant of opposite type follows that compensates the halo and forms the active source and drain region. The resultant structure and its influence on threshold voltage behavior are shown in Figure 18.9. The use of halo implants extends the ability to scale FET gate length to shorter dimensions.

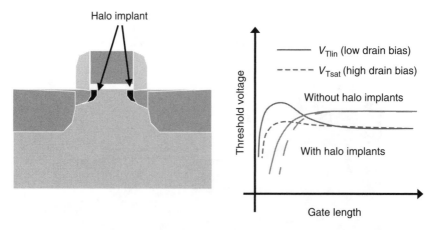

FIGURE 18.9   Schematic illustration of FET halo and its effect on V_T versus gate length.

The halo implant also limits degradation of output conductance (g_{DS}) at shorter gate lengths. An unfortunate side effect of halo implants, which we will discuss later, is an unexpected degradation of g_{DS} for long-channel FETs.

STRESS ENGINEERING

To gain performance benefit of enhanced mobility, advanced CMOS technologies deliberately introduce strain to the silicon lattice. Biaxial strain can be imparted by employing an epitaxial layer, with mismatched lattice constant, grown prior to and below the channel region of the FET [15]. Uniaxial strain can be introduced, after the channel has been formed, using a variety of techniques [16]. These techniques have been used to demonstrate outstanding RF performance in FETs [1]. Some of these techniques, as described in Ref. 16, have dependence on layout details thereby requiring close coordination of the model with the accompanying p-cells and extraction tools so that optimum performance is available to the designer and accurately represented in simulation. Furthermore, physical designers hoping to achieve best matching will now have other design aspects to regulate.

CONTACTS AND WIRING

As contact dimensions shrink, the resistance of individual contacts grows. This is especially worrisome to LNA designers hoping to minimize gate resistance. Placing multiple redundant contacts reduces resistance but introduces parasitic capacitance from the contact landing area. Again, this illustrates another way in which performance optimization will depend on physical design details as we contend with shrinking lithographic dimensions.

The introduction of copper wiring brings lower resistance wires and vias and planar processing that enables better dimensional control and higher yield. The details of planar processing, particularly CMP, demand some constraints be placed on physical design. Maximum and minimum pattern densities on global (chip-wide) and local (hundreds of micrometers) scales are imposed. Furthermore, as part of preparing for mask build, metal fill shapes are added to sparse areas and holes are cut in large metal shapes to assure a uniform pattern density. Effects of these changes need to be accounted for in specifications of wiring properties and in parasitic extraction. The accurate modeling of interconnect effects in RF design is in the critical path to success. At GHz frequencies, shape-based RLC extractors are not always adequate, for example, in the accurate design of matching [17]. Help in addressing these issues comes in the form of p-cells and models for controlled impedance transmission line structures and in applying computer aided design (CAD)-based modeling techniques as described in Ref. 17 with the aim of assuring design success.

DEVICE MENU

Beyond digital FETs, the base CMOS process allows the formation of a suite of passive devices that only require the support of models and design tools. Resistors can be formed from the implanted N-well, the gate polysilicon, or the source/drain active areas. Capacitors and varactors can be formed from the junctions and MOS gates of the FET. Inductors can be formed using the standard wiring. A metal capacitor having very good performance, which we discuss below, can also be formed using the standard wiring. A crude bipolar transistor, formed by the p-type active area, is also created by the base CMOS process. All these devices become usable if they are supported by scalable p-cells and corresponding models. The goal of cost-conscious designs is to implement every function with just these base CMOS devices. However, if application specs require higher performance, a suite of optional high-performance devices should be available.

In the following sections, we will describe devices formed by the base CMOS process and some of the optional high-performance devices made available. Heavy emphasis is placed on IBM's offerings owing to the author's familiarity.

OPTIONAL THICK OXIDE FETS

Additional devices almost ubiquitously included in CMOS technologies and employed in almost every design are input output (I/O) FETs. These are FETs with a thicker gate oxide, made to resemble FETs of previous nodes and used to create IO interfaces that support higher voltages. Figure 18.10 shows the 2.5 and 1.5 V IO FET options from IBM's 90 nm technology compared with the digital FET from the 0.25 μm and 130 nm nodes. A system-on-chip (SoC) design hoping to migrate RF function from an earlier node and combine it with leading-edge digital could, in principle, save time to market by using the IO FETs that resemble the FETs used in the earlier node. Even new designs could find the IO FETs more suitable to their application when higher voltage tolerance or lower leakage is required.

We will discuss how the consequences of dimensional scaling result in lower g_{DS} in the section "Self-Gain." However, an asymmetrically doped FET, with the halo implants removed from the drain, allows lower g_{DS} even down to short-channel lengths. We have employed an approach proposed by

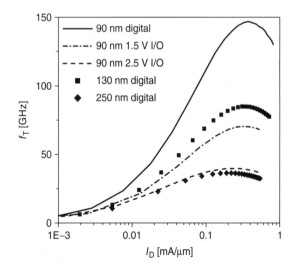

FIGURE 18.10 f_T versus I_D for selected FETs from IBM's 0.25 μm, 130 nm, and 90 nm CMOS nodes illustrating the similarity of IO FETs from newer nodes with digital FETs of earlier nodes. The difference between the 130 nm FET and the 90 nm 1.5 V FET in this example results from a sharing of implant. The properties of the 1.5 V FET could be tuned with independent implants. (From Pekarik, J. et al., *IEEE Custom Integrated Circuits Conference*, IEEE, 2004, p. 217. With permission. © IEEE 2004.)

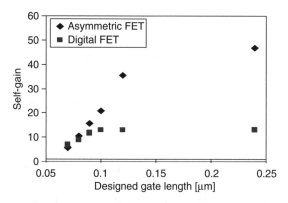

FIGURE 18.11   Shows g_m/g_{DS} comparing the digital FET to the asymmetric FET in 90 nm CMOS. Here g_m and g_{DS} are both measured at ten times the drain current corresponding to threshold. (From Pekarik, J. et al., *IEEE Custom Integrated Circuits Conference*, IEEE, 2004, p. 217. With permission. © IEEE 2004.)

Hook et al. [19] to block the halo implant from the drain side of the FET thereby reducing g_{DS}. Counterintuitively, long-channel digital FETs have high g_{DS}. This results from the halo implant making the V_T of the channel region near the drain a little higher, creating a barrier to current flow. As the V_{DS} is raised, the barrier is lowered allowing more current. Since the asymmetric FET has no drain-side halo, this effect does not occur. Figure 18.11 illustrates the benefit of this asymmetric FET design showing that self-gain is higher even at relatively small gate lengths. We observe that both devices show similar values for g_{DS} and drain induced barrier lowering (DIBL) at short channels, suggesting that the source-side halo controls these short-channel phenomena. Biased at equal current, both devices exhibit the same g_m for a given length. Therefore, we see that g_{DS} decreases by a factor of 5 at three times minimum gate length. Also, because the halo implant is blocked from the drain, we observe lower gate overlap capacitance.

RESISTORS

Two of the most widely used resistors in circuit designs using CMOS technology are the silicide-blocked N-type diffusion resistor and P-type polysilicon resistor. The former is widely used in IO circuits while the latter is employed in most analog circuits where matching and a low temperature coefficient are paramount. The properties of these resistors (sheet resistance, tolerance, temperature coefficient, etc.) are determined by the design of the FET. High-performance resistors with lower overall tolerance can result in low-power designs and improved circuit performance. These devices generally require an additional mask level for improved process control. An added mask level allows the tuning of these properties and results in a silicon resistor with an overall tighter resistance tolerance. Another high-performance resistor can be created by adding a thin metal film embedded in the wiring levels. These resistors have good tolerance and low parasitic capacitance because they are further above the silicon substrate [20].

VARACTORS

Accumulation-mode MOS varactors can be provided in each oxide thickness and have high capacitance density, a large tuning range, and high Q. Reverse-biased junctions can be used as varactors where fine tuning of capacitance is required. However, low-phase-noise, low-power VCOs demand varactors with both high linearity and wide tuning range. IBM has developed [20] an optional hyper-abrupt (HA) junction varactor with a nearly linear C–V tuning ratio of 3.1 and a Q exceeding 100 at 2 GHz. To illustrate the different tuning behavior, Figure 18.12 compares the normalized C–V characteristics of the HA junction varactor and MOS accumulation varactors at the 130 nm node.

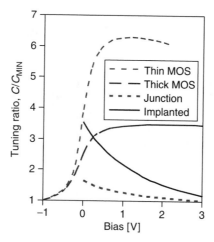

FIGURE 18.12 Normalized *C–V* curves for MOS and HA varactors in IBM's 130 nm CMOS. (From Pekarik, J. et al., *IEEE Custom Integrated Circuits Conference*, IEEE, 2004, p. 217. With permission. © IEEE 2004.)

TABLE 18.2
Inductor Parameters

Spiral Metal	*Q*@5 GHz, *L* = 1.0 nH
2-2X, no M1 ground	9
2-2X, with M1 ground	10
2-2X+Al, no M1 ground	12
2-2X+Al, with M1 ground	14
2-2X+2-4X, no M1 ground	14
2-2X+2-4X, with M1 ground	17
2-2X+2-4X+Al, no M1 ground	15
2-2X+2-4X+Al, with M1 ground	19

INDUCTORS

Thick added metal layers can be employed to build inductors with very high-quality factors [20]. However, inductors suitable for many applications can be formed using standard CMOS wiring levels. In this section we illustrate this with inductors made with standard wiring levels in IBM's 90 nm node. In technologies with copper wiring, low-resistance vias make symmetric inductors feasible. To maintain high *Q*, the spirals consist of parallel combinations of two or more thick Cu metal layers connected together through continuous Cu vias. An additional 1.2 μm final aluminum layer is also available for parallel combination with the Cu levels. By use of a parallel combination of two 2X-pitch (0.5 μm thick) wires and two 4X-pitch (0.81 μm thick) wires and the final Al layer, a very low net sheet resistance of 0.006 Ω/sq. can be achieved. Many digital applications employ these thick wiring levels for power and clock distribution. However, for applications requiring high-*Q* inductors that do not utilize the full wiring stack, an optional single 3 μm thick Cu wire achieves the same performance. A Faraday shield ground plane provides an option to enhance *Q* even further while more effectively isolating the spiral from noise coupling to/from the substrate. Table 18.2 shows *Q* for 1.0 nH spirals in the various combinations of metal stacking and groundplane choice. Figure 18.13 shows representative *Q* versus frequency curves.

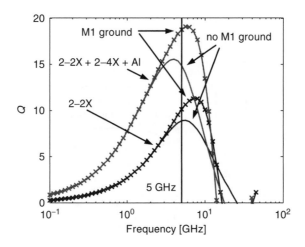

FIGURE 18.13 Symmetric inductor Q (single-ended) for 2-2X and 2-2X, 2-4X+Al options (with and without M1 groundplane). (From Pekarik, J. et al., *IEEE Custom Integrated Circuits Conference*, IEEE, 2004, p. 217. With permission. © IEEE 2004.)

METAL CAPACITORS

The vertical natural capacitor (VNCAP) [21] consists of via-connected stacks of interdigitated metal combs. Details of device layouts should be optimized for reliability and performance. Capacitors including all thin wires and optionally including the 2X-pitch wires have been implemented in scalable p-cells in IBM's 90nm technology. For example, capacitors constructed with four thin wires and two 2X-pitch wires in oxide dielectric have a capacitance density of $1.73\,fF/\mu m^2$. This same structure fabricated in low-K dielectric gives $1.58\,fF/\mu m^2$. Figure 18.14 shows the high-frequency behavior a 90nm, low-K VNCAP with Q of 125 for a 1.6pF capacitor at 4GHz. Frequency characteristics of devices in oxide and low-K dielectric are very similar indicating good integrity of the low-K films.

Parallel-plate metal insulator metal (MIM) capacitors typically require two additional masks. However, they have tighter tolerance limits and can achieve higher capacitance density by either stacking plates or introducing high-K material and they can be optimized for loss giving high Q. For example, in IBM's 90nm CMOS an optional, parallel-plate high-Q MIM with low plate resistance features a capacitance density of $2\,fF/\mu m^2$ and Q in excess of 200 at 5GHz for a $20 \times 20\,\mu m^2$ plate. This device enables the pairing of large capacitors with small inductors in circuits such as VCO tanks, for an overall high Q as well as an area saving. The introduction of high-K dielectric material allows for higher capacitance densities. Capacitors with a density of $4.5\,fF/\mu m^2$ and comparable reliability to oxide- and nitride-based capacitors have been demonstrated [22]. In applications where the amount of capacitance needed is large, it might cost less to add the masks and processing for the planar MIM rather than using the VNCAP at the cost of larger chip area.

FET PERFORMANCE AND SCALING

The scaling of the digital FETs in shrinking CMOS technology nodes presents both advantages and challenges to designers of RF functions. In this section, we will review some FET parameters of interest and how they evolve as digital CMOS shrinks. While the PFET also benefits from scaling and is utilized in some applications such as folded cascode, we focus our discussion details here on the NFET. The parameters, upon which we will base our discussion, are shown in Table 18.3. These are gate length, effective oxide thickness, supply voltage, threshold voltage, transconductance, output conductance, self-gain, cutoff frequency, second-metal pitch, and contact resistance. The values are taken from IBM's general-purpose bulk CMOS technologies but should be representative of industry trends.

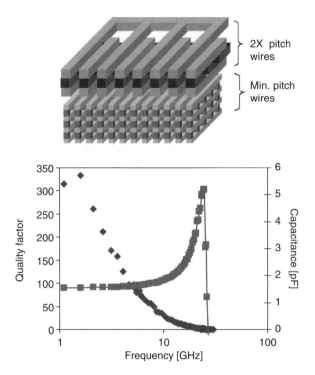

FIGURE 18.14 VNCAP C, Q versus frequency measured on IBM's 90 nm CMOS with low-K dielectric. Capacitor area is 50×20 μm². Structure shown schematically. (From Pekarik, J. et al., *IEEE Custom Integrated Circuits Conference*, IEEE, 2004, p. 217. With permission. © IEEE 2004.)

TABLE 18.3
Parameters of the Digital NFET in IBM CMOS Process

Node (nm)	250	180	130	90	65
L_{GATE} (nm)	180	130	92	63	43
t_{OX}(inv.) (nm)	6.2	4.45	3.12	2.2	1.8
V_{DD} (V)	2.5	1.8	1.5	1.2	1
V_T (V)	0.44	0.43	0.34	0.36	0.24
Peak g_m (μS/μm)	335	500	720	1060	1400
g_{DS}[a] (μS/μm)	22	40	65	100	230
g_m/g_{DS} (−)	15.2	12.5	11.1	10.6	6.1
f_T (GHz)	35	53	94	140	210
M2 pitch (μm)	0.8	0.56	0.4	0.28	0.2
Contact (Ω)	10	11	9	15	20

[a] At peak g_m.

Self-Gain

In the velocity saturation limit, transconductance (g_m) scales with the inverse of gate oxide thickness ($1/t_{OX}$) as shown in the following equation:

$$g_m = \mu \left(\frac{W}{L} \right) C_{OX}(V_{GS} - V_T) \Rightarrow W \frac{\varepsilon v_{SAT}}{t_{OX}} \tag{18.1}$$

For short-channel devices, g_{DS} is related to DIBL that results from electric field at the source side of the FET under conditions of high drain bias. g_{DS} will increase as V_{DS}/L_{GATE} increases. This ratio increases because V_{DD} does not scale as fast as gate length. Furthermore, beyond 90 nm, gate leakage restricts the scaling of the oxide so that L_{GATE} and t_{OX} cannot be reduced proportionally. This results in a precipitous drop in the self-gain, g_m/g_{DS}, for the digital FET at the 65 nm node as shown in Table 18.3. This presents a problem, for example, for low noise amplifier (LNA) designers hoping to realize significant gain. Halo (or pocket) implants are used to counter short-channel effects such as DIBL in digital FETs. Earlier, we discussed an asymmetric FET that, by removing the halo implants from the drain, allows lower g_{DS} even down to short-channel lengths. Ultimately, other technology elements such as high-K gate dielectric, metal gates, and double gate structures will be needed to address this issue [18].

f_T, f_{MAX}

A simple indicator of the scaling of frequency range of device operation is the unity gain frequency of the short-circuit current gain, the cutoff frequency, given by

$$f_T = \frac{g_m}{2\pi C_{in}} \tag{18.2}$$

f_T can be shown to be proportional to the inverse of L_{GATE} in the velocity saturation limit and this is consistent with the data in Table 18.3. A more comprehensive indicator of usable bandwidth is the unity gain point (f_{MAX}) of Mason's unilateral gain (U). Simplified expressions, for example,

$$f_{MAX} = \frac{f_T/2}{\sqrt{g_{DS}(R_G + R_i + R_S) + 2\pi f_T R_G C_{GD}}} \tag{18.3}$$

have been examined by various authors (for example, Ref. 23) as studies of technology scaling. Morifuji's [24] analysis showing the effects of technology scaling and parasitic losses is presented in Figure 18.15. We observe that f_{MAX} has a peak value at some optimum width above which it is limited by physical gate resistance and below which it is limited by parasitic losses in the FET structure. For a given layout, the optimum device width shrinks as the unit gate resistance increases

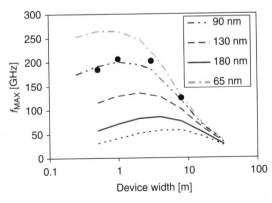

FIGURE 18.15 Calculated f_{MAX} (according to method in Ref. 24) for different technology nodes as a function of FET width. Data from IBM's 90 nm node qualitatively show the predicted dependence. (From Pekarik, J. et al., *European Gallium Arsenide and Other Compound Semiconductors Application Symposium*, EGAAS, 2005, p. 29. With permission. © IEEE 2004.)

with smaller and smaller gate lengths. Care must be taken in FET layout to minimize parasitic gate capacitance that will limit f_T and gate resistance that will limit f_{MAX} and noise figure as discussed in the section "Noise Figure."

While the above analysis indicated that f_{MAX} increases with technology scaling, in practice this is quite difficult to achieve despite the increase in f_T [25]. One reason is the increasing impact of the nonscaling parasitics in the device layout. We will describe a flexible parameterized layout cell that allows the designers to choose layout configurations consistent with their performance requirements.

Noise Figure

The dominant component of broadband noise in FETs is the thermal noise of the channel. Although originating physically as a current noise at the drain, this component may be input referred (in units of V^2) as follows:

$$\frac{\overline{i_d^2}}{g_m^2} = \frac{4kT\Delta f g_{d0}\gamma}{g_m^2} \tag{18.4}$$

where g_{d0} is the channel conductance at zero drain-source bias, Δf the bandwidth of interest, and γ the deviation in noise value from that of an ideal resistor due to both nonuniform charge distribution and high-field effects. The theoretical value for γ in the long-channel approximation is 2/3, but γ can grow significantly greater than 1 when velocity saturation begins to dominate channel transport. In addition to the high-field effects contained within γ, we note from Equation 18.4 that the scaling behavior of channel thermal noise is dominated by that of g_{d0}/g_m^2 and that noise performance should therefore improve with successive technology nodes. Figure 18.16 illustrates this scaling behavior for the digital NFETs measured at 10 and 15 GHz.

Another contribution to the overall noise of an FET is the resistance of the gate, well described by the classical expression for the thermal noise voltage of a resistor, $kTR_G\Delta f$. This contribution is highly layout dependent and may be minimized by dividing the overall gate width into a large number of short gate fingers (again the optimum width scales with node). A well-optimized layout, which minimizes via and contact resistance contributions as well as the resistance along the gate polysilicon, can reduce the relative contribution of gate resistance to the total drain current noise to less than 10%.

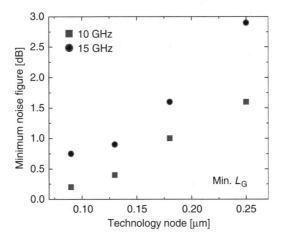

FIGURE 18.16 Minimum noise figure versus technology node showing devices with the best data at minimum L_{GATE} at 10 and 15 GHz. (From Pekarik, J. et al., *IEEE Custom Integrated Circuit Conference*, IEEE, 2004, p. 217. With permission. © IEEE 2004.)

FLICKER NOISE

Although a low-frequency phenomenon, $1/f$ or flicker noise has direct impact on the close-in phase noise of a VCO as well as on baseband noise in circuits such as zero-intermediate frequency (ZIF) receivers. Fundamentally, $1/f$ noise appears in the drain current and arises from fluctuations in the number of channel charge carriers as these carriers are captured and released at random by traps located near the oxide/silicon interface. The magnitude of this noise is determined by the density of these traps, the density of channel charge (and thus by g_m and C_{OX}), and the overall gate area, as described by the following equation:

$$S_{Id} = \left(\frac{1}{f}\right)\frac{M \cdot g_m^2}{WL_{GATE}C_{OX}^2} \qquad (18.5)$$

Alternately, the same noise can be input-referred and described as a voltage noise source on the gate, S_{Vg}, by dividing by g_m^2. This format is particularly useful for comparing noise with the input signal of a baseband amplifier.

As t_{OX} decreases with technology generation, both g_m and C_{OX} increase proportionally. Thus, devices of fixed width W and length L_{GATE} should experience no change in $1/f$ noise, assuming that trap density can be kept constant. If L_{GATE} is allowed to scale to minimum dimensions, however, $1/f$ noise will indeed increase with generation even without a change in trap density, due to both the decrease in L_{GATE} as well as the consequent increase in g_m.

Since the designer can control gate geometry through layout, the key concern for maintaining low $1/f$ performance across generation is preventing trap density from increasing as the result of process-related factors such as strain and changes in oxide composition (e.g., increase in nitrogen content). To explore the trend in trap density across four technology generations of NFET from 0.25 μm to 90 nm, we plot S_{Vg} versus frequency in Figure 18.17. Gate geometry is held constant at 0.25×10 μm^2 while C_{OX} scaling is normalized by measuring each node at a different gate voltage so as to maintain a constant channel surface field and thus a fixed channel charge. Plotted in this manner, differences in S_{Vg} should reflect trap density directly. We observe that each generation approximately overlays the next, suggesting that gate oxide quality is currently being held constant through the 90 nm generation

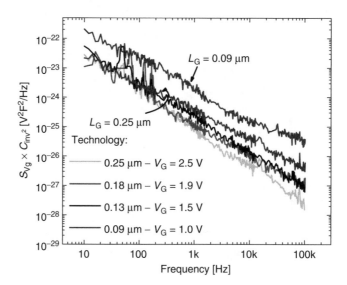

FIGURE 18.17 S_{Vg} versus frequency for four generations of NFET, showing little change in normalized $1/f$ noise and thus in trap density. (From Pekarik, J. et al., *IEEE Custom Integrated Circuit Conference*, IEEE, 2004, p. 217. With permission. © IEEE 2004.)

despite changes in oxide formation processes. However, the spectrum plotted for the minimum gate length at the 90 nm node shows noise power scaling inversely with gate area.

DEVICE MODELS

HIGH-FREQUENCY MODELS

High-frequency behavior is greatly influenced by the impedance of the wiring leading up to the device [26] and the substrate below. To reduce the infinite set of layout variations that contribute these parasitic impedances, we employ p-cells that control layout variations to a few scalable parameters and binary options. The high-frequency model is built to accurately reflect the allowed variations of the p-cell. Figure 18.18a shows the FET p-cell from IBM's 90 nm CMOS. The scalable parameters are gate length, device width, and number of fingers. The binary options are substrate contact ring, one- or two-sided gate contact, and one or two rows of diffusion contacts. Other design aspects, for example, the number and placement of contacts and the spacing of the substrate contact, are controlled to assure a good model fit over the allowed layout variations.

The high-frequency model, which consists of the low-frequency FET model surrounded by a subcircuit network representing parasitics, is shown in Figure 18.18b. The model and p-cell have corresponding scaling variables and options. It is important that the schematic, layout, and parasitic extraction tools all recognize the p-cell, its options, and especially its boundaries in order to avoid double-counting of these parasitics by the model and extraction tools.

Values for components of the high-frequency MOSFET model are determined by measuring FETs in variations of the p-cell configuration and extracting model parameters as a function of geometry. One method to extract the gate resistance term shown in Figure 18.18b from scattering (S-) parameter measurements is by calculating the input resistance as the real part of H_{11} ($Re[H_{11}]$) and comparing this to predicted values. Care must be taken in de-embedding the intrinsic S-parameters of the device because MOSFET data are particularly sensitive to parasitic impedances associated with the probe interface [1]. This is often evident when the low-frequency values of $Re[H_{11}]$ have anomalously low or even negative values. Another approach is to extrapolate S_{11} versus frequency to intercept the real axis on a Smith chart. Both techniques typically give consistent values for input resistance.

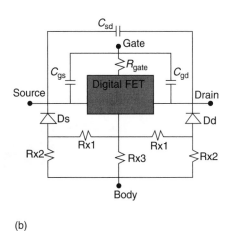

(a)   (b)

FIGURE 18.18 (a) An RF FET p-cell illustrating the optional substrate ring. (b) The high-frequency model as a subcircuit surrounding the low-frequency FET model. (From Pekarik, J. et al., *IEEE Custom Integrated Circuit Conference*, IEEE, 2004, p. 217. With permission. © IEEE 2004.)

MODELING PROCESS VARIATION

The shrinking geometries shown in Table 18.3 also result in increased variation in device performance. As geometries scale down, the impact of dimensional variation to device properties becomes more pronounced and device areas decrease to the point where statistical fluctuations in the count of dopant atoms become evident in electrical characteristics. We introduce a model methodology for capturing process variation in scaled CMOS technologies.

One can visualize process variation as a multidimensional space. Each point in the space represents a particular combination of things that can vary in the process; NFET channel length, PFET mobility, oxide thickness, etc. The Monte Carlo model and simulation technique represents this space by independently varying each of the most important process variables. Figure 18.19 shows three dimensions of this space, NFET channel length, NFET mobility, and NFET base V_T. The Monte Carlo model is represented as a sphere in this picture because any combination of these three parameters (within the process tolerances) can occur in a Monte Carlo simulation. The complete Monte Carlo model has more dimensions, but more than three is difficult to draw. The user-defined corner model is a projection of the Monte Carlo model onto a smaller number of dimensions or corner parameters. In the illustration, NFET channel length is retained as an independent corner parameter (cor_pc), but mobility and V_T are collapsed to a single dimension labeled "cor_nfet." The user-specified corner model is represented by a plane in this picture. Only points on this plane can be specified using the two corner parameters. The foreshortened circle is the portion of this plane containing reasonable process values. Fixed corner models appear as single points in this picture.

For static CMOS logic, fixed corners that represent extremes of gate delay (fast fast (FF) and slow slow (SS)) and extremes of N to P mismatch (fast slow (FS) slow fast (SF)) may provide all the information needed about process variation. However, these points in the process space probably do not represent other extremes of interest such as op amp gain or bandwidth. We recommend that designers use Monte Carlo simulation to understand how critical circuit characteristics vary across the process space and then use corner parameters to define corners that represent interesting extremes of circuit behavior.

Not all of the points within the space defined by the process control limits are physically realizable. This is because some parameters are correlated. For example, poly gates for all FET types are printed together and so L_{GATE} for all FET types are at least partially correlated. If multiple FET types are used in a chip design, the correlations between FET types must be modeled. We have measured the variation of correlation of $I_{D,sat}$ and $V_{T,sat}$ for various FET types and constructed a Monte Carlo model in which correlated L_{GATE} and base V_T distributions reproduce the measured variance

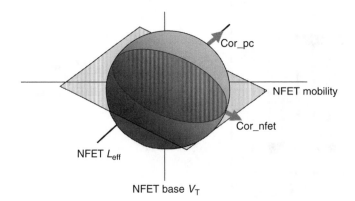

FIGURE 18.19 The Monte Carlo model represented as a sphere provides the most complete coverage of the process space. The user-specified corner models cover only the 2-D plane and the fixed corner models each covers only a single point. (From Pekarik, J. et al., *IEEE Custom Integrated Circuit Conference*, IEEE, 2004, p. 217. With permission. © IEEE 2004.)

and covariance. Using corner parameters specific to each FET type, the user can specify corners that model extremes of circuit behavior.

Similar behavior is seen in the RC delay of different wiring layers, that is, all wires on a given wiring layer will be slow or fast by a similar amount on a given chip. But wires on different layers vary independently of one another. So, for our Monte Carlo model, we provide independent distributions for width of wires on each wiring level and the thickness of each interlay dielectric. For corner analysis, we provide independent corner parameters for each wiring level.

Within a single chip there is also variation among FETs of identical length and width. We model three sources of across-chip FET variation: variation of FET electrical parameters because of random density and placement of dopants under the FET channel, which we will call V_T matching; variation of the length of the FET channel because of variation in poly gate–etched dimensions, which we will call across-chip length variation (ACLV); variation of width of the FET channel, which we will call across-chip width variation (ACWV).

For ACLV, we model variation caused by three layout variables: distance between the devices in question, orientation of the gates (vertical or horizontal), and spacing to the next poly line (poly pitch). ACLV due to line edge roughness and other uncategorized effects is included in the V_T matching term. This and the other four ACLV effects are represented as random distributions in the Monte Carlo model.

Figure 18.20 illustrates the effect of across-chip variation on ring delay. Each group of symbols represents the delay of one particular ring oscillator plotted against the average of the delay of the four rings on that chip. The vertical separation between the groups of symbols is due to the distance part of ACLV and ACWV, which is correlated between FETs within a ring and uncorrelated between FETs in different rings. The vertical spread within one group is due to the other sources of ACLV and ACWV and the dopant mismatch, which are all uncorrelated with a ring. Overall upward trend of each group of data is due to the many sources of chip mean variation.

VARIATION OF FLICKER NOISE

As transistor area is reduced, random telegraph signals (RTS) of individual traps dominate the low-frequency spectrum as can be seen from the Lorentzian signature of RTS in Figure 18.21. This is a significant modeling challenge, since an analytic model can no longer capture the noise spectra due to

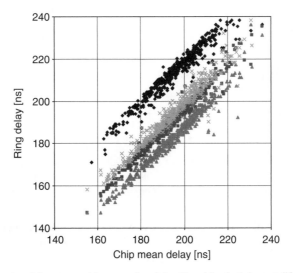

FIGURE 18.20 Ring by ring delay versus chip mean ring delay. Four identical rings at different locations on a chip show both correlated and uncorrelated variations. (From Pekarik, J. et al., *European Gallium Arsenide and Other Compound Semiconductors Application Symposium*, EGAAS, 2005, p. 29. With permission. © IEEE 2004.)

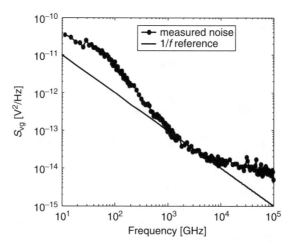

FIGURE 18.21 Low-frequency noise showing the Lorentzian signature of discrete trap RTS (NFET, $W = 10\,\mu m$, $L = 0.08\,\mu m$). (From Pekarik, J. et al., *European Gallium Arsenide and Other Compound Semiconductors Application Symposium*, EGAAS, 2005, p. 29. With permission. © IEEE 2004.)

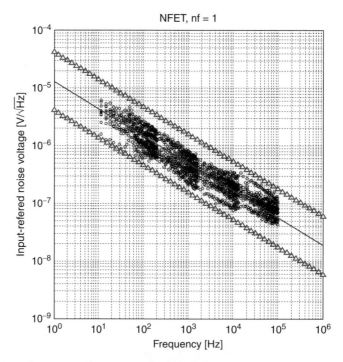

FIGURE 18.22 Low-frequency noise spectra of multiple identical devices with corresponding model and three-sigma corner values (NFET, $W = 10\,\mu m$, $L = 0.08\,\mu m$). (From Pekarik, J.J. et al., in *European Gallium Arsenide and Other Compound Semiconductors Application Symposium*, EGAAS, Paris, 2005.)

the site-to-site variations in effective trap density. A statistical model can enable designers to run Corner and Monte Carlo noise simulations [14]. To implement a statistical noise model, nominal values for three Berkeley short channel IGFET (BSIM) noise parameters NOIA, NOIB, and NOIC are extracted from large area, multifinger devices. In addition to the mean value, a statistical distribution also gets assigned to each parameter. The spread of this distribution is determined from measurements taken from a large sample of single-finger and multifinger devices. Figure 18.22 shows spectra for several identical single-finger devices along with the modeled spectra for nominal and three-sigma corners. Such a model enables accurate statistical phase noise simulations for VCOs and aides in first pass design success of radio frequency integrated circuits (RFICs).

CONCLUSIONS

Design in CMOS is considerably more challenging than with bipolar transistors. Inconvenient device properties, such as flicker noise, low transconductance, and high output conductance, require innovative design techniques. The device characteristics of scaled MOSFETs are difficult to describe with analytic expressions resulting in necessarily complicated compact model and higher risk of disagreement between measured and simulated performance.

It is likely that competitive pressure will dictate that wireless applications, which can be implemented with acceptable performance in CMOS, will be implemented in CMOS. Relative process simplicity and high manufacturing capacity contribute to lower die cost of CMOS when compared to BiCMOS or compound semiconductor technologies. The ability to integrate readily available digital IP, on the same die with a transceiver, provides opportunities to reduce packaging cost, lower power consumption, and simplify system form factor. Therefore, CMOS implementation will be the goal of consumer-oriented and mobile applications.

REFERENCES

1. Lee, S., Wagner, L., Jagannathan, B., Csutak, S., Pekarik, J., Zamdmer, N., Breitwisch, M., Ramachandran, R., and Freeman, G., "Record RF performance of sub-46 nm L_{GATE} NFETs in microprocessor SOI CMOS technologies," *Technical Digest of IEEE International Electron Devices Meeting*, pp. 241–244, 2005.
2. Bennett, H.S., Brederlow, R., Costa, J.C., Cottrell, P.E., Huang, W.M., Immorlica, A.A., Jr., Mueller, J.-E., Racanelli, M., Shichijo, H., Weitzel, C.E., and Zhao, B., "Device and technology evolution for Si-based RF integrated circuits," *IEEE Trans. Electron Devices*, vol. 52, no. 7, pp. 1235–1258, 2005.
3. Pekarik, J., Greenberg, D., Jagannathan, B., Groves, R., Jones, J.R., Singh, R., Chinthakindi, A., Wang, X., Breitwisch, M., Coolbaugh, D., Cottrell, P., Florkey, J., Freeman, G., and Krishnasamy, R., "RFCMOS technology from 0.25 m to 65 nm: The state of the art," *Proceedings of the IEEE Custom Integrated Circuits Conference*, Orlando, FL, pp. 217–224, 2004.
4. Pekarik, J.J., Coolbaugh, D.D., and Cottrell, P.E., "Enabling RFCMOS solutions for emerging advanced applications," *European Gallium Arsenide and Other Compound Semiconductors Application Symposium*, EGAAS, Paris, p. 29, 2005.
5. Semiconductor International Capacity Statistics; http://www.sicas.info/
6. Plummer, J.D., Griffin, P.B., and Deal, M.D., *Silicon VLSI Technology: Fundamentals, Practice, and Modeling*, 1st Ed., Prentice-Hall, 2000.
7. http://www-device.eecs.berkeley.edu/~bsim3/bsim4.html
8. Gildenblat, G., Li, X., Wang, H., Wu, W., Jha, A., van Langevelde, R., Scholten, A.J., Smit, G.D.J., and Klaassen, D.B.M., "Theory and modeling techniques used in PSP model," *Technical Proceedings of 2006 Nanotechnology Conference and Trade Show*, vol. 3, p. 604, 2006 (Chapter 7).
9. Greenberg, D.R., Sweeney, S., Jagannathan, B., Freeman, G., and Ahlgren, D., "Noise performance scaling in high-speed silicon RF technologies," *2003 Topical Meeting on Silicon Monolithic Integrated Circuits in RF Systems*, pp. 22–25, April 9–11, 2003.
10. Giusi, G., "Comparative study of drain and gate low-frequency noise in nMOSFETs with hafnium-based gate dielectrics," *IEEE Trans. Electron Devices*.
11. Razavi, B., "Design considerations for direct-conversion receivers," *IEEE Trans. Circuits Systems II: Analog Digital Signal Process.*, vol. 44, pp. 428–435, 1987.
12. Vandamme, E.P. and Vandamme, L.K.J., "Critical discussion on unified 1/*f* noise models for MOSFETs," *IEEE Trans. Electron Devices*, vol. 47, no. 11, p. 2146, 2000.
13. Scholten, A.J., Tiemeijer, L.F., van Langevelde, R., Havens, R.J., Zegers-van Duijnhoven, A.T.A., and Venezia, V.C., "Noise modeling for RF CMOS circuit simulation," *IEEE Trans. Electron. Devices*, vol. 50, no. 3, pp. 618–632, 2003.

14. Erturk, M., Xia, T., Anna, R., Newton, K.M., and Adler, E., "Statistical BSIM model for MOSFET 1/f noise," *Electron. Lett.*, vol. 41, no. 22, pp. 1208–1210, 2005.
15. Antoniadis, D.A., Aberg, I., Ní Chléirigh, C., Nayfeh, O.M., Khakifirooz, A., and Hoyt, J.L., "Continuous MOSFET performance increase with device scaling: The role of strain and channel material innovations," *IBM J. Res. Dev.*, vol. 50, no. 4/5, 2006 (Available online at: http://www.research.ibm.com/journal/rd/504/antoniadis.html).
16. Horstmann, M., Wei, A., Kammler, T., Hntschel, J., Bierstedt, H., Feudel, T., Frohberg, K., Gerhardt, M., Hellmich, A., Hempel, K., Hohage, J., Javorka, P., Klais, J., Koerner, G., Lenski, M., Neu, A., Otterbach, R., Press, P., Reichel, C., Trentsch, M., Trui, B., Salz, H., Schaller, M., Engelmann, H.-J., Herzog, O., Ruelke, H., Hubler, P., Stephan, R., Greenlaw, D., Raab, M., and Kepler, N., "Integration and optimization of embedded-sige, compressive and tensile stressed liner films, and stress memorization in advanced SOI CMOS technologies," *IEEE Electron. Devices Meeting*, pp. 233–236, 2005.
17. Singh, R., Harame, D.L., and Oprysko, M., *Silicon Germanium—Technology, Models, Design*, Wiley, 2003.
18. Nowak, E.J., Ludwig, T., Aller, I., Kedzierski, J., Leong, M., Rainey, B., Breitwisch, M., Gemhoefer, V., Keinert, J., and Fried, D.M., "Scaling beyond the 65 nm node with FinFET DG-CMOS," *Proceedings of the IEEE Custom Integrated Circuits Conference*, 2003, pp. 339–342, September 21–24, 2003.
19. Hook, T., Brown, J.S., Breitwisch, M., Hoyniak, D., and Mann, R., "High-performance logic and high-gain analog CMOS transistors formed by a shadow-mask technique with a single implant step," *IEEE Trans. Electron. Devices*, vol. 49, no. 9, pp. 1623–1627, 2002.
20. Coolbaugh, D., Eshun, E., Groves, R., Harame, D., Johnson, J., Hammad, M., He, Z., Ramachandran, V., Stein, K., St. Onge, S., Subbanna, S., Wang, D., Volant, R., Wang, X., and Watson, K., "Advanced passive devices for enhanced integrated RF circuit performance," *IEEE Radio Frequency Integrated Circuits Symposium*, pp. 341–344, 2003.
21. Kim, J., Plouchart, J.-O., Zamdmer, N., Sherony, M., Liang-Hung Lu, Yue Tan, Meeyoung Yoon, Jenkins, K.A., Kumar, M., Ray, A., and Wagner, L., "3-Dimensional vertical parallel plate capacitors in an SOI CMOS technology for integrated RF circuits," *2003 Symposium on IEEE VLSI Circuits*, pp. 29–32, June 12–14, 2003.
22. Vaed, K., "A manufacturable high-k MIM dielectric with outstanding reliability and voltage linearity for RF and mixed-signal technologies," *2004 Topical Meeting on Silicon Monolithic Integrated Circuits in RF Systems*, p. 57, 2004.
23. Woerlee, P.H., Knitel, M.J., van Langevelde, R., Klaassen, D.B.M., Tiemeijer, L.F., Scholten, A.J., and Zegers-van Duijnhoven, A.T.A., "RF-CMOS performance trends," *IEEE Trans. Electron Devices*, vol. 48, no. 8, p. 1776–1782, 2001.
24. Morifuji, E., Momose, H.S., Ohguro, T., Yoshitomi, T., Kimijima, H., Matsuoka, F., Kinugawa, M., Katsumata, Y., and Iwai, H., "Future perspective and scaling down roadmap for RF CMOS," *1999 Symposium on IEEE VLSI Circuits*, pp. 163–164, 1999.
25. Dambrine, G., Raynaud, C., Lederer, D., Dehan, M., Rozeaux, O., Vanmackelberg, M., Danneville, F., Lepilliet, S., and Raskin, J.-P., "What are the limiting parameters of deep-submicron MOSFETs for high frequency applications?" *IEEE Electron. Devices Lett.*, vol. 24, no. 3, p. 189–191, 2003.
26. Tiemeijer, L.F., Havens, R.J., de Kort, R., Scholten, A.J., van Langevelde, R., Klaassen, D.B.M., Sasse, G.T., Bouttement, Y., Petot, C., Bardy, S., Gloria, D., Scheer, P., Boret, S., Van Haaren, B., Clement, C., Larchanche, J.-F., Lim, I.-S., Duvallet, A., and Zlotnicka, A., "Record RF performance of standard 90 nm CMOS technology," *IEEE Electron. Devices Meeting*, pp. 441–444, December 13–15, 2004.
27. Hook, T.B., Brown, J., Cottrell, P., Adler, E., Hoyniak, D., Johnson, J., and Mann, R., "Lateral ion implant straggle and mask proximity effect," *IEEE Trans. Electron. Devices*, vol. 50, no. 9, pp. 1946–1951, 2003.

19 Distributed Effects and Coupling in RF Integrated Circuits

Calvin Plett

CONTENTS

Introduction..543
Distributed Effects..544
    Delay and Phase Shift in a Transmission Line...544
    Transmission Line Model...545
    Example: 1 mm Transmission Line 545
    Example: Delay in Power Amplifier...547
    Example: Distributed Capacitance 548
Opportunities with Delay...549
    Opportunities with Transmission Lines 549
    Opportunities with Distributed Amplifiers 549
    Distributed Amplifier Examples..552
Coupling and Its Measurement 554
    Inductors as a Cause of Coupling 554
    LNA Coupling Example 555
    Oscillator-to-Oscillator Coupling...555
Opportunities with Coupling 561
    Transformers with Improved Coupling 561
    Injection-Locked Oscillator..562
    Injection-Locked Oscillator Example...563
Conclusions...563
Acknowledgments...564
References..564

INTRODUCTION

Increasing frequencies and higher levels of integration, including integration of complete systems, results in the increased use of passive components such as transmission lines and inductors. Recent publications of wireless transceivers, for example [1], have shown 15 or more inductors on the same silicon substrate in an area of the order of 10–20 mm². Because of the increased frequencies, long interconnects may no longer behave like an ideal wire with zero delay. Instead it may be more appropriate to think of long interconnect as distributed transmission lines. It will take signals a finite amount of time to travel along such transmission lines. The delay from input to output is equivalent to a phase shift of the output signal with respect to the input signal. Such delay can be problematic in achieving signal alignment, for example, in clock distribution systems, or in the design of combining circuits in power amplifiers. However, the same effects can be exploited, for example, in the design of distributed amplifiers. As well, because of the increased frequency and higher density of components, there can be coupling between components, for example, via conductive coupling through the substrate or via electromagnetic (typically inductive) coupling, for example, from parallel current-carrying wires. Such coupling can present difficulties in achieving isolation between components. However, coupling can also be exploited in the design of transformers, noninvasive testing, and injection-locked oscillators.

In this chapter, we will first look at some of the problems related to distributed effects, specifically delay and signal misalignment. Then we will examine some of the opportunities, such as the use of transmission lines and distributed amplifiers. We will then examine coupling, looking at problems such as noise and change of performance. This will be followed by opportunities such as transformers and injection-locked oscillators.

DISTRIBUTED EFFECTS

Distributed effects result in delay. One of the simplest examples is a series clock distribution system as shown in Figure 19.1. The result of the delay is that clock signals may not be lined up.

DELAY AND PHASE SHIFT IN A TRANSMISSION LINE

The amount of delay t_d can be estimated from

$$t_d = \frac{l\sqrt{\varepsilon_{eff}}}{c} \tag{19.1}$$

where l is the length of the transmission line, ε_{eff} the effective dielectric constant, and c the velocity in free space. The phase shift ϕ across a transmission line can be determined by comparing the time delay to the period T or $1/f$ of the signal, or by comparing the line length to the signal wavelength λ as follows:

$$\phi = \frac{l\sqrt{\varepsilon_{eff}}}{cT} \cdot 360° = \frac{fl\sqrt{\varepsilon_{eff}}}{c} \cdot 360° = \frac{l}{\lambda} \cdot 360° \tag{19.2}$$

where the signal wavelength λ is given by

$$\lambda = \frac{cT}{\sqrt{\varepsilon_{eff}}} = \frac{c}{f\sqrt{\varepsilon_{eff}}} \tag{19.3}$$

As an example, at a frequency of 30 GHz, the period is 33.3 ps. Since c is about 3×10^8 m/s, in free space, the wavelength is 0.01 m or 1 cm. On chip, because the effective dielectric is about 4, the wavelength is reduced to about 5 mm. As a result, for a 30 GHz signal, a 5 mm wire represents a phase shift of 360°. Since phase shift scales with frequency and length, at 5 GHz, and with a 1000 μm line, the phase

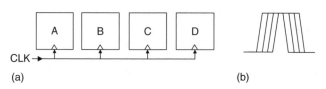

FIGURE 19.1   Delay in a clock distribution system, and pulses arriving progressively later at each stage.

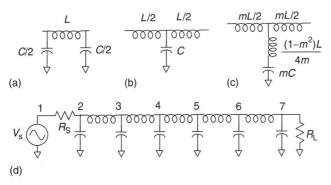

FIGURE 19.2 Transmission line models: (a) simple pi-model for short transmission line, (b) k-section model of a transmission line, (c) m-derived section of a transmission line, typical value for m is 0.6, (d) multisection model valid up to higher frequencies and for longer transmission line lengths.

shift is lowered by a factor of 30 resulting in a 12° phase shift. Equivalently, 1 mm on chip represents a time delay of about 6.7 ps. Large integrated circuits can have signal paths exceeding 10 mm in length; hence, careful design of clock distribution systems is important. In many cases, it may be necessary to align clock with data to reduce the consequences of mismatched delays across a large delay path.

TRANSMISSION LINE MODEL

Delays due to distributed effects can be modeled with a series inductance and a parallel capacitance as shown in Figure 19.2a. An alternative model is given in Figure 19.2b. For a short transmission line, the inductor and capacitor sizes can be estimated from the following equation:

$$L = \frac{Z_0 l \sqrt{\varepsilon_{\text{eff}}}}{c}, \quad C = \frac{l \sqrt{\varepsilon_{\text{eff}}}}{Z_0 c} \tag{19.4}$$

where Z_0 is the characteristic impedance of the line. Breaking a line into a number of shorter lines increases the number of LC sections and results in better modeling accuracy up to a higher frequency and for a total longer length. The characteristic impedance of the transmission line is given by

$$Z_0 = \sqrt{\frac{L}{C}} \tag{19.5}$$

An estimate of the cutoff frequency for a transmission line model is given by

$$f_c = \frac{1}{\pi} \sqrt{\frac{1}{LC}} \tag{19.6}$$

The simple models of the transmission line are useful up to about 45% of the cutoff frequency, as will be illustrated later. At the cost of somewhat more complexity, the m-derived section shown in Figure 19.2c can be used to model a transmission line up to about 85% of the cutoff frequency. Figure 19.2d shows an example of a simple five-stage model of a transmission line terminated in a load resistance.

EXAMPLE: 1 mm TRANSMISSION LINE

As an example, consider a matched 1 mm transmission line with 4 μm oxide, 6 μm width, which using an electromagnetic (EM) simulator results in a transmission line with a characteristic of just over 50 Ω. Using a single stage as in Figure 19.2a to model the line, with this impedance and Equation 19.4

each L is 360 pH and $C = 123.5$ fF. The line is loaded with 50 Ω. As a check, from Equation 19.5, the line impedance is

$$Z_0 = \sqrt{\frac{L}{C}} = \sqrt{\frac{360 \text{ pH}}{123.5 \text{ fF}}} = 54 \, \Omega$$

Note, this is close to the desired 50 Ω. From Equation 19.6, the estimate for model cutoff frequency is

$$f_c = \frac{1}{\pi} \sqrt{\frac{1}{LC}} = \frac{1}{\pi} \sqrt{\frac{1}{360 \text{ pH} \times 123.5 \text{ fF}}} = 47.7 \text{ GHz}$$

For a simple model of the line, the model is valid up to about 40–45% of the cutoff frequency, or, in this example, up to about 21.5 GHz. More sections can be added to increase the valid frequency range. For example, repeating the calculations for two and five stages increases the valid frequency up to 43 GHz and about 107 GHz, respectively. Figure 19.3 shows the simulated gain, phase, and input matching for these transmission line models. The figure shows that while gain is down by 3 dB at roughly the cutoff frequency, both phase and matching deviate from the ideal transmission line at about 45% of the cutoff frequency. In particular, matching is better than about 15 dB up to about 45% of the cutoff frequency.

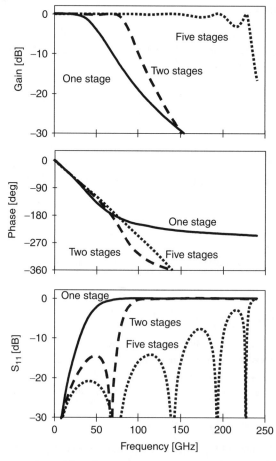

FIGURE 19.3 Gain, phase, and matching (S_{11}) of a 1 mm transmission line, modeled with one, two, and five sections.

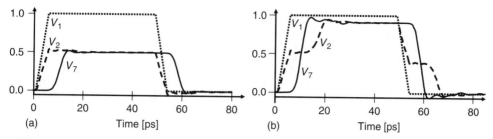

FIGURE 19.4 Pulse response of transmission line model shown in Figure 19.2d, terminated in (a) $50\,\Omega$ and (b) $500\,\Omega$.

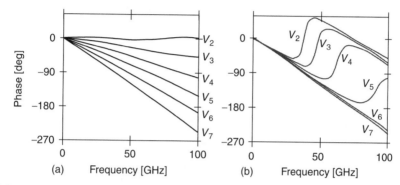

FIGURE 19.5 Phase response of transmission line model shown in Figure 19.2d, terminated in (a) $50\,\Omega$ and (b) $500\,\Omega$.

Note that the m-derived section is another well-known model that extends the usable frequency up to about 85% of the cutoff frequency. The pulse response in Figure 19.4a shows a delay of about 7.0 ns. A simple estimate using Equation 19.1 results in a time delay of 6.7 ps, close to the simulated time. If the same transmission line now is mismatched at its output, for example, connected to $500\,\Omega$, the pulse response appears as in Figure 19.4b. The output now settles to 0.9 of the input voltage, but at the input, the voltage temporarily settles to 0.5 V and then, after the reflection from the mismatch from the far end, settles to the correct final value. Note that if the termination were a short circuit, v_2 would have been a pulse going to half the input voltage, but after the reflection it would have gone back to 0 V. Thus, this is a technique of generating a short pulse.

Figure 19.5 shows the phase response of the 1 mm transmission line model shown in Figure 19.2d. The response is shown up to 100 GHz, roughly the useful limit of the model. This shows that the phase shift is linearly dependent on the line length and frequency for a terminated transmission line. Again, using Equation 19.2 the phase at 100 GHz for a 1 mm line is predicted to be 240°, close to the simulated value. For the incorrectly terminated line shown in Figure 19.5b, due to the reflections, up to about 20 GHz, the phase is nearly the same at each node. This illustrates one of the reasons why it is difficult to design a distributed amplifier if the transmission line is not properly terminated.

EXAMPLE: DELAY IN POWER AMPLIFIER

To deliver high current and for heat dissipation, multiple parallel output transistors are typically used in a power amplifier. The input line connects to bases (gates for MOS) whereas the output line connects to collectors (drains for MOS). Figure 19.6 shows an example of a poor design for delay equalization since the delay is different from the input to each transistor and from each transistor to the output. Thus, each transistor provides a different delay. The exact amount of delay is not easy to predict as illustrated in Figure 19.5b for a mismatched transmission line. Taking advantage of a properly terminated transmission line through the use of a distributed amplifier will be discussed shortly.

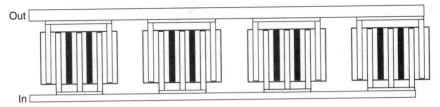

FIGURE 19.6   Power amplifier output transistors with poor delay equalization.

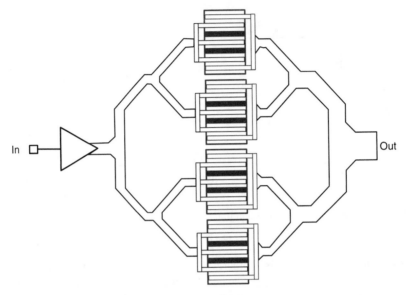

FIGURE 19.7   Power amplifier output transistors with better delay equalization.

A better design is shown in Figure 19.7. In this case, all the input and output lines are of equal length, so while there will be delay, the delay will be equal. As well, this illustrates the principle of widening the lines as the currents get heavier and of avoiding 90° bends in lines (a smoothly flowing electron is a happy electron).

EXAMPLE: DISTRIBUTED CAPACITANCE

A capacitor and its distributed model are shown in Figure 19.8, broken into four sections.

As shown, the capacitor model includes distributed (wanted) capacitance, distributed parasitic capacitance to the substrate, as well as distributed series inductance due to both the capacitor itself and to any leads connecting it to other components. If poorly designed, the distributed components can result in delay across the component and in extreme cases could result in capacitors having self-resonance, and exhibiting inductive reactance beyond the self-resonance frequency. Clearly this could have dire consequences. For example, if such a capacitor is used to decouple a cascode transistor, and if the impedance becomes inductive, instability can be the result. Fortunately, by making the capacitor short and wide, such undesired inductance can be minimized. In the past, the main problem was the substrate capacitance for a nongrounded capacitor as it could result in decreased bandwidth. However, in a modern process, with a very thick oxide between substrate and the bottom plate, the parasitic capacitance is much less important (in some processes, it is lower than the desired capacitor by a factor of more than a hundred).

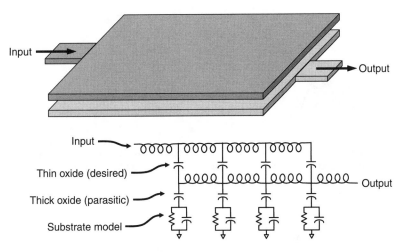

FIGURE 19.8 Capacitor and its distributed model.

OPPORTUNITIES WITH DELAY

Because of the physical distance a signal has to travel, it is not possible to avoid delay. However, if the signal path is a matched transmission line, or if the delay is incorporated into an artificial transmission line, to be discussed later, the frequency response will be high.

OPPORTUNITIES WITH TRANSMISSION LINES

Two types of transmission lines that can be considered are microstrip and coplanar as shown in Figure 19.9. For the microstrip line, field lines can be terminated in the lossy substrate. To avoid the loss, a layer of metal can be used instead; however, because the ground is closer, the lines must also be narrower to maintain the same impedance. As a result, such transmission lines may not be adequate in applications where a lot of current needs to be carried. An alternative to the microstrip line is the coplanar line that has grounds beside the signal line. Because of this ground at the same height as the signal-carrying conductor, ideally a large fraction of the field lines are terminated in metal at the same height. As a result, there is less penetration of field lines into the substrate and hence less loss. Another advantage is that this provides an additional degree of freedom as now both line width and the line separation can be adjusted. As a result, it is possible to achieve the desired impedance while using wider lines. The disadvantage is that the need for the ground lines on either side of the signal lines results in significantly more chip area.

Figure 19.10 shows a comparison of the standard aluminum/SiO_2 to copper/polyimide, with the same thickness of dielectric [2]. As shown in the figure, SiO_2 has a dielectric constant of about 4 whereas polyimide has a dielectric constant of about 2. As a result, with polyimide, the capacitance to the substrate is lower, which minimizes electric field penetration and hence better isolation from the lossy substrate. At the same time, copper is about 30% less resistive than aluminum, so further improvements are expected.

Figure 19.11 shows the measured results with the structures shown in Figure 19.10.

OPPORTUNITIES WITH DISTRIBUTED AMPLIFIERS

A particular application of the artificial transmission line is the distributed amplifier (DA), where such lines are used at both the input and the output. With an artificial transmission line, series inductors (which can be realized with transmission lines or with spiral inductors) and parallel capacitors are used to build a model for a transmission line. Then the parallel capacitors and the parasitic capacitance associated with gain blocks can be incorporated as part of the artificial line as shown in Figure 19.12.

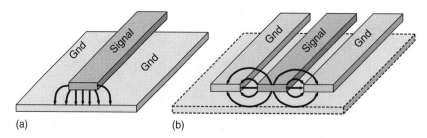

FIGURE 19.9 Transmission lines: (a) microstrip and (b) coplanar. With a small and relatively straightforward amount of postprocessing, additional dielectrics, for example, polyimide, and metals, for example, copper or silver, can be added to any process.

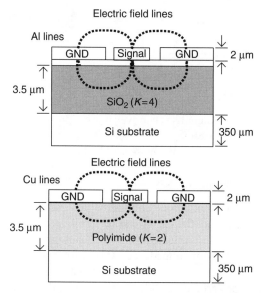

FIGURE 19.10 Structures of Cu/polyimide versus Al/SiO$_2$. (From Amaya, R.E. et al., *International Symposium on Antenna and Applied Electromagnetics*, Ottawa, July 2004, pp. 51–54. With permission. © [2004] IEEE.)

Each gain block is shown as a single MOS transistor in the common-source configuration, but gain blocks can be realized in many different ways, for example, with cascode structures or Darlington pairs, and in different technologies, for example, with bipolar or MESFET transistors. The important thing is that the input and output capacitance is absorbed by the required parallel capacitance of input and output artificial transmission line. Since the gate capacitance C_G is typically significantly higher than the drain capacitance C_D, an additional capacitance C_{add} has been shown connected at the drain. The block labeled as "Match" may contain matching components, coupling capacitors, and biasing circuits. The drain line is typically biased at the positive power supply whereas the gate line is appropriately based to provide the correct V_{GS} bias for the transistors.

At low frequencies ($f_c/10$ or less), the voltage gain of a distributed amplifier, using m-derived sections and terminations, is approximated by

$$A_0 = \frac{g_m \sqrt{Z_d Z_g} \, \sinh\left[N \dfrac{\omega_d}{2\omega_c}\right]}{2 \sinh\left[\dfrac{\omega_d}{2\omega_c}\right]} \tag{19.7}$$

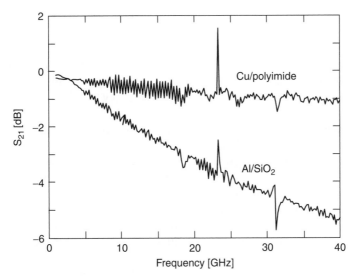

FIGURE 19.11 S_{21} of Cu/polyimide versus Al/SiO$_2$.

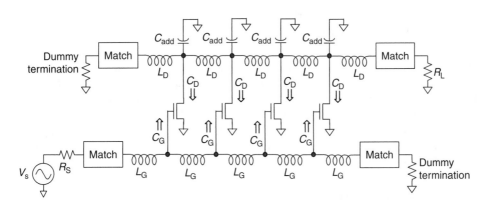

FIGURE 19.12 Principle of distributed amplifier (DA).

where g_m is the transconductance of each gain stage, Z_d and Z_g the characteristic impedances of the drain line and gate line, respectively, N the number of stages, and ω_c the cutoff frequencies for the transmission line. The cutoff frequency for the m-derived termination ω_d is given by

$$\omega_d = \frac{\omega_c}{\sqrt{1-m^2}} \tag{19.8}$$

With a typical value of 0.6 for m, ω_d is equal to $1.25\omega_c$. Substituting these numbers into Equation 19.7 and assuming that both Z_d and Z_g are equal to Z_o, for a four-stage distributed amplifier, gain can be estimated as about $9g_m Z_o$.

The optimum number of stages is given by

$$N_{opt} = \frac{\ln(A_g/A_d)}{A_g - A_d} \tag{19.9}$$

where A_g and A_d are the attenuation of the gate line and the drain line, respectively.

The steps used to design a distributed amplifier depend on what the goals are. If the goal is to maximize the frequency response, the following steps are typically used:

1. The cutoff frequency of the artificial transmission line is first set so that the maximum useful frequency (about 85% of the cutoff frequency for m-derived transmission line sections) is approximately equal to the maximum operating frequency. Note that f_{max} is the upper limit for the maximum operating frequency.
2. The line impedance is typically specified by the need to match source and load. Note that in a fully integrated context such impedances can be higher than $50\,\Omega$, resulting in power savings and making it easier to design into a low-voltage process. As well, up to 30% of mismatch between the drain line and the gate line cutoff frequency is sometimes deliberately introduced to control overshoot [3].
3. This sets both the line inductance and line capacitance.
4. The line capacitance comes from both the transmission line parasitic capacitance and the transistor parasitic capacitance. From the allowed transistor capacitance, the transistors can be sized.
5. The transistor is optimized for highest f_{max} by adjusting the current level and by using multiple fingers to reduce series gate resistance. To optimize f_{max}, gate finger size should be smaller than some specified maximum size, typically a few microns.
6. Simulations and iterations are performed to verify calculations, to include parasitic capacitances, and to optimize performance.

Note in step 1 that if the maximum frequency is set somewhat lower than f_{max}, the line inductors and capacitors will be larger (their ratio is fixed by the line impedance). With a larger capacitance, the transistor size will be larger and the optimal current will be higher. Larger transistor size and higher current result in higher g_m and hence by Equation 19.7 gain will be higher. Thus, there is a trade-off between bandwidth and gain. In fact, if gain is the required parameter, then transistor size is fixed by the need for a particular g_m and this sets the line capacitance. Hence, inductance can be determined from the desired line impedance.

DISTRIBUTED AMPLIFIER EXAMPLES

A series of distributed amplifiers in 0.18 mm CMOS was designed and optimized to explore the maximum achievable performance as a function of the quality factor of the series inductors that make up the artificial transmission line of the passives [4]. The three different gain stages used were a single common-source stage, a cascode stage, and a Darlington stage. The results in Figure 19.13 show that with a cascode gain stage, it is possible to achieve a gain bandwidth product of about 55% of the f_{max} of the process, whereas for GaAs, the unity gain bandwidth can be as high as 80% of the process f_{max}. Silicon is worse due to the lossy substrate in comparison to gallium arsenide processes. The results also show that the simple common-source amplifier and the Darlington stage cannot achieve the same performance as the cascode gain stage.

Using the previously mentioned improved transmission lines as series inductors in an artificial transmission line, distributed amplifiers in 0.35 μm CMOS were designed and simulated, and the results are shown in Figure 19.14.

We can see that the resulting bandwidth is improved from about 7 GHz to about 10 GHz and the gain is also improved.

Finally, to see how far coplanar lines can help to push frequency, a distributed amplifier was designed in standard 0.18 μm CMOS using standard cascode amplifiers. Layout is shown in Figure 19.15.

Measured and simulated results as shown in Figure 19.16 are in reasonable agreement once bond pads are extracted using de-embedding.

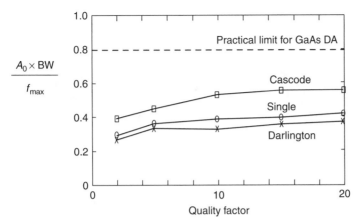

FIGURE 19.13   Distributed amplifier performance limitation.

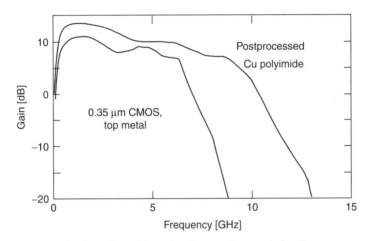

FIGURE 19.14   Improved distributed amplifier using improved transmission lines.

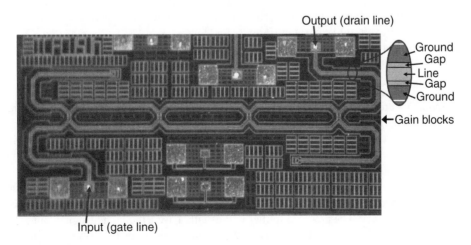

FIGURE 19.15   Photomicrograph of DA in 0.18 μm CMOS.

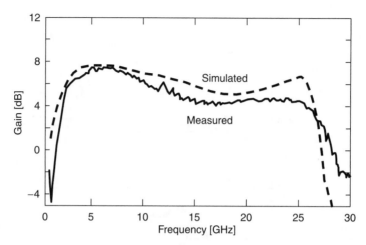

FIGURE 19.16   Measured response of DA in 0.18 μm CMOS.

COUPLING AND ITS MEASUREMENT

Any components can couple to each other, either directly or through common substrate or common biases or supplies, and there can be coupling between bond wires and to the external world. Coupling between digital and analog sections on the same integrated circuit can occur due to switching noise being injected into the common substrate, or on to power supply lines. In this section, the emphasis will be on coupling due to on-chip inductors. Modern integrated radio frequency (RF) circuits can include a large number of inductors, so coupling between them may be significant. This can apply to wireless circuits [1] as well as wireline circuits [5] where inductors could be used in clock and data recovery circuits, laser modulator/driver, or to provide peaking, for example, in limiting amplifiers.

Typical steps taken to reduce coupling are to provide separation and isolation, multiple grounds, and separate supplies. Numerous studies have been done on coupling and its mitigation. Only a few recent references are given here [6–14], but these also refer to numerous earlier publications. Some techniques to reduce coupling are as follows:

1. Place transistors that are particularly sensitive in their own deep n-wells where the process allows it.
2. Use guard rings and guard bands, in some cases connected to separate bond pads.
3. Through circuit design techniques such as the use of differential structures.
4. Through careful circuit layout, for example, for symmetry.
5. Through adjustment of substrate resistivity; and through active noise shaping.

INDUCTORS AS A CAUSE OF COUPLING

On-chip inductors are one of the major causes of coupling. Inductors have large area and thus can inject significant charge into a resistive substrate. In addition, there is EM coupling between inductors and other components, especially other inductors. At the short distance seen on integrated circuits, such coupling is often (not surprisingly) labeled as inductive coupling. On-chip coupling can result in a number of problems. Such problems include noise injection, pulling, injection locking, phase noise, and change of performance. An example application where coupling is especially crucial is in direct conversion transceivers because the power amplifier and a local oscillator could be operating at the same frequency. The power amplifier will usually have a modulated signal on it, whereas the oscillator is typically a single tone; thus, any signal coupled from the power amplifier to the oscillator can cause the frequency to shift and this would be seen as phase noise. It is difficult to

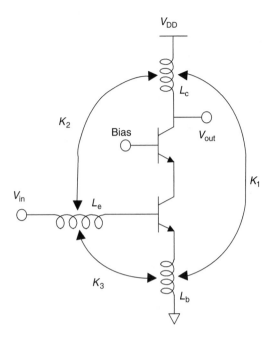

FIGURE 19.17 LNA with coupling between collector, base, and emitter inductors.

incorporate coupling model into system simulation. In this section, we will explore the use of on-chip oscillators to measure coupling.

LNA COUPLING EXAMPLE

A low-noise amplifier (LNA) as shown in Figure 19.17 can have inductors in series with base, collector, and emitter, and coupling can occur between any pair of inductors.

If the LNA is initially designed without consideration of coupling, then adding a coupling coefficient k of 0.015 results in a significant change of performance as shown in Figure 19.18. We note that this high value of coupling coefficient is possible with closely spaced inductors, so the risk is real. Knowing how severe the effect of coupling can be, in this example, a few simple techniques, such as separating the inductors and placing substrate contacts between the inductors, can greatly reduce these effects. It is also possible to design for deliberate coupling between the drain and source inductors. Successful amplifiers have been built where these two inductors are the windings of a transformer. The advantage is that amplifier parameters are less dependent on transistor parameters that are prone to process variations [15].

OSCILLATOR-TO-OSCILLATOR COUPLING

One of the simplest ways to demonstrate and to measure coupling is with two identical voltage-controlled oscillators (VCOs) as shown in Figure 19.19 [16]. The first oscillator VCO_1 is injection locked to an externally applied signal, at some frequency offset from the free-running frequency. The second oscillator VCO_2 operates at its free-running frequency, but the signal from the first oscillator will appear as sidebands around the oscillating frequency and the relative amplitude of the oscillating signal and the sidebands will be an indication of the coupling. In this case, VCO_2 is operating as a high-gain filter, and the outline of the phase noise is also an indication of its gain to an injected signal. To appreciate that oscillators have very high gain, consider that the input is noise whose available power is of the order of kT, or $-174\,dBm$ in a 1 Hz bandwidth. The oscillator outputs could be of the order of 0 dBm; thus, there is about 170 dB of gain to the noise. By design (and verified by various measurements) the VCOs

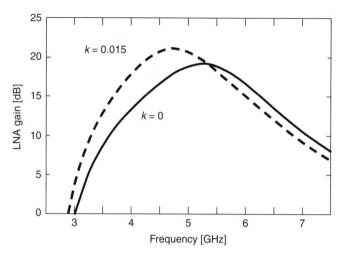

FIGURE 19.18   LNA response with and without coupling.

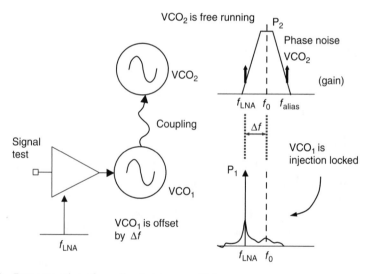

FIGURE 19.19   Demonstration of coupling between oscillators.

are identical and the output powers remained roughly the same, whether injection locked or free running. This is important as it allows measurements with no calibration needed.

It should be noted that injection-locked oscillators can be successfully modeled with a simple transconductance feedback model shown in Figure 19.20. As in all oscillators, to reach a steady amplitude, a nonlinear component is needed and this can be modeled with a nonlinear transconductance. The simplest technique is to combine the fundamental with a third-order component, for example, with transconductance $g_m = i/v_{in}$ defined by $i = 0.005v_{in} + 0.0005v_{in}^3$. Noise can be fed in as a time domain sequence of numbers having the correct power spectral density. The noise can be verified in the time and frequency domain. With the above example transconductance, the oscillator starts up successfully with output voltage going to approximately plus and minus 2.5 V and behaves in many other ways like a real oscillator.

Injection and injection locking can be demonstrated by adding a tonal frequency component into the resonant circuit through the current source labeled i_{inj} in Figure 19.20. Both real oscillators and the model show similar results. For small injected signals, or signals injected far from the free-running

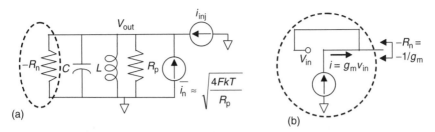

FIGURE 19.20 Oscillator model (a) with a negative resistance, and (b) realization of negative resistance with a transconductor with positive feedback.

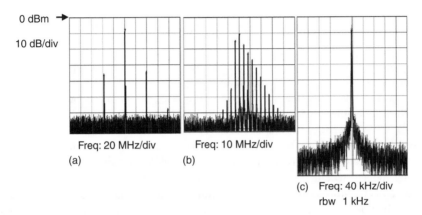

FIGURE 19.21 Measured spectrum of oscillator (a) with injected tone far from injection locking, (b) close to injection locking, and (c) injection locked.

carrier frequency, a single coupled tone and an aliased tone are seen as in Figure 19.21a. If the injected signals are larger, or closer to the free-running frequency, but prior to injection locking, the coupled tone is pulling on the free-running tone and moving it toward the coupled tone. As a result the free-running tone is frequency modulated, resulting in an output spectrum with a number of components as shown in Figure 19.21b. However, it is not a sinusoidal modulation, but is asymmetrically pulled in one direction; hence, there is an asymmetric spectrum of tones. Finally, if the injected signal is sufficiently large and close to the free-running frequency, the oscillator becomes injection locked as shown in Figure 19.21c. Clearly, in all these cases, coupling has resulted in additional signals to the oscillator spectrum. This is typically avoided, except in the special case where the intent is to design an injection-locked oscillator, to be discussed later.

So, what are the levels and frequency offsets required to produce coupling and injection locking? This question has been addressed by Adler [17] and can be explained by treating the oscillator, shown in Figure 19.20, as a high-gain amplifier with a band-pass frequency response shaped by the LC resonator, but because of the active positive feedback circuit in the oscillator, the resonator Q is enhanced to approximately infinity. Thus, the coupled tone will be larger if it is closer in frequency to the free-running carrier frequency, for example, halving the frequency offset will result in double the coupled amplitude, equivalent to 6 dB. Roughly where the amplitude of the coupled signal matches that of the free-running oscillator, the oscillator will follow the injected signal. By a straightforward analysis of Figure 19.20, the amplitude of the coupled tone as a function of frequency can be shown to be

$$v_{out} = \frac{i_{inj}}{\left[\dfrac{1}{R_p} + j\omega C\left(1 - \left(\dfrac{\omega_o}{\omega}\right)^2\right)\right] - \dfrac{1}{R_n}} \tag{19.10}$$

If the offset $\Delta\omega$ from the resonant frequency is small relative to ω_o, a simplified expression for v_{out} in terms of Q_U, the quality factor of the unloaded tank circuit, can be shown to be

$$v_{out} \approx \frac{i_{inj}R_p}{1 - g_m R_p + j2Q_U \dfrac{\Delta\omega}{\omega_o}} \tag{19.11}$$

At resonance, the positive feedback results in the approximate cancellation of the negative and positive resistance; thus, $1 - g_m R_p$ is very small. And since Q_U is equal to $R_p/\omega L$, where ω is approximately equal to ω_o, a very simple expression for the amplitude is given by

$$v_{out} \approx \frac{i_{inj}}{j2C\Delta\omega} \tag{19.12}$$

The onset of injection locking can be estimated as occurring when this coupled amplitude is comparable to the oscillator free-running amplitude.

Coupling current into an oscillator and having it produce output tones is only part of the complete coupling picture. The other part is the EM and substrate coupling between the source of the signal and the oscillator. For this part, EM simulations can be performed. Or, alternatively, measurements can be performed, for example, with two oscillators on the same chip, as suggested in Figure 19.19.

EM simulations can be used to determine a model of coupled inductors as shown in Figure 19.22. The particular inductors being modeled in the figure were symmetrical with an outside diameter of 200 μm and a center-to-center spacing of 175 μm [18]. Such models can be used in a simulator to predict coupling between the two inductors, or by adding capacitors, they can be used to predict coupling between resonators.

As an alternative to circuit simulations, EM simulations (including vias) can be used to predict coupling directly, with results for the same inductor as in Figure 19.22 shown in Figure 19.23. At low frequencies, coupling increases linearly due to mutual inductance, and at high frequencies (beyond the simulated results shown in Figure 19.23), capacitance between the ports and the ground causes the coupling to drop. With oscillators built with these inductors as in the schematic of Figure 19.19, oscillating frequency is predicted to be 3.5 GHz. At 3.5 GHz, the coupling is predicted to be −36 dB.

Simulations were also done for coupling versus separation distance for the previously discussed inductor and for a much simpler inductor, with results shown in Figure 19.24. The large range of results shows the importance of getting the simulation done correctly. For example, the symmetrical differential inductors have underpasses and vias on the sides facing each other, which appears to make a huge difference in the amount of coupling.

To use two on-chip oscillators to measure coupling, the gain of the oscillator as an amplifier must be determined. Starting from Equation 19.12 and noting that the equivalent injected input voltage i_{inj} is multiplied by R_p the equivalent resonator parallel resistance (refer to Figure 19.20), the transfer function can be written as

$$\frac{v_{out}}{v_{inj}} \approx \frac{1}{j2CR_p\Delta\omega} = \frac{1}{j\Delta\omega} \cdot \frac{B_U}{2} \tag{19.13}$$

where B_U is the bandwidth of the resonator and is given by $1/(R_p C)$ or equivalently by ω_o/Q_U. We note that R_p should include the input resistance of the transconductor, and any other parallel resistance. In Figure 19.20, no additional resistance is shown, but in our experiment, it was assumed that the additional resistance was about equal to the parallel resistance of the inductor; hence, the equivalent Q of the inductor was reduced by a factor of 2, to about 3.15. Thus, for the 3.5 GHz oscillator, B_U was 1.1 GHz and the equivalent oscillator gain drops down to unity at an offset of 550 MHz ($B_U/2$).

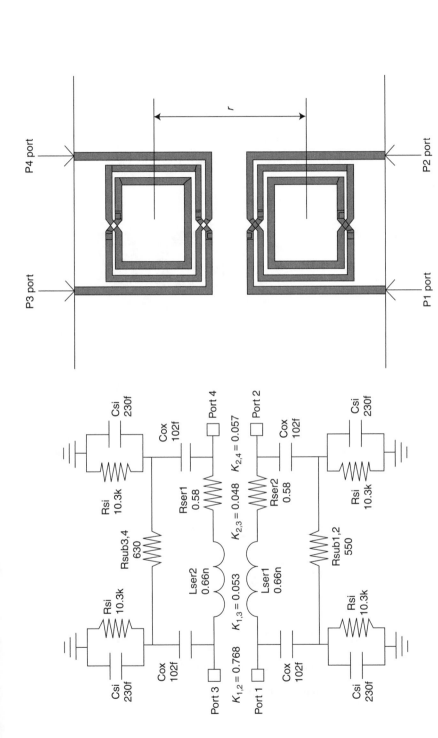

FIGURE 19.22 An inductor model determined by EM simulation of an inductor with outside diameter of 200 μm and center-to-center spacing of 150 μm. (From Amaya, R. E. et al., *IEEE J. Solid-State Circuits*, 40, 1968, 2005. With permission. © [2005] IEEE.)

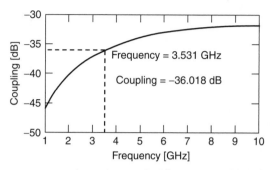

FIGURE 19.23 Simulated inductor coupling as a function of frequency. This is for the inductors shown in Figure 19.22.

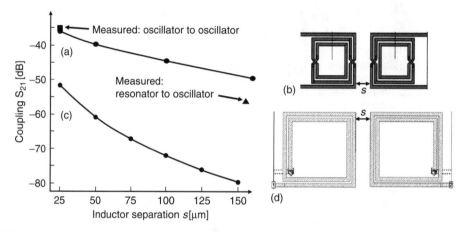

FIGURE 19.24 Simulated inductor coupling versus separation. (a) Simulated coupling curve of symmetrical differential inductor shown in (b). This includes vias. (c) Simulated coupling curve of simple square inductor shown in (d).

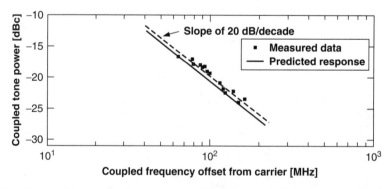

FIGURE 19.25 Measured data and predicted response. (From Amaya, R. E. et al., *IEEE J. Solid-State Circuits*, 40, 1968, 2005. With permission. © [2005] IEEE.)

Thus, at 550 MHz offset, the expected coupling is −36 dBc, increasing at 20 dB/decade closer in to the carrier as shown in Figure 19.25. Measurement results were taken by injection locking VCO_1 (refer to Figure 19.19) to an external signal at an offset of Δf away from VCO_2 at f_0. The measured ratio between the tone at f_0 and the tone at $f_0 - \Delta f$ is indicative of the coupling, and this is also plotted in

Figure 19.25. Measurements with Δf between 64 and 180 MHz agree with the predictions. Measured coupling is ~−35 dBc, 1 dB higher than predicted.

OPPORTUNITIES WITH COUPLING

Coupling can be reduced by separation and by isolation through ground and substrate connections and by deep trenches. However, coupling cannot be completely eliminated. It is also possible to take advantage of coupling, and indeed to have deliberate coupling. Examples include the design of transformers where it is desired to couple signals from a primary to a secondary coil, or in injection-locked oscillators and synthesizers.

TRANSFORMERS WITH IMPROVED COUPLING

Coupling in transformers is achieved by wires that are adjacent to each other. Improved coupling can be achieved through the use of multiturn, multilayer transformers. An example of a fully differential transformer with two layers and two turns is shown in Figure 19.26. In such a case, there is coupling in both horizontal and vertical directions. This reduces the area and the substrate capacitance. However, multilayer transformers (or inductors) mean the bottom layer is closer to the substrate; hence, capacitance is increased. However, if the area reduction is larger than the reduction in distance to the substrate, the total capacitance is reduced. The result is that the self-resonance frequency is increased.

The primary coil, here shown connected to a source, consists of the outer turn of the top layer and the inner turn of the bottom layer whereas the secondary coil, connected to the load resistor, consists of the other two layers. Figure 19.26 shows a transformer with one turn/layer and four layers. As shown, there are two ways to connect the secondary coil. The first, the noninverting configuration, is by grounding the bottom layer, effectively shorting the capacitance from the bottom turn to the substrate. The second configuration is the inverting configuration in which the output is from the bottom plate. Because the bottom plate capacitance is not removed, this configuration is expected to have lower self-resonance frequency.

Transformers such as that shown in Figure 19.26 were fabricated and measured, with results shown in Figure 19.27 and summarized in Table 19.1 [19]. The highest peak Q and self-resonance frequencies were obtained with the two-layer, two-turn transformers. Of these two, the noninverting configuration had higher frequencies due to the bottom capacitance being shorted out, however, with a reduced peak gain (lower S_{21} and coupling coefficient k). The best gain and coupling coefficient was

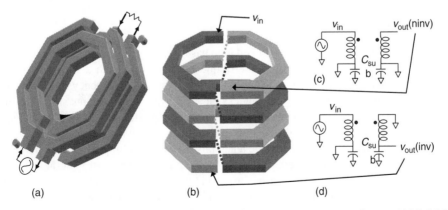

FIGURE 19.26   (a) Two-turn, two-layer transformer. (b) One-turn, four-layer transformer. (c) Model for noninverting configuration with the bottom of the secondary coil grounded. (d) Model for inverting configuration with the output at the bottom of the secondary coil.

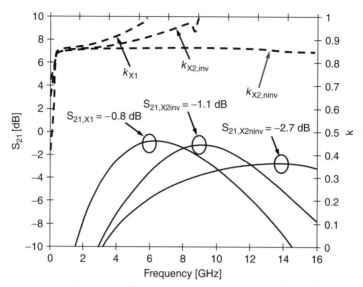

FIGURE 19.27   Measurement results for multiturn, multilayer transformers. X1 is differential two-turn, two-layer transformer. X2 is a one-turn, four-layer transformer. (From Fong, N. et al., *Proc. IEEE Microwave Theory and Techniques Society International Microwave Symposium*, Philadelphia, June 2003, p. 969. With permission. © [2003] IEEE.)

TABLE 19.1
Transformer Performance

Ref cfg.	m	$n_1{:}n_2$	OD (µm)	d (µm)	L_P (nH) (Q)	S_{21} (dB)	k	f (GHz)	f_{res} (GHz)
X1 inv	2	2:2	200	7.8	1 (7.0)	−0.8	0.96	5.0	6.0
X2 inv	4	2:2	108	4.6	0.5 (6.5)	−1.1	0.90	5.4	8.5
X2 ninv	4	2:2	108	4.6	0.5 (10.5)	−2.7	0.86	14	>15

obtained with the two-layer, two-turn differential transformer S_{21} of −0.8 dB measured at 5 GHz. Both inverting and non-inverting four-layer transformers are usable up to 14 GHz. Another important point with these transformers is that the gain is relatively high across several GHz of bandwidth.

INJECTION-LOCKED OSCILLATOR

Most of the power dissipation in a phase-locked loop based frequency synthesizer is due to the high-frequency components, the oscillator, and the frequency divider. In such a system, if the divider divides by N, the oscillator frequency is N times higher than the input frequency. In comparison, it is possible to design an oscillator that injection locks on to a harmonic of the input signal. If this is the Nth harmonic, then this will function in a similar manner to the phase-locked loop with a divide-by-N in the feedback path, but the power dissipation due to the divider will have been eliminated. This is not completely a free ride, as there may be a need for an additional input buffer to make sure the input signal is large enough, or limited, to ensure that there is sufficient energy at the Nth harmonic to lock the oscillator. Ideally, a square wave is used, and in such a case the Nth harmonic component (assuming it is an odd harmonic) has an amplitude that is $1/N$ of the fundamental component. Hence, it may become more and more difficult to lock on to higher harmonics. Figure 19.28 shows an example of an injection-locked oscillator with the input signal being amplified and injected into the tank circuit of the oscillator. With this simple circuit, injection locking will occur as long as the input signal is strong

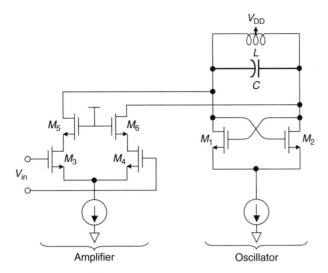

FIGURE 19.28 Injection-locked oscillator.

enough, as determined by Equations 19.12 and 19.13. The equations show that the required input signal level is inversely proportional to the offset frequency. Thus, for the earlier example where the resonator bandwidth was 1.1 GHz ($B_U/2$ is 550 MHz), if the oscillator output is at 0 dBm and the offset frequency is less than 5.5 MHz, the input signal can be smaller than −40 dBm. With a free-running oscillator frequency of 3.5 GHz, with an accuracy of 10%, the frequency can vary by up to 35 MHz; thus, some tuning will likely be necessary if a low-power input amplifier is desired. To provide tunability, the capacitors are typically replaced with varactors, and possibly even switched capacitors.

A number of variations are possible, for example, the amplifier can be ac coupled to the oscillator. It is also possible to inject a signal into the current source rather than directly into the resonant tank circuit. Also, as mentioned earlier, the oscillator can be designed to lock on to a harmonic of the input signal, but care must be taken that the desired harmonic is strong enough and it falls close to the free-running oscillator frequency.

Injection-Locked Oscillator Example

A 1.1 GHz injection-locked oscillator that locks on to the 11th harmonic of a 100 MHz input signal in a 0.18 μm CMOS process is described in Ref. 20. Measurements show that if the free-running frequency is within 300 kHz of the input-injected signal, then −44 dBm is sufficient for injection locking. To ensure that this condition is met, digital pretuning is provided with an on-chip microprocessor. Total layout area is 0.234 mm², and total power dissipation for the injection amplifier and the oscillator is 688 μW. Free-running phase noise is about −80 dBc/Hz at an offset frequency of 50 kHz. After injection locking, phase noise is improved by about 20 dB to about −100 dBc/Hz at 50 kHz offset.

Other injection-locked oscillators have been described in Refs. 21–23. Other applications of injection locking are as injection-locked frequency dividers [24–27], injection-locked frequency multipliers [28], injection-locked oscillators as amplifiers [29–31], injection-locked oscillators as limiters [32], injection-locked oscillator for clock and data recovery [33–35], and finally, injection-locked quadrature oscillators [36–39].

CONCLUSIONS

In this chapter, distributed effects and coupling have been discussed. Distributed effects result in delay. These can be minimized with transmission lines, for which improvements in quality can be achieved with coplanar lines, lower-*k* dielectric, and elevated lines. Furthermore, it is possible to take advantage of distributed effects with distributed amplifiers. Coupling can result in noise and change of performance. Coupling can be analyzed through the use of EM simulators, and can be measured, for example, with on-chip oscillators. With this technique, no calibration is required. Applications of coupling have been shown, which include multiturn, multilayer transformers and injection-locked oscillators.

ACKNOWLEDGMENTS

The author would like to acknowledge the fabrication work by CMC, Nortel, and IBM, funding by Micronet and NSERC, and the direct involvement of students R. E. Amaya, P. H. R. Popplewell, J. Aguirre, N. Fong, and H. Ahmed.

REFERENCES

1. T. Sowlati, D. Rozenblit, R. Pullela, M. Damgaard, E. McCarthy, D.o Koh, D. Ripley, F. Balteanu, and I. Gheorghe, "Quad-band GSM/GPRS/EDGE polar loop transmitter," *IEEE J. Solid-State Circuits*, vol. 39, no. 12, Dec. 2004, pp. 2179–2189.
2. R.E. Amaya, V. Levenets, C. Plett, and N.G. Tarr, "Copper coplanar waveguides on silicon substrates for frequencies up to 40 GHz," *Proc. IEEE International Symposium on Antenna and Applied Electromagnetics*, Ottawa, Canada, July 2004, pp. 51–54.
3. D.G. Sarma, "On distributed amplification," *Proc. Inst. Elect. Eng.*, vol. 102B, 1954, pp. 689–697.
4. R.E. Amaya, J. Aguirre, and C. Plett, "Gain bandwidth considerations in fully integrated distributed amplifiers implemented in silicon," *Proc. IEEE International Symposium on Circuits and Systems*, Vancouver, Canada, May 2004, vol. IV, pp. 273–276.
5. S. Galal and B. Razavi, "10-Gb/s limiting amplifier and laser/modulator driver in 0.18 μm CMOS technology," *IEEE International Solid-State Circuits Conference*, Dig. Tech. Papers, San Francisco, USA, Feb. 2003, pp. 188–189.
6. T. Kadoyama, N. Suzuki, N. Sasho, H. Iizuka, I. Nagase, H. Usukubo, and M. Katakura, "A complete single-chip GPS receiver with 1.6-V 24-mW radio in 0.18 μm CMOS," *IEEE J. Solid-State Circuits*, vol. 39, no. 4, Apr. 2004, pp. 562–568.
7. M.S. Peng and H.-S. Lee, "Study of substrate noise and techniques for minimization," *IEEE J. Solid-State Circuits*, vol. 39, no. 12, Nov. 2004, pp. 2080–2086.
8. M. Pfost, P. Brenner, T. Huttner, and A. Romanyuk, "An experimental study on substrate coupling in bipolar/BiCMOS technologies," *IEEE J. Solid-State Circuits*, vol. 39, no. 10, Oct. 2004, pp. 1755–1763.
9. Y.L. Guillou, O. Gaborieau, P. Gamand, M. Isberg, P. Jakobsson, L. Jonsson, D.L. Déaut, H. Marie, S. Mattisson, L. Monge, T. Olsson, S. Prouet, and T. Tired, "Highly integrated direct conversion receiver for GSM/GPRS/EDGE with on-chip 84-dB dynamic range continuous-time $\Sigma\Delta$ ADC," *J. Solid-State Circuits*, vol. 40, no. 2, Feb. 2005, pp. 403–501.
10. B.E. Owens, S. Adluri, P. Birrer, R. Shreeve, S.K. Arunachalam, K. Mayaram, and T.S. Fiez, "Simulation and measurement of supply and substrate noise in mixed-signal ICs," *IEEE J. Solid-State Circuits*, vol. 40, no. 2, Feb. 2005, pp. 382–391.
11. W. Steiner, H.-M. Rein, and J. Berntgen, "Substrate coupling in a high-gain 30-Gb/s SiGe amplifier— modeling suppression and measurement," *IEEE J. Solid-State Circuits*, vol. 40, no. 10, Oct. 2005, pp. 2035–2045.
12. S. Kristiansson, F. Ingvarson, S.P. Kagganti, N. Simic, M. Zgrda, and K.O. Jeppson, "Surface potential model for predicting substrate noise coupling in integrated circuits," *IEEE J. Solid-State Circuits*, vol. 40, no. 9, Sept. 2005, pp. 1797–1803.
13. W. Xu and E.G. Friedman, "On-chip test circuit for measuring substrate and line-to-line coupling noise," *IEEE J. Solid-State Circuits*, vol. 41, no. 2, Feb. 2006, pp. 474–482.
14. S. Hazenboom, T.S. Fiez, and K. Mayaram, "A comparison of substrate noise coupling in lightly and heavily doped CMOS for 2.4-GHz LNAs," *IEEE J. Solid-State Circuits*, vol. 41, no. 3, Mar. 2006, pp. 574–587.
15. D.J. Cassan and J.R. Long, "A 1-V transformer-feedback low-noise amplifier for 5-GHz wireless LAN in 0.18-μm CMOS," *IEEE J. Solid-State Circuits*, vol. 38, no. 3, Mar. 2003, pp. 427–435.

16. P.H.R. Popplewell, R.E. Amaya, M. Cloutier, and C. Plett, "Calibration-free on-chip inductor coupling experiment with injection lockable VCOs," *Proc. IEEE Bipolar/BiCMOS Circuits and Technology Meeting*, Montreal, Canada, Sept. 2004, pp. 261–264.

17. R. Adler, "A study of locking phenomena in oscillators," *Proc. IRE*, vol. 34, June 1946, pp. 351–357.

18. R.E. Amaya, P.H.R. Popplewell, M. Cloutier, and C. Plett, "EM and substrate coupling in silicon RFICs," *IEEE J. Solid-State Circuits*, vol. 40, no. 9, Sept. 2005, pp. 1968–1971.

19. N. Fong, J.-O. Plouchart, N. Zamdmer, J. Kim, K. Jenkins, C. Plett, and N.G. Tarr, "High-performance and area-efficient stacked transformers for RF CMOS integrated circuits," *Proc. IEEE Microwave Theory and Techniques Society International Microwave Symposium*, Philadelphia, USA, June 2003, pp. 967–970.

20. H. Ahmed, C. DeVries, and R. Mason, "A digitally tuned 1.1 GHz subharmonic injection locked VCO in 0.18 μm CMOS," *Proc. European Solid-State Circuits Conference*, Estoril, Portugal, Sept. 2003, pp. 81–84.

21. S. Ye, L. Jansson, and I. Galton, "A multiple-crystal interface PLL with VCO realignment to reduce phase noise," *IEEE J. Solid-State Circuits*, vol. 37, no. 12, Dec. 2002, pp. 1795–2003.

22. B. Razavi, "A study of injection locking and pulling in oscillators," *IEEE J. Solid-State Circuits*, vol. 39, no. 9, Sept. 2004, pp. 1415–1424.

23. F. Kocer and M.P. Flynn, "A new transponder architecture with on-chip ADC for long-range telemetry applications," *IEEE J. Solid-State Circuits*, vol. 41, no. 5, May 2006, pp. 1142–1148.

24. H.R. Rategh, H. Samavati, and T.H. Lee, "A CMOS frequency synthesizer with an injection-locked frequency divider for a 5-GHz wireless LAN receiver," *IEEE J. Solid-State Circuits*, vol. 35, no. 5, May 2000, pp. 780–787.

25. A. Mazzanti, P. Uggetti, and F. Svelto, "Analysis and design of injection-locked LC dividers for quadrature generation," *IEEE J. Solid-State Circuits*, vol. 39, no. 9, Sept. 2004, pp. 1425–1433.

26. K. Yamamoto and M. Fujishima, "A 44-μW 4.3-GHz injection-locked frequency divider with 2.3-GHz locking range," *IEEE J. Solid-State Circuits*, vol. 40, no. 3, Mar. 2005, pp. 671–677.

27. J. Jeong and Y. Kwon, "A fully integrated V-band PLL MMIC using 0.15-μm GaAs pHEMT technology," *IEEE J. Solid-State Circuits*, vol. 41, no. 5, May 2006, pp. 1042–1050.

28. J.P. Maligeorgos and J.R. Long, "A low-voltage 5.1–5.8-GHz image-reject receiver with wide dynamic range," *IEEE J. Solid-State Circuits*, vol. 35, no. 12, Dec. 2000, pp. 1917–1926.

29. T. Isobe and M. Tokida, "A new microwave amplifier for multichannel FM signals using a synchronized oscillator," *IEEE J. Solid-State Circuits*, vol. 4, no. 6, Dec. 1969, pp. 400–408.

30. E. Marazzi and A. Bellardo, "Thin-film injection-locked oscillators and negative-resistance amplifiers for 2-GHz radio repeater," *IEEE J. Solid-State Circuits*, vol. 7, no. 1, Feb. 1972, pp. 23–32.

31. D.C. Hanson and W.W. Heinz, "Integrated electrically tuned X-band power amplifier utilizing Gunn and IMPATT diodes," *IEEE J. Solid-State Circuits*, vol. 8, no. 1, Feb. 1973, pp. 3–14.

32. J.H. Johnson, R.C. Shaw, and H.L. Stover, "Wide-band oscillating limiters at X-band and L-band frequencies," *IEEE J. Solid-State Circuits*, vol. 3, no. 2, June 1968, pp. 163–165.

33. B.S. Glance, "Minimum required power for carrier recovery at optical frequencies," *IEEE J. Lightwave Technol.*, vol. 4, no. 3, Mar. 1986, pp. 349–355.

34. H. Izadpanah and P.M. Crespo, "A microwave injection-locking technique for direct timing extraction in multigigabit-per-second fiber-optic data links," *IEEE J. Lightwave Technol.*, vol. 8, no. 10, Oct. 1990, pp. 1435–1450.

35. H.-T. Ng, R. Farjad-Rad, M.-J.E. Lee, W.J. Dally, T. Greer, J. Poulton, J.H. Edmondson, R. Rathi, and R. Senthinathan, "A second-order semidigital clock recovery circuit based on injection locking," *IEEE J. Solid-State Circuits*, vol. 38, no. 12, Dec. 2003, pp. 2101–2110.

36. D.K. Ma and J.R. Long, "A subharmonically injected LC delay line oscillator for 17-GHz quadrature generation," *IEEE J. Solid-State Circuits*, vol. 39, no. 9, Sept. 2004, pp. 1434–1445.

37. S.Y. Yue, D.K. Ma, and J.R. Long, "A 17.1–17.3-GHz image-reject downconverter with phase-tunable LO using 3× subharmonic injection locking," *IEEE J. Solid-State Circuits*, vol. 39, no. 12, Dec. 2004, pp. 2321–2332.

38. F. Gatta, D. Manstretta, P. Rossi, and R. Svelto, "A fully integrated 0.18-mm CMOS direct conversion receiver front-end with on-chip LO for UMTS," *IEEE J. Solid-State Circuits*, vol. 39, no. 1, Jan. 2004, pp. 15–23.

39. P. Kinget, R. Melville, D. Long, and V. Gopinathan, "An injection-locking scheme for precision quadrature generation," *IEEE J. Solid-State Circuits*, vol. 37, no. 7, July 2002, pp. 845–851.

20 Substrate Noise Coupling from Digital to Analog Circuits in Mixed-Signal Integrated Circuits

Piet Wambacq, Charlotte Soens, Geert Van der Plas, Mustafa Badaroglu, and Stéphane Donnay

CONTENTS

Introduction..567
Modeling Substrate Noise Generation 569
Substrate Noise Propagation and Impact Model..572
    Propagation of Noise through the Substrate...573
    Entry Points for Substrate Noise in an Analog Circuit..574
    Computation of the Circuit Response due to Substrate Noise..............................574
Experimental Verification of the Model 576
    GHz LC-Tank VCO 576
    Measurement of the Digital Ground Bounce...577
    Impact of Digital Noise on the LC-VCO 578
Reduction of Substrate Noise Impact Using Guard Rings 580
    Guard Ring Modeling...581
    Design Guidelines for Guard Rings 584
Conclusions..586
References..586

INTRODUCTION

An important problem that arises during the design of a single-chip mixed-signal radio is the crosstalk from digital to analog. Simulation of this crosstalk at the circuit level for large systems on chip (SoC) does not provide the necessary insight to resolve it. Moreover, performing circuit simulations to investigate the crosstalk in a large SoC is not a workable solution. First, the number of digital gates is so large that it is not practical to include them in analog circuit simulations. In addition, the crosstalk occurs via the silicon substrate, the PCB, the bonding wires, the package, and the ground and supply lines. All these parasitic paths have to be included in the circuit simulations as well, resulting in extremely large simulation models. Analysis at a higher level of abstraction is more practical and gives much more insight into the coupling mechanisms.

The problem of reducing the impact of substrate noise on analog radio frequency (RF) circuits has already received much attention in literature. Because of the lack of systematic and reliable simulation methods, designers are left with an intuitive and *ad hoc* approach to solve the substrate noise coupling problems. For example, in the single-chip Bluetooth system discussed in Ref. 1, the radio has been carefully isolated with a 300 μm wide p$^+$ guard ring connected to ground with 13 low-impedance bumps. It is estimated that the isolation structure reduces substrate noise coupling by 25 dB at 2.5 GHz. Although the system is functional, it is not clear if this guard ring is performing as intended. Similarly, in Ref. 2, the integration of a radio onto a Pentium die is investigated. A noise transfer function analysis reveals that the use of deep n-well biasing-based isolation in conjunction with differential circuit design is sufficient for radio integration. However, no functional system has been built to verify this analysis.

The largest modeling challenge for substrate noise impact is in lightly doped substrates that are nowadays used in almost any bulk CMOS technology. Unlike the epi-type substrates, which can be easily modeled as a single equipotential node, lightly doped substrates have to be modeled as a three-dimensional RC-mesh. Hence, propagation of noise in a lightly doped substrate is a three-dimensional, layout-dependent phenomenon. Lightly doped substrates are nowadays more widely used than highly doped substrates.

The substrate noise problem can be divided into three aspects: generation of the substrate noise, propagation through the substrate, and impact on the analog/RF circuits.

First, there is the aspect of the noise generation in the digital circuitry. As will be discussed in the section "Modeling Substrate Noise Generation," a prediction of the amount of generated noise with reasonable accuracy requires macromodeling at a higher level than the transistor level, since a detailed circuit-level simulation of the noise generation is only feasible in a reasonable time for very small designs, with a gate count below 1000. Such macromodeling requires a good understanding of the different generation mechanisms of substrate noise. These mechanisms are discussed in the next section. In combination with a model for the assembly parasitics, substrate noise generation macromodels can predict the noise generated by digital circuits of realistic size with accuracy of about 20% compared to measurements. In Refs. 3–6, macromodels for the generation of digital switching noise are presented and validated with measurements, for both epi-type and lightly doped substrates.

Once it is possible to predict the generated noise, different design techniques at the digital side can be evaluated to lower the amount of generated noise [6]. Such techniques will be highly desirable, as scaling will not relieve the problem of substrate noise generation [7].

Next, there is the propagation of the noise through the substrate. To model this, several programs [8–10] exist that generate a three-dimensional resistive or RC network for the substrate.

Finally, there is the aspect of impact of substrate noise on analog/RF circuits. Here, a careful modeling of the assembly characteristics and layout details is required to bring simulations and measurements in agreement. Concerning the impact on analog circuits, mostly circuit-level simulations ([11] on a low-noise amplifier) and interpretation of measurements ([12] on an analog-to-digital converter) have been reported. Refs. 13 and 14 derive an analytical model for substrate noise impact on ring oscillators. The analytical model is validated for epi-type substrates only and does not include a model for the digital noise generation. Moreover, it does not provide much information on the exact coupling mechanisms. As will be shown in the section "Substrate Noise Propagation and Impact Model," substrate noise often impacts the analog circuits via different ways. If these are not well modeled, then it is not clear how this impact can be suppressed during circuit design.

In this chapter, we present a linearized physical model [35] that successfully predicts both the substrate noise generation by a digital modem and the resulting performance degradation for LC-tank voltage controlled oscillators (VCOs) on a lightly doped substrate. This model is valid for both epi-type and lightly doped substrates. The model is validated with measurements on a system containing a functional 40 kgate digital modem and a 3.5 GHz LC-VCO on the same lightly doped substrate. Two designs are used: a reference design and a design containing a p$^+$/n-well guard ring.

In both cases, the model accurately predicts the level of the spurious components appearing at the VCO output due to the digital switching activity. The error remains smaller than 3 dB.

This chapter is organized as follows. The generation of substrate noise and the corresponding macromodeling are discussed in the section "Modeling Substrate Noise Generation." Substrate noise propagation and modeling of the impact of substrate noise on analog circuits are discussed in the section "Substrate Noise Propagation and Impact Model." In the section "Experimental Verification of the Model," the modeling of generation and impact are combined and applied to a mixed-signal chip. The accuracy of the model is also verified with experimental measurements. The section "Reduction of Substrate Noise Impact Using Guard Rings" discusses guard rings that are a means to suppress the impact or generation of substrate noise. A model for the guard ring is constructed. Then this is used to design guard rings.

MODELING SUBSTRATE NOISE GENERATION

Three mechanisms exist that generate substrate noise [7]. The first mechanism is caused by impact ionization: hot electrons that travel from source to drain in a switching n-MOS transistor create hole–electron pairs. The resulting hole current flows into the substrate. The second mechanism is the current injected into the bulk of a transistor via its drain-bulk or source-bulk junctions when the voltage at the drain/source switches. It can be shown [7] that in present and future CMOS technologies (with a high-ohmic substrate) these two mechanisms are negligible compared to the third mechanism, which is bounce on the digital supply lines that is injected into the substrate via many ground contacts.

A circuit-level simulation of the generation of substrate noise would require a transient simulation (e.g., with SPICE or SPECTRE) of the digital circuit extended with a model for the substrate. Such simulations can only be performed in a reasonable CPU time for digital circuits of very limited size (<1000 gates).

To tackle the complexity of the generation problem, different strategies have been presented [5,15–18] to model the generation of substrate noise at the gate level. The idea is as follows: a gate or standard cell is modeled by one or more current sources and some passive components. The current sources model (one of) the generation mechanisms. The passive components are usually linear, and they model the circuit capacitance between the digital V_{DD} and ground, the n-well capacitances, resistances of the epi-layer underneath a standard cell for a low-ohmic substrate, and other parasitic RC components, as they might be relevant for a particular semiconductor process.

For example, the approach discussed in Ref. 5 uses a macromodeling approach to deal with standard cell designs in high-ohmic substrates, as an extension of the modeling with low-ohmic, epi-type substrates [19]. In high-ohmic substrates, only the digital ground bounce needs to be taken into account as substrate noise generation mechanism, as mentioned earlier. This ground bounce is caused by the supply current, which is modeled for each standard cell by a current source between V_{DD} and V_{SS}. The model is completed with admittance in parallel with the current source. This source is only active when the standard cell is switching. In standard cell designs, the cells are usually placed in different a few power domains, i.e., collections of cells that are connected to one single V_{DD} and V_{SS}. To macromodel each power domain, the model of each standard cell of a domain is connected to the same V_{DD} and V_{SS}, which in turn are connected to a package model that includes bond-wire inductances and damping resistances. This model (see Figure 20.1) is completed with a substrate mesh connecting the on-chip power grid to the substrate.

In high-ohmic substrates, the injection of current into the substrate underneath a digital gate can be neglected. One only has to take into account the bounce on the supply lines caused by the switching of a standard cell. This bounce is not zero due to the impedance of the assembly parasitics, which are modeled as well.

A computation of the generation of substrate noise then proceeds as follows (see Figure 20.2): prior to simulation a macromodel is constructed for every standard cell of the library.

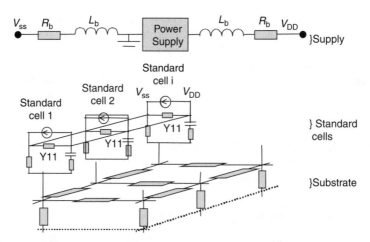

FIGURE 20.1 Macromodel of a power domain in standard cell-based digital circuit. This model is used for simulating substrate noise generation in high-ohmic substrates with the approach discussed in Ref. 5. (From G. Van der Plas et al., *Proceedings of the 41st Annual Conference on Design Automation Conference*, pp. 854–859, June 2004. With permission. © [2004] IEEE.)

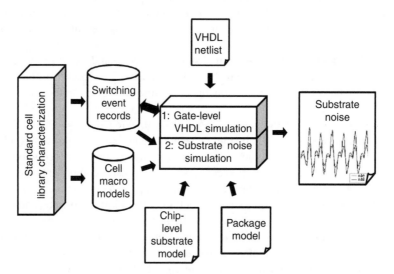

FIGURE 20.2 Substrate noise generation analysis flow. First, a gate-level simulation is performed. This determines for every standard cell the moments at which it is switching. At these moments, the current source of the macromodel of the standard cell is active. Next, this information is used in a time-domain simulation, which comprises the standard cell macromodels as well as a model for the substrate and the package.

This modeling step only needs to be done once. Layout variations, which lead to different capacitances of the interconnections between the standard cells, are added afterward for the circuit under consideration. Next, a gate-level simulation is performed. This gives for each standard cell the exact moment at which it is switching. Finally, a time-domain simulation is performed, where the standard cell models are combined. This yields a waveform for the digital switching current $i_{switching}(t)$ (see Figure 20.3), which is the current drawn by the digital circuit from the digital power supply during switching.

The digital switching current can be approximated by a train of triangular waveforms, a larger one at the clock rising edge and a smaller one at the clock falling edge [3,24]. A part of this current, with spectrum $I_{V_{ss}}(f)$, flows through the digital ground impedance $Z_{V_{ss}}(f)$.

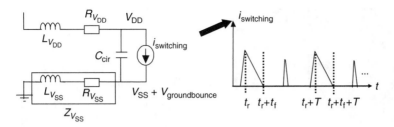

FIGURE 20.3 The ground bounce is modeled as the response of a resonant tank excited by the digital switching current.

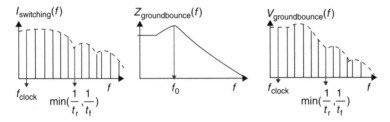

FIGURE 20.4 Representation of the digital ground bounce model in the frequency domain.

Next, the spectrum $V_{\text{groundbounce}}(f)$ of the digital ground bounce voltage and the current $I_{V_{\text{SS}}}(f)$ through the ground lines can be computed from the spectrum of $i_{\text{switching}}$:

$$V_{\text{groundbounce}}(f) = Z_{V_{\text{SS}}}(f) \cdot I_{V_{\text{SS}}}(f) = (R_{V_{\text{SS}}} + j2\pi f L_{V_{\text{SS}}}) \cdot I_{V_{\text{SS}}}(f) \tag{20.1}$$

$$I_{V_{\text{SS}}}(f) = \frac{(j2\pi f C_{\text{cir}})^{-1}}{R_{V_{\text{SS}}} + R_{V_{\text{DD}}} + j2\pi f (L_{V_{\text{SS}}} + L_{V_{\text{DD}}}) + (j2\pi f C_{\text{cir}})^{-1}} I_{\text{switching}}(f) \tag{20.2}$$

Here $R_{V_{\text{SS}}} + j2\pi f L_{V_{\text{SS}}}$ and $R_{V_{\text{DD}}} + j2\pi f L_{V_{\text{DD}}}$ are the interconnect and bonding wire impedances, from, respectively, the on-chip digital ground to the PCB ground and from the on-chip digital supply to the external supply. Further, C_{cir} is the capacitance of the digital circuit. Introducing the impedance $Z_{V_{\text{groundbounce}}}(f)$ that relates the ground bounce voltage for a certain to the switching current,

$$Z_{V_{\text{groundbounce}}}(f) = \frac{V_{\text{groundbounce}}(f)}{I_{\text{switching}}(f)}$$

$$= \frac{R_{V_{\text{SS}}} + j2\pi f L_{V_{\text{SS}}}}{j2\pi f C_{\text{cir}}(R_{V_{\text{SS}}} + R_{V_{\text{DD}}}) - 4\pi^2 f^2 C_{\text{cir}}(L_{V_{\text{SS}}} + L_{V_{\text{DD}}}) + 1} \tag{20.3}$$

the spectrum of the ground bounce voltage $V_{\text{groundbounce}}(f)$ can be found by multiplying $Z_{V_{\text{groundbounce}}}(f)$ with $I_{\text{switching}}(f)$ (see Figure 20.4). The latter has a sinc^2 like shape with notch frequency $f_n = \min(1/t_r, 1/t_f)$ [2], in which t_r is the rise time and t_f is the fall time.

From Equation 20.3, we see that the spectrum of the ground bounce voltage shows a resonance at a frequency $1/(2\pi\sqrt{C_{\text{cir}}(L_{V_{\text{SS}}} + L_{V_{\text{DD}}})})$ with a quality factor given by $\sqrt{(L_s + L_d)}/\sqrt{C_{\text{cir}}}/(R_s + R_d)$.

The accuracy of the approach to macromodel substrate noise generation and ground bounce has been verified with measurements on a 40 kgates telecom circuit (see Figure 20.5 left), fabricated in a 1P6M 0.18 μm CMOS process on a lightly doped substrate with 18 Ωcm resistivity and a wafer thickness of 305 μm. It contains a 20 bit maximum-length-sequence pseudo-random-binary-sequencer (PRBS)

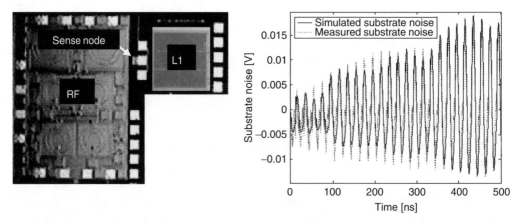

FIGURE 20.5 (Left) Microphotograph of the test circuit, a 40 kgates telecom circuit and the analog victim with the sense node in between; (right) comparison between a simulation of substrate noise using macromodels and measured substrate noise at the start-up of the digital circuit. The noise has been measured at the sense node, left of the digital circuit, at the right border of the VCO. (From G. Van der Plas et al., *Proceedings of the 41st Annual Conference on Design Automation Conference*, pp. 854–859, June 2004. With permission. © [2004] IEEE.)

circuit, driving two cascaded sets of the in-phase and quadrature (IQ) modulator and demodulator chains. The PRBS circuit provides an output signal to synchronize the measurement system with the circuit operation. The operational clock frequency of the IQ modem can be varied from DC to 60 MHz. Its current consumption is 5 mA at a 10 MHz clock and increases linearly with clock frequency.

A 3.5 GHz LC-VCO has been put on the same chip as the digital circuit and serves as a substrate noise victim. The right part of Figure 20.5 shows a comparison of the measured versus simulated substrate noise voltage at a sense node left of the digital circuit at the right border of the VCO. The digital circuit is clocked at 50 MHz, which is close to the resonance frequency of the package (~55 MHz), to maximize the substrate noise voltage (for the experiment's sake). The first 25 clock cycles after end of reset have been plotted. Clearly visible is the start-up of the PRBS that gradually increases the activity of the circuit. The root mean square (RMS) value of the measured substrate noise is 6.5 mV, whereas that of the simulated waveform is 7.7 mV, an error of 20%.

SUBSTRATE NOISE PROPAGATION AND IMPACT MODEL

The digital ground bounce voltage, which has been discussed in the previous section, is injected into the substrate. It propagates through this substrate to the sensitive analog circuits and affects their operation. The way in which the circuit is affected, of course, depends on the specific circuit type, the circuit topology, and its layout. Circuits such as high-frequency narrowband low-noise amplifiers, which operate at frequencies well beyond the clock frequency and its harmonics that contain significant power, are not very sensitive to substrate noise. An example is a narrowband 5 GHz low-noise amplifier, which is on the same chip as a digital system that is clocked at 10 MHz. The harmonics of this clock in the vicinity of 5 GHz do not have much energy and they are well beyond typical values of the resonance frequency of the inductance of the package parasitics with the digital circuit capacitance.

On the contrary, a VCO can be heavily influenced by substrate noise. As the oscillation frequency of a VCO can be tuned by a low-frequency signal at its tuning port, it is also sensitive to substrate noise disturbances at the clock frequency (or its first few harmonics) of a neighboring digital circuit. These disturbances are upconverted by the nonlinear operation of the VCO. In this way, spurious components next to the wanted local oscillator (LO) signal are found in the output spectrum of a VCO (see Figure 20.14 right).

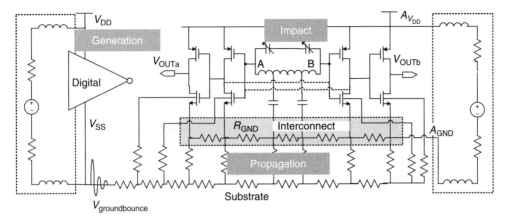

FIGURE 20.6 Schematic representation of generation, propagation, and impact of substrate noise on an analog circuit (here an LC-tank VCO) via different entry points such as the ground interconnect, inductors, and n-MOS bulk connections.

In this section, we describe a model for substrate noise impact on a VCO. The modeling approach for circuits that are nominally linear (such as amplifiers) is similar: the main difference is that the noise is upconverted in a VCO, which is not the case in a linear circuit.

To study substrate noise impact, we split the problem into three parts (see Figure 20.6): the generation of noise ($V_{\text{groundbounce}}(f)$, see the previous section), the propagation ($H^i_{\text{SUB}}(f)$) of noise from the digital noise source through the resistive substrate to the entry points labeled i in the analog circuit, and the impact inside the analog circuit ($H^i_{\text{analog}}(f)$) from entry point i to the analog circuit output [25].

PROPAGATION OF NOISE THROUGH THE SUBSTRATE

As pointed out in the previous section, substrate noise propagates through the substrate before reaching the analog circuit. In epi-type substrates, the highly doped bulk material can be considered as an equipotential node. Hence, the transfer functions $H^i_{\text{SUB}}(f)$, associated to this propagation, are all equal to one. This highly simplifies the model since the substrate can be neglected and all components in the VCO experience the same voltage fluctuation at their substrate terminal. The lightly doped substrate, however, has to be considered as a resistive (R-) mesh and at very high frequencies as an RC-mesh [20].

The relaxation time of the bulk substrate outside the active areas and well diffusions given by $\tau_{\text{SUB}} = \rho_{\text{SUB}}\varepsilon_{\text{SUB}}$ determines the frequency f_{RC} below which the substrate can be treated as purely resistive:

$$f_{\text{RC}} = \frac{1}{2\pi\rho_{\text{SUB}}\varepsilon_{\text{SUB}}} \tag{20.4}$$

For $\rho_{\text{SUB}} = 18\,\Omega\text{cm}$ and $\varepsilon_{\text{rSUB}} = 11.9$, $\varepsilon_0 = 8.85 \times 10^{-12}$ F/m, we find $f_{\text{RC}} = 8.4\,\text{GHz}$. Hence, it is reasonable to neglect intrinsic substrate capacitances for operating speeds of up to a few GHz and switching times of the order of 0.1 ns.

The transfer functions $H^i_{\text{SUB}}(f)$ associated to this propagation decrease as a function of distance. Moreover, they depend on layout details (substrate grounding and capacitance between substrate and metal interconnect lines). Since an analog circuit occupies a certain area, substrate noise is in general larger under parts of the circuit located closer to the noise source and smaller underneath the more remote circuit parts. This has to be taken into account when determining the component in the VCO playing the major role in the impact mechanism. Determining the attenuation by the substrate (the different transfer functions $H^i_{SUB}(f)$) before processing requires a software program such as Substrate Noise Analyst (SNA™) [8] since for real circuits the influence of layout and orientation is extremely complex. The effect of isolation techniques (such as guard rings) is also contained in the substrate transfer functions $H^i_{SUB}(f)$. The role of the metal interconnect parasitics also has to be taken into account for propagation and impact of substrate noise [21]. Extraction of the parasitic resistance in the interconnect can be performed using DIVA [22]. Determining the attenuation by the substrate with these tools, however, is not straightforward since a model for the distributed nature of the interconnect has to be added. This is done by generating contacts distributed over the silicon surface under the interconnect metal. Each contact is connected (resistively or capacitively) to the metal interconnect that is broken down into different resistive parts. If the metal resistivity would not be taken into account, then the substrate noise impact is completely underestimated, as will be shown in the next section.

Entry Points for Substrate Noise in an Analog Circuit

Substrate noise can be coupled into an analog circuit via different ways in an analog circuit: via resistive or capacitive coupling the noise can affect the voltage at different nodes in the circuit. These nodes are called entry points.

An important entry point for resistive substrate noise coupling is the analog ground of the circuit. Typically, a p-type substrate is tied to ground by means of p^+ contacts connected to the off-chip ground with metal interconnect (and bond wires). These contacts and interconnect pick up part of the harmful substrate noise currents from the substrate. Further, the on-chip ground is often implemented as a metal plane over which p^+ contacts are distributed. Contacts located close to the source of noise pick up more current than contacts located further away. In this way, a current can flow through the ground connection. Due to the resistivity of the ground connection, a potential difference occurs, and this can affect the circuit operation. This type of disturbance cannot be suppressed in differential circuits as it is a differential-mode signal, not a common-mode one.

Another example of an entry point via resistive coupling is the bulk of an n-MOS transistor. When substrate noise is generated, a voltage fluctuation can appear on the bulk node of an n-MOS transistor, giving rise to a drain current $i_{ds} = g_{mb}v_{sb}$, with g_{mb} being the body transconductance. This drain current can disturb the circuit operation.

Capacitive coupling of substrate noise into an analog circuit is only important at high frequencies. It can occur via the source-bulk and drain-bulk junction capacitors. However, these capacitors are usually small and this coupling is usually negligible. Another example is capacitive coupling via the n-well underneath a p-MOS transistor or a MOS varactor. Probably the most important capacitive coupling, however, occurs via the passive components of a circuit such as inductors. These components are often very large compared to active elements, and a relatively large parasitic capacitance exists between the substrate and the passive element. For an inductor, this is the capacitance between the substrate and the metal turns.

Computation of the Circuit Response due to Substrate Noise

A substrate noise disturbance that reaches an analog circuit via different entry points can usually be regarded as a small-signal disturbance. In this way, the response of a circuit to substrate noise can in first order be found by small-signal computations. However, a VCO is essentially a strongly non-linear circuit. The small-signal parameters of the active elements in a VCO vary in a periodic way. Hence, with respect to the small substrate noise disturbance, a VCO can be treated as a linear periodically time-varying (LPTV) system. According to Ref. 23, a signal injected into an LPTV system at some frequency yields a spectral component at another frequency. The injection of a single tone disturbance with a small amplitude and frequency f results in two equal power tones around the LO signal at a frequency $f_{LO} \pm f$. The amplitude of these tones depends linearly on the (small) amplitude of the disturbance. For a VCO this frequency translation is the result of the modulation of the voltage over the variable circuit capacitances and thus of the oscillation frequency. In the frequency domain, spurious components V_{spur} that appear around the LO signal can be expressed as a linear sum over the n entry points in the VCO [35]:

$$\left| V_{\text{spur}}(f_{\text{LO}} \pm f) \right| = \left| \sum_{i=1}^{n} H_{\text{analog}}^i(f) \cdot H_{\text{SUB}}^i(f) \cdot V_{\text{groundbounce}}(f) \right| \tag{20.5}$$

The assumption of superposition holds since the substrate noise signals reaching the VCO are typically orders of magnitude smaller than the LO signal. A substrate noise signal that enters the VCO modulates the variable capacitances in the circuit and thus the oscillation frequency. Only frequency modulation (FM) is considered since in a radio amplitude modulation (AM) will most likely be removed by a switching mixer. Since the substrate disturbances are small compared to the LO signal, narrowband FM can be assumed. This assumption is confirmed by measurements. The expression of the amplitude of the spurious tones resulting from narrowband FM shown in Equation 20.5 can be further detailed as follows:

$$\left| V_{\text{spur}}(f_{\text{LO}} \pm f) \right| = \left| \sum_{i=1}^{n} \frac{K_i A_{\text{LO}}}{2f} H_{\text{SUB}}^i(f) \cdot V_{\text{groundbounce}}(f) \right| \tag{20.6}$$

Here A_{LO} is the LO amplitude, the $1/f$ dependency is due to the narrowband FM, and $K_i(V_{\text{tune}})$ is the FM sensitivity function related to the entry point i. It is defined similarly as the VCO gain K_{VCO}:

$$\frac{\partial f_{\text{LO}}}{\partial V_i}(V_{\text{tune}}) = K_i(V_{\text{tune}}) \tag{20.7}$$

The function $K_i(V_{\text{tune}})$ is derived from periodic steady-state simulations with SPECTRE on the circuit model (without substrate). The model of Equation 20.6 is represented schematically in Figure 20.7.

The substrate transfer function $H_{\text{SUB}}^i(f)$ is modeled as a simple voltage division between two resistors: R_{eq}^i and R_{SUB}^i. For example, the coupling from the digital ground to the VCO ground can be expressed as follows:

$$H_{\text{SUB}}^{\text{GND}}(f) = \frac{R_{\text{eq}}^{\text{GND}} + Z_{\text{GND}}^{\text{bond wire}}(f)}{R_{\text{SUB}}^{\text{GND}}} \tag{20.8}$$

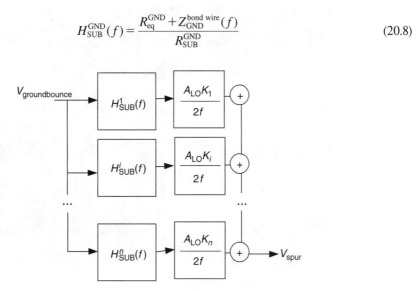

FIGURE 20.7 Model for substrate noise propagation and impact.

Here R_{eq}^{GND} is the equivalent resistance of the VCO ground plane, R_{SUB}^{GND} the resistance between the digital ground and the VCO ground, and $Z_{GND}^{bond\ wire}$ the impedance of the VCO ground bonding wire. Resistance R_{eq}^{GND} models the distributed current flow in both the substrate and the VCO ground plane.

In the next section the proposed model is evaluated and compared to measurements for different situations. The relative importance of the different components of the model gives insight into the substrate noise coupling and the impact problem.

EXPERIMENTAL VERIFICATION OF THE MODEL

The model shown in Figure 20.7 is validated with the mixed-signal integrated circuit (IC) in Figure 20.8. This contains the same circuits as the mixed-signal IC shown in Figure 20.5. However, the measurement setup for these experiments needs to be different from the measurements discussed in the section "Reduction of Substrate Noise Impact Using Guard Rings," leading of course to different package parasitics. Two layouts of the digital circuit have been designed: a reference design without guard ring and a design with a p$^+$/n-well guard ring (see Figure 20.8). The section "GHz LC-Tank VCO" describes the 3.5 GHz LC-tank VCO in more detail. In the section "Measurement of the Digital Ground Bounce" we use the approach discussed in the section "Substrate Noise Propagation and Impact Model" to evaluate the amount of generated noise and we compare the computed noise with measurements. Finally, in the section "Impact of Digital Noise on the LC-VCO" the model for the propagation and the resulting impact on the LC-tank VCO is validated with measurements.

GHz LC-Tank VCO

The VCO that is integrated next to the digital circuit (see Figure 20.8) and that serves as a substrate noise victim is a 3.5 GHz VCO with an LC tank (schematic in Figure 20.9). It uses an n-MOS–p-MOS cross-coupled pair ($W_{n\text{-MOS}} = 50\,\mu m$ and $W_{p\text{-MOS}} = 100\,\mu m$) with an LC tank formed by two on-chip inductors (without patterned ground shield) of 1.2 nH each (realized in metal 6) and two accumulation mode n-MOS varactors (1.2–2.5 pF each). It operates from 3 to 4.4 GHz by changing the tuning voltage at the varactor back gate node from 0 to 1.8 V. The VCO gain K_{VCO} is related to this voltage. The phase noise is −100 dBc/Hz @100 kHz offset at a current consumption of 5 mA (VCO core only) with a supply voltage of 1.8 V. The n-wells and the p-wells have a sheet resistance

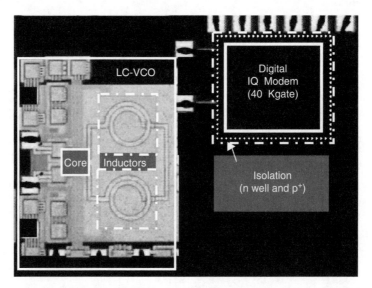

FIGURE 20.8 Microphotograph of LC-VCO and digital IQ modem in 0.18 μm CMOS with a substrate resistivity of 18 Ωcm.

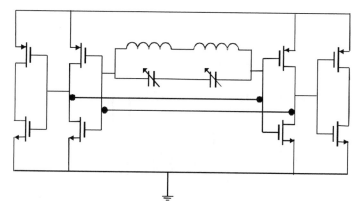

FIGURE 20.9   Schematic of the 3.5 GHz VCO circuit.

of 400 and 800 Ω/sq, respectively. Further, the 20 kA top metal layer and the metal 1 layer have a sheet resistance of 20 and 77 mΩ/sq, respectively.

The chip has been bonded on a PCB using nonconductive epoxy that isolates the die from the PCB ground. Connections between chip and PCB have been realized using bonding wires. The bonding wires have a length of approximately 1.5 mm and a diameter of 25 µm. Simulation with FastHenry™ and calculations according to the formula of Greenhouse [26] both yield an inductance of approximately 1.5 nH. The bonding wire resistance calculated with FastHenry equals approximately 0.1 Ω. Decoupling capacitors on the PCB have been placed as close as possible to the chip.

The following measurements are performed. The digital IQ modem is powered, and the clock and reset signals are applied from a digital pattern generator. While the digital circuit is operational, the power of the spurious tones appearing at both sides of the LO signal are measured and modeled. The LC-VCO is powered from a different source and its tuning voltage V_{TUNE} is set externally to 1 V, which is the value that gives rise to the largest impact. The VCO output spectrum is measured (after limiting to remove AM) with an HP8565ES spectrum analyzer. The ground bounce is measured with an HP8565ES spectrum analyzer and an oscilloscope.

MEASUREMENT OF THE DIGITAL GROUND BOUNCE

The digital ground bounce model is first verified separately on the reference design described in the previous section. The measured digital ground bounce for a clock frequency of 20 MHz is shown in Figure 20.10.

Using the approach explained in the section "Substrate Noise Propagation and Impact Model," this signal is modeled as the response of the resonant circuit formed by the package inductance and the digital circuit capacitance by the digital switching current (Figure 20.1). The Q factor of the resonance circuit equals only 1.2 due to the large resistance in the on-chip interconnect and bonding wires ($R_{V_{SS}} = 1.25\,\Omega$, $L_{V_{SS}} = 1.6\,\text{nH}$, $R_{V_{DD}} = 2.5\,\Omega$, $L_{V_{DD}} = 4\,\text{nH}$, $C_{\text{cir}} = 270\,\text{pF}$). The resonance frequency of $Z_{V_{\text{groundbounce}}}(f)$, which equals $1/(2\pi\sqrt{C_{\text{cir}}(L_{V_{SS}} + L_{V_{DD}})})$, is found to be 127 MHz. The switching current notch frequency is 180 MHz.

In Figure 20.11, a measurement of the digital ground bounce spectrum for a 10 MHz clock frequency is compared to the model computed with the approach explained in the section "Substrate Noise Propagation and Impact Model." At this low clock frequency the spectrum envelope is clearly visible. The ground bounce is measured indirectly. Using the RC-model of the substrate, it is derived from a noise measurement on a substrate contact located between the digital modem and the VCO. The ground bounce spectrum is computed with the model by multiplying the switching current spectrum with the ground impedance. Both the ground impedance and supply current have a finite

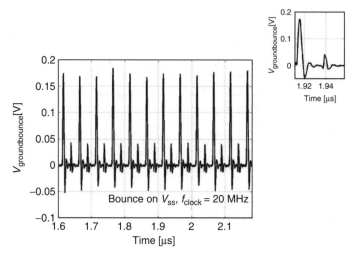

FIGURE 20.10   Measured digital $V_{groundbounce}$ (f_{clock} = 20 MHz).

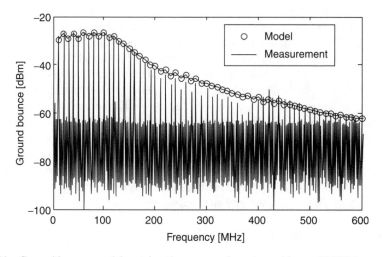

FIGURE 20.11   Ground bounce model matches the measured spectrum (f_{clock} = 10 MHz).

bandwidth and both are approximately equal to 130 MHz. This explains the drop in the ground bounce power spectrum above 130 MHz. For the low-frequency harmonics the deviation from measurements is very small (in the order of 1 dB). It is interesting to note that the power of the even harmonics is slightly higher than the power of the odd harmonics. This is because, besides the large triangular peak of current drawn from the external supply at the rising clock edge, a second smaller current peak occurs at the falling clock edge.

In normal operation a signal is fed to the computational logic at the clock falling edge requiring a small current from the external supply. At the rising clock edge the circuit performs a computation on this input, which requires a larger current from the external supply. After determining the generated substrate noise, we now apply this noise source to the VCO (via the substrate).

IMPACT OF DIGITAL NOISE ON THE LC-VCO

In general, the proposed model of Equation 20.5 is used to locate the main entry points for substrate noise in the analog circuit. When these are identified, isolation techniques can be applied to shield

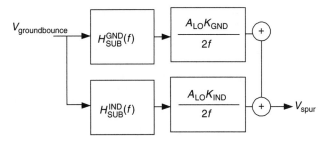

FIGURE 20.12   Impact model of 3.5 GHz LC-VCO: ground and inductor impact are the dominant mechanisms.

these points. In our experiment two coupling mechanisms turn out to be dominant: resistive coupling to the nonideal ground interconnect and capacitive coupling to the inductors (without patterned ground shield in this technology). With these two dominant mechanisms, the level of the spurious tones is found using Equation 20.6:

$$\left| V_{\text{spur}}(f_{\text{LO}} \pm f) \right| = \left| \left(H_{\text{SUB}}^{\text{GND}}(f) K_{\text{GND}} \frac{A_{\text{LO}}}{2f} + H_{\text{SUB}}^{\text{IND}}(f) K_{\text{IND}}(f) \frac{A_{\text{LO}}}{2f} \right) \cdot V_{\text{groundbounce}}(f) \right| \qquad (20.9)$$

Here $H_{\text{SUB}}^{\text{GND}}$ is the transfer in the substrate from the source of noise to the VCO ground interconnect, and $H_{\text{SUB}}^{\text{IND}}$ the transfer of noise to the substrate under the inductors. K_{GND} (defined in Equation 20.7) is the ground FM sensitivity function and K_{IND} the inductor sensitivity function. With these two dominant mechanisms, the impact model shown in Figure 20.6 reduces to the model shown in Figure 20.12.

The FM sensitivity K_{IND} can be expressed as a function of K_{VCO} as follows [34]:

$$K_{\text{IND}}(f) = -K_{\text{VCO}} \frac{2\pi f j C_{\text{IND}}}{g_{\text{m}}} \qquad (20.10)$$

This relation, which has been verified with simulations, can be explained as follows. The substrate noise current couples capacitively through C_{IND}, the capacitance between the substrate and the inductor, and flows into the tank nodes A and B (Figure 20.6) with impedance $1/g_{\text{m}}$, which is the average transconductance of the CMOS cross-coupled pair. The FM sensitivity of these tank nodes equals $-K_{\text{VCO}}$. Combining Equations 20.9 and 20.10, we obtain

$$\left| V_{\text{spur}}(f_{\text{LO}} \pm f) \right| = \left| \left(H_{\text{SUB}}^{\text{GND}}(f) K_{\text{GND}} \frac{A_{\text{LO}}}{2f} - H_{\text{SUB}}^{\text{IND}}(f) K_{\text{VCO}} \frac{\pi j C_{\text{IND}}}{g_{\text{m}}} A_{\text{LO}} \right) \cdot V_{\text{groundbounce}}(f) \right| \qquad (20.11)$$

Figure 20.13 shows the transfer function from the digital ground to the VCO ground ($H_{\text{SUB}}^{\text{GND}}$) and to a point underneath the inductors ($H_{\text{SUB}}^{\text{IND}}$). These transfer functions are determined from AC simulations on the RC substrate model. Whereas $H_{\text{SUB}}^{\text{IND}}$ is flat as a function of frequency, $H_{\text{SUB}}^{\text{GND}}$ increases at higher frequencies. Indeed, at low frequencies, these transfer functions are found as ratios of resistors. For $H_{\text{SUB}}^{\text{GND}}$, there is a voltage division between a distributed resistance between the substrate and the on-chip ground, and the impedance of the off-chip ground connections, which increases as frequency increases due to the inductance of the bond wires.

The output spectrum of the VCO can now be computed with the model of Equation 20.6, using the ground bounce spectrum (see Figures 20.10 and 20.12), the transfer functions $H_{\text{SUB}}^{\text{GND}}$ and $H_{\text{SUB}}^{\text{IND}}$ (Figure 20.13), and the FM sensitivities K_i. The deviation of the computed spectrum with the measured spectrum is smaller than 3 dB for the entire clock range (Figure 20.14). The lower harmonics

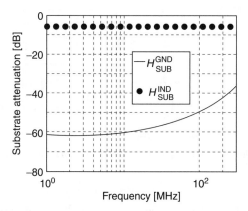

FIGURE 20.13 Substrate transfer functions to dominant entry points on reference design: on-chip VCO ground and inductors.

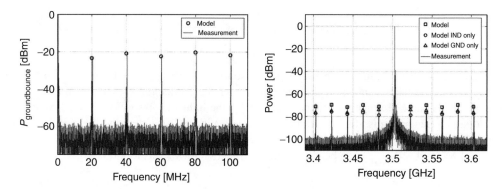

FIGURE 20.14 Left: model of digital ground bounce matches measurement within 1 dB. Right: model for spurious tones at VCO output matches measurements within 3 dB over the full clock range (here: $f_{clock} = 20$ MHz).

of the clock frequency are the most detrimental since they are closest to the carrier signal of the VCO. The spurious level is almost flat because for the impact via the inductors the $1/f$ frequency dependence of the frequency modulation is compensated by the increase of K_{IND} with frequency. For the coupling to the ground, a flat response is obtained as well, because the $1/f$ frequency dependence of the frequency modulation is compensated by the inductive behavior (proportional to f) of the ground network. With this model one can evaluate the efficiency of the guard rings and interpret how they work.

REDUCTION OF SUBSTRATE NOISE IMPACT USING GUARD RINGS

Guard rings (Figure 20.15) provide a way to reduce impact of substrate noise on analog circuits. The operating principle of a guard ring is as follows: a highly doped p-type ring that is connected to a quiet ground can collect the disturbances in the substrate before they reach the analog circuitry. Alternatively, an n-type ring can be put either around the analog circuit or around the digital circuit that generates the noise. The latter ring serves as a barrier for the noisy signals in the substrate rather than as a trap. Another possibility is to put a guard ring around the digital part to prevent substrate noise signals from propagating outside this guard ring. The latter approach will be used as a guarding strategy for a second version of the mixed-signal IC shown in Figure 20.8.

The use of guard rings has been studied extensively [28–32]. In some cases, they can increase the isolation with 40 dB [32]. However, their effectiveness depends on their size, width, operating

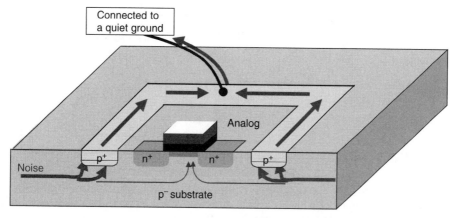

FIGURE 20.15   Principle of a guard ring around an analog circuit to protect this against substrate noise.

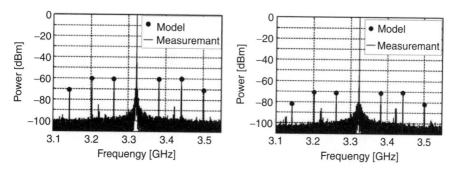

FIGURE 20.16   Level of spurious components when the p^+ guard ring is kept floating (left), and (right) with a grounded p^+ guard ring ($f_{clock} = 60\,MHz$).

frequency, and substrate type. In the following subsections we analyze the effectiveness of guard ring strategies as a function of different parameters such as guard ring size, frequency, and substrate resistivity.

GUARD RING MODELING

In the second version of the mixed-signal IC a standard double isolation structure, an n-well shield and a p^+ guard ring, surrounds the digital circuit (Figure 20.8). Both have a width of 5 μm. With measurements we compare this design in two modes: p^+ guard ring left floating or p^+ guard ring connected to the PCB ground. The n-well shield is always connected to a separate digital supply on the PCB.

The effect of grounding the p^+guard ring is a decrease of the substrate noise impact by approximately 12 dB, which is predicted by the model within 1 dB (Figure 20.16). This decrease is explained by the reduction in noise transfer between the on-chip digital ground and the dominant VCO entry points, namely, the VCO on-chip ground interconnect and the inductors. Both entry points experience this reduction (Figure 20.17).

We now investigate this decrease in the noise transfer function for the coupling to the on-chip VCO ground. For the inductors a similar reasoning applies. Figure 20.18 shows a cross section of the test structure with, from left to right, the analog ground, the p^+guard ring, the n-well shield, and the digital ground. From the layout, using SNA, an equivalent circuit is derived (Figure 20.19).

The equivalent circuit model contains only the nodes that are relevant for the problem: the p^+ guard ring labeled GR, the digital ground labeled V_{SS}, and the VCO ground labeled VCO GND.

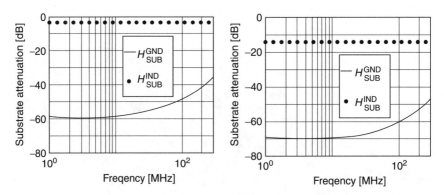

FIGURE 20.17 Transfer to on-chip analog ground interconnect and inductors decreases with 12 dB for a grounded (right) compared to a floating p$^+$ guard ring (left).

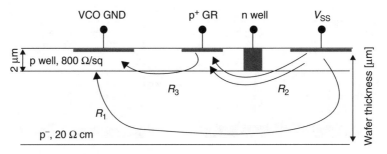

FIGURE 20.18 Cross section of the test structure.

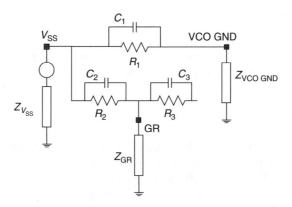

FIGURE 20.19 Equivalent circuit of the test structure containing three external nodes: the p$^+$ guard ring (GR), the digital ground (V_{SS}), and the analog ground (VCO GND).

The equivalent circuit consists of the three substrate impedances between these nodes, Z_1, Z_2, and Z_3 (both resistances and capacitances) completed with the parasitic impedances (both inductance and resistance) of the interconnect and bonding wires connecting them to the PCB ground, Z_{GR}, $Z_{V_{SS}}$, and $Z_{VCO\ GND}$. The bonding wire inductance and resistance are estimated with the FastHenry software [33] and using the Greenhouse formula [26]. Since the studied frequency range is far below $f_{RC} = 8.4\,GHz$, the 18 Ωcm substrate can be modeled as a mesh containing only resistive components [27]. Hence, we can approximate the impedances Z_1, Z_2, and Z_3 by the resistances R_1,

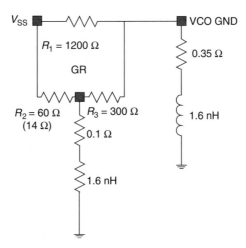

FIGURE 20.20 Extracted substrate and analog ground network. The values in parentheses are for the situation without n-well.

R_2, and R_3 (Figure 20.20). Resistance R_1 is the resistive path for the current that escapes from the influence of the guard ring and reaches the VCO ground by diving deeper into the lightly doped substrate. Resistance R_2 models the resistive path of the current flowing from the digital ground to the guard ring. This current is pushed deeper into the substrate by the presence of the n-well barrier. Resistance R_3 represents the path of the current flowing from the guard ring to the VCO ground via two parallel paths: the lightly doped substrate and the p-well. These resistances are determined using the SNA software and measured. Measured values for R_1, R_2, and R_3 are 1088, 70, and 328 Ω, respectively, and extracted values 1200, 60, and 299 Ω. The error on R_1 (R_{SUB}^{GND} in Equation 20.8), which dominantly determines the coupling, is only 10%, which is within the processing variation of the substrate resistivity.

In the case of a floating p$^+$ guard ring, the ground bounce charge on the on-chip digital ground (V_{SS}) flows both directly to the on-chip analog ground (VCO GND) via R_1, and indirectly through the guard ring node (GR) via R_2 and R_3 (see Figure 20.20). This results in an effective resistive path for the noise current of 277 Ω (1088 Ω in parallel with $328 + 70 \Omega$). In the case of a grounded p$^+$ guard ring the path passing through the guard ring node is shorted to ground through the guard ring bonding wire (both resistive and inductive). The current flows through R_1 only, hence the 12 dB reduction in coupling.

From the model in Ref. 28, a simplified expression is derived for the coupling from the digital to the analog ground, assuming that $Z_{GR}, Z_{VCO\,GND}, Z_{V_{SS}} \ll Z_1, Z_2, Z_3$:

$$\frac{V_{VCO\,GND}}{V_{SS}} = \frac{Z_{VCO\,GND}}{R_1} + \frac{Z_{VCO\,GND}Z_{GR}}{R_3 R_2} \tag{20.12}$$

The first term represents the coupling directly from the digital ground node to the VCO ground when the guard ring is ideally grounded. It results from the voltage division between R_1 and $Z_{VCO\,GND}$ (Figure 20.20). The second term represents the noise transferred from the digital ground via the (not ideally grounded) guard ring to the VCO ground (considered $R_1 \ggg R_3 R_2/Z_{GR}$). It results from the voltage division between R_2 and Z_{GR} followed by the division between R_3 and $Z_{VCO\,GND}$. For our test case the digital modem and VCO have been placed relatively close to each other for reasons of limited chip area, and the direct coupling via R_1 (first term in Equation 20.12) is dominant over the coupling via the guard ring (second term in Equation 20.12). When the coupling via R_1 is reduced, the guard ring ground impedance starts playing an important role as has already been pointed out in Ref. 28.

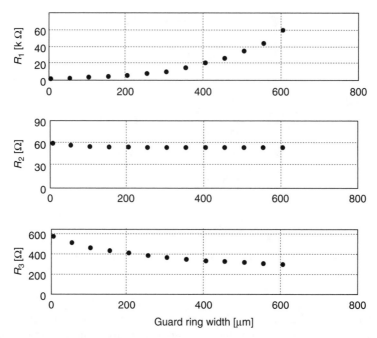

FIGURE 20.21 Resistance values for the equivalent model of the test structure as a function of the guard ring width.

Next, the effect of the n-well shield is studied. The only significant change it brings in the model shown in Figure 20.21 is that R_2 increases from 14 to 60 Ω. R_1 and R_3 remain unchanged. Since in this specific case coupling via R_1 is dominant, the n-well shield does not bring additional shielding when combined with the p$^+$guard ring (below 100 MHz). Only when the coupling via R_1 is negligible and the main coupling occurs via the guard ring, the n-well will help to reduce the impact. Note that the insights presented here can only be obtained with this type of model.

DESIGN GUIDELINES FOR GUARD RINGS

Guarding strategies to reduce substrate noise impact [34] are often based on rules of thumb and intuition, leading to oversized solutions or even system failure. In Ref. 34, a large isolation structure is used to isolate the digital modem from the analog front end. However, the authors do not mention the additional columns of pads placed at both sides of the isolation structure used to connect the analog and digital part with bonding wires on chip. This clearly offers the possibility of a two-chip back-up solution and shows that the effects of substrate noise coupling are not under control. In the next paragraph we demonstrate how the model described in the section "Guard Ring Modeling" allows to apply isolation techniques in a systematic way. The reduction of the impact provided by a digital guard ring is determined as a function of the dimensions of the guard ring.

The model is applied to a test structure containing the digital modem and the 3.5 GHz LC-tank VCO described in the section "GHz LC-Tank VCO," but placed approximately 1000 µm apart. The coupling of digital noise to the VCO ground is considered, but a similar analysis can be made for any circuit node in the VCO that has been identified as an important entry point for substrate noise. The resistances R_1, R_2, and R_3 of the equivalent circuit (Figure 20.20) are determined using SNA for a guard ring width varying from 5 to 600 µm (Figure 20.21). For a grounded guard ring and a given noise voltage V_{noise} on the digital ground the total noise current I_{noise} is

$$I_{\text{noise}} = I_1 + I_2 = \frac{V_{\text{noise}}}{R_1} + \frac{V_{\text{noise}}}{R_2} \tag{20.13}$$

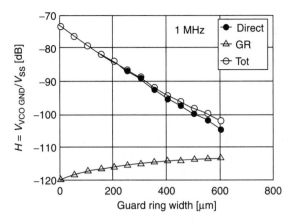

FIGURE 20.22 Comparison of total (tot), direct (direct), and noise transfer via the guard ring (GR) at 1 MHz.

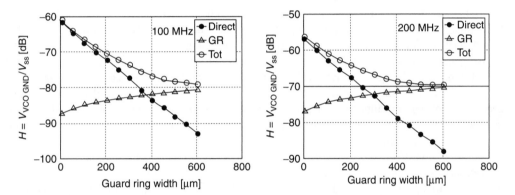

FIGURE 20.23 Total (tot), direct (direct), and noise transfer via guard ring (GR) for a substrate noise disturbance at 100 MHz (left) and 200 MHz (right).

For a width of 5 μm, approximately 97% ($= R_1/(R_1 + R_2) = 1665\,\Omega/(59\,\Omega + 1665\,\Omega)$) of this current flows to the PCB ground via R_2 and the guard ring and only 3% ($= R_2/(R_1 + R_2) = 59\,\Omega/(59\,\Omega + 1665\,\Omega)$) via R_1 and the VCO ground. When the guard ring dimensions are increased, the amount of current taken out of the R_1 path to the VCO ground and led via the R_2 path to the guard ring results in a large relative change of I_1 but only a small relative change of I_2. This is reflected in a large change in R_1 (since $I_1 = V_{noise}/R_1$) and a small change in R_2 (since $I_2 = V_{noise}/R_2$) as shown in Figure 20.21.

From these resistances the coupling from the digital to the analog ground can be estimated according to Equation 20.12. In Figures 20.22 and 20.23 the total coupling, labeled tot and given by Equation 20.12, is plotted, as well as the coupling via R_1, labeled direct and given by the first term in Equation 20.12, and the coupling via the nonideal guard ring labeled GR and given by the second term in Equation 20.12. This is done for noise frequencies of 1, 100, and 200 MHz. For noise frequencies in the 1 MHz range the impedance of the bonding wire connecting the guard ring to the PCB ground is only 0.01 Ω, and the transfer of noise via R_2 and R_3 thus negligible compared to the coupling via R_1. For a noise frequency of 100 and 200 MHz the second term in Equation 20.12 becomes equal to the first term at a guard ring width of 380 and 300 μm, respectively. From this point on increasing the guard ring dimensions is useless. It can be concluded that for large isolation structures, as used by van Zeijl et al. [1] and Kim et al. [34], the bonding wire parasitics can become the bottleneck for substrate noise coupling reduction. Hence, for high-frequency noise it is advised to verify whether the guard ring is made unnecessarily large. To diminish the influence of the guard ring parasitics, multiple bonding wires or bumps can be used for grounding. The required amount of bonding wires can be predicted by the model, taking into account the allowed spurious power and the available chip area (number of ground pads available for bonding).

CONCLUSIONS

Successful integration of analog and digital circuits on one chip can lead to advantages in economics and form factor. However, putting both types of circuits together on a chip introduces an important route for crosstalk from digital to analog, namely, the common substrate. This crosstalk, which we call substrate noise coupling, can seriously harm the analog circuits: comparators in analog-to-digital (A/D) converters can take wrong decisions, RF VCOs can be FM modulated by substrate noise, etc. To master the problem, one should be able to predict the substrate noise during design time. The next step is to develop techniques to reduce this coupling. These techniques can only be validated if substrate noise can be predicted with a reasonable accuracy in a reasonable time.

Since the substrate noise coupling is a complicated problem, it is best split into three different parts: generation in the digital domain, propagation through the substrate, and impact on the analog/RF circuits. To predict the propagation aspect, both academic and commercial tools are available. To predict the digital noise generation, macromodeling strategies are required, since the problem of noise generation by thousands or millions of gates is too complex to face at the circuit level. Gate-level macromodeling techniques with an acceptable accuracy compared to measurements have been demonstrated on digital circuits of reasonable size, but not yet on multimillion gate designs. Nevertheless, research results on noise generation prediction are sufficiently mature to allow for an assessment of digital low-noise design techniques. Such techniques can reduce the generation of noise with about an order of magnitude.

The prediction of impact of substrate noise also needs a high-level modeling approach. The model described in this chapter, which has been validated on a mixed-signal 0.18 μm CMOS IC containing a 40 kgates digital modem and a 3.5 GHz VCO, predicts the level of the spurious components appearing at the VCO output due to the digital switching activity with an error smaller than 3 dB. Finally we demonstrate how the model allows applying isolation techniques as part of a systematic and controlled approach to solve substrate noise coupling problems. As an example we demonstrate for a digital guard ring how the model allows determining suitable dimensions and how the guard ring parasitic inductance lowers its efficiency. The developed model opens the way to solve and understand substrate noise coupling problems for analog/RF circuits.

The next hurdles to be considered in the problem of signal integrity are a study of coupling from analog/RF circuits to other analog/RF circuits and the development of systematic design techniques for low substrate noise impact, instead of a rather blind application of different techniques (differential design, use of guard rings, etc.) as we see today. Only then analog and digital circuits can be put together on a chip without having to resort to conservative designs or without extra design iterations caused by substrate noise coupling. An obvious next step is the construction of one integrated approach that operates on a complete mixed-signal design for the calculation of the substrate noise generated in the digital domain and the impact of that noise on the analog/RF circuits on the same chip.

REFERENCES

1. P.T.M. van Zeijl, J.-W.T. Eikenbroek, P.-P. Vervoort, S. Setty, J. Tangenherg, G. Shipton, E. Kooistra, I.C. Keekstra, D. Belot, K. Visser, E. Bosma, and S.C. Blaakmeer, "A Bluetooth radio in 0.18 mm CMOS," *IEEE J. Solid-State Circuits*, vol. 37, pp. 1679–1687, December 2002.

2. L.M. Franca-Neto, P. Pardy, M.P. Ly, R. Rangel, S. Suthar, T. Syed, B. Bloechel, S. Lee, C. Burnett, D. Cho, D. Kau, A. Fazio, and K. Soumyanath, "Enabling high-performance mixed-signal system-on-a-chip (SoC) in high performance logic CMOS technology," *IEEE VLSI Circuit Symposium*, pp. 164–167, Honolulu, USA, 2002.

3. M. van Heijningen, M. Badaroglu, S. Donnay, G. Gielen, and H. De Man, "Substrate noise generation in complex digital systems: efficient modeling and simulation methodology and experimental verification," *IEEE J. Solid-State Circuits*, vol. 37, no. 8, pp. 1065–1072, August 2002.

4. M. Xu et al., "Measuring and modeling the effects of substrate noise on the LNA for a CMOS GPS receiver," *IEEE Custom Integrated Circuits Conference*, pp. 353–356, Orlando, USA, May 2000.

5. G. Van der Plas, M. Badaroglu, G. Vandersteen, P. Dobrovolny, P. Wambacq, S. Donnay, G. Gielen, and H. De Man, "High-level simulation of substrate noise in high-ohmic substrates with interconnect and supply effects," *IEEE/ACM Design Automation Conference*, pp. 854–859, San Diego, USA, June 2004.

6. M. Badaroglu, P. Wambacq, G. Van der Plas, S. Donnay, G. Gielen, and H. De Man, "Digital ground bounce reduction by supply current shaping and clock frequency modulation," *IEEE Trans. Comput. Aided Design Integr. Circuits Syst.*, vol. 24, no. 1, pp. 65–76, January 2005.

7. M. Badaroglu, P. Wambacq, G. Van der Plas, S. Donnay, G. Gielen, and H. De Man, "Impact of technology scaling on substrate noise generation mechanisms," *IEEE Custom Integrated Circuits Conference*, pp. 501–504, Orlando, USA, 2004.

8. Substrate noise analyst: http://www.cadence.com/products/dfm/substrate_noise_analysis/index.aspx

9. E. Schrik, A.J. van Genderen, and N.P. van der Meijs, "Coherent interconnect/substrate modeling using SPACE—an experimental study," *European Solid-State Device Research Conference*, pp. 585–588, Estoril, Portugal, September 2003.

10. A. Koukab, K. Banerjee, and M. Declercq, "Modeling techniques and verification methodologies for substrate coupling effects in mixed-signal system-on-chip designs," *IEEE Trans. Comput. Aided Design Integr. Circuits Syst.*, vol. 23, pp. 823–836, June 2004.

11. S. Hazenboom, T. Fiez, and K. Mayaram, "Digital noise coupling in a 2.4 GHz LNA for heavily and lightly doped CMOS substrates," *IEEE Custom Integrated Circuits Conference*, pp. 367–370, Orlando, USA, October 2004.

12. M.S. Peng and H.-S. Lee, "Study of substrate noise and techniques for minimization," *IEEE VLSI Circuit Symposium*, pp. 197–200, Kyoto, Japan, June 2003.

13. N. Barton, D. Ozis, T. Fiez, and K. Mayaram, "The effect of supply and substrate noise on jitter in ring oscillators," *IEEE Custom Integrated Circuits Conference*, pp. 505–508, Orlando, USA, May 2002.

14. F. Herzel and B. Razavi, "A study of oscillator jitter due to supply and substrate noise," *IEEE Trans. Circuits Systems II: Analog Digital Signal Process.*, vol. 46, pp. 56–62, January 1999.

15. S. Mitra, R. Rutenbar, L.R. Carley, and D.J. Allstot, "A methodology for rapid estimation of substrate-coupled switching noise," *Proceedings of the IEEE Custom Integrated Circuits Conference*, pp. 129–132, Santa Clara, USA, May 1995.

16. P. Miliozzi, L. Carloni, E. Charbon, and A. Sangiovanni-Vincentelli, "SUBWAVE: a methodology for modeling digital substrate noise injection in mixed-signal ICs," *IEEE Custom Integrated Circuits Conference*, pp. 385–388, San Diego, USA, May 1996.

17. E. Charbon, P. Miliozzi, L. Carloni, A. Ferrari, and A. Sangiovanni-Vincentelli, "Modeling digital substrate noise injection in mixed-signal ICs," *IEEE Trans. Comput. Aided Design Integr. Circuits Syst.*, vol. 18, March 1999.

18. S. Zanella, A. Neviani, E. Zanoni, P. Miliozzi, E. Charbon, C. Guardiani, L. Carloni, and A. Sangiovanni-Vincentelli, "Modeling of substrate noise injected by digital libraries," *IEEE International Symposium on Quality Electronic Design*, pp. 488–492, San Jose, USA, March 2001.

19. M. van Heijningen, M. Badaroglu, S. Donnay, M. Engels, and I. Bolsens, "High-level simulation of substrate noise generation including power supply noise coupling," *IEEE/ACM Design Automation Conference*, pp. 446–451, Los Angeles, USA, June 2000.

20. F. Clement, E. Zysman, M. Kayal, and M. Declercq, "LAYIN: toward a global solution for parasitic coupling modeling and visualization," *IEEE Custom Integrated Circuits Conference*, pp. 537–540, San Diego, USA, May 1994.

21. C. Soens, G. Van der Plas, P. Wambacq, and S. Donnay, "Simulation methodology for analysis of substrate noise impact on analog/RF circuits including interconnect resistance," *Proceedings of the Conference on Design, Automation and Test in Europe*, pp. 270–276, Munich, March 2005.

22. DIVA, http://www.cadence.com/products/dfm/diva

23. T.H. Lee and A. Hajimiri, "Oscillator phase noise: a tutorial," *J. Solid-State Circuits*, vol. 35, pp. 326–336, March 2000.

24. R.C. Frye, "Integration and electrical isolation in CMOS mixed-signal wireless chips," *Proc. IEEE*, vol. 89, pp. 444–455, April 2001.

25. C. Soens, G. Van der Plas, P. Wambacq, S. Donnay, and M. Kuijk, "Performance degradation of LC-tank VCOs by impact of digital switching noise in lightly doped substrates," *IEEE J. Solid-State Circuits*, vol. 40, pp. 1472–1481, July 2005.

26. H.M. Greenhouse, "Design of planar rectangular microelectronic inductors," *IEEE Trans. Parts Hybrids Packag.*, vol. 10, pp. 101–109, June 1974.

27. M. Pfost, H.-M. Rein, and T. Holzwarth, "Modeling substrate effects in the design of high-speed Si-bipolar ICs," *IEEE J. Solid-State Circuits*, vol. 31, pp. 1493–1501, October 1996.

28. T. Blalack, Y. Leclercq, and C.P. Yue, "On-chip RF isolation techniques," *IEEE Bipolar/BiCMOS Circuits and Technology Meeting*, pp. 205–211, Monterey, USA, September 2002.

29. M. Nagata, M. Fukazawa, N. Hamanishi, M. Shiochi, T. Iida, J. Watanabe, Y. Murasaka, and A. Iwata, "Substrate integrity beyond 1 GHz," *IEEE International Solid-State Circuits Conference*, pp. 266–267, San Francisco, USA, February 2005.

30. H.-M. Chen, M.-W. Hu, B.C. Liau, L. Chang, and C.-F. Wu, "The study of substrate noise and noise-rejection-efficiency of guard-ring in monolithic integrated circuits," *IEEE International Symposium on Electromagnetic Compatibility*, vol. 1, pp. 123–128, Washington, August 2000.

31. S. Donnay and G. Gielen (Eds.), *Substrate Noise Coupling in Mixed-Signal ICs*, Kluwer Academic Publishers, Dordrecht, 2003.

32. D. Leenaerts and P. de Vreede, "Influence of substrate noise on RF performance," *European Solid-State Circuits Conference*, pp. 300–304, Stockholm, Sweden, September 2000.

33. M. Kamon, M.J. Tsuk, and J. White, "FASTHENRY: a multipole-accelerated 3-D inductance extraction program," *IEEE/ACM Design Automation Conference*, pp. 678–683, Dallas, USA, June 1993.

34. B.-E. Kim, H.-M. Yoon, Y.-H. Cho, J.-H. Lee, T.-J. Lee, J.-K. Lim, M.-S. Jeong, B. Kim, S.-H. Park, B.-K. Ko, S.-H. Yoon, I. Jung, Y.-U. Oh, and Y.H. Kim, "9 dBm IIP3 direct-conversion satellite broad-band tuner-demodulator SOC," *IEEE International Solid-State Circuits Conference*, pp. 446–507, San Francisco, 2003.

35. C. Soens, G. Van der Plas, M. Badaroglu, P. Wambacq, S. Donnay, Y. Rolain, and M. Kuijk, "A high-level model for substrate noise generation, propagation and resulting RF performance degradation in mixed-signal ICs on a lightly-doped substrate," *IEEE J. Solid-State Circuits*, vol. 41, no. 9, pp. 2040–2051, September 2006.

21 Microelectromechanical Resonators for RF Applications

Frederic Nabki, Tomas A. Dusatko,
and Mourad N. El-Gamal

CONTENTS

Introduction 590
MEM Resonators Basics 590
    Principle of Operation 590
    Quality Factor Definition 591
    Capacitive Transduction and Sensing 591
    Modeling of MEM Resonators 593
    MEM Resonator Nonlinear Effects 596
        Frequency Pulling 596
        Pull-In Voltage 597
        Power Handling 597
    MEM Resonator Energy Loss Mechanisms 597
        Gas Damping 598
        Anchor Loss 598
        Thermoelastic Damping 598
        Q Loading by External Circuitry 598
Applications of MEM Resonators 599
    Filters 599
    Oscillators 599
    Other Applications 600
Evolution of MEM Resonators 600
MEMS-Based Transceivers 605
Mechanical Circuits with MEM Resonators 608
    MEM Resonator–Based Filters 608
    MEM Resonators–Based Arrays 612
Case Studies: Fabricated MEM Resonators 614
    A CMOS-Compatible Tunable Clamped-Clamped Resonator 615
    A Free-Free Beam Resonator 616
    A Radial-Mode Disk Resonator 618
Case Studies: Resonator-Based Systems 620
    MEM Resonator Array–Based Oscillator 620
    Programmable MEM Resonator–Based FSK Transmitter 622
References 624

INTRODUCTION

The introduction of integrated circuits (ICs) in the twentieth century changed the way engineers designed electronic systems. In telecommunications, radio frequency (RF) ICs have performance requirements that are not readily or cheaply attainable using commercial IC technologies. These technologies are usually tailored to digital or more traditional analog designs and, consequently, are limited by the low quality (Q) factors of passive devices such as integrated inductors, capacitors, and filters. This forces engineers to design around performance-limited IC devices, or to resort to cumbersome off-chip elements. Today, the trend toward systems-on-chip (SoC) solutions renders off-chip components undesirable because of their impact on production quality, cost, and size. Microelectromechanical systems (MEMS) offer the opportunity to integrate many RF subcomponents on-chip, which have traditionally been implemented off-chip.

Specifically, microelectromechanical (MEM) devices provide the microelectronics designer with a new toolset of devices and functionalities. These allow for a higher level of integration, which translates into more functionality in the same form factor and enables lighter, lower cost, and more portable wireless systems. Compared to conventional integration-friendly devices such as integrated inductors, MEM components have the potential to offer better performance through enhanced Q factors and lower activation power. MEMS can also allow for improved functionality. For example, MEM resonators can provide flexible on-chip reconfigurability and high filtering performance, while having the potential for integration with electronics.

In this chapter, some conventional transceiver architectures are reviewed, along with their potential for MEMS implementations. MEM resonators are described in detail from their evolution and applications to their operation and use in systems such as MEM resonator–based filters and arrays. Finally, case studies of published works involving MEM resonators are reviewed.

MEM RESONATORS BASICS

PRINCIPLE OF OPERATION

Micromechanical resonators are structures that can be used for filtering or frequency synthesis applications in many subcomponents of RF transceivers, due to their high level of frequency selectivity. Every mechanical structure, such as a disk or a beam, has several natural modes of resonance. At macroscale, these occur at very low frequencies—typically lower than a few kilohertz. By shrinking the dimensions into microscale, higher resonance frequencies can be attained. Mechanical structures are forced into resonance by applying on them mechanical forces at specific frequencies, known as resonant frequencies. Figure 21.1 shows a beam with movement constrained at both ends. This beam configuration is commonly known as "clamped-clamped." Also shown are the deformed shapes for the first flexural mode, which has the lowest resonant frequency and is labeled as "flexural" because the movement is in a direction parallel to the beam thickness. As an example, a clamped-clamped beam that is 40 μm long would have a typical resonant frequency in excess of 8 MHz [1], depending on the structural material used. For smaller beams, the resonant frequency can reach values as high as ∼100 MHz [2].

To provide the mechanical forces necessary to drive the beam into resonance, transducers that convert electrical signals to the mechanical domain are used. Different types of transducers exist. As an example, Figure 21.2 shows the block diagram of a MEM resonator actuated using an electrostatic transducer. A voltage signal is applied to the input of the resonator, which converts it to the mechanical domain as a force.

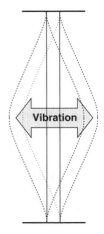

FIGURE 21.1 Clamped-clamped beam with deformed shapes for its first flexural mode of resonance.

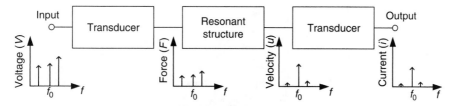

FIGURE 21.2 Electrostatically actuated MEM resonator block diagram.

This force then stimulates a resonant structure by inducing displacement: this filters out the force by attenuating its components that are not at the structure's resonant frequency, f_0. Subsequently, the velocity of the structure, which is the time derivative of the displacement, is sensed by the transducer and is then converted back into the electrical domain as an output current. MEM resonators using other types of transduction mechanisms, such as piezoelectric crystals, operate on different physical quantities, such as mechanical stresses, but their general block diagrams remain the same.

QUALITY FACTOR DEFINITION

The Q factor of a filter is a metric that characterizes its level of selectivity in the frequency domain and is defined as

$$Q = \frac{\omega_0}{\Delta\omega_{-3db}} \tag{21.1}$$

where $\Delta\omega_{-3db}$ is the $-3\,dB$ bandwidth, and ω_0 the center frequency of resonance. In terms of physical quantities, the Q factor is also equal to the ratio of the amount of energy stored in the system to the amount of energy that is lost per cycle:

$$Q = \frac{\Delta E_{stored}}{\Delta E_{lost}} \tag{21.2}$$

Thus, to maximize the Q factor, all sources of energy losses must be minimized. In general, mechanical resonance has a high level of selectivity, as it exhibits very small energy losses. It is therefore ideal for narrowband filtering applications. In the case of a MEM resonator, maximizing the device's Q factor is related to increasing its stiffness (thus the stored energy) and minimizing damping mechanisms (thus the energy lost). When a MEM resonator is used as the frequency-selective tank in an oscillator, its Q factor has a large impact on the phase noise performance of the system and therefore should be as high as possible [3,4].

CAPACITIVE TRANSDUCTION AND SENSING

MEM resonators exploit the high-Q resonance of mechanical structures to perform highly selective filtering for applications in RF-analog electronics. The resonance principle is similar to that used in quartz crystals: the electrical signal is first transformed into a mechanical force that causes the structure to vibrate at specific frequencies. With careful design, the vibrating structure can be used to output an electric current that is a filtered version of the input signal. Thus, the filtering process is composed of three main steps: electromechanical transduction to convert the input electrical signal into a mechanical force, high-Q filtering using mechanical resonance, and lastly mechanical-to-electrical conversion to create an output current. In quartz crystals, these actions are performed

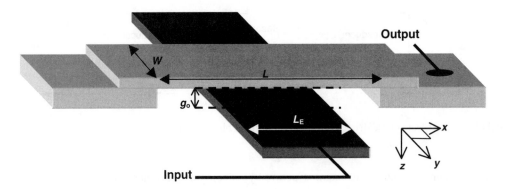

FIGURE 21.3 Simplified beam resonator.

using the piezoelectric properties of the material; however, most MEM resonators are suited to electrostatic (capacitive) excitation. Consider the simple beam resonator shown in Figure 21.3. It is composed of a single mechanical beam with length L and width W. It acts as both the top electrode of a capacitor and a resonant structure. The bottom capacitor electrode is located under the beam at a distance g_o and is connected to the input signal.

When a small-signal voltage $v_i(t)$ is superimposed on a DC voltage V_P at the input electrode, the driving electrostatic force is expressed as

$$F_E(t) = \frac{1}{2}\left|\frac{\partial C}{\partial g}\right|V_P^2 + \left|\frac{\partial C}{\partial g}\right|V_P v_i(t) + \frac{1}{2}\left|\frac{\partial C}{\partial g}\right|[v_i(t)]^2 \qquad (21.3)$$

where C is the overlap capacitance, and g the gap between the beam and the electrode. Since the input signal amplitude is small, the last term of Equation 21.3 can be neglected, yielding an alternating current (AC) force component of

$$F_E(t) = \left|\frac{\partial C}{\partial g}\right|V_P v_i(t) = \eta_e v_i(t) \qquad (21.4)$$

where η_e is known as the electromechanical coupling coefficient. This AC force causes the mechanical structure to vibrate, which in turn modulates the size of the gap between the two electrodes. Note that V_P may be used to effectively turn on and off the resonator and that the structure only vibrates with significant amplitude if the input signal frequency matches its resonant frequency.

When a DC voltage is placed across the gap, the time-varying capacitance generates an output current, $i_o(t)$. This current is proportional to the velocity of vibration and is expressed as

$$i_o(t) = V_P\left|\frac{\partial C}{\partial g}\right|u(t) = \eta_e u(t) \qquad (21.5)$$

where $u(t)$ is the resonator's velocity. By convention, a positive current is defined as the current flowing into the resonator's positive terminal. The exact expression for the capacitance derivative depends on the resonance mode shape [5], but it can be approximated to the following expression by assuming a uniform displacement that is much smaller than the initial gap, g_0:

$$\left|\frac{\partial C}{\partial g}\right| \cong \frac{\varepsilon_0 A_E}{g_0^2} \qquad (21.6)$$

where ε_0 is the permittivity in vacuum, and A_E the electrode overlap area. Equation 21.5 indicates that the magnitude of the output current is directly related to the amplitude of vibration. Thus, the latter must be made as large as possible to increase the power of the output signal. Furthermore, Equation 21.6 shows that the resonator initial gap has a big impact on the output current and should be made small for strong current drive. This small gap, which can be in the order of 100 nm [6], presents one of the biggest fabrication challenges of MEM resonators. The electrode overlap area can be maximized, but higher resonant frequency resonators have small dimensions and hence have limited area overlap. High current drive is essential when MEM resonators are used in reference oscillators, since the phase noise of the system is inversely related to the output power [3,7]. High current drive is also important in resonator-based filters and arrays, where it is directly related to insertion loss [8].

MODELING OF MEM RESONATORS

Since MEM resonators convert signals from the electrical to the mechanical domain, a small-signal model is required that can capture the mechanical resonance of the device along with various electrical effects, such as capacitive feed-through and resistive loading by external circuitry. This section describes how the mechanical resonator can be modeled using a combination of passive electrical components. The resulting models can then be incorporated into circuit simulation packages.

Many analogies can be drawn between mechanical and electrical systems [9]. Since both can be modeled using linear system theory, all concepts and design techniques that are used for circuit design can also be used for mechanical design and vice versa. Like circuits, the response of mechanical systems can be described in terms of the poles and zeros of the system's transfer function. In fact, the linear differential equations that govern the motion of lumped parameter mass-spring-damper systems take the same general form as that of resonant RLC circuits.

To see the similarity between the two systems more clearly, consider the series RLC circuit and the lumped mass-spring-damper mechanical system shown in Figure 21.4. The differential equation governing the RLC is

$$L\frac{d^2i(t)}{dt^2} + R\frac{di(t)}{dt} + \frac{1}{C}i(t) = \frac{dv(t)}{dt} \tag{21.7}$$

where $i(t)$ is the current, L the inductance, R the resistance, C the capacitance, and $v(t)$ the voltage. Similarly, the differential equation governing the mass-spring-damper mechanical system is

$$m\frac{d^2u(t)}{dt^2} + d\frac{du(t)}{dt} + ku(t) = \frac{dF(t)}{dt} \tag{21.8}$$

where $u(t)$ is the velocity, m the mass, d the damping coefficient, k the spring constant, and $F(t)$ the force. By comparing Equations 21.7 and 21.8, it can be noted that there is a clear duality between

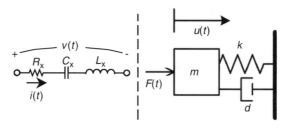

FIGURE 21.4 Lumped electrical RLC circuit and mass-spring-damper system.

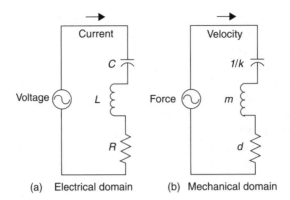

(a)   Electrical domain   (b)   Mechanical domain

FIGURE 21.5   Series RLC mechanical mapping analogy.

TABLE 21.1
Summary of Electrical–Mechanical Analogy

Electrical Variable	Mechanical Variable
Voltage (V)	Force (F)
Current (i)	Velocity (u)
Inductance (L)	Mass (m)
Capacitance (C)	Compliance ($1/k$)
Resistance (R)	Damping (d)

the voltage–current (v–i) response of a series electrical RLC circuit and the force–velocity (F–u) response of a mass-spring-damper system. An RLC circuit is shown in Figure 21.5 along with its dual mechanical equivalent circuit. A summary of the correspondence between the electrical and mechanical domains is provided in Table 21.1.

Using this analogy, the mobility Y_m of the device can be defined as [9]

$$Y_m = \frac{u}{F} \tag{21.9}$$

Mobility, the inverse of mechanical impedance, is the ratio of the velocity of the structure to the driving force. For distributed structures such as MEM resonators, the mobility changes depending on the location at which it is calculated. For example, the mobility of a free end is infinite, whereas the mobility of a clamped end is zero. Using this analogy, it is easy to derive the response of several coupled mechanical systems by simply combining their mobilities [9]. The mobility of a second-order mechanical system such as the one shown in Figure 21.5 can be described as

$$Y_m(\omega) = \frac{U(\omega)}{F(\omega)} = j\omega \frac{1/m}{\left(\omega_n^2 - \omega^2 + j\omega\omega_n/Q\right)} \tag{21.10}$$

where $F(\omega)$ is the input force, $U(\omega)$ the velocity of the mass, Q the Q factor of the mechanical resonance, and ω_n the resonant frequency in radians per second (rad/s). The resonant frequency is defined as

$$\omega_n = \sqrt{\frac{k}{m}} \tag{21.11}$$

Equations to determine the resonant frequencies of different structures are derived in Refs. 9 and 12. The static mass of a structure cannot be used in these models, however, since most resonant structures have complex mode shapes. The resonant structure must be treated as a distributed system, with an *effective* mass, m_N, and an effective spring constant, k_N. This effective mass is usually some fraction of the actual mass of the system and is dependent on the mode shape [10]. A technique to extract the effective mass based on the mode shapes is described in Ref. 9. Once the effective mass is determined, the effective stiffness is found using Equation 21.11.

With the effective mass and spring constant determined, the effective damping factor, d_N, can be shown to be

$$d_N = \frac{\sqrt{m_N k_N}}{Q} \tag{21.12}$$

To map the resonator device to an RLC circuit, the electrostatic transducer needs to be taken into account by using the electromechanical coupling coefficient [5]. The expressions for the RLC components are given as

$$R_x = \frac{v_i(t)}{i_o(t)} = \frac{d_N}{\eta_e^2} \tag{21.13}$$

$$C_x = \frac{\eta_e^2}{k_N} \tag{21.14}$$

$$L_x = \frac{m_N}{\eta_e^2} \tag{21.15}$$

where R_x, C_x, and L_x are the extracted series-RLC circuit parameters and are not physical quantities. R_x is the motional resistance of the resonator and is an important design parameter, as it determines the input–output behavior of the resonator at resonance. As for C_x and L_x, they set the resonant frequency of the RLC circuit, and together with R_x, they set the Q factor such that

$$\omega_n = \sqrt{\frac{1}{L_x C_x}} \tag{21.16}$$

$$Q = \frac{1}{R_x}\sqrt{\frac{L_x}{C_x}} \tag{21.17}$$

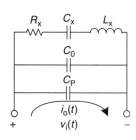

FIGURE 21.6 Resonator small-signal electrical model.

With these three equivalent circuit parameters, an equivalent electrical model can be constructed, as shown in Figure 21.6 [13].

The resonator is represented by its equivalent RLC circuit, whereas the physical overlap capacitance and parasitic feed-through capacitance are modeled with C_o and C_p, respectively. The static electrical overlap capacitance of the resonator input and output port is given by

$$C_o = \frac{\varepsilon_0 A_e}{g_0} \tag{21.18}$$

The parasitic feed-through capacitance from the input to the output port is represented by C_p and depends on the level of coupling between the two ports of the device. It is primarily a result of feed-through across the substrate and of coupling through the resonator's packaging and fixturing. The feed-through capacitance will cause a parallel resonance and, if too large, it can mask the small motional current of the resonator and make the true Q factor difficult to extract. This can be mitigated if the motional resistance of the resonator is low and if care is taken in resonator packaging and interconnecting.

MEM RESONATOR NONLINEAR EFFECTS

So far, small-input signals have been assumed, and nonlinearities could be ignored. However, as the amplitude of the input DC or AC voltages increases, the nonlinear gap capacitances create a number of nonlinear effects. If the resonator electrodes have negligible static bending, and the fringing fields are neglected, the nonlinear capacitance can be modeled using the simple parallel-plate formula, which is given by

$$C = \frac{\varepsilon_0 A_E}{g} = \frac{\varepsilon_0 A_E}{g_0 + x} \tag{21.19}$$

where x is the resonator displacement. As shown previously, the input force and the output current of a resonator are directly related to the derivative of this gap capacitance, given by

$$\left| \frac{\delta C}{\delta x} \right| = \frac{\varepsilon_0 A_E}{g_0^2} \frac{1}{\left(1 + x/g_0\right)^2} \tag{21.20}$$

The derivative of the capacitance exhibits an inverse-square nonlinearity, which not only distorts the output signal but also leads to resonant frequency pulling [6] and Duffing behavior [10,14].

Frequency Pulling

The frequency of a MEM resonator can be tuned to some degree by adjusting the DC bias voltage across the gap capacitance [6]. A positive increase in the bias voltage is accompanied by a decrease in the resonant frequency. This effect, commonly referred to as "spring softening," allows the output frequency to be tuned. A drawback is that the amplitude noise on the bias voltage line directly modulates the output frequency, creating close to carrier noise. Similarly, the resonant frequency also changes as the input signal amplitude grows beyond the small-signal regime, which can have a significant effect on the short-term frequency stability of the resonator.

Frequency pulling is a direct result of the nonlinear gap capacitance. The overall resonant frequency can be described as

$$f_r' = \sqrt{\frac{k_N - k_e}{m_N}} = f_r \sqrt{1 - \frac{k_e}{k_N}} \tag{21.21}$$

where f_r is the resonant frequency with no voltage applied, f_r' the shifted frequency, and k_e the electrostatic effective spring constant, which can be approximated as [13]

$$k_e \approx \frac{\varepsilon_0 A_E V_P^2}{g_0^3} \tag{21.22}$$

The sensitivity of the resonator frequency to changes in the bias voltage can be increased by maximizing the electrode area, decreasing the gap spacing, and reducing the mechanical spring constant. High-frequency resonators have large spring constants and therefore have a smaller tuning range.

For example, a typical 193 MHz disk resonator can have a tuning range as small as 0.01%, and a typical 8 MHz beam resonator can have a tuning range as big as 9.4% [15,16]. Note that increasing the electrical spring constant through gap reduction worsens the overall linearity of the resonator, a tradeoff that should be carefully considered.

Pull-In Voltage

The resonator device must be biased with a DC voltage to generate an output current. As such, it is advantageous to increase the bias voltage to increase the current drive. Unfortunately, because of static instability, there is a limit to how much the bias voltage can be increased before the electrostatic force catastrophically pulls the device into the electrode. The pull-in voltage is highly unpredictable, and in most cases, it is lower than that predicted by analytical equations because of the effects of surface roughness and geometry [17]. The pull-in voltage limits the reduction in motional resistance that can be obtained through biasing and also bounds the tuning range. Balanced resonators, such as disks using *two* symmetric electrodes for input and output, have higher pull-in voltages than beams because the two electrostatic forces oppose each other and cancel. In this case, the bias voltage is limited instead by the breakdown of the air gap across the electrodes.

Power Handling

Nonlinearity in MEM resonators limits the amount of power these devices can handle. As the input amplitude is increased, if the vibration amplitude of the resonator reaches a significant fraction of the electrostatic gap size, the resonator starts to show erratic behavior, or can fail. This limits the power-handling capabilities of the resonator and can constrain performance in systems such as oscillators, in which performance depends not only on the Q factor but also on signal power [6]. To reflect the worst-case scenario, one refers to a point at which the displacement is maximal. The maximum power flowing through a resonator can then be defined as

$$P_{o\max} = \left(R_X i_o^2\right)\bigg|_{\text{worst case}} = \frac{\omega_n}{Q} \cdot k_N \cdot \left(W_{\max}\right)^2 \tag{21.23}$$

where $W_{\max}$ is the maximum allowed displacement expressed as [18]

$$W_{\max} = p \cdot g_0 \tag{21.24}$$

where p is a constant defining how big the displacement of the resonator can be with respect to the gap before device operation is compromised. For example, if pull-in is assumed to be the main limit for the displacement, $p = 0.56$ for a beam resonator [18]. Equations 21.23 and 21.24 show that the power-handling capability of a MEM resonator is increased for stiffer resonators and for bigger gaps. For similar Q factor values and gaps, high-frequency devices such as disk resonators have more power-handling capabilities than beam resonators. For example, a typical 9 MHz clamped-clamped beam resonator has a power-handling capability of −40 dBm, whereas a 60 MHz disk resonator has an improved power-handling capability of −20 dBm [5]. MEM resonator arraying techniques can be used to further improve power handling [19]. Interestingly, a higher Q factor reduces the power-handling capability of the resonator, which can be intuitively understood by the increased displacement-to-input voltage efficiency of a higher-Q resonator.

MEM RESONATOR ENERGY LOSS MECHANISMS

The main sources of energy losses in MEM resonators are through viscous gas damping, thermoelastic damping, and the radiation of acoustic waves through the support of the resonant structure. All these sources can be considered independently and combine to reduce the Q factor of the resonator.

Gas Damping

Gas damping is the energy loss caused by the displacement of air molecules as the resonator vibrates. Because of the small dimensions of the gaps in typical MEM resonators, squeeze-film effects accentuate the effect [20]. Gas damping can be reduced by operating the resonator in vacuum. For high-performance devices, this is becoming the norm, as it is the only way to obtain relatively high Q values ($>$10,000) [21]. For reasonable performance, flexural-mode devices should be operated at pressures that are less than $\sim$100 mTorr [22]. High-frequency resonators such as disks generally exhibit less movement as their spring constants are higher and hence are less affected by gas damping [16,23].

Anchor Loss

Significant energy loss also occurs through the anchors of the resonators. Since the MEM resonator must be affixed to the substrate in some way, this type of loss in unavoidable. As the structure vibrates, it generates a periodic force on the supports that are attached to the substrate. This force, in turn, generates acoustic waves that radiate energy into the substrate. Techniques exist to minimize this loss, for example, locating the supports at stationary nodal points [2] or using different materials for the structure and the anchors to create an acoustic mismatch [24]. Unfortunately, alignment errors and fabrication tolerances ultimately limit the effectiveness of these methods. Anchor loss can also be resonator geometry dependent. For example, in the case of clamped-clamped beam resonators, the geometries of the anchors do not change as the beams get shorter. These shorter beams operate at higher frequencies while exerting bigger moments and forces on the anchors. The anchors undergo more deformation and hence cause added energy loss [2].

Thermoelastic Damping

The last major source of energy loss in MEM resonators is due to thermoelastic damping, which is a characteristic of the resonator material and cannot be avoided. As a result, it sets the upper thermodynamic limit on the Q factor [25]. It is essentially due to the conversion of mechanical strain into heat, which then leads to entropic dissipation. Local adiabatic changes in the stress state result in temperature increases, and therefore to conduction of heat through the material. As shown in Refs. 25 and 26, there are three components of this damping in polycrystalline materials: Zener damping, intercrystalline damping, and intracrystalline damping. Each type of damping contributes to the overall Q factor of the system and is maximized at certain frequencies of vibration. Thus, this should be a major consideration for design, since resonators that vibrate near these frequencies are less efficient.

Q Loading by External Circuitry

Energy loss mechanisms are responsible for determining the *unloaded* Q factor of the resonator; however, when it is used in a system with external circuitry, the effective Q factor of the resonator is in fact much lower because of loading from external components such as resistors. Consider the situation shown in Figure 21.7, in which the terminals of a resonator are connected to external circuitry with output and input resistances equal to R_o and R_i, respectively.

In this case, the loaded Q factor, Q_l, can be expressed as

$$Q_l = \frac{Q_{ul}}{1 + \left((R_i + R_o)/R_x \right)} \tag{21.25}$$

where Q_{ul} is the unloaded Q factor. Equation 21.25 shows that the effective Q factor of the overall system is reduced when loaded by the external circuitry's output and input resistances. To reduce

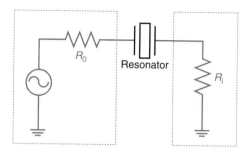

FIGURE 21.7   Resistive loading on resonator by external circuitry.

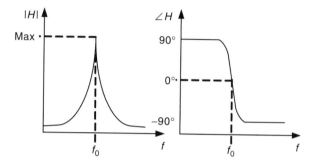

FIGURE 21.8   Typical MEM resonator transfer function.

this Q loading effect, these resistances should be made much smaller than the motional resistance of the resonator, R_x.

APPLICATIONS OF MEM RESONATORS

MEM resonators can be viewed as band-pass filters with small bandwidths that can be designed to resonate at specific frequencies, by appropriate choices of the geometric dimensions and material. With such functionality, many applications are possible through the use of the resonator directly as a filter, or indirectly as the core of an oscillator for frequency generation.

FILTERS

MEM resonators are naturally suitable for band-pass filtering of electrical signals. The magnitude and phase responses of a typical resonant MEM structure are shown in Figure 21.8. At resonance, the MEM resonator essentially acts as a purely resistive element whose value depends on the Q factor of resonance and the actuation transducer efficiency.

MEM resonators can also be combined to create higher order band-pass filters that can be potentially integrated into transceiver front-ends by replacing costly, off-chip, and narrow-bandwidth filters [15]. Furthermore, with the advent of new process technologies, MEM resonators devices can potentially be monolithically fabricated alongside the underlying electronics [1,27].

OSCILLATORS

Oscillators are commonly used in RF systems for high-frequency signal generation in RF voltage controlled oscillators (VCOs) or as low-frequency references for phase-locked loops (PLLs).

The loop topology shown in Figure 21.9 describes the operation of a typical MEM resonator–based oscillator. A wideband amplifier has its frequency response filtered by a MEM resonator. Provided that the amplifier has enough gain to offset the resonator's loss at resonance, and that its bandwidth is wide enough to contribute negligible phase shift to the loop, the circuit oscillates at the resonant frequency of the MEM resonator.

Because of the band-pass nature of the resonator, and the noise shaping caused by the feedback loop, the spectral density of the output is a single tone bounded by an unwanted skirt [28]. This is commonly referred to as phase noise, as it causes jitter in the phase of the output signal. Its magnitude is inversely proportional to the square of the filter's loaded Q factor and to the power of the oscillation [6]. Hence, to reduce this unwanted noise, it is important to use a band-pass filter with a high Q factor and to have a high amplitude of oscillation. The limited power-handling capabilities of MEM resonators restrict phase noise performance improvement through the increase of the oscillation amplitude, but the high Q they provide allows for reasonably good phase noise performance [19].

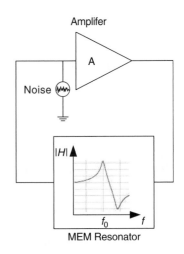

FIGURE 21.9 Typical MEM resonator–based oscillator loop.

In RF VCOs, where full system integration is favored, filters with high center frequencies are used. However, standard LC-based integrated filters can only achieve low Q factors, in the orders of 6–25 [29]. This limits the phase noise performance of integrated VCOs. On the contrary, MEM resonators exhibit Q factors in the order of thousands [2] and have the potential of being integrated with the oscillator electronics.

At low frequencies, an integrated LC-tank has prohibitively large component values. As a result, low-frequency reference oscillators typically use off-chip tanks such as crystals to achieve the required good phase noise performance. MEM resonators are still small enough at low frequencies to be integrated and hence provide a distinct advantage.

OTHER APPLICATIONS

MEM resonators have a resonant frequency that depends on their operating conditions such as temperature, pressure, or ambient chemical content. By capitalizing on these variations, designers can use MEM resonators to measure different physical parameters with high accuracy. MEM resonators have already been considered for use in sensing applications of gas [30,31], vibration [32], ultrasound [33,34] and chemical and biological sensing [35,36]. In other filtering applications, MEM resonators have been investigated for use in such biomedical domains as artificial cochlear implants [37,38].

EVOLUTION OF MEM RESONATORS

MEM resonators have evolved significantly over the last several years. Development has proceeded in two different directions: (1) electrostatically actuated resonators and (2) piezoelectric film, bulk-acoustic-based resonators. Piezoelectric films can be used to generate bulk acoustic waves (BAWs) and, in recent years, have been successfully implemented in high-frequency reference oscillators and filters [39–43]. However, one of the major drawbacks of this technology is that the resonant frequency of the resonators is highly dependent on the thickness of the film—a parameter known to be difficult to control in ICs and systems. Also, it is difficult to create resonators with different frequencies on the same chip, since film thicknesses on a given layer are usually fixed by the process.

This chapter focuses on electrostatically actuated resonators. These allow for more design flexibility, since the resonant frequency is set by geometry parameters, which can be easily modified for different applications. Initial electrostatic resonator designs were mostly based on a comb drive, as shown in Figure 21.10, connected to a large shuttle mass that vibrated laterally on the substrate [6].

Although effective as a proof of concept, these designs had very little practical values for RF systems, as their resonant frequencies were well below 500 kHz—the main reasons for this being the large masses of the structures and the relatively low spring constants. The next generation of designs concentrated on increasing the resonant frequencies of the devices while preserving reasonable Q factors ($Q > 1000$) and were based on the simple clamped-clamped polysilicon cantilever beam, as shown in Figure 21.11 [15]. Although this structure has the potential of generating high-frequency signals, energy loss through the structural anchors makes these designs limited to applications requiring Q factors lower than 10,000. Also, to achieve resonant frequencies above 100 MHz, the length of the cantilever beam becomes very small and subject to variations in processing and mass loading. Another dominant form of energy loss for this structure is squeeze-film air damping. To obtain a reasonable Q factor, these devices must be operated in vacuum at a pressure below $\sim$100 mTorr. This raises issues about packaging and integration.

The second generation of beam resonators was based on a free-free design, as shown in Figure 21.12 [2], where the vibrating structure was a beam that was suspended above the substrate at nodal points.

By anchoring the beams at nodal points, where there is ideally no displacement, and using quarter-wavelength torsional support beams, anchor energy loss was minimized. This significantly increased the Q factor of the device, as it greatly reduced acoustic energy losses to the substrate; however, viscous

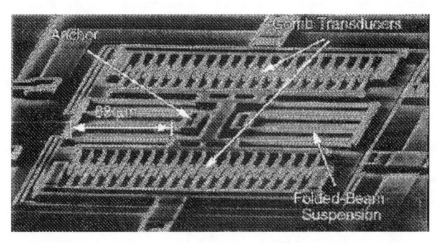

FIGURE 21.10 SEM photograph of a comb resonator. (From Nguyen, C. T. C. and Howe, R. T., *IEEE J. Solid-State Circuits*, vol. 34, no. 4, pp. 440–455, April 1999. With permission. © [1999] IEEE.)

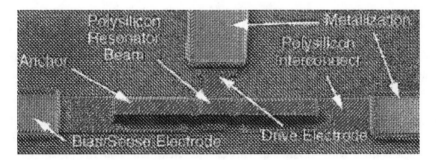

FIGURE 21.11 SEM photograph of a clamped-clamped beam resonator. (From Wang, K., Wong, A.-C., and Nguyen, C. T. C., *J. Microelectromech. Syst.*, vol. 9, no. 3, pp. 347–360, September 2000. With permission. © [2000] IEEE.)

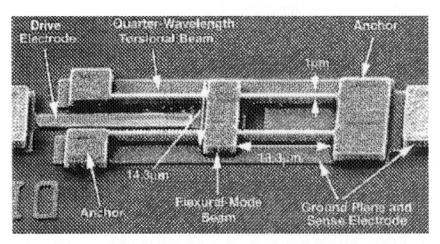

FIGURE 21.12   SEM photograph of a free-free beam resonator. (From Wang, K., Wong, A.-C., and Nguyen, C. T. C., *J. Microelectromech. Syst.*, vol. 9, no. 3, pp. 347–360, September 2000. With permission. © [2000] IEEE.)

gas damping was still an issue. Designs of this type were fabricated and successfully operated at frequencies from 30 to 90 MHz with Q factors of ~8000 [2]. On the basis of this structure, other similar designs using higher order modes were built with resonant frequencies up to 102 MHz [21,44]. Also, because of the complex flexural-mode shapes, differential signals could be generated.

In the last few years, because of the need for resonators that could operate in the ultra high frequency (UHF) range and beyond, a new generation of resonators has been developed. These resonators utilize BAWs instead of flexural movements. A bulk acoustic resonance mode has a very high effective stiffness and thus can be used to generate high frequencies with very high Q factor values. Since the amount of energy stored in the device is related to the stiffness, these resonators store a much larger amount of kinetic energy [16,45]. For example, the spring constant of a typical 1 GHz BAW resonator is on the order of 100 MN/m, whereas that of a flexural-mode beam is on the order of 1500 N/m. Thus, the losses due to gas damping for the BAW resonators are a much smaller percentage of the total energy, which yields a much higher Q factor. For a given frequency, the characteristic dimensions of the devices tend to be much larger than their flexural-beam counterparts. This makes fabrication easier and more reliable.

The most commonly used shape for recent BAW resonators has been the disk structure (Figure 21.13) because of its simplicity and the number of available resonant modes that can be exploited. The first-order contour mode of a disk is illustrated in Figure 21.14.

In this case, the entire diameter of the disk increases and contracts in a way similar to breathing. Recent developments of this design have focused on the support structure, process improvement, and the exploration of new materials. Originally made in polysilicon, the first successful disk resonator was fabricated with a diameter of 34 μm and had a resonant frequency of 160 MHz with a Q factor of over 9000 in vacuum [45].

The main problem with this design was that if the single central support was not placed at the exact center of the device, due to inevitable alignment errors between masking steps, then the Q factor was greatly reduced because of anchor loss. This raised problems regarding the reliability of the design, if it were ever to be implemented in an industrial application. To solve this problem, Nguyen and colleagues pioneered a new process in which the single support stem was self-aligned to the resonator disk [23]. In this process, a stem hole was first etched through the resonator disk and subsequently filled with polysilicon. This ensured that there would be no alignment errors, since the disk and the stem were essentially patterned using the same lithographic mask. With this technique, polysilicon disk resonators were again successfully fabricated with resonant frequencies as high as 1.14 GHz, with a Q factor of higher than 1000 in air. Because of the high-stiffness and

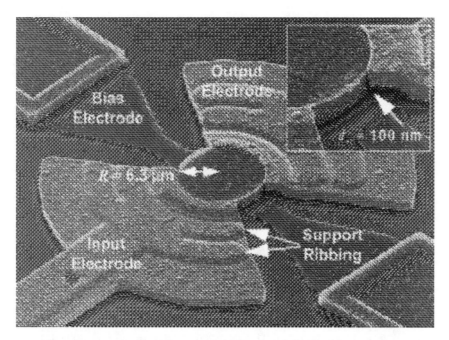

FIGURE 21.13   SEM photograph of a radial contour–mode resonator. (From Clark, J. R. et al., *J. Microelectromech. Syst*, vol. 14, no. 6, pp. 1298–1310, December 2005. With permission. © [2005] IEEE.)

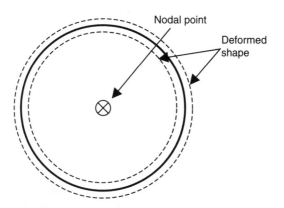

FIGURE 21.14   First radially symmetric resonant mode of a disk.

low-energy loss to the substrate, high Q could still be maintained at atmospheric pressure. Another disk design, the wineglass-mode resonator shown in Figure 21.15, has also been explored. It uses the lower frequency wineglass-mode of operation and has lateral support structures at nodal points. This design operates at a frequency of 73.4 MHz, with an amazing Q factor of 98,000 in vacuum, the largest reported to date for a disk resonator [46].

In 2004, Nguyen's team also unveiled a diamond-disk resonator that was successfully fabricated and resonated at 1.51 GHz with a Q factor of higher than 11,000 in vacuum, the highest frequency of a mechanical resonator, to date [24].

One of the main issues with the disk resonator is its large motional resistance, which makes future use with electronics challenging. The motional resistance of the device is very large because of its high stiffness and small overlap electrode area. If the resonator is used in a filter, the large motional resistance necessitates the use of large terminating resistors to reduce the passband ripples [13]. It also introduces a significant noise component, since the Brownian noise generated by the

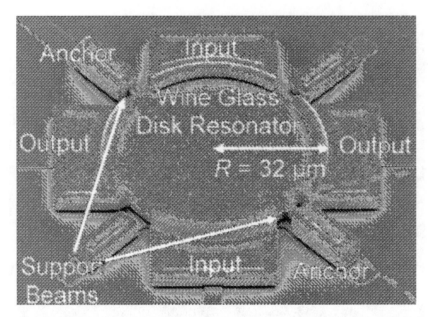

FIGURE 21.15   SEM photograph of a stemless wineglass resonator. (From Lin, Y.-W. et al., *IEEE J. Solid-State Circuits*, vol. 39, no. 12, pp. 2477–2491, December 2004. With permission. © [2004] IEEE.)

device is directly related to the size of the motional resistance. If the resonator is to be used in an oscillator, this large resistance necessitates the use of a transimpedance amplifier (TIA) with an enormous gain. Also, the high motional resistance may limit the highest attainable frequency for a given circuit technology [13]. Although research is still ongoing, there have been several potential solutions to this problem. One solution is to create banks of identical filters that resonate at the same frequency. In this case, the motional resistances of the resonators combine in parallel to reduce the overall resistance. The main difficulty with this solution is that for small variations in the resonant frequencies of the different devices, the combined frequency response will create significant ripples in the passband [47]. However, different mechanical coupling schemes of the resonating elements of such arrays can reduce this problem. More recent approaches increase the efficiency of the transducers through the use of a solid dielectric instead of a hollow gap [48]. This technique is limited to BAW-type resonators as they are not overly affected by the added damping of the dielectric.

Recently, a new circular BAW design has been explored. This new design comprises an annulus, instead of a disk, as shown in Figure 21.16 [49,50]. Its main advantage is that the high-frequency resonant modes are almost completely independent of the average radius of the ring; the resonant frequency of this structure is in fact determined by the width of the ring. Thus, the width of the ring can be used to set the resonant frequency of the device, whereas the average radius can be increased to reduce the effective motional resistance to the desired level by increasing the electrode overlap area. Using this structure, a fabricated prototype has demonstrated a Q factor of 14,600 at a resonant frequency of 1.2 GHz, but with a series resistance that is 12 times smaller than its disk BAW counterpart [50]. The main design challenge for this device is supporting the structure in such a way that energy loss to the substrate is minimized.

An important metric for resonators is the Q factor–frequency (Q–f) product. For high-quality AT-cut quartz crystal oscillators, this value is constant and has a value of approximately 1.6×10^{13} Hz [24]. Whether this empirical relation exists for MEM resonators still remains to be seen. Some general trends can however be observed in the literature. Summarized in Table 21.2 are the highest Q frequency products that have been published to date for polysilicon resonators. Note that the highest Q–f product was obtained using a BAW ring resonator that has high isolation and an extremely high spring constant. Furthermore, this value is at par with those for high-quality quartz crystals,

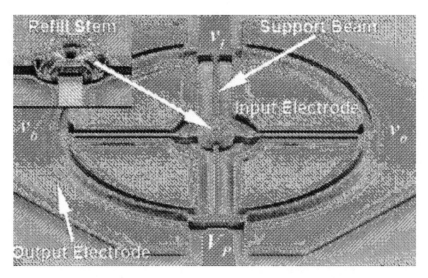

FIGURE 21.16   SEM photograph of a hollow-disk ring resonator. (From Li, S.-S. et al., *IEEE International Conference on Micro-electromechanical Systems*, pp. 821–824, January 2004. With permission. © [2005] IEEE.)

TABLE 21.2
Q-Frequency Product for Several Published Polysilicon Resonators

Type	Frequency (MHz)	Q	Q-Frequency Product ($\times 10^{12}$ Hz)
Annulus BAW [50]	1200	14,600	17.50
Stemless disk BAW [46]	73	98,000	7.15
Self-aligned disk BAW [23]	732	7,330	5.34
Disk BAW [45]	160	9,400	1.50
Higher mode free-free beam [21]	102	11,500	1.17
Free-free beam [2]	92	7,450	0.70
Clamped-clamped beam [15]		8,000	0.07

which is promising. As can be seen from this table, and as expected, the beam designs clearly do not perform as well as the BAW resonators. Although the potential limit for the Q frequency product for polysilicon seems to be on the order of 10^{13} Hz, the development of new materials, such as diamond, will continue to increase this limit.

MEMS-BASED TRANSCEIVERS

A complete MEMS-based transceiver architecture was first proposed by Nguyen [1]. With the successful fabrication of micromechanical resonators with Q values in excess of 10,000, new superheterodyne architectures have become possible, with all large off-chip passive components replaced by RF MEM devices. Also, development of other RF MEM devices such as switches, inductors, and capacitors enables transceiver enhancements [51–54]. Figure 21.17 shows the simplified receive path of a superheterodyne architecture.

In Figure 21.17, all components that are traditionally implemented off-chip are shaded in gray. Specifically, the preselect filter, image-reject filter, and the IF filter are typically implemented using large ceramic or SAW filters, since on-chip LC filters are not able to provide the necessary high Q factor. A similar situation exists for the generation of the reference tone that is used in a channel selection synthesizer. To ensure both long- and short-term stability, it is typically locked to an

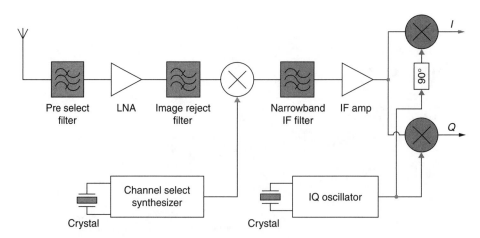

FIGURE 21.17   Simplified superheterodyne receiver architecture.

off-chip quartz crystal that acts as the oscillator's reference. If the Q factor of the crystal is greater than 1000, the reference frequency will not be significantly affected by the temperature dependence of the active electronics, which is typically very large; the reference frequency will depend mainly on the properties of the crystal. Typical uncompensated AT-cut quartz crystals have a frequency drift of around ±50 ppm over the temperature range from -20 to $70°C$ [55], which is orders of magnitude better than that achievable with active electronics.

Off-chip components require much real estate and are costly. MEM devices are easily integrated and hence can be used in greater numbers than off-chip components such as crystals and filters. As has been discussed earlier, MEM devices can offer high performances, which, combined with their integration potential, allow for different approaches to transceiver architectures that were previously not possible because of cost, technical, or power consumption factors.

Replacing external components with their micromachined on-chip equivalents may allow for comparable, if not better, performance. For example, an RF MEM switch can be used as the receive/ transmit relay [56]. The low losses of these switches and their excellent linearity make them excellent candidates for power amplifier (PA) switching or in phased antenna arrays applications [56]. They can also be utilized for higher quality switching of low-noise amplifiers (LNAs) or filters for multiband transceivers. MEM wide-tuning-range capacitors and inductors can be used for better matching networks for PAs, tunable band-select filters, or flexible resonating tanks for VCOs [56].

MEM high-Q resonators, unlike their discrete counterparts, can be used in large numbers, without significantly increasing the overall cost of the system. They can, for example, be arrayed into large filter banks for band, or even channel, selection at RF.

Using these ideas, two transceiver architectures were proposed [1]. The first and most straightforward approach is to simply replace all off-chip components with on-chip MEMS. Although this does not necessarily exploit the full potential of MEM devices, it still allows for monolithic integration, significantly reducing the overall assembly cost. This is illustrated in Figure 21.18, in which all the off-chip components in gray are replaced with their micromachined equivalent devices. Recently, high-Q MEM resonators at frequencies beyond 1 GHz have been demonstrated [50], which indicates that high-quality frequency selectivity can be performed at RF. Similarly, by combining several resonators into arrays, low-insertion-loss band-pass filters have also been demonstrated from high to ultra high frequencies [15,47,57–59]. By combining these MEM resonators with a TIA in a feedback loop, low-phase-noise reference oscillators can also be created to replace the quartz reference crystal. Recently, an array of disk resonators used as a reference oscillator was shown to meet the stringent global system for mobile communication (GSM) phase noise requirements [19].

Another variant of the superheterodyne architecture that makes better use of the large-scale integration afforded by MEM resonators is shown in Figure 21.19 [60]. In this system, the image-reject

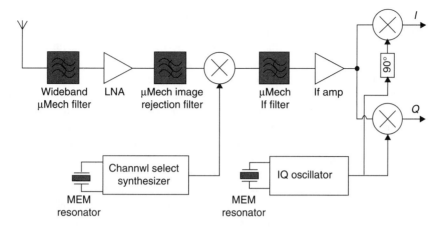

FIGURE 21.18 Superheterodyne architecture with off-chip components replaced by MEM devices.

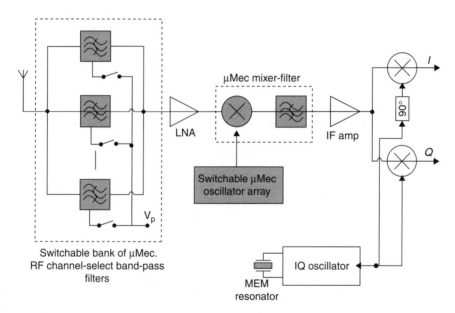

FIGURE 21.19 A MEMS-based receiver architecture.

and the preselect filters are replaced by a bank of switchable high-Q MEM resonators that select the desired channel directly at RF frequencies.

Since MEM resonators can be integrated in large numbers, this bank could contain hundreds of high-Q filters and could be used to implement a truly multiband reconfigurable handset. One very interesting characteristic of MEM resonators is that they can be switched on and off by simply removing their bias voltages. This eliminates the need for lossy series switches in the receive path, which often degrade the overall noise figure of systems [61]. Similarly, the down-conversion to IF can be performed with a programmable bank of micromechanical oscillators. Instead of using a frequency synthesizer for fine channel selection, which consumes a significant amount of power, each channel may use a separate micromechanical oscillator that can be switched on or off. Lastly, the mixer can be implemented using the inherent nonlinearity of a MEM resonator. The MEM resonator simultaneously mixes the RF signal down to intermediate frequency (IF) and filters out unwanted channels [62]. Note that these modifications can also be implemented to a homodyne topology.

Besides the inherent area and cost savings, the architecture shown in Figure 21.19 can be used to trade off high-Q for power consumption [1]. Channel selection directly done at RF yields a substantial advantage: the dynamic range and linearity requirements of the LNA and the mixer in the receive path can be reduced. This is because high-power out-of-band interfering signals are significantly attenuated. For example, in code division multiple access (CDMA) cellular systems, the IIP3 of the LNA is selected to avoid desensitization by a single-tone 900 kHz away from the CDMA signal center frequency and must be greater than +7.6 dBm [63]. As shown in Ref. 1, however, if the MEM channel-select filter can reject the single tone by 40 dB, then the linearity requirement of the LNA relaxes to less than −29 dBm. Another advantage of the reduced level of interfering signals is that the local oscillator phase noise requirements can also be relaxed, further reducing the power-consumption of the system.

Although some MEM devices have become comparable to their macroscopic counterparts in terms of performance, several issues still need to be addressed before they become commercially viable. A significant issue is the stability of these devices with changes in the ambient temperature. One of the main advantages of using quartz crystals is that the resonant frequency is relatively stable with temperature and typically varies by less than 50 ppm over the commercial temperature range. MEM devices, however, are not as stable and therefore require some type of temperature compensation. A recent attempt to mitigate this problem using electronic compensation showed a dramatic improvement in temperature stability [64]; however, the performance is still not comparable to that achievable with compensated quartz crystals.

MECHANICAL CIRCUITS WITH MEM RESONATORS

As MEM resonators are easily integrated, several resonators can be combined to broaden their functionality. For example, resonators may be coupled in arrays to reduce their motional resistance [3,47,59], or they can be interconnected to design higher order series resonator-based filters [8,15,44]. This section gives an overview of MEM resonator-based filters and arrays.

MEM RESONATOR–BASED FILTERS

As was previously discussed, MEM resonators can be modeled using an RLC series-resonant circuit. Using such resonant elements, it is possible to create different types of filters [9]. Figure 21.20 shows a typical model of a three-resonator filter connected together with two coupling networks, along with terminating resistors. Note that feed-through capacitances are omitted here for clarity.

The couplers can be electrical in nature through the use of passive elements such as capacitors or inductors, for example; however, to maintain high quality by minimizing energy loss and improving device fabrication robustness, mechanical coupling using beams acting as springs is often a better choice. An illustration of a triple clamped-clamped resonator filter with flexural couplers is shown in Figure 21.21.

The input voltage is first converted to the mechanical domain through the input transducer. It then gets filtered by the first resonator and subsequently gets mechanically coupled to the second and third resonators, before being finally converted back to the electrical domain through the output transducer. To enable the process of electrostatic transduction, a separate port is required to bias all the resonant beams to a DC voltage different from that of the input and the output, which are biased to ground.

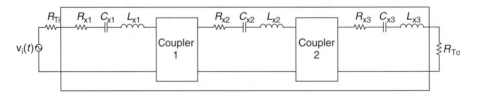

FIGURE 21.20 Three-resonator filter structure.

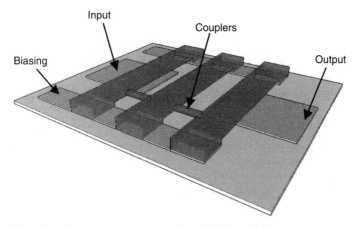

FIGURE 21.21 Three clamped-clamped resonator filter with flexural couplers.

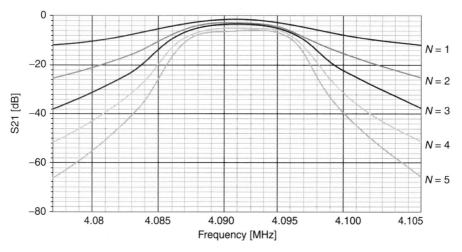

FIGURE 21.22 Transfer functions of different Chebyshev filters made with N 2800-Q resonators.

Figure 21.22 shows the simulated transfer functions of different Chebyshev filters composed of N identical resonators with individual Q factor of 2800. As can be seen, the rejection roll-off is much steeper for filters made of higher number of resonators.

Filters made with more resonators exhibit higher insertion losses and thus require the use of higher Q devices. The insertion loss can be as high as 20 dB for a three-resonator filter with resonator Q factors of 1000 and smaller than 1 dB for Q factor values of 10,000 [8]. Consider a Chebyshev filter composed of four identical resonators. Figure 21.23 shows the simulated transfer functions for different resonator Q factor values. It is clear that filters employing resonators with low Q factors exhibit much higher losses, compared to those employing higher Q resonators.

The coupling beam can ideally be viewed as a spring with stiffness k_{si} and modeled electrically by a capacitor. Because the mass of the coupling beam, m_{si}, cannot be neglected, inductors are added to the capacitor in a T-model formation as shown in Figure 21.24.

Figure 21.25 shows a three-resonator filter structure with couplers assumed to be attached at identical locations on each resonator.

A coupler alters the resonator's effective mass and spring constant, which in turn change the resonator's resonant frequencies. Resonators at the filter's termination ports are loaded by one coupler, whereas other resonators are loaded by two. If all the resonators are identical, the filter structure will be

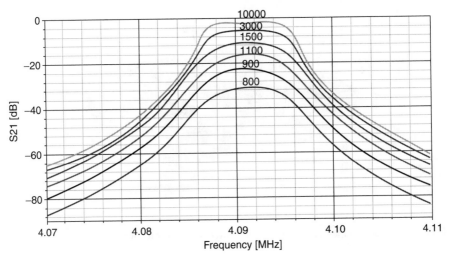

FIGURE 21.23 Transfer functions of four-resonator Chebyshev filters employing resonators with different Q factor values.

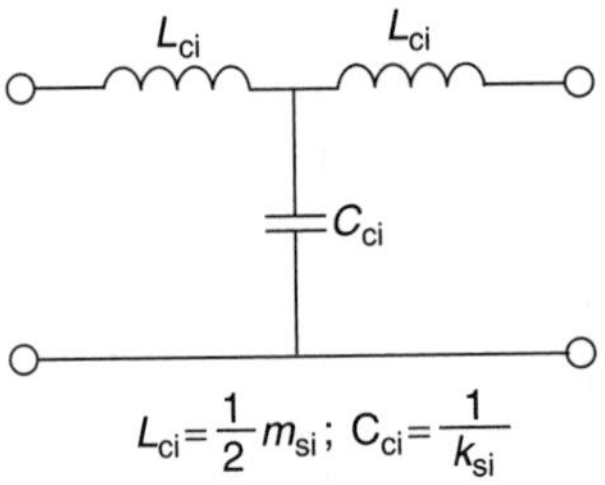

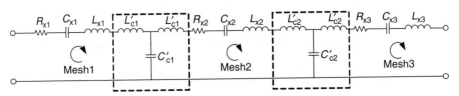

FIGURE 21.24 Lumped coupler T-model.

FIGURE 21.25 Three-resonator filter structure with springs and associated masses.

unbalanced because of the unequal effective resonant frequencies of each mesh [65]. Resonators unbalancing through coupling element loading is illustrated in the transfer functions of Figure 21.26, where a three-resonator filter structure exhibits a center mesh mass variation due to couplers' masses.

To mitigate mass loading and any unbalancing effects, quarter-wavelength supports can be used [9]. If the coupler length is made one-quarter of the acoustic wavelength, it behaves as a free point at its coupling location. It has no loading effect on the filter structure, while allowing for coupling between resonators. The coupler mass elements effectively become negative capacitors that negate the capacitive spring elements and therefore do not affect the resonators' resonant frequencies [15].

The filter structure shown in Figure 21.27 is well known as a mesh-coupled ladder with capacitive coupling. To achieve certain filter types, such as Butterworth or Chebyshev filters, the capacitances C'_{ci} are determined through coupling coefficients (k_i and q_i), which can be found in numerous filter design handbooks (e.g., see Refs. 65–67). These coefficients are shown in Tables 21.3 and 21.4 for a Butterworth filter and a 0.1 dB ripple Chebyshev filter.

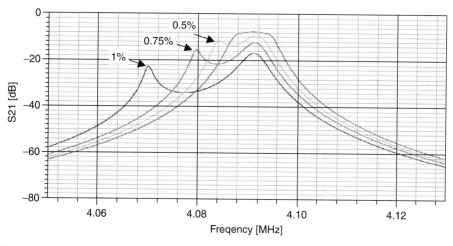

FIGURE 21.26 Three-resonator Chebyshev filter for different center mesh mass mismatch percentages.

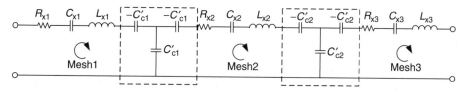

FIGURE 21.27 Three-resonator filter structure with quarter-wave length supports.

TABLE 21.3
Coupling Factors for a Capacitively Coupled Butterworth Filter

Number of Resonators	q_i	q_o	k_{12}	k_{23}	k_{34}	k_{45}
2	1.414	1.414	0.707			
3	1.000	1.000	0.707	0.707		
4	0.765	0.765	0.841	0.541	0.841	
5	0.618	0.618	1.000	0.556	0.556	1.000

Source: Williams, A. and Taylor, F., *Electronic Filter Design Handbook*, McGraw-Hill, New York, 1995.

TABLE 21.4
Coupling Factors for a Capacitively Coupled 0.1 dB Ripple Chebyshev Filter

Number of Resonators	q_i	q_o	k_{12}	k_{23}	k_{34}	k_{45}
2	1.638	1.638	0.771			
3	1.433	1.433	0.662	0.662		
4	1.345	1.345	0.685	0.542	0.685	
5	1.301	1.301	0.703	0.536	0.536	0.703

Source: Williams, A. and Taylor, F., *Electronic Filter Design Handbook*, McGraw-Hill, New York, 1995.

A design methodology for an N-resonator filter with center frequency ω_0 and Q factor Q_{bp} can be outlined using these coupling coefficients. First, the termination resistors R_{Ti} and R_{To} required to achieve the required filter Q factor, Q_{bp}, are determined. Assuming that the couplers are not shifting the resonant frequencies of the resonators, which is the case for quarter-wave supports, these can be determined by [66]

$$R_{Ti,To} = R_x \left(\frac{Q_r}{Q_{bp}q_{i,o}} - 1 \right) \tag{21.26}$$

where Q_r is the resonator's unloaded Q factor. The coupling capacitances may in turn be determined through the following relationship [66]:

$$C'_{Ci} = \frac{Q_{bp}C_x}{k_{i,i+1}} \tag{21.27}$$

where $k_{i,i+1}$ is the ith coupling coefficient. This coupling capacitance value can be used to determine the required spring constant of the support [15].

MEM Resonators–Based Arrays

It is also possible to use the coupling of resonators to reduce the effective motional resistance of a single resonator, which is an important design metric [3,47,59]. If the input signal is distributed across many resonators electrically connected in parallel, the output current is higher, thus making the power handling of the overall array larger than that of a single resonator.

A typical triple clamped-clamped resonator array is shown in Figure 21.28. The structure in Figure 21.28a is similar to that of a filter composed of a resonator cascade (Figure 21.21), but with all electrodes connected together to stimulate a specific mode of resonance. Figure 21.28b adds flexural mechanical couplers that are necessary to better match the resonators' resonant frequencies.

To illustrate a major difference between the two structures shown in Figure 21.28a and 21.28b, the transfer functions of three arrayed resonators with and without mechanical coupling are plotted in Figure 21.29. A 1% resonator-to-resonator mismatch in resonant frequency is assumed to be caused by process variation.

The uncoupled approach does not improve the motional resistance, as the mismatch between the three resonators causes distinct transmission peaks to appear for each resonator. The mechanically coupled approach unifies the resonant frequencies of the resonators. Hence, it exhibits an overall increase in transmission at a single frequency while suppressing the effect of frequency mismatch. This translates into a lower motional resistance. Mechanical coupling is hence paramount in arrays for mitigating mismatches between resonators and ultimately achieving a lower motional resistance.

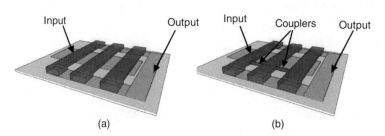

(a)   (b)

FIGURE 21.28 Three clamped-clamped resonator array (a) without mechanical couplers and (b) with flexural couplers.

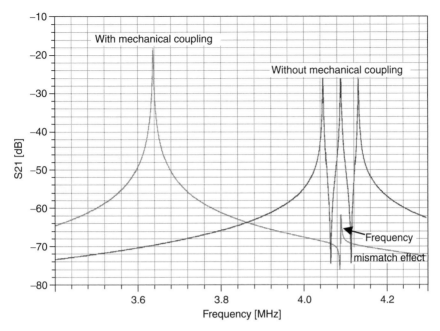

FIGURE 21.29   Transfer functions of three mismatched resonators for different arraying strategies.

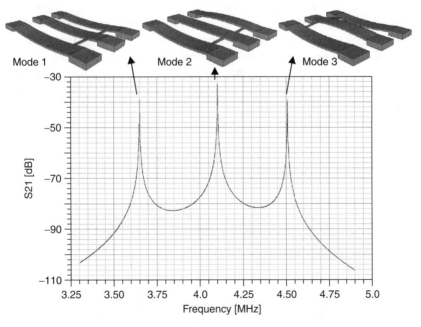

FIGURE 21.30   Resonant modes of a three clamped-clamped resonator cascade.

Figure 21.30 shows the different modes of resonance of a three-resonator filter. Three different flexural modes are possible.

Modes 1 and 3 are similar as they involve the movement of all resonators, whereas mode 2 has one resonator in static equilibrium due to the adjacent resonators' complementary displacements. Small termination resistances are needed in this case to minimize the damping of the resonant modes, which is different than in filter design, where passband ripples need to be minimized. One

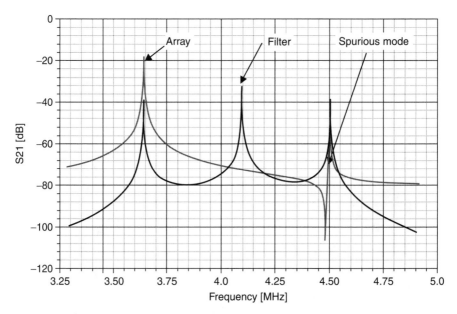

FIGURE 21.31   Transfer functions of a three-resonator array and of a three-resonator filter.

of the modes of operation must be isolated so that the array has a unique resonant frequency and, as such, behaves as a single lower resistance resonating structure. This mode isolation is achieved by electrically stimulating the resonators in such a way as to favor a particular resonant mode of the array. For example, mode 1 of the structure shown in Figure 21.30 may be stimulated by having the electrodes connected to an input signal with uniform phase across all resonators, similarly to the structure shown in Figure 21.28b. The fact that three electrodes stimulate the same mode effectively increases the surface area of the overlap capacitance threefold and hence increases transmission and decreases the motional resistance. Figure 21.31 illustrates the difference in response between three coupled resonators in a filter and the same structure having all resonators connected electrically in parallel to emphasize mode 1.

Due to proper electrode placement and stimulation, transmission at the favored mode of operation is enhanced, whereas it is suppressed at other modes.

It is preferable in an array of resonators to space out the modes away from each other in the frequency domain so that spurious modes lie well out of the band of interest. This can be done through the use of high-stiffness supports and coupling at high-velocity points [47]. This is the opposite of what is required when designing filters where a tight bandwidth is sought after and resonant modes need to be closely spaced. Half-wavelength supports, which have low mobility at the coupling point, can also be used to enhance the effective coupling stiffness of the supports [19].

Finally, Figure 21.32 shows the transfer functions of arrays with different numbers of resonators. Transmission of the favored mode increases as more resonators are used in the array, whereas more spurious modes are present. Spurious modes lower the efficiency with which the dominant-mode motional resistance is reduced. Energy loss in the supports can lower the overall Q factor of the array and hence also reduces the improvement in motional resistance. Eventually, arraying more resonators yields no improvement.

CASE STUDIES: FABRICATED MEM RESONATORS

This section reviews a wide-tuning-range resonator device fabricated in a novel CMOS-compatible process at McGill University and two different resonator designs: a free-free beam resonator [2] and a radial-mode disk resonator [68].

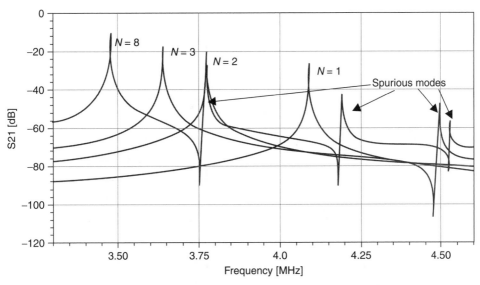

FIGURE 21.32 Transfer functions for different arrays composed of different numbers of resonators.

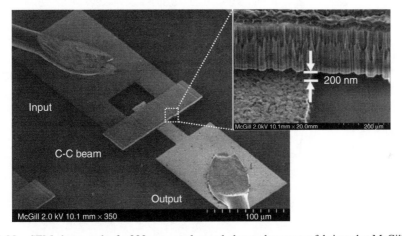

FIGURE 21.33 SEM photograph of a 200 nm gap clamped-clamped resonator fabricated at McGill University.

A CMOS-COMPATIBLE TUNABLE CLAMPED-CLAMPED RESONATOR

At the time of this writing, McGill University has completed development of a novel process that enables the fabrication of CMOS-compatible MEM resonators with a relatively wide tuning range. These resonators are based on a clamped-clamped topology.

A wire-bonded fabricated structure with a 200 nm gap size is shown in Figure 21.33. The response of an 8.3 MHz clamped-clamped resonator is shown in Figure 21.34 with a measured Q factor of 1000. This Q factor performance is in line with comparable work recently done elsewhere. In that work, a similar resonator built on top of a standard bipolar/CMOS (BiCMOS) IC process exhibited a Q factor of 641 at 16 MHz [27].

The resonators utilize a patent-pending tuning scheme that is not based on spring softening bias voltage tuning. The tuning scheme is used to achieve a wider resonant-frequency tuning range with less peak transmission variation. Figure 21.35 shows the transfer functions of a 9 MHz clamped-clamped tunable resonator with an 8.8% tuning range. The peak transmission variation is smaller with the novel tuning method than with the bias voltage tuning method. Wide frequency range tuning, such as demonstrated by this device, is a great asset for offsetting process variation and for offering more flexibility in system design.

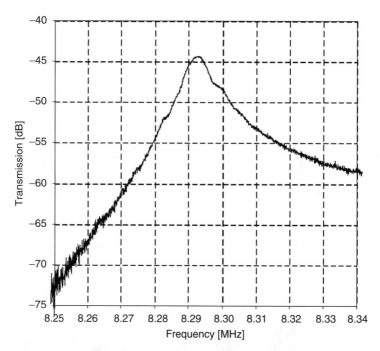

FIGURE 21.34   An 8.3 MHz resonator with a measured Q factor of 1000 fabricated at McGill University.

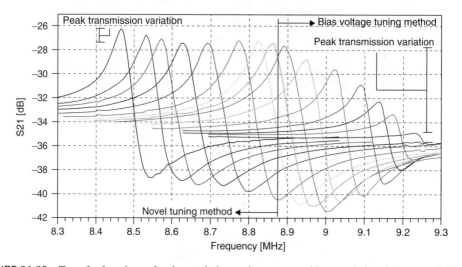

FIGURE 21.35   Transfer functions of a clamped-clamped resonator with extended tuning range (8.8%).

The novel process used to fabricate these resonators allows for post-CMOS integration by involving temperatures and processing steps compatible with standard IC CMOS technologies. It represents a step toward full monolithic integration of MEM resonators with CMOS electronics and toward a fully integrated MEMS/CMOS system.

A Free-Free Beam Resonator

In 2000, Wang et al. published their work on a polysilicon resonator based on a resonant free-free beam structure with length L_r and width W_r. This structure has supports located at the nodal points of a free-free beam resonant mode as shown in Figure 21.36 [2].

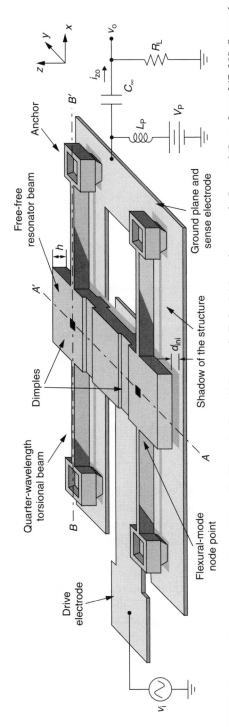

FIGURE 21.36 Free-free beam resonator. (From Wang, K., Wong, A.-C., and Nguyen, C. T. C., *J. Microelectromech. Syst.*, vol. 9, no. 3, pp. 347–360, September 2000. With permission. © [2000] IEEE.)

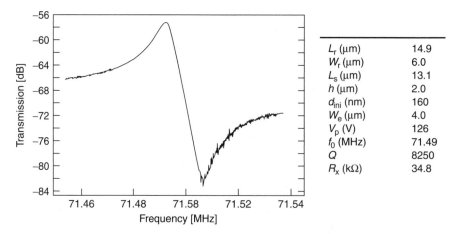

FIGURE 21.37 Summarized fabricated results of 71.49 MHz free-free beam resonator. (From Wang, K., Wong, A.-C., and Nguyen, C. T. C., *J. Microelectromech. Syst.*, vol. 9, no. 3, pp. 347–360, September 2000. With permission. © [2000] IEEE.)

TABLE 21.5
Summary of Different Fabricated Beam Resonator Devices

Resonator Type	Frequency (MHz)	Q	R_x (kΩ)
F-F	31.51	8110	31.1
F-F	50.35	8430	10.7
F-F	71.49	8250	34.9
F-F	92.25	7450	167.0
C-C	54.2	840	8.7
C-C	71.8	300	35.2

Source: Wang, K., Wong, A.-C., and Nguyen, C. T. C., *J. Microelectromech. Syst.*, vol. 9, no. 3, pp. 347–360, September 2000.

Energy loss is reduced by using two strategies. First, the supports are located at nodal points. Second, energy loss due to the finite cross-section of the supports is lowered by having the support length L_s tuned so that its length is one-quarter that of the torsional acoustic wavelength. Dimples provide added reinforcement to the structure and suppress spurious resonant modes caused by the supports. An activation DC voltage, V_P, is applied to the beam through an RF choke, and a signal voltage is applied to the drive electrode of width W_e. A capacitor is also added at the output to DC decouple the load. Figure 21.37 summarizes the transfer function and characteristics of the fabricated device.

The Q factor of the resonator is high, but so is its motional resistance. This is due to the small beam dimensions required to achieve the frequency of operation, which limit the electrode overlap area. Table 21.5 summarizes the performance of different devices published by the authors.

The Q factors of the free-free beam resonators are much higher than similar clamped-clamped beam variants. This is because of anchor energy loss reduction. Also, as anchor loss is resonator length dependent [2,69], reducing it makes the Q factor constant for different resonant-frequency free-free resonators.

A Radial-Mode Disk Resonator

In 2004, Wang et al. presented a polysilicon resonator based on a disk's radial mode of resonance, as shown in Figure 21.38 [68]. The main advantage of this resonator is its higher stiffness. This

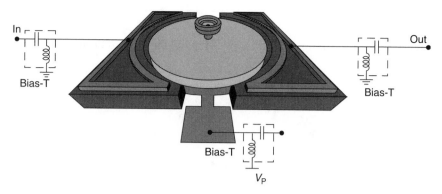

FIGURE 21.38 Radial-mode disk resonator. (From Wang, J., Ren, Z., and Nguyen, C. T. C., *IEEE Trans. Ultrason. Ferroelectr. Freq. Control*, vol. 51, no. 12, pp. 1607–1628, December 2004. With permission. © [2005] IEEE.)

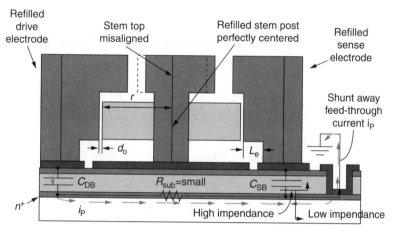

FIGURE 21.39 Disk resonator input–output feed-through. (From Wang, J., Ren, Z., and Nguyen, C. T. C., *IEEE Trans. Ultrason. Ferroelectr. Freq. Control*, vol. 51, no. 12, pp. 1607–1628, December 2004. With permission. © [2005] IEEE.)

translates into higher resonant frequencies for comparable dimensions and a reduced effect of air damping on the resonator's Q factor. A disadvantage of this high stiffness is that the resonator will exhibit a high motional resistance that cannot be considerably reduced electrostatically because of the limited lateral electrode overlap area.

The higher frequency attainable with a disk resonator makes the electrical interconnects to it more sensitive to parasitic capacitances and requires more elaborate measurement methods to extract the actual Q factor [62,68]. As shown in Figure 21.39, feed-through from input to output through the isolating oxide and substrate can be reduced by the addition of a ground connection.

The biasing of this resonator is similar to a beam resonator, with the exception that the input and output ports are not directly capacitively coupled, which mitigates the parallel resonance and reduces feed-through from input to output. The bias voltage, V_P, is applied to the disk structure through an additional electrode. An interesting advantage of this structure is the symmetry of the DC electrostatic forces, which allows for a much higher bias voltage to be applied before pull-in occurs. A disadvantage is that the slightest stem misalignment reduces the Q factor dramatically [45,68]. Therefore, this resonator is fabricated such that the stem is self-aligned by having the disk patterned with a center opening, which is then filled to complete the stem. As only one lithographical mask is

TABLE 21.6
Radial-Mode Resonator Characteristics for the First Three Resonant Modes

	Radial Mode 1	Radial Mode 2	Radial Mode 3
f_0 (MHz)	273	735	1156
Q_{vacuum}	8950	7890	2683
Q_{air}	7500	5160	2655
V_P (V)	30.5	10.5	10.5
R_x (kΩ)	17	521	2442
C_x (F)	3.77×10^{-18}	5.27×10^{-20}	2.10×10^{-20}
L_x (H)	0.896	0.891	0.902

Note: $h = 2.1\,\mu m$, $r = 10\,\mu m$, and $d_0 = 68\,nm$ (extracted R_x, L_x, and C_x are not physical quantities but are used for the RLC model).

Source: Wang, J., Ren, Z., and Nguyen, C. T. C., *IEEE Trans. Ultrason. Ferroelectr. Freq. Control,* vol. 51, no. 12, pp. 1607–1628, December 2004.

effectively used to pattern the disk and stem, no misalignments can occur, even if subsequent patterning of the stem filling is misaligned, as shown in Figure 21.39.

Higher radial modes of resonance can be excited in disk resonators because they do not require a different electrode configuration. By exciting higher modes of resonance, it is possible to push the frequency of operation even further and reach the gigahertz range. Table 21.6 summarizes measurements taken for a radial-mode disk resonator with a radius (r) of $10\,\mu m$ and a thickness (h) of $2.1\,\mu m$ excited at three different modes.

As expected, the frequencies of operation achieved are much higher than that of beam resonators, and the Q factors of the devices are not greatly affected by the vacuum level. The larger effective stiffness of the higher modes increases the motional resistance dramatically and hence makes these devices problematic for use with electronic circuits; however, using arrays, more aggressive voltage biasing, thicker disks, and a smaller gap may help in bringing the impedance levels to as low as $300\,\Omega$ [19,68].

CASE STUDIES: RESONATOR-BASED SYSTEMS

As resonator devices are reaching a certain maturity, they are starting to be integrated with electronic systems such as oscillators [6,19,40], and more complex systems [64]. This section reviews a MEM disk resonator array–based oscillator circuit [19] and a programmable MEM frequency shift keying (FSK) transmitter [64].

MEM RESONATOR ARRAY–BASED OSCILLATOR

In 2005, Lin et al. used the highest Q resonator structure known to date, the disk resonator, to create an oscillator circuit. Stimulating the high-Q compound wineglass mode through specific electrodes placement, an array of identical 60 MHz resonators is used to reduce the high motional resistance of the resonators. This lower motional resistance relaxes the gain requirements of the sustaining amplifier. It thus enabled the design of a TIA in TSMC's (Taiwan Semiconductor Manufacturing Company Ltd.) CMOS $0.35\,\mu m$ process capable of achieving a gain bandwidth product that ensures negligible phase shift at the resonant frequency. A three-disk resonator cascade is shown in Figure 21.40.

Three modes of resonance are possible (Figure 21.41). Using specific electrodes stimulation, similar to what is done for the beam array discussed in the section "Mechanical Circuits with MEM Resonators," the first mode can be favored, and the other modes suppressed, to reduce the overall motional resistance.

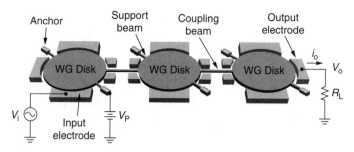

FIGURE 21.40 Wineglass-mode resonator array. (From Lin, Y.-W. et al., *IEEE International Electron Devices Meeting*, pp. 287–290, December 2005. With permission. © [2005] IEEE.)

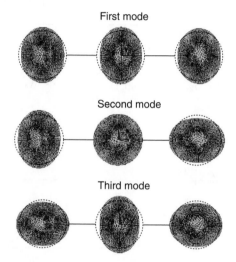

FIGURE 21.41 Resonance modes of wineglass-mode resonator array. (From Lin, Y.-W. et al., *IEEE International Electron Devices Meeting*, pp. 287–290, December 2005. With permission. © [2005] IEEE.)

To spread the modes away from each other in frequency, half-wavelength supports located at high-velocity points are used, which maximizes the effective stiffness of the supports. Figure 21.42 shows the transmission characteristics for different numbers of arrayed resonators. The Q factor decreases slightly because of energy losses due to the supports. Also, transmission does not scale linearly with the number of resonators added because the electrode overlap area is reduced to accommodate mechanical coupling. The motional resistance of a nine-resonator array is 1.25 kΩ.

Along with the array, the TIA circuit used is shown in Figure 21.43. It provides a gain of 8 kΩ and has a bandwidth of 200 MHz, sufficient to sustain oscillation at 60 MHz. The TIA consists of an actively loaded differential pair composed of M_1, M_2, M_3, and M_4. M_{Rf} provides shunt–shunt feedback, which reduces Q factor loading by the amplifier. Finally, M_{11}, M_{12}, M_{13}, and M_{14} provide common-mode feedback to ensure proper DC biasing of the drain nodes at the outputs of the differential pair. The output is taken at the positive terminal of the differential pair so that the phase shift is 0°.

Figure 21.44 shows the output spectrum of the nine-resonator array oscillator and its phase noise plots, with a measured phase noise of −123 dBc at a 1 kHz offset and −136 dBc away from the carrier. As can be seen, the higher power-handling capability of the resonator array makes the oscillator phase noise much smaller than that of the single resonator. Furthermore, thanks to arraying, the $1/f^3$ noise due to the resonator nonlinearity is mitigated to further improve the close-to-carrier phase noise performance. Divided down to 10 MHz, the oscillator meets GSM specifications, which warrants the use of an array structure to improve the resonator performance and bring the technology closer to marketability.

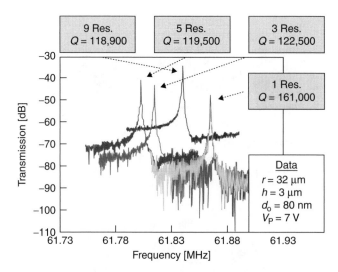

FIGURE 21.42 Transfer characteristics for different disk resonator arrays. (From Lin, Y.-W. et al., *IEEE International Electron Devices Meeting*, pp. 287–290, December 2005. With permission. © [2005] IEEE.)

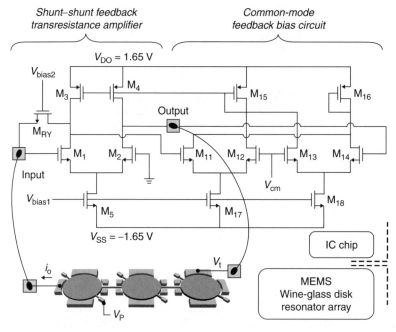

FIGURE 21.43 Circuit schematic of a MEM resonator–based oscillator. (From Lin, Y.-W. et al., *IEEE International Electron Devices Meeting*, pp. 287–290, December 2005. With permission. © [2005] IEEE.)

PROGRAMMABLE MEM RESONATOR–BASED FSK TRANSMITTER

A system-level integration of MEM resonators was presented in 2006 by Hsu et al. in the form of a fully integrated FSK transmitter, shown in Figure 21.45 [64]. The entire system is implemented on a printed circuit board (PCB) in prototype form and on a ball-grid array (BGA) in a more compact final form.

A wide free-free beam resonator is used in an oscillator similar to the work presented in [19] and is packaged in vacuum through a cap wafer as shown in Figure 21.46. The resonator frequency-tuning

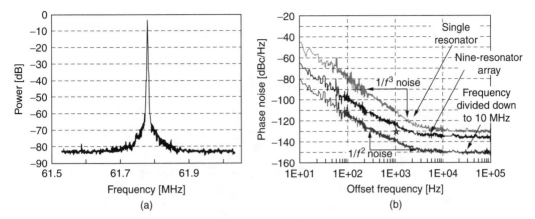

FIGURE 21.44   (a) Output spectrum and (b) phase noise plots of the MEM resonator array–based oscillator. (From Lin, Y.-W. et al., *IEEE International Electron Devices Meeting*, pp. 287–290, December 2005. With permission. © [2005] IEEE.)

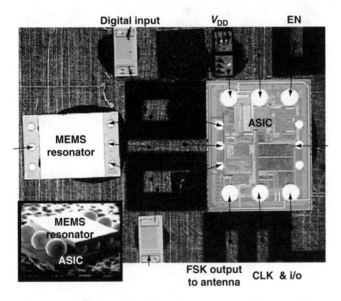

FIGURE 21.45   FSK transmitter photograph of the final product BGA and PCB prototype. (From Hsu, W.-T., Brown, A. R., and Cioffi, K. R., *IEEE International Conference of Solid-State Circuits*, pp. 1111–1120, February 2006. With permission. © [2006] IEEE.)

property through the control of its bias voltage is used to make the oscillator frequency hop between two values. This is controlled by a square pulse that is superimposed to the biasing voltage, V_P. As shown in Figure 21.47, the output of the resonator-based oscillator is fed into a fractional-N sigma-delta PLL, to allow for frequency multiplication of the oscillator signal and electronic temperature compensation of the resonator.

The output center frequency can be programmed from 2 to 437 MHz and is tunable with a 1 ppm accuracy. A frequency variation of 8 ppm from -40 to 85°C is observed. This compensated temperature performance is better than quartz- or SAW-based transmitters [64]. For a resonator biasing of 2.1 V, a 150 kHz frequency deviation is achieved with a 0.9 V square pulse. This application of MEM resonators in a complex system illustrates their potential for higher levels of integration, while providing competitive performances.

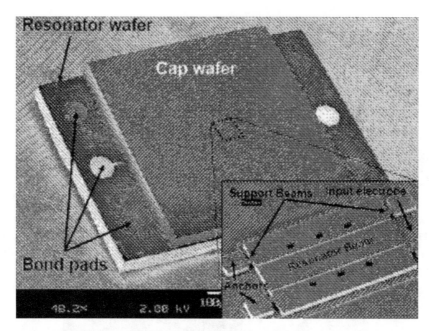

FIGURE 21.46 MEM resonator used in an FSK transmitter with packaging cap wafer. (From Hsu, W.-T., Brown, A. R., and Cioffi, K. R., *IEEE International Conference of Solid-State Circuits*, pp. 1111–1120, February 2006. With permission. © [2006] IEEE.)

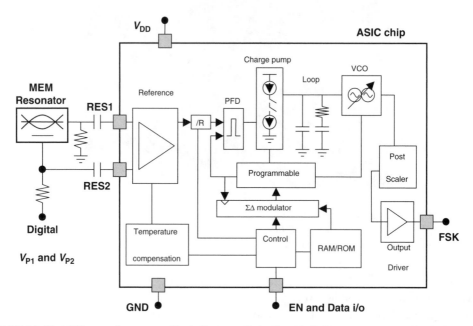

FIGURE 21.47 FSK transmitter system block diagram. (From Hsu, W.-T., Brown, A. R., and Cioffi, K. R., *IEEE International Conference of Solid-State Circuits*, pp. 1111–1120, February 2006. With permission. © [2006] IEEE.)

REFERENCES

1. C. T. C. Nguyen, "Transceiver front-end architectures using vibrating micromechanical signal processors," *Topical Meeting on Silicon Monolithic Integrated Circuits in RF Systems*, Ann Arbor, MI, IEEE, Piscataway, NJ, pp. 23–32, September 2001.

2. K. Wang, A.-C. Wong, and C. T. C. Nguyen, "VHF free-free beam high-Q micromechanical resonators," *J. Microelectromech. Syst.*, vol. 9, no. 3, pp. 347–360, September 2000.

3. S. Lee and C. T. C. Nguyen, "Mechanically-coupled micromechanical resonator arrays for improved phase noise," *IEEE International Frequency Control Symposium and Exposition*, Piscataway, NJ, IEEE, Piscataway, NJ, pp. 144–150, August 2004.

4. V. Kaajakari, T. Mattila, A. Oja, J. Kiihamaki, and H. Seppa, "Square-extensional mode single-crystal silicon micromechanical resonator for low-phase-noise oscillator applications," *IEEE Electron. Device Lett.*, vol. 25, no. 4, pp. 173–175, April 2004.

5. Y.-W. Lin, S. Lee, S.-S. Li, Y. Xie, Z. Ren, and C. T. C. Nguyen, "Series-resonant VHF micromechanical resonator reference oscillators," *IEEE J. Solid-State Circuits*, vol. 39, no. 12, pp. 2477–2491, December 2004.

6. C. T. C. Nguyen and R. T. Howe, "An integrated CMOS micromechanical resonator high-Q oscillator," *IEEE J. Solid-State Circuits*, vol. 34, no. 4, pp. 440–455, April 1999.

7. D. B. Leeson, "A simple model of feedback oscillator noise spectrum," *Proc. IEEE*, vol. 54, no. 2, pp. 329–330, February 1966.

8. K. Wang and C. T. C. Nguyen, "High-order medium frequency micromechanical electronic filters," *J. Microelectromech. Syst.*, vol. 8, no. 4, pp. 534–556, December 1999.

9. R. A. Johnson, *Mechanical Filters in Electronics*, Wiley and Sons, New York, NY, 1983.

10. S. P. Timoshenko, *Vibration Problems in Engineering*, 5th ed., Wiley-Interscience, New York, NY, 1990.

11. S. S. H. Chen and T. M. Liu, "Extensional vibration of thin plates of various shapes," *J. Acoust. Soc. Am.*, vol. 58, no. 4, pp. 828–831, October 1975.

12. M. Onoe, "Contour vibrations of isotropic circular plates," *J. Acoust. Soc. Am.*, vol. 28, no. 6, pp. 1158–1162, November 1956.

13. C. T.-C. Nguyen, *Micromechanical Signal Processors*, PhD Thesis, Berkely, 1994.

14. L. D. Landau and E. M. Lifshitz, *Mechanics vol. 1*, Butterworth-Heinemann, Burlington, MA, 1976.

15. F. D. Bannon, J. R. Clark, and C. T. C. Nguyen, "High-Q HF micro-electromechanical filters," *IEEE J. Solid-State Circuits*, vol. 35, no. 4, pp. 512–526, October 2000.

16. J. R. Clark, W. T. Hsu, M. A. Abdelmoneum, and C. T. C. Nguyen, "High-Q UHF micromechanical radial-contour mode disk resonators," *J. Microelectromech. Syst.*, vol. 14, no. 6, pp. 1298–1310, New York, NY, December 2005.

17. S. Young, D. Weston, B. Dauksher, D. Mancini, S. Pacheco, P. Zurcher, and M. Miller, "A novel low-temperature method to fabricate MEMS resonators using PMGI as a sacrificial layer," *J. Micromech. Microeng.*, vol. 15, pp. 1824–1830, New York, NY, October 2005.

18. S. Lee and C. T. C. Nguyen, "Influence of automatic level control on micromechanical resonator oscillator phase noise," *IEEE International Frequency Control Symposium and Exposition*, Tampa, FL, IEEE, Piscataway, NJ, pp. 341–349, May 2003.

19. Y.-W. Lin, S.-S. Li, Z. Ren, and C. T. C. Nguyen, "Low phase noise array-composite micromechanical wine-glass disk oscillator," *IEEE International Electron Devices Meeting*, Washington, DC, IEEE, Piscataway, NJ, pp. 287–290, December 2005.

20. C. Zhang, G. Xu, and Q. Jiang, "Characterization of the squeeze film damping effect on the quality factor of a microbeam resonator," *J. Micromech. Microeng.*, vol. 14, pp. 1302–1306, July 2004.

21. M. U. Demirci and C. T. C. Nguyen, "Higher-mode free-free beam micromechanical resonators," *IEEE International Frequency Control Symposium*, Tampa, FL, IEEE, Piscataway, NJ, pp. 810–818, May 2003.

22. Y. T. Cheng, W.-T. Hsu, K. Najafi, C. T. C. Nguyen, and L. Lin, "Vacuum packaging technology using localized aluminum/silicon-to-glass bonding," J. Microelectromech. Syst., vol. 11, no. 5, pp. 556–565, October 2002.

23. J. Wang, Z. Ren, and C. T. C. Nguyen, "1.14-GHz self-aligned vibrating micromechanical disk resonator," *IEEE Radio Frequency Integrated Circuits Symposium*, San Francisco, CA, IEEE, Piscataway, NJ, pp. 335–338, June 2003.

24. J. Wang, J. E. Butler, T. Feygelson, and C. T. C. Nguyen, "1.51-GHz nanocrystalline diamond micromechanical disk resonator with material-mismatched isolating support," *IEEE International Conference on Micro-electromechanical Systems*, Maastricht, the Netherlands, IEEE, Piscataway, N.J., pp. 641–644, 2004.

25. V. T. Srikar and S. D. Senturia, "Thermoelastic damping in fine-grained polysilicon flexural beam resonators," *J. Microelectromech. Syst.*, vol. 11, no. 5, pp. 499–504, October 2002.

26. B. H. Houston, D. M. Photiadis, J. F. Vignola, M. H. Marcus, Xiao Liu, D. Czaplewski, L. Sekaric, J. Butler, P. Pehrsson, and J. A. Bucaro, "Loss due to transverse thermoelastic currents in microscale resonators," *Mater. Sci. Eng. A*, vol. 370, no. 1-2, pp. 407–411, April 2003.

27. N. Abele, R. Fritschi, K. Boucart, F. Casset, P. Ancey, and A. M. Ionescu, "Suspended-gate MOSFET: bringing new MEMS functionality into solid-state MOS transistor," *IEEE International Electron Devices Meeting*, Washington, DC, IEEE, Piscataway, NJ, pp. 479–481, December 2005.

28. J. Rogers and C. Plett, *Radio Frequency Intergraded Circuit Design*, Artech House, Norwood, MA, 2003.

29. Y. Koutsoyannopoulos, Y. Papananos, S. Bantas, and C. Alemanni, "Performance limits of planar and multi-layer integrated inductors," *IEEE International Symposium on Circuits and Systems*, vol. 2, pp. 160–163, 2000.

30. J. Zhou, P. Li, S. Zhang, F. Zhou, Y. Huang, P. Yang, and M. Bao, "A novel MEMS gas sensor with effective combination of high sensitivity and high selectivity," *IEEE International Symposium on Applications of Ferroelectrics*, Nara, Japan, IEEE, Piscataway, NJ, pp. 471–474, June 2002.

31. A. Voiculescu, M. Zaghloul, and R. A. McGill, "Design, fabrication and modeling of microbeam structures for gas sensor applications in CMOS technology," *IEEE International Symposium on Circuits and Systems*, Bangkok, Thailand, IEEE, Piscataway, NJ, vol. 3, pp. III-922–III-925, May 2003.

32. D. Scheibner, J. Mehner, D. Reuter, T. Gessner, and W. Dotzel, "A spectral vibration detection system based on tunable micromechanical resonators," *Sens. Actuators A Phys.*, vol. A123-A124, pp. 63–72, 2005.

33. Q. Huang and C. Kuratli, "An ultrasound source based on a micro-machined electromechanical resonator," *IEEE International Symposium on Circuits and Systems*, Atlanta, GA, IEEE, Piscataway, NJ, vol. 4, pp. 348–351, May 1996.

34. M. Hornung, O. Brand, O. Paul, H. Baltes, C. Kuratli, and Q. Huang, "Micromachined acoustic Fabry-Perot system for distance measurement," *IEEE Micro-electromechanical Systems Annual International Workshop*, Heidelberg, Germany, IEEE, Piscataway, NJ, pp. 643–648, January 1998.

35. N. V. Lavrik, M. J. Sepaniak, and P. G. Datskos, "Cantilever transducers as a platform for chemical and biological sensors," *Rev. Sci. Instrum.*, vol. 75, no. 7, pp. 2229–2253, July 2004.

36. J. H. Seo and O. Brand, "Novel high Q-factor resonant microsensor platform for chemical and biological applications," *International Conference on Transducers, Solid-State Sensors, Actuators and Microsystems*, Seoul, Korea, IEEE, Piscataway, NJ, vol. 1, pp. 593–596, June 2005.

37. S. Ando, K. Tanaka, and M. Abe, "Fishbone architecture: an equivalent mechanical model of cochlea and its application to sensors and actuators," *International Conference on Transducers, Solid-State Sensors, Actuators and Microsystems*, Chicago, IL, IEEE, Piscataway, NJ, vol. 2, pp. 1027–1030, June 1997.

38. M. Bachman, F.-G. Zeng, T. Xu, and G.-P. Li, "Micromechanical resonator array for an implantable bionic ear," *J. Audiol. Neurotol.*, vol. 11, no. 2, pp. 95–103, 2006.

39. B. Antkowiak, J. P. Gorman, M. Varghese, D. J. D. Carter, and A. E. Duwel, "Design of a high-Q, low-impedance, GHz-range piezoelectric MEMS resonator," *International Conference on Transducers, Solid-State Sensors, Actuators and Microsystems*, Boston, MA, IEEE, Piscataway, NJ, vol. 1, pp. 841–846, June 2003.

40. B. P. Otis and J. M. Rabaey, "A 300-microwatt 1.9-GHz CMOS oscillator utilizing micromachined resonators," *IEEE J. Solid-State Circuits*, vol. 38, no. 7, pp. 1271–1274, July 2003.

41. A. P. S. Khanna, E. Gane, and T. Chong, "A 2GHz voltage tunable FBAR oscillator," *IEEE MTT-S International Microwave Symposium Digest*, Philadelphia, PA, IEEE, Piscataway, NJ, vol. 2, pp. 717–720, June 2003.

42. M. A. Dubois, C. Billard, C. Muller, G. Parat, and P. Vincent, "Integration of high-Q BAW resonators and filters above IC," *IEEE International Solid-State Circuits Conference*, vol. 1, pp. 392–606, February 2005.

43. J. F. Carpentier, A. Cathelin, C. Tilhac, P. Garcia, P. Persechini, P. Conti, P. Ancey, G. Bouche, G. Caruyer, D. Belot, C. Arnaud, C. Billard, G. Parat, J. B. David, P. Vincent, M. A. Dubois, and C. Enz, "A SiGe:C BiCMOS WCDMA zero-IF RF front-end using an above-IC BAW filter," *IEEE International Solid-State Circuits Conference*, San Francisco, CA, Lisbon falls, Maine: Digital Pub. Inc., 2005/IEEE, Piscataway, NJ, vol. 1, pp. 394–395, 2005.

44. K. Wang and C. T. C. Nguyen, "High-order micromechanical electronic filters," *IEEE Micro-electro-mechanical Systems Annual International Workshop*, Nagoya, Japan, IEEE, Piscataway, NJ, pp. 25–30, January 1997.

45. J. R. Clark, H. Wan-Thai, and C. T. C. Nguyen, "High-Q VHF micromechanical contour-mode disk resonators," *IEEE International Electron Devices Meeting*, San Francisco, CA, IEEE, Piscataway, NJ, pp. 493–496, 2000.

46. M. A. Abdelmoneum, M. U. Demirci, and C. T. C. Nguyen, "Stemless wine-glass-mode disk micromechanical resonators," *IEEE International Conference on Micro-electromechanical Systems*, Kyoto, Japan, IEEE, Piscataway, NJ, pp. 698–701, January 2003.

47. M. U. Demirci, M. A. Abdelmoneum, and C. T. C. Nguyen, "Mechanically corner-coupled square microresonator array for reduced series motional resistance," *International Conference on Transducers, Solid-State Sensors, Actuators and Microsystems*, vol. 2, pp. 955–958, June 2003.

48. Y.-W. Lin, S.-S. Li, Y. Xie, Z. Ren, and C. T. C. Nguyen, "Vibrating micromechanical resonators with solid dielectric capacitive transducer gaps," *IEEE International Frequency Control Symposium and Exposition*, Vancouver, BC, IEEE, Piscataway, NJ, pp. 128–134, 2005.

49. B. Bircumshaw, G. Liu, H. Takeuchi, T.-J. King, R. Howe, O. O'Reilly, and A. Pisano, "The radial bulk annular resonator: towards a 50-Ohm RF MEMS filter," *International Conference on Transducers, Solid-State Sensors, Actuators and Microsystems*, Boston, MA, IEEE, Piscataway, NJ, vol. 1, pp. 875–878, June 2003.

50. S.-S. Li, Y.-W. Lin, Y. Xie, Z. Ren, and C. T. C. Nguyen, "Micromechanical 'hollow-disk' ring resonators," *IEEE International Conference on Micro-electromechanical Systems*, Maastricht, the Netherlands, IEEE, Piscataway, N.J, pp. 821–824, January 2004.

51. A. Dec and K. Suyama, "Micromachined electro-mechanically tunable capacitors and their applications to RF IC's," *IEEE Trans. Microwave Theory Tech.*, vol. 46, no. 12, pp. 2587–2596, December 1998.

52. T. K. K. Tsang and M. N. El-Gamal, "Very wide tuning range micro-electromechanical capacitors in the MUMPs process for RF applications," *Symposium on VLSI Circuits*, Kyoto, Japan, IEEE, Piscataway, NJ, pp. 33–36, June 2003.

53. G. M. Rebeiz and J. B. Muldavin, "RF MEMS switches and switch circuits," *IEEE Microwave Mag.*, vol. 2, no. 4, pp. 59–71, December 2001.

54. J.-B. Yoon, C.-H. Han, E. Yoon, and C.-K. Kim, "High-performance three-dimensional on-chip inductors fabricated by novel micromachining technology for RF MMIC," *IEEE MTT-S International Microwave Symposium Digest*, Anaheim, CA, IEEE, Piscataway, NJ, vol. 4, pp. 1523–1526, 1999.

55 "International crystal manufacturing crystal oscillator and filter products catalog rev. A," *International Crystal Manufacturing Co. Inc.* Oklahoma City, OK, 2006. http://www.icmfg.com/

56. C. T. C. Nguyen, "RF MEMS in wireless architectures," *Proceedings of the Design Automation Conference*, pp. 416–420, June 2005.

57. S.-S. Li, M. U. Demirci, Y.-W. Lin, Z. Ren, and C. T. C. Nguyen, "Bridged micromechanical filters," *IEEE International Frequency Control Symposium and Exposition*, pp. 280–286, August 2004.

58. S.-S. Li, Y.-W. Lin, Y. Xie, Z. Ren, and C. T. C. Nguyen, "Small percent bandwidth design of a 423-MHz notch-coupled micromechanical mixler," *IEEE Ultrasonics Symposium*, Rotterdam, the Netherlands, IEEE, Piscataway, NJ, vol. 2, pp. 1295–1298, September 2005.

59. M. U. Demirci and C. T. C. Nguyen, "A low impedance VHF micromechanical filter using coupled-array composite resonators," *International Conference on Transducers, Solid-State Sensors, Actuators and Microsystems*, Seoul, Korea, IEEE, Piscataway, NJ, vol. 2, pp. 2131–2134, June 2005.

60. C. T. Nguyen, "Vibrating RF MEMS overview: applications to wireless communications," *Proc. Int. Soc. Opt. Eng.*, vol. 5715, pp. 11–25, January 2005.

61. S.-S. Li, Y.-W. Lin, Z. Ren, and C. T. C. Nguyen, "Self-switching vibrating micromechanical filter bank," *IEEE International Frequency Control Symposium and Exposition*, Vancouver, BC, IEEE, Piscataway, NJ, pp. 135–141, August 2005.

62. A.-C. Wong and C. T. C. Nguyen, "Micromechanical mixer-filters ('mixlers')," *J. Microelectromech. Syst.*, vol. 13, no. 1, pp. 100–112, February 2004.

63. W. Y. Ali-Achmad, "RF system issues related to CDMA receiver specifications," *RF Design Magazine*, vol. 22, no. 9, pp. 22–33, September 1999.

64. W.-T. Hsu, A. R. Brown, and K. R. Cioffi, "A programmable MEMS FSK transmitter," *IEEE International Conference of Solid-State Circuits*, San Francisco, IEEE, Piscataway, NJ, pp. 1111–1120, February 2006.

65. A. I. Zverev, *Handbook of Filter Synthesis*, Wiley, New York, NY, 1967.

66. D. S. Humpherys, *The Analysis, Design and Synthesis of Electrical Filters*, Prentice-Hall, Englewood Cliffs: Hemel Hempstead, 1970.

67. A. Williams and F. Taylor, *Electronic Filter Design Handbook*, McGraw-Hill, New York, NY, 1995.

68. J. Wang, Z. Ren, and C. T. C. Nguyen, "1.156-GHz self-aligned vibrating micromechanical disk resonator," *IEEE Trans. Ultrason. Ferroelectr. Freq. Control*, vol. 51, no. 12, pp. 1607–1628, December 2004.

69. Y. Tomikawa, S. Oyama, and M. Konno, "A quartz crystal tuning fork with modified basewidth for a high quality factor: finite element analysis and experiments," *IEEE Trans. Son. Ultrason.*, vol. 29, no. 4, pp. 217–223, July 1982.

22 Membrane-Supported Millimeter-Wave Circuits Based on Silicon and GaAs Micromachining

Alexandru Müller, Dan Neculoiu, George Konstantinidis, and Robert Plana

CONTENTS

Introduction...629
Fabrication and Design 631
    Technological Processes for Membrane-Supported Circuits
        Based on Silicon Micromachining 631
    Design Approach for Membrane-Supported RF MEMS 633
Membrane-Supported Coupled-Line Band-Pass Filters
    Based on Silicon Micromachining 635
Membrane-Supported Antennae Based on Silicon Micromachining.............637
    Membrane-Supported Folded Slot Antennae 637
    Membrane-Supported Yagi–Uda Antennae 640
    Membrane-Supported Yagi–Uda Antennae Characterization 641
Silicon-Based Micromachined Receiving Module.......................................644
GaAs-Micromachined Membrane-Supported Circuits 644
Quasioptical Mixer 647
Conclusions 651
References 652

INTRODUCTION

We have entered the information age with the continuous emergence of numerous wireless applications. Among the objectives are improvement in the quality of daily life (especially of the senior and disabled citizens), upgrading the health care, more efficient air and road traffic control, and more effective environmental monitoring. For the last 20 years, communications were almost exclusively dedicated to military and governmental applications involving only very few users. Today there is a radical change since the trends have totally transformed as most of the applications are devoted to the civil sector. Naturally this resulted (and will continue) in a dramatic increase in the number of customers as well as in the number of applications. All these changes led to an overcrowded frequency spectrum with an ensuing and continuous increase in the value of the allocated frequencies and to a radical modification concerning the required performance of the electronic modules. For example, for general communications systems, there is a continuous tendency to smaller, secure systems that have increased functionality and reduced power consumption. However, these demands place severe constraints on circuit power dissipation and electromagnetic (EM) compatibility and significantly increase the equipment design complexity. At the same time, one key issue is emerging and this is the noise originating from the electronic modules. Today it is mandatory to realize high-frequency receivers featuring very low-noise behavior. This demand, however, is conflicting with the requirements to increase the operating frequency, to decrease the power consumption, and to boost the number of users. Furthermore, the presence of several transmitters operating simultaneously on the same platform with multiple receivers requires receivers with a very high dynamic range and special attention to the overall design of the system for optimum EM compatibility. This usually requires filtering of the transmitters to rule out interference phenomena. Furthermore, the multiplication of the standard requires from the electronic modules to have a degree of agility to optimize the communication. The receiver or the transmitter has to switch to the standard or to the operator featuring the most efficient characteristics. Similar requirements apply to the battery and performance management. Depending on the EM environment, the performance of the receiver/transmitter can be more or less relaxed to save energy, which is a key issue in the case of portable communications, where the system has to be smart enough to choose the best configuration by trading off between the electrical performance (i.e., linearity, noise figure) and the power consumption.

These requirements can be summarized under the motive: "The radio frequency (RF) chip has to work nicely anywhere anytime and has to be as cheap as possible." The task of the system designers is to identify the technology that will meet all the requirements. Concerning the active devices, the situation is relatively clear. For applications up to 30 GHz, silicon will play a major role in the future with the use of silicon germanium bipolar technology that allows for the realization of devices featuring cutoff frequency in the 200 GHz range. For low-noise amplifier and for applications operating at frequencies higher than 30 GHz, the III–V technologies will still be the ideal candidate as they result in active devices exhibiting cutoff frequency in the 400 GHz range. However, concerning the passive elements, the situation is becoming more complex as it is very difficult to deliver passive devices featuring low loss and high quality factor up to millimeter-wave range. Furthermore, the need for smart devices is making the task harder. One emerging solution is the exploitation of the micromachining capabilities of semiconductor materials (i.e., silicon, GaAs, and InP). Silicon is a very promising candidate as it is very easy to micromachine and all the technological processes are very mature. Additionally, the fact that these technologies are compatible with the integrated circuit processes (including digital ones) will make the realization of high-frequency and high-integration level modules possible. This will be a key issue to save wafer area and lower fabrication cost. Another very attractive advantage of the silicon-based micromachined technologies is the use of its mechanical properties. It is possible to realize mobile regions (motion due to electrostatic, magnetic, piezoelectric, or thermal actuation) that can be incorporated into devices featuring tunable behavior. All these concepts are labeled under the name radio frequency microelectromechanical systems (RF MEMS). In many cases, a single MEMS component replaces and outperforms an entire solid-state circuit. In other cases, a thoughtful association of MEMS components with active devices will result in the realization of smart communicating devices. Today everybody is convinced that future telecommunications systems will contain these types of devices. Nevertheless, it is important to assess their performances and to make out the best technological processes in terms of compatibility with integrated circuits, performance, reliability, and finally cost.

A lot of effort has been put into RF MEMS technologies for the last 10 years. A broad range of components and subsystems have been demonstrated, with functions and performances that are significantly improved over conventional microwave and millimeter-wave technologies. Perhaps the most fertile research field for RF MEMS components is the microswitch. It offers small size, good insertion loss and isolation, low power dissipation, and excellent linearity. It can find applications as a standalone component for RF signal routing or as a constitutive element of tuners, phase shifters, reconfigurable filters, and antenna arrays. Switches are devices with moving parts, so they are often called true MEMS. But according to a complete definition, a MEMS is a device or set of devices fabricated with integrated circuit (IC) batch processing techniques combining both electrical and mechanical aspects. As a result, the RF MEMS family includes devices and circuits fabricated essentially using surface micromachining, bulk micromachining, fusion bonding, etc., even they have no moving parts: membrane-supported devices and circuits, three-dimensional integrated circuits, frequency-scaled circuits for terahertz range.

High-resistivity silicon has mechanical, thermal, and electrical properties comparable with best ceramics or other dielectric materials and has been successfully demonstrated as a low-cost substrate for microwave integrated circuits (MICs). The performance of integrated antennae and transmission lines on high-dielectric constant substrates is limited by the power radiation into substrate modes that increases as both frequency and relative dielectric constant increase. Planar components are also subject to dielectric losses, which increase with frequency. All these effects are crucial for the millimeter-wave and higher frequency applications.

One way to solve these problems is to fabricate the MICs on very thin dielectric membranes by removing the bulk silicon underneath the planar circuit. This substrate removal, by the use of micromachining techniques, has as its main effect the substantial reduction of losses in the millimeter-wave range. Additional beneficial effects are the reduction of dispersion effects, the suppression of higher substrate modes, the possibility of using higher transmission line characteristic impedance values in design, and easy scaling of the design to different frequencies. As the membrane supporting the circuits is very thin, the propagating properties are similar to those obtained in the free space. These elements look like being air suspended.

This chapter refers to the design, modeling, technological processes, and characterization techniques developed to manufacture and characterize high-performance millimeter-wave elements and circuits such as membrane-supported low-loss band-pass filters, double-folded slot antennae, and Yagi–Uda end-fire antennae, based on silicon and gallium–arsenide micromachining. The antenna structures are integrated with semiconductor devices into video detection and quasioptical mixing receiver front-ends for millimeter-wave applications.

These new technologies can be employed as an emerging solution for the manufacture of high-performance circuits in the submillimeter frequency range. Most of the presented results were reported by the authors during the last few years.

FABRICATION AND DESIGN

TECHNOLOGICAL PROCESSES FOR MEMBRANE-SUPPORTED CIRCUITS BASED ON SILICON MICROMACHINING

The concept of membrane-supported components was introduced by L. Katehi, G. Rebeiz and colleagues [1–4] by developing a suspended technology from a tri-dielectric layer sandwich ($SiO_2/Si_3N_4/SiO_2$) on high-resistivity (100)-oriented silicon wafers. The membrane has a total thickness of 1.5 μm. Thermal oxidation was used to grow the first 7500Å silicon dioxide layer. The nitride layer (3000Å) was deposed by low pressure chemical vapor deposition (LPCVD) technique. The third silicon dioxide layer (4500Å) was deposited by chemical vapor deposition (CVD) at a temperature of about 400°C. This structure has a slight tensile stress to yield flat and self-supporting rigid membrane characteristics. This kind of technology involves wet backside etching using KOH, ethylene diamine-pyrocatechol (EDP), or tetramethylammonium (TMAH) solutions. Also, MOS-compatible dry backside etching techniques can be used. The membrane formation process has been simplified and optimized with respect to the reliability by Etienne et al. [5] through an investigation of the intrinsic strain between different layers (i.e., dielectrics and metallization). It has been shown that appropriate gas composition and deposition conditions resulted in a minimized overall strain (80MPa tensile) only with two dielectric layers (SiO_2/Si_3N_4: 0.8/0.6 μm).

The improvements obtained by membrane-supported circuits are mainly due to the fact that a very thin dielectric or high-resistivity semiconductor membrane used as support for passive circuit elements (inductors, capacitors, transmission lines, filters, antennae) reduces the effective permittivity to values very close to 1. First experimental measurements of the effective permittivity for $SiO_2/Si_3N_4/SiO_2$ membrane-supported transmission lines were reported in 1994 [6] and it was found to be about 1.08. Experimental determination for the effective permittivity for SiO_2/Si_3N_4 membrane-supported coplanar waveguide (CPW) transmission lines was reported in Ref. 7. A value of about 1.1 was obtained in the 10–65 GHz frequency range.

Three different technological process flows to manufacture membrane-supported circuit elements based on silicon micromachining are presented in Figure 22.1. The top-side process is the same for all three approaches. The initial wafer is a commercially available high-resistivity ($\rho > 5\,k\Omega\,cm$) (100)-oriented silicon wafers with a thickness of 400 μm. The process starts with the bi- or tridielectric layer formation. Then, a 0.55 μm thin TiAu (0.05 μm Ti/0.5 μm Au) seed layer is evaporated over the wafer. Selective electroplating of gold, with a thickness of 3.5 μm, is performed after patterning the top of the wafer with the front mask. A 10 μm thick AZ 4562 type resist is used for the backside processing of the wafer.

For the backside processing, the classical procedure consists of wet etching in KOH (20% at 80°C). A bulk silicon wall will surround the structure on all four edges. Its thickness is determined by the 54.7° angle typical for the anisotropic etching of (100)-oriented silicon in KOH or TMAH solutions and by the thickness of the silicon wafer. The bulk silicon wall is about 1–1.5 mm wide for wafers having usual thickness of 350–525 μm. For lumped elements, filters, and broadside antennae, this topology has no influence on the performance of the circuits. The wet backside etching process has a very good yield and reproducibility.

However, for end-fire membrane-supported antenna structures, a remaining relatively thick bulk silicon wall in the main propagation direction can seriously affect their radiation characteristics. This is the reason that we also developed dry backside etching techniques [8], which were used in manufacturing of membrane-supported end-fire elements. The most important advantage of dry deep reactive ion etching (DRIE) compared to the classical wet etching process consists in the fact

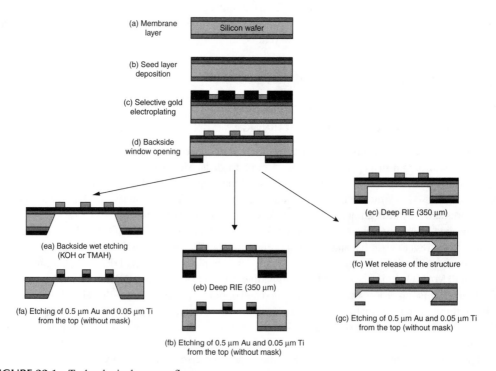

FIGURE 22.1   Technological process flows.

that vertical etch walls can be obtained. This, coupled with the proper design of the backside mask, can reduce the thickness of the silicon etch wall in the direction of the radiation to 50 μm only.

This is very important to minimize the perturbation of the antenna end-fire radiation pattern. The main disadvantage of this technological process is the relatively low yield. The main reason for the reduction of the yield is the nonuniformity of the 400 μm deep reactive ion etching (RIE) process.

We have also developed a new technological process [9], which uses dry followed by wet backside etching of the silicon wafer. Dry RIE (DRIE) is used for the selective removal of 350 μm silicon. The last 50 μm of silicon is removed by wet etching, in KOH solution. Note that no additional mask is needed for the final 50 μm wet etching, as the etching stops at the membrane region. The very spectacular result of this original process was the quasithree edges membrane-supported antenna structures. The bulk silicon wall in the main radiation direction of the antenna is practically removed. At one edge of the structure, the 350 μm high and 50 μm thin silicon wall, remaining after the RIE process, is amputated during the wet etching, in KOH, as depicted in Figure 22.1. The overhanging silicon continues to be etched up to the membrane formation. Only a very thin and narrow silicon layer remains at this end of the membrane. Further wet etching, up to the complete removal of the silicon beak from the structure, has, as effect, the collapse of the membrane due to the excessive compressive stress.

Finally, in all three approaches, the exposed seed layer (in the nonelectroplated regions) is removed from the top of the wafer without any mask by wet etching.

DESIGN APPROACH FOR MEMBRANE-SUPPORTED RF MEMS

Computer-aided design (CAD) tools for membrane-supported microwave components have not yet been adequately developed because of the lack of accurate and efficient models. This modeling problem becomes more important at millimeter-wave frequencies. For planar transmission lines, the design formulas (mostly based on quasistatic approximations) are no longer accurate or are out of their range of validity. One way to solve this problem is to use EM simulators for the characterization of single or coupled transmission lines.

EM simulation techniques for high-frequency structures were developed in the last decade and brought the microwave CAD software to a level of maturity. Generally speaking, there are two large families of commercially available EM solvers. The first one includes general-purpose full-wave simulators that can analyze arbitrarily shaped dielectric and metallic structures, using finite elements or differences, in either the frequency or time domains. The design of the whole micromachined structure using a full-wave EM simulator is prohibitively time-consuming and requires huge computer resources, mainly due to the presence of the very thin dielectric membrane with critical aspect ratio (the ratio between the largest and smallest dimensions involved).

The second one is based on the method-of-moments (MoM) and includes EM simulators specifically optimized for the efficient analysis of printed planar structures with a uniform dielectric layer structure. The spectral domain simulation is based on surface electric currents over infinitely thin conductors (or magnetic currents over gaps) and the information about the surrounding space is embedded within an operator known as the Green function. The efficiency and accuracy are excellent, but only infinitely thin conductors are normally allowed in EM modeling (hence they are indicated as 2.5D methods).

The efficient use of MoM commercial software packages such as Microwave Office (Applied Wave Research Inc.), IE3D (Zeland Inc.), or Sonnet for accurate design of membrane-supported microwave and millimeter-wave circuits was demonstrated for transmission lines (single and coupled), filters, antennae, and receiving front-ends (in the next sections will be presented some representative examples). These EM simulators perform EM analysis for arbitrary 3D planar geometry (e.g., microstrip, coplanar, stripline, etc.), maintaining full accuracy at all frequencies. They have extensive postprocessing capabilities (displaying of [S] parameters, the characteristic impedance, and the effective dielectric constant of the section of line leading up to a port, current density distribution, antennae radiation pattern). The EM analysis includes dispersion, discontinuities, surface waves, higher order modes, metallization loss, dielectric loss, and radiation loss.

The IE3D software package is based on open-boundary Green's function formulation [10]. This approach is well matched for the modeling of antenna structures without metallic enclosures. The IE3D software includes an optimization engine that allows the use of multiple objective functions such as return losses and antenna gain/directivity. The optimization variables are the main layout dimensions. The EM simulation results are saved in Touchstone-type files and the whole circuit is analyzed and optimized using commercial nonlinear circuit simulation software based on harmonic balance method.

The design methodology of the membrane-supported circuits is based on the following design rule: "Divide the microwave circuit into membrane-supported components, bulk-supported components, and membrane-to-bulk transition regions. Design every type of components using appropriate design methods and techniques. Determine the performances of the whole circuit using linear/nonlinear circuit techniques using the partial analyses results." This rule is illustrated in the flowchart in Figure 22.2.

Because of the advantages of the membrane-supported components, all the critical millimeter-wave circuitry (antennae, selective networks for band-pass filters) must be placed on the micro-machined dielectric membrane. The membrane area must be as small as possible. The rest of the elements (feed transmission lines for filters, the Schottky diodes, and the low-pass filter that connects the microwave circuit with the video output) are situated on bulk silicon.

The membrane-supported components can be modeled and optimized using IE3D simulation software. Although IE3D package includes an optimization engine, it is mandatory in the design to start from an initial solution for layout dimensions obtained by means of an analytical design approach. The bulk-supported components can be modeled and designed using both EM simulations and closed-form formula (analytic design). Every circuit block is finally EM simulated and the results as S-parameters are stored in *.sxp Touchstone-type files.

Membrane-to-bulk transition regions cannot be simulated using IE3D (or other MoM simulators) because of the nonuniform dielectric layer (see Figure 22.1). One solution can be the use of a general-purpose full-wave simulator. But this approach is very time-consuming and requires huge computer resources. Another solution is to use a circuital approach: the nonuniform CPW transmission line (because of the substrate thickness variation) is divided into small length sections of uniform transmission line with constant substrate thickness. Every section is simulated using IE3D, and the main transmission line parameters (characteristic impedance, effective permittivity, loss

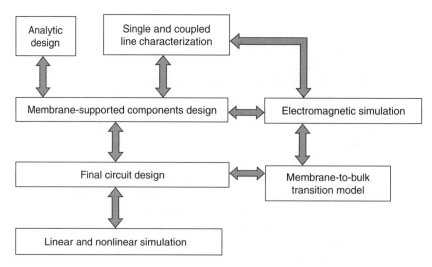

FIGURE 22.2 Design flowchart.

coefficient) are extracted for a wide frequency range. The equivalent circuit model consists in a chain (cascade) of small-length transmission lines, which is analyzed for [S] parameters.

The final circuit design uses a linear and nonlinear circuit simulator and [S] parameter data files. Each region is modeled in terms of multiport linear parameters and the results are put together to obtain network parameters of the whole receiver structure.

There is a problem with the integration in the same design environment of EM simulation and nonlinear analyses using the harmonic balance technique.

The harmonic balance solution requires the solution of the linear components in the circuit at DC, the fundamental frequency, and the specified number of harmonics of the fundamental frequency. The range of frequencies used to simulate the structure with an EM simulator should be wide enough to cover the frequency band of the circuit simulation. If the EM simulation is over a narrower band of frequencies, then the system must extrapolate the frequency response of the EM structure. The obtained results can be incorrect. On the contrary, it is not possible to simulate using an EM simulator (like IE3D) the microwave circuit at high-order harmonics (because of computing resources and simulation time). Also, the EM simulators can produce unusable results at low frequencies and for DC analyses. The integration of nonlinear and EM simulation is an important topic of further research.

MEMBRANE-SUPPORTED COUPLED-LINE BAND-PASS FILTERS BASED ON SILICON MICROMACHINING

The first membrane-supported filter structures were reported in 1995. The two filters proposed by Chi and Rebeiz [11] were coupled-line type structures. They had a bandwidth of 40 and 5% with losses of 0.7 and 2 dB, respectively, at 14–15 GHz. A 250 GHz microshield band-pass filter was proposed by Weller et al. [12]. An open-end series stub configuration was used. Weller and Katehi presented in 1996 [13] a low-pass filter using lumped elements obtaining losses of about 0.5 dB in the Ka band. High-performance 37 and 60 GHz planar micromachined filters on silicon were presented by Blondy et al. [14].

The use of micromachining techniques offers clear advantages in comparison with traditional technologies used in the millimeter-wave region, such as (1) lower insertion losses than any other existing technology available for millimeter-waves; (2) large bandwidth of the devices; and (3) easy fabrication and chip integration.

We have developed a new design for membrane-supported coupled-line filter structures. The filter design is based on cascade connection of identical symmetric elementary cells and the image parameters concept. The design strategy is based on the image parameter representation of two port networks. This approach, originally used only for the design of low-pass microstrip filters, is modified and applied to micromachined band-pass filters based on CPW, working at millimeter-wave frequencies. The band-pass filter is composed of a number of identical symmetric elementary cells connected in cascade. Image parameters are utilized to obtain a simple analytical model of the entire structure. For such a network, the image impedance of the filter (Z_{if}) is the same for each basic cell (Z_{ic}). In this way, all conditions regarding the EM performances of the filter can be analytically imposed directly on Z_{ic}, i.e., on one elementary cell. The general circuit theory states that a frequency-dependent, lossless two-port network has a band-pass behavior if its image impedance Z_{ic} is real and a band-stop behavior if Z_{ic} is purely imaginary. Therefore, two equations are provided by the conditions $Z_{ic} = 0$ or $Z_{ic} = \infty$ at the two cutoff frequencies. One more equation is obtained imposing the matching condition at the central frequency of the passband.

The elementary cell consists of three series-connected two-port networks made by means of ideal transmission lines: (1) two single transmission line sections characterized by the parameters Z_c (the characteristic impedance) and θ_p (the electrical length) and (2) one coupled line section characterized by Z_{oe} (the even-mode impedance), Z_{oo} (the odd-mode impedance), and θ (the coupling electrical length). Since the image parameter approach provides three analytic conditions and the

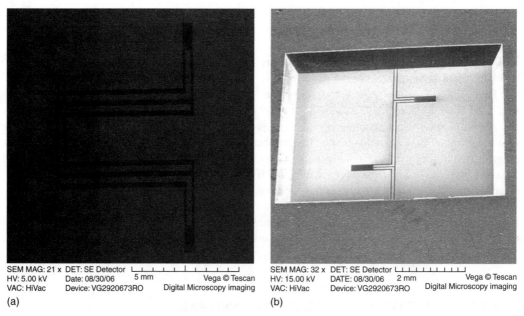

(a)

(b)

FIGURE 22.3   Top- (a) and bottom-side (b) photos of the 38 GHz filter structure supported on $SiO_2/Si_3N_4/$ SiO_2 membrane.

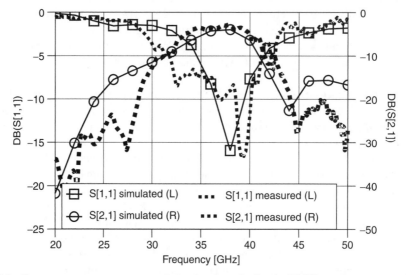

FIGURE 22.4   S-parameter measurements and simulated results for the 38 GHz coplanar waveguide filter.

basic cell is characterized by five parameters, there are two degrees of freedom to satisfy the technological constraints, in particular, to have suitable values for the characteristic impedance. The above approach represents the analytical design step for the design approach presented in the section "Design Approach for Membrane-Supported RF MEMS."

A first example is a coupled-line filter for 38 GHz. Top- and bottom-side SEM photos of these filter structures are presented in Figures 22.3a and 22.3b. The comparison between the measured values of the S-parameters and the modeled ones is presented in Figure 22.4 (for the 38 GHz receiver). Minimum losses are around 1.5 dB [15].

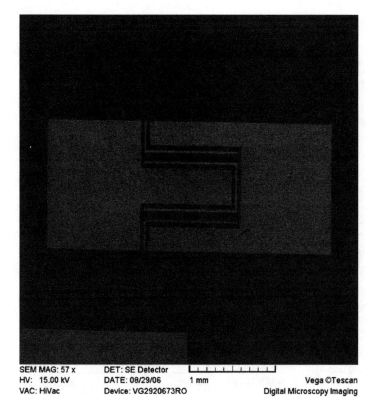

SEM MAG: 57 x   DET: SE Detector   ⌐ı ı ı ı ı ı ı ı ⌐
HV: 15.00 kV   DATE: 08/29/06   1 mm   Vega ©Tescan
VAC: HiVac   Device: VG2920673RO   Digital Microscopy Imaging

FIGURE 22.5   Topside photos of the 45 GHz filter structure supported on $SiO_2/Si_3N_4/SiO_2$ membrane.

The second example is a compact filter structure for 45 GHz operating frequency [16]. Only the coupled line section was changed in comparison with Figure 22.4. Top- and bottom-side SEM photos of this structure are presented in Figures 22.5 and 22.6, and the measured S-parameters are presented in Figures 22.7a and 22.7b. An excellent agreement is observed between the computed and measured values. Losses as low as 0.9 dB are obtained. This new filter configuration is also tested on micromachined GaAs substrate and the results will be presented in the section "GaAs-Micromachined Membrane-Supported Circuits."

MEMBRANE-SUPPORTED ANTENNAE BASED ON SILICON MICROMACHINING

MEMBRANE-SUPPORTED FOLDED SLOT ANTENNAE

The first membrane-supported antennae were proposed by Rebeiz et al. [18]. This antenna was a dipole suspended in an etched pyramidal cavity on a 1 μm thick silicon oxinitride membrane. Micromachined microstrip antennae were first presented in Refs. 19 and 20. The thick silicon substrate of high dielectric constant under the antennae is first removed and then a low dielectric constant substrate is locally synthesized. In Ref. 21, a double-folded slot antenna is presented. The antenna is placed on an extended hemispherical high-resistivity silicon lens to achieve a high directivity. The double-folded slot antenna has a broadside, symmetric bidirectional radiation pattern and can be easily manufactured on thin membranes obtained by silicon micromachining. Membrane-supported folded slot antenna test structures for 77 and 94 GHz were reported by our group in Ref. 22.

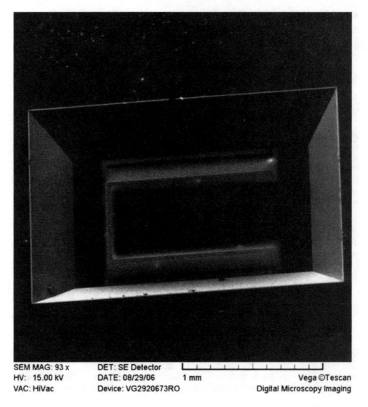

FIGURE 22.6   Bottom-side photos of the 45 GHz filter structure supported on $SiO_2/Si_3N_4/SiO_2$ membrane.

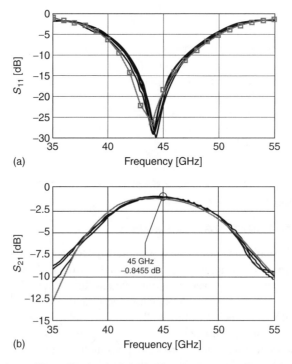

FIGURE 22.7   (a) Band-pass filter reflection losses: simulated (squares) and experimental results for several structures. (b) Band-pass filter transmission: simulated and experimental results for several structures. (From Neculoiu, D. et al., *Electron. Lett.*, 40, 180–182, 2004. With permission. © IEEE 2004.)

A folded slot antenna for 38 GHz was designed, manufactured, and characterized, and results will be presented in this section.

In the first stage of the design, the characteristic impedance and effective permittivity of the membrane-supported CPW lines are determined. The second stage addresses the estimation of the slot length S and slot separation D (Figure 22.8) through the calculation of the half wavelength at the operating frequency. Because the antenna is supported on a very thin dielectric membrane (the effective permittivity is close to 1), both S and D are chosen to be equal to the half of the free-space wavelength at the frequency of operation. After the optimization of the antenna performance, the new layout dimensions are: $L = 3.000\,\mu m$, $D = 3.600\,\mu m$, and $S = 1.550\,\mu m$. The simulated radiation pattern of the demonstrator is presented in Figure 22.9 (input-matching conditions at the input port

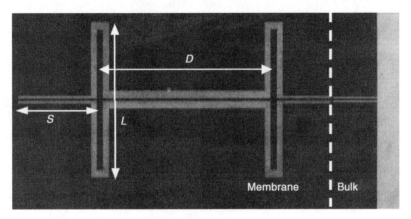

FIGURE 22.8 The layout of the 38 GHz folded slot antenna.

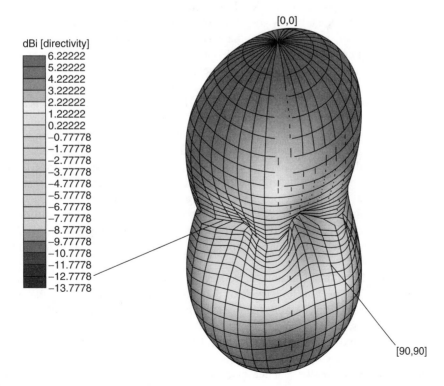

FIGURE 22.9 The radiation pattern for the 38 GHz folded slot antenna.

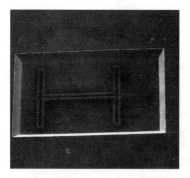

FIGURE 22.10 Bottom-side SEM photo of the membrane-supported folded slot antenna for 38 GHz.

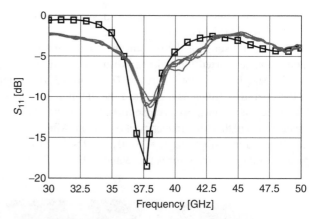

FIGURE 22.11 Experimental (dotted line) and simulated (continuous line) measurements return losses for the 38 GHz antenna.

are assumed). The moderate gain of about 6 dBi of the antenna can be increased by almost 2.5 dB by placing a reflector at the backside of the antenna. Several structures of the membrane-supported double-folded slot antennae are fabricated and tested. Bottom-side SEM photo of the membrane-supported folded slot antenna for 38 GHz is presented in Figure 22.10. The structure is visible by transparency through the 1.4 μm thin membrane. The experimental measurements for the return losses are compared with that of the simulated (Figure 22.11). The agreement between simulated and measured data is very good and validates the design approach.

MEMBRANE-SUPPORTED YAGI–UDA ANTENNAE

For the millimeter-wave range, recent studies developed by our group [23,24] have demonstrated that the Yagi–Uda configuration is a very good solution for millimeter-wave frequencies and above. The Yagi–Uda antenna is an end-fire antenna that has the radiation pattern in the same plane with the antenna structure in contrast with the folded slot antenna, which is a broadside antenna (the radiation pattern is perpendicular to the antenna plane).

The Yagi–Uda antenna was invented in 1926 by Yagi and Uda and has become a standard for multielement antenna arrays from the HF through the UHF range. Yagi–Uda antenna is a traveling wave structure that, as the number of elements increases, has improved directivity, gain, and front-to-back ratio. The Yagi–Uda array is based on principles of parasitic elements that are not directly fed by an energy source. These elements focus the radiation pattern by the currents induced in them by radiation from the driving element. According to their length and spacing from the driving element, the parasitic elements become either directors or reflectors. Directors are shorter than the driving element and result in the radiation pattern being sharply distorted in the direction of the director. Reflectors are longer than the driving element and result in a radiation pattern directed away from them. A novel concept named "quasi-Yagi" antenna has been proposed in Ref. 25, which takes advantage from both the generation of surface wave and the truncated CPW ground plane as the reflecting element.

It is possible to fabricate Yagi–Uda antennae on a very thin dielectric membrane by using micromachining techniques. The overall dimensions of the antenna are comparable with the free-space wavelength, so this approach is very well suited for millimeter-wave and sub–millimeter-wave frequency range, up to the terahertz region. The advantages of the membrane-supported components are preserved. Using the Zeland IE3D software, the Yagi–Uda antenna was designed by optimizing the layout dimensions (driver, directors, and reflector parameters, in terms of spacing, length, and width). The main target parameter was the antenna gain, which is in close connection with the antenna

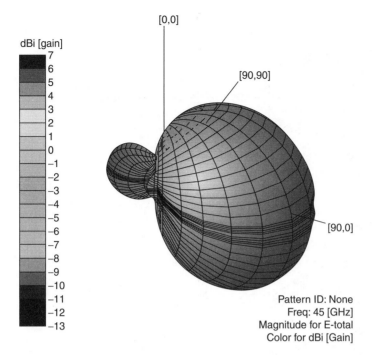

FIGURE 22.12 Simulated 3D radiation pattern of the Yagi–Uda antenna.

reflection losses and radiation pattern. Antenna gain must be maintained at reasonable values across the entire operating bandwidth centered on 45 GHz (for this specific case). The final optimization of the antenna layout includes the CPW slot-line transition parameters (the length of the slots, the length of the slot line, etc.). The optimized antenna radiation pattern is presented in Figure 22.12, and the final layout dimensions are presented in Figure 22.13 (all dimensions are in micrometers).

The manufacturing of membrane-supported end-fire antennae (like Yagi–Uda antennae) is a difficult task and this was the reason to develop dry and dry followed by wet etching techniques for the membrane formation presented in the section "Introduction." Yagi–Uda antennae were manufactured using dry etching techniques [8]. An only 50 μm wide silicon wall remained in the main radiation direction of the antenna with no observable influence on the radiation characteristics. Dry followed by wet etching techniques [9] were also used to manufacture high-performance Yagi–Uda antennae. An SEM photo of a quasithree-edge-supported Yagi–Uda antenna structure is presented in Figure 22.14.

Membrane-Supported Yagi–Uda Antennae Characterization

The antennae experimental characterization is a difficult task. The target parameters can be divided into two categories:

1. Circuit parameters (antenna impedance, voltage standing wave ratio (VSWR), reflection loss, etc.)
2. Radiation parameters (radiation pattern, antenna gain and directivity, main direction of radiation, etc.)

The parameters from the first category can be measured using an on wafer experimental setup. Free-space conditions must be provided for the antenna radiation. This can be done by placing the antenna on a dielectric slab with low loss and a dielectric permittivity close to 1, by covering the metallic parts of the experimental setup with appropriate absorber material. This approach is suitable only at

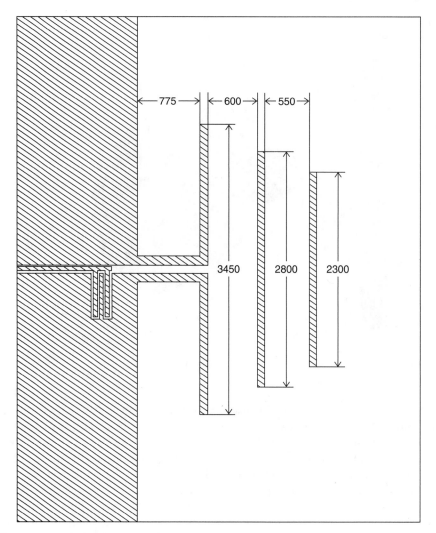

FIGURE 22.13   The layout of the membrane-supported Yagi–Uda antenna.

millimeter-wave range and above when it is possible to place the building blocks of the experimental setup far away from the antenna structure (at least several wavelengths).

The parameters from the second category can be measured using special experimental setup. The end-fire antennae (like Yagi–Uda, tapered slot antennae, etc.) can still be evaluated for antenna gain using an on wafer experimental setup. The accuracy of the experimental results is affected due to the interference from the metal platform and other metal parts of the probe station that are close to the antenna structures.

The characterization technique presented in Ref. 26 was used in membrane-supported Yagi–Uda antenna measurements. This characterization method is accurate, fast, and inexpensive when automated for repeated measurements. The main drawback of the method is the fact that it does not allow to determine the antenna radiation pattern. However, using the on wafer characterization the technological processes can be validated, to evaluate the repeatability and the possible integration of the antenna structure with active components in emitting/receiving front-ends, for wireless communication microsystems.

Two identical antennae have been placed face to face separated by a distance R (Figure 22.15). One antenna operates as an emitter, the other as a receiver. From transmission measurements and

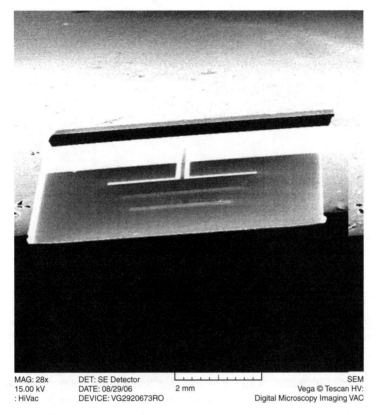

MAG: 28x   DET: SE Detector   SEM
15.00 kV   DATE: 08/29/06   2 mm   Vega © Tescan HV:
: HiVac   DEVICE: VG2920673RO   Digital Microscopy Imaging VAC

FIGURE 22.14   SEM photo of a quasithree-edge-supported Yagi–Uda antenna structure.

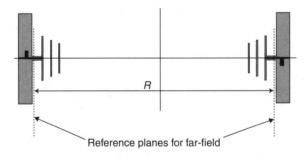

FIGURE 22.15   Experimental setup for end-fire antenna characterization. (From Neculoiu, G. et al., *IEE Microwave Antennae Propagation*, 151, 311–314, 2004. With permission. © IEE 2004.)

assuming that the antennae have similar feature characteristics, the gain G can be calculated using the Friis' formula:

$$G^2 = |S_{21}|^2 \left(\frac{4\pi R}{\lambda_0} \right)^2 \tag{22.1}$$

where S_{21} is the transmission S-parameter and λ_0 is the free-space wavelength.

The reflection losses for the manufactured Yagi–Uda antenna structures are presented in Figure 22.16 while the gain is presented in Figure 22.17.

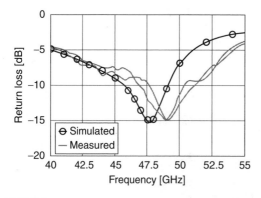

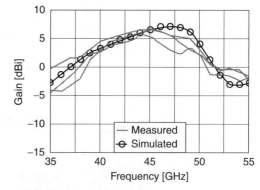

FIGURE 22.16 The reflection losses for the manufactured Yagi–Uda antenna structures. (From Neculoiu, G. et al., *IEE Microwave Antennae Propagation*, 151, 311–314, 2004. With permission. © IEE 2004.)

FIGURE 22.17 The gain of the manufactured Yagi–Uda antenna structures. (From Neculoiu, G. et al., *IEE Microwave Antennae Propagation*, 151, 311–314, 2004. With permission. © IEE 2004.)

SILICON-BASED MICROMACHINED RECEIVING MODULE

Membrane-supported antennae can be readily used for the manufacturing of hybrid integrated direct (video-type) receiving modules. The receiver consists of a membrane-supported antenna, a matching network, a flip chip Schottky detector diode bonded on bulk material, and a low-pass filter. Such a structure was first reported by us in Ref. 27. The receiver is based on a membrane-supported folded slot antenna hybrid integrated with GaAs flip chip Schottky detector diode, DMK 2790. The data sheet electrical specifications are series resistance $R_S = 4\,\Omega$, total capacitance $C_T = 0.05\,\text{pF}$ (the overlay capacitance $C_p = 0.02\,\text{pF}$), ideality factor $n = 1.05$, and saturation current $I_S = 0.5\,\text{pA}$. The Schottky diode is placed in series between the matching network and the low-pass filter and is mounted using conductive epoxy. A bridge circuit biases the detector device and, using a second Schottky diode, allows bridge balance and temperature effect compensation. The membrane to bulk transition is a nonuniform CPW line (the substrate thickness changes from membrane thickness to 525 µm) and it is modeled as a cascade of very short transmission lines with constant effective permittivity and characteristic impedance. The transmission line parameters are found by means of EM simulations. The matching network is fabricated on bulk material and consists of a 1000 µm long 50/50/50 µm CPW line followed by a 450 µm long 175/350/175 µm CPW line (350 µm is the width and 175 µm is the gap of the CPW line). The low-pass filter presents a short circuit at the input at the frequency of 38 GHz. The filter consists of three sections: two low-impedance sections (25/350/25 µm CPW line) and one high-impedance section (175/50/175 µm CPW line). All the sections are 775 µm long. The membrane-supported antenna is a folded slot type. The layout of the receiver module is presented in Figure 22.18, and the photo of the receiver structure with the mounted Schottky diode placed on the PCB is presented in Figure 22.19. The setup used to measure the radiation pattern is shown in Figure 22.20 while the voltage sensitivity of the receiver module is presented in Figure 22.21.

GaAs-MICROMACHINED MEMBRANE-SUPPORTED CIRCUITS

Micromachining of GaAs is an exciting less-explored alternative for the manufacturing of components and modules for high-performance communication systems. GaAs micromachining is very interesting for the RF MEMS field also due to the potential for easy monolithical integration of micromachined passive circuit elements with active devices manufactured on the same chip.

The molecular beam epitaxy (MBE) or metal-organic CVD (MOCVD) epitaxy grown layers of III–V compound semiconductor heterostructures provide flexibility and precision in micromachining. These layers have sharp interfaces, and due to different composition, can be etched by

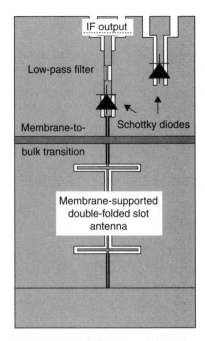

FIGURE 22.18 The layout of the silicon micromachined receiver for 38 GHz.

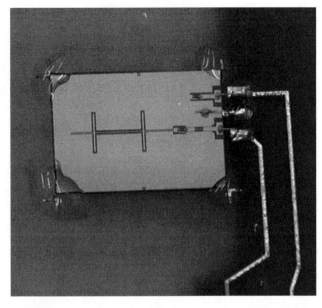

FIGURE 22.19 Photo of the receiver structure with the mounted Schottky diode.

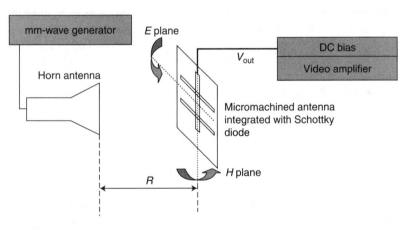

FIGURE 22.20 Experimental setup for radiation pattern measurements.

wet or dry techniques with excellent selectivity. The GaAs/Al$_x$Ga$_{1-x}$As system can be used for this purpose. Some dry and wet etching systems exhibit etching rates of GaAs orders of magnitude higher than those for Al$_x$Ga$_{1-x}$As and vice versa (if $x \geq 0.5$). Most commonly, AlGaAs is used as an etch-stop layer for GaAs. A lot of selective wet etching solutions for the GaAs/AlGaAs system were reported [28–30], but the best and reproducible results were obtained using dry etching systems. Using the GaAs/AlGaAs heterostructure and the etch-stop properties of AlGaAs with respect to GaAs, pressure, power, and thermoelectric sensors were reported [31–33]. The typical thickness of the GaAs membranes for this type of sensing structures is about 1 μm and the thickness of the etch-stop AlGaAs layer is about 0.2 μm. Millimeter-wave filter structures supported on GaAs membranes were manufactured in the last years by groups involving the authors of this chapter [34,35]. Substrateless Schottky diodes for applications in the terahertz range were manufactured in the last few years [36,37]. These diodes have as main advantages the reduction of series resistance, less

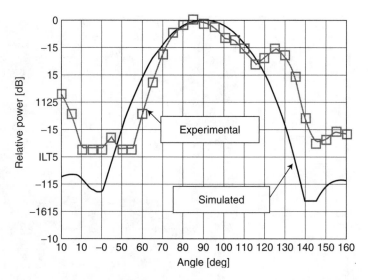

FIGURE 22.21 Experimental and simulated radiation patterns at 38 GHz (E plane) for the double-folded slot receiver antenna.

influence of skin effect, quasivertical current flow, and increase of power handling capabilities. A monolithical integration of an antenna with a Schottky diode, on the same thin GaAs membrane, in a millimeter-wave receiver module was recently reported by the authors of this chapter [38]. Some circuits for applications in the sub–millimeter-wave range based on GaAs micromachinig were also recently reported and are described in Ref. 39.

Conventional and low-temperature (LT) III–V MBE growth was used to fabricate the GaAs/AlGaAs/GaAs heterostructure. Semi-insulating GaAs wafers ($\rho = 10^7\,\Omega$ cm), with a thickness of 460 μm, were used as substrate. The MBE process started with a very thin (50 nm) buffer GaAs layer deposition. Over this layer, a 0.2 μm thin $Al_xGa_{1-x}As$ etch-stop layer (with $x = 0.6$) was deposited. Over the AlGaAs layer, an LT semi-insulating 2 μm thin GaAs layer ($\rho > 10^6\,\Omega$ cm) was deposited. Over this layer, GaAs layers of various thicknesses and doping levels are grown depending on the application and the frequency of operation. The growth experiments were performed in a VG80 horizontal MBE chamber with a background pressure of 10^{-10} mbar. During growth, the chamber pressure was 10^{-7} mbar.

Contact lithography, e-gun evaporation, and liftoff techniques were used to define filter or antenna structures. A 500 Å Ti/7000 Å Au metallization was used, and then the wafers were mounted facedown on special glass plates and the GaAs substrate was thinned down to 150 μm by lapping technique.

The etching pattern for the membranes was defined by backside alignment contact photolithography. The membranes were fabricated in a Vacutec 1350 RIE chamber using CCl_2F_2. End-point detection and optical (visual) detection were used during the RIE etching. After the selective etching, the thickness is approximately 2.2 μm. The photo of the quasivertical etching profile obtained using selective DRIE process is presented in Figure 22.22.

The effective permittivity of the GaAs-membrane-supported structures was determined using two different lengths of CPW lines. The S-parameters were measured and the effective permittivity was extracted. The value was 1.45 in the range 10–65 GHz.

A coupled-line filter structure for 45 GHz operating frequency (similar to the structure presented in the section "Membrane-Supported Antennae Based on Silicon Micromachining") was designed and manufactured on a GaAs membrane [40].

In Figure 22.23a, top view of the filter structure is presented. Figure 22.23b presents a bottom view of a GaAs-membrane-supported filter structure; details from the filter structure are visible through the transparency of the membrane. The comparison between the measured and simulated results for the filter structure are presented in Figure 22.24.

FIGURE 22.22   SEM photo of the 2 μm thin GaAs/AlGaAs membrane and of the selective dry etching profile.

The most exciting circuits we have developed using GaAs micromachining are two fully mono-lithic integrated direct receiver modules on thin GaAs membrane. The first one is based on a folded-slot membrane-supported antenna [38]; the second one uses a Yagi–Uda antenna [41]. For the two receivers, both the antennae as well as the monolithic integrated Schottky diodes are supported on the same 2 μm thin GaAs membrane. The heteroepitaxial structure parameters are presented in Figure 22.25. A photo of the monolithic integrated receiver module for 38 GHz is presented in Figure 22.26, and a detail from the Schottky diode region is presented in Figure 22.27. The mea-sured voltage sensitivity for the 38 GHz receiver structure is presented in Figure 22.28.

The Yagi–Uda monolithic integrated receiver module for 45 GHz was recently manufactured by our group [42]. Photo of the receiver structure is presented in Figure 22.29. The microwave mea-surements are in progress.

The monolithic integration of an antenna with a Schottky diode on the same GaAs membrane is important due to future possibilities to employ them in receiving modules working in the submil-limeter frequency range up to the terahertz region.

QUASIOPTICAL MIXER

The receiving front-end presented in the section "Silicon-Based Micromachined Receiving Mod-ule" can be used as a quasioptical mixer. The radiation pattern of the double-folded slot antenna is bidirectional and almost symmetric. One lobe of the antenna is used to receive the RF signal while the opposite lobe is used for the quasioptical coupling with the local oscillator (LO) signal. The experimental setup used for the characterization of the micromachined heterodyne receiver is pre-sented in Figure 22.30 [43]. The horn antennae, the quasioptical mixer, and the mechanical system for positioning were shielded inside a wooden box covered with suitable absorber material for 38 GHz frequency range.

Since the circuit combines the function of antenna and mixer, the isotropic conversion loss L_{iso} parameter is used to characterize the efficiency of the receiver. L_{iso} is defined as the ratio between the power level at intermediate frequency (IF) and the incident isotropic power at RF. The output power at intermediate frequency was measured using a spectrum analyzer.

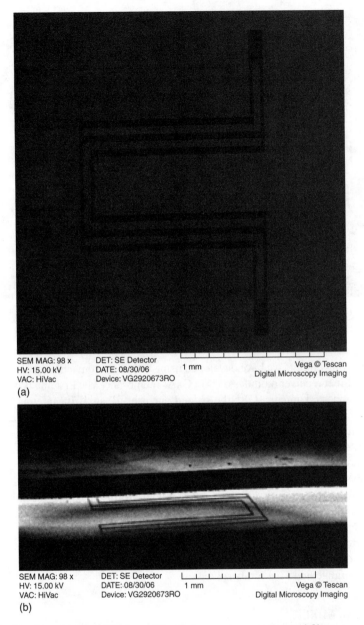

FIGURE 22.23 Top (a) and bottom (b) view of the GaAs-membrane-supported filter.

The following formula is used to calculate the isotropic power at the membrane-supported antenna plane:

$$P_{iso} = P_{gen} \, G_{Horn} \left(\frac{\lambda}{4\pi R} \right)^2 \tag{22.2}$$

where λ is the free-space wavelength, R the distance between horn antenna and double-folded slot antenna (defined in Figure 22.5), G_{Horn} the gain of the horn antenna, and P_{gen} the power level at the output of the cables connecting the horn antennae and the signal sources.

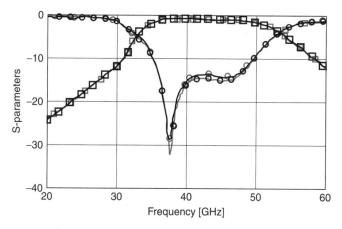

FIGURE 22.24 The measured and simulated results of the 45 GHz filter structure (circles and squares represent measured S_{11} and S_{21} respectively).

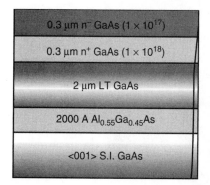

FIGURE 22.25 The receiver heterostructure.

FIGURE 22.26 Optical photo of the monolithic integrated receiver module for 38 GHz.

FIGURE 22.27 A detail from the Schottky diode region of the monolithic integrated receiver module for 38 GHz.

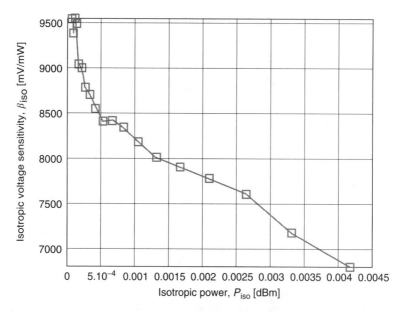

FIGURE 22.28   The measured voltage sensitivity for the 38 GHz receiver structure.

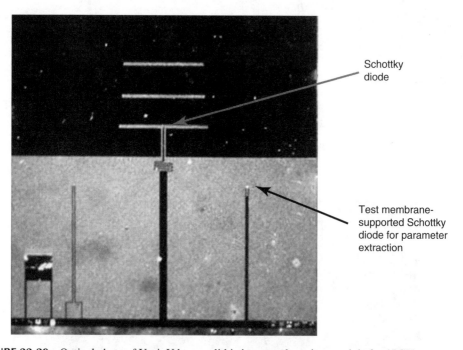

FIGURE 22.29   Optical photo of Yagi–Uda monolithic integrated receiver module for 45 GHz.

The isotropic conversion losses of upper sideband versus IF frequency are presented in Figure 22.31 (almost identical results were obtained for the lower sideband). The isotropic conversion losses have a constant value of about 10 dB for a wide band of the IF frequency range. The increase of L_{iso} as the IF frequency increases over 1 GHz is due to the mismatch between the quasioptical mixer and the IF chain.

The mixer was also measured for RF frequencies between 35 and 40 GHz. The experimental results for L_{iso} at 100 MHz intermediate frequency value are shown in Figure 22.32. Good experimental

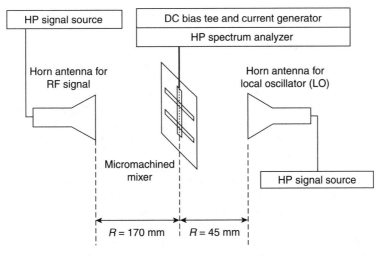

FIGURE 22.30 The experimental setup used for the characterization of the micromachined quasioptical mixer.

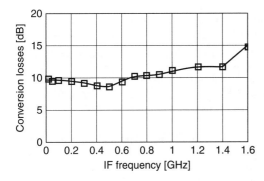

FIGURE 22.31 The isotropic conversion losses of upper sideband versus IF frequency.

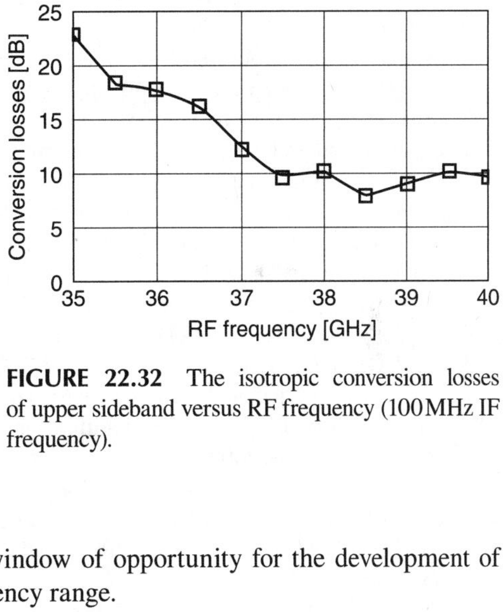

FIGURE 22.32 The isotropic conversion losses of upper sideband versus RF frequency (100 MHz IF frequency).

results obtained using this approach open a new window of opportunity for the development of circuits and systems operating up to terahertz frequency range.

CONCLUSIONS

The membrane-supported circuits have demonstrated high performance for millimeter-wave filter, antennae, and receiver applications. When fabricated using bulk micromachining of high-resistivity silicon and semi-insulated GaAs substrates, these circuits exhibit low losses, low dispersion, and can lead to innovative subsystem topologies like in the quasioptical heterodyne receiver.

The main task for the RF MEMS researchers in the next decade will be the integration of micromachining technologies with advanced silicon germanium (SiGe) and CMOS processes developed for millimeter-wave applications. This will result in sophisticated front-ends for short-range collision-avoidance radars, imaging, and indoor telecommunication systems for frequencies up to 100 GHz and beyond.

REFERENCES

1. Dib N. I., Harokopus W. P., Katehi L. P. B., Ling C. C., and Rebeiz G. M., "Study of a novel planar transmission line," *IEEE MTT-S Digest*, Boston, MA, pp. 623–626, June 1991.
2. Weller T. M., Rebeiz G. M., and Katehi L. P. B., "Experimental results on microshield transmission line circuits," *IEEE MTT-S Digest*, Atlanta, Georgia, pp. 827–830, June 1993.
3. Drayton R. F. and Katehi L. B., "Development of self-packaged high frequency circuits using micromachining techniques," *IEEE Trans. MTT*, vol. 43, no. 9, pp. 2073–2080, September 1995.
4. Katehi L. B. and Rebeiz G. M., "Novel micromachined approaches to MMIC's using low-parasitic, high performance transmission media and environments," *IEEE MTT-S Digest,* San Francisco, pp. 1145–1148, June 1996.
5. Saint Etienne E., Pons P., Blasquez G., Temple P., Conedera V., Dilhan M., Chauffleur X., Ménini Ph., Plana R., Parra T., Guillon B., and Lalaurie J. C., "A dedicated micromachined technology for high aspect ratio millimeter-wave circuits," *Sensors Actuators A*, vol. 68/1–3, pp. 435–441, 1998.
6. Robertson S. V., Katehi L. P. B., and Rebeiz G. M., "W-band microshield low-pass filters," *IEEE MTT-S Digest,* pp. 625–628, 1994.
7. Neculoiu D., Plana R., Pons P., Müller A., Vasilache D., Petrini I., Buiculescu C., and Blondy P., "Microwave characterization of membrane supported coplanar waveguide transmission lines—Electromagnetic simulation and experimental results," *Proceedings of International Conference CAS 2001,* Sinaia, Romania, pp. 151–154, October 2001.
8. Neculoiu D., Pons P., Bary L., Saadaoui M., Vasilache D., Grenier K., Dubuc D., Müller A., and Plana R., "Membrane supported Yagi-Uda antennae for millimeter-wave applications," *IEE Microwave Antennae Propagation*, vol. 151, no. 4, pp. 311–314, August 2004.
9. Saadaoui M., Pons P., Plana R., Bary L., Dureuil P., Bourrier D., Vasilache D., Neculoiu D., and Muller A., "Dry followed by wetbackside etching processes for micromachined endfire antennae," *J. Micromech. Microeng.*, vol. 15, no. 7, pp. S65–S71, 2005.
10. IE3D User's Manual, www.zeland.com, Zeland Software Inc., Freemont, CA, 2004.
11. Chi C. Y. and Rebeiz G. M., "Planar microwave and millimeter-wave lumped elements and coupled-line filters using micro-machining techniques," *IEEE Trans. MTT,* vol. 43, no. 4, pp. 730–738, 1995.
12. Weller T. M., Katehi L. P. B., and Rebeiz G. M., "A 250-GHz microshield bandpass filter," *IEEE Microwave Guided Lett.,* vol. 5, no. 5, pp. 153–155, 1995.
13. Weller T. M. and Katehi L. P. B., "A millimeter-wave micromachined lowpass filter using lumped elements," *IEEE MTT-S Digest,* pp. 631–634, 1996.
14. Blondy P., Brown A., Cross D., and Rebeiz G. M., "Low-loss micromachined filters for millimeter wave communications systems," *IEEE Trans. MTT,* vol. 46, no. 12, pp. 2283–2288, 1998.
15. Bartolucci G., Neculoiu D., Dragoman M., Giacomozzi F., Marcelli R., and Muller A., "Modeling, design and realisation of micromachined millimeter wave band-pass filters," *Int. J. Circuit Theory Appl.*, vol. 31, no. 5, pp. 529–539, 2003.
16. Neculoiu D., Bartolucci G., Pons P., Bary L., Vasilache D., Buiculescu C., Vladoianu F., Dragoman M., Petrini I., Muller A., and Plana R., "Low-Losses coupled-lines silicon micromachined band-pass filters for the 45GHz frequency band," *Proceedings of International Conference CAS 2003*, Sinaia, Romania, pp. 109–112.
17. Neculoiu D., Bartolucci G., Pons P., Bary L., Vasilache D., Muller A., and Plana R., "Compact membrane-supported bandpass filter for millimeter-wave applications," *Electron. Lett.,* vol. 40, no. 3, pp. 180–182, February 2004.
18. Rebeiz G. M., Kasilingam D. P., Guo Y., Stimson P. A., and Rutledge D. B., "Monilithic millimeter-wave two-dimensional horn imaging arrays," *IEEE Trans. AP*, vol. 38, no. 9, pp. 1473–1482, 1990.
19. Gauthier G. P., Courtay A., and Rebeiz G. M., "Microstrip antennae on synthesized low dielectric-constant substrates," *IEEE Trans. AP*, vol. 45, no. 8, pp. 1310–1314, 1997.
20. Gauthier G. P., Raskin J. P., Katehi L. P. B., and Rebeiz G. M., "A 94-GHz aperture-coupled micromachined microstrip antenna," *IEEE Trans. AP*, vol. 47, no. 12, pp. 1761–1766, 1999.
21. Gauthier G. P., Raman S., and Rebeiz G. M., *IEEE Trans. MTT,* vol. 48, pp. 1416–1419, 2000.
22. Neculoiu D., Pons P., Plana R., Blondy P., Muller A., and Vasilache D., "MEMS antennae for millimeter wave applications," *Proceedings of SPIE*, San Francisco, vol. 4559, pp. 66–73, 2001.
23. Neculoiu D., Pons P., Vasilache D., Bary L., Muller A., and Plana R., "Membrane-supported Yagi–Uda antennae for millimeter-wave applications," *Proceedings of 3rd ESA Workshop*, Espoo, Finland, pp. 603–608, May 2003.

24. Muller A., Saadaou M., Pons P., Bary L., Neculoiu D., Giacomozzi F., Dubuc D., Grenier K., Vasilache D., and Plana R., "Fabrication of silicon based micromachined antennae for millimeter-wave applications," *Proceedings of MEMSWAVE Workshop,* Toulouse, pp. D15–D18, July 2003.

25. Sor J., Qian Y., and Itoh T., "Coplanar waveguide fed quasi-Yagi antenna," *Electron. Lett.,* vol. 36, pp. 1–2, 2000.

26. Simions R. N. and Lee R. Q., "On-wafer characterization of millimeter-wave antennae for wireless application," *IEEE Trans. MTT,* vol. 47, pp. 92–96, 1999.

27. Muller A., Neculoiu D., Giaccomozzi F., Petrini I., Buiculescu C., Vasilache D., Dragoman M., Avramescu V., Dascalu D., and Zen M., "Silicon based micromachined receiving module for 38 GHz with bonded GaAs Schottky detector diode," *Workshop Digest of MME'01—Micromechanics Europe,* Cork, Ireland, pp. 115–119, September 2001.

28. Hjort K., "Gallium arsenide micromechanics—A comparation to silicon and quartz," *Proceedings of GaAs Applications Symposium,* Torino, pp. 65–72, 1994.

29. Hjort K., "Micromachining of non-silicon materials," *NEXUS Workshop on Micro-Machining,* IMSAS, University of Bremen, pp. 1–18, May 1995.

30. Hjort K., "Sacrificial etching of III-V compounds for micromechanical devices," *J. Micromech. Microeng.,* no. 6, pp. 370–375, 1996.

31. Dehe A., Pavlidis D., Hong K., and Hartnagel H. L., "InGaAs/InP thermoelectric infrared sensor utilizing surface bulk micromachining technology," *IEEE Trans. Electron. Devices,* vol. 44, pp. 1052–1056, July 1997.

32. Frike K., Dehe A., Schussler M., Lee W. Y., and Hartnagel H. L., "Micromechanical sensors based on GaAs:AlGaAs," *Proceedings of GaAs Applications Symposium,* Torino, pp. 65–72, 1994.

33. Lalinsky T., Haseik S., Mozolova Z., Drzik M., and Hatzopoulos Z., "Micromachined power sensor microsystem," *Proceedings 9th Micromechanics Europe Workshop MME'98,* Ulvik in Hardanger, Norway, pp. 139–142, June 1998.

34. Muller A., Konstantinidis G., Giaccomozzi F., Lagadas M., Deligeorgis G., Iordanescu S., Petrini I., Vasilache D., Marcelli R., Bartolucci G., Neculoiu D., Buiculescu C., Blondy P., and Dascalu D., "Micromachined filters for 38 and 77 GHz supported on thin membranes," *J. Micromech. Microeng.,* vol. 11, pp. 1–5, 2001.

35. Konstantinidis G., Muller A., Deligiorgis G., Petrini I., Vasilache D., Neculoiu D., Lagadas M., Buiculescu C., Avramescu V., Iordanescu S., and Blondy P., "GaAs membrane supported millimeter wave filters," *Proceeding of SPIE on MEMS Components and Applications for Industry, Automobiles, Aerospaces and Communication,* vol. 4559, pp. 157–161, 2001.

36. Siegel P. H., Smith R. P., Martin S., and Gaidans M., "2.5-THz GaAs monolithic membrane-diode mixer," *IEEE Trans. Microwave Theory Tech.,* vol. 47, pp. 596–604, May 1999.

37. Ichizli V., Rodriguez-Girones M., Lin C.-I., Szeliga P., and Hartnagel H. L., "The effect of gas plasma on the deposition quality of Schottky metals and interconnect metallisation for planar diodes structure for THz applications," *9th International Conference on THz Electronics,* Charlottesville, Virginia, pp. 54–57, October 2001.

38. Konstantinidis G., Neculoiu D., Lagadas M., Deligiorgis G., Vasilache D., and Muller A., "GaAs membrane supported millimeter-wave receiver structures," *J. Micromech. Microeng.,* vol. 13, pp. 353–358, 2003.

39. Siegel P., "Terahertz technology," *IEEE Trans. Microwave Theory Tech.,* vol. 50, pp. 910–928, 2003.

40. Muller A., Konstantinidis G., Neculoiu D., and Plana R., "GaAs MEMS for millimeter wave communications," *Proceedings of EuMW-GaAs, Munich,* pp. 337–340, October 2003.

41. Neculoiu D., Muller A., and Konstantinidis G., "Electromagnetic modelling of GaAs membrane supported mm-wave receivers," *J. Phys. Conf. Series,* vol. 34, pp. 28–33, 2006.

42. Stavinidris A., Muller A., Neculoiu D., Konstantinidis G., Vasilache D., Dragoman M., Petrini I., Buiculescu C., Chatzopoulos Z., Bary L., and Plana R., "GaAs membrane-supported Yagi-Uda antenna 45 GHz receiver," *Proceedings of the MEMSWAVE Conference,* Orvieto, Italy, pp. 16–20, 2006.

43. Neculoiu D., Petrini I., Buiculescu C., Dragoman M., Muller A., Giacomozzi F., Bartolucci G., and Marcelli R., "Design and characterization of a quasi-optical mixer fabricated using silicon micromachining," *Proceedings of IEEE MELECON,* Malaga, Spain, pp. 210–213, May 2006.

Index

A

across-chip length variation (ACLV), CMOS technologies, 539
across-chip width variation (ACWV), CMOS technologies, 539
active area implants, CMOS technology, 527–528
active postdistortion (APD) techniques, low-noise amplifier linearity, 322
active transmitters, short-distance wireless communication, 165
actuator design, health care monitoring, 236–237
additive white Gaussian noise (AWGN)
  differential chip detection, crystal elimination, single-chip radios, 253–255
  multiple input, multiple output systems, 123–124
  ultra-low-power radio frequency transceivers
    spread-spectrum systems, 193
    wireless microsensor nodes, 189
ADS®, ultra-low-power radio frequency transceivers, simulation results, 215
advanced mobile phone systems (AMPS), power amplifiers, 373
aging population demographics, health care technology and, 222
all-digital phase-locked loop (ADPLL) system, digital radiofrequency processor (DRP™)
  evolution of, 266
  local oscillator generation, 271–275
  transmitter specifications, 275–279
alternating current (AC) voltage, MEM resonator, capacitive transduction and sensing, 592–593
ambient intelligence (AmI)
  defined, 186
  future research applications, 218–219
ambulatory multiparameter monitoring, characteristics and applications, 224–229
amplitude modulation (AM)
  power amplifiers, back-off applications, 367–368
  single-chip radios, node-to-node communication, 249–251
"AM suppression test," global system for communication (GSM) receivers, 57–58
analog circuitry
  passive CMOS technology, 10–11
  substrate noise impact, 568
    entry points, 574
  substrate noise impact reduction, guard ring modeling, 581–586
analog demodulation, Bluetooth receivers, 65
analog-double quadrature sampling (A-DQS), frequency-hopping spread-spectrum synthesizer, 194
analog filtering, cognitive radio spectrum-sharing technology, suboptimal noncoherent detection, 148–149

analog-to-digital conversion (ADC)
  cognitive radio spectrum-sharing technology
    dynamic range reduction, 144–146
    impulse radio ultra wide band (IR-UWB) technology, 135–136
  multimode/multistandard receivers, 74
  power amplifiers, 32–33
  single-chip radios, digital implementation and verification, 257–258
  ultra-low-power transceivers
    intermodulation distortion characterization, 216–218
    link budget analysis, 211–212
    ultra wideband design, 190
  ultra wideband systems, body area networks, 230–232
  wireless standards and, 5
anchor loss, MEM resonators, 598
annulus-designed resonator, 604–605
antenna array
  membrane-supported millimeter-wave circuits
    folded slot antennae, 637–640
    Yagi-Uda antenna, 640–644
  mm-microwave CMOS systems, 45–48
  multimode/multistandard receivers, 72–73
  short-distance wireless systems, 161–164
  single-chip radios, unwanted signal coupling, 259–261
antenna reference point (ARP), ultra wideband systems, 85
antialiasing (AA) filter
  ultra-low-power radio frequency transceivers, link budget analysis, 211–215
  ultra-low-power transceivers, intermodulation distortion characterization, 216–218
application-specific integrated circuits (ASIC), health care monitoring sensor and actuators, 236–237
arbitrary weighted Gaussian noise channel (AWGN), ultra wideband systems, receiver specifications, 83–84
array architecture, MEM resonators, 612–614
  oscillator system, 620–622
automatic amplitude control (AAC) loop, multiband voltage-controlled oscillator design, 445–447
automatic frequency control (AFC), phase-locked loop bandwidths, 385–386
automatic gain control (AGC), single-chip radios, digital implementation and verification, 257–258

B

back-off applications, power amplifiers, 367–372
  Doherty PA, 371–372
  envelope elimination and reconstruction, 369–370
  envelope following, 369
  envelope tracking, 369
  out-phasing, 370–371

back-off applications, power amplifiers (*contd.*)
  requirements and efficiency penalties, 367–368
  subranging, 368–369
backside processing, membrane-supported millimeter-wave
          circuits, 632–635
band-pass filter (BPF)
  ultra-low-power radio frequency receiver
          architectures, 191
  ultra-low-power radio frequency transceivers, link
          budget analysis, 210–211
bandwidth gain
  active CMOS technology, 11–14
  ultra wideband (UWB) systems, 108
baseband filters, multimode/multistandard receivers, 74
behavioral modeling and simulation, digital radiofrequency
          processor (DRP™), 295–298
Berkeley short channel IGFET (BSIM) noise parameters,
          CMOS technologies, 540–541
bipolar CMOS technology
  complete phase noise analysis, 447–452
  global system for communication (GSM)
          receivers, 58–59
  multimode/multistandard receivers, 71–72
  overview, 522–524
  ultra wideband systems, 83–103
bipolar voltage pulse, short-distance wireless systems,
          174–178
bit-error rate (BER)
  Bluetooth receivers, 64–65
  dynamic radio principles and, 7
  global system for communication (GSM)
          receivers, 56–58
  short-distance wireless systems
    dense networks, 175–178
    wideband capacitive transmission, 171–172
  ultra-low-power radio frequency transceivers
    spread-spectrum systems, 193
    wireless microsensor nodes, 188–193
Bluetooth technology
  digital radiofrequency processor (DRP™)
    all-digital phase-locked loop system, 275–279
    overview, 266–270
  future research issues, 65–66
  low-end extension, 186–187
  overview of, 64
  personal area networks, 4–5
  ΣΔ synthesizer
    basic requirements, 465–466
    blocks characterization, 476–482
    charge pump design, 480–482
    dither signal contribution, 475
    divider chain, 478–479
    finite image rejection, 461–462
    fractional divider, 465
    future research issues, 483–484
    input noise sources, 473–474
    loop filter design, 482
    modulator characterization, 476
    momentary frequency ripple, RF output signal,
          463–465
    overview, 456

phase frequency detector, 479–480
phase noise contribution, 474–475
residual FM calculation integrals, 471–472
residual FM contributions, 472–475
simulation results, 482–483
spurious modulation, voltage-controlled oscillator,
          460–461
third-order distortion, quadrature low-frequency
          paths and carrier feed-through, 462–463
transfer derivations, 466–471
transmitter imperfections, "modulation
          characteristics" test, 458–465
voltage-controlled oscillator design, 476–478
zero-IF transmitter architecture, 456–458
Bode plots, phase-locked loop bandwidths, loop stability,
          400–402
body area networks (BAN), 160, 224–225
bond-wire inductance, silicon integrated circuit W-band
          low-noise amplifiers, 336–338
bridge-based inductive driver, short-distance wireless
          systems, 166–167
broadband low-noise amplifier, basic principles, 15–18
broadband low single-frequency mixers, low-noise
          amplifier, 21–24
broadband radio access networks (BRAN), IEEE 80211a
          standard, 68
Bruccoleri low-noise amplifier, 20–21
BSIM3 model, mm-microwave CMOS systems, 36–37
built-in self-test, digital radiofrequency processor
          (DRP™), 294–295
bulk acoustic wave (BAW)
  MEM resonators, 602–605
  ultra-low-power radio frequency receiver architectures,
          191
Butterworth filter, MEM resonator coupling, 610–612

C

capacitive antennas, short-distance wireless systems,
          163–165
  wideband capacitive transmission, 171–172
capacitive coupling
  MEM resonators, coupled Butterworth filter, 610–612
  radio frequency integrated circuits, delay opportunities,
          552–553
  substrate noise propagation entry point, 574
capacitive transduction and sensing, MEM resonator,
          591–593
capacitor models, radio frequency integrated circuits
          (RFICs), distributed capacitance, 548–549
capacitor multiplication, PLL circuit implementation,
          415–417
carrier feed-through, third-order distortion, 462–463
carrier frequency, short-distance wireless systems, passive
          receivers, 179
cascaded multitap direct sampling mixer (MTDSM)
          filtering, digital radiofrequency processor
          (DRP™), 288–289
cascode amplifiers
  CMOS W-band LNA designs, 343–345
  intrastage matching, 338–339

cascode transistor stages
  low-noise amplifiers, narrowband topology, 309–315
  microwave CMOS systems, 34–35
  mm-microwave CMOS systems, 43–44
  predistortion-based FHSS transmitter design, 203–205
cellular communication networks, wireless transceivers, 4
channel design and components
  arbitrary weighted Gaussian noise channel, 83–84
  CMOS technology, 11–14
  control channels, cognitive radio spectrum-sharing
      technology, 140–141
  fading wireless channels, multiple input, multiple out
      systems, 110–111
  multiple input, multiple output systems
    fading wireless channels, 110–111
    transmitter channel knowledge, 115–116
  short-distance wireless systems, 160–164
    capacitive vs. inductive, 163–164
    electrically small antennas, 160–161
    reactive vs. radiative design, 161–163
channel knowledge at transmitter, multiple input, multiple
    out systems, 115–116
channel reciprocity, multiple input, multiple out systems,
    115–116
characteristic wave impedance, electrically small antenna
    transmission, short-distance wireless
    systems, 161–163
charge-pump phase-locked loops (PLL) bandwidths
  circuit implementation, 412–414
  complete phase noise analysis, 447–452
  continuous time modeling, 393
  digital radiofrequency processor (DRP™), local
      oscillator generation, 271–275
  discrete-time modeling, 403–404
  fractional N-synthesizers, programmable
      multimodulus divider phase frequency
      detector, 442–445
  linear modeling, 392–393
  noise models, 435
  phase-error response, 399–400
  phase frequency detector, 393–395
  $\Sigma\Delta$ modulators, 480–482
Chebyshev filter
  low-noise amplifier, wideband design, 319–321
  MEM resonators, 609–612
chemical vapor deposition (CVD), membrane-supported
    millimeter-wave circuits, 631–635
chip architecture and design, multiple input, multiple
    output systems, 121–122
Chireix power amplifier, basic principles, 370–371
chronic disease epidemiology, health care technology and,
    222–223
circuit implementation, phase-locked loops (PLL) building
    blocks, 411–423
  charge pump, 412–414
  loop filter, 414–417
  phase-frequency detector, 411–412
  prescalers (dividers), 418–423
  voltage-controlled oscillator, 417–418
circuit response computation, substrate noise propagation,
    574–576

circuit simulations, RFIC parametric converters, 508–516
clamped-clamped beam design, MEM resonator
  array architecture, 612–614
  CMOS-compatible tunable clamped-clamped
      resonator, 615–616
  design evolution, 601–605
  mechanical circuits, 609–612
  operating principles, 590–591
class A, B, and C voltage and current waveforms, power
    amplifiers, 354–358
class AB operation, power amplifiers, 358–359
  envelope tracking, 369
  g_m ratio biasing, 376–377
class D operation, switching power amplifiers, 362
class E operation, switching power amplifiers, 362
class F operation, switching power amplifiers, 362
clock distribution system, radio frequency integrated
    circuits (RFICs), transmission line delay,
    544–545
clock frequency spectrum, substrate noise generation
    modeling, digital ground bounce
    measurement, 577–578
closed-loop frequency transfer function
  digital radiofrequency processor (DRP™), all-digital
      phase-locked loop system, 278–279
  $\Sigma\Delta$ converters, 466–471
closed-loop transmit diversity, multiple input, multiple out
    systems, 111–113
CMOS technology
  active devices, 11–14
  basic process flow, 524–537
    active area implants, 527–528
    contacts and wiring, 528
    device menu, 529
    FET performance and scaling, 532–533
    flicker noise, 536–537
    f_t, f_{max} scaling, 534–535
    gate properties, 526–527
    inductors, 531–532
    isolation process, 525–526
    metal capacitors, 532
    noise figure, 535
    optional thick oxide FETs, 529–530
    resistors, 530
    self-gain, 533–534
    stress engineering, 528
    varactors, 530–531
    well implants, 526
  Bluetooth systems, 64–66
  device models, 537–541
    flicker noise variation, 539–541
    high-frequency models, 537
    process variation, 538–539
  digital radiofrequency processor (DRP™)
    evolution of, 265–270
    radiofrequency circuits, 270–271
  evolution of, 521–524
  future research issues, 541
  global system for mobile communications, 56–60
  low-noise amplifiers
    evolution of, 305

CMOS technology (*contd.*)
    future applications, 325–326
    high-frequency noise mechanisms, MOSFETs, 306–308
    millimeter-wave layout techniques, 339–343
    radio astronomy applications, 322–325
    W-band designs, 343–345
    wideband design, 320–321
  MEM resonators, tunable clamped-clamped resonator, 615–616
  microwave systems, 33–35
  mm-microwave systems, 35–49
    active elements, 35–37
    antenna array, 45–48
    key building blocks, 41–45
    passive elements, 37–41
    radio architecture, 45–49
    transceiver architecture, 48–49
  multistandard receiver development, 71–74
  passive devices, 9–11
  power amplifiers, 373
  RFIC parametric converters
    elastance varactor model, 501–502
    varactor structures, 504–508
  short-distance wireless systems, narrowband sinusoidal transmission, 171
  $\Sigma\Delta$ converters, precalculated seeds, 442
  universal mobile radio, overview, 54–56
  universal mobile telecommunication system, 60–64
  wireless local area networks, 66–71
cochannel interference, multiple input, multiple output systems, 128
code division multiple access (CDMA), 4
  MEM resonators, 607–608
  multimode/multistandard receivers, 71–74
  power amplifiers, 374–375
  wireless standards for, 5
cognitive radio spectrum-sharing technology
  cognitive radio systems architecture, 139–141
  frequency-agile wideband transmission, 153–156
  future research issues, 156
  overlay approach, 131–132, 137–138
  overview, 131–132
  signal processing, 146–153
    feature detection, 150–153
    optical coherent detection: matched filter, 146–148
    suboptimal noncoherent detection: radiometer, 148–149
  spectrum sharing principles, 132–133
  ultra wideband transmission, 133–138
    impulse detection, 136–137
    impulse radio architecture, 135–136
    system architecture, 134–135
  wideband spectrum-sensing architecture, 141–146
    dynamic range reduction, 144–146
    frequency agility regimes, 141–144
cognitive universal dynamic radio (COGUR)
  analog and digital baseband, 32–33
  basic principles, 8–9
  receiver architecture, 14–15

Colpitts topology, predistortion-based FHSS transmitter design, 203–205
comb design, MEM resonators, 601–602
common-gate noise-canceling low-noise amplifier, 20–21
  cascode amplifier intrastage matching, 338–339
  ultra wideband systems, 88–89
  wideband design, 317–321
common-source transistors, cascode amplifier intrastage matching, 338–339
communication systems operating regions, short-distance wireless systems, 162–163
complementary code keying (CCK), wireless local area networks, 66
computer-aided design (CAD), membrane-supported millimeter-wave circuits, 634–635
conduction angles, power amplifier classification, 354
contact resistance, CMOS technology, 528
continuous time modeling, phase-locked loop bandwidths, 393
control channels, cognitive radio spectrum-sharing technology, 140–141
coordinate rotation digital computer (CORDIC), cognitive radio spectrum-sharing technology, impulse detection, 137
cost issues
  ambulatory multiparameter monitoring, 227–229
  CMOS technology, 523–524
coupled-line band-pass filters, membrane-supported millimeter-wave circuits, silicon micromachining, 635–637
coupling effects
  radio frequency integrated circuits
    measurement, 554–561
      inductors, 554–555
      LNA example, 555
      oscillator-to-oscillator, 555–561
      transformer coupling improvement, 561–562
  substrate noise propagation and impact, guard ring modeling, 585–586
cross-coupled oscillators, predistortion-based FHSS transmitter design, 203–205
crosstalk
  mixed-signal integrated circuits, 567–568
  multiple input, multiple output systems
    performance evaluation, 125–127
    radio frequency integrated circuit design, 116–119
crystal elimination, single-chip radios, 251–258
  differential chip detection, 252–253
  digital implementation issues and verification, 257–258
  modulation format, 253
  noise analysis, 253–255
  simulation results, 255–257
crystal reference noise, phase-locked loop bandwidths, building block phase noise models, 434
cubic sensor nodes, health care monitoring systems, 237–240
current density, silicon integrated circuit W-band low-noise amplifiers, 331–333
current feedback, ultra wideband systems, low-noise amplifiers, 88–89

current mode logic (CML)
  complete phase noise analysis, 447–452
  low-noise amplifier, 24
  phase-locked loop systems, dual-mode dividers,
    420–423
current-mode passive mixers, low-noise amplifier, 22–24
current power amplifier, classification, 354
cutoff frequency
  radio frequency integrated circuits, delay
    opportunities, 552–553
  RFIC parametric converters, 502–508
cyclic delay diversity (CDD), multiple input, multiple out
    systems, 112–113
cyclostationary detectors, cognitive radio spectrum-sharing
    technology, feature detection, 151–153

D

data recovery, predistortion-based hopping-frequency
    synthesizers, 200–201
deep reactive ion etching (DRIE), membrane-supported
    millimeter-wave circuits, 632–635
degenerate amplifier, RFIC parametric converters, 490–492
delay effects, radio frequency integrated circuits (RFICs)
    opportunities with, 549–544
      distributed amplifiers, 549–554
      transmission lines, 549
    power amplifier, 547–548
    transmission line, 544–545
$\Delta\Sigma$ synthesizer. *See* $\Sigma\Delta$ synthesizer
dense networks, short-distance wireless systems, 172–178
derivative superposition technique, low-noise amplifier
    linearity, 322
desired-to-undesired (D/U) ratios, cognitive radio
    spectrum-sharing technology, frequency
    agility regimes, 153–156
deterministic errors, predistortion-based hopping-
    frequency synthesizers, 196–197
device models, CMOS technology, 529, 537–541
  flicker noise variation, 539–541
  high-frequency models, 537
  process variation, 538–539
device size, silicon integrated circuit W-band low-noise
    amplifiers, simultaneous noise and input
    impedance matching, 331–333
diamond-disk resonator, 603–605
  array architecture, 603–605, 620–622
differential binary phase shift keying (DBPSK)
  crystal elimination, single-chip radios, 256–257
  wireless local area networks, 66–68
differential chip detection (DCD), single-chip radios,
    crystal elimination, 252–257
differential phase shift keying (DPSK), crystal elimination,
    single-chip radios, differential detection,
    252–253
differential quadrature phase shift keying (DQPSK)
  crystal elimination, single-chip radios, differential
    detection, 252–253
  wireless local area networks, 66–68
digital compensation, digital radiofrequency processor
    (DRP™), 293–294

digital enhanced cordless telecommunications (DECT)
    standard, integer-*N* PLL, 387
digital flip-flops (FF), phase-locked loop systems,
    dual-mode dividers, 420–423
digital frequency smoothing algorithms, cognitive
    radio spectrum-sharing technology, feature
    detection, 153
digital ground bounce, substrate noise generation modeling,
    570–572
  measurement, 577–578
digital implementation and verification, single-chip radios,
    257–258
digital noise impact, GHz LC-tank voltage controlled
    oscillator (LC-VCO), 578–580
digital radiofrequency processor (DRP™)
  all-digital transmitter, 275–279
  basic components, 266–270
  behavioral modeling and simulation, 295–296
    VHDL RX simulations, 296–298
    VHDL TX simulations, 296
  built-in self test, 294–295
  CMOS scaling technology, 270–271
  digital compensation, 293–294
  discrete-time receiver, 279–292
    cascaded MTDSM filtering, 288–289
    direct sampling mixer, 282
    high-rate IIR filtering, 284–285
    lower-rate IIR filtering, 286–288
    MTDSM feedback path, 290–292
    near-frequency interferer attenuation, 289
    receiver architecture, 279–282
    signal processing, 289–290
    spatial MA filtering zeros, 285–286
    temporal moving-average, 282–284
  energy management, 295
  evolution of, 265–266
  frequency response, 275–279
  local oscillator properties, 271–275
  script processor, 293–295
  summary of applications, 298–301
digital signal processing (DSP) technique
  dual-loop PLL archtecture, 409
  multiple input, multiple out systems, 116
  power amplifiers
    envelope elimination and reconstruction, 369–370
    out-phasing power amplifier, 370–371
digital-to-analog conversion (DAC)
  frequency-hopping spread-spectrum synthesizers,
    194–195
  mixed-signal integrated circuits, substrate noise
    coupling
    future research, 586
    guard ring noise impact reduction, 580–586
    model verification, 576–580
    noise generation modeling, 569–572
    noise propagation, 572–576
    overview, 567–569
  predistortion-based FHSS transmitter design,
    201–205
  predistortion-based hopping-frequency synthesizers,
    197–200

digital-to-RF-amplitude converter (DRAC), digital radiofrequency processor (DRP™), all-digital phase-locked loop system, 275–279

digitally controlled oscillator (DCO), digital radiofrequency processor (DRP™), 271–275

digitally controlled power amplifier (DPA), digital radiofrequency processor (DRP™), all-digital phase-locked loop system, 275–279

dimensional variation, CMOS technologies, device models, 538–539

dipole antennas, short-distance wireless systems, 163–166

direct analog synthesis (DAS), phase-locked loop bandwidths, 384–386

direct and reciprocal mixing, integer-N PLL synthesizers, 391–392

direct conversion (DC) receivers
global system for communication, 58–60
power amplifiers
DC-to-RF mathematics, 362–363
fundamental-to-DC ratios, 363–365
universal mobile telecommunication systems, 63–64
wireless local area networks, 70–71

direct current (DC) voltage, MEM resonator, capacitive transduction and sensing, 592–593

direct digital synthesizer (DDS), ultra wideband systems, 100

direct sampling mixer, digital radiofrequency processor (DRP™), 282

direct sequence spread spectrum (DSSS) system
single-chip radios, crystal elimination, 252–258
ultra-low-power radio frequency transceivers, 192–193
wireless local area networks, 66

discrete-time receiver
digital radiofrequency processor (DRP™), 279–292
cascaded MTDSM filtering, 288–289
direct sampling mixer, 282
high-rate IIR filtering, 284–285
lower-rate IIR filtering, 286–288
MTDSM feedback path, 290–292
near-frequency interferer attenuation, 289
receiver architecture, 279–282
signal processing, 289–290
spatial MA filtering zeros, 285–286
temporal moving-average, 282–284
phase-locked loop bandwidths, 403–404

distortion cancellation, low-noise amplifiers, 20–21

distributed active transformer (DAT) power combining, power amplifiers, 26–30

distributed amplifiers
radio frequency integrated circuits, delay opportunities, 549–553
ultra wideband systems, 88–89

distributed capacitance, radio frequency integrated circuits (RFICs), 548–549

distributed effects, radio frequency integrated circuits (RFICs), 544
capacitance, 548–549
one millimeter transmission line, 545–547
power amplifier delay, 547–548

transmission line delay and phase shift, 544–545
transmission line model, 545–549

distributed multiple input, multiple output systems, research and development of, 127–128

dither signal, $\Sigma\Delta$ synthesizer, 475

divider chain, $\Sigma\Delta$ modulators, 478–479

Doherty power amplifier design, 371–372

double-sideband phase noise, multiband wireless networks, 430–431

double-transistor architecture, switching power amplifiers, 359–362

down-converter design
global system for communication (GSM) receivers, 59
24-GHz receiver, single-chip radios, 247
ultra wideband systems, 89–92

drain and source metallization
substrate noise generation, 569
W-band CMOS circuits, 343

drain induced barrier lowering (DIBL), CMOS technology
optional thick oxide FETs, 529–530
self-gain, 534

drain voltage and current
power amplifiers
class A, B, and C operations, 354–358
class AB operations, 358–359
switching power amplifiers, 359–362

drawn minimum channel length, active CMOS technology, 11–14

dual-antenna phased array transceiver, multiband orthogonal frequency division multiplexing, 103

dual-IF architecture, multimode/multistandard receivers, 71–72

dual-loop PLL architecture, 409

dual-mode dividers (DMDs)
fractional-N synthesizers, multiband wireless networks, frequency synthesis, 431–434
phase-locked loop systems, 419–423

dual modulus divider architecture, $\Sigma\Delta$ modulators, 478–479

dual-path technique, PLL loop filter circuit implementation, 414–417

dynamic biasing power amplifier, 30

dynamic radio, basic principles, 6–7

dynamic random access memory (DRAM), CMOS technology process, 525–526

dynamic range
active CMOS technology, 12–14
cognitive radio spectrum-sharing technology, reduction techniques, 144–146

E

ECMA-368 standard, MB-OFDM systems, 82–83

Eco node, development of, 187

effective damping factor, MEM resonator modeling, 595–596

effective number of bits (ENOB) calculations, ultra-low-power radio frequency transceivers, link budget analysis, 213–215

efficiency parameters, power amplifiers, 352–353
back-off applications, 367–368
class A, B, and C operations, 355–358

Doherty power amplifier design, 372
finite bandwidth, 366–367
switching configuration, 359–362
E-GSM blocking mask, global system for communication (GSM) receivers, 57–58
elastance varactor model, RFIC parametric converters, 501–502
electrally small antennas, short-distance wireless systems, 160–163
electrical-mechanical analogy, MEM resonator modeling, 594
electroencephalography/electrocardiography/electromyography systems, ambulatory multiparameter monitoring, 224–229
electromagnetic simulations, radio frequency integrated circuits, oscillator-to-oscillator coupling, 558–561
electromechanical coupling coefficient, MEM resonator, capacitive transduction and sensing, 592–593
electronic-health systems (e-health)
body area networks, 224
evolution of, 223–224
electrostatic discharge (ESD) structures
MEM resonator, 591
ultra wideband systems, low-noise amplifier and, 87–89
electrostatic force, MEM resonator, capacitive transduction and sensing, 592–593
electrostatically actuated resonators, evolution of, 600–605
emitter length, W-band radio, 335–336
end-fire processing, membrane-supported millimeter-wave circuits, 632–635
energy gain behavior, RFIC parametric converters, 491–943
energy loss mechanisms, MEM resonators, 597–599
energy management, digital radiofrequency processor (DRP™), 295
energy-scavenging techniques
health care monitoring, micropower generation and storage, 232–235
ultra-low-power radio frequency receivers, 186
enhanced data for GSM evolution (EDGE), 4
digital radiofrequency processor (DRP™), 266–270
power amplifiers, 374
enhanced GSM bands (EGSM), power amplifiers, 373
envelope elimination and reconstruction (EER), power amplifiers, 369–370
envelope following, power amplifiers, 369
envelope tracking, power amplifiers, back-off applications, 369
equivalent circuit model, substrate noise generation, guard ring modeling, 581–586
erasable programmable read only memory (EPROM), predistortion-based FHSS transmitter design, 203–205
error vector magnitude (EVM)
multiple input, multiple out systems
performance evaluation, 124–127
transmitter specifications, 120–121
wireless local area networks, 66–68
external circuitry, Q loading, MEM resonators, 598–599
external frequency divider, single loop architectures, 408

F

fabrication techniques, MEM resonators, 614–620
fading wireless channels, multiple input, multiple out systems, 110–111
fast Fourier transform (FFT)
multiband orthogonal frequency division multiplexing, transceiver specifications, 100–104
multiple input, multiple out systems, fading wireless channels, 110–111
fast-hopping synthesizer, ultra wideband systems, 95–100
integer-N PLL approach, 95–96
single PLL and single-side-band mixer, 98–99
two PLLs and single-side-band mixer, 96–98
feature detection, cognitive radio spectrum-sharing technology, 149–153
feedback division ratio, reference frequency and feedback division ratio, 387
feedback elements, low-noise amplifier, wideband design, 320–321
feedback path, digital radiofrequency processor (DRP™), multitap direct sampling mixer, 290–292
feed-forward architecture
cognitive radio spectrum-sharing technology, dynamic range reduction, 145–146
phase-locked loop bandwidths, 397–398
field effect transistor (FET)
CMOS technologies, process variation, 538–539
CMOS technology
layout, 526
optional thick oxide FETs, 529–530
overview, 521–524
performance and scaling, 532–533
stress engineering, 528
well implants, 526
fractional-N synthesizers, programmable multimodulus divider phase frequency detector, 444–445
short-distance wireless systems, narrowband sinusoidal transmission, 171
field programmable gate array, single-chip radios, digital implementation and verification, 258
figure of merit (FOM) characterization
RFIC parametric converters, varactor structures, 502–508
ultra-low-power radio frequency transceivers, link budget analysis, 215
film bulk acoustic resonator (FBAR), ultra-low-power radio frequency transceivers, 187–188
link budget analysis, 210–211
filter systems, MEM resonators, 599
mechanical circuits, 608–612
finger width, W-band CMOS circuits, millimeter-wave layout, 340–341
finite bandwidth signals, power amplifiers, 365–367
finite image rejection, Bluetooth technology, 461–462
fixed dividers, integer-N systems, 419–423
flicker noise, CMOS technology, 536
metal oxide semiconductor (MOS) transistor, 526–527
variation, 539–541
folded slot antennae, membrane-supported millimeter-wave circuits, 637–640

Fourier transforms, power amplifiers, DC-to-RF mathematics, 362–363

fractional-*N* synthesizers
multiband wireless networks, frequency synthesis, 431–434
ΣΔ converters
Bluetooth technology, 465–471
multiband implementation, 440–447
multimodulus divider phase frequency detector and charge pump, 442–445
phase noise models, 436–438
precalculated seeds, 440–442
voltage-controlled oscillator designs, 445–447
ultra-low-power radio frequency systems, 194

frame error rate (FER), wireless local area networks, 66–68

free-free design, MEM resonators, 601
fabrication, 616–618
programmable MEM-based FSK transmitter, 622–624

frequency accuracy, reference frequency and feedback division ratio, 387–388

frequency agility regimes, cognitive radio spectrum-sharing technology
transmission specifications, 153–156
wideband spectrum-sensing architecture, 141–144

frequency command word (FCW), digital radiofrequency processor (DRP™), local oscillator generation, 271–275

frequency dividers
Bluetooth technology, 466–471
phase-locked loop bandwidths, 397
building block phase noise models, 435

frequency division duplexing (FDD) systems, universal mobile telecommunications, 60–61

frequency domain duplexed (FDD) systems, integer-*N* PLL synthesizers, 388–389

frequency domain filtering, cognitive radio spectrum-sharing technology, dynamic range reduction, 144

frequency domain processing, cognitive radio spectrum-sharing technology, frequency agility regimes, 156

frequency-hopping spectrum spreading (FHSS)
Bluetooth standard and, 64
cognitive radio spectrum-sharing technology, ultra wideband systems, 134–135
integer-*N* PLL synthesizers, 389
micro-Watt node development, 218–219
predistortion-based transmitter design, 201–205
ultra-low-power radio frequency synthesizers, 194–201
data recovery, 200–201
deterministic errors, 196–197
experimental results, 204–205
predistortion-based design, 195–196, 202–204
stochastic errors, 197–200
ultra-low-power radio frequency transceivers
DSSS *vs.*, 192–193
wireless microsensor nodes, 189, 192
wireless local area networks, 66

frequency modulation (FM)
GHz LC-tank voltage controlled oscillator (LC-VCO), digital noise impact, 578–580

substrate noise generation modeling, circuit response computation, 574–576

frequency offsets, phase-locked loop systems, 408–409

frequency pulling, MEM resonators, 596–597

frequency reference (FREF) clock, digital radiofrequency processor (DRP™), local oscillator generation, 271–275

frequency response, digital radiofrequency processor (DRP™), all-digital phase-locked loop system, 275–279

frequency synthesis
multiband wireless networks
base phase noise concepts, 428–431
fractional-*N* ΣΔ synthesizer, 440–447
in-band/out-of-band phase noise, 438–440
noise analysis and measurement comparisons, 447–452
overview, 427–428
phase-locked loop architectures, 431–434
building block phase noise models, 434–438
multimode/multistandard receivers, 74
phase-locked loop bandwidths, 384–386

FSK transmitter, programmable MEM resonator and, 622–624

fundamental-component-to-DC-component (FDC) ratios, power amplifiers, 363–365

G

GaAs micromachining, membrane-supported millimeter-wave circuits
antenna fabrication, 637–644
coupled-line band-pass filters, 635–637
fabrication and design, 631–635
future research issues, 651
GaAs membrane-supported circuits, 644–647
overview, 629–631
quasioptical mixer, 647–651

gain achievements
low-noise amplifiers
narrowband topology, 314–315
wideband design, 320–321
ultra-low-power radio frequency transceivers, link budget analysis, 214–215

gain coefficients, multiple input, multiple out systems, channel reciprocity, 116

gain optimization, silicon integrated circuit W-band low-noise amplifiers, 334

gas damping, MEM resonators, 598

gate contact
substrate noise generation, 569–572
W-band CMOS circuits, millimeter-wave layout, 340–343

gate delay, CMOS technologies, process variation, 538–539

gate pitch, W-band CMOS circuits, 343

gate-source capacitance, low-noise amplifiers, narrowband topology, 309–315

Gaussian low-pass filter (LPF), Bluetooth technology, ΣΔ synthesizer, 457–458

Gaussian minimum shift keying (GMSK), power amplifiers, 373

generalized gain, short-distance wireless systems, narrowband sinusoidal transmission, 168–169 general packet radio service (GPRS), power amplifiers, 373

GHz LC-tank voltage controlled oscillator (LC-VCO), substrate noise propagation, 576–577

    digital noise impact, 578–580

Gilbert cell mixer

    mm-microwave CMOS systems, 41–42

    ultra wideband systems, 89–92

global positioning systems (GPS), wideband low-noise amplifiers, 322–325

global roaming applications, multimode/multistandard receivers, 71–74

global system for communication (GSM) standard, 4

    direct conversion receiver construction, 59–60

    multimode/multistandard receivers, 71–74

    power amplifiers, 373

    receiver evolution and development, 58–59

    receiver requirements, 56–58

Greenhouse formula, substrate noise propagation and impact, guard ring modeling, 582–586

group control channels (GCCs), cognitive radio spectrum-sharing technology, 140–141

guard rings, substrate noise impact reduction, 580–586

    design guidelines, 584–586

    modeling protocols, 581–584

H

halo implants, CMOS technology, 527–528

    optional thick oxide FETs, 529–530

    self-gain, 534

health care monitoring

    aging demographics and, 222

    ambulatory multiparameter monitoring, 224–229

    body area networks, 160, 224

    chronic disease epidemiology and, 222–223

    economic burden of, 221–222

    electronic health systems, 223–224

    future research issues, 240

    integration technology, 237–240

    micropower generation and storage, 232–235

    patient-centric approaches, 223

    sensors and actuators, 236–237

    ultra-wideband communications

        analog receiver, 230–232

        pulse generator, 229–230

    wireless communication systems, 229

heatsink very-thin quad flat-pack no leads (HVQFN) package, multiband orthogonal frequency division multiplexing, 100–104

heterodyne receiver architecture, ultra-low-power radio frequency transceivers, 207–209

heterostructure bipolar transistors (HBTs), SiGe technology, 522–524

high dynamic front-end receiver, 14–24

    broadband noise $1/f$ noise mixers, 22–24

    noise and distortion cancellation, 20–22

    receiver specifications, 14–15

    shunt feedback LNA, 18–20

    wideband LNA design, 15–18

high electron mobility transistors (HEMTS), wideband low-noise amplifiers, 322–325

high-frequency devices, CMOS technology, 537

high-frequency noise mechanisms, low-noise amplifiers, 306–308

high-ohmic substrates, noise generation, 569–572

history sampling capacitor, digital radiofrequency processor (DRP™), high-rate infinite impulse response filtering, 284–285

hole current, substrate noise generation, 569

hollow-disk ring resonator, 604–605

homodyne receiver architecture, ultra-low-power radio frequency transceivers, 208–209

Human++ research program

    aging demographics and, 222

    ambulatory multiparameter monitoring, 224–229

    body area networks, 160, 224

    chronic disease epidemiology and, 222–223

    economic burden of, 221–222

    electronic health systems, 223–224

    future research issues, 240

    integration technology, 237–240

    micropower generation and storage, 232–235

    patient-centric approaches, 223

    sensors and actuators, 236–237

    ultra-wideband communications

        analog receiver, 230–232

        pulse generator, 229–230

    wireless communication systems, 229

I

IEEE 802.11 standards

    cognitive radio spectrum-sharing technology, ultra wideband systems, 134–135

    MB-OFDM systems, 82–83

    medium-range LAN networks, 4

    power amplifiers, 375

IEEE 802.11a standard, wireless local area networks, 68–69

IEEE 802.11/.11b standard, wireless local area networks, 66–68

IEEE 802.11g standard, wireless local area networks, 69

IEEE 802.15 standard, cognitive radio spectrum-sharing technology, ultra wideband systems, 135

IEEE 802.16 standard, evolution of, 5

IEEE 802.22 standard, cognitive radio spectrum-sharing technology, 139–141

IIP2 specification

    global system for communication (GSM) standard, 57–58

    universal mobile telecommunication receivers, 63

IIP3 performance

    active CMOS technology, 14

    ultra-low-power transceivers, intermodulation distortion characterization, 217–218

    ultra wideband systems, linearity and filter requirements, 85–86

    universal mobile telecommunication receiver specifications, 61–63

image rejection ratio (IRR)

    Bluetooth technology, finite image rejection, 462

    phase-locked loop bandwidths, 392

IMD sinking circuit, low-noise amplifier linearity, 322

impact ionization, substrate noise generation, 569

implementation protocols, short-distance wireless systems, 179–181

impulse detection, cognitive radio spectrum-sharing technology, ultra wideband systems, 136–137

impulse radio ultra wide band (IR-UWB) technology, 82
cognitive radio spectrum-sharing technology, 135–136

in-band noise, phase-locked loop systems, 438–440

inductive-coupled antennas, short-distance wireless systems, 163–166

inductive inner-chip communication topology, short-distance wireless systems, 173–178

inductive in-plane antennas, short-distance wireless systems, 173–174

inductive transceiver, short-distance wireless systems, dense networks, 176–178

inductor parameters
CMOS technology, 531–532
GHz LC-tank voltage controlled oscillator (LC-VCO), digital noise impact, 578–580
radio frequency integrated circuits, coupling effects, 554–555

industrial-scientific-medical (ISM) band
Bluetooth standard and, 64
MB-OFDM systems, 82–83
$\Sigma\Delta$ modulators, divider chain, 479
ultra-low-power transceivers, 187

infinite impulse response (IIR), digital radiofrequency processor (DRP™)
discrete-time receiver architecture, 280–282
high-rate filtering, 284–286
lower-rate filtering, 286–288
MTDSM feedback path, 290–292
open-loop z-domain transfer function, 277–279

injection-locked frequency division (ILFD), phase-locked loop systems, 420–423

injection-locked oscillators (ILO)
phase-locked loop systems, 420–423
radio frequency integrated circuits
coupling improvement, 562–563
oscillator-to-oscillator coupling, 556–561

input impedance
low-noise amplifiers, narrowband topology, 309–315
wideband low-noise amplifier, 16–18
radio astronomy applications, 323–325

input impedance matching, silicon integrated circuit W-band low-noise amplifiers, 331–335
device size, 331–333
gain optimization, 334
multistage designs, 334–335
optimum biasing, 331

input/output field effect transistor (I/O FET), CMOS technology, 529–530

integer-N PLL system
architecture, 407–408
frequency accuracy, 387–388
multiband wireless networks, frequency synthesis, 431
prescalers (dividers), 418–423
reference frequency and feedback division ratio, 387

settling time, 388–389
ultra wideband systems, 95–96

integration density, CMOS technology, 523–524

integration technologies
health care monitoring systems, 237–240
power amplifier design, 25
single-chip radios, 241–242

intelligent objects, current research on, 159–160

interference scenario
digital radiofrequency processor (DRP™), near-frequency interferer attenuation, 289
ultra-low-power radio frequency transceivers, intermodulation distortion characterization, 216–218
ultra wideband systems, 84–85

intermediate frequency/baseband filter
cognitive radio spectrum-sharing technology, impulse detection, 137
lower sideband down-converter, 497
ultra wideband systems, 92–93

intermodulation distortion characterization, ultra-low-power radio frequency transceivers, 216–218

intermodulation testing, wireless local area networks, 67–68

intersymbol interface (ISI), IEEE 80211a standard, 68

intrastage matching techniques, cascode amplifiers, 338–339

inverse fast Fourier transforms (IFFT)
cognitive radio spectrum-sharing technology, frequency agility regimes, 153–156
multiple input, multiple out systems, spatial diversity, 112–113

I/Q accuracy, integer-N PLL synthesizers, 392

isotropic power calculations, membrane-supported millimeter-wave circuits, quasioptical mixer, 648–651

J

jitter deviation, short-distance wireless systems, dense networks, 177–178

Johnson counter, phase-locked loop systems, dual-mode dividers, 421–423

L

Leeson's phase noise model, voltage-controlled oscillators, 418–419

linear MIMO crosstalk, radio frequency integrated circuit design, 116–118

linear modeling, phase-locked-loop systems, 392–393

linear periodically time-varying (LPTV) time system, substrate noise propagation, 574–576

linearization techniques
low-noise amplifiers, 321–322
power amplifiers, 353–354, 372

line impedance, radio frequency integrated circuits, delay opportunities, 552–553

link budget analysis, ultra-low-power radio frequency transceivers
discrete parts, 210–211
integrated parts, 210–212

local oscillator (LO) generation

cognitive radio spectrum-sharing technology, feature detection, 153

digital radiofrequency processor (DRP™), 271–275

low-noise amplifier, 23–24

multiple input, multiple systems, radio frequency integrated circuit design, 119

substrate noise generation modeling, circuit response computation, 574–576

local oscillator (LO) phase noise (PN), global system for communication (GSM) standard, 57–58

local oxidation of silicon (LOCOS) process, CMOS technology, 525–526

loop antennas, short-distance wireless systems, 163–164

loop delay, phase-locked loop bandwidths, 397

loop filter (LPF)
  phase-locked loop bandwidths, 395
    noise models, 435–436
    phase-error response, 398–400
    type and order, 398
  PLL circuit implementation, 412–417
  ΣΔ synthesizer, 482
    Bluetooth frequency dividers, 468–471

loop stability, phase-locked loop bandwidths, 400–402

loop topology, MEM resonators, 599–600

Lorentzian signature, random telegraph signals, CMOS technology, 539–540

lossless parametric amplifier, upper sideband up-converter, 495–497

lower sideband down-converter (LSBDC), RFIC parametric converters
  basic properties, 497
  circuit simulations, 514–516
  transmit/receive chains, 497–500

low-frequency noise spectra
  CMOS technology, 540–541
  substrate noise generation modeling, digital ground bounce measurement, 577–578

low-IF-architecture
  Bluetooth receivers, 65
  cognitive universal dynamic radio, 8–9
  passive CMOS technology, 10–11
  wireless local area networks, 69–71

low-jitter loops, PLL architecture, 410–411

low-noise amplifier (LNA)
  broadband low single-frequency mixers, 21–24
  classical optimization, 306–308
  CMOS technology
    contact resistance, 528
    self-gain, 534
  digital radiofrequency processor (DRP™), discrete-time receiver architecture, 280–282
  dynamic radio and, 7
  evolution of, 305
  frequency specificity of, 4
  future trends, 325–326
  high-frequency noise mechanisms, CMOS MOSFETS, 306–308
  linearity, 321–322
  MEM resonators, 607–608
  multiband orthogonal frequency division multiplexing, transceiver specifications, 100–104

multimode/multistandard receivers, 73–74

narrowband topology, 309–315
  gain cascade, 314–315
  noise figure signal source impedance, 315

noise and distortion cancellation, 20–21

power-matched topologies, 308–309

radio frequency integrated circuits, coupling effects, 555

receiver specification, 14–15

short-distance wireless systems, 166

shunt feedback topology, 18–20

silicon integrated circuit W-band design
  bond-wire inductance, 336–338
  cascode intrastage matching, 338–339
  design philosophy, 330–331
  future trends, 345–347
  matching results, 345
  millimeter-wave CMOS layout techniques, 339–343
    drain and source metallization and gate pitch, 343
    finger width, 340–341
    gate contacts, 340
    optimization techniques, 341–343
  simultaneous noise and input impedance matching, 331–335
    device size, 331–333
    gain optimization, 334
    multistage designs, 334–335
    optimum biasing, 331
  V-band and W-band design, 330
  wideband design
    CMOS technology, 343–345
    matching methodology, 335–336

24-GHz receiver, single-chip radios, 245–247

ultra-low-power transceivers
  intermodulation distortion characterization, 216–218
  link budget analysis, 211–215
  ultra wideband transceiver design, 190

ultra wideband systems, 87–89

wideband design, 15–18, 315–321
  gain case, 320–321
  radio astronomy applications, 322–325

low-noise transconductance amplifier (LNTA), digital radiofrequency processor (DRP™), direct sampling mixer, 282

low-power wireless systems, trends in, 186–187

low pressure chemical vapor deposition (LPCVD) technique, membrane-supported millimeter-wave circuits, 631–635

lumped parameter mass-spring-damper systems, MEM resonator modeling, 593–594

M

macromodeling, substrate noise generation, 569–572

Manley-Rowe relations, RFIC parametric converters, 494–495
  circuit simulations, 508–514

M-ary orthogonal signaling, single-chip radios
  digital implementation and verification, 257–258
  format, 253
  noise analysis, 253–255
  simulation techniques, 255–257

Mason's unilateral gain, CMOS technology, 534–535

matched filters, cognitive radio spectrum-sharing technology, optical coherent detection, 146–148

maximal ratio combining (MRC), multiple input, multiple out systems, receive diversity, 111–112

maximum likelihood (ML) decoding, multiple input, multiple out systems, 114–115

maximum output power matching, power amplifiers, 353

maximum stable gain (MSG) measurements, mm-microwave CMOS systems, 35–37

maximum uncorrected frequency errors, predistortion-based hopping-frequency synthesizers, 196–197

measurement noise analysis, multiband wireless networks, frequency synthesis, complete phase noise analysis *vs.*, 447–452

mechanical circuits, MEM resonators, 608–614

medium access control (MAC)
    cognitive radio spectrum-sharing technology, ultra wideband systems, 135
    low-power autonomous devices, 186–187
    multiple input, multiple out systems, channel reciprocity, 115–116

medium-spectrum-scarcity regime, cognitive radio spectrum-sharing technology, wideband spectrum-sensing architecture, 141–143

membrane-supported millimeter-wave circuits, silicon/ GaAs micromachining
    antenna fabrication, 637–644
    coupled-line band-pass filters, 635–637
    fabrication and design, 631–635
    future research issues, 651
    GaAs membrane-supported circuits, 644–647
    overview, 629–631
    quasioptical mixer, 647–651
    silicon-based receiving module, 644

metal capacitors, CMOS technology, 532

metal-insulator-metal (MIM) capacitors
    mm-microwave CMOS systems, 37–41
    passive CMOS technology, 10–11

metal-organic chemical vapor deposition (MOCVD), GaAs micromachined membrane-supported millimeter-wave circuits, 644–647

metal oxide semiconductor field-effect transistor (MOSFET)
    high-frequency devices, 537
    low-noise amplifiers
        future applications, 326
        high-frequency noise mechanisms, 306–308
    W-band CMOS circuits, millimeter-wave layout, 339–343

metal oxide semiconductor (MOS) transistor
    active CMOS technology, 12–14
    gate architecture, 526–527
    RFIC parametric converters, varactor structures, 504–508

method-of-moments (MoM) design, membrane-supported millimeter-wave circuits, 634–635

microelectromechanical systems (MEMS) technology
    evolution of, 159
    low-noise amplifiers, 15
    resonators

applications, 599–600
array systems, 612–614
basic principles, 590
capacitive transduction and sensing, 591–593
CMOS-compatible tunable clamped-clamped resonator, 615–616
energy loss mechanisms, 597–599
evolution of, 600–605
fabrication case study, 614–620
filter systems, 608–612
free-free beam resonator, 616–618
mechanical circuits, 608–614
modeling, 593–596
nonlinear effects, 596–597
operating principles, 590–591
programmable FSK transmitter, 622–624
quality factor definition, 591
radial-mode disk resonator, 618–620
system case study, 620–624
    array-based oscillator, 620–622
transceiver architecture, 605–608
RFIC parametric converters, 489
ultra-low-power transceivers, 187–188

micropower generation and storage, health care monitoring, 232–235

micro-Watt nodes
    defined, 186
    future research applications, 218–219

microwave CMOS systems, 33–35

millimeter wave layout
    membrane-supported systems, silicon/GaAs micromachining
        antenna fabrication, 637–644
        coupled-line band-pass filters, 635–637
        fabrication and design, 631–635
        future research issues, 651
        GaAs membrane-supported circuits, 644–647
        overview, 629–631
        quasioptical mixer, 647–651
        silicon-based receiving module, 644
    microwave CMOS systems, 35–49
        active elements, 35–37
        antenna array, 45–48
        key building blocks, 41–45
        passive elements, 37–41
        radio architecture, 45–49
        transceiver architecture, 48–49
    W-band CMOS circuits, 339–343

milli-Watt nodes, defined, 186

minimum mean square error (MMSE) algorithm, multiple input, multiple out systems, spatial multiplexing, 114–115

minimum noise figure, low-noise amplifiers, 306–308

minimum shift keying (MSK), 24-GHz transmitter, single-chip radios, 248–249

mixed-signal integrated circuits, substrate noise coupling, digital-to-analog conversion
    future research, 586
    guard ring noise impact reduction, 580–586
    model verification, 576–580
    noise generation modeling, 569–572

noise propagation, 572–576
overview, 567–569
modulation ratio
differential chip detection, crystal elimination,
single-chip radios, 253
RFIC parametric converters, 502–508
modulator characteristics, $\Sigma\Delta$ synthesizer, 476
molecular beam epitaxy (MBE), GaAs micromachined
membrane-supported millimeter-wave
circuits, 644–647
Monte Carlo model, CMOS technologies, process variation,
538–540
moving-average coefficients, digital radiofrequency
processor (DRP™), 283–284
spatial filtering zeros, 285–286
multiband orthogonal frequency division multiplexing
(MB-OFDM)
interference scenario, 84–85
synthesizer requirements, 86–87
transceiver specifications, 100–104
ultra wide band technology and, 81–83
multiband wireless networks, frequency synthesis
base phase noise concepts, 428–431
fractional-N $\Sigma\Delta$ synthesizer, 440–447
in-band/out-of-band phase noise, 438–440
noise analysis and measurement comparisons, 447–452
overview, 427–428
phase-locked loop architectures, 431–434
building block phase noise models, 434–438
multimode power amplifiers, 25
multimode radio, component count explosion in, 5
multimodulus divider (MMD)
complete phase noise analysis, 447–452
fractional-N synthesizers, multiband wireless networks,
frequency synthesis, 433–434
fractional N-synthesizers, programmable
multimodulus divider phase frequency
detector, 442–445
multiple antenna systems, multiple input, multiple output
systems, 128
multiple-antenna transceiver architecture, mm-microwave
CMOS systems, 47–49
multiple-finger varactor, RFIC parametric converters,
505–508
multiple input, multiple output (MIMO) systems
chip architecture and design, 121–122
evolution of, 108–109
fading wireless channels, 110–111
future research issues, 127–128
programmable $\Sigma\Delta$ modulator, precalculated seeds,
440–442
radio frequency integrated circuit design
crosstalk, 116
linear crosstalk, 116–118
nonlinear crosstalk, 118–119
overview, 108–109
shared local oscillator generation, 119
transmitter error vector magnitude, 120–121
variable gain amplifier control, 119–120
spatial diversity, 111–113
spatial multiplexing, 113–115

testing procedures and measured results, 122–127
theoretical background, 109–116
transmitter channel knowledge, 115–116
multiple input, single output (MISO) systems, defined, 109
multistage noise shaping (MASH)
$\Sigma\Delta$ converters
phase noise models, 436–438
precalculated seeds, 440–442
silicon integrated circuit W-band low-noise
amplifiers, 334–335
multistandard receiver development, CMOS technology,
71–74
multitap direct sampling mixer (MTDSM), digital
radiofrequency processor (DRP™)
cascaded filtering, 288–289
discrete signal processing, 289–290
discrete-time receiver architecture, 280–282
feedback path, 290–292
lower-rate IIR filtering, 286–288

N

narrowband sinusoidal transmission, short-distance wireless
systems, 165, 167–171
narrowband topology, low-noise amplifier, 309–315
gain cascade, 314–315
noise figure signal source impedance, 315
n-channel MOS (N-MOS) device
low-noise amplifiers, wideband design, 321
PLL circuit implementation, 413–414
voltage-controlled oscillators, 417–418
RFIC parametric converters, varactor structures, 504–508
single-chip radios, unwanted signal coupling, 259–261
substrate noise propagation entry point, 574
GHz LC-tank voltage controlled oscillator, 576–577
"near-far" scenario
receiver blockage and, 7
ultra-low-power radio frequency transceivers,
spread-spectrum systems, 192–193
near-frequency interferer attenuation, digital radiofrequency
processor (DRP™), 289
node-to-base-station communication, single-chip radios, 251
node-to-node communication, single-chip radios, 249–250
noise analysis
differential chip detection, crystal elimination,
single-chip radios, 253–255
residual FM calculations, Bluetooth technology, 473–474
noise figure (NF)
active CMOS technology, 12–14
CMOS technology, 535
dynamic radio principles and, 7
global system for communication (GSM) standard, 56–58
low-noise amplifiers, 14–15
distortion cancellation and, 20–21
high-frequency noise mechanisms, 306–308
narrowband topology, 310–315
radio astronomy applications, 324–325
signal source impedance, 315
mm-microwave CMOS systems, 37–41
multiple input, multiple output systems, 124–125
phase-locked loop bandwidths, modeling of, 402–403

noise figure (NF) (*contd.*)
  silicon integrated circuit W-band low-noise amplifiers, 331–335
    device size, 331–333
    gain optimization, 334
    multistage designs, 334–335
    optimum biasing, 331
  24-GHz receiver, single-chip radios, 246–247
  ultra-low-power radio frequency transceivers
    link budget analysis, 211–215
    wireless microsensor nodes, 188–193
  ultra wideband systems, 83–84
  universal mobile telecommunication systems, 63–64
  wireless local area networks, 66–68
noise management, single-chip radios, 258–261
noise transfer function (NTF), phase-locked loop systems, in-band/out-of-band noise, 438–440
nonconstant envelope modulation, power amplifier design, 24
nonlinear MIMO crosstalk, radio frequency integrated circuit design, 118–119
nonlinearity
  MEM resonators, 596–597
  power amplifiers, 353–354
no-spectrum-scarcity regime, cognitive radio spectrum-sharing technology, wideband spectrum-sensing architecture, 141–142
Nyquist theorem, ultra-low-power radio frequency receiver architectures, 191

O

on-chip analog ground, substrate noise propagation and impact, guard ring modeling, 582–586
on-chip antennas, single-chip radios, 242–244
on-chip components, MEM resonators, 605–608
on-chip power combining, power amplifiers, 27–30
on/off keying (OOK)-modulated data, ultra-low-power radio frequency receivers, 191
1 dB compression point
  low-noise amplifier linearity, 321–322
  W-band radio matching, 335–336
one-port negative-resistance amplifier, RFIC parametric converters, 488–489
op-amp design topology, low-noise amplifier, 23–24
open-loop transmit diversity
  multiple input, multiple out systems, 111–113
  phase-locked loop bandwidths, loop stability, 400–402
open-loop *z*-domain transfer function, digital radiofrequency processor (DRP™), all-digital phase-locked loop system, 277–279
operational amplifier (OPAMP) system, ultra-low-power transceivers, intermodulation distortion characterization, 217–218
optical coherent detection, cognitive radio spectrum-sharing technology, matched filter, 146–148
optimization techniques
  low-noise amplifiers, 306–308
    radio astronomy applications, 325
    wideband design, 317–321
  W-band CMOS circuits, finger width and gate contacts, 341–343

optimum biasing, silicon integrated circuit W-band low-noise amplifiers, simultaneous noise and input impedance matching, 331
orthogonal frequency division multiplexing (OFDM)
  Bluetooth standard and, 4–5
  cognitive radio spectrum-sharing technology
    frequency agility regimes, 153–156
    ultra wideband systems, 134–135
  multiple input, multiple out systems
    fading wireless channels, 110–111
    spatial diversity, 112–113
oscillator-to-oscillator coupling, radio frequency integrated circuits, 555–561
out-of-band noise, phase-locked loop systems, 438–440
out-phasing power amplifier, 370–371
output *vs.* input power, power amplifiers, class A, B, and C operations, 357–358
overlay approach, cognitive radio spectrum-sharing technology
  basic components, 137–138
  overview, 131–132
  systems architecture, 139–141
oversampled parametric amplifier, RFIC parametric converters, 492–493

P

packet error rate (PER)
  multiple input, multiple output systems, testing and evaluation, 124–125
  ultra wideband systems, 83–84
paintable electronics, development of, 160
pair gain, on-chip antennas, single-chip radios, 243–244
parametric converters, radio frequency integrated circuits
  amplifier configuration, 493–500
  basic properties, 487
  circuit simulations, 508–516
  degenerate amplifier, 490–492
  future research issues, 516–517
  historical background, 488–489
  lower sideband down-converter, 497, 514–516
  operating principles, 490–493
  oversampling amplifier, 492–493
  resistive and transresistive amplification, 488
  transmit and receive chains, 497–500
  upper sideband up-converter, 495–497, 508–514
  varactor structures, 500–508
    elastance model, 501–502
    figures of merit, 502–508
parametric dividers, phase-locked loop systems, 420–423
passive filter transfer function, phase-locked loop bandwidths, 394–396
passive receivers, short-distance wireless systems, 178–179
passive transmitters, short-distance wireless communication, 165
passive-voltage-mode CMOS ring mixer, low-noise amplifier, 21–24
patch antennas, short-distance wireless systems, 161–163
patient-centered health care, wireless technology for, 223
p-channel MOS (PMOS) device
  low-noise amplifier, wideband design, 318–321

multiband voltage-controlled oscillator design, 445–447
RFIC parametric converters, varactor structures, 504–508
substrate noise propagation entry point, 574
    GHz LC-tank voltage controlled oscillator, 576–577
peak-to-average ratio (PAR), power amplifiers, back-off
    applications, 367–368
personal area networks (PANs)
  cognitive radio spectrum-sharing technology, ultra
      wideband systems, 134–135
  evolution of, 4–5
  ultra wide band technology and, 81–83
phase detector (PD)
  charge-pump phase-locked loops (PLL) bandwidths, 393
  PLL synthesizer noise models, 435
phase-error response, phase-locked loop bandwidths,
    398–400
phase fluctuation terminology, phase noise concepts,
    428–431
phase-frequency detector (PFD)
  complete phase noise analysis, 447–452
  fractional N-synthesizers, programmable multimodulus
      divider phase frequency detector, 442–445
  phase-locked loop circuit implementation, 411–412
  phase-locked loop modeling, 393–397
    frequency divider, 397
    loop delay, 397
    loop filter, 395
    voltage-controlled oscillator, 395–397
  phase-locked loop systems, in-band/out-of-band noise,
    438–440
  $\Sigma\Delta$ modulators
    basic requirements, 479–480
    charge pump, 480–482
phase-locked loop (PLL) systems
  architectures, 405–411
    adaptive loops, fast settling, 409–410
    dual-loop architecture, 409
    integer-N-architecture, 407–408
    low-jitter loops, 410–411
    phase noise, 406
    reference feed-through, 406–407
    settling time, 406
    single loop with external frequency divider, 408
    single loop with external frequency offset, 408–409
  building block circuit implementation, 411–423
    charge pump, 412–414
    loop filter, 414–417
    phase-frequency detector, 411–412
    prescalers (dividers), 418–423
    voltage-controlled oscillator, 417–418
  continuous-time modeling, 393
  control dynamics, 397–402
    loop stability, 400–402
    phase-error response, higher order systems origin,
      398–400
    type and order systems, 398
  discrete time modeling, 403–405
  future research issues, 424
  linear modeling, 392–393
  multiband wireless networks, frequency synthesis
    architectures, 431–434

building block phase noise models, 434–438
  overview, 428
noise modeling, 402–403
overview, 384–386
PFD/QP modeling, 393–397
  frequency divider, 397
  loop delay, 397
  loop filter, 395
  voltage-controlled oscillator, 395–397
power amplifier, universal frequency synthesizer, 31
short-distance wireless systems, dense networks, 176–178
system specification, 386–392
  frequency accuracy, 387–388
  reference frequency and feedback division ratio, 387
  settling time, 388–389
  spectral purity, 389–392
  synthesis (tuning) range, 387
ultra-low-power transceivers, 187–188
ultra wideband systems
  integer-N PLL technique, 95–96
  single PLL/single-side-band (SSB) mixer, 98–99
  two PLLS/single-side-band mixer, 96–98
phase noise
  circuit design theory, voltage-controlled oscillators,
    417–418
  integer-N PLL synthesizers, 390–392
  multiband wireless networks, frequency synthesis
    basic principles, 428–429
    measurement comparisons, 447–452
    overview, 428
  phase-locked loop systems, 406
    in-band/out-of-band noise, 438–440
    models, 434–438
  residual FM calculations, Bluetooth technology, VCO
    contribution, 474
  $\Sigma\Delta$ converters, 436–438
  voltage-controlled oscillators, 476–478
phase shift keying (PSK)
  radio frequency integrated circuits (RFICs),
    transmission line, 544–545
  wireless local area networks, 66–68
phase transfer function, phase-locked loop bandwidths,
    loop stability, 400–402
physical layer (PL), Bluetooth technology, 458
piezoelectric acoustic-base resonators, evolution of, 600–605
pole-zero location, phase-locked loop bandwidths, 395–396
Pospiesalski noise model, active CMOS technology, 13–14
  mm-microwave CMOS systems, 37
power-added efficiency (PAE), power amplifiers, 352–353
  class A, B, and C operations, 355–358
  finite bandwidth, 366–367
  switching configuration, 359–362
power amplifiers (PAs)
  AMPS, GSM, and GPRS applications, 373
  analog and digital baseband, 32–33
  architecture, 25–26
  back-off applications, 367–372
    Doherty PA, 371–372
    envelope elimination and reconstruction, 369–370
    envelope following, 369
    envelope tracking, 369

power amplifiers (PAs) (*contd.*)
    out-phasing, 370–371
    requirements and efficiency penalties, 367–368
    subranging, 368–369
  basic properties, 351–352
  CDMA/WCDMA technologies, 374–375
  class A, B, and C operations, 354–358
  class AB operation, 358–359
    g_m ratio biasing, 376–377
  classification and conduction angles, 354
  direct current-to-radiofrequency conversion
      mathematics, 362–363
  dynamic biasing, 30
  EDGE technology, 374
  efficiency, 352–353
    finite bandwidth, 366–367
  evolution of, 350
  finite bandwidth signals, 365–366
  frequency specificity of, 4
  fundamental-to-direct current ratios, 363–365
  future design and application, 24–33
  historical perspective, 377–379
  IEEE 802.11a/b/g standard, 375
  linearization, 353–354, 372
  maximum output power matching, 353
  mm-microwave CMOS systems, 45
  multimode amplifiers, 25
  power-combining technologies for, 26–30
  power control, 372
  radio frequency integrated circuits (RFICs), delay
      effects, 547–548
  real-world wireless systems, 372–373
  relative signal bandwidth, 350–351
  switching PA systems, 359–366
  target specification, 25
  terminology, 351–352
  24-GHz transmitter, single-chip radios, 248–249
  ultra-low-power radio frequency transceivers, wireless
      microsensor nodes, 188–193
  ultra wideband systems, transmitter specifications, 94–95
  universal frequency synthesizer, 30–31
  voltage and current signal pairing, 365
  voltage-controlled oscillator, spurious modulation,
      460–461
  wide tuning range voltage-controlled oscillator, 31–32
  zero harmonic power, 363
power and area demands, integer-N PLL synthesizers, 392
power-constrained optimization, low-noise amplifiers,
      narrowband topology, 311–315
power control
  MEM resonators, 597
  power amplifiers, 372
    flow and balance diagram, 351–352
power domain modeling, substrate noise generation,
      569–572
power gain
  RFIC parametric converters, upper sideband
      up-converter, 510–514
  24-GHz receiver, single-chip radios, 246–247
  wideband LNA design, 16–18
power-matched topologies, low-noise amplifier, 308–309

matching requirements, wideband design, 318–321
  narrowband systems, 309–315
power spectral density (PSD)
  dither signal, $\Sigma\Delta$ synthesizer, 475
  integer-N PLL synthesizers, phase noise mathematics,
      390–392
  phase noise principles, 429–431
  residual FM calculations, Bluetooth technology,
      472–475
  $\Sigma\Delta$ converters, phase noise models, 437–438
Poynting vector, electrically small antenna transmission,
      short-distance wireless systems, 161–163
precalculated seeds, programmable $\Sigma\Delta$ modulator, 440–442
predistortion-based FHSS transmitter design, 201–205
predistortion-based hopping-frequency synthesizers, 195–204
  data recovery, 200–201
  deterministic errors, 196–197
  stochastic errors, 197–200
prescaler (divider) topology, integer-N systems, 418–423
process variation, CMOS technologies, 538–539
programmable MEM resonator-based FSK transmitter,
      622–624
programmable multimodulus divider phase frequency
      detector, fractional N-synthesizers, 442–445
propagation-link budget analysis, ultra-low-power radio
      frequency transceivers, 209–215
pseudo-random-binary-sequencer (PRBS), substrate noise
      generation modeling, 571–572
pseudorandom noise, crystal elimination, single-chip
      radios, differential detection, 252–253
P-type polysilicon resistor, CMOS technology, 530
pull-in voltage, MEM resonators, 597
pulse-based wideband links, short-distance wireless systems,
      wideband inductive transmission, 172
pulse generator, ultra wideband systems, body area
      networks, 229–230
pump signal, RFIC parametric converters, 489

Q

Q-factor-frequency product, resonator design, 604–605
Q loading, external circuitry, MEM resonators, 598–599
  mechanical circuits, 609–612
quadrature amplitude modulation (QAM)
  mm-microwave CMOS systems, 45–48
  power amplifier design, 24
quadrature low-frequency paths, third-order distortion,
      462–463
quadrature mixer design, multimode/multistandard
      receivers, 73–74
quadrature phase shift keying (QPSK)
  cognitive radio spectrum-sharing technology, 149
  MB-OFDM systems, 82–83
  universal mobile telecommunication system, 60
quality factor definition, MEM resonator, 591
quarter-wave transformer, Doherty power amplifier design,
      371–372
quasioptical mixer, membrane-supported millimeter-wave
      circuits, 647–651
quasithree-edge-supported Yagi-Uda antenna, membrane-
      supported millimeter-wave circuits, 642–644

R

radial contour-mode resonator
  evolution of, 602–605
  fabrication of, 618–620
radiative transmission, short-distance wireless systems, 161–163
radio architecture
  mm-microwave CMOS systems, 45–47
  ultra-low-power radio frequency transceivers, wireless microsensor nodes, 189
radio astronomy, wideband low-noise amplifier design, 322–325
radio frequency bands, power amplifiers, 350–351
  DC-to-RF mathematics, 362–363
radio frequency filters, multimode/multistandard receivers, 72–73
radio frequency identification (RFID)
  evolution of, 159
  short-distance wireless systems, passive transmitters, 179
  ultra-low-power radio frequency receivers
    energy-scavenging techniques, 186
    system architecture, 190–191
radio frequency integrated circuits (RFICs)
  coupling measurement, 554–561
    inductors, 554–555
    LNA example, 555
    oscillator-to-oscillator, 555–561
  coupling opportunities, 561–563
    injection-locked oscillator, 562–563
    transformer coupling improvement, 561–562
  delay opportunities, 549–554
    distributed amplifiers, 549–554
    transmission lines, 549
  distributed effects, 544
    capacitance, 548–549
    one millimeter transmission line, 545–547
    power amplifier delay, 547–548
    transmission line delay and phase shift, 544–545
    transmission line model, 545–549
  evolution of, 107–108
  future research, distributed effects and coupling, 563–564
  multiband wireless networks, frequency synthesis, 427–428
    complete phase noise analysis vs., 451–452
  multiple input, multiple output systems
    crosstalk, 116
    evolution, 108–109
    linear crosstalk, 116–118
    nonlinear crosstalk, 118–119
    overview, 108–109
    shared local oscillator generation, 119
    transmitter error vector magnitude, 120–121
    variable gain amplifier control, 119–120
  overview, 543
  parametric converters
    amplifier configuration, 493–500
    basic properties, 487
    circuit simulations, 508–516
    degenerate amplifier, 490–492

future research issues, 516–517
historical background, 488–489
lower sideband down-converter, 497, 514–516
operating principles, 490–493
oversampling amplifier, 492–493
resistive and transresistive amplification, 488
transmit and receive chains, 497–500
upper sideband up-converter, 495–497, 508–514
varactor structures, 500–508
  elastance model, 501–502
  figures of merit, 502–508
passive CMOS technology, 10–11
radio frequency-microelectromechanical systems (RF-MEMS) technology
  cognitive radio spectrum-sharing technology, dynamic range reduction, 144–146
  short-distance wireless systems, narrowband sinusoidal transmission, 169–171
radiometry, cognitive radio spectrum-sharing technology, suboptimal noncoherent detection, 148–149
random phase noise, integer-N PLL synthesizers, 390–392
random telegraph signals (RTS), CMOS technology, flicker noise variation, 539–541
Rauch filter, ultra wideband systems, 92–94
reactive ion etching (RIE)
  CMOS technology process, 525–526
  GaAs micromachined membrane-supported millimeter-wave circuits, 646–647
reactive transmission, short-distance wireless systems, 161–163
  narrowband interference, 176–178
read-only memory (ROM)
  frequency-hopping spread-spectrum synthesizers, 194–195
  predistortion-based FHSS transmitter design, 201–205
  predistortion-based hopping-frequency synthesizers, 197–200
real-world wireless systems, power amplifiers, 372–373
receive chains, RFIC parametric converters, 497–500
received signal strength indicator (RSSI)
  dynamic radio principles and, 7
  ultra-low-power radio frequency receivers, 186
receiver diversity, multiple input, multiple out systems, 111–113
receiver specifications
  Bluetooth standard, 64–65
  digital radiofrequency processor (DRP™), discrete-time receiver, 279–282
  global system for communication (GSM) standard, 56–59
  membrane-supported millimeter-wave circuits, silicon-based micromachined module, 644
  MEM resonators, 605–608
  multimode/multistandard receivers, 71–72
  short-distance wireless systems, 166
  24-GHz CMOS RF circuits, single-chip radios, 245–247
  ultra-low-power radio frequency transceivers, 205–209
    architecture, 207–209
  ultra wideband systems
    down-converter mixer, 89–92
    IF/baseband filter, 92–94
    linearity and filter requirements, 85–86

receiver specifications (*contd.*)
    low-noise amplifier, 87–89
    specificity, 83–84
   universal mobile telecommunications, 61–63
   wireless local area networks, 69–70
reciprocity, multiple input, multiple out systems, 115–116
reference feed-through, phase-locked loop systems, 406–407
reference frequency, phase-locked loop bandwidths, 387
reference spurs, integer-*N* PLL synthesizers, 389–390
regenerative dividers, phase-locked loop systems, 420–423
relative signal bandwidth, power amplifiers, 350–351
residual FM calculations, Bluetooth characteristics
   important integrals, 471–472
   synthesizer characteristics, 472–475, 483
residual offset cancellation, predistortion-based hopping-
       frequency synthesizers, data recovery,
       200–201
resistance values, substrate noise propagation and impact,
       guard ring modeling, 582–586
resistive amplification, RFIC parametric converters, 488
resistors, CMOS technology, 530
resonators, microelectromechanical systems technology
   applications, 599–600
   array systems, 612–614
   basic principles, 590
   capacitive transduction and sensing, 591–593
   CMOS-compatible tunable clamped-clamped resonator,
      615–616
   energy loss mechanisms, 597–599
   evolution of, 600–605
   fabrication case study, 614–620
   filter systems, 608–612
   free-free beam resonator, 616–618
   mechanical circuits, 608–614
   modeling, 593–596
   nonlinear effects, 596–597
   operating principles, 590–591
   programmable FSK transmitter, 622–624
   quality factor definition, 591
   radial-mode disk resonator, 618–620
   system case study, 620–624
      array-based oscillator, 620–622
   transceiver architecture, 605–608
ring delay, CMOS technology, 539–540
ripple contributions
   MEM resonator coupling, 610–612
   transmitter imperfections, RF output signal, 463–465
root locus plots, phase-locked loop bandwidths, loop
      stability, 401–402

S

sample-reset loop filter, PLL circuit implementation, 415–417
scatter plots, power amplifiers, 377–379
scattering parameters, on-chip antennas, single-chip radios,
      243–244
Schottky diodes, membrane-supported millimeter-wave
      circuits, quasioptical mixer, 648–651
script processor architecture, digital radiofrequency
      processor (DRP™), 293–295
seamless system assembly, development of, 160

second-order intermodulation (IM2) products, global system
      for communication (GSM) standard, 57–58
self-gain, CMOS technology, 533–534
sensing applications, MEM resonators, 600
sensitivity levels, wireless local area networks, direct
      conversion receivers, 70–71
sensor node power specifications, ambulatory
      multiparameter monitoring, 225–229
sensor systems, health care monitoring, 236–237
settling time
   Bluetooth frequency dividers, 468–471
   integer-*N* PLL systems, 388–389
   phase-locked loop systems, 406
      core requirement, 409–410
shallow trench isolation structure, CMOS technology,
      525–526
Shannon's theorem, radio frequency integrated circuits,
      107–108
short-channel effect (SCE), CMOS technology, active area
      implants, 527–528
short-distance wireless
   channel design and components, 160–164
      capacitive *vs.* inductive, 163–164
      electrically small antennas, 160–161
      reactive *vs.* radiative design, 161–163
   dense networks, 172–178
   development and applications, 159–160
   future research issues, 181
   implementation comparisons, 179–181
   narrowband sinusoidal transmission, 167–171
   passive receivers, 178–179
   receiver design, 166
   transceiver classification, 165–167
   transmitter design, 166–167
   wideband capacitive transmission, 171–172
   wideband inductive transmission, 172
short-time discrete Fourier transform (ST-DFT),
      predistortion-based hopping-frequency
      synthesizers, data recovery, 200–201
shunt feedback topology, low-noise amplifier, 18–20
   wideband design, 317–321
$\Sigma\Delta$ synthesizer
   Bluetooth transmitters
      basic requirements, 465–466
      blocks characterization, 476–482
      charge pump design, 480–482
      dither signal contribution, 475
      divider chain, 478–479
      finite image rejection, 461–462
      fractional divider, 465
      future research issues, 483–484
      input noise sources, 473–474
      loop filter design, 482
      modulator characterization, 476
      momentary frequency ripple, RF output signal,
         463–465
      overview, 456
      phase frequency detector, 479–480
      phase noise contribution, 474–475
      residual FM calculation integrals, 471–472
      residual FM contributions, 472–475

simulation results, 482–483
spurious modulation, voltage-controlled oscillator, 460–461
third-order distortion, quadrature low-frequency paths and carrier feed-through, 462–463
transfer derivations, 466–471
transmitter imperfections, "modulation characteristics" test, 458–465
voltage-controlled oscillator design, 476–478
zero-IF transmitter architecture, 456–458
multiband implementation, 440–447
multimodulus divider phase frequency detector and charge pump, 442–445
phase noise models, 436–438
precalculated seeds, 440–442
voltage-controlled oscillator designs, 445–447
signal characteristics, global system for communication (GSM) receivers, 56–58
signal coupling management, single-chip radios, 258–261
signal noise and distortion ratio (SNDR), active CMOS technology, 14
signal processing, cognitive radio spectrum-sharing technology, 146–153
feature detection, 150–153
optical coherent detection: matched filter, 146–148
suboptimal noncoherent detection: radiometer, 148–149
signal-processing applications, digital radiofrequency processor (DRP™), 289–290
signal source impedance, low-noise amplifiers, noise figure effects, 315
signal-to-noise-and-distortion ratio (SNDR), ultra-low-power radio frequency transceivers, intermodulation distortion characterization, 217–218
signal-to-noise ratio (SNR)
differential chip detection, crystal elimination, single-chip radios, 254–255
integer-N PLL synthesizers
random phase noise, 390–392
reference spurs, 389–390
multiple input, multiple out systems, receive diversity, 111–112
phase-locked loop bandwidths, 385–386
short-distance wireless systems, 175–178
ultra-low-power radio frequency transceivers
intermodulation distortion characterization, 216–218
link budget analysis, 209–215
simulation results, 215
wireless microsensor nodes, 188–193
ultra wideband systems, 83–84
significant-spectrum-scarcity regime, cognitive radio spectrum-sharing technology, wideband spectrum-sensing architecture, 141, 143–144
silicide-blocked N-type diffusion resistor, CMOS technology, 530
silicon-germanium technology
global system for communication (GSM) receivers, 58–59
heterostructure bipolar transistors, 522–524
silicon integrated circuit W-band low-noise amplifiers
bond-wire inductance, 336–338
cascode intrastage matching, 338–339
design philosophy, 330–331

future trends, 345–347
matching results, 345
millimeter-wave CMOS layout techniques, 339–343
drain and source metallization and gate pitch, 343
finger width, 340–341
gate contacts, 340
optimization techniques, 341–343
simultaneous noise and input impedance matching, 331–335
device size, 331–333
gain optimization, 334
multistage designs, 334–335
optimum biasing, 331
V-band and W-band design, 330
wideband design
CMOS technology, 343–345
matching methodology, 335–336
silicon micromachining, membrane-supported millimeter-wave circuits
antenna fabrication, 637–644
coupled-line band-pass filters, 635–637
fabrication and design, 631–635
future research issues, 651
overview, 629–631
quasioptical mixer, 647–651
silicon-based receiving module, 644
simulated division ratio, $\Sigma\Delta$ converters, precalculated seeds, 440–442
simulated inductor coupling, radio frequency integrated circuits, 558–561
simulation techniques
crystal elimination, single-chip radios, 255–257
$\Sigma\Delta$ synthesizer, 482–483
sine wave detection, cognitive radio spectrum-sharing technology, 149
single-chip radios
crystal elimination, 251–258
differential chip detection, 252–253
digital implementation issues and verification, 257–258
modulation format, 253
noise analysis, 253–255
simulation results, 255–257
digital radiofrequency processor (DRP™), 266–270
evolution of, 241–242
future research issues, 261
noise and unwanted signal coupling management, 258–261
on-chip antennas, 242–244
24-GHz CMOS RF circuits, 244–251
integrated receiver, 245–247
integrated transmitter, 247–249
node-to-base-station communication, 251
node-to-node communication, 249–251
single-finger varactor, RFIC parametric converters, 505–508
single-frequency noise, global system for communication (GSM) receivers, 58–59
single-input, multiple-output (SIMO), defined, 109
single-input, single-output (SISO) systems
defined, 109
performance evaluation, 124–127

single-loop PLL architectures
  external frequency divider, 408
  frequency offsets, 408–409
single-side-band (SSB) mixer
  integer-*N* PLL synthesizers, 390–392
  phase noise principles, 429–431
  $\Sigma\Delta$ converters, phase noise models, 437–438
  24-GHz receiver, single-chip radios, 246–247
  ultra wideband systems, 96–99
single-transistor architecture, switching power amplifiers, 359–362
slow-wave transmission lines, mm-microwave CMOS systems, 41–44
small-signal models, MEM resonators, 593–594
smart surfaces, development of, 160
software-defined radio (SDR), development of, 6
source-degenerated low-noise amplifier
  input impedance matching, 333–334
  narrowband topology, 309–315
source-sink current, fractional *N*-synthesizers, programmable multimodulus divider phase frequency detector, 443–445
space-time codes, multiple input, multiple out systems, spatial diversity, 111–113
S-parameter measurements
  CMOS W-band LNA designs, 345–347
  membrane-supported millimeter-wave circuits, 635–637
  Yagi-Uda antennas, 642–644
spatial differential line inductor, mm-microwave CMOS systems, 37–41
spatial diversity, multiple input, multiple out systems, 108–109, 111–113
spatial filtering, cognitive radio spectrum-sharing technology, 145–146
spatial moving average, digital radiofrequency processor (DRP™), additional zeros, 285–286
spatial multiplexing, multiple input, multiple out systems, 108–109, 113–115
spectral correlation function (SCF), cognitive radio spectrum-sharing technology, feature detection, 150–153
spectral mask, cognitive radio spectrum-sharing technology, ultra wideband systems, 134
spectral purity, integer-*N* PLL synthesizers, 389
spectrum analyzer, phase noise, 429–431
spectrum sharing, cognitive radio spectrum-sharing technology
  basic principles, 132–133
  signal processing for, 146–153
  systems architecture, 139–141
spectrum utilization measurement, cognitive radio spectrum-sharing technology, 132–133
spiral inductors
  passive CMOS technology, 10–11
  wideband LNA design, 16–17
spreading factor (SF), universal mobile telecommunication system, 60–61
spread-spectrum (SS) systems, ultra-low-power radio frequency receiver, 191–192
spurious modulation

substrate noise impact reduction, guard ring modeling, 581–586
  voltage-controlled oscillator, 459–460
square kilometer array (SKA) radio telescope, wideband low-noise amplifiers, 322–325
stagger-tuned stages, silicon integrated circuit W-band low-noise amplifiers, 334–335
stemless wineglass resonators, 603–605, 620–622
stochastic errors, predistortion-based hopping-frequency synthesizers, 197–200
stress engineering, CMOS technology, 528
subharmonic pumping, RFIC parametric converters, 499–500
suboptimal noncoherent detection, cognitive radio spectrum-sharing technology, radiometers, 148–149
subranging architecture, power amplifiers, back-off applications, 368–369
subsampling architecture, ultra-low-power radio frequency receivers, 190–191
substrate noise impact, digital-to-analog mixed-signal integrated circuits
  future research, 586
  guard ring noise impact reduction, 580–586
  model verification, 576–580
  noise generation modeling, 569–572
  noise propagation, 572–576
  overview, 567–569
superheterodyne front-end architecture
  cognitive universal dynamic radio, 8–9
  MEM resonators, 605–608
super-regenerative architecture, ultra-low-power radio frequency receivers, 190–191
surface acoustic wave systems
  low-noise amplifiers, 14–15
  multimode/multistandard receivers, 73–74
sweet spot biasing, switching power amplifiers, 360–362
switching power amplifiers, 359–362
switch-related errors, phase-locked loop bandwidths, charge pump circuit implementation, 412–414
synthesis range, phase-locked loop bandwidths, 387
synthesizer requirements
  Bluetooth technology, 465–471
  ultra wideband systems, 86–87
system transfer function, phase-locked loop bandwidths, 397–398

T

tank inductors, multiband voltage-controlled oscillator design, 445–447
telescope design, wideband low-noise amplifiers, radio astronomy applications, 322–325
Telos node, development of, 187
temporal moving-average operation, digital radiofrequency processor (DRP™), 282–284
  additional spatial filtering zeros, 285–286
  discrete signal processing, 289–290
thermal channel noise, CMOS technology, 535
thermal energy scavengers (thermal electric generators/thermopiles), health care monitoring, 233–235

thermoelastic damping, MEM resonators, 598
Thevenin transformation, loop antennas, short-distance wireless systems, 163–164
third-order distortion, Bluetooth technology, $\Sigma\Delta$ synthesizer, 462–463
third-order intercept point (IP3), low-noise amplifier linearity, 321–322
three-dimensional system-in-a-package (3D SiP) technology, health care monitoring systems, 237–240
three-resonator filter structure, MEM resonators, 608–612
TI BRF6100 single-chip radio, 241–242
time division duplexing (TDD)
    Bluetooth standard and, 64
    phase-locked loop bandwidths, 386–387
time domain cancellation, cognitive radio spectrum-sharing technology
    dynamic range reduction, 144–146
    frequency agility regimes, 156
time-to-digital converter (TDC), digital radiofrequency processor (DRP™), local oscillator generation, 273–275
timing deviation (TDEV), digital radiofrequency processor (DRP™), all-digital phase-locked loop system, 276–279
top-side fabrication, membrane-supported millimeter-wave circuits, 632–635
total gate width
    low-noise amplifiers, 331–333
    W-band radio, 335–336
transceiver specifications
    count explosion for, 5
    MEM resonators, 605–608
    mm-microwave CMOS systems, 47–49
    multiband orthogonal frequency division multiplexing, 100–104
    multiple input, multiple out systems, performance evaluation, 124–127
    short-distance wireless systems
        energy efficiency, 180–181
        narrowband sinusoidal transmission, 167–171
    ultra-low-power transceivers, 187–188
    ultra wideband design, 190
        ultra wideband systems, 83–87
        interferer scenario, 84–85
        MB-OFDM UWB, 100–103
        receiver linearity and filter requirements, 85–86
        receiver specificity, noise figure, and signal-to-noise ratio, 83–84
        synthesizer requirements, 86–87
        transmitter requirements, 86
transconductance amplifier (TA), digital radiofrequency processor (DRP™), discrete-time receiver architecture, 280–282
transfer functions
    MEM resonators, mechanical circuits, 609–612
    substrate noise propagation, 572–576
        circuit response computation, 575–576
        digital noise impact, 578–580
transformer coupling
    power amplifiers, 26–30

radio frequency integrated circuits, improvement in, 561–562
transimpedance amplifier (TIA), resonator design, 604–605
transistor size, silicon integrated circuit W-band low-noise amplifiers, 331
transmission antennas, short-distance wireless systems, 163–164
transmission line, radio frequency integrated circuits
    delay and phase shift, 544–545
    delay opportunities, 549
    model, 545
    one-millimeter transmission line, 545–547
transmit chains, RFIC parametric converters, 497–500
transmit diversity, multiple input, multiple out systems, 111–113
transmitter imperfections
    Bluetooth "modulation characteristics" test, $\Sigma\Delta$ synthesizer, 458–465
    momentary frequency ripple contributions, 463–465
transmitter leakage, universal mobile telecommunication receiver specifications, 61–63
transmitter specifications
    cognitive radio spectrum-sharing technology, 139–141
        frequency agility regimes, 153–156
    digital radiofrequency processor (DRP™), all-digital phase-locked loop system, 275–279
    multiple input, multiple out systems
        channel knowledge, 115–116
        error vector magnitude, 120–121
        testing and verification, 123–127
    predistortion-based FHSS transmitter design, 201–205
    short-distance wireless systems, 166–167
    24-GHz receiver, single-chip radios, 247–249
    ultra wideband systems, 86, 94–95
transresistive amplification, RFIC parametric converters, 488
true single phase clock (TSPC) register, phase-locked loop systems, dual-mode dividers, 420–423
tuned radiofrequency (TRF) architecture, short-distance wireless systems, narrowband sinusoidal transmission, 170–171
24-GHz CMOS RF circuits, single-chip radios, 242, 244–251
    integrated receiver, 245–247
    integrated transmitter, 247–249
    node-to-base-station communication, 251
    node-to-node communication, 249–251
two-port CMOS transistor
    low-noise amplifiers, 306–308
    RFIC parametric converters, upper sideband up-converter, 508–514

U

ultrahigh frequency (UHF) range, MEM resonators, 602–605
ultra-low-power radio frequency transceivers
    development and applications, 159–160
    energy-scavenging techniques, 186
    frequency-hopping spread-spectrum synthesizers, 194–205
        data recovery, 200–201
        deterministic errors, 196–197
        experimental results, 204–205

ultra-low-power radio frequency transceivers (*contd.*)
  predistortion-based design, 195–196, 202–204
    stochastic errors, 197–200
  future research issues, 218–219
  industrial research, 187
  intermodulation distortion characterization, 216–218
  low-power wireless systems trends, 186–187
  overview of, 185–186
  receiver planning, 205–209
    architecture, 207–209
    simulation results, 215
  2.4 GHz receiver link budget analysis, 209–215
  university research, 187–188
  wireless microsensor nodes, 188–193
    DSSS *vs.* FHSS, 192–193
    RFID, subsampling, and super-regenerative
      architectures, 190–191
    spread-spectrum systems, 191–192
    system architecture, 189
    ultra wideband transceivers, 190
ultra wideband (UWB) systems
  body area networks, 229–232
    analog receiver, 230–232
    pulse generators, 229–230
  cognitive radio spectrum-sharing technology, 133–138
    impulse detection, 136–137
    impulse radio architecture, 135–136
    system architecture, 134–135
  cognitive radio systems, 8
  evolution of, 4, 108
  fast-hopping synthesizer, 95–100
    integer-*N* PLL approach, 95–96
    single PLL and single-side-band mixer, 98–99
    two PLLs and single-side-band mixer, 96–98
  low-noise amplifier design, 315–321
  overview, 81–83
  radiofrequency transmitter specifications, 94–95
  receiver specifications, 87–94
    down-converter mixer, 89–92
    IF/baseband filter, 92–94
    low-noise amplifier, 87–89
  short-distance wireless communication, 165
  transceiver specifications, 83–87
    interferer scenario, 84–85
    MB-OFDM UWB, 100–103
    receiver linearity and filter requirements, 85–86
    receiver specificity, noise figure, and signal-to-noise
      ratio, 83–84
    synthesizer requirements, 86–87
    transmitter requirements, 86, 94–95
  ultra-low-power radio frequency transceivers, 190
  wideband LNA design, 15–18
underlay approach, cognitive radio spectrum-sharing
      technology
  overview, 131–132
  ultra wide band systems, 133–138
unilateral gain, mm-microwave CMOS systems, 35–37
unity gain frequency
  CMOS technology, 521–522
    scaling, 534–535
  passive CMOS technology, 10–11

universal control channel (UCC), cognitive radio
      spectrum-sharing technology, 140–141
universal frequency synthesizer, power amplifier, 30–31
universal mobile telecommunication system (UMTS), 60–64
  multimode/multistandard receivers, 71–74
  receiver specifications, 61–63
universal software-defined radio
  cognitive radio concept, 8–9
  development of, 6
university research, ultra-low-power transceivers, 187–188
upper sideband up-converter (USBUC), RFIC parametric
      converters
  basic properties, 495–497
  circuit simulations, 508–514
  transmit/receive chains, 497–500

V

varactor structures
  CMOS technology, 530–531
  RFIC parametric converters, 500–508
    elastance model, 501–502
    figures of merit, 502–508
    subharmonic pumping, 499–500
    upper sidband up-converter, 509–514
  substrate noise propagation entry point, 574
variable clock (CKV) period, digital radiofrequency
      processor (DRP™), local oscillator
      generation, 273–275
variable gain amplifier (VGA)
  dynamic radio principles and, 7
  multimode/multistandard receivers, 74
  multiple input, multiple systems, radio frequency
      integrated circuit design, 119–120
  ultra wideband systems, 92–94
V-band radios, silicon integrated circuit W-band low-noise
      amplifiers, 330
vector signal generators (VSGs), multiple input, multiple
      output systems, 123–124
verification methodology, digital radiofrequency processor
      (DRP™), behavioral modeling and
      simulation, 295–298
vertical natural capacitor (VNCAP), CMOS technology, 532
VHSIC hardware description language (VHDL), digital
      radiofrequency processor (DRP™)
  behavioral modeling and simulation, 296–298
  evolution of, 266
  RX simulations, 296–298
  TX simulations, 296
virtual multiple input, multiple output systems, research
      and development of, 127–128
voltage-controlled oscillator (VCO)
  Bluetooth receivers, 65
    design criteria, 476–478
  charge-pump phase-locked loops (PLL) bandwidths,
      continuous time modeling, 393
  CMOS technology, varactor structure, 530–531
  flicker noise, 527
  fractional-*N* synthesizers
    multiband wireless networks, frequency synthesis,
      433–434

programmable multimodulus divider phase frequency detector, 444–445

MEM resonators, 599–600
    array-based oscillator, 620–622

mm-microwave CMOS systems, 43–45

multiband wireless networks, frequency synthesis
    complete phase noise analysis, 448–452
    design criteria, 445–447
    integer-N PLL system, 431

multimode/multistandard receivers, 74

multiple input, multiple output systems, chip architecture and design, 121–122

phase-locked loop systems
    building block phase noise models, 434
    charge pump circuit implementation, 412–414
    design issues, 417–418
    in-band/out-of-band noise, 438–440
    low-jitter loops, 410–411
    mathematics of, 395–396
    noise modeling, 402–403
    phase-error response, 399–400
    synthesis range, 387

power amplifiers
    universal frequency synthesizer, 30–31
    wide tuning range systems, 31–32

predistortion-based FHSS transmitter design, 201–205

radio frequency integrated circuits, oscillator-to-oscillator coupling, 555–561

residual FM calculations, Bluetooth technology, phase noise contribution, 474

RFIC parametric converters, elastance varactor model, 501–502

simulation techniques, 483

single-chip radios, unwanted signal coupling, 259–261

spurious modulation, 459–460

substrate noise propagation and impact, 571–576
    circuit response computation, 574–576
    digital noise impact, 578–580
    GHz LC-tank voltage controlled oscillator, 576–577
    guard ring modeling, 581–586

transmitter imperfections, RF output signal, 463–465

ultra-low-power radio frequency receiver architectures, 191

ultra wideband systems, two PLLs/single-side-band (SSB) mixer, 96–98

wireless standards and, 5

voltage-current signal pairing, power amplifiers, 365

voltage-switched charge pump, circuit implementation, 412–414

W

Watt nodes, defined, 186

W-band radio
    CMOS W-band LNA designs, 343–345
    silicon integrated circuit W-band low-noise amplifiers, 330
    matching methodology, 335–336

well implants, CMOS technology, 526

wide tuning range voltage-controlled oscillator, power amplifiers, 31–32

wideband capacitive transmission, short-distance wireless systems, 171–172

wideband code division multiple access (WCDMA), power amplifiers, 350–351, 374–375

wideband digital wireless communication systems, power amplifiers, back-off applications, 367–368

wideband inductive transmission, short-distance wireless systems, 172

wideband low-noise amplifier design, 15–18, 315–321
    basic principles, 15–18
    gain case, 320–321
    radio astronomy applications, 322–325

wideband spectrum-sensing architecture, cognitive radio spectrum-sharing technology, 141–146
    dynamic range reduction, 144–146
    frequency agility regimes, 141–144

Wiener-Khinchin-Einstein theorem, integer-N PLL synthesizers, 390–392

wireless local area networks (WLANs)
    direct conversion receiver, 70–71
    evolution of, 4
    IEEE 802.11a standard, 68–69
    IEEE 802.11/.11b standard, 66–68
    IEEE 802.11g standard, 69
    multimode/multistandard receivers, 71–74
    multiple input, multiple output technology, 128
    overview, 66

wireless microsensor nodes, ultra-low-power radio frequency transceivers, 188–193
    DSSS *vs.* FHSS, 192–193
    RFID, subsampling, and super-regenerative architectures, 190–191
    spread-spectrum systems, 191–192
    system architecture, 189
    ultra wideband transceivers, 190

wireless sensor networks, evolution of, 159

wireless standards, evolution of, 5

wireless transceivers, evolution of, 4–9

wiring systems, CMOS technology, 528

Y

Yagi-Uda antenna, membrane-supported millimeter-wave circuits, 640–644

Z

zero harmonic power, power amplifiers, 363

zero-IF architecture
    Bluetooth technology, $\Sigma\Delta$ synthesizer, 456–458
    CMOS technology, flicker noise, 536
    micro-Watt node development, 219
    multiband orthogonal frequency division multiplexing, 102–104
    universal mobile telecommunication receiver specifications, 62–63
    wireless local area networks, 69–71

zero synthesis, PLL loop filter circuit implementation, 415–417

Zigbee standard, low-power autonomous devices, 186–187

zigzag dipole antenna, single-chip radios, 242–244